D0857306

Mobile Genetic Elements

Mobile Genetic Elements

EDITED BY

James A. Shapiro
Department of Microbiology
University of Chicago
Chicago, Illinois

1983

ACADEMIC PRESS

A Subsidiary of Harcourt Brace Jovanovich, Publishers

New York London

Paris San Diego San Francisco São Paulo Sydney Tokyo Toronto

ACADEMIC PRESS, INC.
111 Fifth Avenue, New York, New York 10003

United Kingdom Edition published by
ACADEMIC PRESS, INC. (LONDON) LTD.
24/28 Oval Road, London NW1 7DX

Library of Congress Cataloging in Publication Data

Main entry under title:

Mobile genetic elements.

Includes index.
1. Extrachromosomal DNA. 2. Translocation (Genetics)
3. Cytoplasmic inheritance. I. Shapiro, James Allen,
Date
QH462.T7M63 1982 574.87'3282 82-11624
ISBN 0-12-638680-3

PRINTED IN THE UNITED STATES OF AMERICA

83 84 85 86 9 8 7 6 5 4 3 2 1

Contents

4. Prokaryotic IS Elements

SHIGERU IIDA, JÜRG MEYER, AND WERNER ARBER

5. Tn*3* and Its Relatives

FRED HEFFRON

6. Transposon Tn*10*

NANCY KLECKNER

7. Transposable Elements in Yeast

G. SHIRLEEN ROEDER AND GERALD R. FINK

12. Phase Variation and Related Systems

MICHAEL SILVERMAN AND MELVIN SIMON

13. Mating-Type Genes of *Saccharomyces cerevisiae*

JAMES E. HABER

14. Antigenic Variation in Trypanosomes

P. BORST

Contributors

Numbers in parentheses indicate the pages on which the authors' contributions begin.

WERNER ARBER (159), Department of Microbiology, Biozentrum, University of Basel, Basel, Switzerland

P. BORST (621), Section for Medical Enzymology and Molecular Biology, Laboratory of Biochemistry, Jan Swammerdam Institute, Amsterdam, The Netherlands

JEAN-CLAUDE BREGLIANO (363), Laboratorie de Génétique, Université de Clermont-Ferrand II, Aubière, France

ALLAN CAMPBELL (65), Department of Biological Sciences, Stanford University, Stanford, California 94305

NINA V. FEDOROFF (1), Department of Embryology, Carnegie Institution of Washington, Baltimore, Maryland 21210

GERALD R. FINK[1] (299), Section of Biochemistry, Molecular and Cell Biology, Cornell University, Ithaca, New York 14853

HOWARD M. GOODMAN (505), Department of Molecular Biology, Massachusetts General Hospital, Boston, Massachusetts 02114

JAMES E. HABER (559), Department of Biology, Rosenstiel Basic Medical Sciences Research Center, Brandeis University, Waltham, Massachusetts 02254

FRED HEFFRON (223), Cold Spring Harbor Laboratory, Cold Spring Harbor, New York 11724

SHIGERU IIDA (159), Department of Microbiology, Biozentrum, University of Basel, Basel, Switzerland

MARGARET G. KIDWELL (363), Division of Biology and Medicine, Brown University, Providence, Rhode Island 02912

NANCY KLECKNER (261) Department of Biochemistry and Molecular Biology, Harvard University, Cambridge, Massachusetts 02138

JÜRG MEYER (159), Department of Microbiology, Biozentrum, University of Basel, Basel, Switzerland

[1]Present address: Department of Biology, Massachusetts Institute of Technology, Cambridge, Massachusetts 02139

ANNE RÉSIBOIS (105), Laboratoire d'Histologie, Université Libre de Bruxelles, Bruxelles, Belgium

G. SHIRLEEN ROEDER (299), Department of Biology, Yale University, New Haven, Connecticut 06511

GERALD M. RUBIN (329), Department of Embryology, Carnegie Institution of Washington, Baltimore, Maryland 21210

JEFF SCHELL (505), Max-Plank-Institut für Züchtungsforschung, Köln, Federal Republic of Germany, and Laboratorium voor Genetika, Rijksuniversiteit Gent, Gent, Belgium

MICHAEL SILVERMAN (537), Department of Biology, University of California at San Diego, La Jolla, California 92093

MELVIN SIMON (537), Agouron Institute, La Jolla, California 92037

ARIANE TOUSSAINT (105), Laboratoire de Génétique, Université Libre de Bruxelles, Bruxelles, Belgium

MARC VAN MONTAGU (505), Laboratorium voor Genetische Virologie, Vrije Universiteit Brussel, Brussel, Belgium

HAROLD E. VARMUS (411), Department of Microbiology and Immunology, University of California, San Francisco, California 94143

PATRICIA ZAMBRYSKI[1] (505), Department of Biochemistry and Biophysics, San Francisco Medical School, University of California, San Francisco, California 94143

[1]Present address: Laboratorium voor Genetika, Rijksuniversiteit Gent, Gent, Belgium

Genomic Reorganization in Cell Lineages

The chapters of this book describe several well-studied cases in which genetic determinants—often identified as specific nucleic acid sequences—repeatedly change their positions within or between cellular genomes. Because their genomic positions are not fixed, these determinants may conveniently be classed together under the rubric of mobile genetic elements. Because all genomic configurations change over evolutionary time, this term has no rigorous definition. Nonetheless, certain identifiable components of cellular genomes (such as proviruses and transposable elements) display marked variability when compared to the Mendelian factors whose usually reliable behavior permits the construction of genetic maps.

It is a great temptation to try to place the discovery of mobile elements in the development of our concepts about heredity. Certainly, their complex recombinational activities and sometimes non-Mendelian behavior pose fundamental problems in understanding the organization of chromosomes and extrachromosomal hereditary nucleic acids. They also open many possible mechanisms leading to the formation of novel genomic configurations. However, our knowledge is still too immature to permit fruitful generalizations. There is no organism—not even *Escherichia coli,* in which intriguing new phenomena appear almost monthly—in which the mobile elements have been cataloged and their effects on phenotype and karyotype described with any completeness, and we do not yet understand in sufficient detail the relationship between mobile elements that are amenable to genetic analysis and the multiple repetitive elements that comprise a major fraction of eukaryotic nuclear DNA. I shall therefore simply describe the rationales that underlie the following collection of chapters and mention a few points that I find particularly thought-provoking.

From an editorial point of view, this book has two general objectives. The first is to introduce the nonspecialist to the biology and genetics of mobile elements. If we are successful, the chapters will make the biochemistry of DNA rearrangements more accessible to embryologists and evolutionists and illuminate the related developmental cycles to the biochemist. The second objective is to show

how natural the activity of mobile elements can be in diverse biological situations. This is an effort to counter the tendency to regard mobile elements as interesting curiosities in a background of genomic stability. Although a century of cytogenetic literature has already shown that chromosomes are dynamic structures, it is still useful to see the different ways that mobile elements can fit specific adaptive needs and respond to elaborate control systems.

The copious illustrations in Chapter 1 by Fedoroff on maize controlling elements show how well suited corn is to clonal analysis of genetic change. This made it possible for McClintock and others to uncover several systems of nuclear differentiation within plant tissues. These systems can alter chromosome structure, change specific cellular phenotypes, and create new developmental patterns. They utilize intranuclear communication between related but often distinct mobile elements and reveal a wealth of regulatory phenomena linked to plant development and cell lineages. Controlling elements were the first mobile elements to be described with precision, and they demonstrate best of all the power of mobile elements to remodel hereditary programs. Perhaps the most remarkable property of controlling elements is their capacity for generating patterns, as seen in the color illustrations of kernels that have developed during controlling element activity.

Chapters 2 and 3 on bacteriophages λ and Mu by Campbell and by Toussaint and Résibois discuss two prokaryotic mobile elements that have been analyzed in considerable detail. In large measure, so much information is available because they can be isolated in virus particles and are thus easier to study physically and genetically than elements that are always incorporated into more complex genomic structures. More than two decades of attention by molecular biologists and biochemists to the specific recombination system associated with λ insertion and excision make it possible *(1)* to define the specific proteins and DNA-substrate sequences involved and *(2)* to describe with considerable precision the sophisticated regulatory system that controls the specific recombination events. Interestingly, a part of this regulation depends on DNA structures, so the recombination events are self-regulatory. For λ, insertion and excision are not essential for reproduction as a virus, but this is not the case for phage Mu. Mu's reproduction (including both replication and morphogenesis) is completely dependent on its ability to move through the bacterial genome and recombine with many regions of the host DNA. Thus Mu illustrates how genetic mobility can be totally integrated into a biological cycle. In addition to their existence as viruses (which still depend on cells for reproduction), λ and Mu form prophages and are capable of incorporating and mobilizing nonviral genetic material intracellularly and intercellularly. In this way the distinctions between host and viral genomes become less rigid, and it is apparent that the hereditary determinants of one cell can communicate directly with those of another cell (sometimes of a completely different species).

Three chapters deal with nonviral mobile elements in bacteria. Iida, Meyer, and Arber (Chapter 4) review IS elements comprehensively, and Heffron (Chapter 5) and Kleckner (Chapter 6), focus, respectively, on the Tn*3* family of transposons and on Tn*10*, which have been objects of more detailed analysis. Although there are certain basic similarities between all the transposable elements discussed in these chapters—in particular, the use of recombination mechanisms that involve DNA replication—they appear to be functionally distinguishable into at least two classes. The distinction relates to the biochemical pathways for transposition and other DNA rearrangements, the biological significance of each element, and the origins of related transposable elements within each class. The Tn*3* family of transposons uses biochemical and regulatory mechanisms that limit DNA reorganizations almost entirely to the dispersal of sequences that are *internal* to the element among the various replicons within a bacterial cell. Thus we understand why Tn*3* and its relatives are efficient agents for the rapid spread of specific phenotypic determinants. What is not yet understood is how these determinants become associated with the recombination sequences of this class of element. Many IS elements (including the IS*10* extremities of Tn*10*) present a contrast. They utilize replicative recombination pathways which frequently lead to the stable reorganization of DNA sequences that are *external* to the element. Because IS elements do not determine specific cellular phenotypes, their role appears to reside chiefly in restructuring and recombining bacterial replicons. It sometimes happens that these DNA reorganizations create structures in which specific determinants are flanked by duplicate IS elements, thereby creating a composite mobile element or transposon encoding a particular phenotype. In other words, the genetic activity of this second class of transposable element means that any sequence can be part of a mobile element.

Recombinant DNA technology has made it possible to analyze genetic phenomena in eukaryotic cells with the same precision at the molecular level as in bacteria. Chapter 7 on transposable Ty elements in brewer's yeast *(Saccharomyces cerevisiae)* by Roeder and Fink illustrates how molecular analysis leads to a clear understanding of the mobile element events underlying phenotypic changes. Because Ty elements form a major proportion of the repeated DNA sequences in yeast nuclei, it is possible to identify the role of repeat sequences in chromosome reorganization. The results show that related but distinguishable Ty elements are in physical communication with each other and that an important source of variability results from recombination between different Ty elements. Although Ty elements can move to new positions in the yeast genome by mechanisms that may resemble those for prokaryotic transposable elements (and thereby alter the regulation of coding-sequence expression), these events are rare. The most frequent events are exchanges between repeated sequences at particular locations, and this means that variability is channeled in certain directions by the distribution of elements in the chromosome complement. It is also intriguing to note that Ty elements have never been observed to excise precisely from a chromosomal site and leave it as it was before insertion. Thus transposition of a Ty element to a particular site initiates a particular set of chromosome genealogies.

The genome of *Drosphila melanogaster* is more complex than that of *Saccharomyces cerevisiae,* and *Drosophila* contains a greater diversity of mobile elements. The catalog of these elements is far from complete, but there is already a sizable corpus of molecular and genetic data about them. The connections have become much clearer between mobile elements that have been identified as repeated DNA sequences and those identified by genetic analysis. The two complementary aspects of this research are discussed in Chapter 8 by Rubin on *Drosophila* transposable elements and in Chapter 9 by Bregliano and Kidwell on hybrid dysgenesis. The phenomenology of hybrid dysgenesis (which is not yet widely known by nonspecialists) has a great deal to teach us about several fundamental aspects of mobile genetic element biology. These include *(1)* the cellular and environmental controls that govern when mobile elements become active, *(2)* the spread of mobile elements through large populations (on a global scale in this case because *D. melanogaster* is a cosmopolitan species), and *(3)* the role of mobile elements in creating reproductively isolated populations (in other words, their role in speciation).

It is ironic that the most intensively studied of all groups of mobile genetic elements—the vertebrate retroviruses described in Chapter 10 by Varmus—are not generally considered genetic elements at all. The formation of a chromosomal provirus by reverse transcription in retrovirus reproduction is most often viewed as a curiosity of viral nucleic acid replication, to be explained and understood in terms of virus adaptation rather than as a fundamental genetic process. This situation is analogous to that of temperate bacteriophages like λ and Mu, and it arises because we have not yet learned how to incorporate the basic, ubiquitous phenomena of virus–cell and parasite–host interactions into a more comprehensive picture of heredity. There are alternatives to the strictly virological point of view, and these are increasingly compelling as the parallels between retroviruses and other mobile and repeated elements come into sharper focus. In their DNA form (as proviruses), retroviruses display much the same behavior (including insertional mutagenesis, alteration of developmental programs, and mobilization of external chromosomal sequences) as mobile elements in other organisms. Moreover, sequence analogies between retroviruses and other repeated elements in vertebrates as well as the abundant transcription of transposable elements in yeast and *Drosophila* have suggested that reverse transcription may be a rather widespread mechanism of genome reorganization.

Chapter 11 on *Agrobacterium* oncogenesis in plants by Zambryski *et al.* describes the most thoroughly investigated situation in which mobile elements play a role in intergeneric genome communication and host–parasite interaction.

There are several fascinating aspects to this phenomenon. First, it involves a highly evolved mechanism to induce a particular type of neoplastic change. Second, under the control of a transmissible hereditary element (a plasmid), one organism (a bacterium) directly modifies for its own benefit the hereditary apparatus of another organism (a plant) by directing the insertion of a specific DNA sequence. Third, analysis of this system of natural genetic engineering may eventually permit human modification of plant genomes by piggybacking specific determinants onto the bacterial transforming DNA. Indeed, most techniques of genetic engineering utilize or mimic processes that occur naturally with mobile genetic elements, and it may be that these parallels can teach us a great deal about the modification of hereditary programs during evolution.

Not all genomic changes are limited to the germ line or have significance only in organismal lineages. It has been known since at least the 1890s that cellular differentiation during embryonic development involves nuclear and chromosomal changes. So it would be surprising if DNA reorganization did not play a role in some of the highly programmed and specific changes that underlie differentiation. Chapter 12 by Silverman and Simon on flagellar phase variation in *Salmonella* describes the first case in which the new science of molecular biology revealed (in 1956) that a change in DNA structure could control the state of cellular differentiation. One of the intriguing aspects of the site-specific recombination system that changes flagellar phases is its close biochemical relationship to other prokaryotic recombination systems discussed in this book. One (the inversion of host-range determinants in phage Mu) performs a similar regulatory function, but the other (Tn*3* resolvase) has a different purpose. It seems that bacterial cells have been able to adapt one basic biochemical recombination mechanism to meet different needs in DNA metabolism. Two other cases from lower eukaryotes are discussed in Chapter 13 by Haber on yeast mating type and in Chapter 14 by Borst on surface antigenic variation in trypanosomes. In both cases, predictable movements of coding sequences to new genomic locations lead to particular changes in cellular phenotypes. At least three features of these systems merit special emphasis. One is the importance of coding-sequence location for expression of the information it contains. In the yeast mating-type system, in which the details of transcription are known, this change is more complex than simply the placing of a protein-coding sequence downstream of appropriate transcription signals; it must involve regulation by a higher order of chromosome organization. Both systems, in fact, illustrate a situation complementary to that seen with other mobile elements in which the movement of noncoding sequences alters the expression of specific phenotypes. A second noteworthy feature of mating-type and surface antigen changes is the limitation of DNA movement to the appropriate sequence. This means that the underlying biochemistry—which, at least in yeast, involves general recombination functions—can be recruited for more specific changes. The third feature that should be highlighted is the connection between DNA reorganization and the history of the cell in which a genetic switch is to occur. Not all cells of a particular genotype have the same capacity for change. In order to switch mating type, a homothallic yeast cell must have budded at least once; a trypanosome recently introduced into a mammalian bloodstream from an insect vector has a potential antigenic repertoire that is different from one transferred by injection from one mammal to another. So there must exist regulatory mechanisms operating on DNA reorganization with "memories" of the previous cell lineage.

In 1977 the Cold Spring Harbor Laboratory published the first book dedicated to mobile genetic elements, *DNA Insertion Elements, Plasmids, and Episomes.* Up to the time of the meeting that formed the basis for the volume, I think it is fair to say that only a handful of geneticists perceived how various and widespread are the hereditary processes that involve mobile elements. Then the major events in hereditary variation seemed to most of us to be changes in protein-coding sequences or unexplained sudden rearrangements of chromosome structure. Since that time, there has been tremendous progress in defining the molecular parameters of genomic change and elucidating the complexities of chromosome organization. Some of this progress is described in the pages that follow. Nonetheless, I think it is true even today that a majority of geneticists would argue that mobile elements are fascinating but only have significance as part of the "background noise" of random variability needed to provide the raw material of evolutionary change. I will use the editor's privilege to express a different view. In my opinion, we will only integrate mobile elements fully into our picture of heredity when we have formulated entirely new mechanisms for cellular differentiation in both development and evolution. When Barbara McClintock demonstrated that specific heritable nuclear elements were responsible for changing the structure of maize chromosomes, she made a discovery comparable to observing spontaneous atomic decay. Like scientists in other fields at various periods in the histories of their disciplines, we geneticists now have to come to terms with unanticipated levels of structure and organization.

The following chapters represent extensive summaries of these fourteen mobile element systems. Although each chapter contains current and until now unpublished information, the authors have also taken great pains to place current observations within the historical development of their systems. It is pleasing to think that so many diverse phenomena can unexpectedly reveal common patterns of genetic control. In my opinion, these thoughtful reviews will not quickly become dated. Through their efforts, the authors have created a unique statement of a particularly exciting stage in the history of genetics. I hope the reader will share the exhilaration of our voyage toward a richer appreciation of heredity.

JAMES A. SHAPIRO

CHAPTER 1

Controlling Elements in Maize

Nina V. Fedoroff

I. Introduction

Controlling elements in maize (*Zea mays*) were the first transposable elements discovered and investigated. Their ability to transpose, affect gene expression, and cause chromosomal rearrangements had already been documented in substantial detail through genetic analyses by the early 1960s. Serious work on the molecular basis of controlling element phenomenology began more than a decade later and is only now beginning to yield important information. In the intervening time, transposable elements were discovered in bacteria, lower eukaryotes, and other higher eukaryotes. Analysis of these elements progressed

Mobile Genetic Elements

ISBN 0-12-638680-3

quickly, because it was undertaken when it was already possible to do molecular and genetic studies simultaneously. Although the fundamental properties of maize controlling elements can readily be represented diagrammatically and although they resemble those of prokaryotic and other eukaryotic transposable elements, there are many features that are even now unique to the maize elements. Perhaps the most striking feature of controlling elements is their ability to sense developmental parameters. Controlling elements transpose and activate other types of genetic change at a predetermined time and with a characteristic frequency in the development of the organism. The nature, frequency, and timing of element-associated changes are genetically specified and can be altered by mutation. There is even some evidence that controlling elements can be preprogrammed to alter the timing and frequency of element-associated genetic events at a precise time in development. In addition, elements have been isolated that show differences in expression as a function of their location, either within the organism as a whole or within a particular tissue. Because a controlling element can insert in or near a gene and alter its ability to be expressed, excision of the element or other types of element-associated genetic events occurring during development can result in a regular and characteristic pattern of gene expression in the organism. The pattern is determined by the properties of the element and is therefore subject to heritable alterations that affect the element. It was McClintock's awareness that elements distinct from the genes themselves could modify or control the pattern of a gene's expression in the organism that led her to designate them *controlling* elements (McClintock, 1956a,b).

The genetic behavior of the maize controlling elements is understood in much greater detail than is the behavior of any other eukaryotic transposable element. This is, in part, because of the unique attributes of maize as an experimental organism. I will therefore begin with a general introduction to maize and a sufficiently detailed consideration of maize reproductive biology and certain aspects of its development to provide a basis for understanding the phenotypes of the kernels that I have used to illustrate controlling element phenomena. I will then give a brief overview of controlling element behavior, making an effort to simplify and clarify some of the terminology that has been used. In subsequent sections, I will give a detailed description of the two controlling element families that have received the most extensive study, the *Ac–Ds* (*Mp*) family and the *Spm* (*En–I*) family. In structuring my presentation, I have endeavored, whenever possible, to include illustrations of the phenotypes that underlie the interpretations presented in the text.

The maize ear, with its hundreds of kernels, can be likened to the microbiologist's petri dish. Each kernel represents the outcome of a separate mating event. The bulk of the kernel belongs to the next generation, comprising the diploid plant embryo and the triploid endosperm tissue that will nourish the progeny plant during germination (Fig. 1a). Because of the double-fertilization event (described in detail in a subsequent section) that initiates kernel development, the embryo and the endosperm generally have the same genetic constitution, except

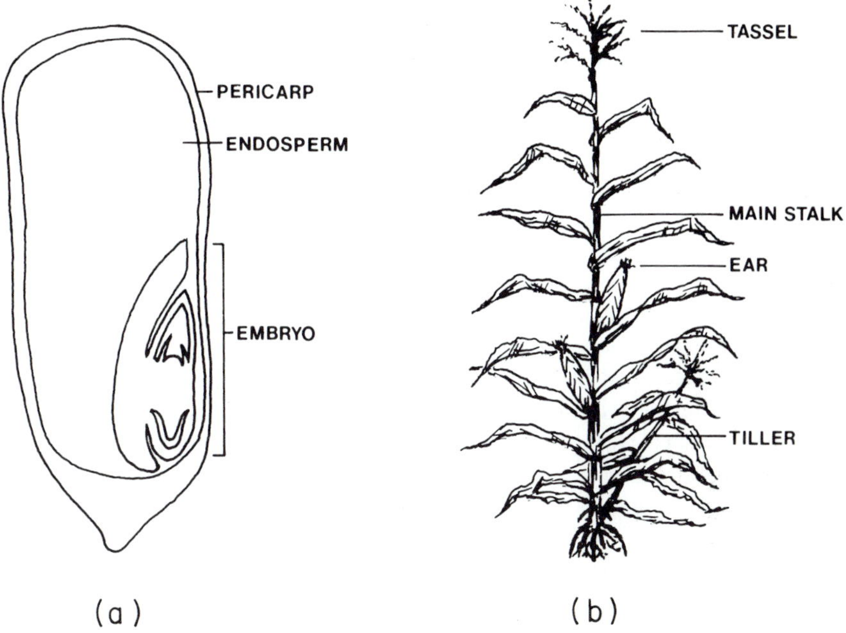

Fig. 1 (a) Cross-sectional drawing of the mature kernel, showing the outer pericarp layer, the underlying endosperm, and the embryo. (b) Drawing of a maize plant.

for a difference in ploidy. The endosperm is a bulky, nutritive storage tissue, and many mutations have been identified that affect its phenotype. Because the embryo and endosperm generally receive the same genic complement from each parent, the genetic behavior of mutations affecting endosperm phenotypes can be scored directly in kernels. Mutations and other types of genetic change that occur premeiotically in germ-line tissue can be expressed in the endosperm, as well as represented in the embryo, substantially facilitating the detection and analysis of even relatively rare genetic events. Moreover, because plants do not segregate the germ line in the same way animals do, genetic events happening early in the development of a plant can both be detected sectorially in the plant and reflected in a large fraction of the gametes. Thus, the relationship between somatic and germinal genetic changes can be determined, at least in those cases in which the events occur sufficiently early and in the right cell lineages to be included in the germ line.

A factor that has been extremely important in the analysis of controlling elements is that there are many genes that are expressed in the maize kernel the inactivation of which has little or no deleterious effect on viability and fertility. These include genes affecting starch and storage protein biosynthesis in the bulk of the endosperm, as well as genes affecting the biosynthesis of the red and purple anthocyanin pigments present in the surface (aleurone) layer of the endosperm and other plant parts. Mutations affecting pigment biosynthesis have proved especially rewarding in this connection, not only because the pigment pathway is dispensable, but also because pigment expression is easily scored visually and relatively subtle changes in levels of gene expression can be reflected in the production of intermediate levels of pigment. Thus, genes involved in pigment biosynthesis have served the maize geneticist in much the same capacity that the *lac* operon has served the microbial geneticist.

The clonal character of maize development has also been important in understanding the types of genetic change associated with transposable elements. This, too, will be described in some detail and is introduced here only in a general sense. Mitotic descendants of a given cell give rise to contiguous tissue sectors, the shape of which is determined by the pattern of cell division in a given tissue and the size of which is determined by the number of subsequent cell divisions to tissue maturity. This has several implications. First, it is possible to trace the fate of both daughter cells produced at a single mitotic division. Because events happening at early mitotic divisions can be reflected in the germ line, it has been possible, for example, to analyze both of the chromosomes involved in a single controlling element transposition event. In addition, successive genetic events occurring in a single cell lineage can be reflected in subsectors of phenotypically distinct tissue within larger sectors. The order of successive events can be deduced from the hierarchical ordering of phenotypically different subsectors. By the same token, the developmental timing of a genetic change determines the size of the sector exhibiting the phenotypic consequences of that change. A genetic change occurring during the first embryonic cell division can, for example, give a plant with genotypically and phenotypically distinct halves, each half producing correspondingly distinct gametes. Similar changes occurring at mitotic divisions early in the development of the ear can give rise to twinned sectors of genetically and phenotypically distinct kernels. Genetic changes that occur late in development may be confined to somatic tissue and expressed purely as a somatic variegation pattern. Thus, changes in the size of phenotypically distinct sectors reflect changes in the developmental time of origin of the sectors and therefore the developmental timing of the responsible genetic events.

A. Maize Reproductive Biology

The haploid (gametophytic) generation is much reduced in maize, as it is in other angiosperms, comprising the pollen grain on the male side and the embryo sac on the female side (Weatherwax, 1923). Pollen is produced in the anthers of the brush-like tassel at the top of the maize plant (Figs. 1b and 2b). Pollen grains are derived from microspore mother cells within the developing anther. The microspore mother cell undergoes meiosis to give rise to four haploid spores, each of which gives rise to a pollen grain. Microsporogenesis is depicted in Fig.

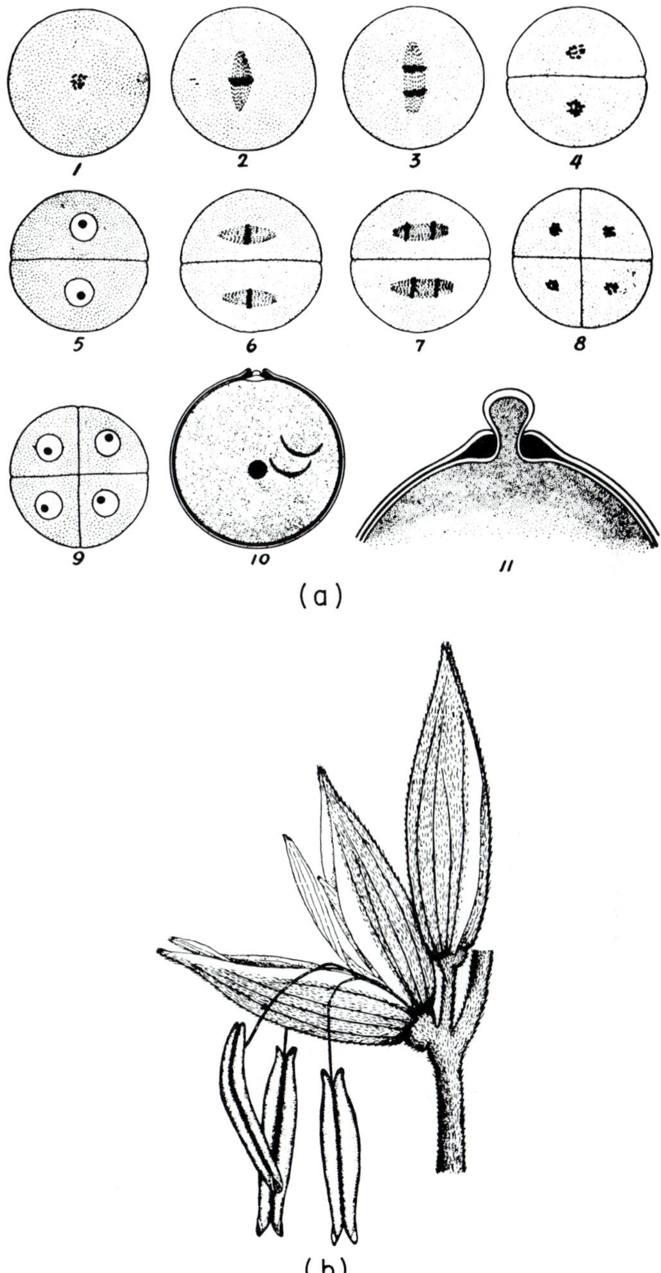

Fig. 2 (a) Microsporogenesis or meiotic stages from the microspore mother cell (1) to the four-spore stage (9); magnification 75–100×. All four spores give rise to pollen grains (10) with a vegetative nucleus and two sperm cells. Diagram 11 shows the germinating pollen grain. (b) A pair of staminate spikelets, showing the emerging anthers. [(a) Reproduced from Kiesselbach, 1951; (b) reproduced from Weatherwax, 1923]

2a, which is reproduced from Kiesselbach (1951). The mature pollen grain comprises a large haploid cell within which the two small crescent-shaped sperm cells are nested [Fig. 2a(10)]. The three nuclei present in the pollen grain are mitotic descendants of the haploid nucleus represented in the spore. When the pollen is mature, a pore opens at the end of the anther, releasing the pollen.

The female gametophyte, the embryo sac, is derived from the diploid megaspore mother cell. Megasporogenesis is depicted in Fig. 3a (from Kiesselbach, 1951). The megaspore mother cell undergoes meiosis, giving rise to four haploid spores, three of which disintegrate. The innermost haploid spore divides mitotically to form the embryo sac (Fig. 3b). The haploid unfertilized egg cell and two adjacent haploid synergid cells are at the base of the embryo sac [Fig. 3b(5)]. At the apex are the haploid antipodals, which either disintegrate or persist as a patch of haploid tissue during kernel development (Randolph, 1936). The large central cell is occupied by two haploid prmary endosperm nuclei, which lie at the base of the cell near the egg (Randolph, 1936). It is the large central cell that gives rise, after fertilization, to the nutritive endosperm tissue that comprises the bulk of the kernel. The embryo sac, inside its ovule, is contained within the ovary (Fig. 3c), from the wall of which grows a long style, termed the silk. The silks grow upward, emerging from the husk leaves at the top of the ear and serve to entrap pollen.

The pollen grain germinates within minutes of falling or being placed on a silk, sending a long pollen tube through the silk. The two sperm cells migrate down the pollen tube and are deposited within the embryo sac (Fig. 3d). One of the sperm cells fuses with the egg cell, followed by fusion of the nuclei to produce the diploid zygote that develops into the embryo. The fusion of the second sperm nucleus with the two primary endosperm nuclei of the central cell gives the triploid primary endosperm nucleus. Because the embryo sac develops from a single haploid, postmeiotic cell, as does the pollen grain, the sperm cells contain sister haploid nuclei, as do the egg cell and the central cell of the embryo sac. Thus, the embryo that develops from the fertilized egg and the endosperm that develops from the central cell normally have the same genetic constitution, except that the endosperm contains two copies of the maternal haploid genome represented in the embryo sac. The importance of this from the geneticist's standpoint is that the same maternal and paternal haploid genomes are present both in the endosperm and in the embryo that will develop into the progeny plant. This means that in an appropriately structured cross, many progeny kernels can be screened for rare mutant phenotypes, and the phenotypically mutant kernels will generally carry the corresponding mutant allele in the embryo that is expressed in the endosperm. This, in turn, makes it possible to detect and recover rare mutations fairly efficiently, because each ear produces as many as several hundred kernels, each of which will give a progeny plant.

Additional genetic advantages of maize lie in the fact that it can readily be crossed by or to either itself or another plant. Crossing is accomplished by covering developing ears before the silks emerge, then cutting back and pollinat-

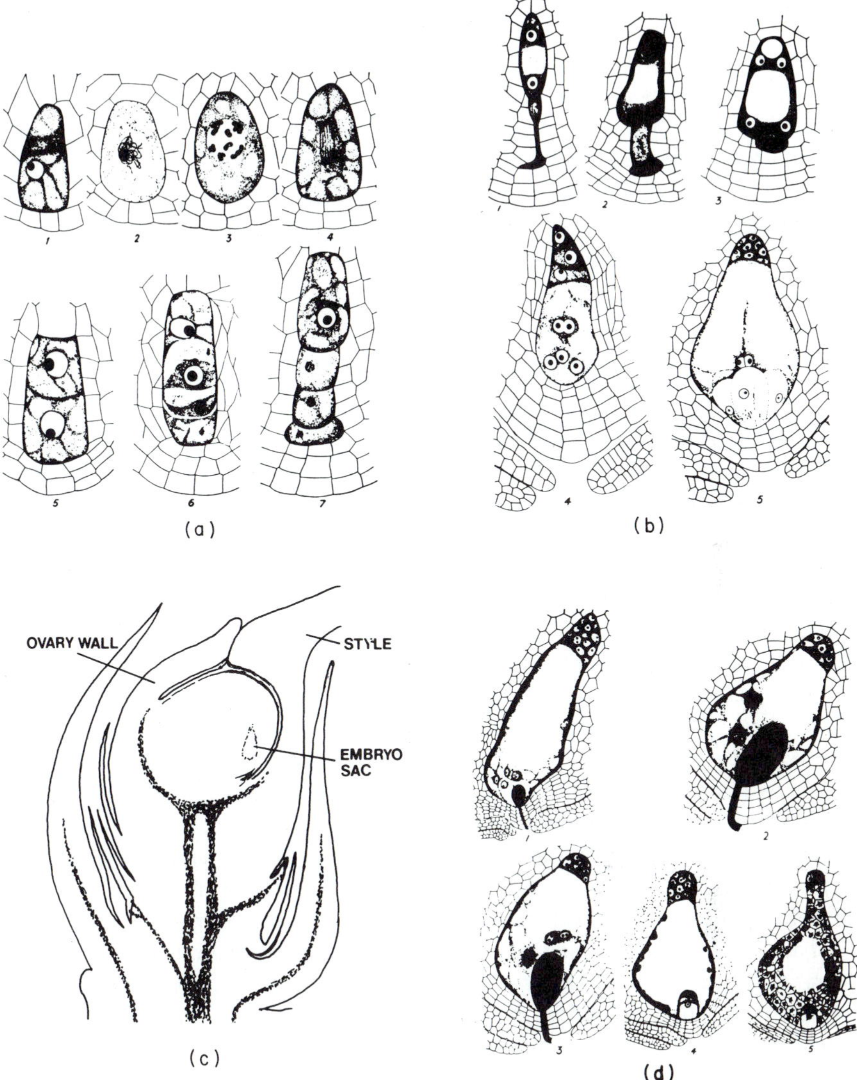

Fig. 3 (a) Megasporogenesis or meiotic stages from the megaspore mother cell (1) to the four-spore stage (7), magnification 125×. (b) Development of the embryo sac from the innermost spore; magnification 50–100×. The mature embryo sac is shown in (5). The egg cell at the base is flanked by the synergids. The large central cell contains the polar nuclei. The antipodals are at the top. (c) The pistillate inflorescence in cross section, showing the embryo sac within the ovule, surrounded by the ovary wall from which the style (silk) grows. (d) The pollen tube penetrates the embryo sac (1), discharging the two sperm cells. Double fertilization occurs (2), followed by division of the primary endosperm nucleus (3). The free-nuclear stage in endosperm development and the three-cell stage in the development of the embryo are shown in (4). A slightly more advanced stage in embryo development and cellularization of the endosperm are shown in (5) [(a, b, d) reproduced from Kiesselbach, 1951]

ing the silks with pollen collected from either the same or a different plant. Depending on their genetic constitution, plants can produce two or more fertile ears, each of which can be crossed by a different pollen parent. Additional stalks, called tillers, can produce fertile ears as well (Fig. 1b). Moreover, a single plant produces millions of pollen grains, shedding them gradually over a period of approximately a week. This permits a single plant to be used as a pollen parent repeatedly in a variety of genetic tests.

B. Development of the Endosperm and Pericarp

Endosperm development is clonal, and cells give rise to clones of cells the shape of which reflects the pattern of cell division within the tissue (Randolph, 1936; Coe, 1978; McClintock, 1978a). As noted earlier, genetic events occurring at mitotic divisions can give rise to genetically distinct daughter cells, which can develop into phenotypically distinguishable twinned sectors in the mature endosperm. The first division of the primary endosperm nucleus occurs 2–4 hr after fertilization, and the sister nuclei migrate to positions opposite each other on either side of the zygote (Randolph, 1936). The first nuclear division plane is vertical, and the planes of the subsequent early nuclear divisions are diagonal (Coe, 1978). Free nuclear divisions continue in synchrony for 7–8 divisions, and the resulting 128–256 nuclei line the central cell (Randolph, 1936). Cell walls begin to form gradually, progressing from the region of the embryo toward the antipodals (Fig. 3d). During this early period, there is no specialized epidermal layer, and cell division occurs throughout the tissue. Subsequently, cell division becomes localized in the periphery, and the internal cells increase in size and become polyploid. Cell walls form periclinally in the epidermal and underlying cell layers repeatedly, generating rows of cells that extend inward toward the central region of the endosperm. Enlargement of cells in the central region is compensated for by anticlinal divisions at the surface. This generates irregularly conical sectors of cells in mature endosperm that are descended from the earliest cells (see kernel 8, in which there are deeply stained, fan-shaped sectors of cells that are genetically distinct from the adjacent colorless cells).

About midway through endosperm development, the epidermal layer of cells begins to differentiate into the aleurone layer (Randolph, 1936; Kyle and Styles, 1977). Subsequent growth of the aleurone is by cell enlargement and anticlinal wall formation (Randolph, 1936). The final divisions occur in alternating perpendicular planes (Coe, 1978), generating relatively symmetrical surface clones of cells. Many examples of phenotypically distinct aleurone clones are shown in Plates 1 and 2 (between pp. 12 and 13). Developing endosperm cells are characterized by high rates of starch and storage protein synthesis. Differentiation of the surface aleurone layer involves many biochemical changes, among which are the cessation of starch and storage protein synthesis and expression of the anthocyanin pigment biosynthetic pathway. Many loci[1] have been identified genetically by mutations that affect starch, storage protein, and anthocyanin biosynthesis (see Coe and Neuffer, 1977, for a succinct summary of maize genetics). At least 14 different loci are known to affect the intensity, quality, and tissue distribution of anthocyanin pigments in maize plants (Coe and Neuffer, 1977). The loci discussed in the present context are listed in Table I, which includes a brief summary of what is known about each locus and the phenotypes resulting from certain types of mutations at the loci. Briefly, the *R* and *C* loci are believed to be regulatory loci (Styles *et al.*, 1973; Chen and Coe, 1977), whereas the *C2*, *A*, and *A2* loci are either known or surmised to encode enzymes directly involved in anthocyanin biosynthesis (Reddy and Coe, 1962; Kirby and Styles, 1970; Styles and Ceska, 1977; Dooner, 1980a). The *Bronze* (*Bz*) locus encodes an enzyme that glycosylates the pigment molecule (Larson and Coe, 1977; Dooner and Nelson, 1977a), whereas the *Pr* locus is believed to be involved in its hydroxylation (Coe, 1955; Harborne and Gavazzi, 1969). Expression in the aleurone of the deep purple pigment evident in kernel 1 (Plate 1) is dependent on the presence of the *R*, *A*, *A2*, *C*, *C2*, *Bz*, and *Pr* alleles, as well as the appropriate allele of the *Bz2* locus not discussed here. Pigment expression in the pericarp, but not in the aleurone, depends on the additional presence of the *P-RR* allele of the *P* locus. The phenotypes observed in the aleurone layer of strains homozygous for certain mutant alleles of these loci are illustrated in Plate 1.

Many mutations are known to affect the quantity or quality of the starch synthesized in endosperm tissue. Only two loci identified by such mutations will be discussed here. These are the *Shrunken* (*Sh*) locus and the *Waxy* (*Wx*) locus. The *Sh* locus encodes sucrose synthetase, an enzyme required for starch biosynthesis (Chourey and Nelson, 1976; Chourey and Schwartz, 1971; Chourey, 1981a). Endosperm tissue in strains homozygous for recessive mutations (*sh*) at this locus contain less than the normal amount of starch and have the shrunken phenotype for which the locus was named (kernel 4 shows a cross section of a kernel with the deeply indented crown characteristic of the *sh* allele). The *Wx* locus encodes a starch granule-bound glucosyltransferase that catalyzes the synthesis of the straight-chain amylose starch in endosperm tissue (Nelson and Rines, 1962; Chaleff *et al.*, 1981; Echt and Schwartz, 1981). Endosperm tissue containing amylose stains a deep blue-black with iodine–potassium iodide reagent and is somewhat translucent (kernel 5). Endosperm tissue in a strain homozygous for a recessive mutation at the *Wx* locus contains only the branched starch amylopectin, has an opaque appearance, and stains a brownish color that fades to yellow or becomes colorless on heating (kernel 6).

Surrounding the endosperm and embryo is a layer of maternal origin termed

[1]The term locus denotes a genetic unit defined by mutations having a similar phenotypic effect and a common map position. Its imprecision is appropriate to the paucity of molecular information we have about maize genes. A locus is presumed to include a coding sequence or sequences, as well as closely associated regulatory sequences.

Table I Information about the Loci Discussed in the Text[a]

Chromosome and map position[b]	Allele	Expression
1–26	*P–RR* or *p*	Red pericarp and cob
	P–WW or *p*	Colorless pericarp and cob; recessive to *P*
10–57	*R*	Anthocyanin produced in aleurone and plant; a complex locus that determines the tissue specificity of pigment expression; alleles are known to affect 12 tissue types differentially (Styles *et al.*, 1973)
3–111	*A*	Anthocyanin produced in aleurone and plant; may encode an enzyme involved in pigment synthesis (Kirby and Styles, 1970); certain alleles appear to be internally duplicated (Laughnan, 1952)
	a	Colorless aleurone, green or brown plant; recessive to *A*
4–123	*C2*	Anthocyanin produced in aleurone and plant; may encode flavanone synthetase (Dooner, 1980a)
	c2	Colorless aleurone, reduced plant color; recessive to *C*
5–15	*A2*	Anthocyanin in aleurone and plant; may encode an enzyme involved in pigment synthesis (Kirby and Styles, 1970)
	a2	Colorless aleurone, green or brown plant; recessive to *A2*
5–46	*Pr*	Purple aleurone due to presence of cyanidin glucoside; may encode an enzyme involved in hydroxylation of pigment molecules (Harborne and Gavazzi, 1969)
	pr (kernel 2)	Red aleurone due to presence of pelargonidin glucoside; recessive to *Pr*
9–26	*C*	Anthocyanin produced in aleurone; may be compound (McClintock, 1948)
	C–I	Colorless aleurone; dominant to *C*
	c	Colorless aleurone; recessive to *C*
9–29	*Sh*	Normal starch synthesis in endosperm; encodes sucrose synthetase (Chourey and Nelson, 1976)
	sh (kernel 4)	Shrunken kernels due to reduced starch content, recessive to *Sh*
9–31	*Bz*	Anthocyanin produced in aleurone and plant; encodes UDPglucose: flavonoid 3–*0*–glucosyltransferase (UFGT) (Larson and Coe, 1977; Dooner and Nelson, 1977a)
	bz (kernel 3)	Aleurone brownish or bronze due to absence of UFGT activity; recessive to *Bz*
9–59	*Wx* (kernel 5)	Endosperm contains amylose; encodes starch–particle bound glucosyltransferase (Nelson and Rines, 1962); endosperm stains blue–black with I_2–KI solution
	wx (kernel 6)	Endosperm waxy, contains only amylopectin due to absence of glucosyltransferase; recessive to *Wx;* endosperm stains reddish–brown (fades to yellow or colorless with heat) with I_2-KI solution

[a] An additional locus that affects aleurone pigment expression is the *Bz2* locus. Purple aleurone (kernel 1) is observed in plants that are genotypically *A*, *A2*, *C*, *C2*, *Bz*, *Bz2*, *Pr*, and *R*. The *P–RR* allele must be present for pericarp, but not for aleurone pigmentation.

[b] The chromosome number is given, followed by the map position, expressed in recombination units relative to one end of the chromosome (Coe and Neuffer, 1977).

the pericarp. The pericarp is derived from the ovary wall (Fig. 3c) and therefore has the genetic constitution of the maternal plant. The ovary wall develops into a thin, extensively lignified protective layer (Randolph, 1936). The pattern of late cell division in the pericarp is strikingly different from that in the aleurone layer and results in the formation of wedge-shaped sectors that terminate at the silk-attachment point on the crown of the kernel. Red phlobaphene pigments are synthesized in the pericarp tissue in strains having the appropriate genetic constitution (Styles and Ceska, 1977). The *P* locus is involved in pigment expression in the pericarp and cob, and strains homozygous for certain recessive alleles at this locus have an unpigmented pericarp layer (Anderson, 1924). The pericarp layer has been mechanically removed from the illustrative kernels shown in Plates 1 and 2 to permit clear visualization of the aleurone layer.

Because the aleurone constitutes the peripheral layer of the endosperm, clonal lineages in the endosperm and aleurone are directly related and their genetic constitution is the same. The endosperm and aleurone layers are nevertheless biochemically distinct, and mutations are known that affect loci expressed in one or the other, but not both tissues. As a consequence, it has been possible not only to study the simultaneous effects of genetic changes on expression of several genes but to infer the existence of certain kinds of interactions between genetic elements. As described previously, the pericarp tissue has a different origin from endosperm tissue, consisting of diploid maternal tissue. However, it should be noted that the lineages giving rise to the gametophytic and ovary wall tissue become distinct rather late in the development of the ear. As a consequence, genetic changes that occur even at quite late premeiotic divisions can be reflected phenotypically in the pericarp layer and represented genotypically in the embryo.

II. Overview of Maize Controlling Elements

Controlling elements are transposable elements capable of causing unstable insertion mutations as well as a variety of chromosomal rearrangements, including deletions, duplications, inversions, and translocations (McClintock, 1951a, 1956b, 1965a). Insertion of a controlling element can give rise to an unstable allele of a locus. Alleles of this type were originally designated *mutable* loci or alleles of a locus. This terminology has persisted, and unstable alleles are designated by an *m* that follows the allele designation (e.g., *bz-m*). Controlling element alleles of a locus can show both somatic and germinal instability. Somatic reversion of an unstable recessive allele can occur in many cell lineages during development, yielding multiple clones of phenotypically dominant cells in the organism. The somatic variegation patterns generated by reversion of unstable alleles of this type are illustrated using the *bz-m1*[2] allele of the *Bz* locus (kernel 7)

[2]The first number following the mutable designation is the number of the isolate in order of discovery. Additional numbers in parentheses refer to further derivatives of the same mutant allele. The mutable alleles discussed here are listed and briefly described in Table III.

and the *wx-m7* allele of the *Wx* locus (kernel 8). The deeply pigmented sectors apparent in the aleurone layer of kernel 7 constitute clones of cells, each of which was derived from a single cell in which the unstable *bz-m1* allele reverted to the dominant *Bz* allele (the colorless areas have a different origin and will be considered later). Kernel 8 has been cut in half, and the endosperm tissue has been stained with iodine. The darkly staining sectors of cells constitute clones derived from single cells in which the *wx-m7* allele has reverted to the dominant *Wx* allele.

Unstable controlling element insertion mutations each give a characteristic pattern of gene expression in the organism. It will become evident, as the discussion unfolds, that the pattern is determined by the properties of the transposable element and its interaction with the gene. McClintock's designation of these elements as controlling elements followed from her realization that these elements were distinct from the genes themselves, could alter the expression of genes reversibly, and acted in the nucleus to modulate gene expression (McClintock, 1956a,b). McClintock showed that the properties of the controlling elements could be independent of their chromosomal location, and that an element could have a similar effect on the expression of genes involved in different biosynthetic pathways and located in different places in the genome (McClintock, 1956a). Moreover, it became apparent that a single element could exert control over the expression of more than one gene simultaneously.

Among the mutant strains in which a controlling element has become associated with a known locus, two basic categories of controlling-element alleles have been distinguished. These were designated *autonomous* and *nonautonomous* (McClintock, 1950a). An autonomous allele is inherently unstable because the element at the locus is capable of autonomous excision and transposition. I will therefore refer to such an element as an autonomous element. The *wx-m7* allele shown in kernel 8 represents a mutation caused by insertion of an autonomous element at the *Wx* locus. Alleles of the second type, termed nonautonomous, are unstable only when an autonomous element is also present in the genome. The element causing this type of insertion mutation does not appear to be capable of autonomous excision and transposition and will therefore be referred to as a nonautonomous element. An autonomous element provides a nonautonomous

---

Plate 1 Phenotypic expression in the aleurone and endosperm of several loci and their stable and unstable mutant alleles. The relevant genotype and controlling element constitution are given for each kernel. The phenotypes are explained in the text and in Tables I and III. **1.** *R, Pr, A, A2, C, C2,* and *Bz* (expression of the anthocyanin pigment also require the *Bz2* allele, not discussed here. See the footnote to Table I). **2.** *pr*. **3.** *bz*. **4.** *sh* (the kernel has been cut longitudinally). **5.** *Wx* (the kernel has been cut and stained with iodine); the aleurone is colorless due to the presence of the *C–I* allele. **6.** *wx* (the kernel has been cut and stained with iodine); the aleurone is colorless due to the presence of the *C–I* allele. **7.** *C bz–ml/c bz (Pr), Ac* present. **8.** *wx–m7/wx, Ac* at the *Wx* locus. **9.** *C–I Bz Wx Ds/C bz wx (pr), Ac* present. **10.** *C–I Ds Bz/C bz (Pr), Ac* present. **11.** *C–I Ds Bz Wx/C bz wx (pr), Ac* present (the kernel has been cut and stained with iodine). **12.** *sh–m(5933)/sh, Ac* present (the kernel has been cut longitudinally). **13.** *C bz–m4/C bz (Pr), Ac* present. **14.** *C bz–ml/C bz (Pr),* two *Ac* present. **15.** *a–ml (5719A1)/a (pr),* no *Spm* present.

1
pr
2
bz
3
sh
4
Wx
5
wx
6
Bz
bz
c
7
Wx
wx
8
wx
Cbz
Wx
C-IBz
9
bz
Bz
bz
Bz
10
C-IBz
Cbz
Wx
wx
CBz
11
Sh
12
Bz
c
bz
13
2 Ac
14
-Spm
15

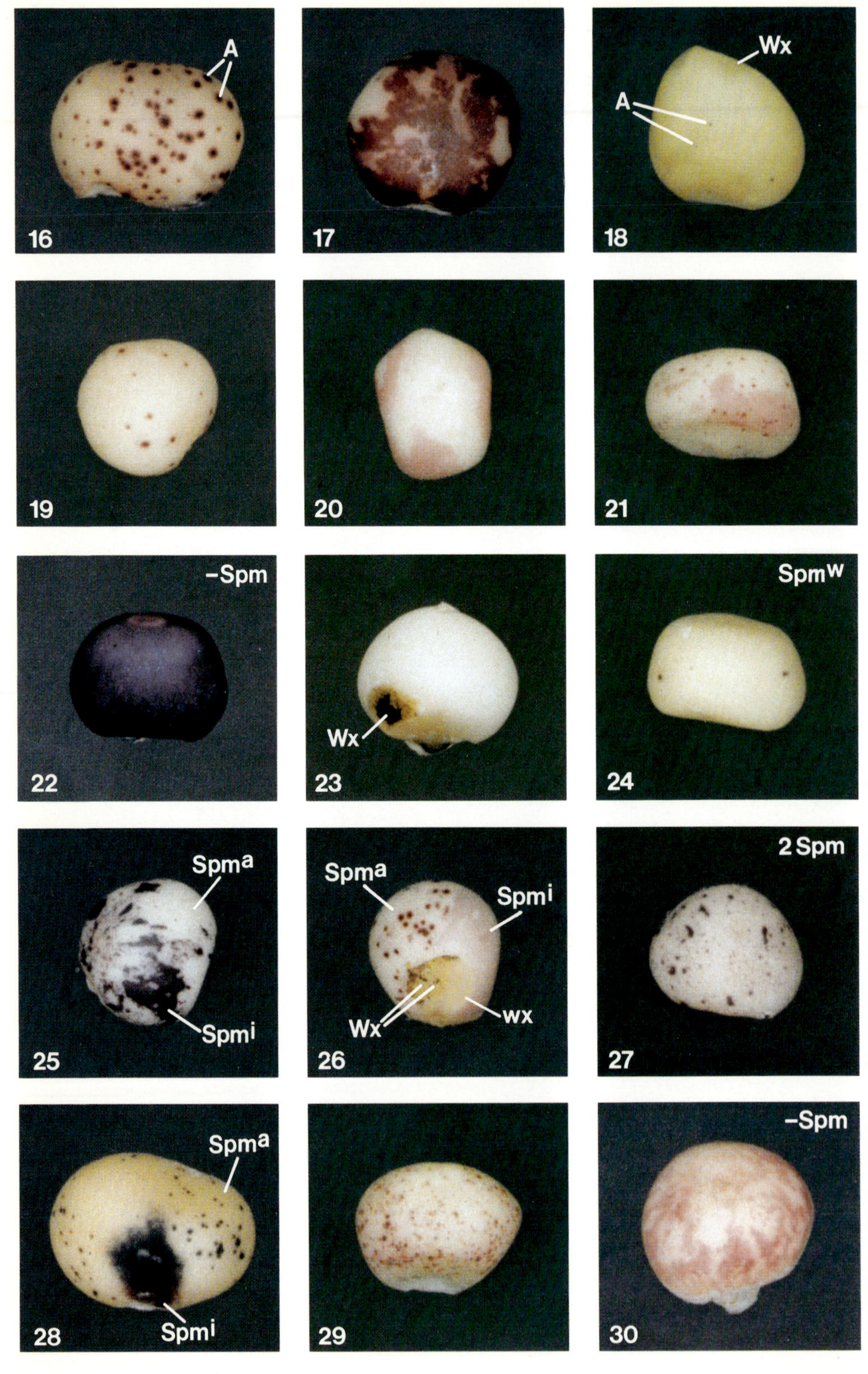
A
16
17
Wx
A
18
19
20
21
-Spm
22
Wx
23
Spm^w
24
Spm^a
Spm^i
25
Spm^a
Spm^i
Wx
wx
26
2 Spm
27
Spm^a
Spm^i
28
29
-Spm
30

element with a function or functions necessary to activate excision and transposition, as well as the other types of genetic rearrangements associated with controlling elements. A recessive mutation caused by insertion of a nonautonomous element at a locus is stable in the absence of the cognate autonomous element. Kernel 7 is from a strain of this type (*bz-m1*). In the absence of the appropriate autonomous element, the kernel aleurone layer shows the uniform pale brown pigmentation characteristic of a stable *bz* mutant strain (kernel 3). Somatic instability is observed only when a second, autonomous controlling element is present together with the *bz-m1* allele. Kernel 7 shows the deeply pigmented sectors that result from somatic mutation of the *bz-m1* allele to the *Bz* allele activated by the autonomous element. The ability of an autonomous element to trans-activate transposition and other genetic changes involving a nonautonomous element is diagrammatically represented in Fig. 4. Because nonautonomous elements can be activated to transpose in the presence of an autonomous element, they may represent elements with genetic defects affecting expression of the trans-acting function or functions required for transposition. The possibility that nonautonomous elements are derived from autonomous elements is represented in Fig. 4 by an arrow that is interrupted by a question mark to indicate that there is, as yet, no direct evidence for the implied sequence relationship.

The existence of nonautonomous elements that can be activated in trans by autonomous elements has provided a basis for the classification of elements. A nonautonomous element can be activated by certain autonomous elements, but not by all (Peterson, 1980). Elements that interact with each other by this criterion are defined here as belonging to the same *family* of elements. There are three relatively well-characterized families of controlling elements in maize and several more that have received less attention (Peterson, 1980). Both autonomous and nonautonomous elements have been identified within each of the well-characterized families. Elements of all families cause unstable insertion mutations, but elements belonging to different families differ from each other in their mode of action.

The three controlling element families that have received the most extensive

---

Plate 2 Phenotypic expression in the aleurone and endosperm of several unstable alleles of the *A*, *A2*, and *Wx* loci. The relevant genotype and controlling-element constitution are given for each kernel. The phenotypes are explained in the text and in Table III. **16.** *a–m1 (5719A1)/a (pr), Spm* present. **17.** *a–m1 (5996)/a (pr), Spm* present. **18.** *a–m1 (5996–340)/a, wx–m8 (pr), Spm* present (note the translucent sectors of *Wx* endosperm). **19.** *a–m1 (5718)/a (pr), Spm* present. **20.** *a–m1 (5720)/a (pr), Spm* present. **21.** *a–m1 (5720)/a–m1 (5718) (pr), Spm* present. **22.** *a2–m1(A2)/a2 (Pr)*, no *Spm* present. **23.** *a2–m1(A2)/a2, wx–m8/wx, Spm* present (the kernel has been cut and stained with iodine). **24.** *a–m1 (5719A1)/a (Pr), Spm–w* present. **25.** *a2–m1(A2)/a2 (Pr), Spm–c* present; the areas in which the *Spm* is active (*Spm*[a]) and inactive (*Spm*[i]) are indicated. **26.** *a–m1 (5719A1)/a, wx–m8/wx (pr), Spm–c* present; the areas in which the *Spm* is active (*Spm*[a]) and inactive (*Spm*[i]) are indicated. **27.** *a2–m1(A2)/a2 (Pr)*, two *Spm–c* present. **28.** *a–m1(5719A1)/a (Pr), Spm* present; the areas in which the *Spm* is active (*Spm*[a]) and inactive (*Spm*[i]) are indicated. **29.** *a–m2(7977B)/a (pr), Spm* present. **30.** *a–m2(7977B)/a (pr)*, no *Spm* present; preset pattern.

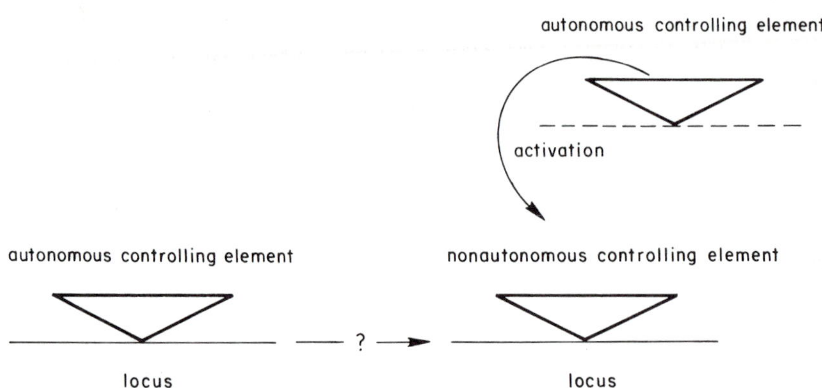

Fig. 4 A diagrammatic representation of the trans-activation of a nonautonomous element by an autonomous element. The possibility that nonautonomous elements are derivatives of autonomous elements is represented by an arrow interrupted by a question mark.

analysis are the *Ac–Ds, Spm,* and *Dt* families. The family designations used here are those given by McClintock for the *Ac–Ds* and *Spm* families (McClintock, 1947, 1954). By the activation test defined above, the independently isolated *Mp* element was shown to belong to the *Ac–Ds* family (Barclay and Brink, 1954), and independently isolated *En–I* elements were shown to belong to the *Spm* family (Peterson, 1960, 1961, 1965). The *Dt* family was identified by Rhoades (1936, 1938, 1941, 1945) and has also been studied by Neuffer (Nuffer, 1955, 1961; Neuffer, 1965), and Doerschug (1968a,b,c, 1973). Several other apparently unrelated controlling element families have been identified as responsible for unstable mutations at the *R* locus, as well as at a locus that affects storage protein synthesis in the endosperm (Kermicle, 1970, 1973; Gonella and Peterson, 1977, 1978; Friedemann and Peterson, 1980; Salamini, 1980). Certain element families, such as *Bg* (*Sd*) and *Ub,* have been shown to differ from each other and from the more thoroughly studied families (Salamini, 1980; Gonella and Peterson, 1977, 1978; Friedemann and Peterson, 1980; Peterson, 1980). The I^R factor associated with the unstable *R-st* allele of the *R* locus has been reported to be transposable, as is a modifier of *R-st* instability (Ashman, 1959, 1960; Kermicle, 1973). The paramutational properties of the fascinating *R-st* allele have received more attention than has the associated transposable element, which is not well understood and whose relationship to other elements is not known (Brink, 1973; Kermicle, 1970, 1973). There are, in addition, numerous reports in the literature of unstable alleles at a variety of loci. Whether these are caused by elements representing additional controlling element families or independent isolates of elements belonging to families that have already been identified is not known in most cases.

In the following sections, I will describe two controlling element families in detail, the *Ac–Ds* (*Mp*) and *Spm* (*En–I*) families. This choice was governed by

Table II Controlling Element Families

Family	Autonomous elements	Nonautonomous elements
Ac–Ds	*Activator (Ac)*	*Dissociation (Ds)*
	Modulator (Mp)	
Spm	*Suppressor–mutator (Spm)*	Unnamed
	Enhancer (En)	*Inhibitor (I)*
Dt	*Dotted (Dt)*	Unnamed

the amount of information available concerning their mode of action and by the existence of substantial differences between them. The element designations are summarized in Table II. I have not devoted a separate section to a description of the *Dt* family because it is less well studied than the other families and strongly resembles them in its mode of action. I have, however, listed it in Table II and have included descriptions of important differences between *Dt* and the other controlling-element families in the text. Although the element designations used are predominantly those given by McClintock, I have endeavored to use the relevant alternative designations when discussing the work of other authors. I have done this because a given pair of elements, such as *Ac* and *Mp*, that are indistinguishable by the trans-activation test need not be identical in all respects. Indeed, it will become evident that controlling elements are quite unstable genetically. Altered forms of controlling elements arise frequently and have, in some cases, received separate designations. Heritable changes that affect a nonautonomous element pose a special problem, because it is not always possible to determine whether a change in the sequence of the element or in its location is responsible for the observed alteration. Nonetheless, a given element may comprise a group of subtly different elements, only some of which have received separate names.

McClintock's emphasis on the ability of the transposable elements to change or regulate the pattern of gene expression in development led her to give the designation of controlling element to both autonomous and nonautonomous elements. This term appears to have been accepted widely and is used in the present discussion. For a time, McClintock made a further distinction between them, calling the activating autonomous element the *regulator* and the nonautonomous element the *operator* (McClintock, 1961a). Several authors have used *regulator* or *regulatory element* to designate an autonomous element and *receptor* to designate a nonautonomous element (Fincham and Sastry, 1974; Peterson, 1965). Because I believe that the molecular analyses currently in progress in several laboratories may soon provide a basis for a better understanding of the relationship between these elements and their effect on gene expression, I have used the somewhat cumbersome but more general and neutral terms autonomous and nonautonomous element in the present discussion.

III. The Ac–Ds Controlling Element Family

The first transposable element family investigated in detail by McClintock was *Activator–Dissociation* (*Ac–Ds*). Unstable mutations caused by an autonomous element belonging to the same family were described by Emerson (1914, 1917, 1929). The element was later named *Modulator* (*Mp*) by Brink and Nilan (1952) and was shown to resemble the *Ac* element (Barclay and Brink, 1954). *Dissociation* (*Ds*) is the generic term used to designate nonautonomous elements belonging to this family. The name *Dissociation* derives from the early observation that this element can provide a specific site of chromosome breakage or dissociation into two fragments (McClintock, 1946, 1947). Chromosome breakage at *Ds* requires the presence of an autonomous element belonging to the same family (McClintock, 1947; Barclay and Brink, 1954). *Activator* (*Ac*) is an autonomous element of this family, and its name derives from its ability to activate chromosome breakage at *Ds*. *Ds* can also transpose in the presence, but not in the absence, of *Ac* and can insert into or become associated with a locus in such a way that it affects expression of the locus (McClintock, 1948, 1949). A mutation caused by insertion of *Ds* is unstable only in the presence of *Ac*. The *Ac*-activated somatic instability of one such mutant allele (*Bz-m1*) has already been illustrated (kernel 7) (McClintock, 1951b). The autonomous element *Ac* can itself insert at a locus to produce an unstable mutant allele. The phenotypic expression in endosperm tissue of one such unstable *wx* allele (*wx-m7*) is illustrated by kernel 8 (McClintock, 1964). In the subsequent discussion, *Ds* will be addressed first, with the trans-acting *Ac* element treated as a constant. The properties of *Ac* will then be discussed, followed by a consideration of their structural relationship.

A. The Nonautonomous *Ds* Controlling Element

1. Chromosome Breakage at *Ds* and Its Consequences

Ds was first identified by its ability to act as a specific site of chromosome breakage in the presence of *Ac* (McClintock, 1945, 1946). This property was recognized in a strain in which *Ds* was located proximal to the *Wx* locus on the short arm of chromosome 9. There are several loci on the short arm of chromosome 9 that are expressed in kernels and for which recessive mutations are known. These include the *C*, *Sh*, *Bz*, and *Wx* loci (Table I). In strains of the appropriate genetic constitution, chromosome breakage at *Ds* is clearly manifested as a variegation pattern in the kernel. The kernel used to illustrate this phenomenon (kernel 9) is heterozygous at the *C*, *Bz*, and *Wx* loci, having the constitution *C-I Bz Wx*/*C bz wx*, and also carries *Ac*. Chromosome breakage at *Ds* gives rise to an acentric fragment carrying the markers distal to *Ds* (*C-I*, *Bz*,

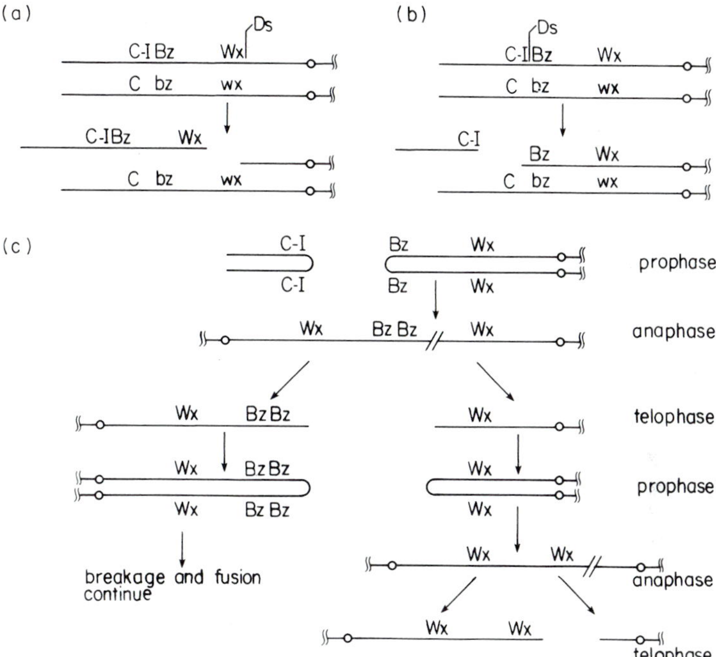

Fig. 5 A diagrammatic representation of chromosome breakage and acentric–dicentric formation at *Ds*. (a) Chromosome breakage at *Ds* proximal to the *Wx* locus. Loss of the acentric fragment carrying the *C-I Bz* and *Wx* alleles gives sectors of *C bz wx* tissue. (b) Chromosome breakage at *Ds* located between the *C* and *Bz* loci gives sectors of cells that are *C*, *Bz*, and *Wx* as a consequence of the loss of the *C-I* allele carried on the acentric fragment. (c) The chromatid-type of BFB cycle. Terminal fusion of sister chromatids following breakage at *Ds* yields a U-shaped acentric fragment and a dicentric chromosome that breaks at subsequent mitotic divisions. Fusion of newly broken ends after chromosome replication perpetuates the cycle of breakage. The BFB cycle is illustrated using the markers on chromosome 9 that are present in illustrative kernels 10 and 11. Sequential loss of the dominant alleles on the chromosome undergoing breakage during endosperm development results in the formation of sectors of endosperm tissue in which the recessive alleles borne on the unbroken homolog are expressed.

and *Wx*) and a broken chromosome terminating at *Ds* (McClintock, 1947, 1948, 1951a). The acentric fragment is lost at the subsequent mitotic division, giving rise to sectors in the kernel that have only those markers proximal to *Ds* and on the unbroken homolog. Chromosome breakage and the ensuing marker loss are diagrammed in Fig. 5a. Breaks occurred during the development of kernel 9, producing sectors that have the phenotype determined by the *C bz wx* homolog not carrying *Ds*. Parts of the kernel are colorless and *Wx*, the phenotype observed in kernels of a *C-I Bz Wx*/*C bz wx* heterozygote. A border of intense anthocyanin pigmentation appears between the genetically different sectors of the kernel. This is due to complementation between tissues that are genotypically *C-I Bz* and *C bz*. The *Bronze*-encoded glycosylating enzymatic activity is not present in *bz* tissue but is present in *C-I Bz/C bz* tissue, which is colorless because of an early block in anthocyanin biosynthesis (Dooner and Nelson, 1979). Thus, the pigment that is synthesized, but not glycosylated in *bz* tissue, is glycosylated by the *Bz*-encoded enzyme synthesized in the *C-I Bz/C bz* tissue as a result of diffusion at the border between the sectors. The small pigmented spots appearing in predominantly *C-I Bz Wx* areas of the kernel represent smaller sectors of *C bz wx* tissue resulting from chromosome breakage and loss later in the development of the kernel. Because the substance diffusing from the *C-I Bz/C bz* tissue penetrates several cell layers, very small *C bz* sectors often become completely pigmented. A section of the kernel that crosses the border between the *C-I Bz* and *C bz* sectors has been cut away and stained with iodine. It can be seen that the border between the *Wx* and *wx* tissues coincides with the border between the *C-I Bz* and *C bz* sectors. Thus, all three markers are lost simultaneously as a consequence of chromosome breakage at *Ds*. Hence, the site of chromosome breakage is proximal to the *Wx* locus. In cytological studies on the sporogenous cells in anthers, it was established that the physical site of chromosome breakage correlated with its genetic position (McClintock, 1946, 1947). It was further established that there was a correlation between the presence of a terminally deficient chromosome 9 and a deeply staining chromatin body in the cytoplasm (McClintock, 1946). Thus, the somatic loss of markers distal to *Ds* on chromosome 9 is correlated with the physical loss of the distal end of the chromosome.

As studies on *Ds* progressed, it became clear that the marker loss just illustrated was consequent on the formation of sister chromatids attached to each other at *Ds* and that other types of chromosomal changes could occur as well (McClintock, 1947, 1948, 1949, 1950b, 1951a, 1956a). These include transpositions of *Ds* either unaccompanied by visible chromosomal abnormalities or accompanied by direct or inverted duplications or translocations having *Ds* as one endpoint, as well as deletions, chromosome losses, and the formation of ring chromosomes. Analysis of *Ds* behavior was facilitated by the isolation of mutants in which *Ds* had transposed to sites sufficiently close to chromosome 9 endosperm markers to affect their expression, producing unstable mutant phenotypes (McClintock, 1948, 1949, 1950b, 1951a,b). Relationships among the various *Ds*-associated chromosomal events were discovered in studies on such mutants and will be considered in subsequent sections.

2. Acentric and Dicentric Chromosome Formation at *Ds*

It was established in cytological and genetic studies that chromosome breakage at *Ds* in the presence of *Ac* is associated with the formation of a U-shaped acentric fragment consisting of sister chromatids distal to *Ds* and a U-shaped dicentric chromosome consisting of the sister chromatids proximal to *Ds* (Fig. 5c). The site of attachment of the chromatid fragments to each other coincides with the chromosomal position of *Ds* (McClintock, 1948). The loss of markers distal to *Ds* illustrated in the preceding section results from the loss of the acentric fragment at mitosis. The consequences of dicentric chromosome formation at *Ds* can be observed phenotypically in a strain of the appropriate genetic constitution in which *Ds* is located distal to one or more genetic markers. This is illustrated using a kernel that has the chromosome 9 constitution *C-I Ds*[3] *Bz Wx*/*C bz wx* (Fig. 5b). The consequences of chromosome breakage and acentric–dicentric formation at *Ds* are diagrammed in Fig. 5c. The fused sister chromatids form a bridge connecting the centromeres during mitotic disjunction. The bridge breaks along the length of the dicentric chromosome. When a break occurs between a centromere and the fusion site, one daughter cell receives a broken chromosome with a duplication of the genetic material between the break point and the fusion site. The other daughter cell receives a chromosome deleted for the corresponding chromosomal segment, and mitotic descendants of this cell can express only the alleles represented on the corresponding segment of the unbroken homolog. The processes of chromosome fusion, bridge formation, and breakage continue in both daughter cells receiving broken chromosomes. McClintock (1942a) called this the *chromatid* type of breakage–fusion–bridge (BFB) cycle, and it is diagrammatically represented in Fig. 5c for the loci on chromosome 9 used in illustrative kernels 10 and 11. Breakage of the dicentric chromatid between the *Bz* and *Wx* loci results in the segregation to the daughter cells of broken chromatids, one of which has a *Bz* duplication and the other of which is missing the *Bz* locus. The descendants of these two genotypically different daughter cells give rise to adjacent twin sectors in the mature kernel that are deeply pigmented (*Bz Bz Wx*/*C bz wx*) and bronze (*Wx*/*C bz wx*), respectively. Such twin sectors are evident in kernel 10.[4] The BFB cycle continues, generating *bz* and *wx* subsectors in the *Bz Wx* sector and *wx* subsectors in the *bz Wx* tissue. A portion of kernel 11 has been cut away and stained with iodine to reveal the *Wx* phenotype.

The chromatid type of BFB cycle initiated by acentric–dicentric formation at *Ds* ceases in the embryo but can persist in the endosperm (McClintock, 1946, 1947, 1948, 1951a). The cessation of the BFB cycle involving sister chromatids had been noted earlier by McClintock (1941), although occasional exceptions showing persistence of the BFB cycle in the developing embryo were also observed (McClintock, 1944, 1951a). Using chromosome-9 markers affecting both plant and kernel morphology, McClintock (1946, 1947, 1948, 1949) established that *Ac* activates acentric–dicentric formation at *Ds* relatively late in plant development but at virtually any stage in endosperm development.

[3]*Ds* is located the *Sh* locus in this strain. Its effect on the *Sh* locus will be considered subsequently and is not apparent in the illustrative kernels (10 and 11) due to the presence of the *Sh* allele on the homolog.

[4]Differences between the red and purple pigments in the kernels used as illustrations are due to differences at the *Pr* locus on chromosome 5 (*Pr*-purple, *pr*-red). These differences are not important in the present context. Those kernels that have been cut and stained with iodine are generally homozygous for the *pr* allele because the red pigment provides the best contrast between the blue-black color of the stained *Wx* sectors and the colored sectors appearing in the aleurone layer.

3. TRANSPOSITION OF *Ds*

It is evident from the foregoing illustrations that when *Ds* is proximal to the *Wx* locus (kernel 9), the pattern of *Ac*-activated kernel variegation is strikingly different than when *Ds* is between the *C* and *Bz* loci (kernels 10 and 11). One method used for detecting *Ds* transposition events is therefore the selection of individual kernels showing an altered pattern of marker loss in a multiply marked strain of appropriate genetic constitution (McClintock, 1950b). Plants grown from many such kernels were examined cytologically and tested genetically to determine the new location of *Ds* (McClintock, 1950b). Other *Ds* transposition events were initially detected as a consequence of newly arising spontaneous, unstable mutations at loci affecting plant or kernel phenotypes (McClintock, 1948, 1949, 1951a,b, 1956a). The origin of the *c-m1* allele of the *C* locus was, for example, shown to be associated with the disappearance of *Ds*-associated acentric–dicentric formation at the original site proximal to the *Wx* locus and its appearance at a site coinciding with the chromosomal location of the *C* locus (McClintock, 1948, 1949). In many strains, *Ds* transposition was not accompanied by a visible chromosomal alteration. Because *Ds* is stable in the absence of *Ac*, its chromosomal position can be determined by recombinational analysis. A rough map of the several sites on the short arm of chromosome 9 at which *Ds* has been detected is given in Fig. 6 (B. McClintock, personal communication). In addition to sites depicted in Fig. 6, *Ds* has been identified as causing mutations at the *C*, *Sh*, *Bz*, and *Wx* loci on chromosome 9, as well as loci on other chromosomes (see Table III for examples).

Transposition of the *Ds* element can also be accompanied by the production of chromosomal rearrangements (McClintock, 1948, 1949, 1950b). The rearrangements that have occurred in two particularly well-studied strains are diagrammed in Fig. 7. *Ds* transposed from its original position proximal to the *Wx* locus to a new site near the *C* locus in both strains. In both strains, transposition of *Ds* was associated with removal of *Ds* from its original position and the origin of a duplication extending from the original (donor) site to the new (recipient) site of *Ds* insertion. In one strain (Fig. 7b), the portion of the chromosome between the donor and recipient sites was duplicated in direct order, whereas in the other strain, the duplicated chromosome segment was inverted (Fig. 7a). Because the duplication endpoints precisely coincide with the donor and recipient sites, it appears likely that the duplication accompanied the transposition event, occurred during chromosome replication, and resulted in the concomitant deletion of the corresponding chromosome region from the sister chromatid. Moreover, the presence of two *Ds* elements, both at the recipient site in both duplicated chromosomes, suggests a mode of transposition in which the element replicates while maintaining a physical association with both donor and recipient sites. Although the transposition of *Ds* appears to be associated with chromosome replication, it is not replicative in the same sense as transposition of a bacterial transposon (Chapter 6). In both the rearranged chromosomes illustrated in Fig. 7 and in

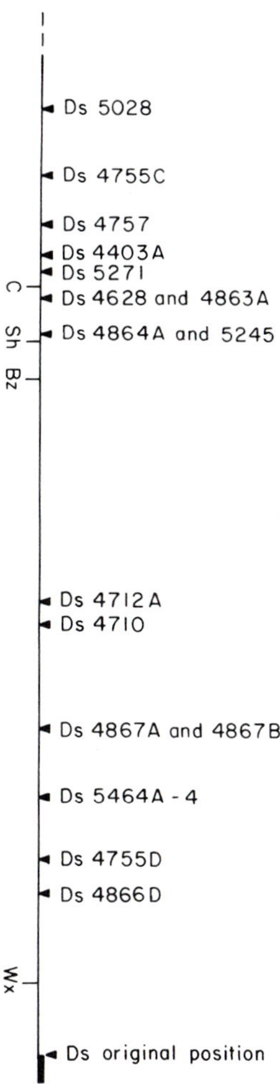

Fig. 6 A diagrammatic representation of the approximate position occupied by several *Ds* elements on the short arm of chromosome 9. Each element has been designated by the number of the culture in which it was isolated. The map positions are based on recombination analysis and are approximate. The original position of *Ds* was determined cytologically and coincides with the junction between the euchromatic and heterochromatic segments of the short arm of chromosome 9 in the pachytene chromosome complement. *Ds* elements have also been mapped at the *C*, *Sh*, *Bz*, and *Wx* loci (Table III). (McClintock, unpublished observations)

other chromosomes with transposed *Ds* elements, *Ds* is no longer detectable at the donor site (McClintock, 1949, 1950b, 1951a).

Although chromosome 9 is morphologically normal in most of the strains that have been analyzed in which *Ds* has undergone an intrachromosomal transposition, the infrequent detection of transposition-associated rearrangements may

Table III Summary of the Properties of the Unstable Alleles Discussed in the Text

Controlling element family	Element at locus	Allele	Kernel phenotype
Ac–Ds	*Ds*	*a–m3*	Pale aleurone, with *A* sectors when *Ac* present (McClintock, 1951b)
		bz–m1	*bz* aleurone, with *Bz* sectors when *Ac* present (kernels 7, 14) (McClintock, 1951b)
		Bz–wm	Pale *Bz* aleurone, with *Bz* and *bz* sectors when *Ac* present (McClintock, 1956b)
		c–m1	*c* aleurone, with *C* sectors when *Ac* present (McClintock, 1948)
		c–m2	*c* or intermediate aleurone pigmentation with sectors ranging from very pale to very deeply pigmented when *Ac* present (McClintock, 1948)
		wx–m1	*wx* endosperm or intermediate, with sectors showing lower and higher *Wx* expression when *Ac* present (McClintock, 1948)
	Nontransposing *Ds*	*bz–m4*	*bz* aleurone, with *Bz* sectors when *Ac* present (kernel 13) (McClintock, 1953)
		sh–m	*sh* endosperm, with *Sh* sectors when *Ac* present (kernel 12) (McClintock, 1952)
	Ac	*bz–m2*	*bz* aleurone, with *Bz* sectors (McClintock, 1951b)
		wx–m7	*wx* endosperm, with *Wx* sectors (kernel 8) (McClintock, 1964)
	Mp	*P–RR–Mp*	*p* pericarp, with *P* sectors (Brink and Nilan, 1952)

(*Continued*)

Table III (*Continued*)

Controlling element family	Element at locus	Allele	Kernel phenotype
Spm	Nonautonomous element activated by *Spm*	*a–m1*	*a* or pale aleurone in absence of *Spm;* colorless with *A* sectors when *Spm* present; (kernels 15–21, 24, 26, 28) (McClintock, 1951b)
		a2–m1(A2)	*A2* aleurone; colorless when *Spm* present (kernels 22–23) (McClintock, 1957)
		wx–m8	*wx* endosperm, with *Wx* sectors when *Spm* present (kernels 23, 26) (McClintock, 1961b)
	Spm	*a–m2*	*a* aleurone when *Spm* inactive; pale with *A* sectors when *Spm* active (kernel 29) (McClintock, 1951b)
		a–m5	*a* aleurone, with pale and dark *A* sectors (McClintock, 1961b)

reflect the poor transmission of rearranged chromosomes rather than the actual frequency with which transposition of *Ds* is associated with chromosomal rearrangement (McClintock, 1950b). McClintock (1950b, 1956a) noted that many kernels that show a variegation pattern indicative of a *Ds* transposition fail to germinate, suggesting dominant lethality of some insertions or gross chromosomal imbalances. Despite the fact that the methods by which *Ds* transpositions were detected preclude a random sampling of such events, it is evident that transpositions can occur with or without major chromosomal rearrangements.

4. SOMATIC AND GERMINAL REVERSION OF *DS* INSERTION MUTATIONS: THE *BZ-M1* AND *C-M1* ALLELES

The *bz-m1* allele of the *Bz* locus and the *c-m1* allele of the *C* locus carry unstable recessive mutations caused by insertion of *Ds* (McClintock, 1948, 1949, 1951b). Both alleles revert somatically in the presence, but not in the absence, of *Ac*. Somatic reversion of the allele during endosperm development results in the development of clones of deeply pigmented cells in the aleurone

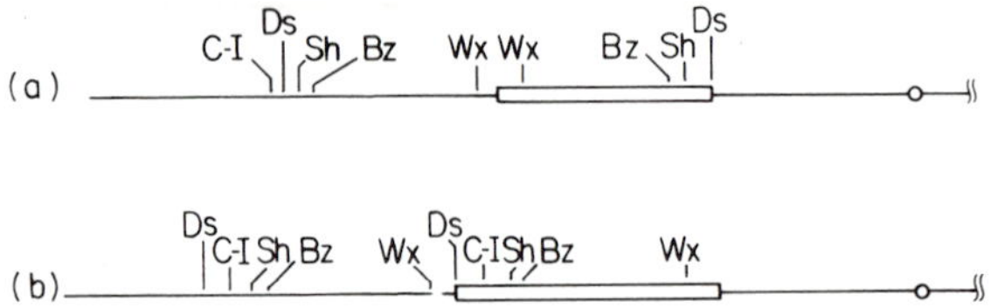

Fig. 7 A diagrammatic representation of two chromosomes 9 with internal duplications and transposed *Ds* elements isolated and analyzed by McClintock (1950b). The open bars represent the duplicated segment of the short arm of chromosome 9. The original position of *Ds* in the strain from which these chromosomes were derived was proximal to the *Wx* locus and coincides with one of the duplication endpoints in both strains.

layer (kernel 7, *bz-m1*). Chromosome breakage and acentric–dicentric formation also occur at the site of *Ds* in these strains. The colorless sectors in kernel 7, which has the chromosome 9 constitution *C bz-m1/c bz,* arise as a consequence of chromosome breakage and acentric–dicentric formation at *Ds*. The acentric fragment carrying the *C* locus is lost from some cell lineages, giving rise to colorless (*c bz*) sectors in the kernel aleurone. Germinal reversion of these alleles is mediated by the excision of *Ds* from the locus. This conclusion is based on the observation that *C* and *Bz* strains derived from *c-m1* and *bz-m1* strains by spontaneous mutation no longer show chromosome breakage at the respective loci and are subsequently stable in the presence of *Ac* (McClintock, 1949, 1951b). That somatic reversion is also mediated by excision of the element from the locus is suggested by the observation that the phenotypic consequences of *Ds*-mediated acentric–dicentric formation are not observed within somatic revertant sectors. Thus, for example, the colorless sectors arising from loss of the *C* allele following acentric–dicentric formation are observed within the *bz* areas in kernel 7 but not in the *Bz* areas. Although the transposition of *Ds* to a new site concomitant with its excision has been detected in the study of germinal revertants derived from *c-m1* (McClintock, 1949), the frequency with which excision and transposition are linked cannot readily be determined because transpositions of *Ds* to unmarked chromosomes cannot be detected.

5. THE RELATIONSHIP BETWEEN ACENTRIC–DICENTRIC FORMATION AND TRANSPOSITION (OR EXCISION) OF *DS*

In studies on the *c-m1* allele, McClintock (1949) noted that genetic changes producing alterations in the frequency of somatic reversion to *C* resulted in an inverse change in the frequency of acentric–dicentric formation. The original *c-m1* allele showed a low frequency of reversion and a high frequency of acentric–dicentric formation (McClintock, 1949). It was relatively unstable and produced derivatives that ranged from ones that showed a high reversion frequency and no detectable chromosome breaks to ones that did not revert or reverted only occasionally and showed a very high frequency of chromosome breakage (McClintock, 1951a). McClintock inferred that acentric–dicentric formation and removal of *Ds* were alternate consequences of the same breakage events (McClintock, 1949). Acentric–dicentric formation, as well as the origin of the inverted duplication depicted in Fig. 7a, may involve the ligation of the 3′- and 5′-ends of complementary DNA strands. A biochemical precedent for this type of reaction is provided by the observation that the A* protein of φX174 generates hairpin structures from the circular replicative form I of the phage DNA under appropriate conditions (Van der Ende *et al.*, 1981). It may be that the formation of acentric and dicentric chromatids at the site of *Ds* is a consequence of trans-esterification reactions carried out by element-associated or element-induced recombination enzymes involved in *Ds* transposition.

Genetic changes that alter the relative frequencies of *Ds* excision and acentric–dicentric formation occur only in the presence of *Ac* (McClintock, 1949, 1951a). They appear to affect the *Ds* element, because the behavior of a given derivative is the same with different *Ac* elements. McClintock noted the occurrence of such changes early in the study of *Ds*, designating them changes in the *state* of *Ds* (McClintock, 1947). Derivatives that show little or no acentric–dicentric formation are more stable genetically than those with a high frequency of acentric–dicentric formation (McClintock, 1949, 1950b). The underlying genetic changes can occur at mitotic divisions, producing somatic sectors that show an altered frequency of chromosome breakage (McClintock, 1947, 1949). Although a derivative showing a high frequency of acentric–dicentric formation can mutate in a single step to one showing a low frequency, the converse is not true (McClintock, 1949, 1950b). Repetitive selection through three or four generations of individual kernels showing breakage frequencies slightly higher than that in the parental strain was required to reisolate a derivative showing a high frequency of breakage from one that had mutated to a low frequency (McClintock, 1950b). These observations were interpreted as suggesting the existence of *Ds*-associated repetitive elements (McClintock, 1949).

6. Other Types of *Ds* Mutations

There are strains with *Ds* mutations that show a more complex pattern of phenotypic change in the presence of *Ac* than the simple reversion described for *c-m1* and *bz-m1*. The *c-m2* and *wx-m1* alleles produce somatic sectors showing phenotypes intermediate between the recessive and dominant *C* and *Wx* alleles, equivalent to the dominant allele or more intense than expected for a single copy of the dominant allele (McClintock, 1948, 1949, 1951a,b). Derivatives have been obtained from both that show a phenotype intermediate between the recessive and dominant alleles in the absence of *Ac*. In the presence of *Ac*, the intermediate alleles are unstable, mutating toward either the dominant or the recessive phenotype. The expression of the *c-m2* allele appears to be further complicated by a complementation reaction that intensifies pigment expression at the interface between two sectors of intermediate phenotype, suggesting the possibility of more than a single coding sequence at the locus. The behavior of these alleles is in some respects similar to that of certain transposable element insertion mutations in yeast, which result in the under- or overexpression of the gene adjacent to the inserted transposable element (Fink *et al.*, 1980; Errede *et al.*, 1980). The germinal mutations at *c-m2* and *wx-m1* appear to result in an altered level of gene expression without loss of *Ds* from the locus. An additional observation is that mutations at *wx-m1* are, in some cases, accompanied by a dominant or semidominant mutation that affects the growth of endosperm tissue, particularly if the mutations are toward the recessive *wx* phenotype (McClintock, 1951a). Whatever its nature, the mechanism by which these alleles mutate does not appear to involve simple excision of the *Ds* element or even a major portion of it, because the derivatives continue to be unstable.

7. NONTRANSPOSING *Ds* ELEMENTS

The *Ds* elements in two independently derived strains, *Ds4864A* and *Ds5245* (Fig. 6), cause mutations in the immediate vicinity of the insertion site, but rarely or never transpose to other sites (McClintock, 1952, 1953, 1954, 1955, 1956b). *Ds* is located just distal to the *Sh* locus in both strains. Recombination between *Ds* and *Sh* is less than 0.5% in both (McClintock, 1951a, 1952). Mutations that affect one or more markers on either side of the *Ds* element occur in these strains in the presence, but not in the absence, of *Ac* (McClintock, 1952). Among these are mutations that affect the *C* locus to the left of *Ds* and mutations that affect the *Sh* locus or both the *Sh* and *Bz* loci to the right of *Ds*, but none that extend to markers on both sides of the element simultaneously (McClintock, 1952, 1953). Of the 37 *sh* mutations examined in some detail, all were viable as homozygotes, showed normal transmission, and contained *Ds* at a position that was not detectably altered (McClintock, 1953). Nine of 37 *sh* mutations reverted to *Sh;* reversion was *Ac*-activated and occurred at a high frequency in two cases. Recombination frequencies between *sh* and nearby loci were reduced in several mutants and increased substantially in the *sh-m(5933)* mutant. The endosperm phenotype of the *sh-m(5933)* allele in the presence of *Ac* is illustrated in kernel 12. The kernel has been cut longitudinally. Sectors of *Sh* tissue appear as irregular areas of translucent starchy tissue in the crown of the kernel. In the absence of *Ac*, this strain shows the endosperm phenotype characteristic of a stable *sh* strain (kernel 4).

Mutations affecting both the *Sh* and *Bz* loci proximal to *Ds* have occurred in the *Ds4864A* and *Ds5245* strains, as have mutations that affect the *C* locus, but do not result in loss of the entire end of the chromosome distal to *Ds* (McClintock, 1953, 1954). These mutations, like the *sh* mutations described above, arise by a mechanism that does not involve the loss or transposition of *Ds*. Among the *sh bz* mutants derived from the *Ds4864A* and *Ds5245* strains, most proved to have stable mutations at both loci (McClintock, 1953). The doubly mutant strains differed from each other in the extent to which the mutations were transmissible and adversely affected the development of homozygous and heterozygous kernels (McClintock, 1953). One mutant, designated *bz-m4,* showed a stable *sh* phenotype and an unstable *bz* phenotype that reverted to *Bz* in the presence of *Ac* (McClintock, 1953, 1956c). The position of the *Ds* element was not detectably altered in any of the *sh bz* strains analyzed, as judged by the site of *Ds*-associated chromosome breakage (McClintock, 1953). The *C* locus mutations derived from the *Ds4864A* and *Ds5245* strains also behaved as stable recessive mutations (McClintock, 1953, 1954). The position of *Ds* did not appear to have changed, and the derivatives differed from each other in the transmissibility of the mutations as well as the viability of the homozygotes. None of the mutations caused visible alterations in chromosome morphology. Thus, the *Ds* elements in strains *Ds4864A* and *Ds5245* appear to promote unidirectional rearrangements that extend from the site of *Ds* insertion but do not involve either excision or transposition of *Ds*. Although the genetic behavior of several of the mutant strains is consistent with the interpretation that the mutations are due to deletions of various lengths, direct evidence of deletion exists only for the stable *sh* mutation in the *bz-m4* mutant (Burr and Burr, 1980, 1981; Döring *et al.*, 1981; Chaleff *et al.*, 1981). Because the *Ds* elements in the two independently derived strains *Ds4864A* and *Ds5245* are either at the same or nearly the same site, it is not clear whether the behavior of these elements is dictated by their chromosomal position or by their sequence.

Reversion of the several unstable *sh* alleles and the unstable *bz-m4* allele to or toward the dominant *Sh* and *Bz* alleles of these loci may occur by a rather different mechanism than the simple excision or transposition event that occurs most frequently in *c-m1* or *bz-m1* strains. The *bz-m4* and the unstable *sh-m* alleles revert both somatically and germinally in the presence of the *Ac* element (McClintock, 1952, 1953, 1956c). Neither germinal nor somatic reversion results in the loss or transposition of *Ds* (McClintock, 1953, 1956c). In the presence of *Ac,* the revertants show *Ds*-mediated breakage of chromosome 9 at the *Bz* locus and give rise to new unstable mutants. That *Ds* is also retained at the locus in somatic revertant sectors in the *bz-m4* mutant is evident in kernel 13, which has the constitution *C bz-m4/c bz* (*Ac* present). Colorless sectors resulting from loss of the *C* locus are evident within both the *bz* and *Bz* areas of the aleurone. The behavior of the *Ds* element in the *bz-m4* strains contrast markedly, in this respect, to its behavior in the *bz-m1* (kernel 7) and *c-m1* strains described earlier. Chromosome breakage is generally not observed within somatic revertant sectors in *bz-m1* and *c-m1* strains, and germinal revertants show neither *Ds*-mediated chromosome breakage at the locus nor further instability in the presence of *Ac* (McClintock, 1949, 1951b). Like those derived from *bz-m4,* germinal revertants derived from the *sh-m* strains retain *Ds* in the immediate vicinity of the *Sh* locus

(McClintock, 1953). Such revertants are unstable and yield *sh-m* alleles in the presence of *Ac*. Reversion of the *sh-m* and *bz-m4* alleles may therefore occur by secondary *Ds*-associated chromosomal rearrangements.

8. EFFECT OF *DS* ON GENE STRUCTURE AND EXPRESSION

Detailed information on how *Ds* can affect the structure and expression of a locus is beginning to accumulate. There is evidence that *Ds* mutations can alter the level of gene expression, the structure of the gene product, and the developmental profile of gene expression as a consequence of sequence changes inside and outside of the transcription unit at a locus. Evidence that *Ds*-mediated structural alterations can occur within and near a transcription unit has come primarily from the analysis of three *sh-m* strains, designated *sh-m(5933), sh-m(6258),* and *sh-m(6233)* and derived from the *Ds4864A* and *Ds5245* strains described in the preceding section. The *sh-m(5933)* and *sh-m(6233)* strains both show alterations in the location of restriction endonuclease sites near the 5′-end of the sucrose synthetase coding sequence at the *Sh* locus, but it is not yet known whether they fall inside or outside of it (Burr and Burr, 1980, 1981; Döring *et al.*, 1981; Chaleff *et al.*, 1981). When *Ac* is not present, immature endosperm tissue from both strains contains a low level (0.5–1.0%) of a poly-A+ mRNA that is homologous to the *Sh*-encoded mRNA and appears to be the same length as the mRNA for the *Sh*-encoded sucrose synthetase (Chaleff *et al.*, 1981). The coding sequence at the *Sh* locus appears to be rearranged in both the original *sh-m(6258)* strains and an *sh-m* strain, designated *sh-m(6795)*, derived from an *Sh* revertant of *sh-m(6258)*. Although the nature of the rearrangements present in these strains is not entirely clear, both strains contain a copy of the coding sequence that is interrupted near the 3′-end either by an insertion or by a rearrangement (Chaleff *et al.*, 1981; Fedoroff *et al.*, 1983). Both strains contain abundant aberrant transcripts slightly shorter than the 3-kb (kilobase) sucrose synthetase mRNA. The aberrant transcripts appear to contain the 5′-end of scurose synthetase mRNA but are missing the sequence distal to the insertion or inversion break point (Fedoroff *et. al.*, 1983). When *Ac* is absent, none of the mutant strains produce detectable levels of *Sh*-encoded sucrose synthetase (Schwartz, 1960; Chaleff *et al.*, 1981; Chourey, 1981; Fedoroff *et al.*, 1983). It has been observed, however, that two polypeptides antigenically related to sucrose synthetase, but having a higher electrophoretic mobility, can be synthesized *in vitro* from mRNA purified from *sh-m(6258)* and *sh-m(6795)* endosperm tissue (Fedoroff *et al.*, 1983). Because the aberrant transcripts produced in strain *sh-m(6258)* and its derivative *sh-m(6795)* diverge from the normal transcript at the point where the coding sequence is disrupted, the structural alteration in these strains is clearly related to the mutant phenotype.

Studies have been done on the proteins encoded by the *Bz* and *Wx* loci in several strains homozygous for *Ds* alleles of these loci (Dooner and Nelson, 1977b; Chaleff *et al.*, 1981; Echt and Schwartz, 1981). Dooner and Nelson

(1977b) reported that in the absence of *Ac*, little or no *Bz*-encoded UFGT enzyme was produced in aleurone tissue from several strains with *Ds* mutations at the *Bz* locus. They further reported that a small amount of heat-labile enzyme was produced in a strain homozygous for the *Bz-wm* allele (Table III), which supports the synthesis of almost half of the normal amount of glycosylated pigment. They also reported that a small amount of an apparently normal UFGT enzyme was produced in a strain homozygous for the *bz-m4* allele (in the absence of *Ac*), but that it was synthesized abnormally early, for a brief period, and in endosperm rather than in aleurone cells (Dooner and Nelson, 1977b; Dooner, 1980b). Studies on the protein encoded by the *Wx* locus in strains with *Ds* mutations at this locus likewise suggest the existence of different categories of mutants (Chaleff *et al.*, 1981; Echt and Schwartz, 1981). These include mutants that produce no detectable *Wx* protein or low levels of a normal *Wx* protein, as well as mutants that produce an altered *Wx* protein. McClintock (1965b, 1978a) described derivatives of the *wx-m7* allele, the phenotype of which suggests the existence of a gradient of gene expression during kernel development. There is evidence from studies carried out by Nelson (1968) that *Wx* recombinants can be obtained from strains that are heterozygous for two different *wx* insertion mutations, indicating that the mutations affect different sites within the locus. Kermicle (1980) also found that *Ds* mutations mapped at different sites within the *R* locus.

Taken together, these studies indicate that *Ds* mutations can alter the structure and expression of a locus in quite different ways. *Ds* mutations can reduce the level of transcription, shift the developmental timing and tissue specificity of gene expression and alter the primary structure of the transcription unit. From the known genetic properties of the mutants described in preceding sections, *Ds*-associated mutations are unlikely to have a single origin and will probably include both simple insertion mutations and element-associated rearrangements that alter gene expression.

B. The Autonomous Controlling Element *Ac*

In the preceding discussion, the *Ac* controlling element has been treated as the genetic entity responsible for the behavior of the nonautonomous *Ds* element. The *Ac* element is genetically separable from *Ds* and need only be in the same nucleus as the *Ds* element to destabilize the *Ds* element. What is known about *Ac* will be explored in some detail below, followed by a consideration of the evidence that nonautonomous *Ds*-like elements can be derived from an autonomous *Ac* element.

1. THE MUTATOR FUNCTION OF *AC*

Ac is an autonomous element capable of causing unstable mutations. The phenotype associated with one such insertion mutation of the *Waxy* locus is shown in kernel 8. The other unstable mutations caused by the *Ac* or the related

Mp element that are discussed here are listed in Table III. In addition to its ability to cause unstable mutations, *Ac* can be detected by its effect on a *Ds* element located elsewhere in the genome, as described in detail in previous sections. Because *Ac* need not be at the same site or on the same chromosome as *Ds, Ac* must have a *trans*-acting function that is necessary for chromosome breakage at *Ds* and the ensuing excision, transposition, or acentric–dicentric chromatid formation. Because it appears to be quite analogous to the mutator function of the *Spm* controlling element, I have used the same designation of *mutator* in discussing this function of the *Ac* element (McClintock, 1954). The properties of the mutator function are consistent with the production of a diffusible, element-encoded, sequence-specific enzyme necessary for transposition. Although the *Ac* mutator has so far not been further subdivided genetically, mutator function may well reside in more than one diffusible gene product. Moreover, the possibility cannot presently be discarded that mutator function is mediated by cellular enzymes or involves direct contact between elements. There are, for example, observations that show coincident timing of genetic changes at distant *Ds* elements (McClintock, 1949, 1956a). In a strain containing *Ds* proximal to the *Wx* locus on one chromosome 9 and *Ds* at the *C* locus (*c-m1*) on the homolog, *Ac*-activated changes at the two *Ds* elements were frequently observed in precisely the same somatic sectors (McClintock, 1949). That is, the borders of sectors showing loss of the *Wx* locus precisely coincided with those of sectors in which the *c-m1* allele had reverted to the *C* allele. It follows that the original genetic changes at both *Ds* elements occurred in precisely the same progenitor cell. Similar coincidences have been observed in the cell lineages in which *Ac* mutates or transposes and *Ds* sustains a heritable change in the relative frequency of transposition and acentric–dicentric formation (McClintock, 1949, 1956a). McClintock (1956a) noted that such coincidences were sufficiently numerous so that a change affecting the *Ds* element could serve as a useful selection for changes in *Ac* location. The occurrence of coincident changes at *Ac* and *Ds* elements at disparate chromosomal locations raises the possibility of physical interactions between elements. Indeed, McClintock (1956a) suggested that *Ds*-like elements could arise from an *Ac* element by a direct genetic exchange, as has recently been observed with yeast transposable elements (Chapter 7, this volume).

2. The Effect of *Ac* Dosage

When more than one *Ac* element is present, there is a delay in the developmental timing and a reduction in the apparent frequency of *Ac*-activated somatic mutation and acentric–dicentric formation at *Ds* (McClintock, 1948; Brink and Nilan, 1952; Wood and Brink, 1956). This is illustrated in kernel 14, which has the genetic constitution *C bz-m1/C bz* and has two *Ac* elements present. When the phenotype of this kernel is compared with the phenotype of kernel 7, which has the same *bz-m* allele and one *Ac*, it is evident that the average size of the somatic *Bz* sectors is smaller. This means that the somatic mutations occur later in development, when there are two *Ac* elements present, and that the size of the average revertant clone is therefore smaller. The same is true of acentric–dicentric formation, which occurs later in the presence of more than one *Ac* element than it does in the presence of one element. When a strain contains more than one *Ds* element or more than one *Ds* insertion mutation, all of the *Ds* elements respond similarly to the *Ac* dosage. At high *Ac* dosages, the apparent frequency of somatic mutation decreases; however, the reduction may be due to a delay in the timing of mutation beyond cessation of cell division in a given tissue (McClintock, 1948). That is, a delay in the timing of excision events can have the same phenotypic consequences as a reduction in frequency, so that factors affecting timing cannot always be distinguished from factors affecting frequency. A decrease in the germinal, as well as the somatic, reversion frequency accompanying an increase in dosage of the controlling element has been described for *Mp,* which is similar or identical to *Ac* (Emerson, 1929; Brink and Nilan, 1952). Again, this may be attributable to a delay in the time of reversion, because the timing of premeiotic events critically influences the fraction of gametes carrying altered chromosomes.

In contrast to the delay in the developmental timing (and possible decrease in frequency) of *Ac*-activated genetic changes with increasing numbers of the element, the frequency of the *Dt*-activated somatic mutation increases with the number of elements present (Rhoades, 1936, 1938; Nuffer, 1955, 1961). *Dt,* like *Ac,* activates somatic reversion of mutations, presumed to be caused by a related nonautonomous element (Table II). As the number of *Dt* elements present in a strain carrying such a mutation increases, the number of revertant sectors increases exponentially, without discernible change in sector size (Rhoades, 1936, 1938; Nuffer, 1955, 1961). Somatic reversion of mutations belonging to the *Spm* family, discussed in detail below, is insensitive to the dosage of the *Spm* element (McClintock, 1955, 1965a).

Analysis of the relationship between the dosage of the *Ac* element and the developmental timing of somatic mutation is complicated by the genetic lability of both *Ds* and *Ac*. Heritable changes have already been described that affect the relative frequencies of *Ds* excision and acentric–dicentric formation (McClintock, 1948, 1949). *Ac* also sustains heritable changes that affect the timing of *Ac*-activated mutations (McClintock, 1948). Such changes can mimic a change in *Ac* dosage or give a mutation pattern intermediate between those obtained with one and two doses of the parent *Ac* (McClintock, 1948). McClintock (1948) believed these to be changes that affect the *Ac* element and designated them as states of *Ac*. Analogous differences between *Mp* insertion mutations have been noted and will be considered below (Orton and Brink, 1966; Brink and Williams, 1973).

3. Mechanism of *Ac* Transposition

There is evidence that *Ac* transposition is correlated with chromosome replication and that it results in the removal of the element from its original site on one of the two sister chromatids. *Ac*-bearing plants of the appropriate genetic constitution often show twin somatic sectors whose phenotype is suggestive of complementary changes in *Ac* dosage (McClintock, 1948). The most extensive studies of the transposition mechanism were carried out by Brink, Greenblatt, and their co-workers on the *Mp* element. *Mp* was identified by Brink and Nilan (1952) as a factor causing the variegation in pericarp and cob pigmentation studied initially by Emerson (1914, 1917, 1929). *Mp* behaves as an autonomous element tightly linked to the *P* locus, and the unstable allele of the locus is designated *P-RR-Mp*. Strains heterozygous for *P-RR-Mp* and the recessive *p* allele exhibit wedge-shaped revertant sectors of red pigmentation in an otherwise colorless pericarp tissue (Emerson, 1914). *Mp* shows a dosage effect like that described for *Ac* (Emerson, 1929; Brink and Nilan, 1952). Hence, the size and number of revertant red sectors decrease with increasing numbers of *Mp* elements, giving a phenotype that is designated *light-variegated* to distinguish it from the *medium-variegated* phenotype observed with one *Mp* element (Brink and Nilan, 1952). Transposition of *Mp* away from the *P* locus almost always restores the dominant *P* allele, which gives a uniformly red-pigmented pericarp (Brink, 1958a). The frequency of kernels with red pericarp (no *Mp* at the *P* locus) and with light-variegated pericarp (two *Mp*) is roughly equivalent among the progeny of a *P-RR-Mp/p* heterozygote grown from a medium-variegated kernel, suggesting that the *Mp* element lost by transposition from some chromosomes is gained by others (Brink and Nilan, 1952). The number of *Mp* elements recovered per transposition event is 1.44, indicating that replication of the element is associated with transposition (Greenblatt, 1966).

The observation that large twinned sectors signaling changes in *Mp* dosage appear on medium-variegated ears provided the basis for a precise and elegant analysis of the mechanism of *Mp* transposition (Brink and Nilan, 1952; Brink and Barclay, 1954; Greenblatt and Brink, 1962, 1963; Greenblatt, 1968, 1974a,b,c). As noted earlier, genetic changes occurring at mitotic divisions during ear development can give rise to adjacent, phenotypically distinct twinned sectors on an ear, each comprising several to many kernels and descended from one of the two genotypically distinct daughter cells. Twinned red (no *Mp* at the *P* locus) and light-variegated (two *Mp*) sectors are observed on medium-variegated (one *Mp*) ears (Fig. 8), suggesting that changes in *Mp* constitution occur at mitotic divisions, giving rise to daughter cells that have, respectively, lost and gained a copy of the element. The validity of this inference was amply confirmed by growing plants from the kernels in twinned sectors and analyzing them for the number and position of *Mp* elements by three-point linkage tests and by scoring for chromosome breakage at *Ds* by *Mp* (Greenblatt and Brink, 1962, 1963). It was found that kernels with red pericarp produced plants that no longer had *Mp* at the *P* locus, whereas light-variegated kernels had an *Mp* element at the *P* locus, as well as a second *Mp* element, often located near the first one on the same chromosome. However, when plants grown from kernels with red pericarp (no *Mp* at the *P* locus) were examined for the presence of *Mp*, a transposed element

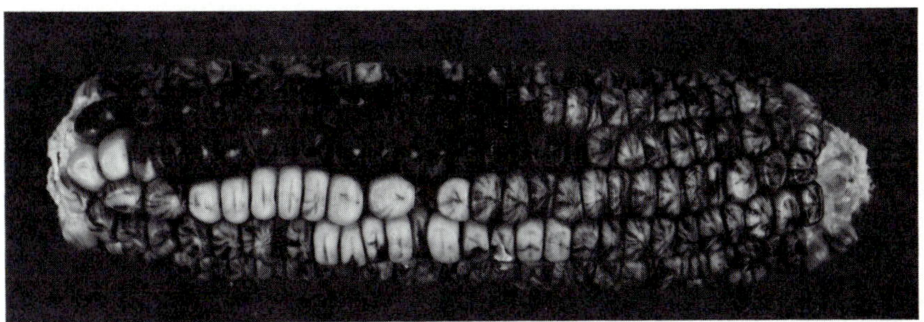

Fig. 8 Twinned red and light-variegated sectors on a medium-variegated ear with *Mp* at the *P* locus. (Courtesy of I. Greenblatt)

was present in some cases and not in others. In 38% of the twinned sectors analyzed, there was a transposed *Mp* element present in the light-variegated sector and no *Mp* element in the red sector, implying that the element had transposed from the *Mp* locus on one replicated sister chromatid to a replicated recipient site on the other sister chromatid, as diagrammed in Fig. 9a. An *Mp* element was present in plants grown from the red kernels in 62% of twinned sectors analyzed, and its chromosomal position coincided with that of the transposed *Mp* element in the adjacent light-variegated twinned sector, suggesting that *Mp* had transposed after replication from one sister chromatid to an unreplicated recipient site elsewhere on the chromosome, as diagrammed in Fig. 9b. These observations clearly establish that *Mp* transposition can occur at a mitotic division and can result in the removal of the element from the donor site on only one sister chromatid, accompanied by its appearance at a new site.

The recipient site is commonly near the donor site on the same chromosome (Van Schaik and Brink, 1959; Greenblatt and Brink, 1962). The location of the recipient site relative to the donor site was determined in 87 independently isolated strains having an *Mp* element at the *P* locus and a second, transposed *Mp* element (Van Schaik and Brink, 1959). This was done by determining the recombination frequency between the two elements in plants that had undergone an apparent doubling of the *Mp* dosage as judged by the light-variegated phenotype of the pericarp. The recipient site showed no linkage to the donor site in only one-third of the cases analyzed. No further tests were carried out to distinguish between distant sites on the same chromosome and sites on different chromosomes; hence, the fraction of intrachromosomal transpositions remains uncertain. Moreover, *Mp* increases recombination frequencies in its immediate vicinity, possibly further reducing apparent linkage (Greenblatt, 1974c, 1981). In 29% of the isolates analyzed, no recombination was detected between the *P* locus and the transposed *Mp* element presumed to be present. This group of isolates may well have sustained mutations in the *Mp* element that affect the timing and frequency of transposition, although tandem duplication and very short-range transpositions are not excluded. The donor and recipient sites were

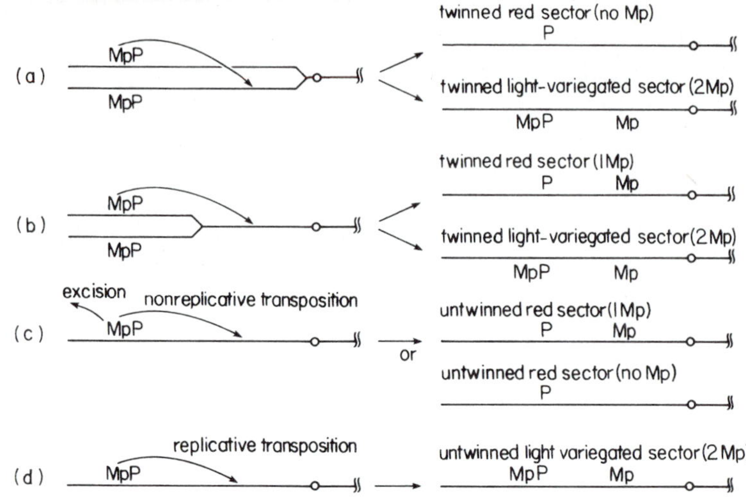

Fig. 9 A diagrammatic representation of various types of intrachromosomal transpositions. The transpositions diagrammed in (a) and (b) occur during or after replication of the *Mp* element at the *P* locus and result in its removal from the *P* locus on one sister chromatid and its insertion either in a replicated site on the other sister chromatid (a) or in an unreplicated site on the same chromosome (b). The chromosomes received by the daughter cells derived from the cell in which each transposition has occurred are indicated on the right, as are the expected phenotypes of kernels on ears of plants in which such transpositions have occurred at premeiotic mitoses. As described in the text, transpositions of the type diagrammed in (a) and (b) have been documented. Other possible types of transpositions, as well as simple excision of the element, are diagrammed in (c) and (d), and their genotypic and phenotypic consequences are indicated at the right. For reasons discussed in the text, it is unclear whether such transposition and excision events occur. Although intrachromosomal events are diagrammed, transposition to unlinked sites is also known to occur.

linked in 64% of the strains analyzed, with 43% of the recipient sites mapping within five recombination units of the *P* locus. In recent studies, I. M. Greenblatt (personal communication) has shown that among 54 independent intrachromosomal transpositions of *Mp*, 31% of the recipient sites mapped within an interval of four to five recombination units just distal to the *P* locus. In contrast, no recipient sites were found in the same interval proximal to the *P* locus, indicating that there is polarity in the transposition mechanism, in addition to a strong preference for short-range transposition events. The more distant recipient sites on the same chromosome were found proximal and distal to the *P* locus at about equal frequency and showed no tendency toward clustering.

Although these studies establish that the transposition events depicted in Figs. 9a and 9b occur, they leave open the possibility that simple excision and other types of transpositions, including intrachromatid and replicative transpositions, also occur. As diagrammed in Figs. 9c and 9d, such events would give rise to untwinned sectors in which there has been a change in the number or position of *Mp* elements. Greenblatt (1968) analyzed many untwinned red sectors and found that the fraction containing a transposed *Mp* element was the same as among twinned red sectors. He presented evidence that untwinned red and light-variegated sectors, which constitute a small fraction of all of the sectors in his material, occur at equal frequencies and argued that they are identical in origin to those present as twinned sectors, but are untwinned because occasionally only one of the two original daughter cells with an altered *Mp* constitution develops into a visible sector of kernels, the other cell having given rise to underlying cob tissue in the ear (Greenblatt, 1974c). He also reported that red and light-variegated pericarp kernels occur at equal frequencies in the progeny of plants homozygous for the *P-RR-Mp* allele (1974c). This is an issue of central importance to the transposition mechanism. If Greenblatt's argument is correct, it leads to the conclusion that *Mp* undergoes only those transpositions diagrammed in Figs. 9a and 9b. That is, *Mp* is removed from one sister chromatid after DNA replication and inserted at an unreplicated site elsewhere on the chromosome or into the sister chromatid from which the *Mp* element was not removed. Simple excision, intrachromatid transpositions, and replicative transpositions do not occur, according to this interpretation. However, if untwinned red and light-variegated sectors are accepted as representing the total progeny derived from a cell in which transposition or excision has occurred, it follows that some or all of the types of excision and transposition events depicted in Figs. 9c and 9d can occur. It has been observed in some studies that even when all relevant crossover products are recovered, the frequency of untwinned red sectors slightly exceeds that of untwinned light-variegated sectors (Brink and Nilan, 1952). A slight excess of untwinned red sectors is expected even in the absence of intrachromatid transpositions, because 38% of recipient sites are replicated, and of these, some (one-third or less) are on another, independently assorting chromosome (Van Schaik and Brink, 1959; Greenblatt and Brink, 1962). Segregation of the recipient site with the donor site at the mitotic division following transposition should result in the formation of untwinned red sectors. However, if intrachromatid transpositions (Fig. 9c) are as frequent as interchromatid transpositions (Fig. 9a), untwinned red sectors should constitute a large fraction of all sectors. In careful studies, the proportion of twinned sectors has been found to be more than 80% of all sectors (Greenblatt, 1974c, personal communication), and the number of untwinned red sectors has not greatly exceeded the number of untwinned light-variegated sectors (Brink and Nilan, 1952; Greenblatt, 1974c). Thus, it is clear that transpositions of the type depicted in Fig. 9a and 9b predominate, although there remains the possibility that other types of excision and transposition events occur at a low frequency.

The results of studies on transposition of *Mp* from the *P* locus indicate that when *Mp* is removed from the locus, expression of the locus almost always returns to the original level characteristic of the dominant *P* allele. Colorless stable mutants occur at a very low frequency (Brink, 1958a). The results of McClintock's (1956b, 1962) studies on the *bz-m2* allele, which has an *Ac* element at the *Bz* locus (Table III), stand in contrast to these findings. The *bz-m2* allele yielded stable *bz* mutations almost five times more frequently than mutations to the fully pigmented, stable *Bz* phenotype (McClintock, 1956b, 1962). A majority of both the null and *Bz* mutants analyzed no longer had *Ac* at the *Bronze* locus, although many had either one or two *Ac* elements located elsewhere in the genome. The difference in the frequency with which null mutations arise when the controlling element transposes from these two unstable alleles (*P-RR-Mp* and *bz-m2*) may represent a difference in the frequency of precise excision by these two elements (*Mp* and *Ac*), the sensitivity of the insertion sites to sequence changes, or the effect of the insertion site on the types of excision or transposition events that occur.

4. A Transposition-Defective *Mp* Element

Brink (1958a) described an *Mp* element that appears to be altered in its ability to transpose. The *Mp* element is located at the *P* locus and stably inhibits expression of the locus. The mutator function of this element is unaltered, as judged by its ability to cause chromosome breakage at *Ds* and to contribute to the dosage effect on somatic mutation observed when more than one copy of the *Mp* element is present. Strains carrying this element at the *P* locus, however, show a very low frequency of somatic mutation and no germinal mutation of the allele to the dominant *P* allele. This raises the possibility that there is a cis-dominant mutation in the element itself that alters its capacity to transpose.

5. Reconstitution of an Unstable Allele

Orton and Brink (1966) designed a study to detect and investigate unstable mutations of the *P* locus resulting from transposition of *Mp* to the locus. The study was carried out using a strain in which *Mp* had originally been located at the *P* locus and had subsequently been removed by transposition, yielding a strain with uniformly deeply pigmented pericarp. A total of 146 independent unstable mutations was recovered, and in the vast majority of them, the *Mp* element was shown to have returned to the *P* locus. The striking outcome of this study was, however, that less than 20% of the mutants showed the variegation pattern characteristic of the original unstable mutant. The remainder differed from the original mutant in the frequency of somatic and germinal mutation, the background pigmentation level of the pericarp and cob, or both. The possibility that mutants showing a low frequency of somatic mutation contained two *Mp* elements was tested and eliminated in several cases. The results of this study showed not only that the *Mp* element could transpose back to the *P* locus but also that different insertion mutations of the same locus differ in stability and in their effect on expression of the locus. In a similar study of mutants in which *Mp* had transposed away from and back to the *R* locus, Brink and Williams (1973) also observed that the reconstituted unstable alleles of the *R* locus differed widely in stability. Based on tests of the ability of the *Mp* element to contribute to the dosage effect observed with the *P-RR-Mp* allele and to trans-activate chromosome breakage at *Ds,* they concluded that the *Mp* elements were alike in their capacity to interact with other elements in trans (Brink and Williams, 1973). However, it is unclear whether the results of these experiments were analyzed in sufficient detail to reveal shifts in the developmental timing of trans-activation of chromosome breakage at *Ds* or transposition of *Mp* from the *P* locus.

6. Control of *Ac*-Activated Genetic Events

Early in the study of the *Ac* element, McClintock noted heritable differences among *Ac* elements in the developmental timing and frequency of *Ac*-activated chromosome breakage and acentric–dicentric formation at *Ds* (McClintock, 1948). Elements showing such differences were designated as being in different states, signifying heritable but relatively unstable forms of the element (McClintock, 1948). McClintock also noted that changes in state, as well as changes in *Ac* dosage consequent on transposition (as described for the *Mp* element in preceding sections), could occur at a very precise time in development. Thus, for example, McClintock isolated one *Ac* element that transposed or changed in state during the first or second nuclear division in endosperm development in over 90% of the kernels examined (McClintock, 1948, 1949, 1951a). Because different *Ac* isolates undergo changes and transpositions at different developmental times, McClintock concluded that *Ac* controlled the developmental timing of such changes and transpositions (McClintock, 1949, 1951a). In the studies on independent *Mp* insertion mutations at the *P* and *R* loci described above, Orton and Brink (1966) and Brink and Williams (1973) entertained the alternative or, perhaps, additional possibility that the effect of the insertion mutation on expression of the locus and the stability of the insertion mutation are influenced by its position within the locus. The suggestion that the element can insert at different sites within the locus is quite reasonable and is supported by recombinational analysis of insertion mutations at the *Wx, A,* and *R* loci (McClintock, 1965c; Nelson, 1968; Kermicle, 1980). In all of these studies, it was found that certain controlling element insertion mutations recombined with each other, indicating that they affect different sites within the locus. What is not known is whether the elements occupying different positions within a locus are identical in sequence; it may be that the elements are especially subject to genetic change during transposition. Kedharnath and Brink (1958) provided some evidence that transposed *Mp* elements fall into two categories when tested for their contribution to the *Mp* dosage effects and that the categories differ from each other with respect to the frequency of somatic reversion but not developmental timing. In view of the evidence, to be discussed in a subsequent section, that cryptic controlling elements are regular residents of the maize genome, it seems quite possible that expression of the element, as reflected in its transposition and its ability to activate transposition and acentric–dicentric formation in trans, can be subject to local control. That is, the element may in some cases come under the influence of the insertion site's developmental program or may be affected by the sequence at the insertion site. Peterson (1976a,b, 1977) has proposed such a position hypothesis to explain differences in expression of *En* elements at different locations.

Whatever the underlying causes, the genetic instability of the factors controlling the developmental timing and the apparent frequency of *Ac* and *Mp* transposition complicate estimates of transposition frequency. Moreover, because plants do not have a segregated germ line in the sense that animals do, the developmental timing of a mitotic event critically determines its representation in the population of gametes produced by the plant. Thus, if a single transposition event occurs early enough in the development of the plant and in a cell lineage that will give rise to germ cells, the transposed element can be represented in a large fraction of the gametes. What is strikingly characteristic of controlling elements is that a particular isolate trans-activates somatic reversion and acentric–dicentric formation within a narrowly defined developmental interval and can undergo heritable changes within an equally restricted developmental window (McClintock, 1948, 1949, 1951a, 1956a,b, 1965a).

As discussed in previous sections, the extent to which the behavior of an element is influenced by the sequence at the site of insertion is not known. Nonetheless, the substantial evidence that elements exhibit similar properties at different chromosomal locations (McClintock, 1956a) supports the view that element-associated sequence differences are involved in determining the genetic and developmental behavior of a particular element. Thus, *Ac* (and *Mp*) may constitute a group of closely related elements, each of which has its own characteristic, genetically determined developmental profile of expression, transposition, and mutation. In view of this complexity, estimates of transposition frequency are not especially meaningful with most of the elements that have been studied. Nuffer (1961) isolated a *Dt*-activated unstable allele of the *A* locus that appeared to lack strict developmental control of genetic change. It differs from previously isolated unstable alleles that respond to *Dt* in showing no uniformity in revertant sector size. Sectors range in size from a single cell to the entire kernel, with no tendency toward grouping at an average size. By contrast, the average sector size observed with a previously isolated *Dt*-responsive mutant allele of the same locus was rather uniform, averaging 78 cells and ranging from 25 to 560 cells. Nuffer (1961) estimated the frequency with which the newly isolated allele reverted to the *A* allele at various stages of development and arrived at estimates ranging from 1.1×10^{-4} to 17×10^{-4} per cell division in the presence of a single *Dt* element.

7. Reversible Inactivation of *Ac*

Another type of change that *Ac* has been observed to undergo is a reversible inactivation (McClintock, 1964, 1965b). An inactive *Ac* can activate neither somatic reversion of a *Ds* insertion mutation nor acentric–dicentric formation at *Ds* (McClintock, 1964, 1965b). McClintock first observed such a change in a plant carrying the *wx-m7* allele, which has *Ac* at the *Wx* locus (kernel 8), and the *a-m3* allele, which has *Ds* at the *A* locus (McClintock, 1964). Neither mutable allele showed somatic mutation in most of the kernel progeny. Return of the *Ac* to the active phase did occur in a few cells, as inferred from the simultaneous destabilization of both alleles in the same cell lineage (McClintock, 1965a,b). It was further established that the inactive *Ac* did not contribute to the *Ac* dosage effect and that the *wx* mutation could mutate somatically if an active *Ac* element was introduced together with the inactive *Ac* at the locus (McClintock, 1964, 1965b). Changes of the type just described were designated changes in *phase* of activity. Reversible changes in phase were first identified and have been studied in more detail in the *Spm* controlling element system discussed in subsequent sections.

8. Derivation of Nonautonomous Elements

There is evidence that a nonautonomous, *Ds*-like element can be derived directly from an autonomous *Ac* element. Three independent strains with non-autonomous elements were derived from *bz-m2*, which has *Ac* at the *Bz* locus (McClintock, 1955, 1956b, 1962). All three resemble the many *Ac*-activated *Ds* insertion mutations that have been studied. Two were derived from the original *bz-m2* allele. Of these, one strain was identified as a single kernel that showed a higher frequency of somatic mutation than was typically observed with *bz-m2* (McClintock, 1955). The other strain was originally identified as a possible null mutant (McClintock, 1956b). The *bz* phenotype by which it was originally identified was evident in the absence of *Ac*, but the characteristic pattern of somatic mutation was seen when *Ac* was reintroduced into plants carrying this allele. The third mutant (*Bz-wm*) was derived from a revertant of *bz-m2* that had a *Bz* phenotype but still had *Ac* at the locus (McClintock, 1956b, 1962). The derivative did not have an autonomous *Ac* at the locus and showed a somewhat recessive phenotype in the absence of *Ac*. Introduction of *Ac* resulted in the production of both more deeply and less deeply pigmented somatic sectors. All three mutants showed the delayed timing of somatic mutation observed with increasing doses of *Ac* in other *Ds* insertion mutants. All three mutations, therefore, show the behavior characteristic of *Ds* insertion mutations. Because they occurred in a strain having an autonomous *Ac* element at the *Bronze* locus, it seems likely that the nonautonomous, *Ds*-like element at the locus is, in fact, a mutant *Ac* element. Unlike the transposition-defective *Mp* element described earlier, the nonautonomous element in these strains appears to lack mutator function but can be activated in trans by a nondefective autonomous element.

An additional indication that *Ds* resembles an *Ac* element lacking mutator function comes from the observation that an inactive *Ac* can be activated in trans by an active *Ac* element (McClintock, 1964, 1965a). Thus, the properties of the *bz-m2* derivatives just described are rather similar to those of the *wx-m7* allele with an inactive *Ac* element (McClintock, 1964), except that the inactive *Ac* element is capable of returning to its previously active state. Whether the underlying changes differ in stability or are fundamentally different is not known. Although there is no direct evidence bearing on the mechanisms by which *Ds*-like elements arise, it is perhaps worth pointing out that these could be varied and include mutations in element-encoded functions for transposition, permanent inactivation resulting from a defect in the phase-change mechanism, and interconversion or recombination between elements at different chromosomal locations. The last possibility was suggested by McClintock (1956a) and has recently been observed to occur among transposable elements in yeast (Chapter 7, this volume).

IV. The *Spm* Controlling Element Family

Like the *Ac–Ds* family, the *Spm* controlling element family comprises autonomous and nonautonomous elements. The autonomous element was designated *Suppressor-mutator* (*Spm*) by McClintock (1954) and *Enhancer* (*En*) by Peterson (1953, 1960), who initially described mutations caused by this family of elements. The nonautonomous element was left unnamed by McClintock and was termed *Inhibitor* (*I*) by Peterson (1953). *Spm* elements have in common with the other controlling elements the capacity to cause unstable mutations by insertion at a locus. The *Spm* and *En* elements were identified as members of the same family by the activation criteria described earlier (Peterson, 1965). The *Spm* family of elements differs from the *Ac-Ds* family in several ways. Although some *Spm*-mediated chromosome breakage may occur, no systematic acentric–dicentric formation of the kind characteristic of *Ds* has been observed. *Spm* is further distinguished from *Ac* in its effect on certain alleles of a locus containing a nonautonomous element belonging to the same family (McClintock, 1953, 1954; Fowler and Peterson, 1974). In certain strains, insertion of the nonautonomous element at a locus results in a phenotype that is intermediate between dominant and recessive. Kernel 15 illustrates the phenotype associated with one such mutation at the *A* locus. The aleurone layer in a kernel that is heterozygous for the allele designated *a-m1(5719A1)* and the recessive *a* allele is palely pigmented when *Spm* is absent (kernel 15). When *Spm* is present, the aleurone layer is colorless and shows deeply pigmented revertant sectors (kernel 16). The function of the element responsible for inhibiting pigmentation was designated the *suppressor* (*sp*) function, whereas the function responsible for somatic reversion was designated the *mutator* (*m*) function (McClintock, 1954). The *sp* and *m* functions are genetically distinguishable but co-transpose (McClintock, 1955). The *m* function is required for both somatic and germinal instability

(McClintock, 1957, 1961b, 1965a). Thus, the autonomous and nonautonomous elements of the *Spm* family exhibit two distinct types of interactions, as represented diagrammatically in Fig. 10. McClintock (1965a,b) also used the designations *component I* for the suppressor and *component II* for the mutator. Although there is virtually no molecular information about the effect of elements belonging to the *Spm* family on the structure and expression of a locus, there is a wealth of genetic information about this complex and interesting family of elements.

A. Mutations that Affect the Behavior of the Nonautonomous Element of the *Spm* Family

I will first address mutations that affect the response of a locus having an integrated, nonautonomous element to a *standard Spm*. It will become evident later that the *Spm* element itself is so changeable that there may be little meaning in defining a standard. However, much of the early work done by McClintock in this system involved an element that she occasionally termed standard (McClintock, 1957). This element is active throughout the developmental cycle of the plant and has a mutator function that is expressed relatively early in development. In illustrating the types of mutations that affect the nonautonomous element, I have chosen kernels from strains carrying the standard *Spm*.

1. Mutations that Affect the Pattern of Somatic Mutation

A locus with an integrated, nonautonomous controlling element belonging to the *Spm* family can sustain changes that affect the pattern of somatic mutation (McClintock, 1955; Peterson, 1961, 1970b). The changes are heritable, giving new alleles that differ from the parent allele in such respects as the frequency of somatic mutation, the time in development when somatic mutations occur, and the level of gene expression in mutant cell clones. McClintock (1955) designated new alleles of this type states of a locus (McClintock, 1955, 1957, 1958, 1968). I have used the more familiar term *allele* in the present description because the various states of a locus indeed behave as alleles, albeit often unstable. Kernels from strains heterozygous for one of several additional unstable alleles of the *A* locus and a recessive allele that is stable in the presence of the *Spm* element are shown in kernels 17-20.[5] Unlike the *a-m1(5719A1)* allele, which produces substantial aleurone pigmentation in the absence of the *Spm* element, the alleles represented in kernels 17–20 produce very palely pigmented or unpigmented aleurones in the absence of *Spm*. The alleles the phenotypes of which are displayed by kernels 18 and 19 [*a-m1(5996-340)* and *a-m1(5718)*] resemble the *a-*

[5]Except when explicitly stated, all of the kernels used here to illustrate the *Spm* system are heterozygotes of this type.

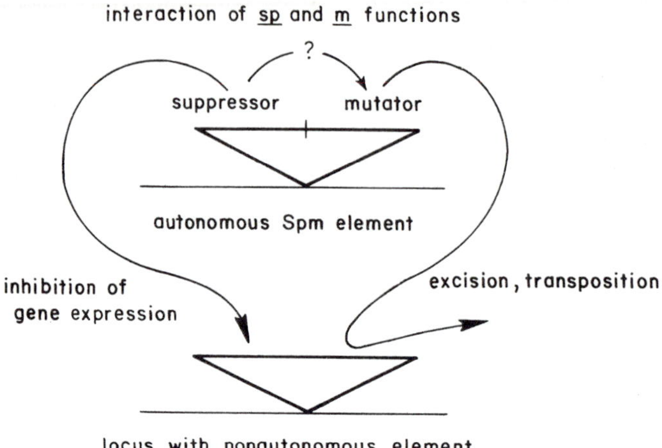

Fig. 10 A diagrammatic representation of the interactions between autonomous and non-autonomous elements belonging to the *Spm* family. The suppressor function of the autonomous *Spm* element inhibits expression of a locus with an integrated nonautonomous element. The mutator function of the *Spm* acts in trans to promote excision and transposition of the nonautonomous element. The mutator function is required for transposition of the *Spm* element and is expressed or functional only when the suppressor is expressed. The dependence of mutator expression on suppressor expression is indicated by an arrow. As discussed in the text, the nature of this dependence is not known, and the arrow does not imply a direct regulatory interaction.

m1(5719A1) allele in the small size of the pigmented sectors observed in the aleurone layer. Somatic mutation occurs late in the development of kernels carrying these alleles and an *Spm* element. The alleles differ from each other in the frequency of somatic mutation. The *a-m1(5719A1)* allele shows the highest frequency, and the *a-m1(5996-340)* shows the lowest frequency. Strains carrying the alleles *a-m1(5996)* and *a-m1(5720)* show strikingly different patterns of aleurone pigmentation when *Spm* is present (kernels 17 and 20). In strains carrying either of these two alleles, the kernel aleurone layer shows large areas of pigmentation, indicating that somatic mutations occur early in kernel development in these strains. The alleles differ from each other, however, in the intensity of the pigmentation observed in the mutant sectors. A strain that carries the *a-m1(5996)* (kernel 17) allele and *Spm* produces intensely pigmented somatic sectors, whereas a strain that carries the *a-m1(5720)* allele produces almost exclusively the palely pigmented sectors observed in kernel 20.

Mutations that alter the pattern of somatic mutation occur frequently in some strains (McClintock, 1955). Strain *a-m1(5996)* is one example, and the strain designated *a-m1(5996–340)*, which exhibits a pattern of late somatic mutation (kernel 18), was derived from it by spontaneous mutation. Strains in which somatic mutations occur early in development give rise to new alleles more frequently than do strains in which somatic mutations occur late in development (McClintock, 1955). The higher frequency of germinal mutations in strains that show early somatic mutation supports the view that similar mechanisms are involved in premeiotic (germinal) and postmeiotic (somatic) mutations.

Mutations that alter the pattern of somatic mutation are cis-dominant, and alleles that produce different patterns of somatic mutation are codominant (McClintock, 1955, 1965a). Codominance of the two alleles the phenotypes of which are displayed in kernels 19 and 20 is illustrated by kernel 21, which is an *a-m1(5718)/a-m1(5720)* heterozygote and also contains an *Spm* element. Thus, a nonautonomous element affects the expression only of the allele with which it is associated. The autonomous element, on the other hand, is trans-acting and is required for somatic mutation of both alleles. Mutations that alter the pattern of somatic reversion occur in the presence, but not in the absence, of the trans-acting *Spm* element (McClintock, 1955). McClintock (1955) attributed the changes to the nonautonomous element at the locus. What is not known is whether the element itself undergoes internal sequence changes or whether the changes are due to local, short-range transpositions and, perhaps, rearrangements that leave the internal sequence of the element unaltered. McClintock (1965c) has obtained recombinational evidence that elements can be located at different sites within the *A* locus, and Nuffer (1962) has suggested that short-range transpositions may occur between components of the complex *A* locus. Moreover, McClintock (1963, 1965a, 1968) isolated two unstable alleles of the *A* locus in which somatic mutation produced colorless sectors that develop pigmented rims at the boundary with surrounding tissue, suggesting the production of two inactive gene products capable of complementation. Such alleles might also mutate somatically by short-range transposition of the element within the locus. It seems a reasonable conjecture that the behavior of an element upon trans-activation is determined both by the sequence of the element and the insertion site.

Perhaps the most striking feature of the genetic changes that affect the response of a nonautonomous element to trans-activation is that they alter developmental parameters. Each allele or state (in McClintock's terminology) has its own characteristic response, definable in terms of the frequency, developmental time of origin, and level of gene expression characteristic of somatic revertant sectors in the presence of *Spm,* as well as the level of gene expression in its absence. Each allele is therefore responsible for a characteristic pattern of gene expression in the organism.

2. Mutation of Unstable Alleles to Stable Alleles

The study of germinal mutations that show no further somatic instability in the presence of *Spm* has provided evidence that such mutations arise by transposition of the element away from the locus (McClintock, 1954, 1955; Peterson, 1961; Fowler and Peterson, 1974; Reddy and Peterson, 1976). Stable alleles are derived from unstable alleles with an integrated nonautonomous element in the presence, but not in the absence, of the autonomous *Spm* element (McClintock, 1954, 1955). Stable mutants derived from unstable mutants of the *A* locus vary in aleurone color from completely colorless (null mutants) to very deeply pigmented (*A*). The aleurone phenotype of most such mutants is the same in the presence and absence of the *Spm* controlling element. The frequency with which stable mutations arise, as well as their characteristic phenotype, varies from strain to strain in a way that is consistent with the somatic mutation pattern of the strain (McClintock, 1955). The *a-m1(5996)* (kernel 17) and *a-m1(5720)* (kernel 20) alleles, for example, produce early somatic mutations and a high frequency of stable mutations. The *a-m1(5996)* allele produces deeply pigmented somatic sectors in aleurone tissue and frequently produces stable mutants that show intense aleurone pigmentation (although stable, colorless and intermediate, palely pigmented mutants also occur). The *a-m1(5720)* allele generally produces palely pigmented aleurone sectors, as well as many stable mutants that show pale aleurone pigmentation. Very occasionally, small, deeply pigmented sectors are seen on kernels carrying the *a-m1(5720)* allele and *Spm,* suggesting that the locus is capable of reverting to a form that is expressed at a higher level. The *a-m1(5719A1), a-m1(5718),* and *a-m1(5996-340)* alleles, all of which show very late somatic mutation, have a low germinal mutation frequency. As noted earlier, such differences among strains support the hypothesis that germinal and somatic mutations occur by the same mechanism, differing only in whether they are pre- or postmeiotic.

There is a second type of stable genetic change that is not associated with transposition of the element away from the locus, but instead appears to affect the capacity of the element of undergo transposition. Such a mutation can affect the ability of the element to respond to the *m* function of *Spm* without altering its response to the *sp* function (McClintock, 1957, 1958, 1961a,b). Kernels 22 and 23 illustrate the phenotypes displayed by one such mutant allele of the *A2* locus. The aleurone layer of kernels that have the genetic constitution *a2-m1(A2)/a2,*[6] but have no *Spm,* is deeply pigmented (kernel 22). By contrast, kernels with the same genetic constitution and containing a standard *Spm* element have a completely colorless aleurone layer (kernel 23). That the *Spm* element present in kernel 23 is capable of eliciting somatic mutation is evident from the endosperm starch variegation due to the presence of the *wx-m8* allele, which is an unstable allele of the *Wx* locus the reversion of which is activated by *Spm* (Table III). The kernel has been cut and stained with iodine, and *Wx* sectors arising by somatic mutation of the *wx-m8* allele are evident. The mutant allele of the *A2* locus can therefore respond only to the *sp* function of the activating *Spm* element; this phenotype suggests the existence of a cis transposition defect in the element at the *A* locus. Similar derivatives of the *a-m5* allele have been isolated (McClintock, 1961b). The *a2-m1(A2)* allele proved valuable in analyzing the autonomous *Spm* element.

[6]McClintock designated mutants such as *a2-ml(A2)* as *class II* states of a locus and referred to this particular allele as the *A2* state of *a2-ml* (McClintock, 1957, 1958, 1971).

B. Autonomous Controlling Elements Belonging to the *Spm* Family

There are several different autonomous elements that belong to the *Spm* family, as defined by the activation test. The elements differ from each other in their interactions with nonautonomous elements of the same family. Because they are autonomous elements, sometimes of independent origin, it cannot always be established with certainty that they are mutational derivatives of each other. In discussing these elements, I will generally adhere to the terminology used by the authors who first described them, using new abbreviations or terms only when I believe that they will clarify or facilitate the discussion.

The mutations considered in the previous section are cis-dominant mutations affecting the nonautonomous element at a locus. In this section, I will be concerned with the trans-acting *sp* and *m* functions of autonomous *Spm* elements. I will describe factors that affect the ability of an *Spm* element to activate any one of a number of different mutations caused by insertion of the cognate nonautonomous element. Changes affecting the trans-acting functions of *Spm* are most readily identified in double mutants containing two different *Spm*-responsive unstable alleles, such as the *wx-m8* allele and an unstable allele of either the *A* or the *A2* locus, because a change in the *Spm* element simultaneously alters the response of both mutant loci. In addition, the *Spm* itself can cause unstable insertion mutations, and the analysis of such mutations has provided insight into the mechanism of *Spm* transposition.

1. *Spm-s*, *Spm-w*, and *Modifier*

Spm-w is an element belonging to the *Spm* family that activates somatic mutation later in development and at a lower frequency than does the standard *Spm* element. The interaction between a standard or *Spm-s* (McClintock, 1955, 1957, 1965a) element and several alleles of the *A* and *A2* loci was described and illustrated in the preceding section. *Spm-s* and *Spm-w* elements differ from each other in the frequency and timing of the somatic mutations that they elicit (McClintock, 1956c, 1957, 1961a, 1963). The pattern of small, deeply pigmented sectors that results when an *Spm-s* is present in a kernel having the genetic constitution *a-m1(5719A1)/a* at the *A* locus is illustrated in kernel 16. The pattern obtained with the same allele of the *A* locus in the presence of an *Spm-w* is illustrated in kernel 24. The deeply pigmented sectors are smaller and fewer in the aleurone layer of kernel 24 (*Spm-w*) than in that of kernel 16 (*Spm-s*). Thus, somatic reversion occurs later and less frequently when *a-m1(5719A1)* is activated by *Spm-w* than when it is activated by *Spm-s*. *Spm-s* is dominant to *Spm-w* in a heterozygote (McClintock, 1957). Peterson noted a similar "weak" *En* element, which he designated *En'* (Peterson, 1966).

It is primarily the *m* function that differs between an *Spm-s* and an *Spm-w* element, although there is some indication that the *sp* functions differ as well.

Spm-w elements transpose at a much lower frequency than do *Spm-s* elements, but they can be mobilized by an *Spm-s* element in trans (McClintock, 1965a,b). As is evident in kernels 16 and 24, both *Spm-s* and *Spm-w* elements are capable of suppressing the expression of the *A* locus in aleurone tissue. McClintock (1957) observed, however, that plants having an *a-m1(5719A1)* allele and an *Spm-w* showed a deeper color than plants of the same genetic constitution but having an *Spm-s* element. This observation suggests that the *sp* function is somewhat attenuated in *Spm-w*. Thus, although *Spm-w* and *Spm-s* elements appear to differ primarily in the developmental timing of *m* function, there is also some evidence of a difference in *sp* function. At least one of the *Spm-w* elements identified by McClintock appeared to mutate in its expression to *Spm-s* both somatically and germinally (McClintock, 1961a, 1963, 1965b). Peterson (1976a) observed that such a change from late to standard timing of somatic mutation accompanied the transposition of an *En* element, suggesting either that expression of the element can be influenced by its position or that the element is especially labile during transposition.

McClintock identified an element, which she designated *Modifier,* that interacts with both *Spm-s* and *Spm-w* (McClintock, 1956c, 1957, 1958, 1965a). The properties of the *Modifier* element suggest that it represents a defective element that can be rescued by complementation. The *Modifier* element itself does not affect the expression of a locus having an integrated, nonautonomous element belonging to the *Spm* family, nor does it transpose autonomously. That is, it shows neither the *sp* nor the *m* function of an *Spm* element. However, when present together with such a locus and either an *Spm-s* or an *Spm-w* element, *Modifier* increases the rate of somatic mutation of alleles that show a relatively low frequency, but not that of alleles that show a high frequency (McClintock, 1958). *Modifier* itself is a transposable element and belongs to the *Spm* family by the criteria applied here. The mechanism by which it affects the *m* function of an *Spm* element is not understood, nor is its origin clear. One interpretation is that this element represents a defective *Spm* element. The high somatic mutation frequency elicited by *Spm-w* and *Modifier* together may be the result of genetic complementation between these elements. The observation that *Modifier* can enhance somatic mutation even with an *Spm-s* element suggests that it may either be responsible for or promote the overproduction of enzymes involved in excision or transposition. Peterson reported the presence of a factor, termed *Rst,* that appears to be similar to *Modifier,* although its transposability has not been investigated (Peterson, 1976b). *Modifiers* have also been identified in other controlling element families (Rhoades, 1938; Ashman, 1959, 1960).

2. *SPM* ELEMENTS THAT UNDERGO CYCLIC CHANGES IN ACTIVITY

There are *Spm* elements that undergo cyclic changes in expression of both the *sp* and *m* functions of the element (McClintock, 1957, 1958, 1959, 1961b, 1962,

1971). Cycling derivatives have also been described for the related *En* element (Peterson, 1966) and the unrelated *Ac* (McClintock, 1964) and *Dt* (Doerschug, 1968c) elements. I have used the designation *Spm-c* for such derivatives of the *Spm* element. There are *Spm-c* elements that undergo reversible changes in activity rather frequently, at intervals of a few cell divisions during plant development, and others that change infrequently (McClintock, 1958, 1959). The phenotypic effect of an *Spm-c* that undergoes frequent changes to an inactive state late in development is illustrated by kernel 25. The genetic constitution of the kernel is *a2-m1(A2)/a2*, and it contains an *Spm-c* element. This allele produces deeply pigmented aleurone tissue in the absence of *Spm* (kernel 22) and colorless aleurone tissue in the presence of *Spm-s* (kernel 23), because it is unable to respond to the *m* function. The sectorial pattern of pigment expression seen in the aleurone layer of kernel 25 is therefore a consequence of the inactivation of the *Spm* element. The pigmented areas correspond to sectors in which the *sp* function is not expressed (inactive *Spm*, or Spm^{i}), whereas the colorless areas correspond to sectors in which the element's *sp* function is expressed (active *Spm*, or Spm^{a}). Within the large pigmented sector are several colorless subsectors that reflect the return of the *Spm* to an active state. Thus, the *Spm-c* element present in this strain undergoes frequent reversible changes that alter its capacity to suppress expression of the *a2-m1(A2)* allele of the *A2* locus.

The *sp* and *m* functions of an *Spm-c* are activated and inactivated as a unit (McClintock, 1958). The coincident activation and inactivation of both functions become evident when an *Spm-c* is combined with an unstable allele that responds to both the *sp* and *m* functions of the *Spm* element. Coincident inactivation of the two functions is illustrated in kernel 26, the genetic constitution of which is *a-m1(5719A1)/a, wx-m8/wx,* and that also contains an *Spm-c* element. Half of the kernel shows the pattern of small, deeply pigmented aleurone sectors characteristic of this allele in the presence of an *Spm-s*. The other half of the kernel's aleurone layer shows the uniform pale pigmentation characteristic of the *a-m1(5719A1)* allele in the absence of *Spm,* except for a few small areas of colorless tissue that represent sectors in which the *Spm-c* has returned to the active form. A portion of the kernel has been cut away, and the underlying endosperm has been stained with iodine to reveal the *Wx* phenotype. *Wx* sectors resulting from somatic reversion of *wx-m8* appear only in endosperm tissue underlying the variegated portion of the aleurone tissue. The phenotype of this kernel illustrates the coincident change in expression of both the *sp* and *m* functions of the *Spm*. Thus, it appears that expression of the *m* function is contingent on expression of the *sp* function. Whether this is because *sp* regulates *m* gene expression in some way or because both gene products are required for excision and transposition cannot be discerned from the existing information.

Spm-c elements appear to undergo cyclic changes in expression independently of each other (McClintock, 1959, 1961a). The autonomy of the cyclic changes in different elements has been inferred from the observation that the size and number of pigmented sectors observed with the *a2-m1(A2)* allele (as well as with other alleles) decrease with increasing numbers of *Spm-c* elements (McClintock, 1957, 1958, 1959, 1961a). The *Spm-c* dosage can readily be changed in the endosperm because the female parent contributes two identical haploid genomes to the primary endosperm nucleus, whereas the male parent contributes one. Thus, if the *Spm-c* is present only in the female parent, the endosperm will have two copies of the element, whereas if it is present in the male parent, the endosperm will have only one copy. Kernel 25 came from a cross in which the male parent contributed the *Spm-c,* whereas kernel 27 came from an identical cross in which the female parent contributed the *Spm-c*. When the two kernels are compared, it is evident that the average size of the pigmented sectors is smaller in kernel 27 than in kernel 25. Kernels of the same genetic constitution but carrying three *Spm-c* elements have few small, pigmented areas, and similar kernels with four *Spm-c* elements are almost completely colorless (McClintock, 1971). Because the *sp* function of the element is trans-acting, only one element need be in the active phase for the expression of the *a2-m1(A2)* allele to be suppressed, resulting in a colorless aleurone phenotype. In order for the *a2-m1(A2)* allele to be expressed and the aleurone tissue to be pigmented, all of the resident *Spm-c* elements must be inactive. It follows that if the elements undergo changes in activity independently, then the average size of the pigmented sectors, containing all resident *Spm-c* elements in an inactive form, should decrease with an increase in the number of such elements present. The frequency and developmental timing of reversals in the activity phase are genetically controlled and subject to heritable changes. McClintock (1959) noted that phase reversals were associated with changes in duration of the active and inactive phases and suggested that the modification responsible for changing the phase likewise resets the duration of each phase.

There is evidence that a frequently cycling *Spm-c* can transiently activate an *Spm-c* element in an inactive phase of long duration. Such transient activation has been inferred from the extremely interesting observation that an *Spm* element in a prolonged inactive phase contributes to the *Spm-c* dosage effect in the same way as an element undergoing frequent changes in the phase of activity (McClintock, 1958, 1959). *Spm* elements have been identified that have inactive phases extending over several plant generations. A plant with an *Spm*-responsive allele of the *A* or *A2* locus and an inactive *Spm* of this type has the phenotype of a similar plant lacking *Spm.* Evidence that an *Spm* element is still present is shown by the occasional appearance of plant sectors or kernel sectors exhibiting the phenotype characteristic of the allele in the presence of an active *Spm.* When an active *Spm-c* is introduced into such a strain, however, a fraction of the kernels exhibits the phenotype observed when an active *Spm-c* element is present in each parent. Kernels exhibiting this phenotype have received both the inactive *Spm* and the active *Spm-c*. Thus, the *Spm-c* appears to activate the inactive *Spm* element so that it undergoes changes in the activity phase at a similar time in development as the active *Spm-c*. This change in the expression of the inactive

Spm element is a transient effect expressed in the *Spm* heterozygote, because the two elements segregate unchanged (McClintock, 1959).

3. *SPM* ELEMENTS THAT ARE EXPRESSED DIFFERENTIALLY WITHIN A PLANT OR A TISSUE

Spm and *En* elements have been identified the phase of activity of which depends on their location within a plant or a single tissue (McClintock, 1965a, 1971; Peterson, 1966; Fowler and Peterson, 1978). Kernel 28 is an example of the phenotypic effect of such an element. This kernel has the genetic constitution *a-m1(5719A1)/a* and carries an *Spm* element that is active in the base of the kernel and inactive in its crown. Unlike the *Spm-c* elements described above, which undergo inactivation sectorially throughout the kernel, this *Spm* element is consistently inactive only in a small area at the top of each kernel. Peterson (1966) identified such an altered form of the *En* controlling element and designated it *En-flow*. He also identified an *En* element (designated *En-crown*) that showed an inverse pattern of expression, being consistenly inactive in the sides of the kernel and active only in the crown. McClintock (1971) has isolated strains in which the *Spm* element is consistently inactive in the kernels of ears on the main stalk of the plant and active in the kernels on tiller ears (Fig. 1b). Fowler and Peterson (1978) reported an *En* element that reproducibly caused later somatic mutations in main stalk ears than in tiller ears. The behavior of such elements suggests that they can be subject to regulation by factors that are differentially distributed within a plant or a single tissue. That this type of element is not unique to the *Spm* family is indicated by Doerschug's (1973) description of a *Dt* element resembling *En-crown*.

C. Presetting

In all of the examples given so far to illustrate the interactions between autonomous controlling elements and loci with nonautonomous elements, it is implicit that the loss or inactivation of an element is immediately reflected by a change in the phenotype of a plant or plant part. The inactivation of the *Spm-c*, for example, is manifested in changes in the pigmentation of sectors of aleurone tissue in a single kernel. The close correlation between genotype and phenotype is a general rule that *presetting* violates. Briefly, presetting is a unique pattern of gene expression observed with certain *Spm*-responsive alleles that have been exposed to an active *Spm* element transiently during plant development (McClintock, 1963, 1964, 1965b, 1967b). *Spm* presets a gene to be expressed in a unique way after the *Spm* has been removed either by somatic loss or by meiotic segregation. Presetting at the *A* locus has been observed with two similar derivatives of the *a-m2* allele, designated *a-m2(7977B)* and *a-m2(7995)*. Before discussing presetting, however, a brief description of the *a-m2* allele is necessary, because its behavior is distinctly different from that of other *Spm* alleles.

The properties of the *a-m2* allele suggest that *Spm* is inserted at the locus in such a way that the locus is coordinately expressed with the element itself (McClintock, 1951b, 1961b, 1962). *Spm* is located at the *A* locus in *a-m2*, and expression of the locus is directly correlated with the activity phase of *Spm*. When the *Spm* element is inactive, the *a-m2* allele is not expressed and produces a colorless aleurone layer. By contrast, certain derivatives of the *a-m1* and *a2-m1* alleles are expressed and produce a moderately or deeply pigmented aleurone layer when *Spm* is absent or inactive (see kernels 15 and 22). When the *Spm* located at the *A* locus in *a-m2* is active, the locus is expressed at a low level, giving the kernel aleurone layer a uniform pale pigmentation on which deeply pigmented somatic mutant sectors appear. By contrast, the aleurone layer is colorless when an active *Spm* is present together with *a-m1*, and deeply pigmented somatic sectors appear on the colorless background. Thus, in the *a-m2* allele, the expression of the *A* locus is directly correlated with the expression of the *Spm* element, suggesting that the relationship between the controlling element and the coding sequence at the locus may be different in this mutant than in other strains with mutations caused by elements belonging to the *Spm* family.

In the two derivatives of the original *a-m2* allele that exhibit presetting, *a-m2(7977B)* and *a-m2(7995)*, the autonomous *Spm* at the *A* locus has been converted to or replaced by a nonautonomous element belonging to the same family (McClintock, 1963, 1964, 1965b). Both alleles yield a pigmentation pattern resembling that of the *a-m2* allele in the presence of an autonomous *Spm* element (kernel 29). No aleurone pigmentation is observed with either allele in the absence of the *Spm* element. If a plant that carries either of these alleles and one active *Spm* element located on a different chromosome (e.g., *a-m2(7977B)/a*, *Spm* present) is crossed by a plant that is homozygous for the recessive allele of the *A* locus and lacks *Spm* (*a/a*, no *Spm*), only half of the kernels that have the genetic constitution *a-m2(7977B)/a* will receive *Spm*, because the element is assorting independently of the *A* locus. Those that receive *Spm* and the *a-m2(7977B)* allele exhibit the variegated phenotype evident in kernel 29. Those that receive the *a-m2(7977B)* allele, but no *Spm*, are expected to be colorless; however, among them are examples of kernels showing the irregularly mottled pattern of pigmentation evident in kernel 30 (McClintock, 1964). The presence or absence of *Spm* in kernels exhibiting such a pigmentation pattern can be judged either by incorporating a second *Spm*-responsive allele, such as *wx-m8*, into the strain or by growing mottled kernels and testing the resulting plants for the presence of *Spm* (McClintock, 1961b, 1963, 1964). Irregular pigmentation patterns of the type exhibited in kernel 30 are observed only in kernels that do not contain an active *Spm*. However, such patterns are detected only if the *Spm* is initially present in the same plant as the *a-m2(7977B)* or the *a-m2(7995)* allele. Thus, for example, if the *Spm* element in the cross described above resided in the *a* parent rather than the *a-m2(7977B)* parent, no such irregular pigmentation patterns would occur in progeny kernels, even though the crosses are formally equivalent (McClintock, 1963, 1964). Such patterns are also not observed if *Spm* is present in neither parent (McClintock, 1963, 1964). It follows that the *Spm* element must be present initially in a plant together with either the *a-m2(7977B)* or the *a-m2(7995)* allele in order for the pattern to be expressed in kernels receiving the responsive allele, but not the element, following meiotic segregation. Thus, the premeiotic exposure of either allele presets it for later expression in the absence of the element. The preset pattern clearly differs from the pattern of somatic reversion obtained with either allele in the presence of an active *Spm* (compare kernels 29 and 30). The pigment intensity varies from kernel to kernel among those showing preset patterns and can generally be characterized as an irregular, mottled pattern (McClintock, 1964).

Further studies on the presetting phenomenon have revealed that such patterns are, to a large extent, not heritable. The results of genetic tests on plants derived from irregularly pigmented kernels showed that *Spm* was not present in them and that the patterns were reproduced sporadically in a few progeny kernels with the *a-m2(7995)* allele and not at all in plants carrying the *a-m2(7977B)* allele (McClintock, 1964, 1965b). The unstable alleles themselves did not appear to be altered by having undergone presetting. Upon reintroduction of an active *Spm* element, alleles that have been preset showed an unaltered capacity to respond to the element. Thus, the preset patterns were "erased" in subsequent generations and had no permanent effect on the locus. Moreover, the preset pattern could be reestablished by a cross of the same type in which presetting was originally observed (McClintock, 1963, 1964).

Presetting appears to be independent of the developmental stage and requires only the initial presence of an active *Spm* element in a plant carrying either the *a-m2(7977B)* or *a-m2(7995)* allele (McClintock, 1964). Plants can undergo sectorial losses of *Spm* early in development, either by virtue of excision of the element or as a consequence of transposition, followed by mitotic segregation. Sectorial losses can occur sufficiently early in development to eliminate the *Spm* element from an entire tiller and its ear, or later in development to produce subsectors of kernels lacking *Spm* on ears in which it is present (McClintock, 1964). Preset patterns are observed within both large and small *Spm*-free sectors, suggesting that it is only necessary for the active *Spm* element to be present together with the appropriate allele early in plant development (McClintock, 1964). *Spm-w* and *Spm-s* elements are equally effective in presetting, but an inactive *Spm* is incapable of presetting either allele for later expression, suggesting that the *sp* component of the element is responsible for presetting (McClintock, 1965b).

McClintock has observed analogous presetting phenomena with the unstable alleles of the *C2* locus caused by insertion of elements belonging to the *Spm* family (McClintock, 1967b). As early as 1917, Emerson reported the sporadic occurrence of "freak" kernels or sectors of kernels in strains carrying the *Mp* element at the *P* locus. Such kernels showed nonheritable changes in pericarp and cob pigmentation, suggesting that analogous presetting may occasionally occur in the *Ac–Ds* family (Emerson, 1917). In both of these studies, it was observed that the phenotype of the underlying cob and pericarp tissue did not correspond to the genotype of the kernels, despite a common developmental origin. McClintock suggested that an allele of this type could be preset for expression in one type of tissue but that the setting could be erased in germ-line tissue or early in the development of the embryo (McClintock, 1967b).

D. Mechanism of *Spm* Transposition

Because *Spm* does not show dosage-dependent changes in its ability to trans-activate genetic changes, the lineage analyses that provided insight into the mechanism of *Mp* transposition cannot readily be performed with elements of this family. Nonetheless, analysis of stable mutants derived from the *a-m2* and *a-m5* alleles that harbor *Spm* at the *A* locus yielded information that is consistent with a nonreplicative transposition mechanism analogous to that deduced for the *Mp* element. In studies on the original *a-m2* allele described earlier, McClintock (1962) reported that there was a good correlation between transposition of the *Spm* element away from the locus and the production of stable mutants. Among the stable mutants were alleles that resembled the original *A* allele in expression, as well as alleles that gave a rather irregular expression of the locus. Approximately 40% of the stable mutants analyzed had one or more *Spm* elements in the genome. Less than 10% of such mutants retained an *Spm* element either at the locus or nearby on the same chromosome, suggesting that transposition is generally accompanied by the removal of the element from the donor site and that donor and recipient sites are commonly unlinked. In a similar study on the *En* element at the *A* locus, Peterson (1970a) reported transposition of the element to a site on the same chromosome in 25% of the stable mutants tested.

The observation that plant and kernel variegation patterns reflect the dosage of an *Spm-c* permits some further parallels to be drawn between the mechanisms of *Spm* and *Mp* transposition. Twinned sectors revealing changes in *Spm-c* dosage appear in plants, ears, and individual kernels carrying the *a2-m1(A2)* allele and an *Spm-c* element. One-half of such a twinned sector shows the deeply pigmented phenotype observed in the absence of *Spm*, whereas the other half shows the sparser variegation observed when the *Spm-c* dosage is increased (McClintock, 1971). The phenotypes of such twinned sectors resemble those observed on ears of plants carrying the *Mp* elements at the *P* locus (Greenblatt and Brink, 1962) and suggest that the *Spm* element lost at a mitotic division from one daughter cell is gained by the other daughter cell, increasing the number of elements present in that cell and its progeny. Although this phenomenon has not been analyzed in detail, as it has been for *Mp*, it appears reasonable that the twinned sectors arise as a result of transposition of the element from only one of the two sister chromatids, followed by its segregation with the other sister chromatid at cell division.

A consideration of the different kinds of stable mutants obtained during transposition permits some further inferences to be made about the *Spm* transposition mechanism. Among the stable mutations already discussed, mutations that restore full gene expression generally occur less frequently than do those that result in a null or moderate level of gene expression (McClintock, 1962; Peterson, 1970a), suggesting that imprecise excision occurs frequently. However, as noted earlier, the types of stable mutants recovered and their relative frequencies are strain specific. These observations reflect differences among the inserted elements, the relative sensitivity of the insertion site to sequence changes occurring as a consequence of the excision process, or both.

Nonautonomous elements have repeatedly been derived from (or replaced by) autonomous elements, and there is some evidence that the origin of a nonautonomous element is associated with transposition of the element (Peterson, 1961, 1968; McClintock, 1962; Reddy and Peterson, 1976). Typical two-element systems have been derived by spontaneous mutation in strains with an autonomous *Spm* element at the *A* locus (McClintock, 1962). In some instances, the mutation was originally identified by a change in the somatic reversion pattern in a single kernel. A strain derived from such a kernel has both an autonomous *Spm* element no longer located at the *A* locus and an element that behaves as a typical nonautonomous element of the *Spm* family at the *A* locus. Not surprisingly, mutant strains of this type can also lack an *Spm* element, and the presence of a nonautonomous element at the *A* locus is discovered only after the reintroduction of an autonomous *Spm* element (McClintock, 1962). These observations suggest the existence of a mode of transposition in which a copy of the element transposes without concomitant excision of the resident copy, resembling that observed in bacteria (Shapiro, 1979; Chapters 4 and 5, this volume). However, because mutations that give rise to nonautonomous elements occur at a much lower frequency than mutations that give rise to stable alleles, they may represent independent mutations that affect the *m* function of an autonomous element and that occur either in the absence of a transposition event or subsequent to a transposition event in the same cell lineage. Some evidence in support of the hypothesis that the origin of nonautonomous elements is linked to transposition was provided by the analysis of plants derived from the *a-m5* strain having an autonomous *Spm* element at the *A* locus (McClintock, 1962). Plants were grown from kernels exhibiting a somatic mutation pattern unlike that of the parent strains. The change in the somatic mutation pattern proved heritable in all 10 of the plants analyzed, and seven of the plants had a typical two-element system. Although the sample size is small, these results suggest a correlation between transposition and the origin of a nonautonomous element.

V. The Origin of Controlling Elements

It is likely that controlling elements or their components are regular residents of the maize genome. McClintock's studies on transposable elements originated in experiments on plants having broken chromosomes undergoing the BFB cycle (McClintock, 1942a,b). What emerged unexpectedly was an array of unstable mutations affecting many different loci (McClintock, 1946, 1950a). The study began with mutations that affect plant characters and later focused on mutations that affect kernel phenotypes because such characters are more easily analyzed. It was subsequently established in independent experiments by McClintock (1950b, 1951b, 1965a) and Doerschug (1973) that chromosome breakage was an important factor in the emergence of the phenotype associated with the presence of a transposable element. In both experiments, a recessive allele of the *A* locus responsive to *Dt* was present in plants that received a broken chromosome 9 from each parent, as well as in controls that did not. The genetic stocks used had never exhibited behavior indicating the presence of *Dt*. Somatic mutation of the *a* allele to the dominant *A* allele was observed in 0.1–0.2% of the kernels, but only when chromosome breakage was also occurring in the strain. Doerschug (1973) isolated two new *Dt* elements by this means, one of which was similar to a previously isolated *Dt* element and the other of which resembled the *En-crown* element belonging to the *Spm* family (Peterson, 1966).

The consequences in progeny plants of introducing a broken chromosome from each parent have been studied in detail (McClintock, 1942a,b, 1943, 1944, 1951a, 1978b). The broken homologs fuse early in kernel development, initiating the BFB cycle. McClintock (1942a, 1951a) designated this the *chromosome* type of BFB cycle. Unlike the chromatid type of BFB cycle, which occurs in the endosperm but generally not in the embryo, the chromosome type of BFB cycle can persist in the developing plant, giving rise to inversions and both terminal and internal deletions (McClintock, 1941, 1942a, 1951a, 1978b). In addition, broken chromosomes participate in the formation of grossly modified translocation products comprising sections of a broken chromosome and a nonhomologous chromosome (McClintock, 1978b). Thus, the introduction of the broken chromosomes initiates widespread genomic change. It is under these conditions that unstable mutations occur at a high frequency (McClintock, 1946). Neuffer (1966) and Bianchi *et al.* (1969) further showed that physical agents such as X and ultraviolet rays evoked somatic expression of transposable elements, including *Dt, Ac,* and *Spm*.

Because autonomous controlling elements affect nonautonomous elements in trans, their presence on any chromosome can be detected. In addition, an inactive *Spm* controlling element can be detected by its ability to contribute to the *Spm-c* dosage effect (McClintock, 1958, 1959), and modifiers, which may represent a class of defective element (McClintock, 1956c, 1957, 1958), can be detected by their ability to interact with an *Spm* element. Thus, the interactions between elements permit the detection of several otherwise cryptic elements. The very fact that controlling elements can be subjected to rather standard genetic analyses in which they behave as single Mendelian factors, albeit transposable, implies that the number of active copies is small. Although the components may exist separately without being detected by the genetic criteria employed, the striking observation is that the properties of independently isolated controlling elements are rather similar. The complex *Spm* element, for example, strongly resembles the *En* element identified by Peterson (Peterson, 1961, 1965, 1966; Fowler and Peterson, 1974). It seems improbable, therefore, that controlling elements are separately assembled from undetectable components and is more likely that such elements exist in a cryptic form in the genome.

VI. Discussion

The similarities between maize controlling elements and the transposable elements described in other chapters of this volume are evident. Controlling elements can transpose with or without concomitant chromosomal rearrangements, cause unstable mutations at a variety of loci, and produce local chromosomal aberrations and rearrangements without transposing. In one particularly well-investigated case, it has clearly been shown that transposition is closely associated with chromosome replication and results in the removal of the element from one of the two sister chromatids, with concomitant insertion at another site, often nearby, either on the nondonor sister chromatid or in an unreplicated region of the chromosome. It is suspected that simple excision, as well as replicative and nonreplicative intrachromatid transpositions, also occur at a low frequency.

A wide variety of controlling element variants has been discovered and studied. Included are elements that appear to have cis-defects in transposition, elements that can be trans-activated to transpose, and elements that undergo reversible changes in their ability to transpose and trans-activate the transposition of nonautonomous elements. Perhaps the most striking property of maize controlling elements is their ability to sense developmental time, catalyzing transpositions and other genetic changes within a restricted developmental period. This capacity is manifested in the fairly uniform size of independent somatic sectors arising from cells in which a phenotypically detectable genetic change caused by a controlling element has occurred. Controlling elements can undergo heritable alterations in the nature, frequency, and developmental timing of the genetic changes that they evoke. Moreover, some isolates behave as though they are programmed to undergo such changes at a specific time in development. There are, in addition, elements the behavior of which is determined by their location within the organism as a whole or within a particular tissue. Thus, each controlling element family comprises a substantial group of related elements, some of which appear to have major functions that others lack. In addition, family members show many heritable differences that affect their developmental profiles of expression, as well as the nature and timing of the genetic changes and transpositions that they undergo. It is the rich subtlety of their ability to respond to and control developmental parameters that constitutes one of the most fascinating aspects of controlling elements in maize.

Equally extraordinary is the capacity of controlling elements to reprogram genes. Insertion of a controlling element at a locus brings expression of that locus under the control of the element. The developmental pattern of expression for the locus is subsequently determined by the nature of the insertion mutation and the properties of the inserted element. Insertion mutations can alter both the developmental profile and the level of gene expression. The interaction between the autonomous and nonautonomous elements belonging to the *Spm* family imposes a level of regulation consequent on the capacity of the element's suppressor function to inhibit expression of a locus having an inserted nonautonomous element belonging to this family. Transposition or loss of the element from the locus during development gives rise to a pattern of gene expression in the organism that is dictated by the properties of the controlling elements. Because an autonomous element can affect insertion mutations at different loci, a single element can simultaneously determine the pattern of expression of more than one locus. Despite control by a common autonomous element, a nonautonomous element at a locus can determine the particular pattern of expression for the locus, with the consequence that the different loci under common control can each be expressed in a unique way.

McClintock focused on the interactions between transposable elements and genes, attributing a regulatory function to the elements (McClintock, 1956a,b, 1965a, 1967a, 1978a). Although it is evident that controlling elements have genetic mechanisms for sensing developmental time and position, the view that they are fundamental to gene regulation in development has not been widely accepted (Brink, 1958b; Fincham and Sastry, 1974; Campbell, 1980). Current notions on the nature and function of such elements vary from postulates of a role for transposable elements in evolution to the suggestion that such elements are parasitic sequences the ability of which to outreplicate chromosomes prevents their elimination (Cohen, 1976; Nevers and Saedler, 1977; Kleckner, 1981; Doolittle and Sapienza, 1980; Orgel and Crick, 1980; Campbell, 1981). There is a growing appreciation that mechanisms for rapid restructuring of genomes must participate in the long-term genetic change we call evolution and that transposable elements are commonly involved in chromosomal rearrangements in eukaryotes, as well as prokaryotes (Chapters 4, 5, 8, and 9). There are tantalizing indications from both genetic and biochemical studies that developmental patterns of gene expression can be determined by both cis-acting elements at a locus and trans-acting elements located elsewhere in the genome (Tan, 1946; McClintock, 1967a; Paigen, 1979; Rabinow and Dickinson, 1981) and that transposable elements can alter gene regulation (Schon *et al.*, 1981; Errede *et al.*, 1980). Two well-studied cases in which DNA rearrangements are known to regulate gene expression are reviewed in Chapters 13 and 14. On balance, however, it must be acknowledged that our understanding of the importance of transposable elements in genome structure and gene expression is still quite limited. The insight that inherent in the transposition mechanism is a means by which transposable elements can outreplicate the genome as a whole has unfortunately drawn attention away from other important implications of their behavior. But even proponents of the notion that transposable elements are molecular parasites acknowledge that it is difficult to imagine that such sequences have, and have had, no impact on gene structure and function (Doolittle and Sapienza, 1980; Orgel and Crick, 1980). Indeed, the ability of transposable elements to reprogram, restructure, and relocate genes is no longer in doubt. What we do not know is how important transposable elements have been in shaping and reshaping genes and genomes historically. And we have not yet begun to come to terms with the implications of genetic mechanisms that control the tissue specificity, timing, and nature of element-associated genetic changes in development, the existence of which is so elegantly revealed by the studies on maize controlling elements.

Acknowledgments

I am grateful to Drs. B. McClintock and I. Greenblatt for materials and information and to Drs. D. Botstein, D. Chaleff, E. Coe, G. Rubin, J. Shapiro, A. Spradling, and S. Wessler for their helpful criticism of this manuscript. I take full responsibility for the interpretations presented herein.

References

Anderson, E. G. (1924). Pericarp studies in maize. II. The allelomorphism of a series of factors for pericarp colors. *Genetics* **9,** 442–453.

Ashman, R. B. (1959). Transposability of M^{st}, a modifier of stippled aleurone. *Maize Genet. Coop. News Lett.* **33,** 122–123.

Ashman, R. B. (1960). Stippled aleurone in maize. *Genetics* **45,** 19–34.

Barclay, P. C., and Brink, R. A. (1954). The relation between Modulator and Activator in maize. *Proc. Natl. Acad. Sci. USA* **40,** 1118–1126.

Bianchi, A., Salamini, F., and Restaino, F. (1969). Concomitant occurrence of different controlling elements. *Maize Genet. Coop. News Lett.* **43,** 91.

Brink, R. A. (1958a). A stable somatic mutation to colorless from variegated pericarp in maize. *Genetics* **43,** 435–447.

Brink, R. A. (1958b). Mutable loci and development of the organism. *J. Cell. Comp. Physiol.* **52,** Suppl. 1, 169–196.

Brink, R. A. (1973). Paramutation. *Annu. Rev. Genet.* **7,** 129–152.

Brink, R. A., and Barclay, P. C. (1954). An accessory chromosomal element associated with variegated pericarp in maize. *Science (Washington, D.C.)* **119,** 577–578.

Brink, R. A., and Nilan, R. A. (1952). The relation between light variegated and medium variegated pericarp in maize. *Genetics* **37,** 519–544.

Brink, R. A., and Williams, E. (1973). Mutable R-Navajo alleles of cyclic origin in maize. *Genetics* **73,** 273–296.

Burr, B., and Burr, F. A. (1980). Detection of changes in maize DNA at the *Shrunken* locus due to the intervention of *Ds* elements. *Cold Spring Harbor Symp. Quant. Biol.* **45,** 463–465.

Burr, B., and Burr, F. A. (1981). Controlling-element events at the *Shrunken* locus in maize. *Genetics* **98,** 143–156.

Campbell, A. (1980). Some general questions about movable elements and their implications. *Cold Spring Harbor Symp. Quant. Biol.* **45,** 1–9.

Campbell, A. (1981). Evolutionary significance of accessory DNA elements in bacteria. *Annu. Rev. Microbiol.* **35,** 55–83.

Chaleff, D., Mauvais, J., McCormick, S., Shure, M., Wessler, S., and Fedoroff, N. (1981). Controlling elements in maize. *Carnegie Inst. Washington Year Book* **80,** 158–174.

Chen S.-M., and Coe, E. H., Jr. (1977). Control of anthocyanin synthesis by the *C* locus in maize. *Biochem. Genet.* **15,** 333–346.

Chourey, P. S. (1981a). Electrophoretic analysis of sucrose synthetase proteins in the complementing heterozygotes at the *Shrunken* locus in maize. *Theor. Appl. Genet.* **59,** 231–234.

Chourey, P. S. (1981b). Genetic control of sucrose synthetase in maize endosperm. *Mol. Gen. Gent.* **184,** 373–376.

Chourey, P. S., and Nelson, O. E. (1976). The enzymatic deficiency conditioned by the *shrunken*-1 mutations in maize. *Biochem. Genet.* **14,** 1041–1055.

Chourey, P. S., and Schwartz, D. (1971). Ethyl methanesulfonate-induced mutations of the *Sh*1 protein in maize. *Mutat. Res.* **12,** 151–157.

Coe, E. H., Jr. (1955). Anthocyanin synthesis in maize, the interaction of *A*2 and *Pr* in leucoanthocyanin accumulation. *Genetics* **40,** 568.

Coe, E. H., Jr. (1978). The aleurone tissue of maize as a genetic tool. *In* "Maize Breeding and Genetics" (D. B. Walden, ed.), pp. 447–459. Wiley, New York.

Coe, E. H., Jr., and Neuffer, M. G. (1977). The genetics of corn. *In* "Corn and Corn Improvement" (G. F. Sprague, ed.), pp. 111–223. Amer. Soc. Agronomy, Madison, Wisconsin.

Cohen, S. N. (1976). Transposable genetic elements and plasmid evolution. *Nature* (*London*) **263,** 731–738.

Doerschug, E. (1968a). Transpositions of Dt_1. *Maize Genet. Coop. News Lett.* **42,** 22.

Doerschug, E. (1968b). Stability of Dt_1. *Maize Genet. Coop. News Lett.* **42,** 24–26.

Doerschug, E. (1968c). Activation cycles of Dt_1^B. *Maize Genet. Coop. News Lett.* **42,** 26–28.

Doerschug, E. B. (1973). Studies of Dotted, a regulatory element in maize. *Theor. Appl. Genet.* **43,** 182–189.

Doolittle, W. F., and Sapienza, C. (1980). Selfish genes, the phenotype paradigm and genome evolution. *Nature (London)* **284,** 601–603.

Dooner, E. B. (1980a). Genetic regulation of anthocyanin biosynthetic enzymes in maize. *Genetics* **94,** s29.

Dooner, H. K. (1980b). Regulation of the enzyme UFGT by the controlling element *Ds* in *bz-m*4, an unstable mutant in maize. *Cold Spring Harbor Symp. Quant. Biol.* **45,** 457–462.

Dooner, H. K., and Nelson, O. E. (1977a). Genetic control of UDPglucose: flavonol 3-*O*-glucosyltransferase in endosperm of maize. *Biochem. Genet.* **15,** 509–519.

Dooner, H. K., and Nelson, O. E. (1977b). Controlling element-induced alterations in UDPglucose: flavonoid glucosyltransferase, the enzyme specified by the *bronze* locus in maize. *Proc. Natl. Acad. Sci. USA* **74,** 5623–5627.

Dooner, H. K., and Nelson, O. E. (1979). Interaction among *C, R,* and *Vp* in the control of the *Bz* glucosyltransferase during endosperm development in maize. *Genetics* **91,** 309–315.

Döring, H.-P., Geiser, M., and Starlinger, P. (1981). Transposable element *Ds* at the shrunken locus in *Zea mays. Mol. Gen. Genet.* **184,** 377–380.

Echt, C. S., and Schwartz, D. (1981). Evidence for the inclusion of controlling elements within the structural gene at the *waxy* locus in maize. *Genetics* **99,** 275–284.

Emerson, R. A. (1914). The inheritance of a recurring somatic variation in variegated ears of maize. *Am. Nat.* **48,** 87–115.

Emerson, R. A. (1917). Genetical studies of variegated pericarp in maize. *Genetics* **2,** 1–35.

Emerson, R. A. (1929). The frequency of somatic mutation in variegated pericarp of maize. *Genetics* **14,** 488–511.

Errede, B., Cardillo, T. S., Wever, G., and Sherman, F. (1980). Studies on transposable elements in yeast. I. ROAM mutations causing increased expression of yeast genes: their activation by signals directed toward conjugation functions and their formation by insertion of Ty1 repetitive elements. *Cold Spring Harbor Symp. Quant. Biol.* **45,** 593–602.

Fedoroff, N., Mauvais, J., and Chaleff, D. (1983). Molecular studies on mutations at the *Shrunken* locus in maize caused by the controlling element *Ds. J. Mol. Appl. Genet.* (in press).

Fincham, J. R. S., and Sastry, G. R. K. (1974). Controlling elements in maize. *Annu. Rev. Genet.* **8,** 15–50.

Fink, G., Farabaugh, P., Roeder, G., and Chaleff, D. (1980). Transposable elements (Ty) in yeast. *Cold Spring Harbor Symp. Quant. Biol.* **45,** 575–580.

Fowler, R., and Peterson, P. A. (1974). The *a2-m* (r-pa-pu) allele of the *En*-controlling element system in maize. *Genetics* **76,** 433–446.

Fowler, R. G., and Peterson, P. A. (1978). An altered state of a specific *En* regulatory element induced in a maize tiller. *Genetics* **90,** 761–782.

Friedemann, P. F., and Peterson, P. A. (1980). The *Ub* controlling element system. *Maize Genet. Coop. News Lett.* **54,** 2–3.

Gonella, J. A., and Peterson, P. A. (1977). Controlling elements in a tribal maize from Colombia: *Fcu,* a two-unit system. *Genetics* **85,** 629–645.

Gonella, J. A., and Peterson, P. A. (1978). The *Fcu* controlling-element system in maize. II. On the possible heterogeneity of controlling elements. III. On the variable dilute pigmenting capacity of r-cu. *Mol. Gen. Genet.* **167,** 29–36.

Greenblatt, I. M. (1966). Transposition and replication of modulator in maize. *Genetics* **53,** 361–369.

Greenblatt, I. M. (1968). The mechanism of modulator transposition in maize. *Genetics* **58,** 585–597.

Greenblatt, I. M. (1974a). Proximal–distal polarity of Modulator transpositions upon leaving the *P* locus. *Maize Genet. Coop. News Lett.* **48,** 188–189.

Greenblatt, I. M. (1974b). Modulator: a modifier of crossing over. *Maize Genet. Coop. News Lett.* **48,** 189–191.

Greenblatt, I. M. (1974c). Movement of Modulator in maize: a test of an hypothesis. *Genetics* **77,** 671–678.

Greenblatt, I. M. (1981). Enhancement of crossing-over by the transposable element, Modulator, in maize. *Maydica* **26,** 133–140.

Greenblatt, I. M., and Brink, R. A. (1962). Twin mutations in medium variegated pericarp maize. *Genetics* **47,** 489–501.

Greenblatt, I. M., and Brink, R. A. (1963). Transpositions of Modulator in maize into divided and undivided chromosome segments. *Nature (London)* **197,** 412–413.

Harborne, J. B., and Gavazzi, G. (1969). Effect of *Pr* and pr alleles on anthocyanin biosynthesis in *Zea mays. Phytochemistry* **8,** 999–1001.

Kedharnath, S., and Brink, R. A. (1958). Transposition and the stability of Modulator in maize. *Genetics* **43,** 695–704.

Kermicle, J. L. (1970). Somatic and meiotic instability of R-stippled, an aleurone spotting factor in maize. *Genetics* **64,** 247–258.

Kermicle, J. L. (1973). Organization of paramutational components of the *R* locus in maize. *Brookhaven Symp. Biol.* **25,** 262–280.

Kermicle, J. L. (1980). Probing the component structure of a maize gene with transposable elements. *Science (Washington, D.C.),* **208,** 1457–1459.

Kiesselbach, T. A. (1951). A half-century of corn research. *Am. Sci.* **39,** 629–655.

Kirby, L. T., and Styles, E. D. (1970). Flavonoids associated with specific gene action in maize aleurone, and the role of light in substituting for the action of a gene. *Can. J. Genet. Cytol.* **12,** 934–940.

Kleckner, N. (1981). Transposable elements in prokaryotes. *Annu. Rev. Genet.* **15,** 341–404.

Kyle, D. J., and Styles, E. D. (1977). Development of aleurone and subaleurone layers in maize. *Planta* **137,** 185–193.

Larson, R. L., and Coe, E. H., Jr. (1977). Gene-dependent flavanoid glucosyltransferase in maize. *Biochem. Genet.* **15,** 153–156.

Laughnan, J. R. (1952). The action of allelic forms of the gene *A* in maize. IV. On the compound nature of the A^b and the occurrence and action of its A^d derivatives. *Genetics* **37,** 375–395.

McClintock, B. (1941). The stability of broken ends of chromosomes in *Zea mays. Genetics* **26,** 234–282.

McClintock, B. (1942a). The fusion of broken ends of chromosomes following nuclear fusion. *Proc. Natl. Acad. Sci. USA* **28,** 458–463.

McClintock, B. (1942b). Maize genetics. *Carnegie Inst. Washington Year Book* **41,** 181–186.

McClintock, B. (1943). Maize genetics. *Carnegie Inst. Washington Year Book* **42,** 148–152.

McClintock, B. (1944). Maize genetics. *Carnegie Inst. Washington Year Book* **43,** 127–135.

McClintock, B. (1945). Cytogenetic studies of maize and *Neurospora. Carnegie Inst. Washington Year Book* **44,** 108–112.

McClintock, B. (1946). Maize genetics. *Carnegie Inst. Washington Year Book* **45,** 176–186.

McClintock, B. (1947). Cytogenetic studies of maize and *Neurospora. Carnegie Inst. Washington Year Book* **46,** 146–152.

McClintock, B. (1948). Mutable loci in maize. *Carnegie Inst. Washington Year Book* **47,** 155–169.

McClintock, B. (1949). Mutable loci in maize. *Carnegie Inst. Washington Year Book* **48,** 142–154.

McClintock, B. (1950a). The origin and behavior of mutable loci in maize. *Proc. Natl. Acad. Sci. USA* **36,** 344–355.

McClintock, B. (1950b). Mutable loci in maize. *Carnegie Inst. Washington Year Book* **49,** 157–167.

McClintock, B. (1951a). Chromosome organization and genic expression. *Cold Spring Harbor Symp. Quant. Biol.* **16,** 13–47.

McClintock, B. (1951b). Mutable loci in maize. *Carnegie Inst. Washington Year Book* **50,** 174–181.

McClintock, B. (1952). Mutable loci in maize. *Carnegie Inst. Washington Year Book* **51,** 212–219.

McClintock, B. (1953). Mutation in maize. *Carnegie Inst. Washington Year Book* **52,** 227–237.

McClintock, B. (1954). Mutations in maize and chromosomal aberrations in *Neurospora*. *Carnegie Inst. Washington Year Book* **53,** 254–260.

McClintock, B. (1955). Controlled mutation in maize. *Carnegie Inst. Washington Year Book* **54,** 245–255.

McClintock, B. (1956a). Intranuclear systems controlling gene action and mutation. *Brookhaven Symp. Biol.* **8,** 58–74.

McClintock, B. (1956b). Controlling elements and the gene. *Cold Spring Harbor Symp. Quant. Biol.* **21,** 197–216.

McClintock, B. (1956c). Mutation in maize. *Carnegie Inst. Washington Year Book* **55,** 323–332.

McClintock, B. (1957). Genetic and cytological studies of maize. *Carnegie Inst. Washington Year Book* **56,** 393–401.

McClintock, B. (1958). The suppressor–mutator system of control of gene action in maize. *Carnegie Inst. Washington Year Book* **57,** 415–429.

McClintock, B. (1959). Genetic and cytological studies of maize. *Carnegie Inst. Washington Year Book* **58,** 452–456.

McClintock, B. (1961a). Some parallels between gene control systems in maize and in bacteria. *Am. Nat.* **95,** 265–277.

McClintock, B. (1961b). Further studies of the suppressor–mutator system of control of gene action in maize. *Carnegie Inst. Washington Year Book* **60,** 469–476.

McClintock, B. (1962). Topographical relations between elements of control systems in maize. *Carnegie Inst. Washington Year Book* **61,** 448–461.

McClintock, B. (1963). Further studies of gene-control systems in maize. *Carnegie Inst. Washington Year Book* **62,** 486–493.

McClintock, B. (1964). Aspects of gene regulation in maize. *Carnegie Inst. Washington Year Book* **63,** 592–602.

McClintock, B. (1965a). The control of gene action in maize. *Brookhaven Symp. Biol.* **18,** 162–184.

McClintock, B. (1965b). Components of action of the regulators *Spm* and *Ac*. *Carnegie Inst. Washington Year Book* **64,** 527–534.

McClintock, B. (1965c). Restoration of A_1 gene action by crossing over. *Maize Genet. Coop. News Lett.* **39,** 42–45.

McClintock, B. (1967a). Genetic systems regulating gene expression during development. *Dev. Biol. Suppl.* **1,** 84–112.

McClintock, B. (1967b). Regulation of pattern of gene expression by controlling elements in maize. *Carnegie Inst. Washington Year Book* **65,** 568–578.

McClintock, B. (1968). The states of a gene locus in maize. *Carnegie Inst. Washington Year Book* **66,** 20–28.

McClintock, B. (1971). The contribution of one component of a control system to versatility of gene expression. *Carnegie Inst. Washington Year Book* **70,** 5–17.

McClintock, B. (1978a). Development of the maize endosperm as revealed by clones. *In* "The Clonal Basis of Development" (S. Subtelny and I. M. Sussex, eds.), pp. 217–237. Academic Press, New York.

McClintock, B. (1978b). Mechanisms that rapidly reorganize the genome. *Stadler Genet. Symp.* **10,** 25–48.

Nelson, O. E. (1968). The *Waxy* locus in maize. II. The location of the controlling element alleles. *Genetics* **60,** 507–524.

Nelson, O. E., and Rines, H. W. (1962). The enzymatic deficiency in the *waxy* mutant of maize. *Biochem. Biophys. Res. Commun.* **9,** 297–300.

Neuffer, M. G. (1965). Crossing over in heterozygotes carrying different mutable alleles at the *A*1 locus in maize. *Genetics* **52,** 521–528.

Neuffer, M. G. (1966). Stability of the suppressor element in two mutator systems at the *A*1 locus in maize. *Genetics* **53,** 541–549.

Nevers, P., and Saedler, H. (1977). Transposable genetic elements as agents of gene instability and chromosome rearrangements. *Nature (London)* **268,** 109–115.

Nuffer, M. G. (1955). Dosage effect of multiple *Dt* loci on mutation of *a* in the maize endosperm. *Science (Washington, D.C.),* **121,** 399–400.

Nuffer, M. G. (1961). Mutation studies at the *A*1 locus in maize. I. A mutable allele controlled by *Dt. Genetics* **46,** 625–640.

Nuffer, M. G. (1962). Transposition of mutability between components of the A_1 locus. *Maize Genet. Coop. News Lett.* **36,** 59–60.

Orgel, L. E., and Crick, F. H. C. (1980). Selfish DNA: the ultimate parasite. *Nature (London)* **284,** 604–607.

Orton, E. R., and Brink, R. A. (1966). Reconstitution of the variegated pericarp allele in maize by transposition of Modulator back to the *P* locus. *Genetics* **53,** 7–16.

Paigen, K. (1979). Acid hydrolases as models of genetic control. *Annu. Rev. Genet.* **13,** 417–466.

Peterson, P. A. (1953). A mutable pale green locus in maize. *Genetics* **38,** 682–683.

Peterson, P. A. (1960). The pale green mutable system in maize. *Genetics* **45,** 115–133.

Peterson, P. A. (1961). Mutable a_1 of the *En* system in maize. *Genetics* **46,** 759–771.

Peterson, P. A. (1965). A relationship between the *Spm* and *En* control systems in maize. *Am. Nat.* **99,** 391–398.

Peterson, P. A. (1966). Phase variation of regulatory elements in maize. *Genetics* **54,** 249–266.

Peterson, P. A. (1968). The origin of an unstable locus in maize. *Genetics* **59,** 391–398.

Peterson, P. A. (1970a). The *En* mutable system in maize. III. Transposition associated with mutational events. *Theor. Appl. Genet.* **40,** 367–377.

Peterson, P. A. (1970b). Controlling elements and mutable loci in maize: their relationship to bacterial episomes. *Genetica (The Hague)* **41,** 33–56.

Peterson, P. A. (1976a). Change in state following transposition of a regulatory element of the enhancer system in maize. *Genetics* **84,** 469–483.

Peterson, P. A. (1976b). Basis for diversity of states of controlling elements in maize. *Mol. Gen. Genet.* **149,** 5–21.

Peterson, P. A. (1977). The position hypothesis for controlling elements in maize. *In* "DNA Insertion Elements, Plasmids and Episomes" (A. I. Bukhari, J. A. Shapiro, and S. L. Adhya, eds.), pp. 429–435. Cold Spring Harbor Lab., Cold Spring Harbor, New York.

Peterson, P. A. (1980). Instability among the components of a regulatory element transposon in maize. *Cold Spring Harbor Symp. Quant. Biol.* **45,** 447–455.

Rabinow, L., and Dickinson, W. J. (1981). A cis-acting regulator of enzyme tissue specificity in *Drosophila* is expressed at the RNA level. *Mol. Gen. Genet.* **183,** 264–269.

Randolph, L. F. (1936). Developmental morphology of the caryopsis in maize. *J. Agric. Res. (Washington, D.C.)* **53,** 881–916.

Reddy, A. R., and Peterson, P. A. (1976). Germinal derivatives of the *En* controlling-element system in maize: characterization of colored, pale and colorless derivatives of *a*2-*m*. *Theor. Appl. Genet.* **48,** 269–278.

Reddy, G. M., and Coe, E. H., Jr. (1962). Inter-tissue complementation: a simple technique for direct analysis of gene-action sequence. *Science (Washington, D.C.)* **138,** 149–150.

Rhoades, M. M. (1936). The effect of varying gene dosage on aleurone colour in maize. *J. Genet.* **33,** 347–354.

Rhoades, M. M. (1938). Effect of the *Dt* gene on the mutability of the a_1 allele in maize. *Genetics* **23,** 377–397.

Rhoades, M. M. (1941). The genetic control of mutability in maize. *Cold Spring Harbor Symp. Quant. Biol.* **9,** 138–144.

Rhoades, M. M. (1945). On the genetic control of mutability in maize. *Proc. Natl. Acad. Sci. USA* **31,** 91–95.

Salamini, F. (1980). Controlling elements at the *Opaque*-2 locus of maize: their involvement in the origin of spontaneous mutation. *Cold Spring Harbor Symp. Quant. Biol.* **45,** 467–476.

Schon, E. A., Cleary, M. L., Haynes, J. R., and Lingrel, J. B. (1981). Structure and evolution of goat γ-, β^C- and β^A-globin genes: three developmentally regulated genes contain inserted elements. *Cell* **27,** 359–369.

Schwartz, D. (1960). Electrophoretic and immunochemical studies with endosperm proteins of maize mutants. *Genetics* **45,** 1419–1427.

Shapiro, J. A. (1979). Molecular model for the transposition and replication of bacteriophage Mu and other transposable elements. *Proc. Natl. Acad. Sci. USA* **76,** 1933–1937.

Styles, E. D., and Ceska, O. (1977). The genetic control of flavonoid synthesis in maize. *Can. J. Genet. Cytol.* **19,** 289–302.

Styles, E. D., Ceska, O., and Seah, K.-T. (1973). Developmental differences in action of *R* and *B* alleles in maize. *Can. J. Genet. Cytol.* **15,** 59–72.

Tan, C. C. (1946). Mosaic dominance in the inheritance of color patterns in the lady-bird beetle, *Harmonia axyrodis. Genetics* **13,** 195–210.

Van der Ende, V., Langeveld, S. A., Teertstra, R., Van Arkel, G. A., and Weisbeck, P. J. (1981). Enzymatic properties of the bacteriophage ØX174 A* protein on superhelical ØX174 DNA: a model for the termination of the rolling circle DNA replication. *Nucleic Acids Res.* **9,** 2037–2053.

Van Schaik, N. W., and Brink, R. A. (1959). Transpositions of Modulator, a component of the variegated pericarp allele in maize. *Genetics* **44,** 725–738.

Weatherwax, P. (1923). "The Story of the Maize Plant." Univ. of Chicago Press, Chicago.

Wood, D. R., and Brink, R. A. (1956). Frequency of somatic mutation to self color in maize plants homozygous and heterozygous for variegated pericarp. *Proc. Natl. Acad. Sci. USA* **42,** 514–519.

CHAPTER 2

Bacteriophage λ

ALLAN CAMPBELL

Mobile Genetic Elements

ISBN 0-12-638680-3

I. The λ Life Cycle

Viewed as an inserting element, bacteriophage λ may be classified as a circular extrachromosomal replicon capable of inserting its DNA into the host chromosome by reciprocal, site-specific recombination. Like most aspects of λ biology, the insertion reaction is remarkable in the degree to which its specfic attributes are well suited to its function in the life of the phage. The insertion reaction is therefore best understood in the context of the normal λ life cycle. This chapter begins with an outline of that life cycle. Although it will not be necessary to cover all aspects of viral replication and maturation, an account of early gene function and its regulation is essential.

Figure 1 depicts the major features of the λ life cycle. The free virus particle contains a double-stranded DNA molecule about 47,000 nucleotide pairs in length, with complementary projecting single-stranded 5′-ends 12 bases long. This molecule is encapsidated in an isometric head attached to a long-flexible tail. Both the head and the tail contain several different phage-coded proteins. When a phage particle encounters a cell of a susceptible host such as *Escherichia coli* K12, the DNA is injected through the tail into the bacterial cell. Once the entire λ DNA molecule has entered the cell, the sticky ends are sealed by the host enzyme polynucleotide ligase to form a covalently closed circular molecule.

A. Early Gene Expression

An ordered program of gene expression from the intracellular λ chromosome then commences. This program is so organized that it has two possible outcomes, only one of which is realized in a particular infected cell. In some infected cells, replication of viral DNA is followed by assembly of progeny viral particles, which are liberated from the cell by lysis. Other cells survive infection. In these cells insertion usually occurs, so that the survivors and their descendants harbor an inserted, repressed prophage. These two alternative pathways are called *lytic* and *lysogenic,* respectively. The decision as to which pathway to follow is not made immediately upon infection. Rather, cells destined for both pathways proceed initially along a common course.

The program begins with transcription by the host RNA polymerase from two phage promoters, p_L and p_R. Transcription from p_L proceeds leftward, through gene *N,* whereas transcription from p_R proceeds rightward, through gene *cro.* In the absence of the protein product of gene *N* (gp*N*), transcription stops at termination sites shortly beyond *N* and *cro,* respectively. Once formed, gp*N* confers a special property on transcription initiated at p_L and p_R. This transcription now becomes resistant to normal termination signals and continues leftward through gene *int,* across gene *att,* and rightward through gene *Q.* The mechanism of

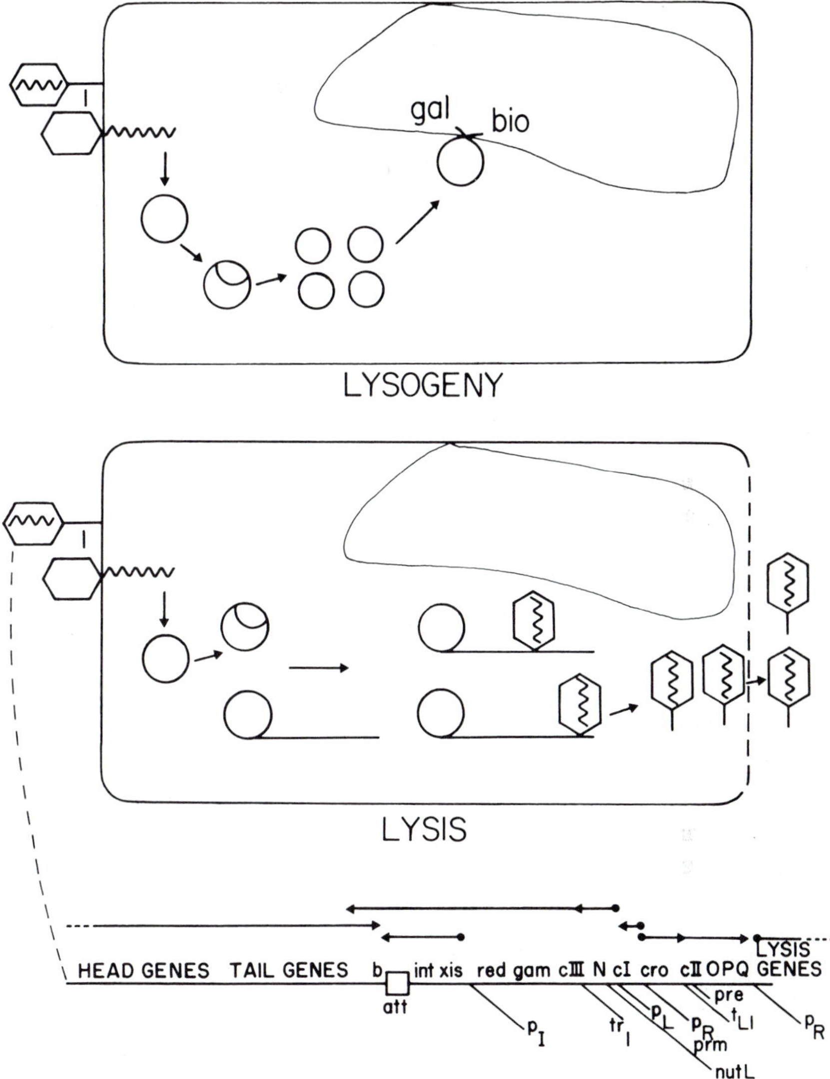

Fig. 1 Life cycle of bacteriophage λ. (Top frame) Infected cell headed for lysogeny. A phage particle attaches to the cell surface and injects its DNA, which becomes circular through the action of the host enzyme polynucleotide ligase and then replicates. A circular phage DNA molecule then becomes inserted into the bacterial chromosome between the *gal* and *bio* genes. λ repressor is synthesized, turning off other viral genes, thus allowing the lysogenic cell to survive and multiply. (Middle frame) Infected cell headed for lysis. The initial steps are the same as in the top frame. Insertion and repression do not occur. Viral proteins eventually assemble around DNA products of rolling circle replication and package unit lengths of DNA into viral particles, which are liberated when the cell lyses. (Lower frame) Abbreviated map of phage λ, showing genes (above the line) and control sites (below the line) relevant to the lysis–lysogeny decision.

antitermination is poorly understood, but it is known to be specific for transcription originating at p_L and p_R and to require the integrity of recognition sequences (*nutL* and *nutR*) located early in the p_L and p_R operons.

Most of the genes transcribed on these extended messages from p_L and p_R (which are expressed at their maximal rate early in infection, before the cell is firmly committed to either the lytic or the lysogenic pathway) can be divided into two groups according to their function. Several genes (*O, P, redX, redB, gam*) function in DNA replication. Of these, only *O* and *P* are essential for replication. The *red* and *gam* genes are frequently classified as recombination genes because the effect of *red* mutations on recombination is more severe than their effect on replication. However, their most obvious consequence for viral growth is that the rate of viral replication, and therefore the yield of viral progeny, is reduced. Other genes (*N, cro, cII, cIII, Q*) encode products the functions of which are purely regulatory. At least the first four of these regulatory genes influence the decision between lytic and lysogenic pathways.

B. Regulatory Genes

The product of gene *cII* (gp*cII*) has three known regulatory effects:

1. It turns on leftward transcription of the *cI* gene from the promoter pre. The product of the *cI* gene is a repressor that binds to operator sites that overlap p_L and p_R, respectively. At the terminal stage of the lysogenic pathway, in an established lysogenic cell, repressor binds to these operator sites and prevents transcription. Repressor bound to the operator on the right has a second effect: At the same time that it turns off rightward transcription of genes *cro, cII,* and so on, repressor turns on leftward transcription of the *cI* gene from the promoter prm, thus stabilizing the repressed state.
2. gp*cII* turns on transcription of the *int* gene from the p_I promoter.
3. By an unknown mechanism, gp*cII* bound to pre delays expression of the late genes *S-J* (Court *et al.*, 1975). In a wild-type K12 host, these effects of gp*cII* are enhanced by gp*cIII*. Mutants of both the phage and the host have been isolated in which the need for gp*cIII* is reduced or eliminated (Belfort and Wulff, 1971; Jones and Herskowitz, 1978). These and other results suggest an ancillary role for gp*cIII*. Whenever we discuss *in vivo* effects of gp*cII*, it should be understood that under ordinary circumstances gp*cIII* is required as well.

By its antiterminating action, gp*N* serves as a positive regulator of all other early genes except *cro*. The gp*cro* protein is a negative regulator. It binds to the same sites as the repressor does, and it prevents transcription leftward from p_L and rightward from p_R. It also represses leftward transcription of the *cI* gene initiated from prm. The effect of gp*cro* is thus qualitatively similar to that of repressor, except that repressor can stimulate *cI* transcription from prm. The role of gp*cro* in the λ life cycle is quite different from that of repressor, however. Repressor formation is confined to cells that have chosen the lysogenic pathway. The *cI* gene is either stably turned on (in lysogenic cells), in which case repression at p_L and p_R is virtually complete, or stably turned off (early in infection or in lytically infected cells). On the other hand, gp*cro* gradually accumulates during the lytic cycle and inhibits transcription from p_L and p_R to an ever greater degree, so that early gene expression is progressively reduced as phage development proceeds. Because the *cro* gene itself belongs to an early operon, the reduction never becomes complete. The last regulatory gene, gp*Q*, turns on transcription of the late genes *S-J*, which encode enzymes for lysis and protein components of the virus particle. The mechanism, like that of gp*N* action, is antitermination of a specific message—in this case, late message initiated at $p_{R'}$ (Forbes and Herskowitz, 1982).

C. Decision between Lytic and Lysogenic Pathways

The exact manner in which a cell becomes committed to the lytic or the lysogenic pathway has not yet been identified with certainty. The immediate effectors appear to be gp*cII* and gp*cro*. If the concentration of gp*cII* reaches a level sufficient to turn on transcription from pre, the resulting amount of repressor formed is adequate to shut off p_L and p_R and to initiate the self-sustaining transcription of *cI* from prm. Because of the progressive buildup of gp*cro*, a cell that does not attain the necessary gp*cII* concentration early enough during the cycle loses its chance to do so forever.

Rapid, effective commitment is most easily understood if the action of gp*cII* is strongly cooperative, so that it functions maximally above some critical concentration and minimally below that concentration. Positive feedback of gp*cII* on its own synthesis has also been suggested as an additional means of imparting an all-or-none character to the switch (Herskowitz and Hagen, 1980). Good data on the *in vitro* effects of gp*cII* are just becoming available and will not be reviewed here. There is a well-known *in vivo* effect of the multiplicity of infection. Multiply infected cells have a higher probability of choosing the lysogenic pathway than do singly infected cells. Also, under the experimental conditions generally used to follow integrase production, significant *cII*-stimulated integrase synthesis is observed only at high multiplicity. The result seems reasonable according to the above discussion, because the several phage DNA molecules present from the outset might produce adequate gp*cII* before much gp*cro* has accumulated. A complete kinetic description should include the rates of production and decay of gp*cro* and gp*cII* as well as the number of DNA copies (and therefore of binding sites). Other factors affecting the frequency of lysogenization (such as the concentration of divalent cations in the medium or of cyclic AMP in the cell) may operate by influencing the production of gp*cro* or gp*cII*.

Later in this chapter, we will consider the coordination of the control of the integrase and excisionase genes with the effectors of the switching event, to which both the mechanism of *cII* action and the antiterminating effect of gp*N* are especially relevant.

Once a cell has become committed to the lysogenic pathway, repressor turns off transcription from p_L and p_R, with the result that all phage gene transcription except the *cI* message is reduced or eliminated. Normal cell growth and division resume. Generally, phage DNA becomes inserted into the chromosome prior to the first cell division, so that each daughter cell harbors an inserted prophage as a permanent part of its chromosome.

The lysogenic condition can persist through an indefinite number of cell divisions. If repressor is inactivated (either spontaneously or through induction by various experimental treatments), transcription from p_L and p_R is initiated, and a new lytic cycle ensues. Transcription from p_L results in expression of the *int* and *xis* genes, the products of which excise the viral DNA from the host chromosome. This reverses the insertion event and regenerates a circular viral chromosome, which then multiplies to generate progeny that can be packaged into extracellular particles.

The major means of spontaneous repressor inactivation is cleavage of the repressor protein by the *recA* protease. This suggests that *recA* undergoes occasional spontaneous activation, perhaps by accidental breakdown of the controls normally operating on it. Deliberate activation of *recA*, through induction of the SOS repair system, causes derepression of almost the entire cell population. The *recA* protein is a very specific protease, with only one known cellular substrate—the *lexA* protein of the SOS induction system. The fact that λ repressor is sensitive to such a specific protease strongly indicates that the *recA*–λ interaction plays an important role in allowing phage production from lysogenic bacteria under natural conditions.

It is important to appreciate that, although our knowledge of λ regulation may be extensive, it is unlikely to be complete. Several new regulatory effects, previously unsuspected, have been discovered in the last few years. We may expect that an equal number remain undiscovered. It is therefore worth considering what sorts of undetected circuitry remain possible and what others are precluded by existing data. For example, nothing in our current knowledge eliminates the possibility that some negative feedback on *cII* function from *Q* or the genes it activates serves to reinforce the commitment to the productive cycle. On the other hand, the possibility that insertion of phage DNA into the bacterial chromosome feeds back on the components of the regulatory switch seems unlikely. The frequency of insertion can be drastically reduced by various manipulations. These can be either environmental (insertion is much less frequent at 43°C than at 37°C) or genetic (insertion is virtually eliminated by mutations inactivating the integrase gene or the insertion sites of either the bacterium or the phage). Under these conditions, the failure to insert has no significant influence on the decision to survive infection and establish repression. Where repression is

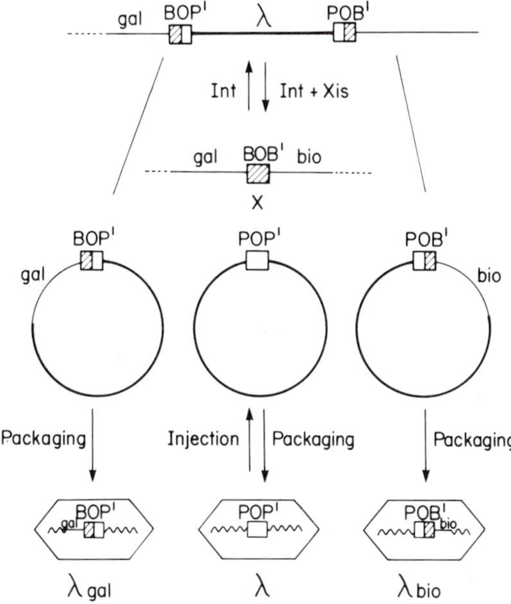

Fig. 2 Normal and abnormal excision from lysogenic bacteria. The center column shows the normal pathway of excision and packaging and (in the reverse direction) injection and insertion. Abnormal excision (rare, not catalyzed by integrase) generates λ*gal* and λ*bio* particles, as shown in the side columns.

established without insertion, infected cells grow and divide, so that the non-multiplying phage genomes they harbor are eventually diluted out with growth (Ogawa and Tomizawa, 1967).

D. Deviations from the Normal Cycle

On rare occasions, λ development deviates from the normal life cycle shown in Fig. 1. Two deviations of interest in the present context are the generation of specialized transducing phages and insertion at secondary sites on the bacterial chromosome.

The mode of origin of specialized transducing phages is indicated in Fig. 2. Whereas the normal excision reaction precisely reverses the results of insertion, excision can be imprecise, generating a new phage variant in which some bacterial DNA that was adjacent to the insertion site replaces some of the phage DNA from the opposite end of the prophage. The ratio of such specialized transducing phages to normal phages in a lysate prepared by induction of lysogenic bacteria is typically about 10^{-5}. Imprecise excision does not require the phage enzymes (integrase, excisionase) needed for precise excision. Although the enzymology of imprecise excision is unknown, its net result is breakage and rejoining of heterologous DNA in the same manner as in deletion mutations. To a first approximation, the exact locations of phage and bacterial endpoints are random. Within the size limits prescribed by the packaging ability of the phage, the bacterial DNA segment may be either longer or shorter than the phage DNA segment it replaces. When the replaced phage segment includes any genes essential for viral development, the specialized transducing phage is defective and can grow only in a cell coinfected with a nondefective λ.

During the lysogenic response, λ ordinarily inserts at a specific site on the chromosome between the *gal* and *bio* operons. With appropriate selective techniques, insertions at other chromosomal sites can be detected among survivors from λ infection of bacterial deletion mutants that have lost the primary site (Shimada *et al.*, 1972). There are several major secondary sites at which insertions have recurred, as well as other sites at which insertion has been observed only once. Imprecise excision from secondary sites generates specialized transducing phages carrying genes from the adjacent bacterial DNA. The proportion of specialized transducing phages in lysates prepared by induction of secondary site lysogens is frequently much higher than that of λ*gal* or λ*bio* in a normal lysogen, because the frequency of precise excision from the secondary sites is lower than that from the primary site.

Specialized transducing phages and secondary site insertions have been useful for studying λ insertion and excision. They may also be important in bacterial evolution, because they allow the phage to move bacterial genes from one cell to another or from one chromosomal location to another.

II. Mechanism of Insertion and Excision

A. The Insertion Reaction: *In Vivo* Assays

The insertion reaction comprises breakage of phage DNA at a specific site (designated *attPOP'*) and of bacterial DNA at site *attBOB'*, and rejoining to create hybrid sites *attPOB'* and *attBOP'* at the termini of the inserted prophage (Fig. 2). The reaction can be described by

$$attPOP' + attBOB' \rightarrow attPOB' + attBOP' \quad (1)$$

This terminology for the four sites, which can now be replaced where desired by the corresponding four base sequences, was introduced to indicate that a hybrid site such as the left prophage terminus consisted of a crossover point (O) flanked by bacterial sequences to the left (B) and phage sequences to the right (P').

Such a terminology was needed to describe the apparent difference in catalytic requirements for the insertion reaction Eq. (1) and its reversal. Insertion required only the product of the *int* gene, whereas excision also required the *xis* gene product. Soon after the discovery of these genes, it was shown that the genetic requirements depend on the chemical nature of the interacting sites, not on whether the interaction results in excision rather than insertion. For example, when a cell is infected with the two specialized transducing phages λ*gal* and λ*bio*, recombination between them at the insertion site follows the rule

$$\underset{(attBOP')}{\lambda gal} + \underset{(attPOB')}{\lambda bio} \underset{\text{Int}}{\overset{\text{Int + Xis}}{\rightleftarrows}} \underset{(attBOB')}{\lambda gal\ bio} + \underset{(attPOP')}{\lambda} \tag{2}$$

The concentration of integrase in cells can be monitored by measuring the extent of the insertion reaction under standardized conditions. The relationship between the measured numbers and the actual concentration of integrase is unknown, but such assays have provided most of the available qualitative information on the regulation of integrase. Determining the fraction of surviving cells that become lysogenized is the most straightforward method, but frequently not the most convenient one.

A widely used *in vivo* assay employs specially constructed λ derivatives (called λatt^2) that contain, in a single phage genome, both *att* sites, *attBOB′* and *attPOP′*, separated by a segment of nonessential DNA (Fig. 3). Integrase can excise the intervening DNA, leaving a shortened phage genome with an *attBOP′* site. The shortened phage genome can be distinguished from its att^2 parent by the greater resistance to EDTA of phage particles with the shortened genome. Intracellular integrase concentrations can then be investigated by infecting the cell under study with an att^2 phage (generally an int^- mutant so that it will generate no integrase itself) and measuring the fraction of progeny that are EDTA-resistant.

Another method employs specially constructed hosts in which the gene for an easily scorable bacterial trait such as Gal^+ is bracketed by *attPOP′* and *attBOB′* sites (Fig. 4). Site-specific recombination between these two sites excises the gal^+ gene. The extent of the reaction is determined by scoring the fraction of treated cells that form Gal^- colonies. The method has the special virtue of being extremely sensitive and thus suitable for detecting low rates of integrase expression, such as that of the p_I promoter in a repressed prophage.

B. The Insertion Reaction: *In Vitro* Assays

The first successful *in vitro* assays employed supercoiled DNA from att^2 phage as substrate. At the end of the reaction, the DNA was used for transfection, and the EDTA-resistant fraction of the phage yield was measured. Later

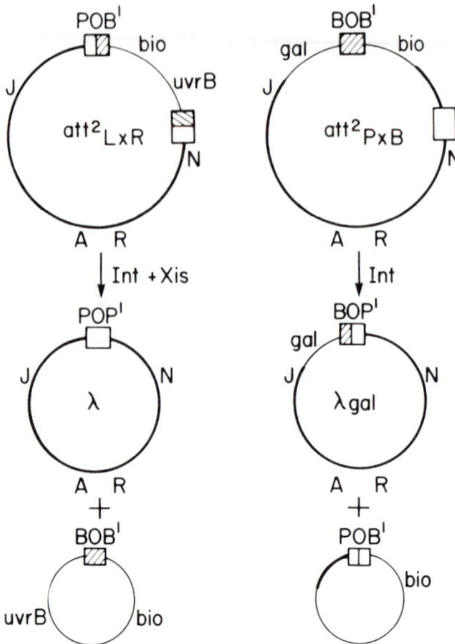

Fig. 3 Breakdown of λ*att*² phage DNA by integrase and excisionase. (Left column) An EDTA-sensitive λ*att*²L×R phage (with *att* sites that are normally involved in excision) breaks down to normal EDTA-resistant λ plus a fragment of bacterial DNA (not recovered). (Right column) An EDTA-sensitive λ*att*²P×B phage (with *att* sites that are normally involved in insertion) breaks down to an EDTA-resistant λ*gal* and a circular DNA fragment (not recovered). In the presence of a DNA packaging system (*in vivo* or *in vitro*), the *att*² phages are distinguished from their smaller derivatives by the greater EDTA sensitivity of the virus particles. *In vitro*, the products can also be distinguished from the *att*² phages by digestion with restriction enzymes.

assays followed the change in DNA substrate directly, by examining the restriction patterns of *att*² DNA before and after integrase action (Mizuuchi *et al.*, 1978).

Use of *att*² DNA simplifies the reaction from an intermolecular to an intramolecular one. Subsequently, material and conditions were developed in which the intermolecular reaction could also be studied *in vitro* (Mizuuchi and Mizuuchi, 1979). The intermolecular reaction offers the advantage that the *attPOP'* and *attBOB'* substrates can be manipulated individually. For example, either of the two may be individually labeled with isotopes or provided in a specific physical state (such as supercoiled, relaxed, or linear).

C. Enzymes: Phage-Coded

The only λ-coded protein known to be needed for *in vitro* insertion is λ integrase. This protein has been highly purified and has a monomer molecular weight of about 40,000. This is the expected size for the product of the *int* gene

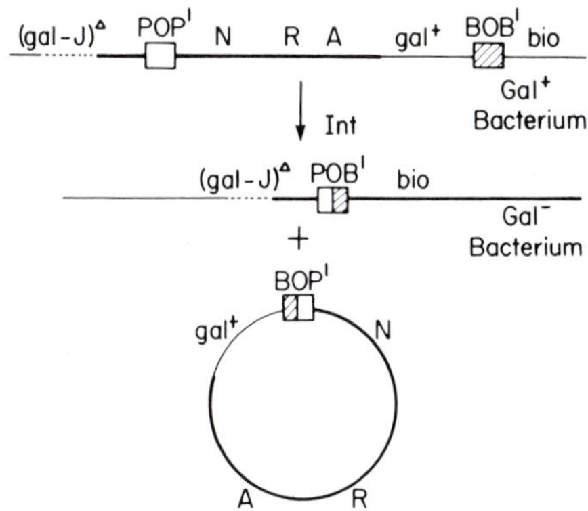

Fig. 4 Tester strain for measuring integrase by the production of Gal$^-$ derivatives of Gal$^+$ lysogen by excision of λ*gal*$^+$. This strain (RW309), constructed by R. A. Weisberg, was used by Shimada and Campbell (1974).

as calculated from the DNA sequence, and the N-terminal amino acid sequence of the protein is that predicted from the DNA sequence (Nash, 1981a). Studies with λ mutants indicate that *int* is the only λ gene the product of which is required for insertion *in vivo*. Some mutations (*hen*) cause a defect in insertion but not in excision (Echols *et al.*, 1974). The *hen* defect is cis-specific and results from a single base change in the integrase binding site of P′ (H. Echols, personal communication).

D. Enzymes: Host-Coded

The *in vitro* reaction also requires two host polypeptides, which are subunits of a dimeric protein called integration host factor (IHF). The larger subunit has been identified as the product of the *himA* gene, which is also required for insertion *in vivo*. Other host genes needed for insertion *in vivo* that have been identified in mutant studies are *hip, himD* (which could be the same gene as *hip*), and *gyrB* (Miller *et al.*, 1979). The *gyrB* gene encodes a subunit of DNA gyrase. Although gyrase was discovered through its contribution to the *in vitro* insertion reaction, it is not needed for *in vitro* insertion if the substrate is supplied in supercoiled form (Mizuuchi *et al.*, 1978).

E. DNA Substrates

The nucleotide sequences of *attPOP′*, *attBOB′*, and the hybrid attachment sites *attBOP′* and *attPOB′* (as represented in λ*gal* and λ*bio* phages) have been determined (Fig. 5). These sequences localize the crossover point to a 15-base sequence that is identical for all four sites (and thus can be equated to the "O" of

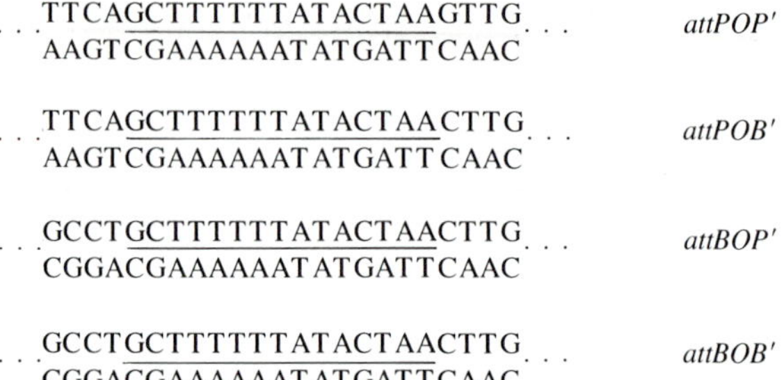

Fig. 5 Nucleotide sequence in the vicinity of the crossover point. The 15 bases common to all four sites are underlined. Based on results of Landy and Ross (1977).

attPOP') and verify that the hybrid sites have the sequences expected for a simple crossover event. By themselves, these sequences shed no light on how much of the flanking DNA is directly involved in the insertion reaction. Information relevant to that question comes from the study of sites with altered base sequences (reviewed in Nash, 1981a, and Gottesman, 1981). Alterations have been introduced by various methods: (*1*) Non-complementable mutations that reduce insertion ability have been isolated and sequenced. Such mutations include a single base deletion within the 15-base common core and deletion mutations that remove all or part of *attPOP'* or *attBOB'*. Of special interest are deletion mutations the genesis of which is *int*-dependent, which appear to arise by site-specific recombination between *attPOP'* and secondary *att* sites located on λDNA (Davis and Parkinson, 1971). (*2*) The nucleotide sequences of non-transducing or specialized transducing phages that have excised from secondary insertion sites on the bacterial chromosome have been correlated with their properties as substrates for insertion or excision *in vivo*. (*3*) Plasmids bearing cloned segments of *attPOP'* DNA have been subjected to enzymatic digestion to see how much of the flanking DNA can be removed without impairing substrate ability. (*4*) The sequences flanking λ*attPOP'* have been compared with those of the λ-related phage 434, which has the same insertion specificity as λ. Study of the *in vitro* intermolecular reaction showed that a supercoiled *attPOP'* molecule can recombine with a linear *attBOB'* molecule, whereas the reverse combination is ineffective. In recognition of this asymmetry, the *attPOP'* partner is termed the *donor* and the *attBOB'* partner the *recipient*. Full donor activity requires the presence of at least 150 bases to the left and 75 bases to the right of the crossover point. On the other hand, the recipient needs only the 15-base pair common sequence and, at most, four nucleotides on either side (Nash, 1981a; Mizuuchi *et al.*, 1981).

Figure 6 compares the nucleotide sequences of *attPOP'* from phages λ and 434, and shows the integrase binding sites in λDNA determined by DNase footprinting. The experiments just described indicate that all four binding sites

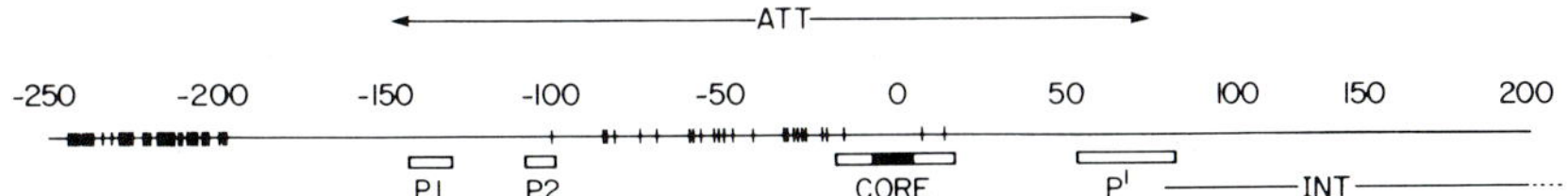

Fig. 6 Structure of *attPOP'*. The extent of *att* from −150 to +75 is based on resectioning experiments (Mizuuchi *et al.*, 1981). P1, P2, "core," and P′ are integrase binding sites determined by footprinting. The shaded part of the core binding site is the 15-base common sequence shown in Fig. 5. Vertical bars indicate positions at which the sequence of phage 434 differs from that of λ (Mascarenhas *et al.*, 1981). Thus, the 120 bases at the 3′-end of the *int* gene are identical in the two phages, whereas the sequence to the left of −200 is quite different.

are needed for normal activity. The λ–434 comparison shows that the binding sites are generally conserved, whereas the long spacer region to the left of the crossover point has diverged to a much greater degree.

F. The Reaction Mechanism

The *in vitro* reaction requires the DNA substrates, λ integrase, IHF, and a favorable ionic environment. Perhaps as noteworthy as the requirements are some of the things that are apparently not required: The reaction does not require added ATP or any equivalent energy source, deoxyribonucleotide triphosphates, DNA polymerase, *recA* protein, or polynucleotide ligase (Mizuuchi *et al.*, 1978). The implication is that the integrase and IHF do the entire job of breaking all four polynucleotide chains of the two partners and reconnecting the eight broken ends in a manner that conserves the energy of the original phosphodiester bonds.

Additional evidence for the absence of any repair synthesis at the crossover point, as well as insight into the reaction mechanism itself, comes from *in vitro* studies on the origin of the phosphorus atoms in the products of *in vitro* recombination. By examining the results of chain cleavage due to phosphorus decay in recombinant molecules derived from parents that had been labeled separately in each of the four possible polynucleotide chains, Mizuuchi *et al.* (1981) showed that the two chains of each parent are broken at different positions, as though they had been cut by an enzyme that made staggered nicks 5–7 base pairs apart.

These elegant *in vitro* studies were paralleled by some equally refined investigations on *in vivo* insertion at secondary sites that indicate that the staggered cuts are in fact 7 base pairs apart. Several secondary sites have now been sequenced. The 15 base pairs at the crossover site are similar but not identical to those of λ. If λ insertion into such secondary sites proceeds by staggered cuts in both partners, any single insertion event will produce a hybrid site that switches the information source at one end of the 7-base overlap in one polynucleotide chain and the other end in the complementary chain. If this primary insertion product then replicates without mismatch correction, each of the two daughter molecules will have its information switch at one end of the 7-base overlap. Therefore, study of different lysogens independently derived by insertion at a given secondary site should reveal two and only two classes of hybrid sites at each prophage terminus, where the switch in the information source is at the two ends of the same 7-base overlap segment that was identified by *in vitro* labeling. Additionally, if excision from the secondary site lysogens proceeds by the same staggered cut mechanism, excision may not only regenerate wild-type λ, leaving the original secondary site in the chromosome, but may also generate a λ in which the 7 base pairs of λ are replaced by the corresponding bases from the secondary site, and a cured bacterium the secondary site of which has experienced the reciprocal replacement. All these predictions have in fact been fulfilled.

To emphasize that λ insertion is a pure polynucleotide exchange involving neither new synthesis nor DNA degradation, this type of site-specific recombination has been classified as a *conservative* reaction (as contrasted to *duplicative* reactions such as transposition of IS elements). Resolution of Tn*3* cointegrates is likewise a conservative reaction in the same sense, as shown by similar *in vitro* experiments (Reed and Grindley, 1981).

An additional feature of the integrase reaction *in vivo* is the occasional generation (in crosses in which both parents are *attPOP'*) of recombinants between point mutations in genes such as *int* and *xis* that are close to the *att* site. The *int*-dependent generation of such recombinants suggests that there is occasional branch migration into adjacent DNA (extending a distance of 1000 nucleotides or more) from recombinational intermediates generated at *att* (Enquist *et al.*, 1979b; Echols and Green, 1979).

In addition to its specific binding to *att* DNA, purified integrase has a non-site-specific topoisomerase activity; its relation to the integrase reaction is unknown. Like integrase, IHF has several binding sites in *attPOP'*. Some of these overlap but do not coincide with integrase binding sites.

G. The Excision Reaction

The excision reaction can be followed *in vivo* by breakdown of an *att*^{2} phage with *attPOB'* and *attBOP'* sites, or by measuring the fraction of surviving lysogenic cells that have lost the prophage (*prophage curing*) following treatments such as superinfection or transient induction. A highly sensitive indicator for excisionase activity is a strain carrying a defective prophage inserted into the *galT* gene. Excision restores the *galT*$^+$ gene. Even though the excision frequency from the secondary site is lower than that from the primary site, only the excisants form colonies on a galactose minimal medium and therefore can be enumerated even when very rare. In principle, excision from any secondary site within a gene the phenotype of which is selectable can be used in this manner.

The *gal* insertion is especially handy for scoring the *int* or *xis* phenotype of a phage under test, because only an int^+ xis^+ phage will generate Gal^+ colonies within a plaque on a test plate. The defective prophage in the test strain bears a large deletion removing the *cI* gene, so that λ can form plaques on it (Enquist and Weisberg, 1976). The excision reaction can also be studied by breakdown of an att^2 phage bearing the appropriate sites (Fig. 3). The first att^2 phage to be constructed was of this type (Shulman and Gottesman, 1971). Because the excision reaction requires both integrase and excisionase, these att^2 phages, modified by mutation in the appropriate gene, can be used to assay either excisionase or integrase *in vivo*.

The excision reaction has been studied *in vitro*, using λatt^2 DNA as substrate and extracts of induced or infected cells as sources of integrase and excisionase (Gottesman and Gottesman, 1975). Starting with an xis^+ infection, excisionase can be isolated separately from integrase and shown to be needed for the *in vitro* reaction. It is a small protein of molecular weight about 8000, as expected from the size of the excisionase gene. Integrase and IHF are also needed for the *in vitro* reaction (reviewed in Nash, 1981a). *In vivo*, extensive excisive recombination can occur in the *himA1* mutant when excisionase is abundant, but when excisionase is limiting, much more excision is observed in a $himA^+$ host than in the *himA* mutant (Gottesman and Abremski, 1980).

The requirement for excisionase is not absolute. A low rate of excisive recombination (absolutely dependent on *himA*) is observable *in vivo* with all xis^- mutants. This result is corroborated by *in vitro* studies. Under appropriate *in vitro* conditions, extensive excisive recombination is observable with integrase alone (Abremski and Gottesman, 1981). The concentration of integrase required for the excision reaction *in vivo* appears to be much lower than that needed in the insertion reaction (Enquist *et al.*, 1979a).

H. Physical Basis of Directional Specificity

The genetic control of insertion and excision is summarized

$$\text{Phage} \underset{\text{Int + Xis}}{\overset{\text{Int}}{\rightleftarrows}} \text{prophage} \tag{3}$$

Or, in more molecular detail,

$$attPOP' + attBPB' \underset{\text{Int + Xis}}{\overset{\text{Int}}{\rightleftarrows}} attPOB' + attBOP' \tag{4}$$

Although these equations embody the major features of insertion and excision, they cannot adequately represent their chemistry. Enzymes written above arrows customarily indicate catalysts that accelerate the reaction without themselves undergoing any net change. A catalyst cannot shift the position of an equilibrium; it can only hasten the approach to an equilibrium state. If the sole net change in the system were the polynucleotide exchange, a difference in catalytic specificities for the forward and reverse reactions would be a thermodynamic impossibility.

The thermodynamic problem was first addressed by Dove (1970). Dove postulated that the equilibrium of Eq. (4) is far to the right (which could be explained if there were stacking interactions between nucleotides closely flanking the crossover point) and that the function of excisionase is to pump energy into the reaction, thus reversing its direction. The prediction of Dove's hypothesis that seemed most readily testable at the time was that integrase alone (without excisionase) should be able to cause a limited amount of excision. The prediction has turned out to be correct.

Subsequent work has made Dove's question seem even more cogent than it did a decade ago. Until the *in vitro* assay had been perfected, there was no strong reason to imagine that Eq. (4) even approximated a summary of the complete reaction, rather than simply indicating the fate of the macromolecular reactants. It seemed at least as reasonable to suppose that the reaction proceeded through an energy-wasting step, in which DNA chains were cut, to an energy-requiring ligation. Under those circumstances, the net reaction when the two steps were summed would consume ATP or some equivalent energy source. The actual insertion and excision reactions might then have been something like the following:

$$n\text{ATP} + n\text{H}_2\text{O} + attPOP'\ attBOB' \xrightarrow{\text{Int}} attPOB' + attBOP'\ n\text{ADP} + \text{P}_i \tag{5}$$

$$n\text{ATP} + n\text{H}_2\text{O} + attPOB' + attBOP' \xrightarrow{\text{Int + Xis}} attPOP' + attBOB' + n\text{ADP} + \text{P}_i \tag{6}$$

The different catalytic specificites of these two reactions present no thermodynamic problem, because the reactions are not true chemical reversals of each other (Campbell, 1972).

In fact, however, the *in vitro* studies have not revealed a need for any added energy source, such as ATP. The reaction proceeds with purified substrates and enzymes. The enzyme appears to break polynucleotide chains in such a manner that the energy of the phosphodiester bonds is stored in the enzyme–substrate complex and reutilized for ligation rather than being released to the surroundings. This compels us to reexamine possible explanations for the question that Dove raised.

Although underscoring the relevance of Dove's question, the knowledge gathered during the last decade has effectively weakened the specific solution he offered. The existence of the 15-base segment common to reactants and products at the crossover point makes a significant difference in energy content unlikely. (Dove explicitly recognized this point, and he postulated that no such common sequence would be found.) Furthermore, the absence of a required energy source *in vitro* means that excisionase cannot *catalytically* pump energy into the reaction.

Before going into the possible nontrivial explanations for the observed results, I should stress that it remains possible that no real problem exists. Although it is true that the *in vitro* assays display the catalytic specificities observed *in vivo,* that fact by itself provides no assurance that the *in vitro* systems retain all the important attributes of the *in vivo* reactions. Furthermore, the *in vitro* insertion and excision reactions have generally not been performed in such a manner that one is truly a reversal of the other. We are not discussing a single experiment in which reactants are converted quantitatively to products with integrase present and then reappear from the product when excisionase is added. More typically, the insertion reaction is the breakdown of one type of *att*[2] phage, and the excision reaction is the breakdown of another (Fig. 3). One could explain away such results by assuming that in both cases the design of the experiment allows both reactions to proceed with a net loss of free energy. The *in vivo* studies suffer from similar limitations, as well as offering other possible driving forces, such as removal of products by replication.

Such trivial (but perhaps correct) explanations offer one the option of continuing to ignore the issue rather than addressing it. If we reject that course, we are left with only two available energy sources that might provide the driving force for the reaction: Either the energy is present in the DNA substrates, or else it comes from the specific protein factors added (Int, Xis, and IHF). The simplest way in which internal energy of the DNA substrate could be contributed to the reaction is through relaxation of supercoils. This possibility seems unlikely from experiments in which the state of supercoiling has been deliberately varied. Although supercoiled molecules are frequently superior substrates, the state of supercoiling changes little during the reaction (Mizuuchi *et al.*, 1980). Also, the *in vitro* excision reaction, although accelerated by the use of supercoiled substrates, proceeds to a considerable extent without supercoiling (Pollock and Abremski, 1979). A possible energy contribution from any of the proteins cannot be excluded, because in the *in vitro* assays used to date, all the required proteins are present in molar excess over the DNA. This leaves open the possibility that they function as reactants rather than as catalysts and undergo some permanent change as the reaction proceeds. This change might be a loss of internal energy by the protein itself. However, the simplest possibility, which invokes no unknown properties, is that the binding energy of the protein to the DNA drives the reaction.

The importance of these energetic considerations is emphasized because it seems doubtful that much real headway can be made in understanding the determination of directionality until the energetic driving force has been clearly identified. For example, the multiple integrase binding sites on *attPOP′* might allow integrase to bind cooperatively to sites that become separated during the reaction (although binding studies conducted to date fail to reveal any strong cooperativity, as reviewed in Nash, 1981b), and excisionase might then act by altering the pattern of cooperativity. Such cooperative binding seems to be assumed, although not always explicitly stated, in some current models of directional control. The implication that cooperative binding to the substrate should accelerate the reaction is reasonable if the reaction is driven by some energy source other than integrase binding. However, if the binding energy itself drives the reaction, then the affinity for the product should exceed that for the substrate. In order to drive the reaction to the right, the binding to the substrate should then be anticooperative rather than cooperative.

Up to this point, we discussed only the forward and reverse reactions of Eq. (4). These are not the only reactions that integrase can mediate. One can test the genetic requirements for site-specific recombination between two partners bearing any of the four *att* sites appearing in Eq. (4), including identical ones. Most of the 16 possible pairwise combinations have been tested *in vivo*. The observed differences in the behavior of the four types of sites in such tests were the original basis for introducing the use of different symbols, such as *attPOP'* and *attBOB'*, to distinguish among them (Guerrini, 1969). The only combination that requires excisionase is the natural excision reaction, *attPOB'* + *attBOP'*. Most other combinations (including all those in which *attPOP'* is one partner) require integrase alone. The reaction *attBOB'* + *attBOP'* is catalyzed poorly, if at all. One reaction, *attBOP'* × *attBOP'*, requires only integrase, but unlike the normal insertion reaction, it proceeds as well at 43°C as at 36°C (Guarneros and Echols, 1973). Any satisfactory explanation for the determination of directionality must explain all of these facts.

III. Regulation of Integration Genes

A. Relation to the λ Life Cycle

Whatever the chemical reason for the directional specificity of the insertion–excision reaction may be, its biological utility is so apparent that we may expect many other site-specific recombinases to exhibit similar properties. By making those enzymes suitable to the occasion, the phage can control the direction of the reaction so as to coordinate it with the λ life cycle. To take full advantage of this potentiality, the phage must exert some differential control over the production, activity, or stability of integrase and excisionase.

At least four states in the life of λ can be distinguished that differ in their requirements for integrase or excisionase: (*1*) In infected cells that have become committed to the lysogenic pathway, integrase is needed and excisionase is undesirable. (*2*) In infected cells before commitment, or after commitment to the lytic pathway, neither integrase nor excisionase is needed. Integrase could be a liability, because it might sequester the phage DNA in the chromosome. (*3*) In established lysogenic cells, neither enzyme is needed. Integrase alone would not hurt anything, but integrase and excisionase together could cause spontaneous prophage loss. (*4*) In lysogenic celis that have been derepressed, both enzymes are needed.

B. Transcriptional Control: Dual Promoters

Differential control of transcription accomplishes much of the desired regulation (Fig. 1). The integrase gene can be transcribed from two promoters, p_I and p_L. Transcription from p_I generates integrase alone, whereas transcription from p_L produces both integrase and excisionase. Transcription from p_I is turned on by gp*cII*. Transcription from p_L is repressible by repressor and gp*cro*, and extends as far as *int* and *xis* only when antiterminated by gp*N*. In conjunction with the regulation of *cI* transcription, described earlier, these controls on integrase formation have the following consequences: (*1*) In infected cells that have become committed to the lysogenic pathway, gp*cII* turns on transcription of *cI* and *int*. The repressor produced by the *cI* gene then turns off transcription from p_L. Therefore, integrase, but not excisionase, is made. (*2*) In infected cells committed to the lytic pathway, both *int* and *xis* are transcribed from p_L. (*3*) In established lysogenic cells, repressor turns off transcription from p_L and p_R. Therefore, *cII* and *cIII* are not transcribed. Because neither the p_I nor the p_L transcript is formed, neither *int* nor *xis* is transcribed. (*4*) In derepressed lysogenic cells, as in cells committed to the lytic pathway, both *int* and *xis* are transcribed from p_L.

In three of these four conditions, transcriptional control produces the desired result. Only the infected cell committed to the lytic pathway is out of line. Although no integrase is needed, *int* is transcribed from p_L. In this case, integrase is also regulated at another level, to be described later.

C. The p_I Promoter

As shown in Fig. 7, the −10 region of the p_I promoter overlaps the initiation codon of the *xis* gene. The p_I message thus includes most of *xis* as a leader that precedes the initiation codon of the *int* gene. The −35 region of the p_I promoter is upstream from the *xis* gene and strongly resembles the −35 region of the pre promoter. Definitive identification of this promoter sequence as the *cII*-controlled promoter was made possible by the study of deletions removing DNA from the right. Of the deletions indicated in Fig. 7, Δ29 allows *cII*-promoted transcription of *int*, but Δ303 does not. These deletions not only localize the *cII*-stimulated promoter to this spot, but also speak against the hypothesis propounded at one time that gp*cII* acts by extending upstream leader sequences rather than by initiating new messages. Δ29 fuses λ genes to the *trp* leader

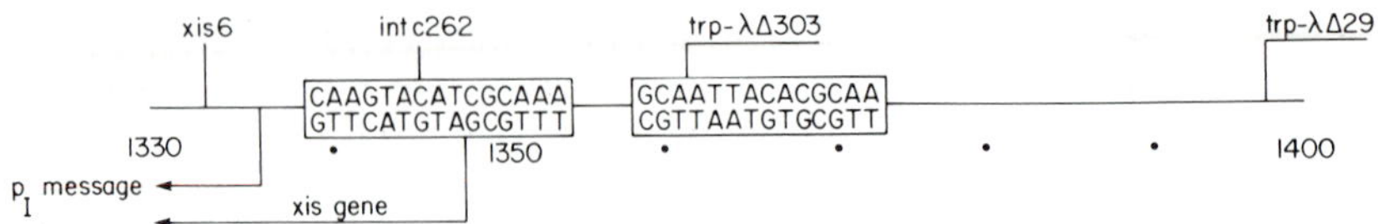

Fig. 7 Structure of p_I. Of the two *trp*–λ fusions shown, Δ303 destroys p_I function, whereas Δ29 does not. Sequences are shown of the −10 region (which includes the first three codons of the *xis* gene) and the −35 region (presumed to be the site of *cII* interaction because of its similarity to the −35 region of the pre promoter). Based on Abraham *et al.* (1980). The indicated start site of the p_I message is based on results of Abraham and Echols (1981) and Schmeissner *et al.* (1981).

sequence upstream from the attenuator site, yet allows good *cII* stimulation under conditions of tryptophan repression (Abraham *et al.*, 1980). Analogous deletion studies on pre have led to the same conclusion (Jones *et al.*, 1979).

A point mutation (*int-c*226) in the −10 region allows a high rate of constitutive transcription (and integrase production) in the absence of gp*cII*. Because it changes the initiation codon of the *xis* gene, it also causes a Xis$^-$ phenotype. The *int-c*226 mutation increases the similarity between the −10 region and the "consensus" sequence for bacterial and phage promoters (Rosenberg and Court, 1979). Mutants with a phenotype similar to that of λ*int-c*226 (and encountered in the same mutant selections) can also arise by insertion of IS2 into the *xis* gene, so that *int* is transcribed by readthrough from the IS2 promoter (Zissler *et al.*, 1977).

The basal level of transcription for the wild-type p_I promoter in the absence of gp*cII* is low but not zero. Two results of this low rate of transcription, which persists even in a repressed lysogenic cell, are (*1*) the occasional *int*-dependent occurrence of prophage losses in strains so constructed that a prophage is bracketed by sites the recombination of which requires integrase but not excisionase; and (*2*) readthrough transcription from the prophage into adjacent bacterial DNA, first studied in a strain with λ inserted at a secondary site in the *trpC* gene. Both results could include contributions from unidentified weak promoters within *xis* (or, in the second case, still further downstream) as well as from p_I itself. There is no doubt that transcription from p_I can read through into bacterial DNA, because the transcription from λ*int-c*226 prophage does so (the basis for its original selection). Whether the low rate of integrase expression in lysogenic cells serves any purpose is unknown (Campbell, 1976).

An interesting teleological question about p_I concerns its location. Why should the promoter lie at the beginning of the *xis* gene rather than farther downstream? There are several possible explanations, which are not mutually exclusive: (*1*) Perhaps the resulting long leader region on the *int* message plays some special role in the regulation of translation or the stabilization of the message. (*2*) Perhaps there is no other location for the p_I promoter that would be less disruptive of other requirements. The logical location for the promoter might be between *int* and *xis*, but these two genes overlap each other. If the overlap itself serves some useful function, that function might not permit a promoter sequence to be interspersed there. (*3*) Perhaps the location of the promoter is determined more by historical reasons than by optimization of function. If the *att-int-xis* segment joined the viral genome as a unit the recombinase functions of which had already evolved (as discussed in the final section of this chapter), the p_I promoter might be descended from the ancestral promoter of the previral segment.

D. Gene Overlap

The *int* and *xis* genes overlap by 20 base pairs. This raises the possibility that there is some competition between the translation of these two genes from the p_L message. Such competition could determine the relative amounts of integrase and excisionase made by the p_L message.

E. Control of Messenger Stability: Retroregulation

Control of transcription initiation provides an adequate explanation for the observed and expected regulation of integrase and excisionase, with one exception: In infected cells destined to lyse, the transcriptional controls are the same as in lysogenic cells that have been derepressed. However, integrase and excisionase are needed in the latter case, but not in the former. Moreover, little integrase activity is found in infected cells that form only the p_L transcript and not the p_I transcript (for example, cells infected with λ*cII*).

The important breakthrough in understanding this deficiency in integrase production was the discovery that much higher rates of integrase production were achieved in cells infected by λ*gal* phages or by λ phages with deletions of DNA immediately to the left of the prophage site, rather than by wild-type λ. The results of mixed infection showed that the effect is cis-specific: An integrase gene on a phage with wild-type phage DNA sequences to the left of *att* is poorly expressed, but the presence of such a phage in the same cell does not inhibit expression of an *int* gene on a phage from which this downstream DNA is deleted. The effect is unusual in that most prokaryotic regulation results from interactions with sites that are upstream rather than downstream from the affected genes. This effect was therefore given a special name: *retroregulation* (Epp *et al.*, 1981; Guarneros and Galindo, 1979; Schindler and Echols, 1981).

Retroregulation achieves the result we might consider ideal from the phage's viewpoint: In an infected cell, where integrase is not needed, the downstream inhibitory site (called *sib* for *si*tio *in*hibitor in the *b* region of λ) prevents its formation. In a derepressed lysogenic bacterium, however, the *sib* site is at the opposite end of the prophage and hence is not downstream from the *int* gene.

Hypotheses concerning the mechanism of retroregulation must account for the fact that retroregulation does not (and, teleologically, should not) prevent integrase production from the p_I transcript. It is thus special, not only as a downstream effect but also in its specificity for the origin of the transcript.

Although several questions about retroregulation remain unresolved, the following mechanism accounts for the known facts: The p_L message carrying the *sib* site is degraded by RNase III in a manner that drastically reduces *int* expression. Susceptibility to RNase III degradation is conferred on the p_L message by the presence of a stem and loop structure formed by bases −154 to −193. The p_I message, which terminates at a chain of U's that follow the stem and loop structure, is not degraded by RNase III; but the p_L message, because of the antiterminating activity of gp*N*, reads through the terminator and thereby acquires the proper secondary structure. The principal evidence for this hypothesis is (*1*) the absence of retroregulation in RNase III^- mutant hosts; and (*2*) the location in the −154 to −193 region of *sib* mutations that relieve retroregulation. It is not yet clear exactly how cleavage at the *sib* site leads to inactivation of the *int* gene, or how localized the degradation of the p_L message is.

Although retroregulation clearly operates to make λ closer to the ideal, its major significance to λ biology is still open to discussion. It avoids wasteful production of integrase where it is not needed, and perhaps prevents some infected cells from losing their phage DNA into the chromosome. However, whatever problems the lack of retroregulation might create should be encountered by *sib* mutants and perhaps even more strongly by *int-c* mutants, which produce integrase under all conditions, including repression. In fact, however, neither of these mutants exhibits any obvious deficiencies in phage growth as normally studied in the laboratory.

It remains to be seen whether retroregulation is a common method for controlling the activity of prokaryotic genes or whether its use is restricted to special circumstances. Certainly the particular version of retroregulation used by λ, which requires an antiterminated message, may have a limited distribution. Retroregulation appears to be especially suited to movable elements that resemble λ in that site-specific recombination results in the apposition of different sequences downstream from the gene. Retroregulation, along with site recognition such as that afforded by *int* and *xis*, allows the system to discriminate between its own state before recombination and its state afterward, and to modulate the regulation of the genes responsible for recombination according to the state of the system.

F. Metabolic Instability of Excisionase

Although retroregulation drastically reduces integrase production (as estimated from *in vivo* assays), it does not prevent the formation of excisionase from the same p_L message (36). However this result comes about mechanistically, the phage uses some logic in making excisionase under circumstances in which insertion is not desired, because it will offset the effect of any integrase that is accidentally produced. However, there is also the potential disadvantage that excisionase made in an infected cell before commitment may remain present after the cell has become committed to the lysogenic pathway, where excision is unwanted.

This disadvantage is reduced by the fact that excisionase is unstable *in vivo*. Weisberg and Gottesman (1971) transiently derepressed lysogenic bacteria, so that integrase and excisionase were formed during a limited time period, and then measured the decay of integrase and excisionase activity by superinfecting with a tester att^2 phage bearing either an int^- or an xis^- mutation. By that test, excisionase decays rapidly, whereas integrase is stable. This was the first regulatory mechanism discovered that affects the insertion and excision pathways differentially.

G. Adequacy of Current Models

The major controls known to operate on integrase and excisionase can be put into a tidy package in which everything makes teleological sense. It is prudent to question how complete our knowledge is likely to be, and also whether the raw facts fit the scheme as tidily as we have indicated.

Some perspective on both questions may be gained by recalling how the subject appeared to lambda workers about a decade ago (Hershey, 1971). At that time, it was commonly believed that the only mode of transcription for *int* and *xis* was from the p_L message, that the proteins were made coordinately, and that the metabolic instability of integrase provided a sufficient mechanism for directional control *in vivo*.

If someone had suggested at that time that λ needed a mechanism such as retroregulation, it is unlikely that the suggestion would have received much serious attention. This was not only because of the absence of experimental support, but because a substantial body of data had been collected that seemed to run counter to such a suggestion. These were data of a type I have already mentioned in passing: In a mixed infection between two λ phages (or between a combination such as λ*gal* × λ*bio*), the progeny include individuals that have undergone recombination at *att*, and the genetic requirements for the production of such recombinants are determined by the nature of the recombining sites, just as they are in insertion and excision.

If all the controls we have described operated with 100% efficiency, no such recombinants should be observed. We have implied that integrase is seldom made from the p_L message, and that the p_I message is turned on abruptly when the cells become committed to the lysogenic pathway. This would leave no opportunity for appreciable *int*-promoted recombination in a phage cross, because almost all such recombinants should arise after commitment to the lysogenic pathway, in cells that will never lyse and produce progeny. Conscious of this problem, current workers generally use att^2 phages with a cI^- mutation, so that cells whose gp*cII* concentration directs them toward the lysogenic pathway will nevertheless lyse and produce progeny. However, there are many earlier crosses in which no such precautions were taken, and in which *int*-promoted recombinants were nevertheless observed.

These observations present no serious discrepancy with respect to experimental facts. They simply indicate that the controls we have described are not always 100% effective, although well-controlled experiments show that they are quantitatively quite strong. I mention them here to give the nonlambdologist a feeling for the present state of the subject. First, we know little about the major sources of deviations from the simplistic application of the on–off rules of regulation—for example, whether the integrase that contributes phage recombinants in a cross between two cI^+ phages comes mainly from p_L or p_I. Second, we are pleased that the rules themselves accord so well with what we might expect if λ is a highly refined product of natural selection. This fact provides encouragement, but no assurance that our current perspective is at all comprehensive.

IV. Comparative Properties of Phages that Insert by Reciprocal Site-Specific Recombination

Although no other temperate phage's insertion system has been studied so intensively as that of λ, some interesting variations on a common theme can be identified. The phages of interest include λ-related phages such as coliphages 434, 81, 82, 21, and 80 and *Salmonella* phage P22, and λ-unrelated phages, such as coliphage P2.

A. Coliphages 81, 82, and 434

These phages insert at the same site on the bacterial chromosome as λ. Their integrase and excisionase proteins have been found to be functionally equivalent to those of λ in complementation studies (Balandina *et al.*, 1977). The DNA sequence of phage 434 *attPOP′* has been determined from −153 to +247. Comparison of the λ and 434 sequences (Fig. 6) reveals some highly conserved properties: (*1*) The 165 nucleotides at the 3′-terminus of the *int* gene are completely conserved. (*2*) The integrase binding sites are more highly conserved than the largest of the spacer regions separating them. (*3*) The 15-base sequence at the crossover point is completely conserved. (*4*) The length of the spacer regions is conserved. (*5*) The sequence to the left of the *sib* terminator (beyond −197) is totally different in the two phages (Mascarenhas *et al.*, 1981).

The last finding is of interest because it had been suggested that complementarity between the sequence distal to *sib* and a segment of the *int* gene might play some role in retroregulation (Epp *et al.*, 1981; Schindler and Echols, 1981). Phage 434 lacks this complementarity but shows retroregulation, like λ (Mascarenhas *et al.*, 1982). At a more general level, the finding suggests that, although the message must extend beyond the *sib* site in order to be processed, the nature of the distal sequence is not critical.

A DNA segment including the p_I promoter of 434 has also been sequenced (D. Mascarenhas, unpublished). The critical regions of the promoter do not differ from those of λ. No sequence data are available for phages 81 and 82.

B. Coliphages 21 and 80 and *Salmonella* Phage P22

These phages can all form viable recombinants with λ, but they insert at chromosomal sites different from that used by λ. The int^+ and xis^+ genes of phage 21 will not complement a λint^- or λxis^- mutant in mixed infection (Liedke-Kulke and Kaiser, 1967). Whereas phages 21 and 80 are similar to λ in their life cycles and genome organization, P22 differs not only in its normal host species but also in its mode of DNA packaging and its use of an ancillary regulation system for repressor control.

The chromosomal insertion sites for phages 21 and 80 are close to the *trp* operon and may be identical to each other. Little information, either genetic or molecular, is available about the insertion sites and genes of these phages. Such information should be of interest because the phages are related to λ and have a similar overall genome organization, yet the 21 insertion site is not even an identified secondary site for λ insertion. If λ and 21 diverged from a common ancestor, without independent intercalation of insertion genes from foreign sources, then attachment sites and integrase genes must have changed drastically but coordinately.

The integrase gene of phage P22, which was the first integrase gene to be identified in any phage, is located close to the *att* site (Smith and Levine, 1967). The chromosomal insertion site of P22, near the *proB* gene, is distinct from that of λ or known related coliphages (Susskind and Botstein, 1978).

C. Coliphage P2

Coliphage P2 is completely unrelated to λ, as judged by DNA hybridization. It inserts by reciprocal site-specific recombination at a preferred chromosomal site near the *his* operon. As in λ, the integrase gene of P2 is adjacent to the insertion site. Mutations affecting excision map in another gene (*cox*) on the other side of the repressor gene from *att*. Because the frequency of generalized recombination in P2 is very low, *int*-promoted recombination at the *att* site is more frequent than the total amount of recombination observed in the rest of the P2 chromosome during a phage cross (Bertani and Bertani, 1971; Lindahl and Sunshine, 1972).

The most profound difference between P2 insertion and λ insertion is in the genetic control of integrase expression. Not only is P2 insertion less affected (perhaps unaffected) by repressor than λ insertion, but, more strikingly, the expression of P2 integrase from a derepressed prophage is very small compared to that from an infecting phage in the same cell. Thus (as with retroregulation in λ), the activity of an insertion gene is altered by the act of insertion. In P2, the effect is just the opposite of that in λ. P2 insertion prevents, rather than enhances, integrase expression. It is hypothesized that either the *int* gene itself or the operon to which it belongs is split by the insertion event. The direction of transcription of P2*int* being unknown, a split in the operon might either disconnect the gene from its major promoter or, alternatively, might remove a downstream site that protects the RNA from degradation. In any event, some limited expression of *int* must be possible from an inserted prophage because spontaneous phage production from lysogenic cells requires an int^+ prophage. As in many lysogenic systems, such spontaneous phage production is a rare event, and spontaneous excision to produce viable cells cured of the prophage is even rarer. Both the P2 mode of integrase regulation and the λ mode stabilize the system against such spontaneous loss.

One consequence of the nonrepressible expression of P2 integrase is that, unlike λ, P2 is unable to form tandem double lysogens, in which the *int* gene adjacent to the central *att* site would be expressed like that of an infecting phage and would therefore cause one of the two prophages to be excised. Such excision comprises an *attPOP′* × *attBOP′* or *attPOP′* × *attPOB′* reaction that (if P2 follows the same rules as λ) should require only integrase, not excisionase. We might expect a λ*int-c* mutant to behave like P2 in this respect. Tandem double lysogens of P2 can be formed only if the central *int* gene carries an int^- mutation (Bertani, 1971). Because of this feature, P2 forms double lysogens less readily than λ does, and those rare double lysogens that are formed carry the P2 prophage at one of various possible secondary sites.

Another noteworthy property of P2 is its ability to undergo occasional *int*-dependent imprecise excision, leaving behind a chromosome that is not only cured of P2 but has also suffered a deletion of DNA adjacent to the P2 insertion site. Such deletion (termed *eduction*) excises the prophage together with adjacent DNA segments of various lengths (Sunshine, 1972). Eduction has some similarity to adjacent deletion formation by transposons.

V. Reciprocal Recombination as a Transposition Mechanism

The final two sections of this chapter depart from a factual account of λ and similar phages to discuss more general issues that have been raised concerning the properties of movable elements.

Because λ insertion was already well documented at the time bacterial insertion sequences were discovered, it was natural to entertain the hypothesis that the latter elements employed an insertion mechanism similar to that of λ. In particular, it was hypothesized that bacterial insertion sequences generally moved from one site to another by excising as a circle from their original location and reinserting elsewhere. The major difference between an episome such as λ or F and typical transposable elements would then be that, although both classes of elements can insert and excise, only the episomes can replicate autonomously. The fact that precise excision of transposons almost invariably resulted in loss rather than in relocation, the general continued presence of a copy of the transposed element in its old location as well as its new one, and especially the example of Mu phage, the replication of which is closely coupled to its transposition, made it obvious that different explanations were required (Shapiro *et al.*, 1977).

These facts have stimulated the development of various ingenious models to explain replicative transposition. They also generate another question that has received less attention. At the current state of knowledge about transposition and λ insertion, can we see any reason why some transposable elements should not employ a λ-like mechanism for excision and reinsertion? If not, may we anticipate the discovery that some transposable elements insert by reciprocal, site-specific recombination? The fact that new transpositions of maize controlling elements can be found among individuals that have undergone excision from the original site (McClintock, 1956; Greenblatt, 1968) suggests that excision and transposition can be correlated in some cases. At any rate, at our present state of knowledge, it seems at least as important to discuss the possible reasons for any general rules that we perceive as it is to recount the rules that apply within the limited domain that has been examined carefully.

Regarding this general question, perhaps the most relevant advance in our understanding of λ insertion is the demonstration that it comprises precise reciprocal recombination between identical 15-base pair sequences. Such a mechanism implies a high degree of specificity for the chromosomal site into which λ inserts. The specificity is not absolute. Insertion can take place into secondary sites that are not identical to λ in all 15 base pairs (or even in all of the 7 base pairs between the staggered nicks), and such insertions can be precisely reversed by excision. The rates of insertion and excision at secondary sites are much lower than those at the primary site. Transposons and insertion sequences generally exhibit a much lower specificity for the target sites at which they insert. There are conspicuous exceptions, such as IS*4*. It is reasonable to suppose that the insertion–excision mechanism exemplified by λ is better suited to an element that inserts at a unique site than is the mechanism of replicative transposition of IS*1*, although we do not know how readily either system can evolve in the direction of increased or decreased specificity for the target site.

We might expect that a mechanism with low target site specificity would be chosen whenever the purpose of insertion is pure mobility—such as the ability of an antibiotic resistance determinant to hitchhike its way into other cells by hopping onto a plasmid or phage. On the other hand, whenever transpositions take place in a controlled manner where they are directed toward specific target sites (as they might if transposable elements carry out some of the determinative changes during the development of eukaryotes), excision and reinsertion by a λ-like mechanism should serve the purpose. In terms of the survival of the element itself (viewed as potentially "selfish DNA" of no use to the host), a replicative mechanism with low target specificity may facilitate the increase and dispersal of the element.

One could ask how many mutational steps would be required to change λ itself to a nonreplicating element capable of transposing from one secondary site to another (or back to the primary site). So far, as we know, only two steps (both known) are required: an internal deletion that eliminates the replication origin, and an *int-c* or o_L-*c* mutation that causes a high rate of constitutive production of integrase. An o_L-*c* mutation also causes constitutive production of excisionase, which facilitates excision from secondary sites as well as from the primary site. To my knowledge, no one has actually verified that such a phage derivative can transpose; nor is it obvious that experimental verification is worth the effort. However, the prediction that such a phage derivative should transpose by excision and reinsertion is based entirely on known mechanisms. If it failed to transpose as predicted, we would be forced to search for an explanation of this failure, and such an explanation would necessarily invoke new phenomena of which we are currently unaware.

To summarize, there are clearly some circumstances in which replicative transposition should be a superior strategy to excision and insertion. There are no compelling reasons to believe that it should always be so, even for elements incapable of autonomous replication.

VI. Evolution of λ and Its Relation to Host Evolution

The generation of λ*gal* from λ lysogens was the first unequivocal example of a host gene becoming stably incorporated into the genome of a virus. Consequently, it played an important role in enlarging our perspectives on viral evolution. Whereas the potentialities of λ for gene incorporation initially appeared to be limited to host genes in the immediate vicinity of its chromosomal insertion site, it soon became clear that, from secondary site lysogens or lysogens of hosts that had experienced genetic rearrangements, λ derivatives could be obtained bearing any gene of the host. This raised the question of whether λ as we now know it might have acquired much of its present genome by picking up host genes and adapting them to its own purposes. A related question is whether λ has had any significant impact on the genetic constitution of its host species by transferring host genes among strains, by translocating genes from one chromsomal location to another, or by the separation of viral genes from the rest of the viral genome and their reprogramming to serve host functions.

A. Possible Evolution by Accretion of Host Modules

Once we consider the possibility that λ evolution has proceeded by the acquisition of genes or blocks of genes borrowed from the host or from other viruses or plasmids in the host (followed, of course, by mutations that increase the adaptive value of the new complex), it is natural to wonder whether the strong tendency for genes with related functions to be clustered together on the λ genetic map might not be a consequence of the progressive acquisition of blocks of genes that were already functioning as units before they became a part of λ. Thus, we might imagine that λ was derived from a small plasmid that contained the replication genes *O* and *P* and the replication origin, a bacterial operon that functioned in the turnover of cell envelope components that included the lysis genes, a DNA packaging system (perhaps functional in the genetic exchange between bacteria) consisting of the head and tail genes, an insertion sequence that includes *int, xis,* and *att,* and so forth. The implication is that each of these functional modules that comprises a segment of the viral genome has a more ancient pedigree than λ itself does, and that it was already performing a function similar to its present one in λ before it ''joined'' the λ genome.

There is no evidence against such a scheme. Evidence of several types is superficially supportive, but none is really convincing.

First, λ can rather easily lose, by deletion, large segments of its genome, thus generating derivatives that resemble its postulated progenitors. It can, for example, shed 80–90% of its genome, including the *N* gene, but leaving genes *cro* through *P* to become a multicopy plasmid (λ*dv*) that can be propagated indefinitely in that form. Likewise, deletions of DNA from an inserted prophage can remove almost all the λDNA except the genes for assembly of viral particles. Such deletions can occur spontaneously, but also under the combined agency of ultraviolet light and insertion sequence IS2.

One strain of λ that harbors an IS2 just upstream from *xis* (λ*crg*) repeatedly generates specific deletions that remove all DNA between the IS2 and a point just to the left of the *cos* site. Interestingly, such deletions can also be induced by ultraviolet light (though less frequently) in wild-type λ, perhaps under the influence of other IS2 sequences in the cell (Marchelli *et al.*, 1968). Such events show that objects analogous to the postulated progenitors can be stably maintained, and that the transition between wild-type λ and such objects can occur fairly readily.

The relationship between λ and IS2 and their interactions leading to specific deletion formation, though still poorly understood, have several intriguing aspects that deserve further study: (*1*) In addition to the most common deletion type, which removes genes *N* through *R* and leaves *cos-J,* there are other rarer types, such as one that removes genes *N-P* and leaves genes *Q-J*. (*2*) Deletion is frequently, but not always, accompanied by chromosomal rearrangement that destroys the genetic linkage between the *gal* and *bio* genes flanking the λ prophage. Different rearrangements are found in different isolates (Zavada and Calef, 1969). (*3*) The phage, as well as the IS2, plays an active role in cryptic formation. For example, replacement of the λ immunity region with the 434 immunity region eliminates cryptic formation (Adhya and Campbell, 1970). (*4*) IS2 has a specific preference for insertion into λDNA in the immediate vicinity of the *crg* insertion. Two spontaneous IS2 insertions (*bi2* and *crg*) appeared in λ stocks without deliberate selection for them (Zissler *et al.*, 1977). An artificial IS2 made selectable by intercalation of a gene for kanamycin resistance repeatedly inserted at a specific site located between *crg* and *bi2,* in the opposite orientation from *crg* (Saint-Girons *et al.*, 1981). It is not known whether the resulting phages undergo specific deletions.

Second, λ-related phages such as 434, 21, and so on collectively exhibit considerable diversity in their specificities of repression, insertion, surface receptor recognition, replication, and positive regulation. Recombinants bearing various combinations of these specificities can be generated in laboratory crosses but are also found in natural strains. This functional diversity among phages that are sufficiently related to recombine is mirrored in heteroduplex studies. Heteroduplexes between any two natural lambdoid phages show an interspersion of homologous segments (that include the crossover points) and heterologous segments (that include the heterospecific genetic determinants). One possible manner in which such a pattern might have arisen is if some of the various lambdoid phages have evolved by the independent recruitment of distinct functional modules from different sources. The extreme alternative is that all known lambdoid phages are descended from a common ancestor of the same general size and complexity as λ, and that the homologous segments have been conserved while the heterologous ones diverged. In my judgment, the actual data on lambdoid phages fit much better with descent from a common ancestor.

Third, *E. coli* and related bacteria harbor in their DNA various segments (that differ among bacterial isolates) that are either homologous to parts of the λ genome by Southern hybridization, or can recombine with λ mutants to generate functional phages, or both. *E. coli* K12, for example, harbors genes that can replace the *red* and *gam* genes of λ, and also the late regulation and lysis genes *Q, S, R,* and *Rz,* with heterologous but functionally equivalent DNA. Some of these segments might be related to the postulated precursors from which λ originated. Alternatively, they could be the degenerate remains of lambdoid prophages that suffered large deletions. In my judgment, the actual data fit much better with the idea that they are deletion derivatives of lambdoid phages.

B. Distinction between Host Modules and Defective Prophages

I have not tried to justify my evaluation of the data on λ-related phages and phage-related DNA segments. These issues have been discussed in more detail elsewhere (Campbell, 1977). However, in the case of phage-related segments, it is important to explain what question is really being asked, and why it seems necessary to ask that question rather than some other one.

Most scientists distrust concepts that cannot be translated rather directly into experimental operations. If we ask whether portions of a viral genome can originate from host genes, we might hope to settle the matter by showing that a DNA segment that consists of host genes at the beginning of an experiment can become part of a viral genome during the experiment. However, to define the question in that manner would require us to conclude that viruses originate from host genes every time a lysogenic bacterium is induced, because the genes of the prophage behave as host genes, operationally defined. Few virologists would want to equate induction of a lysogen with the origin of a virus. Because bacteria can become lysogenic in the laboratory by infection, we suspect (although we cannot prove) that most naturally lysogenic bacteria likewise acquired the viral genome through infection. The prophage is therefore treated as a stage in the phage life cycle, rather than as an evolutionary precursor of the phage. To anyone taking the opposite view, the problem of the origin of viruses is of course already solved.

Although the distinction just made may appear obvious and trivial, it needs to be stated explicitly before we consider the subject of natural strains carrying defective or incomplete prophages. Suppose that we take two natural strains of bacteria, neither of which is lysogenic, cross them together, and observe that a recombinant bacterium produces a phage the genome of which is derived in part from the chromosome of one of these strains and in part from the other. Are we not then witnessing the origin of a virus from host genes? And if that observation does not settle the matter, what observation would?

The problem is that, just as we can make bacteria lysogenic by laboratory infection, we can also obtain defective prophages from complete ones. We can therefore easily construct precisely the result described above, in which two strains, neither overtly lysogenic but harboring complementary remnants of a phage that infected them earlier, can generate a complete phage by genetic recombination. The laboratory study of defective prophages generally entails infection of one such bacterial strain by a mutant phage, which is a minor variation on the same theme.

In asking whether viruses have originated from host genes or host modules, what we would like to see is the incorporation into a viral genome of genes that not only were present in the host chromosome but that also served some function in the economy of the host cell (such as the incorporation of *gal* genes into λ to give λ*gal*), where the incorporated genes (unlike the *gal* genes) come to serve some useful function in the life of the virus. As soon as we introduce "useful function" as a criterion (especially if we stipulate that the relevant utility must be observable in the natural habitat of the phage and bacterium), we have gone beyond the limits of any simple operational definition. Furthermore, once we accept as real the distinction between a genuine host module and a degenerate prophage, we appreciate that recombination among degenerate prophages should be much more frequent than the evolutionarily successful incorporation of host genes, so that most of the apparent examples of the latter process are expected to be spurious.

C. Tentative Conclusions

Nothing that has been said here contradicts the idea that λ originated by compounding host modules in the manner postulated at the beginning of this section. The tentative conclusion is that, regardless of how λ may have originated, the organization of the entire gene complex as now constituted preceded the diversification of the different lambdoid phages and the dissemination of fragments of λ-derived DNA among natural enterobacteria. The implication is that the process of gene acquisition by which a virus as complex as λ may originate from host modules is not an ongoing process that is happening at a frequency detectable in ordinary laboratory-scale experiments, but rather consists of natural events that very rarely occur in such a manner that their products are successful in evolution.

The results also indicate that λ has moved around a lot in nature, occasionally leaving as souvenirs defective prophages scattered about in the chromosomes of various natural strains. This underscores the potentiality of λ for effecting gene translocations of functional significance to the host, although direct evidence is unavailable.

D. Sample Pathway for the Possible Evolution of the Insertion Region of λ

The modular hypothesis of λ evolution implies that *int* and *xis* should be more closely related to the analogous genes of simpler inserting elements, such as transposons, than they are to other genes of λ, such as *red* or *cIII*. Of the known transposon genes, those with the greatest functional similarity to λ*int* are genes encoding resolvases for cointegrates, such as the *tnpR* gene of Tn*3*.

Although there is no detectable homology between *tnpR* and *int*, either in DNA sequence or in amino acid sequence, it may be worth considering briefly the steps that would be needed for a phage such as λ to have acquired its present structure by incorporation of an element such as Tn*3*. The purpose of the exercise is not to suggest that any particular course of events offers unique advantages, but rather to provide a framework for examining the implications and problems that are common to various schemes. For this purpose, there are advantages to working through one possible scheme from beginning to end without trying to enumerate all the possible alternatives at each step. To avoid any misunderstanding about the literal identification of the ancestral transposon with Tn*3*, I will consider a hypothetical element called Tnλ, the genome organization of which resembles that of Tn*3*. Its resolvase gene (analogous to *tnpR*) will be named *tin*.

A possible sequence of steps is diagrammed in Fig. 8. The first step is the insertion of Tnλ into an ancestral noninserting phage. The phage, which can now insert into the chromosome by reciprocal recombination with a preinserted Tnλ, evolves a repression mechanism that allows the prophage to persist without harming the cell. One day, the phage infects a cell the chromosome of which contains no Tnλ and inserts instead into a chromosomal site between *gal* and *bio* the 15-base pair sequence of which resembles the resolution site of Tnλ to some extent (like a secondary insertion site of λ). Successive steps that follow are (*1*) inversion of the resolvase gene *tin* so that it can be transcribed from the p_L promoter; (*2*) base changes in the original *tin* promoter to make it activable by *cII;* (*3*) differentiation of the N-terminus of the *tin* polypeptide so that this part of the protein specifically promotes excision; (*4*) evolution of a new translation-initiation codon in the middle of the *tin* gene, allowing production of the Int protein that does not cause efficient excision; (*5*) evolution of a termination signal for the *xis* gene, generating a separate Xis polypeptide; (*6*) deletion of base sequence changes at the 5′-end of the *xis* gene, so that Xis is no longer produced from the p_I message; (*7*) evolution of the *sib* site to create the RNase III sensitivity of the P_L message.

The possible alternative hypotheses at each step can be illustrated by considering as an example the manner in which the difference between *attPOP′* and *attBOB′* may have arisen. Any hypothesis on this subject must be consistent with three facts: (*1*) In a genome of about 5×10^3 kb, such as that of *E. coli,* a specific 15-base pair sequence is unlikely to recur by chance. Applied to the 15 base pairs at the crossover points of phage and bacterial DNA (or, equivalently, of the two prophage ends), this might mean either that the phage sequence is a replica of DNA originally at the bacterial site or that the bacterial sequence arose by a duplication of DNA introduced by the phage or its ancestral transposon. It could, of course, also be derived in part from both sources. (*2*) There is a bacterial insertion site for λ not only in cured derivatives of the *E. coli* K12 strain (which is naturally lysogenic for λ) but also in other enterobacterial strains, such as *E. coli* B and *Salmonella typhimurium* (Zichichi and Kellenberger, 1963; Smith, 1971). (*3*) The flanking sequences of *attBOB′* and *attPOP′* are unrelated.

In the above scheme, the two prophage ends are derived from the bacterial

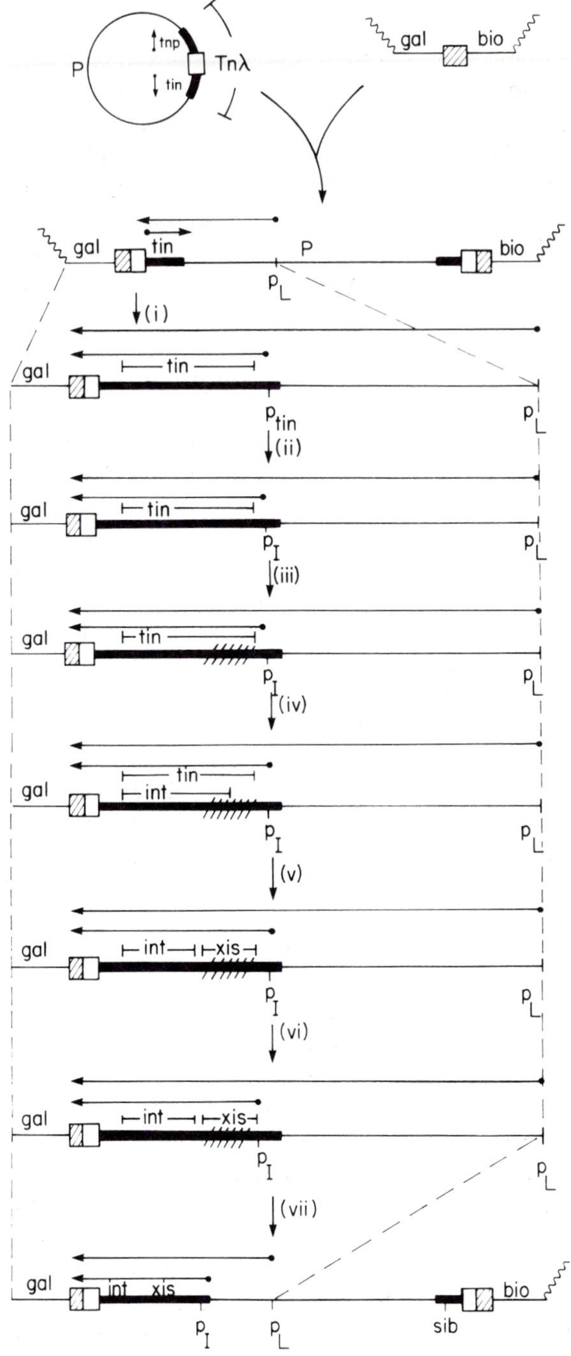

Fig. 8 Possible scheme for evolution of the insertion genes of λ. See the text for details.

insertion site. Although insertion into a secondary site results initially in imperfect matching, it is expected that repeated cycles of insertion into the same bacterial site would favor mutation or correction that would change the phage site in the direction of the bacterial site rather than the reverse. No additional assumptions are needed to account for the second and third facts listed above.

A reasonable alternative is that the ancestral lysogen, where λ was inserted into a Tnλ, lost the distal segments of the Tnλ by *tin*-promoted deletion and then went on to evolve *int* and *xis*. In this case, both phage and bacterial sites would be derived from the Tnλ.

One difficulty with making a definite choice between these alternatives comes from the fact that λ-related phages insert at several different sites that are not obviously related to one another. If such site preferences have evolved by divergence from a common ancestor after the *int* and *xis* genes were already present, then the scheme of Fig. 8 is simplistic in assuming that the ancestral phage was inserted at the present λ site rather than at the site of phage 21 or P22.

At any rate, this particular scheme calls attention to some general features that are common to most alternative schemes: (*1*) Although it is tempting to rearrange the order in which the indicated steps occur, there seems to be no possible order that would generate a better-adapted element at each step. For example, the development of a fully effective excisionase function prior to the elimination of *xis* from the p_I message would generate a phage incapable of stable insertion. It is unlikely that evolution has often proceeded very far in an antiadaptive direction before reaching the next adaptive peak. It is more probable that a step such as the development of fully effective excisionase is really the sum of a series of gradual changes that went hand in hand with changes in the p_I promoter. In general, we may expect that components such as *xis, sib,* and p_I acquired their present properties gradually and concurrently, and that some adjustments in the rest of the phage genome took place simultaneously. (*2*) There may well be existing elements that diverged from λ by branching off from the scheme at various points. For example, a split operon control such as that exhibited by phage P2 is an alternative that might be reached by a phage that did not undergo the initial inversion of the *tin* gene. (*3*) As with all schemes for viral evolution, changes in the reverse direction are equally plausible. If viruses and transposons are related at all, there is an equal chance that existing transposons are descendants of temperate phages, rather than the other way around.

We may note the contrast between the evolutionary question we posed in the previous section of this chapter (Could λ change into an element that transposes among secondary sites?) and the one we have been considering here. The former question can be given a definite answer: Two known changes should do the job. Here, on the other hand, we have postulated numerous steps, with little direct evidence for any of them. We hope that awareness of the question and its possible answers may expedite the identification of the specific steps that are largely hypothetical at present.

References

Abraham, J., and Echols, H. (1981). Regulation of *int* gene transcription by bacteriophage λ. Location of the RNA start generated by an *int* constitutive mutation. *J. Mol. Biol.* **146,** 157–165.

Abraham, J., Mascarenhas, D., Fischer, R., Benedik, M., Campbell, A., and Echols, H. (1980). DNA sequence of regulatory region for integration gene of bacteriophage λ. *Proc. Natl. Acad. Sci. USA* **77,** 2477–2481.

Abremski, K., and Gottesman, S. (1981). Site-specific recombination: *xis*-independent excisive recombination of bacteriophage lambda. *J. Mol. Biol.* **53,** 67–78.

Adhya, S., and Campbell, A. (1970). Crypticogenicity of bacteriophage λ. *J. Mol. Biol.* **50,** 481–490.

Balandina, L. A., Sineokii, S. P., and Krylov, V. (1977). Isolation and study of Int-mutants of φ 81. *Genetika (Moscow)* **13,** 1103–1107. [*Sov. Genet. (Engl. transl.)*].

Belfort, M., and Wulff, D. L. (1971). A mutant of *Escherichia coli* that is lysogenized with high frequency. *In* "The Bacteriophage λ" (A. D. Hershey, ed.), pp. 739–742. Cold Spring Harbor Lab., Cold Spring Harbor, N.Y.

Bertani, L. E. (1971). Stabilization of P2 tandem double lysogens by *int* mutations in the prophage. *Virology* **46,** 426–436.

Bertani, L. E., and Bertani, G. (1971). Genetics of P2 and related phages. *Adv. Genet.* **16,** 199–237.

Campbell, A. (1972). Episomes in evolution. *In* "Evolution of Genetic Systems" (H. H. Smith, ed.), *Brookhaven Symp. Biol.* **23,** 534–562.

Campbell, A. (1976). Significance of constitutive integrase formation. *Proc. Natl. Acad. Sci. USA* **73,** 887–890.

Campbell, A. (1977). Defective bacteriophages and incomplete prophages. *In* "Comprehensive Virology" (H. Fraenkel-Conrat and R. R. Wagner, eds.), Vol. 8, pp. 259–328. Plenum, New York.

Campbell, A. (1981). Some general questions about movable elements and their implications. *Cold Spring Harbor Symp. Quant. Biol.* **45,** 1–9.

Court, D., Green, L., and Echols, H. (1975). Positive and negative regulation by the cII and cIII products of bacteriophage λ. *Virology* **63,** 484–493.

Davis, R., and Parkinson, J. (1971). Deletion mutants of phage lambda. III. Physical structure of attφ. *J. Mol. Biol.* **56,** 403–423.

Dove, W. F. (1970). An energy-level hypothesis for λ prophage insertion and excision. *J. Mol. Biol.* **47,** 1115–1120.

Echols, H., and Green, L. (1979). Some properties of site-specific and general recombination inferred from int-initiated exchanges by bacteriophage lambda. *Genetics* **93,** 297–307.

Echols, H., Chung, S., and Green, L. (1974). Site-specific recombination: genes and regulation. *In* "Mechanisms in Recombination" (R. F. Grell, ed.), pp. 69–77. Plenum, New York.

Enquist, L. W., and Weisberg, R. A. (1976). The red plaque test: a rapid method for identification of excision defective variants of bacteriophage λ. *Virology* **72,** 147–153.

Enquist, L. W., Kikuchi, A., and Weisberg, R. A. (1979a). The role of λ integrase in integration and excision. *Cold Spring Harbor Symp. Quant. Biol.* **43,** 1115–1120.

Enquist, L. W., Nash, H., and Weisberg, R. A. (1979b). Strand exchange in site-specific recombination. *Proc. Natl. Acad. Sci. USA* **76,** 1363–1367.

Epp, C., Pearson, M. L., and Enquist, L. (1981). Downstream regulation of *int* gene expression by the *b2* region in phage lambda. *Gene* **13,** 327–337.

Forbes, D., and Herskowitz, I. (1982). Polarity suppression dependent on the *Q* gene product of phage λ. *J. Mol. Biol.* (in press).

Gottesman, S. (1981). Lambda site-specific recombination: the *att* site. *Cell* **25,** 585–586.

Gottesman, S., and Abremski, K. (1980). The role of himA and xis in lambda site-specific recombination. *J. Mol. Biol.* **138,** 503–512.

Gottesman, S., and Gottesman, M. (1975). Excision of prophage λ in a cell-free system. *Proc. Natl. Acad. Sci. USA* **72,** 2188–2192.

Greenblatt, I. M. (1968). The mechanism of modulator transposition in maize. *Genetics* **58,** 585–597.

Guarneros, G., and Echols, H. (1973). Thermal asymmetry of site-specific recombination by bacteriophage λ. *Virology* **52,** 30–38.

Guarneros, G., and Galindo, J. M. (1979). The regulation of integrative recombination by the *b2* region and the *cII* gene of bacteriophage λ. *Virology* **95,** 119–126.

Guerrini, F. (1969). On the asymmetry of λ integration sites. *J. Mol. Biol.* **46,** 523–542.

Hershey, A. D. (ed.) (1971). "The Bacteriophage Lambda." Cold Spring Harbor Lab., Cold Spring Harbor, N.Y.

Herskowitz, I., and Hagen, D. (1980). The lysis–lysogeny decision of phage λ: explicit programming and responsiveness. *Annu. Rev. Genet.* **14,** 399–446.

Jones, M. O., and Herskowitz, I. (1978). Mutants of bacteriophage λ which do not require the *cIII* gene for efficient lysogenization. *Virology* **88,** 199–212.

Jones, M. O., Fischer, R., Herskowitz, I., and Echols, H. (1979). Location of the regulatory site for establishment of repression by bacteriophage λ. *Proc. Natl. Acad. Sci. USA* **76,** 150–154.

Landy, A., and Ross, W. (1977). Viral integration and excision: structure of the lambda *att* site. *Science* (Washington, D.C.) **197,** 1147–1160.

Liedke-Kulke, M., and Kaiser, A. D. (1967). Genetic control of prophage insertion specificity in bacteriophages λ and 21. *Virology* **32,** 465–474.

Lindahl, G., and Sunshine, M. (1972). Excision-deficient mutants of bacteriophage P2. *Virology* **49,** 180–187.

Marchelli, C., Pica, L., and Soller, A. (1968). The cryptogenic factor in λ. *Virology* **34,** 650–663.

Mascarenhas, D., Kelley, R., and Campbell, A. (1981). DNA sequence of the *att* region of coliphage 434. *Gene* **15,** 151–156.

Mascarenhas, D., Trueheart, J., Benedik, M., and Campbell, A. (1982). Retroregulation: control of integrase expression by the *b2* region of bacteriophage λ and 434. *Virology* (in press).

McClintock, B. (1956). Controlling elements and the gene. *Cold Spring Harbor Symp. Quant. Biol.* **21,** 197–216.

Miller, H. I., Kikuchi, A., Nash, H. A., Weisberg, R. A., and Friedman, D. I. (1979). Site-specific recombination of bacteriophage λ: the role of host gene products. *Cold Spring Harbor Symp. Quant. Biol.* **43,** 1121–1126.

Miller, H. I., Abraham, J., Benedik, M., Campbell, A., Court, D., Echols, H., Fischer, R., Galindo, J. M., Guarneros, G., Hernandez, T., Mascarenhas, D., Montanez, C., Schindler, D., Schmeissner, V., and Sosa, L. (1981). Regulation of the integration–excision reaction by bacteriophage λ. *Cold Spring Harbor Symp. Quant. Biol.* **45,** 439–445.

Mizuuchi, K., and Mizuuchi, M. (1979). Integrative recombination of bacteriophage λ: *in vitro* study of the intermolecular reaction. *Cold Spring Harbor Symp. Quant. Biol.* **43,** 1111–1114.

Mizuuchi, K., Gellert, M., and Nash, H. A. (1978). Involvement of supertwisted DNA in integrative recombination of bacteriophage lambda. *J. Mol. Biol.* **121,** 375–392.

Mizuuchi, K., Gellert, M., Weisberg, R. A., and Nash, H. A. (1980). Catenation and supercoiling in the products of bacteriophage λ integrative recombination *in vitro*. *J. Mol. Biol.* **141,** 485–494.

Mizuuchi, K., Weisberg, R., Enquist, L., Mizuuchi, M., Buraczynska, M., Foeller, C., Hsu, P.-L., Ross, W., and Landy, A. (1981). Structure and function of the phage *att* site: size, Int binding sites and location of the crossover point. *Cold Spring Harbor Symp. Quant. Biol.* **45,** 429–437.

Nash, H. A. (1981a). Integration and excision of bacteriophage λ: the mechanism of conservative site specific recombination. *Annu. Rev. Genet.* **15,** 143–168.

Nash, H. A. (1981b). Site-specific recombination protein of phage λ. *In* "The Enzymes" (P. Boyer, ed.), Vol. 14A, pp. 471–480. Academic Press, New York.

Ogawa, T., and Tomizawa, J. (1967). Abortive lysogenization of bacteriophage λ *b2* and residual immunity of non-lysogenic segregants. *J. Mol. Biol.* **23,** 225–245.

Pollock, T. J., and Abremski, K. (1979). DNA without supertwists can be an *in vitro* substrate for site-specific recombination of bacteriophage λ *J. Mol. Biol.* **131,** 651–654.

Reed, R. R., and Grindley, N. D. F. (1981). Transposon-mediated site-specific recombination *in vitro:* DNA cleavage and protein-DNA linkage at the recombination site. *Cell* **25,** 721–728.

Rosenberg, M., and Court, D. (1979). Regulatory sequences involved in the promotion and termination of RNA transcription. *Annu. Rev. Genet.* **13,** 319–353.

Saint-Girons, I., Fritz, H.-J., Shaw, C., Tillman, E., and Starlinger, P. (1981). Integration specificity of an artificial kanamycin transposon constructed by the *in vitro* insertion of an internal Tn*5* fragment into IS2. *Mol. Gen. Genet.* **183,** 45–50.

Schindler, D., and Echols, H. (1981). Retroregulation of the *int* gene of bacteriophage λ: control of translation completion. *Proc. Natl. Acad. Sci. USA* **78,** 4475–4479.

Schmeissner, V., Court, D., McKenney, K., and Rosenberg, M. (1981). Positively activated transcription of λ integrase gene initiates with UTP *in vivo. Nature* (*London*) **292,** 173–175.

Shapiro, J. A., Adhya, S. L., and Bukhari, A. I. (1977). Introduction: new pathways in the evolution of chromosome structure. *In* "DNA Insertion Elements, Plasmids, and Episomes" (A. I. Bukhari, J. A. Shapiro, and S. L. Adhya, eds.), pp. 3–12. Cold Spring Harbor Lab., Cold Spring Harbor, N.Y.

Shimada, K., and Campbell, A. (1974). Lysogenization and curing by *int*-constitutive mutants of phage λ. *Virology* **60,** 157–165.

Shimada, K., Weisberg, R. A., and Gottesman, M. E. (1972). Prophage lambda at unusual chromosomal locations. I. Location of the secondary attachment sites and the properties of the lysogens. *J. Mol. Biol.* **63,** 483–503.

Shulman, M., and Gottesman, M. (1971). Lambda *att*^{2}*:* a transducing phage capable of intramolecular *int-xis* promoted recombination. *In* "The Bacteriphage Lambda" (A. D. Hershey, ed.), pp. 477–489. Cold Spring Harbor Lab., Cold Spring, N.Y.

Smith, G. R. (1971). Specialized transduction of the *Salmonella hut* operons by coliphage λ: deletion analysis of the *hut* operons employing λ p*hut*. Virology **45,** 208–223.

Smith, H. O., and Levine, M. (1967). A phage P22 gene controlling integration of prophage. *Virology* **31,** 207–216.

Sunshine, M. G. (1972). Dependence of eduction on P2 *int* product. *Virology* **47,** 61–67.

Susskind, M., and Botstein, D. (1978). Molecular genetics of bacteriophage P22. *Microbiol. Rev.* **42,** 385–413.

Weisberg, R. A., and Gottesman, M. E. (1971). The stability of Int and Xis functions. *In* "The Bacteriophage Lambda" (A. D. Hershey, ed.), pp. 489–500. Cold Spring Harbor Lab., Cold Spring Harbor, N.Y.

Zavada, V., and Calef, E. (1969). Chromosomal rearrangements in *Escherichia coli* strains carrying the cryptic lambda prophage. *Genetics* **61,** 9–22.

Zichichi, M. L., and Kellenberger, G. (1963). Two distinct functions in the lysogenization process:

the repression of phage multiplication and the incorporation of the prophage in the bacterial genome. *Virology* **19,** 450–460.

Zissler, J., Mosharaffa, E., Pilacinski, W., Fiandt, M., and Szybalski, W. (1977). Position effects of insertion sequences I*S2* near the genes for prophage λ insertion and excision. *In* "DNA Insertion Elements, Plasmids, and Episomes" (A. I. Bukhari, J. A. Shapiro, and S. L. Adhya, eds.), pp. 381–388. Cold Spring Harbor Lab., Cold Spring Harbor, N.Y.

CHAPTER 3

Phage Mu: Transposition as a Life-Style

Ariane Toussaint
Anne Résibois

I. Introduction

The temperate and mutator phage Mu was the first prokaryotic transposable element to be described (Taylor, 1963). Since then, a second mutator phage very closely related to Mu, D108, has been discovered and characterized (Mise, 1971; Hull *et al.*, 1978; Gill *et al.*, 1981; Gaelen and Toussaint, 1980b). As with other temperate bacteriophages, when Mu infects a sensitive host, it can enter either

Mobile Genetic Elements

ISBN 0-12-638680-3

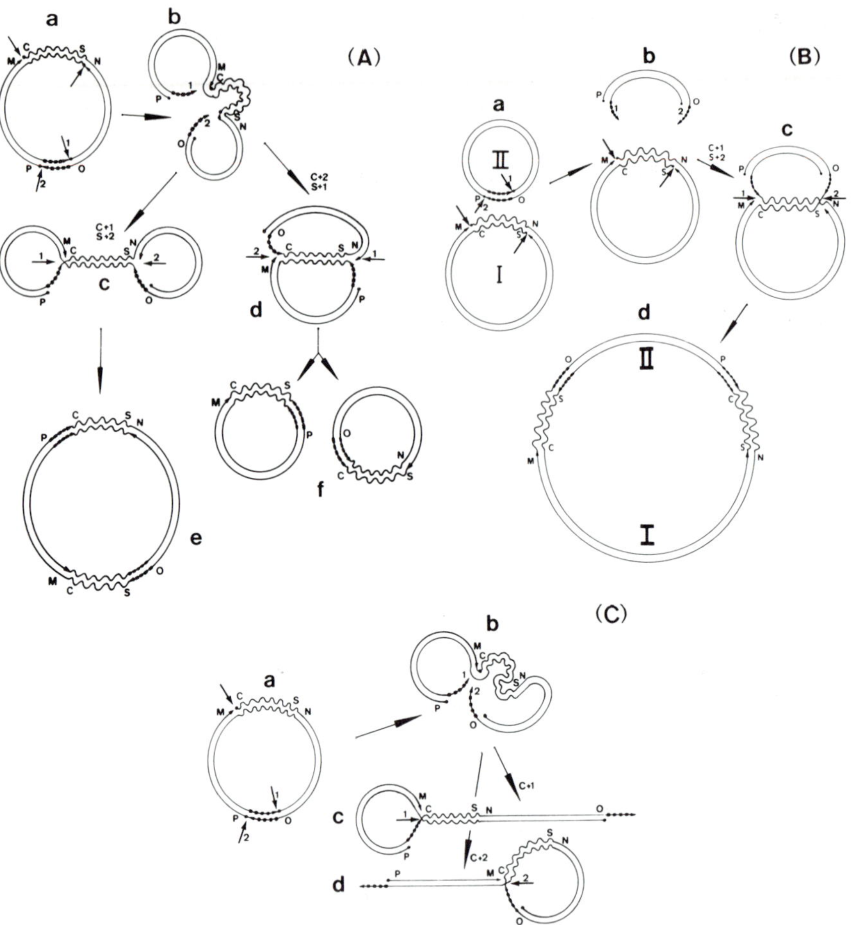

Fig. 1 Symmetrical cleavage-ligation model (Shapiro, 1979). (a) The bacterial chromosome containing one Mu prophage and the target site. M and N and O and P are bacterial sequences that flank the prophage and the traget site, respectively. The arrows in (a) indicate sites where nicks occur to generate the structure shown in (b). In (c) the 3'-end (1) of the target has bound to the Mu *c* end and the 3'-end (2) of the target has bound to the Mu *S* end. This generates the dumbbell structure with Mu connecting the two circles. In (d) the 3'-end (1) has bound to the Mu *S* end and the 3'-end (2) has bound to the Mu *c* end, generating a ribosome-like structure where the Mu connects the two circles. Replication of the Mu in the dumbbell generates a chromosome with two Mus in opposite orientation flanking an inverted segment ON (e). Replication of the Mu in the ribosome-like structure generates two circles containing one Mu each, one of which has only the bacterial origin of replication. The other will thus be lost, generating a deletion (f). (B) Replicon fusion according to the symmetrical model. The Mu prophage is on replicon I, the target site on replicon II. The same sequence of events as in (A) generates a unique molecule with two Mu prophages in the same orientation, joining replicons I and II. (C) Staggered cleavage-ligation model (Harshey and Bukhari, 1981). In (a) a nick occurs at the Mu *c* end and a staggered cut occurs at the target site. This generates the structure shown in (b). In (c) the 3'-end at the target, labeled (1) in (b), has bound to the 5' Mu *c* end, generating a key structure in which the Mu lies in the tail at the circle–tail joint. In (d) the 3'-end (2) of the target has bound to the 5' Mu *c* end, generating a key where the Mu is located in the circle at the circle–tail joint. The free 3'-OH end of the target have been separated to make the drawings clearer. Upon

the lytic cycle or the lysogenic state. In the first case, all the phage functions are expressed, the virus replicates and synthesizes its coat proteins, the DNA is processed (i.e., packaged in the coat), and the host cell membrane is lysed, releasing 50–100 phage particles. In the second case, a repressor is synthesized that blocks the expression of most viral functions, and the viral DNA forms a stable association with the host. In that state, the phage genome is called a *prophage* and the bacterium that propagates the prophage is called a *lysogen*. The repressor synthesized by the prophage also blocks the expression of a superinfecting Mu, so that the lysogen is immune to superinfection by Mu. In a lysogenic population some bacteria ($\sim 10^{-4}$) are spontaneously induced. That is, their prophage is no longer repressed; it expresses all its functions, and phage progeny is released. If the repressor is inactivated—for instance, by raising the temperature in the case of prophages that have a mutation making the repressor thermosensitive—the whole population of lysogens is induced and lyses, releasing 50–100 phages per bacterium.

A striking difference between Mu and other temperate phages is that the Mu genome integrates in the host chromosome whether it enters the lytic cycle or the lysogenic state. Moreover, in both cases it integrates at a random location in the host chromosome, inactivating the target gene, so that about 3% of the lysogens have acquired a mutation (Taylor, 1963). During the lytic cycle, Mu transposes to new locations and generates chromosomal rearrangements including deletions, inversions, duplications, and transpositions of host DNA segments, and fuses independant circular DNA molecules such as two plasmids, or a plasmid and a chromosome, or a circular phage DNA and a plasmid or the chromosome. Mutations resulting from Mu insertion are strongly polar (Jordan *et al.*, 1968). Moreover, Mu insertion results in the duplication of 5 bp of the target site (Kahmann and Kamp, 1979; Allet, 1979). Excision of an integrated Mu genome is a very rare event and can be either imprecise or precise (Bukhari, 1975; Khatoon and Bukhari, 1981). All the properties just described have now been found to be shared by all the bacterial transposable elements, IS sequences, and transposons, so that Mu is now considered to be a gigantic transposon (Tn) 39 kb long, disguised as a virus.

As we shall discuss later, chromosomal rearrangements and transposition are most probably a direct consequence of the mode of Mu replication. The easiest way to visualize how this happens is to look at the different models proposed to explain transposition, which are summarized in Fig. 1. In the first class of models (Shapiro, 1979; Arthur and Sherratt, 1979), the ends of a Mu prophage integrated in the host chromosome are nicked (most probably by a Mu-encoded enzyme) and bound to a new target site where a staggered cut of opposite polarity

---

completion of replication of the Mu and binding of the Mu *S* end strand, opposite to that cut at the *c* end in (a), a deletion and an inversion will be generated from (d) and (c), respectively, as in [A(c)] and [A(d)]. Wavy lines, Mu DNA; filled squares, the target site; normal lines, host DNA. Arrowheads indicate the 3' hydroxyl ends; solid circles indicate 5' phosphate ends; C and S are the *c* and *S* ends of Mu.

has occurred, as proposed by Grindley and Sherratt (1978). Structures very similar to replication forks are thus generated at the ends of Mu. Replication could then proceed from either or both ends of the Mu. Upon completion of Mu replication, depending on how binding of Mu ends to the target occurred (see Fig. 1A), the host chromosome has suffered either a deletion or an inversion, and a new transposed copy of Mu is present at the target site. If the Mu and the target sites had been located on two different replicons, the same set of events would have led to the fusion of the replicons (see Fig. 1B).

In the second class of models (Harshey and Bukhari, 1981; Galas and Chandler, 1981), only one end of Mu would bind to the target at a time, so that only one replication fork is created from which the Mu genome would replicate. After completion of replication, binding of the second Mu end to the target could generate the same rearrangements as above (see Fig. 1C).

The models also make different predictions about how simple transposition events (i.e., transpositions occurring in the absence of concomitant rearrangements of the chromosome) could occur. It is not yet clear to what extent these specific predictions are relevant to the mode of Mu replication. As we shall discuss later, they may, however, be relevant to the mechanism by which an infecting Mu first integrates in the host genome (see Section III,A).

Mu has two very useful advantages over IS and Tn: It is not a normal constituent of the bacterial genome, so that isogenic bacteria with and without a Mu can be compared; most importantly, Mu is inducible. This allows one to provoke at will transposition events at high frequencies and to create experimental conditions far more advantageous than those that can be produced with the other transposable elements. The disadvantage of Mu is the complexity resulting from the presence of a whole set of functions related to its phage personality. In this chapter, we will review what is known about Mu as a phage and as a Tn, the two aspects being, as we shall see, intimately linked.

II. Mu as a Temperate Phage

A. The Phage Particles

A Mu particle (Fig. 2) consists of an icosahedral head 600 Å long, bound to a contractile tail (1000 Å long when extended, 600 Å long when contracted, 200 Å wide) terminating in a base plate supporting spikes and fibers (To *et al.*, 1966). The density of the phage particle is 1.468 (To *et al.*, 1966), and the phage DNA has a GC content of 51% (Martuscelli *et al.*, 1971). Mu can infect many enterobacteria, including strains of *Escherichia coli, Citrobacter freundii, Erwinia,* and *Salmonella typhimurium*. The adsorption of the phage tail on a lipopolysac-

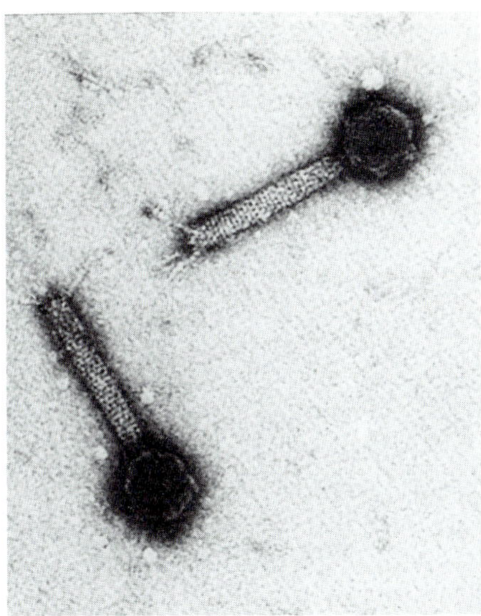

Fig. 2 Electron micrograph of the Mu phage particle. (Kindly supplied by A. L. Taylor.)

charide component of the bacterial outer membrane (R. Sandulache *et al.*, unpublished) is followed by the injection of the viral DNA into the host cytoplasm.

B. The Phage Genome

Lysates of Mu phage particles can be obtained either by infecting sensitive bacteria or by inducing lysogenic bacteria, which, in the case of Mu, is always performed by shifting the culture of a lysogen for a thermoinducible mutant to high temperature (42°C). The most widely used mutant is Mu*cts*62 (Howe, 1973a). Unexpectedly, the content of Mu lysates prepared by infection and by induction is different. Mu DNA, as extracted from phage particles purified from lysates prepared upon induction of a lysogen, consists of a 39-kb-long, linear, double-stranded DNA. (This value is the average of values obtained in different laboratories both by the analysis of Mu restriction fragments and by electron microscopy.) After denaturation and renaturation, Mu DNA displays a particular structure (Fig. 3) that defines the primary physical map of the virus. 50% of the homoduplex molecules have a long stretch of double-stranded segment (the α segment, 33 kb long) followed by a single-stranded bubble (the G loop, 3 kb long), a second double-stranded segment (the β segment, 1.7 kb long), and single-stranded "split ends" (SE) of variable length, with an average length of 1.7 kb.

In some renaturation conditions, the G loop shows an internal renaturation bubble (see Fig. 3) (Bade, 1972; Hsu and Davidson, 1972; Daniell *et al.*,

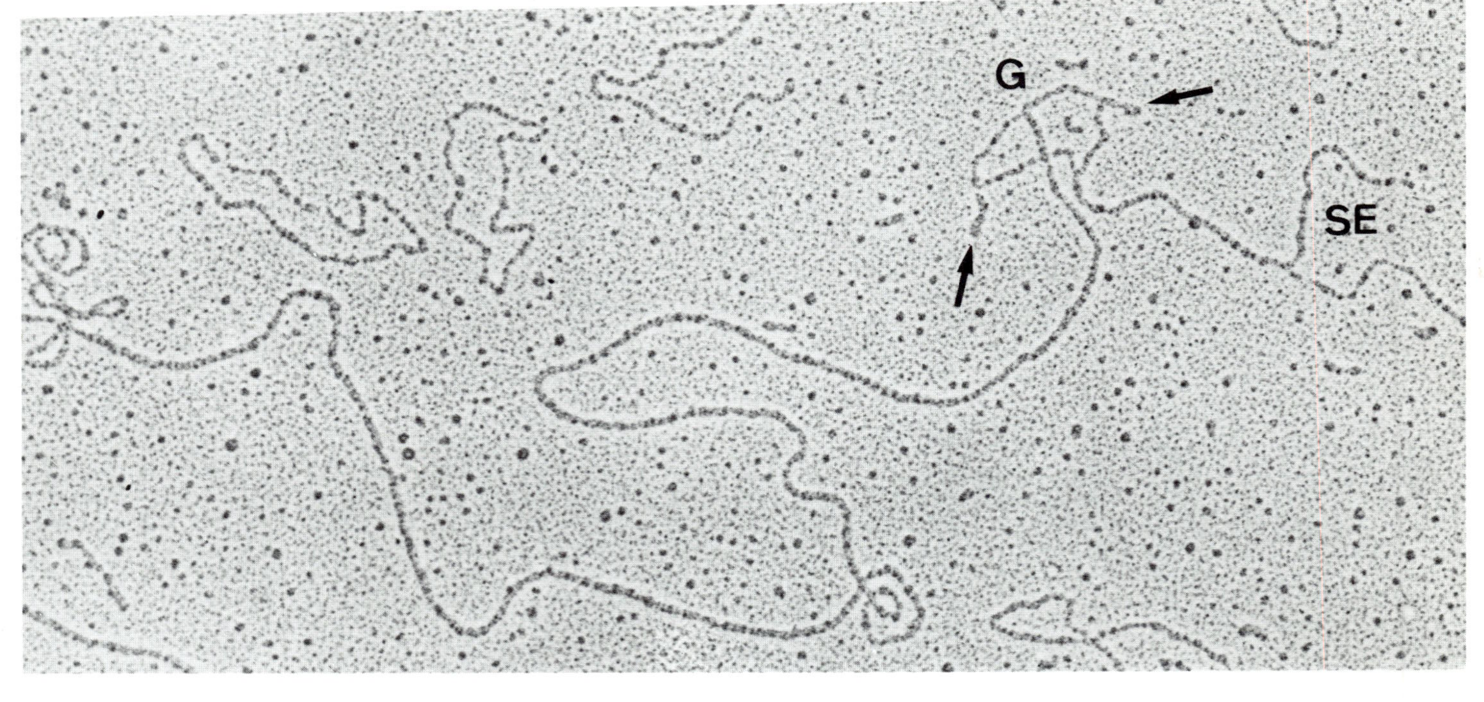

Fig. 3 Electron micrograph of a Mu homoduplex DNA molecule. The long double-stranded DNA segment on the left is part of the α region. G is the G loop. The arrows point toward the internal renaturation bubble in the G region. The β region is the short double-stranded segment included between the G loop and the split ends (SE).

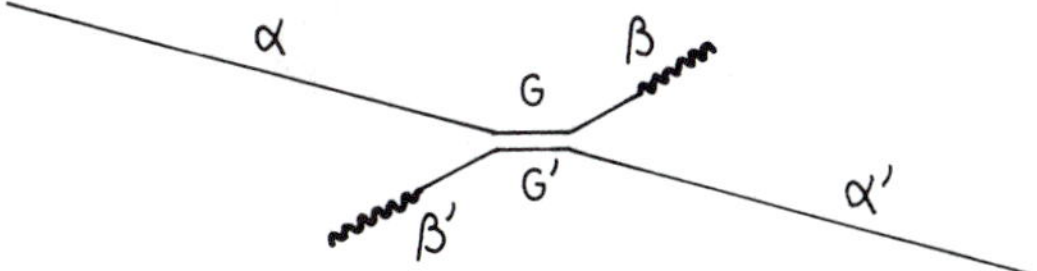

Fig. 4 Schematic representation of a Mu homoduplex DNA molecule where two complementary G regions in opposite orientation have reannealed. See the text for more details.

1973a). These four segments correspond to four discrete regions on the Mu genome, α, G, β, and the variable ends. The other 50% of the molecules do not show the G loop: The G region is double-stranded, so that α, G, and β are indistinguishable from each other and form a continuous double-stranded DNA, 37.5 kb long, ending in SEs of the same length as those mentioned above. Hsu and Davidson (1974) found rare homoduplex molecules showing only a short, double-stranded segment having the length of G, flanked at each side by asymmetrical single-stranded segments, one of which had the length of α and the other the length of β plus the SE (see Fig. 4). They also found single-stranded DNA, that had self-reannealed on a very short segment flanking the G region. From these two observations, they concluded that the G region is flanked by inverted repeats, that it can invert by reciprocal recombination between these short sequences, and that the G loops are formed by annealing of complementary strands carrying G segments with opposite orientation.

Mu DNA extracted from phage particles purified from lysates prepared upon infection of sensitive bacteria is different in that it very rarely shows a G loop on the homoduplex DNA. We shall come back to this point later.

C. Physical and Genetic Maps of Mu

A number of conditional defective mutants of Mu were isolated (mostly *amber* and a few thermosensitive mutants). Complementation tests grouped them in 22 functional units corresponding to 22 genes (*A, B, C–W;* see Table I). In addition, early (*kil, cim, gam, sot, arm,* and *lig*) and late (*gin* and *mom*) nonessential functions have been defined. These phenotypes are summarized in Table I.

The genetic map of Mu has been established mostly by deletion mapping. Bacterial strains carrying deletions starting in a host gene located near the Mu prophage and extending to different positions in the Mu genome were superinfected with defective mutants. The genes remaining on the cryptic prophages were detected by their ability to recombine with the superinfecting phages (marker rescue test, Franklin *et al.*, 1965). Joint efforts (Boram and Abelson, 1973; Bukhari and Metlay, 1973; Faelen and Toussaint, 1973; Howe, 1973a; Wijffelman *et al.*, 1973, summarized by Abelson *et al.*, 1973) determined that the prophage map is always the same, whatever the site of prophage insertion. Moreover, it was found that the prophage can be inserted in either orientation with respect to the host chromosomal markers. This first map was recently

Table I Characteristics of the Mu Genes and Functions

Gene	Localization with respect to the *c* end (kb)	Function	Reference[a]	Molecular weight of the protein
c	0–0.9	Repressor		26,000
ner (negative regulation)	—	Shutoff of early transcription	(1)	6,000
A	1.3–3.3	Transposase		70,000
B	3.3–4.3	Replication		33,000
cim (control of immunity)	—	When inactivated in prophage lacking the S end, no repressor synthesis	(2)	7,000
kil (killing)	—	Killing of the host even in the absence of replication	(3)	8,000
gam	5–6	Protects DNA from exoV digestion by binding to DNA	(4)	14,000
sot (stimulation of transfection)	6–7	When present and expressed in a bacterium, stimulates the efficiency of transfection with Mu DNA	(5)	
arm (amplified replication of Mu)	—	When deleted, less replication of Mu DNA	(6)	
lig (''ligase'')	—	Expression of an activity that can substitute for *E. coli* and T4 ligase	(7)	
C	9–10	Positive regulator of late gene expression		15,500
lys (lysis)	—	Lysis of the host cell		
D	—	Head protein		
E	—	Head protein		

H	—	Head protein		64,000
F	—	Head protein		54,000
G	—			
I	—			
T	—	Major head protein	(8)	
J	—	Head protein		
K	—	Tail protein		
L	—	Tail protein		55,000
M	—	Tail protein		
Y	—	Tail protein		12,500
N	—	Tail protein		60,000
P	—	Tail protein		43,000
Q	—	Tail protein		
V	—	Tail protein		
W	—	Tail protein		
R	—	Tail protein		
S	α and G	Tail protein		56,000
S'	α and G	Tail protein		48,000
U	In G	Tail protein		22,000
U'	In G	Tail protein		22,000
gin (G inversion)	Left of β	Inversion of the G region	(9)	21,500
mom (modification of Mu)	Middle of β	Modifies Mu DNA protecting against restriction. Is not a methylase	(10)	37,000 (?)

[a] (1) Wijffelman *et al.* (1974); (2) van de Putte *et al.* (1978); (3) van de Putte *et al.* (1977a); (4) Giphart-Gassler and van de Putte (1979); van Vliet *et al.* (1978); Williams and Radding (1981); (5) van de Putte *et al.* (1977b); (6) Waggoner *et al.* (1981); Goosen *et al.* (1982); (7) Ghelardini *et al.* (1979, 1980); Paolozzi *et al.* (1980); (8) Scott and Howe (1982); (9) Chow *et al.* (1977); (10) Toussaint, 1976; Hattmann (1979, 1980). All other genes were described in Howe *et al.* (1979a,b) and O'Day *et al.* (1979).

[b] Most of the molecular weights come from Giphart-Gassler *et al.* (1981a,b). The question marks in parentheses indicate uncertain molecular weights.

refined after isolation of many more conditional defective mutants and Mu-deleted prophages, using the same technique (O'Day *et al.*, 1979). By recombination between pairs of conditional defective mutants, the vegetative map of Mu was established and found to be the same as that of the prophage (Wijffelman *et al.*, 1972; Couturier and van Vliet, 1974).

The genetic and physical map of phage Mu shown in Fig. 5 was assembled by M. Faelen. This map was derived from those described by Allet and Bukhari (1975) and Kahmann *et al.* (1977). It was corrected and refined using unpublished information communicated by Howe, van de Putte, and their coworkers and by Kahmann and Kwoh.

The molecular weights of the products encoded by different Mu genes were characterized using the mini-cell system (Giphart-Gassler *et al.*, 1981a,b; Magazin *et al.*, 1978; Coelho and Symonds, 1981) or *in vitro* systems (Magazin and Allet, 1978). Mini-cells are produced by *E. coli min* mutants and are devoid of chromosomal DNA but contain all the elements necessary for transcription and translation (Reeve, 1977). They can be infected or transfected with phage DNA and will then synthesize only phage proteins (although sometimes not with the proper regulation). Table I gives the molecular weight of Mu proteins where they are known.

1. THE α REGION

The α region contains the repressor gene, all the early functions, all the head functions, and most of the tail functions. The properties of the early functions are multiple, being involved in integration, transposition, replication, and regulation of late gene expression. They are summarized in Table I. The late functions of Mu include the head and tail genes, which were differentiated both by electron microscopic analysis of defective lysates and by head–tail complementation tests in which lysates of pairs of defective mutants were mixed in conditions allowing head-tail joining and analysis for the formation of active phages (Giphart-Gassler *et al.*, 1981b).

2. TRANSCRIPTION AND REGULATION OF GENE EXPRESSION IN THE α REGION

As shown by hybridization of Mu-specific mRNA with separated strands of Mu DNA, the induced Mu genome is transcribed asymmetrically from the left end of the α region (called *c* end) toward the right end of the β region (called the *S* end), that is, from the *r* strand (Bade, 1972; Wijffelman *et al.*, 1974). The early mRNA, which is 8.5 kb long (J. Engler, unpublished results), hybridizes with the *c* end. It is first synthesized 0–8 min after induction or infection. Late RNA synthesis starts 25 min after induction or infection and rises sharply thereafter (Wijffelman *et al.*, 1974). Between 8 and 25 min, RNA synthesis remains

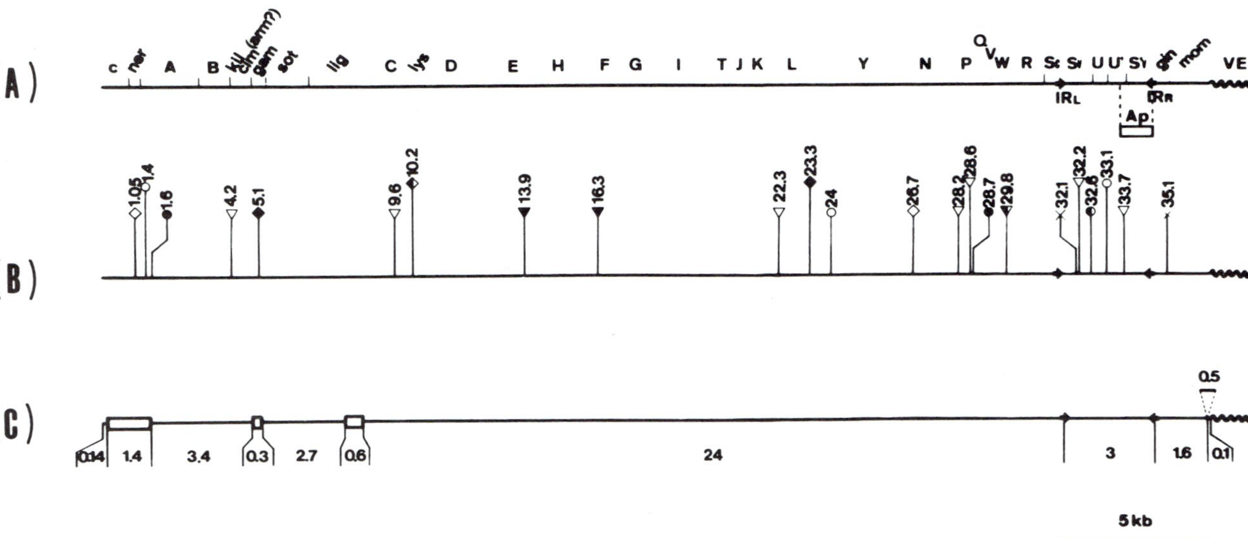

Fig. 5 (A) Genetic map of phage Mu. Additional information about the different genes can be found in Table I. The exact order of the *cim*, *kil*, *arm*, *gam*, *lig*, and *sot* functions has still to be confirmed. IRL and IRR represent the 34-bp-long inverted repeats flanking the G region. Ap is the substitution in the MupAp1 derivative (Leach and Symonds, 1979). VE represents the *S* variable end. (B) Restriction map of phage Mu. The sites for cleavage by the different restriction enzymes are represented as follows. *Hind*III, ◇; *Eco*RI, ◆; *Bgl*II, ◇; *Bal*I, ○; *Pst*I, ●; *Kpn*I, ◐; *Hpa*I, ▽; *Bam*HI, ▼; *Sal*I, ▲; *Hind*II, X. (The HindII sites located in the α region are not indicated.) Only few recognition sites have been located on D108. It carries two *Eco*RI sites, 1.3 and 5.0 kb from the *c* end (only the second one corresponds to a site on Mu); three *Pst*I sites at 1.6, 13.9, and 28.7 kb from the *c* end; one *Bal*I site, one *Hind*III, and one *Bam*HI site at 33.1, 26.1, and 16.8 kb from the *c* end, respectively. (C) Schematic representation of a heteroduplex between the Mu and D108 DNAs. The three regions of nonhomology that are detected in the α region are represented by boxes. The β regions of the two phages differ only by an insertion of 0.5 kb in the D108 DNA.

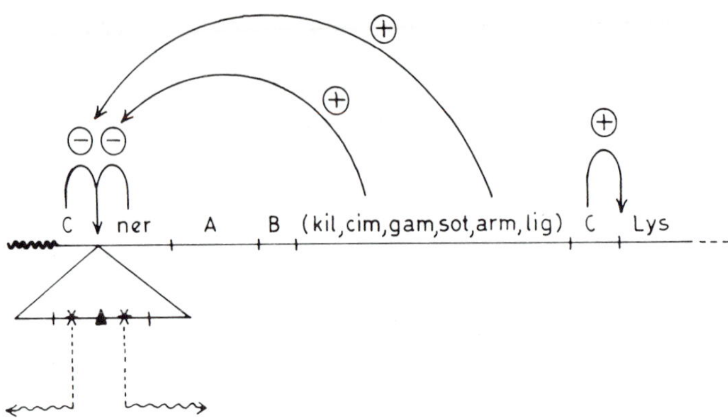

Fig. 6 Schematic representation of regulatory mechanisms of the Mu early and late functions. ▲, *hind*III recognition site located ~1 kb from the *c* end; ∗, Pribnow boxes flanking the *hind*III site; |, Shine-Dalgarno sequences presumably used for translation initiation of *ner* and *c*, respectively; ⇝; mRNA. It is not known whether the *c* and *ner* mRNA overlap or not; ⊕, positive regulatory mechanism; ⊖, negative regulatory mechanism.

low, and the exact nature of the RNA made has not been analyzed. 1% of the total late RNA hybridizes to the early region (van Meeteren, 1980). Both early and late RNA synthesis are repressed in lysogens while the *c* gene encoding the reprssor is transcribed from the opposite (*l*) strand (van Meeteren *et al.*, 1980; van Meeteren, 1980). The repressor binds to an operator most probably located near the *Hind*III restriction site located 1 kb from the *c* end (Kwoh and Zipser, 1979). By looking at the immunity of strains that carry either a Mu prophage deleted from various portions of the *S* end or a high copy number plasmid in which various portions of the *c* end had been cloned, van de Putte *et al.* (1980a) concluded that repressor synthesis is positively controlled by the *cim* function (see Table I). Indeed, strains that carry the *cts*62 mutation but lack *cim* and the *S* end do not recover immunity when grown at 30°C after they have been shifted to 42°C in order to inactivate the repressor. The exact function of *cim*, its substitution by the *S* end, and the level at which it acts are not yet known. Synthesis of early mRNA reaches a maximum at 4 min after induction or infection and decreases thereafter. The shutoff is the result of negative regulation by the *ner* function (see Table I) (Wijffelman *et al.*, 1974) which was shown by the analysis of plasmids carrying cloned *c* end fragments to map between the *c* and *A* genes (Giphart-Gassler and van de Putte, 1979; H. Priess *et al.*, 1982). The product of the *ner* gene seems, in addition, to regulate transcription of the early region at later times at a constant low level (van Meeteren, 1980) and to regulate negatively repressor synthesis (van Leerdam *et al.*, 1982). Mutants in the *A* and *B* genes, as well as prophages deleted from the *S* end, synthesize only early RNA (Wijffelman *et al.*, 1974; E. G. Bade, unpublished results). The product of the *C* gene is also required for late RNA synthesis (van Meeteren, 1980). This suggests that late RNA synthesis is positively controlled by the *C* gene product and requires replication, because A^- and B^- mutants and prophages deleted from the *S* end do not replicate (see Section III,C). The first 1500 bp of the *c* end have been sequenced (Priess *et al.*, 1982; B. Allet *et al.*, unpublished results). As expected, promoter and translation initiation sequences were found flanking the *Hind*III restriction site located 1 kb from the end. Three promotor mutations that also lead to insensitivity to the shutoff of early transcription by *ner* have been sequenced and found to have the same single base-pair substitution in that region (N. Goosen and B. Allet, unpublished), confirming that the promoter for early RNA synthesis and the *ner* recognition site are located in that region. The *c* and early transcripts do not seem to overlap (Priess *et al.*, 1982).

As mentioned later, the *arm* function (see Table I may positvely regulate expression of the *A* and *B* genes (Waggoner *et al.*, 1981; Goosen *et al.*, 1982). All these regulatory mechanisms, which have been reviewed recently by van de Putte *et al.* (1980a), are summarized in Fig. 6.

3. The G Segment and the Phage Host Range

As pointed out earlier, the electron microscopic analysis of homoduplex Mu DNA revealed the characteristic G loop with its internal and flanking inverted repeats. Moreover, G loops were found to be much more abundant on homoduplex DNA prepared from induced lysates than on those prepared from lytically grown lysates (50% vs. 1%) (Daniell et al., 1973 a,b).

The relationship between the G segment and the phage host range was established through a set of converging experiments. Symonds and Coelho (1978) found that in cultures of Mu lysogens only half of the bacteria gave infective centers on a lawn of *E. coli*. However, when the nonproducing cells were allowed to grow for 4 hr or more, they acquired the ability to produce phage on the same strain. Knowing that inversion of the G segment occurs in the prophage state (Kamp *et al.*, 1978a,b), they suggested that the capacity to produce phage was related to the orientation of the G segment on the prophage. It was not clear whether those samples that did not produce infective centers actually produced no phage particles or whether they produced defective phages unable to infect the *E. coli* strain used as an indicator. At the same time, Bukhari and Ambrosio (1978) found that when a phage lysate grown upon induction and containing 50% Mu with the G region in each orientation was used to infect *E. coli* K12, only those Mu with G in one orientation were able to inject their DNA into the host cytoplasm. Finally, using a mutant defective for G inversion, Kamp *et al.* (1978a) generated phages with the G region frozen in either orientation and found that only those with G in one orientation (defined as +) could infect *E. coli* K12.

All these data suggest that (*1*) induced lysates contain a portion of phages that cannot adsorb or inject their DNA on *E. coli* K12, most probably G(−) phages; (2) half of the cells in a culture of a Mu lysogen carry a prophage with their G region in the (+) orientation, and half carry a G(−) prophage; and (*3*) recovery of the ability to produce phage active on *E. coli* by nonproducing bacteria corresponds to inversion of the G segment from the (−) to the (+) orientation. Chow and Bukhari (1977) found that Mu and phage P1 DNA can reanneal along the Mu G region, showing that the two unrelated phages share the invertible segment that is called *C* in P1 (Yun and Vapnek, 1977). The P1 C segment inverts and is flanked by longer inverted repeats (~600 bp) than the Mu G region. In addition to this physical homology, Mu and P1 are related through their host range; *E. coli* strains resistant to P1 are also resistant to Mu, and 50% of the strains selected for Mu resistance are also resistant to P1 (Howe, 1972). Toussaint *et al.* (1978) found that Mu could acquire the P1 host range; that Mu mutants in genes *S* and *U* could be rescued by the wild-type alleles from P1 and, in contrast, that P1 could acquire the Mu host range; and that mutants of P1 in genes 33, 72, and 66 could be rescued by the wild-type alleles from Mu by homologous recombination. This strongly suggests that Mu genes *S* and *U* and P1 genes 33, 72, and 66 map very near or in the G and C segments, respectively, and are involved in specifying the phage's host ranges.

Using marker rescue analysis with deleted lysogens that carry only the α or the G and β regions of the Mu genome, Howe *et al.* (1979b) showed that the *S* and *U* genes map in the G region. Although it has not been proven it is generally assumed that these genes specify Mu host range. Using induced lysates of Mu that contained MuG(+) and MuG(−) in equivalent amounts, attempts were made to find strains, that, in contrast to *E. coli* K12, would be sensitive to MuG(−) phages. Such strains were found, including *E. coli C, C. freundii, Shigella sonneii, Enterobacter cloacae, Erwinia carotovora, Erwinia uredovora, Erwinia amylovora,* and *Erwinia herbicola.* On the other hand, *E. coli B, S. typhimurium,* and *Arizona* are sensitive to MuG(+) as is *E. coli* K12 (van de Putte *et al.*, 1980b; Kamp, 1981; Faelen *et al.*, 1981 a,b; N. Lefèbvre, unpublished observations). Because induced lysates contain MuG(+) and MuG(−) phages in equivalent amounts, when such an induced lysate is used to infect *E. coli* K12, only MuG(+) phages inject their DNA and multiply, so that MuG(−) phages become a minor fraction of the phage population. Conversely, if the induced lysate is used to infect *Erwinia,* for instance, only MuG(−) phages multiply and MuG(+) phages become a minority.

More recently, the isolation of point and insertion mutations in and around the G segment, the cloning of Mu DNA segments covering that region, the identification of the molecular weight of the proteins encoded by that region in minicells (Kwoh and Zipser, 1981; M. Giphart-Gassler *et al.*), and the sequencing of a DNA segment encompassing 500 bp of α, all of G, and 700 bp of β (R. Kahmann and D. Kamp, unpublished) have provied a consistent picture of the Mu G segment (see Fig. 7). Part of the *S* and *S'* genes seem to map in the right end of the α region, so that the NH_2-terminal regions of both proteins are encoded by the same regions. The C-terminal portion of *S* (*Sv*) would be encoded

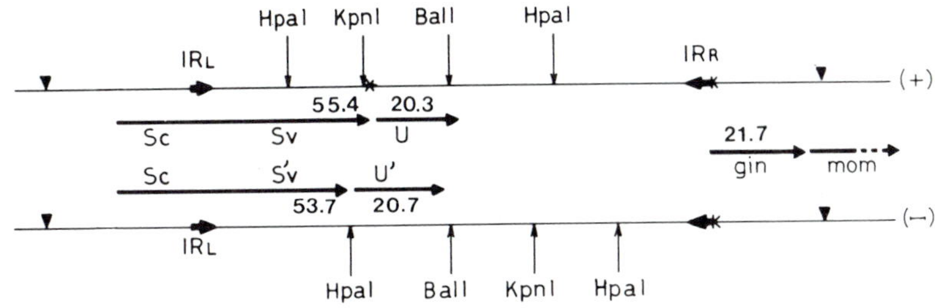

Fig. 7 The stars indicate Pribnow boxes. Short arrows represent the 34-bp inverted repeats flanking the G region that are the sites for reciprocal recombination catalyzed by the *gin* protein leading to G inversion. The long arrows indicate the longest open reading frames. The numbers on these arrows are the molecular weights of the different proteins calculated from the nucleotide sequences of these reading frames. The molecular weights (kilodaltons) are those found in the mini-cell system: S, 56; S′, 48; U, 22; U′, 18 or 26; gin, 21.5. The vertical arrows show recognition sites for restriction enzymes *Hpa*I, *Kpn*I, and *Bal*I located in the G segment. (+) and (−) refer to the G orientation providing a host range suitable for *E. coli* K12 and *E. coli* C, respectively.

by the left end of G in the (+) orientation, whereas that of *S′* (*S′v*) would be encoded by the left end of G in the (−) orientation. The *U* and *U′* genes map in the central part of G and are expressed from the (+) and (−) orientations, respectively. They have similar sequences because they cover the part of G that contains the internal inverted repeat. The flanking inverted repeats are 34 bp long. The molecular weights calculated from the base sequence giving the largest open reading frames are 64,000, 58,000, 23,000, and 29,000 for *S*, S′, *U*, and *U′*, respectively, compared to 56,000–64,000, 48,000–58,000, 22,000–23,000, and 18,000–24,000 found in the mini-cell systems (see Table I). However, there are several other shorter open reading frames in the sequenced region, and their possible physiological role remains to be elucidated. The fact that all known point mutations in *S* map in G also remains to be explained.

4. The β Region: G Inversion and DNA Modification

Two nonessential genes—*gin*, which is required for inversion of the G segment (Chow *et al.*, 1977; Kamp *et al., 1978b)* and *mom* which is responsible for modification of Mu DNA and protection against restriction (Allet and Bukhari, 1975; Toussaint, 1976)—have been mapped in the β region (Chow *et al.*, 1977; Toussaint *et al.*, 1980). By electron microscopic analysis of homoduplex and heteroduplex molecules of Mu DNA, Chow *et al.* (1977) characterized deletions in the G region that abolish G inversion; one of them leaves the G segment intact and is complemented for inversion by phage P1 (Kamp *et al.*, 1978b) and by the invertible DNA segment controlling phase variation in *Salmonella* (see Chapter

12) (Kamp and Kahmann, 1981). This allowed the characterization of the gene *gin*, which maps at the left end of β, the product of which is required for G inversion, The sequencing of that region (D. Kamp and R. Kahmann, unpublished) reveals a possible promoter sequence at the beginning of β, possibly straddling the right inverted repeat (Fig. 7). It is followed by an open reading frame compatible with a protein of 25 K. Comparison of the sequences of G and *gin* with that of the invertible segment determining phase variation in *S. typhimurium* and its inversion gene shows that both systems are related, although inverted with respect to each other.

Although no detailed study of the regulation of *gin* expression has been done so far, it is known that *gin* is expressed by the repressed Mu prophage (Kamp *et al.*, 1978a). It is assumed that it is expressed at the same level during the lytic cycle and that inversion of G is a slow process, so that it occurs only rarely during one lytic cycle. This explains the almost total absence of inversion in lytically grown lysates.

Mu DNA is unexpectedly resistant to many restriction enzymes. This is seen both *in vitro* when one tries to digest Mu DNA with purified restriction enzymes (Allet and Bukhari, 1975) and *in vivo* by the fact that the lysates grown on *E. coli* grow well on other bacteria that synthesize their own host-specific restriction-modification enzymes that can digest any foreign DNA as it enters the bacterial cell (Toussaint, 1976). This resistance is due to the expression of a phage function, called *mom*, that maps in the middle of the β region (Toussaint *et al.*, 1980). *mom*$^-$ mutants have been isolated that are normally restricted both *in vivo* and *in vitro*. The *mom* function can also modify other DNAs, as shown by the fact that other phages grown in induced Mu lysogens are resistant to restriction (Toussaint, 1976). The fact that λ prophages present in a Mu lysogen are not modified strongly suggests that the *mom* gene is repressed on the Mu prophage (Toussaint, 1976). However, expression of the *mom* gene is very peculiar in that it seems to occur at a much higher level after induction than after lytic infection. Indeed, phages grown upon induction are fully resistant to restriction, whereas phages grown upon infection are partially sensitive to restriction (Toussaint, 1976).

Both MuG(+) and MuG(−) are modified more strongly after induction than after infection, suggesting that the G orientation has no effect on *mom* expression (A. Toussaint and E. Schoonejans, unpublished results). The second peculiarity of *mom* regulation is that the modification directed by *mom* does not occur in *E. coli dam*$^-$ mutants that lack an adenine methylase enzyme (Toussaint, 1977; Khatoon and Bukhari, 1978). DNA extracted from Mu *mom*$^+$ phages grown upon induction of a *dam*$^+$ lysogen contain ~15% modified adenine residues. The precise nature of the modification remains to be elucidated. However, it is not a methylation, showing that the *mom* function is not a methylase (Hattman, 1980). The region just in front of the *mom* gene has been sequenced (R. Kahmann and D. Kamp, unpublished), revealing two successive overlapping inverted repeated sequences in which each of the three segments contains a GATC site for methylation by *dam*. This suggested that *dam* may regulate *mom* expression rather than interact with *mom* at a functional level. Hattman (1982) has shown that *dam* is required for transcription of the *mom* gene. In addition to being positively regulated by *dam*, the expression of *mom* also requires a Mu-encoded function, most probably the product of gene *C* (M. Goradia, S. Hattman, R. Plasterk, personal communication).

5. THE VARIABLE ENDS AND THE PACKAGING OF MU DNA

As mentioned earlier, Mu homoduplex and heteroduplex DNA always show SEs of about 1.7 kb next to their β region (*S* end). Electron microscopic analysis of Mu homoduplex DNA bound to phage T4 protein 32, which binds to single-stranded DNA, revealed the presence of very short SEs attached to the α region (*c* end) (H. Delius, unpublished observation). Analysis of Mu DNA with appropriate restriction enzymes revealed the presence of two fragments that, due to size variation, form fuzzy bands upon separation by electrophoresis on gels; they correspond to the *c* and *S* ends (Allet and Bukhari, 1975). Daniell *et al.* (1975) showed that Mu DNA extracted from purified phage particles grown on *E. coli* hybridizes with *E. coli* DNA. The fact that only the restriction fragments corresponding to the ends of Mu hybridize with host DNA (Bukhari *et al.*, 1976) nicely confirmed that the SEs correspond to random host DNA sequences. Mu DNA carrying an insertion in α or in β have SEs at the *S* end shortened by a length equivalent to the length of the insertion. Where the insertion is longer than 1.7 kb, Mu DNA also lacks the right end of the β region (Bukhari and Taylor, 1975; Chow *et al.*, 1977). Similarly, large deletions in the Mu genome are compensated for by lengthening the host DNA segment at the *S* end (Faelen *et al.* 1978a). The length of the host DNA segment at the *c* end, however, is not affected by either a deletion or an insertion in Mu DNA (Résibois *et al.*, 1978; Résibois *et al.*, 1981).

All these observations confirm those made by Bukhari and Taylor (1975), who proposed that both host DNA segments present at the ends of Mu DNA are there because Mu DNA is packaged from Mu copies randomly integrated in the host genome, and that packaging proceeds from the *c* end toward the *S* end, terminating when the phage head is full (head full packaging). Because the head is able to accommodate about 39 kb of DNA and the genome is only 37.5 kb long, packaging proceeds ~2 kb farther than the actual Mu *S* end, generating the *S* variable end. The *c* variable ends must be generated by another mechanism that seems to be related to the double-helix structure because their lengths vary by multiples of 11 bp (George and Bukhari, 1981).

As we shall see later, although it may be surprising to consider that maturation of Mu phage DNA occurs from randomly integrated Mu phage genomes, it is perfectly relevant in the case of Mu, which transposes while it replicates and remains integrated throughout its whole lytic cycle. It is worth mentioning here that Mu phage DNA has a structure very similar to that of the Mu prophage because both have the same map and are flanked by random DNA sequences.

III. Mu as a Transposable Element

So far, we have described Mu functions that are involved with its phage properties. We have avoided discussing Mu early functions and their regulation because these functions are necessary simultaneously for integration, for transposition, for induction of chromosomal rearrangements, for replication, and even for the expression of the late functions. It is best to discuss their properties by considering Mu both as a phage and as a transposable element.

Experimentally, the separation between the phage behavior and the transposable element behavior of Mu has been achieved by the isolation of the so-called mini-Mus. These derivatives carry large internal deletions and retain intact ends of Mu and sometimes one or several early genes. Some were constructed *in vitro* by cloning the left and right ends of a Mu prophage in a plasmid or in a λ vector (Schumann and Bade, 1979: Maynard-Smith *et al.*, 1980; Chaconas *et al.*, 1981a; van Leerdam *et al.*, 1981; Schumm and Howe, 1980,1981; Howe and Schumm, 1980; D. Kamp *et al.*, unpublished). Others were isolated *in vivo,* screening for deletions induced by a Tn inserted near either end of the Mu genome (Faelen *et al.*, 1978a; Maynard-Smith *et al.*, 1980; Résibois *et al.*, 1981). These mini-Mus were very helpful in studying some aspects of Mu replication and transposition. All the mini-Mus isolated so far are defective, but they replicate and are matured in the presence of a helper Mu. Matured mini-Mu DNA consists of a double-stranded mini-Mu DNA linked to ~100 bp of host DNA at the *c* end and to a very long segment of host DNA at the *S* end that is equal to the normal length of Mu (39 kb) minus the length of the mini-Mu (Faelen *et al.*, 1978a; Résibois *et al.*, 1981).

It has been impossible experimentally to separate the transposition and integration of Mu from its replication. All the mutants that are defective for integration and transposition are also defective for replication; this is true for both phage and host mutants. Moreover, although it has not been definitely proven, the first event occurring after infection with Mu is usually considered the integration of the Mu DNA in the host chromosome, and after this first event has occurred, there are no further differences between the lytic cycles after infection and induction.

A. The First Event after Mu Infection

Most viruses replicate and integrate from a circular form of their genome. Upon entry into the host cytoplasm, their DNA, which is usually linear in the viral particle, circularizes either by the rejoining of sticky ends (complementary single-stranded ends) or by reciprocal recombination between terminally redundant ends. In the case of adenoviruses and phage ϕ29, the circular forms are maintained by a protein that is covalently linked to both 5′-termini (Ito *et al.*, 1978; Coombs *et al.*, 1978; Salas *et al.*, 1978).

The ends of Mu have been sequenced (Kahmann and Kamp, 1979; Allet, 1979). They are not circularly permuted, nor do they contain complementary single strands. Ljungquist and Bukhari (1979) analyzed the fate of ^{32}P-labeled Mu DNA early after infection of either a sensitive or a Mu lysogenic host. In the lysogen, the superinfecting phage is repressed. In both infections, the incoming DNA is converted to a form that, in neutral sucrose gradients, sediments twice as fast as the linear Mu DNA used as a marker. In the sensitive host the fraction of fast-sedimenting material is small and transient, whereas in the lysogen it is more abundant and detectable as long as 40 min after infection. After phenol extraction, the DNA cosediments with a linear Mu DNA marker in either CsCl–EtBr or alkaline sucrose gradients, showing that the fast-sedimenting fraction is not covalently closed circular DNA (ccc). After infection of a Mu-immune host or of a *himA* strain in which Mu does not replicate (Miller and Friedman, 1980; M. Pato, unpublished results), a circular form of Mu DNA can be isolated and visualized with the electron microscope and by electrophoresis on agarose gels. These circles are linearized upon treatment with either pronase, SDS, high salt, high temperature (75°C), or phenol, showing that Mu DNA can circularize and that circularization is achieved by means of one or several proteins (R. Harshey and A. I. Bukhari, unpublished). It is not yet clear whether the proteins(s) are covalently linked to the DNA and whether the circular form is actually achieved after infection of a sensitive host. It could very well be a short-lived intermediate of the first integration event, which is known to occur very soon after the onset of Mu development (Ljungquist and Bukhari, 1979). Integration seems to occur primarily through a replicative event. Ljungquist *et al.* (1978) showed that 10 min after infection of a sensitive strain with ^{32}P-labeled Mu, 60% of the labeled DNA sedimented with the host DNA and 40% sedimented as free DNA; in immune strains, no Mu DNA was incorporated into the bacterial genome. The bacterial sequences that flank Mu DNA within the phage particle are lost when Mu DNA inserts in the host genome, as shown by the electron microscopic analysis of heteroduplex molecules made between Mu phage and prophage DNA (Hsu and Davidson, 1974). The first integration therefore resembles a transposition event. However, it differs from other transpositions in that DNA is transposed from a linear molecule, possibly maintained as a circle by proteins(s), as mentioned above, to a circular target replicon; other transpositions proceed from one circular replicon to another. As a consequence, if the infecting Mu DNA integrates into the host chromosome via the mechanisms shown in the models outlined in Fig. 1, the chromosome will become linear (see Fig. 8) unless a further step occurs. This step may be a reciprocal recombination between the two Mu copies (Shapiro, 1979). Such a recombination would regenerate a linear Mu genome and a circular host chromosome that carries an insertion of Mu flanked by a 5-bp repeat. Alternatively, if only one end of Mu binds to the target at a time, this step could consist of nicking at the opposite end of the same strand that was nicked in the first step (see Fig. 8d,e) and binding of that strand and its complementary strand to the target (Galas and Chandler, 1981).

Linearization of the chromosome may be of no importance when the infecting phage begins lytic development that leads to cell death, but in order for a lysogen to survive, the host chromosome must be resealed. It is therefore tempting to speculate that lysogenization may require one more enzymatic activity than normal transposition. So far, there is no experimental evidence for the existence of such a function, but it has been shown that plasmids that have acquired a Mu prophage by lysogenization usually have simple insertions, whereas those that received a Mu prophage by transposition from the chromosome are rearranged, suggesting that the two mechanisms of integration may be at least partially different (Chaconas *et al.*, 1980a). On the other hand, there are weak indications that the first integration event may be conservative (L. Paolozzi and P. Liebart, unpublished results; R. Bourret and M. Fox, unpublished results). If this turns out to be true, it would lead us to believe that the first integration proceeds through different mechanisms than those shown in Fig. 1.

1. THE INTEGRATION SITE OF MU IN THE BACTERIAL CHROMOSOME

Bukhari and Zipser (1972) and Daniell *et al.* (1972) studied the distribution of Mu insertions in the *lacZ* gene of *E. coli* K12. Both groups mapped the Mu prophages by recombination between Mu-induced Z^- mutations and known deletions in the *Z* gene. They also looked at Mu insertions in the *lacY* gene. The ratio between the number of insertions in *lacY* and *lacZ* was roughly equal to the ratio between the length of the two genes, suggesting a random distribution of Mu insertions in that region of the bacterial chromosome. From the number of insertion sites found by Bukhari and Zipser (1972) (more than 60), one can calculate that if Mu recognizes a specific sequence on the bacterial chromosome, it is not longer than 2 bp (Couturier, 1976). Daniell *et al.* (1972) located 40% of the insertions they had isolated in the *Z* gene in the same deletion interval, which would suggest a nonrandom distribution. Unfortunately, in no case was a physical map of the *lacZ* gene and of the insertions available, making interpretation of those results difficult. More recently, Mu insertions have been isolated in the

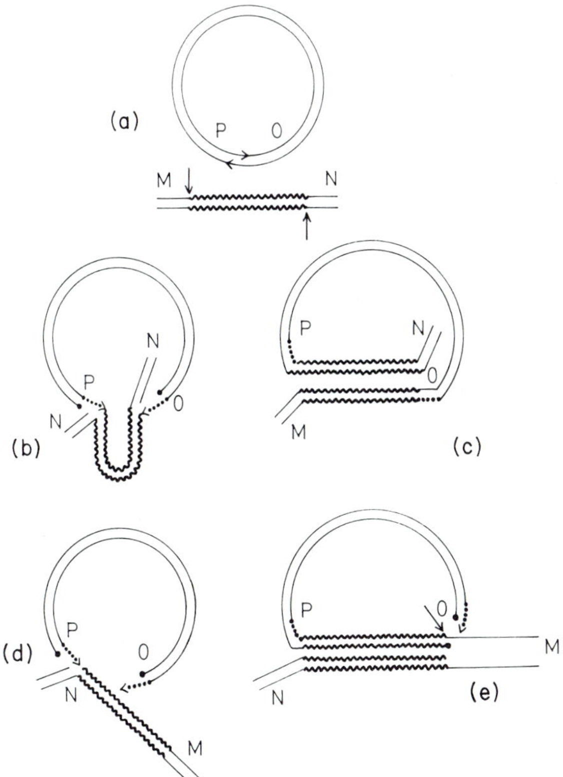

Fig. 8 Integration of infecting Mu DNA. (a) The circle represents the host chromosome; the arrowheads mark the staggered cut at the target site for integration. M and N are the variable bacterial sequences flanking the Mu DNA (zigzag line) in the viral DNA. (b) After nicking at the two Mu–host DNA junctions indicated by the arrows in (a), the free Mu ends have been covalently lined to the target on the host chromosome. (c) The Mu DNA has been replicated; the host chromosome is now linear. It could be recircularized by a reciprocal recombination between the two Mu copies. This would regenerate a linear Mu genome. The Mu prophage will be flanked by a 5-bp repeat. (d) After nicking of one Mu–host DNA junction, the free end of Mu has bound to the target site. (e) The Mu DNA has been replicated, and the chromosome has become linear. Nicking at the opposite end of Mu marked by the arrow, and binding of that Mu strand and its newly synthesized complementary strand to the target, will generate an integrated Mu prophage flanked by a direct repeat of 5 bp.

malK–lamB region of the bacterial chromosome; they were found to be non-randomly distributed. Raibaud *et al.* (1979) and Silhavy *et al.* (1979) found that 20% of the insertions occurred in a region that represents 2-8% of the gene, and Emr and Silhavy (1980) located 12 out of 25 insertions in *lamB* in a 12-bp interval. These results could be considered as showing a regional preference for integration, as is the case for the other prokaryotic transposable elements (Kleckner, 1981).

2. THE GENETIC CONTROL OF MU INTEGRATION

Although many independently isolated Mu prophages were found to be integrated in either orientation in the host chromosome, the relative order of the genes was found to be the same on all of them (Abelson *et al.*, 1973; Boram and Abelson, 1973; Bukhari and Metlay, 1973; Faelen and Toussaint, 1973; Howe, 1973a; Wijffelman *et al.*, 1973), showing that a unique site of the Mu genome is used for integration. Moreover, the order of the markers is the same on the phage and on the prophage, as shown both by genetic analysis (Wijffelman *et al.*, 1973; Couturier and van Vliet, 1974) and by electron microscopic analysis of heteroduplexes between phage and prophage DNA (Hsu and Davidson, 1974), suggesting that the integration site on the Mu genome is either located near one end or is formed by the joining of the two ends. Although there are no conclusive experimental data, it is universally accepted (Howe and Bade, 1975; Couturier, 1976; Bukhari, 1976; Toussaint *et al.*, 1977) that both ends of Mu are necessary for the first integration after infection. As we shall see later, it is known that both ends of Mu are required for transposition of its DNA. Because the first integration event is considered a transposition, it is also believed to require the two ends. There is experimental evidence that the *S* end is indeed required. Mu phage particles lacking the *S* end are easily obtained through the packaging of a mutant that carries an insertion longer than the size of the *S* variable end, that is, > 2 kb. Résibois *et al.* (1978) and Chow and Bukhari (1978) showed that such deleted phages were unable to go into a lytic cycle after infection. It is also known that 116 bp of the *S* end are sufficient because the phage MudII301, in which most of the G–β region has been substituted for by a segment of bacterial DNA and that retains only the last 116 bp of β, integrates normally after infection (M. Casadaban, unpublished results).

Integration of infecting Mu DNA does not occur in a bacterium already lysogenic for Mu and therefore immune. This shows that the synthesis of at least some of the functions required for integration is directed by Mu. O'Day *et al.* (1978) showed that the only mutants of Mu unable to lysogenize are the A^- mutants. Because the product of the *A* gene is absolutely required for transposition (see later), it is considered the main enzyme involved in integration and transposition and is now usually called *transposase*.

B. Mu-Induced Chromosomal Rearrangements

As summarized in the introduction, Mu can mediate the formation of chromosomal rearrangements such as deletions, inversions, duplications, and transpositions of host DNA segments as well as replicon fusions. All these rearrangements always involve transposition of a Mu genome and require the presence of the *A* gene product and of intact ends of the Mu genome. This means, for instance, that mini-Mus that lack the *A* gene or prophages deleted from the right end are unable to induce chromosomal rearrangements, whereas A^+ mini-Mus are able to do so. The *B* gene product, although not essential for the occurrence of rearrangements, increases their frequencies (see below).

Mu-mediated chromosomal rearrangements never occur when Mu is repressed. Because the phage normally kills its host within an hour after induction or infection, the detection of rearrangements by genetic methods requires techniques to ensure either the survival of the bacteria or the maintenance of the plasmids that have been rearranged by the phage. This was achieved by several means. In some cases, the Mu prophages were only partially induced by growing the lysogens at 37°C instead of 42°C (Faelen *et al.*, 1975). Such conditions seem to allow expression of the Mu functions necessary for transposition and induction of rearrangements without allowing completion of the lytic cycle, and most of the induced bacteria survive the induction. Mu*X* mutants were also used. They carry an insertion in the *B* gene that inactivates all the phage functions required for the lytic cycle except *A* (Bukhari, 1975; Bukhari and Taylor, 1975). Finally, mini-Mus that are deleted for all the killing and maturation functions were used (Faelen *et al.*, 1978a).

1. MU-MEDIATED REPLICON FUSION

The ability of Mu to generate replicon fusion was first described in 1971. Faelen *et al.* (1971) showed that integration of a λ*pgal* phage, mutated in such a way that it can neither replicate nor integrate, can be detected after mixed infection with this phage and Mu. The presence of the λ*pgal* was detected using bacteria deleted for the *gal* operon—therefore Gal^-, which became Gal^+ once they integrated the transducing λ*pgal* phage. All the Gal^+ transductants obtained are stably Gal^+, contain λ as an integrated prophage, and are lysogenic for Mu. No integration of λ*pgal* occurs upon infection of a Mu lysogen, indicating that some of the functions required for Mu-mediated integration of the λ*pgal* at least are encoded by Mu. Moreover, mutations inactivating gene *A* or gene *B* prevent integration of the λ*pgal* (Toussaint and Faelen, 1973).

Van de Putte and Gruijthuijsen (1972) obtained similar results. Using a strain carrying an F′ plasmid thermosensitive for its replication, they found Mu-mediated integration of the plasmid in the host chromosome after infection with Mu. It soon became obvious that the *E. coli* chromosome and the λ*pgal* genome in one case, and the F plasmid in the other case, are fused through two entire Mu genomes (Toussaint and Faelen, 1973; P. van de Putte and M. Gruijthuijsen unpublished observations). Indeed, in bacteria that have a normal recombination system (Rec^+), the Gal^+ transductants are unstable and segregate about 1% colonies that have lost the λ*pgal* but that keep a Mu prophage. This suggested that in the Gal^+ transductants the λ*pgal* prophage is flanked by two extensive regions of homology. By the analysis of the Gal^- segregants and by deletion mapping of the integrated Mu and λ*pgal* prophages, it has been shown that the two prophages are adjacent on the host chromosome and that the λ*pgal* prophage is flanked by two entire Mu prophages similarly oriented. The target sites for the fusions are randomly located both on the host chromosome and on the λ*pgal* phage because auxotrophic mutants are found that have the λ genes in an unusual circular permutation (Toussaint and Faelen, 1973).

A further insight into the possible mechanism of that replicon fusion was provided by the following observation. After partial induction of a strain that carries two genetically distinguishable Mu prophages, one located on the chromosome and the other on an F′ episome, the frequency of insertion of a superinfecting integration-defective λ*pgal* phage is greatly enhanced, as originally observed after mixed infection with the λ*pgal* and Mu. Hoeever, in this case, although always present in an unusual circular permutation, the λ*pgal* is virtually always located at the site previously occupied by a Mu prophage either on the chromosome or on the F′. Moreover, the λ*pgal* is flanked by two entire Mu genomes in the same orientation and with the same genotype as the Mu originally present at the site now occupied by the Mu–λ*gal*–Mu structure (Faelen *et al.*, 1975). This strongly suggests that, at least in the partial inducing conditions used in those experiments, the Mu prophages are not excised and reintegrated but rather duplicated *in situ* to generate the two Mu copies flanking the λ*pgal*. These results led Faelen *et al.* (1975) to propose a model in which, after this *in situ* duplication of the prophage, the opposite ends of the two Mu copies interact with a target site on the λ*pgal* to generate the cointegrate structure.

The model predicted (*1*) that the same process should lead to the formation of deletions and inversions adjacent to the prophage; (*2*) that deletion formation should be accompanied by the excision of a circular DNA molecule containing at least one Mu genome and host DNA; (*3*) that the inverted host segment should be surrounded by two Mu genomes in opposite orientation; and (*4*) that, again by the same process, after duplication of the Mu in the circular molecule, this DNA could reintegrate at a random location in the chromosome. This would lead to the transposition of the host DNA carried on the circular DNA, which in its new location would be flanked by two copies of Mu in direct repeat.

Mu-mediated fusions were later found between a plasmid and the bacterial chromosome (Chaconas *et al.*, 1980b; Bialy *et al.*, 1980) between two plasmids (Maynard-Smith *et al.*, 1980; Chaconas *et al.*, 1980b), and mini-Mu-mediated fusions were observed between a phage and the chromosome (Schumm and Howe, 1980, 1981; Howe and Schumm, 1980) or a phage and a plasmid (Toussaint and Faelen, 1981). Fusions between two plasmids have been analyzed genetically by mating experiments and physically by digestion with restriction endonucleases (Maynard-Smith *et al.*, 1980; Chaconas *et al.*, 1980). In the matings (Fig. 9), the donor bacteria contain two plasmids, each of which carries a selectable marker such as an antibiotic resistance marker. One plasmid is transferrable; the second is not and cannot be mobilized by the first one. A Mu or a mini-Mu prophage that also carries a marker is inserted in one of the two plasmids. The recipient strain is lysogenic for Mu in order to prevent any ex-

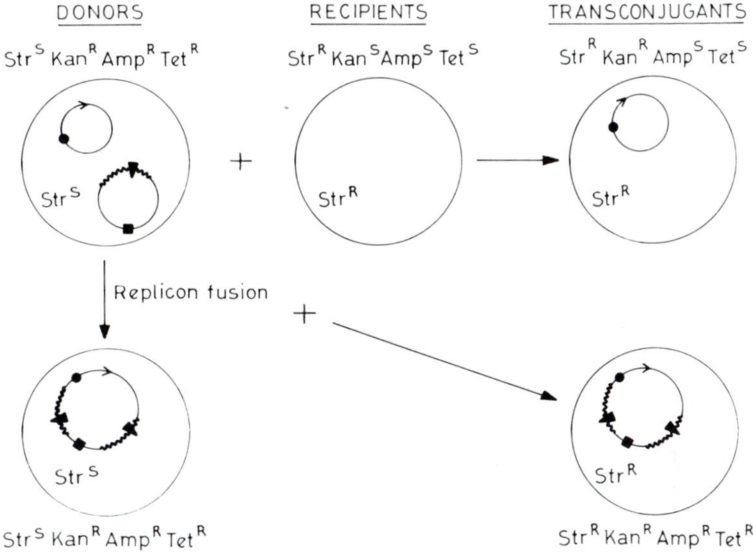

Fig. 9 Detection of replicon fusion mediated by Mu. Only plasmids present in the donors and the transconjugants are represented. The donor is Str^S and the recipient is Str^R. The transmissible plasmid, the nontransmissible and nonmobilizable plasmid, and Mu carry a Kan^R, a Tet^R, and an Amp^R marker, respectively. The fused replicons formed in the donor after Mu induction are easily detected after transfer in the Str^R recipient where they confer resistance to ampicillin and tetracycline, whereas transfer of the transmissible plasmid alone confers resistance only to kanamycin. The donor is killed by the streptomycin added in the medium used to select the tranconjugants. Origin of transfer, >; Mu DNA, ⁓; Kan^R marker, ●; Tet^R marker, ■; Amp^R marker, ▼.

pression of the Mu lytic functions by the Mu received from the donor. It contains either a normal recombination system (Rec^+) or a defective one (Rec^-). The donor strain is induced at 42°C and is mated at 42°C with the recipient. When the two plasmids are fused in the donor through the action of Mu, they can be transferred to the recipient and generate transconjugants that can be selected for having the selective markers carried by the plasmids and the Mu (or mini-Mu). Such fused plasmids, known as *cointegrates,* are recovered at a frequency of 10^{-3}–10^{-4} (in the absence of Mu, the frequency is less by at least 10^{-2}). The cointegrates segregate the original plasmid, which carries the Mu or the mini-Mu, and the second plasmid, which carries a transposed copy of the Mu or the mini-Mu. This segregation, which occurs in Rec^+ bacteria, is believed to proceed through a reciprocal recombination between the two Mus or mini-Mus that fuses the two plasmids.

Whether this segregation is strictly dependent on the host recombination system remains to be proven. Toussaint and Faelen (1981) reported one case in which segregation from the fused replicons occurred in the absence of the host recombination system and was dependent on the presence of Mu functions, most probably the *A* and *B* gene products.

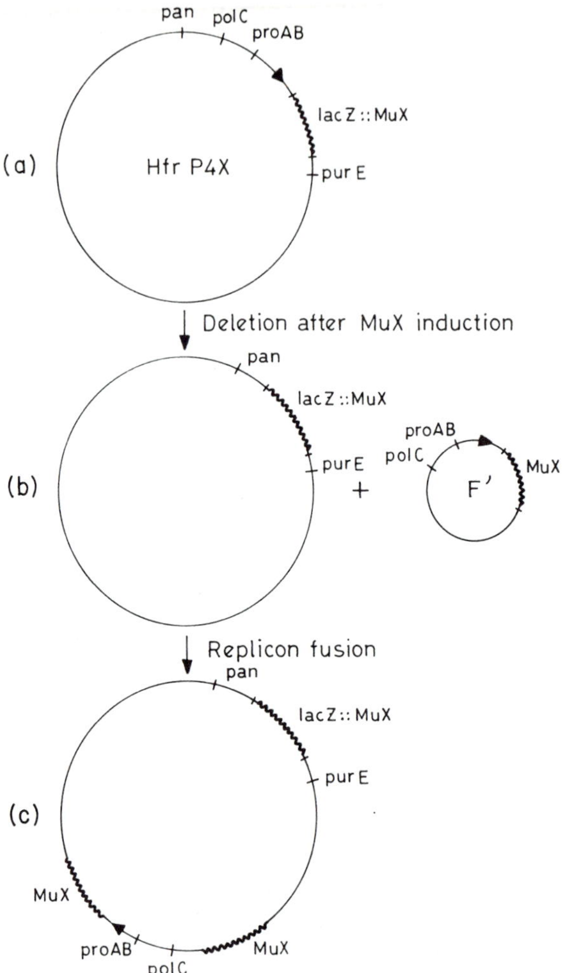

Fig. 10 Mu-induced deletion. (a) The strain Hfr P4X, which has a Mu*X* prophage inserted near the Hfr origin of transfer (indicated by the arrow), was induced at 42°C. Derivatives that had acquired an F′ *pro* episome were sought. These could not survive in the absence of their episome, which was shown to carry the *polC* gene by the fact that it could complement a *polC*$^-$ mutant. This strongly suggests that *polC* was deleted from the chromosome upon formation of the F′*pro, polC* episome (b). After a second round of induction, Hfr derivatives could be found that had the *polC* gene at a new location (c).

The fusions mediated by Mu or mini-Mu require a normal gene *A* product and intact ends of the Mu genome. They occur in the absence of the *B* gene product, although at a reduced frequency. Mu or mini-Mu-mediated fusions can occur even when either the donor or the recipient replicon cannot replicate. Fusions occur when the replicon that carries the Mu does not replicate (Schumm and

Howe, 1980, 1981; Howe and Schumm, 1980), and conversely, a Mu inserted in the bacterial chromosome stimulates the integration of a nonreplicating λ DNA (Faelen *et al.*, 1975). It is not known if fusion can occur if neither the donor nor the recipient DNA replicates, although the models described in the introduction would predict that it should occur.

2. MU-MEDIATED DELETIONS

Mu-mediated deletions are found either after partial induction of a Mu prophage or after infection with Mu (Howe and Zipser, 1974; Cabezon *et al.* 1975; Faelen and Toussaint, 1978). Two types of Mu-mediated deletions are detected in partially induced lysogens: those that occur adjacent to the integration site of the parental prophage (cis deletions) and those that are formed far from that site (trans deletions).

The cis deletions leave behind an entire Mu genome, eliminate the host DNA linked either to the *c* or to the *S* extremity of the prophage, and have one variable endpoint; trans deletions and deletions induced by an infecting Mu have two variable endpoints and a new prophage substituting for the deleted segment. The absence of the *B* gene product has little effect on the formation of cis deletions, but it considerably reduces the number of trans deletions (Faelen and Toussaint, 1978; Faelen *et al.*, 1978b).

The phage probably induces deletions by splitting the chromosome into two circular segments that are complementary and each of which contains a Mu prophage. Indeed, from a strain that carries an F plasmid integrated in its chromosome (Hfr) and a Mu*X* prophage integrated in the nearby *lac* operon (Fig. 10), Faelen isolated a derivative in which the F plasmid has been reexcised, most probably under the influence of the induced Mu*X*. This episome carries a Mu*X* prophage, the *proAB* and the *polC* genes from the host (*polC* is the structural gene for the *E. coli* DNA polymerase III that is essential for replication of the chromosome and therefore for the survival of the cell). The chromosome of that strain still carries a Mu*X* prophage at its original location but is deleted for the *proAB* and *polC* genes. The strain can survive only if it carries the plasmid either in an autonomous state or reintegrated in the chromosome. This reintegration occurs by Mu-mediated replicon fusion, generating strains in which the *proAB* and *polC* genes are transposed into a new location and, with the F flanked by two copies of Mu*X* similarly oriented.

3. MU-MEDIATED INVERSIONS

Hfr bacteria transfer their DNA into a recipient bacterium in a unique and oriented fashion from the F origin of transfer. The induction of a Mu prophage inserted in the chromosome of an Hfr bacterium generates derivatives that transfer their genes to a recipient in an inverted order as compared to the parental Hfr strain (Faelen and Toussaint, 1980a). The analysis of one of these derivatives has indicated that the inversion of the gradient of gene transfer results from the inversion of a DNA sequence including the F genome and that the inverted DNA fragment is bracketed by two Mu prophages in opposite orientations. The fact that the prophage is duplicated during the inversion suggests that the Mu DNA physically participates in the inversion process and that Mu functions are required because inversion is not observed under Mu immunity. Mu*X* prophages do induce inversion of adjacent host DNA (inversion in cis) but not of DNA segments that are not linked to the prophage (inversion in trans). By analogy with what is observed in the other Mu-induced rearrangements, this strongly suggests that the induction of inversions in trans requires the presence of the *B* gene product, whereas inversion in cis occurs even in its absence.

4. Mu-Mediated Duplications

There is strong evidence that Mu can stimulate the duplication of host DNA segments. M. Faelen and N. Glansdorff (unpublished results) found that partial induction of a Mu*cts*62 prophage inserted in the *glpK* gene increases the number of clones that synthesize twice as much arginosuccinate lyase as the parental strain. Transduction experiments have indicated that these clones have acquired a second Mu prophage and a second copy of *argH* (the structural gene of the arginosuccinate lyase) and that all these DNAs are closely linked (see Fig. 11). Rec^+ bacteria cured of the parental prophage inserted in the *glpK* gene segregate derivatives that simultaneously lose one of the *argH* genes and the second Mu genome. The simplest interpretation of this result is that the two copies of the *argH* gene are tightly linked, similarly oriented, and separated by a Mu prophage (Fig. 11).

5. Mu-Mediated Transpositions

In strains that contain a nonintegrated plasmid, Mu stimulates the formation of plasmid derivatives that carry any given gene of the host chromosome (Faelen and Toussaint, 1976; Dénarié *et al.*, 1977). The bacterial DNA that is transposed onto the plasmid is inserted in different regions of the plasmid and surrounded by two Mu prophages that have the same orientation (Faelen and Toussaint, 1976; Faelen *et al.*, 1977). The phage transposes host DNA sequences of variable sizes. Linked genes are very often simultaneously transposed by Mu (cotransposition: Faelen and Toussaint, 1976), and the frequency of cotransposition between two linked markers has been shown to increase, whereas the distance that separates them decreases. More rarely, unlinked genes are transposed onto the same F molecule (coincident transposition: A. L. Taylor, unpublished), and the frequency of coincident transposition of unlinked genes does not depend

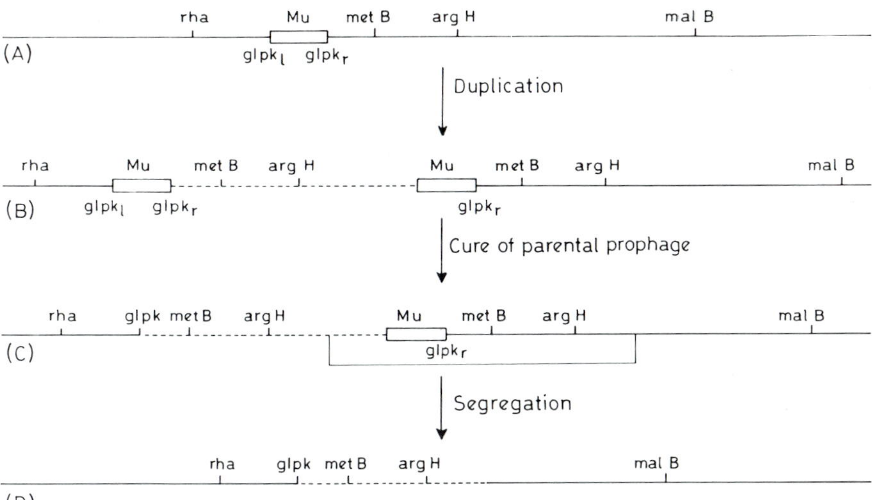

Fig. 11 Mu-induced duplication. The starting strain carries a Mu*cts*62 prophage within the *glpK* gene (A). Incubation of that strain at 37°C stimulates the formation of bacteria containing two copies of the *argH* and *metB* genes. In these derivatives, the DNA fragment corresponding to the duplicated DNA is surrounded by two Mu*cts*62 prophages that have the same orientation, one of which is located within *glpK* (B). When made Rec$^+$ and then *glpK*$^+$ by P1 transduction, these bacteria lose the Mu*cts*62 prophage that was inserted in the *glpK* gene but retain the second Mu*cts*62 prophage located at the junction of the tandem duplication (C). *Rec*-mediated recombination between the two copies of the duplicated region leads to the loss of one of these copies and eliminates the second Mu prophage (D).

on the distance that separates them. In Rec$^+$ conditions, the plasmids that have received linked genes segregate derivatives that have simultaneously lost all the cotransposed genes: Those that received two unlinked genes segregate derivatives that have lost one or both genes. All these segregants retain at least one Mu genome. The simplest interpretation of such segregation patterns is that the linked genes are inserted between two Mu prophages similarly oriented, whereas each of the unlinked genes is flanked by different pairs of Mu genomes that all have a common orientation.

However, the structure of many of these episomes seems to be more complex. One easy way to measure the number of Mu copies in a strain is to use a Mu phage that carries an ampicillin resistance marker (MupAp), that is, a β-lactamase gene *bla*. The amount of β-lactamase synthesized, which is easily measured, is directly related to the number of *bla* genes, and hence to the number of Mu prophages. Using such a MupAp phage, A. L. Taylor (unpublished) determined that some plasmids that underwent a coincident transposition induced by MupAp carry six or more MupAp prophages and that the number of Mu genomes is usually progressively reduced to one in Rec$^+$ bacteria. He also showed that Mu-mediated transposition of a single host DNA region onto F often generates plasmids that contain more than two Mu genomes. In some cases, he observed segregation of the chromosomal markers independent of the host recombination system. The mechanism of that segregation remains to be elucidated.

Mu stimulates the transposition of host genes either adjacent to or distant from the original prophage location. Figure 12 shows different possible pathways that can be envisaged to generate Mu-mediated transposition of adjacent and distant host genes using Mu-mediated deletions and replicon fusions. The Y gene adjacent to the prophage could be excised on a circular DNA containing one copy of Mu by Mu-mediated deletion (Ia). This circle could then be integrated into the plasmid by Mu-mediated replicon fusion (IIa). These two events could just as well occur in the reverse order. The X gene is distant from the prophage. In that case, the first event could be (Ib) the transposition of Mu to a site near X. The second event could be either the excision of a circular DNA containing X and one copy of Mu by Mu-mediated deletion (IIba) or the integration of the plasmid near X by Mu-mediated replicon fusion (IIbb). The third event could be Mu-mediated replicon fusion between the plasmid and the circle in the first case (IIIba) and Mu-mediated deletion-excision of the plasmid, the X host DNA, and two Mus in the second case (IIIbb). Both possibilities generate a plasmid that carries the X host gene flanked by two entire Mu genomes in the same orientation. As mentioned earlier the *B* gene product stimulates all Mu-induced DNA rearrangements. Because, as shown in Fig. 12, it seems most likely that Mu-mediated transposition of host DNA requires at least two Mu-mediated events and most often three, it is not surprising that the product of the *B* gene is required for Mu-mediated transposition of host DNA (Faelen *et al.*, 1978 a,b).

Transpositions of host genes onto a plasmid are frequently isolated upon induction of lysogens that carry the Mu in the chromosome but are not detected when the prophage is located on the episome (M. Faelen and F. van Gijsegem, unpublished results). Mini-Mu MuΔ26, which replicates upon induction, although it is deleted for most Mu late functions, efficiently stimulates the transposition of host genes whether it is integrated in the bacterial DNA or in the plasmid (Toussaint *et al.*, 1981). Mini-Mu Mu3A (which is A^+B^-) stimulates the transposition of host genes at a detectable frequency only when it is inserted in a plasmid (Toussaint *et al.*, 1981; van Gijsegem and Toussaint, 1982); however in the presence of a B^+ helper, it induces transposition very efficiently wherever it is located (Faelen *et al.*, 1978a; Toussaint *et al.*, 1981).

There is no definitive explanation for these last observations. However, taking into account another set of data, one can make some hypotheses. After induction of cells carrying plasmids with a Mu prophage, the plasmids can no longer be conjugally transferred (Schröder and van de Putte, 1974; Razzaki and Bukhari, 1975), most probably because they become integrated into the chromosome through Mu-mediated fusion. After integration, they most probably still undergo several rearrangements induced by Mu, and they could finally be destroyed by maturation of the Mu DNA. This would prevent reappearance of transferrable and maintainable F or F′ plasmids (i.e., F plasmids that carry host DNA) and

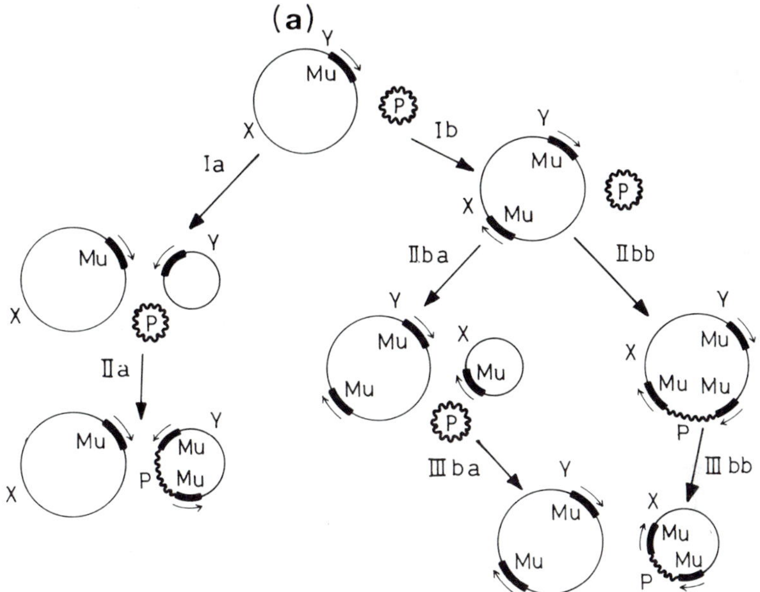

Fig. 12 Different pathways for Mu-mediated transposition of chromosomal genes onto a plasmid. (a) The bacterial chromosome carries a Mu (or mini-Mu) insertion next to the Y gene. P is a plasmid. Ia, Mu-mediated deletion of the Y gene generates a closed circular DNA containing one Mu and the Y gene. One Mu prophage is left behind at the original location during the process of deletion. IIa, Mu-mediated replicon fusion between the circle and the plasmid leads to transposition of the Y gene onto the plasmid, where it is flanked by two similarly oriented copies of Mu. Ib, the Mu prophage has transposed near the X chromosomal gene. IIba, the new Mu copy has generated a deletion of the X gene generating a circle containing X and one Mu copy. IIIba, Mu-mediated replicon fusion between the circle and the plasmid leads to transposition of the X gene onto the plasmid. IIbb, Mu-mediated replicon fusion leads to integration of the plasmid near the X gene. As a consequence, the plasmid DNA is flanked by two similarly oriented copies of Mu. IIIbb, Mu-mediated deletion has led to the formation of a plasmid carrying X flanked by two Mu copies in the same orientation. Ia and IIa could also have occurred in the reverse order, that is, Mu-mediated integration of P in the chromosome followed by Mu-mediated deletion of the plasmid carrying Y flanked by two Mu copies. Transposition of Y adjacent to the original Mu prophage has required two steps; transposition of X, which is away from the original Mu, has required three steps. ———, bacterial DNA; ▬, Mu DNA; 〰, plasmid DNA. The arrows indicate the orientation of the Mu prophages.

therefore the detection of Mu-mediated transposition. MuΔ26 has no maturation functions. This could be the reason why some F or F′ plasmids can be regenerated during MuΔ26 replication and recovered. Finally, Mu3A, when alone, induces rearrangements at low frequency because it is B^-. If present on the plasmid, two Mu3A-mediated rearrangements (replicon fusion and deletion) would be sufficient to generate an F′ carrying a segment of transposed host DNA. The same process would require three Mu3A-mediated events if Mu3A was originally located on the chromosome (transposition, fusion, deletion) (see Fig. 12). For unknown reasons, plasmids that carry a mini-Mu do not integrate in the chromosome after induction as efficiently as those carrying a Mu, even when the mini-Mu is complemented by a Mu located on the chromosome (Chaconas *et al.*, 1980b). The fact that plasmids carrying Mu3A do generate F′s when the mini-Mu is complemented by Mu is consistent with that observation.

Mu-mediated transposition that requires intensive Mu replication seems to be a much more complex event than Mu-mediated deletions, inversions, and replicon fusions.

6. Mini-Muductions

Besides possessing the ability to induce all these rearrangements, Mu is a generalized transducing phage. It can be used as such only in Rec^+ recipient strains (Howe, 1973b). The frequencies of transduction are very low, and markers from the 0–2-min region of the *E. coli* K12 map are transduced about 10 time more frequently than the other markers (Howe, 1973b; Bade *et al.*, 1978). The phage contransduces closely linked genes. Analysis of the DNA present in the transducing particles has shown that it consists only of host DNA (Teifle and Schmieger, 1981).

The ability of Mu to transduce in Rec^+ bacteria is sharply increased when the transducing lysate originates from a strain that contains a mini-Mu prophage in addition to the Mu prophage (Faelen *et al.*, 1979). The high transduction frequencies by the mini-Mu/Mu mixed lysate are probably due to the large number of phage particles that contain a mini-Mu genome linked to giant variable ends.

Moreover, when the mini-Mu has a functional *A* gene, the mini-Mu/Mu mixed lysate can transduce any gene into a *recA* recipient (Faelen *et al.*, 1978a, 1979). Contrary to what happens in Rec^+ bacteria, the transductants recovered in *recA* strains do not result from the substitution of a segment of chromosomal DNA by homologous DNA of the transducing phage; in *recA* recipients, the transduced DNA is integrated at a random location in the host chromosome between two mini-Mu genomes that have the same orientation. This mode of transduction is called *mini-muduction*. From a molecular point of view, the mini-Mu/transduced DNA/mini-Mu structures that do transpose en bloc can be considered as Tns with a structure similar to that of Tn*9* (see Chapter 4) where the two mini-Mus replace the IS sequences.

7. Mu Transposition

Two types of transpositions may be considered. The first one has already been described. It occurs at the same time as bacterial chromosomal rearrangements such as transpositions of bacterial genes, replicon fusions, deletions, inversions, and deletions of markers located far from the original Mu insertion. In each case, a Mu copy remains at its original location, and one or several additional copies are found closely linked to the rearranged bacterial DNA. Such transpositions are always accompanied by a Mu-specific replication.

The second category consists of Mu transpositions that occur without any concomitant host gene rearrangements. Whether such simple transpositions do occur is not clear. Chaconas *et al.*, (1981b) have shown that F plasmids that received a copy of Mu transposed from the chromosome never have the size expected for an F (Mu). However, Maynard-Smith *et al.* (1980) and Coelho *et al.* (1980) found rare R388 plasmids that received Mu by transposition from the chromosome and had the expected size for an R388 (Mu), although very short deletions or insertions would not have been seen by the methods used in these two last cases. In the case of mini-Mu, the frequencies of simple transposition versus cointegrate formation have been measured in many different experimental conditions and are summarized in Table II. Each system has its own characteristics, and the results differ. In some systems the frequency of cointegrate formation is 100%, but in others it is not, suggesting that in at least some conditions, transposition is always accompanied by replicon fusion. In most instances, the cointegrate structures are stable unless they are in a Rec$^+$ host, in which case they resolve by reciprocal recombination between the two mini-Mus using the host recombination enzymes. Where simple transpositions of the mini-Mu are recovered, it is difficult to decide whether they actually reflect the existence of simple transposition events per se or result from the resolution of cointegrate structures. Indeed, as mentioned earlier, there are some experiments suggesting that resolution could occur even in the absence of the host recombination enzymes, although the mechanism of such a resolution remains completely unknown.

8. MU EXCISION

Precise and imprecise excisions are characteristic of prokaryotic transposable elements. Mu shows that behavior only in particular conditions, and differences exist with the excision of other transposable elements. Excisions of a Mu prophage inserted in the *lacZ* gene have been detected only using Mu*X* mutants (Bukhari, 1975), probably because they do not kill the host but do express the *A* gene after induction.

Excision requires intact ends and an active *A* gene (Bukhari, 1975; Khatoon and Bukhari, 1981). The frequency of excision of Mu*X* prophages is low (10^{-6}) as compared to their frequency of transposition (10^{-3}), supporting the idea that excision is not required for transposition. Imprecise excision of Mu from the *lacZ* gene is detected by the fact that although *lacZ* is still inactivated after excision, the polar effect of the Mu insertion on the distal gene *lacY* is relieved. Such *lacZ*$^-$, *lacY*$^+$ derivatives are 10–100 times more frequent than the *lacZ*$^+$, *lacY*$^+$ derivatives that result from precise excision. Many of the *lacZ*$^-$, *lacY*$^+$ revertants do not seem to contain any Mu DNA; the majority of these clones do

Table II Frequency of Simple Transposition and Cointegrate Formation

Donor replicon		Recipient replicon				Mu functions provided		Amount of time of induction or *A* and *B*	Time of induction versus time of mating	% Cointegrates	Cointegrate	Reference[a]
Nature	Replication	Nature	Transfer	Recipient bacterium	Element transposed	In cis	In trans					
PML2	+	R388	+	*recA* *i*Mu$^+$Mu^R	Mu	All	None	Large	0–30-min induction before mating	50	Mostly as expected: some longer or shorter	(1,2)
PML2	+	R388	+	*recA* *i*Mu$^+$Mu^R	Mu*B*$^-$	All except *B*	None	Small	Idem	50	Shorter than expected	
λ	–[b]	Chromosome	–	*recA* *i*Mu$^-$*i*λ$^+$	Mini-Mu	*A B* NEE	None	Large	—	100	As expected	(3,4)
λ*i*21	–[b]	F*pro,lac*	+	*recA* *i*Mu$^+$*i*21$^+$	Mini-Mu	None	*AB*	Low	Simultaneous	95	Measured genetically; roughly as expected	(5)
						A	None	Low		100		
						A	*AB*	Low		100		
				recA *i*21$^+$		None	*AB*	Low		80		
						A	None	Low		50		
						A	AB	Low		15		
Chromosome	+	F*pro,lac*	+	*recA* *i*Mu$^+$	Mu	All	None	Large	50-min induction before mating	0[c]	Variables sizes, shorter and longer than expected	(6)
pSC101	+	Chromosome	–	*recA*	Mu	All	None	Large	—	100	As expected; analyzed blot hybridization	(7)
pSC101	+	F*pro,lac*	+	*recA* *i*Mu$^+$	Mini-Mu	None	All	Large	50-min induction before mating	100	As expected	(8)

[a] (1) Maynard-Smith *et al.*, 1980; (2) Coelho *et al.*, 1980; (3) Schumm and Howe, 1980, 1981; (4) Howe and Schumm, 1980; (5) Toussaint and Faslen, 1981; (6) Chaconas *et al.*, 1981a,b; (7) Chaconas *et al.*, 1980a,b; (8) Chaconas *et al.*, 1981b.

[b] The λ or λ*i*21 does not replicate because the recipient bacterium is λ or λ*i*21 immune.

[c] No cointegrate is expected in this experiment because the plasmid has to transfer to be recovered. The interesting point here is that none of the F′ recovered have the expected size for an F*pro,lac* that received a Mu as a result of a simple transposition event.

[d] Only a fraction of pSC101 :: mini-Mu associates with the host chromosome, and only part of those associations are cointegrates. The other type of association is not understood so far.

not have detectable deletions in the *lacZ* gene, and most of them can revert further to *lacZ*$^+$. Most of the other *lacZ*$^-$, *lacY*$^+$ clones, which retain at least part of the prophage genome, seem to have a complete *lacZ* gene and can usually revert to Lac$^+$ (Khatoon and Bukhari, 1981). At least part of the imprecise excisions must occur by a different mechanism than the precise excisions because the frequencies of both types of excisions are not similarly affected by various factors: Precise excision is completely blocked under immunity or in the absence of the *A* gene product, whereas imprecise excision is lowered by a factor of only 10–100 in the same conditions. Products synthesized by the host are thus sufficient to carry out the imprecise excision but not the precise excision, and Mu A protein somehow stimulates the frequency of the imprecise excision but is absolutely essential for the precise excision. This situation is different from that observed with elements such as Tn*10*, in which precise and imprecise excisions do occur in the absence of Tn-encoded functions, albeit at reduced frequencies (Foster *et al.*, 1981). Mu and Tn*10* excisions also differ in that *recA* mutations have no effect on the excisions of Tn*10* but lower the precise and imprecise excisions of Mu by a factor of 10–100. This further confirms the fact that the mechanisms controlling the excisions of Mu and Tn*10* are not identical, but it does not rule out the possibility that these mechanisms involve similar host components.

The ability of Mu to induce all the rearrangements that are summarized in Fig. 13 makes it a very useful tool for genetic analysis, particularly for bacteria into which the *E. coli* F plasmid cannot be transferred and that will not allow growth of the transducing phage P1. This aspect of Mu will not be discussed here because it will be extensively described elsewhere (A. Toussaint, unpublished).

C. Mu Replication and DNA Synthesis

Mu transposition and Mu-induced chromosomal rearrangements seem always to involve replication of the Mu genome, for after a rearrangement has occurred there are always additional copies of Mu flanking the rearranged DNA segment. One should, however, make a distinction between intensive replication, that is, replication that can be detected by biochemical methods such as DNA–DNA hybridization and replication that is seen by genetic analysis of clones that underwent Mu-induced rearrangements and cannot be detected biochemically. One example is provided by *B*$^-$ mutants that do induce genetically detectable rearrangements involving replication at reduced frequencies but are considered as completely replication deficient because after induction no increase in Mu-specific DNA synthesis can be detected by hybridization.

As seen previously, excision of Mu from the host chromosome does not seem to be a usual step in the lytic development of Mu prophage. The scarcity of extensive excision of the prophage was emphazised by Ljungquist and Bukhari (1977). At different times after induction, they isolated the total DNA from a

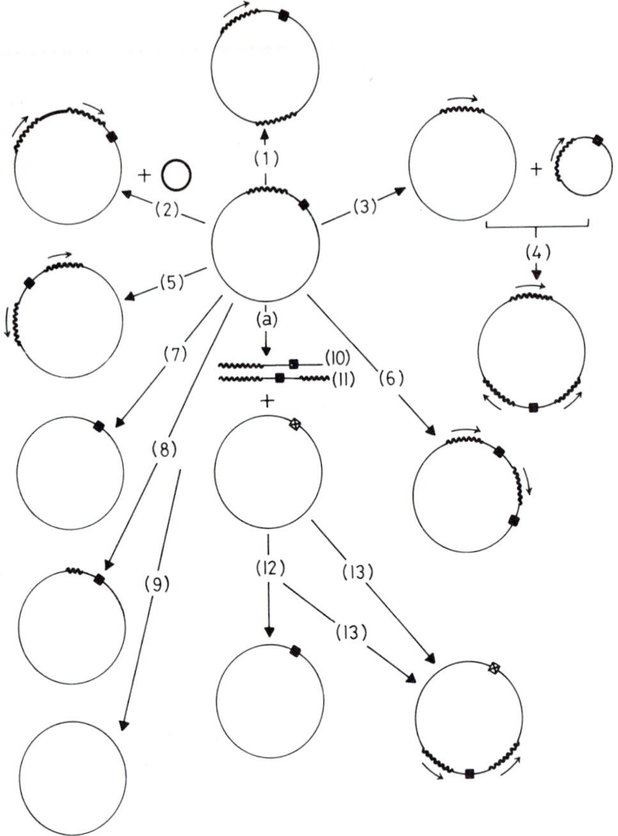

Fig. 13 Chromosomal rearrangements induced by Mu and mini-Mu. The arrows indicate the orientation of the Mu or mini-Mu prophages. Rearrangements include (1) transposition; (2) replicon fusions where the two replicons (here, the chromosome and a plasmid) are joined by two Mus in the same orientation; (3) deletion where a circle containing one Mu and a chromosomal segment is excised, leaving one Mu copy behind; (4) transposition of a chromosomal segment, for instance by replicon fusion between the chromosome and a circle formed upon the mediated deletion (see also Fig. 12); the transposed segment is flanked by two Mu copies in the same orientation; (5) inversion where the inverted segment is flanked by copies of Mu in opposite orientation; (6) duplication where the duplicated segment is flanked by two Mus in the same orientation; (7) precise excision; (8) imprecise excision leaving a fragment of the Mu prophage behind; (9) imprecise excision removing a segment of host DNA adjacent to the Mu prophage; (10–13) induction of a strain lysogenic for mini-Mu and Mu providing lysates with phage particles containing mini-Mu DNA linked to long stretches of host DNA (10) or host DNA segments flanked by two copies of mini-Mu (11). Upon infection of a sensitive Rec$^+$ strain with such a lysate, two modes of transduction are observed: generalized transduction (12) where homologous recombination catalyzed by host recombination enzymes leads to substitution of the mutated (defective) allele by the wild-type allele carried by the phage, and mini-muduction (13) where the mini-Mu host DNA mini-Mu structure is transposed at random on the chromosome using the Mu transposition enzymes. If the recipient is Rec$^-$ only mini-muduction takes place. —, bacterial DNA; ∿∿, Mu or mini-Mu; ■, bacterial marker located near the Mu or mini-Mu prophage and moved around under the influence of Mu; ⊠, defective allele of the bacterial marker; ▬, plasmid DNA.

strain carrying one Mu prophage. The DNA was digested with restriction enzymes, and the fragments were separated by electrophoresis on an agarose gel, transferred to a nitrocellulose filter by the blotting technique (Southern, 1975), and hybridized with ^{32}P-labeled Mu DNA. Two characteristic fragments corresponding to the junction between the *c* and *S* ends of the original Mu prophage and the host chromosome were detected with the same intensity up to the time of onset of Mu DNA maturation. This shows that the original prophage copy is not excised from host DNA.

DNA synthesis is initiated 6–8 min after induction or infection (Waggoner *et al.*, 1977; Wijffelman and Lotterman, 1977), as shown by the hybridization of total labeled DNA extracted from the induced or infected cells with Mu DNA or RNA. No form of free Mu DNA can be detected, whereas the number of Mu copies increases, first slowly (two copies at 9 min, four at 12 min, six at 15 min) and then more quickly after 20-min induction (Waggoner *et al.*, 1977; Wijffelman and Lotterman, 1977). There is no amplification of host sequences adjacent to either end of the original prophage (Waggoner and Pato, 1978). These observations were important in the formulation of the models summarized in Fig. 1. It was later found that the Mu copies remain associated with the host nucleoid, which tends to become more compact as the number of Mu copies increases (Pato and Waggoner, 1981). Looking at Fig. 1, one can see clearly that the Mu prophage is not excised while it binds to the new target and that adjacent host DNA is not replicated during replication of the Mu genome. Moreover, each new round of replication involves additional covalent binding between the end(s) of Mu and a distant target site, which would lead to the alteration of the tertiary structure of the chromosome and consequently to an increase in the sedimentation rate of the nucleoid. As also predicted by the models, circular forms, nicked or supercoiled, and with or without a tail, start to appear 10–12 min after induction or infection (Schröder *et al.*, 1974; Waggoner *et al.*, 1974, 1977; Harshey and Bukhari, 1981). Their length varies, showing no size classes, and they contain host sequences covalently linked to one or several Mu genomes. At most, they represent 3% of the total DNA content of the cell.

Most of the Okazaki fragments purified at different times after induction of Mu hybridize with the same strand of Mu (the *r* strand), which led Wijffelman and van de Putte (1977) and Goosen (1978) to conclude that Mu replication is unidirectional, starting at the *c* end and proceeding toward the *S* end. However, these results are not easy to interpret because up to 20% of the Okazaki fragments hybridize to the opposite (*l*) strand, which could be taken to mean that some replication proceeds in the opposite direction. Moreover, even if the DNA replication starts exclusively from the *c* end, the right end must be involved in some way in the initiation of DNA synthesis because prophages deleted from their right end do not replicate (van de Putte *et al.*, 1978; Waggoner *et al.*, 1981).

Mu replication requires intact ends and the products of genes *A* and *B*. Prophages deleted from either end, A^- and B^- mutants, do not replicate (Wijffelman *et al.*, 1974; Wijffelman and Lotterman, 1977; van de Putte *et al.*, 1978;

Waggoner *et al.*, 1981) and do not induce formation of closed circular DNA molecules (Waggoner *et al.*, 1977). Mini-Mu Mud123 (Maynard-Smith *et al.*, 1980), which kept 5.3 kb from the *c* end, and thus genes *A* and *B* and part of the nonessential early (NEE) region, replicates extensively after induction (Waggoner et al, 1981). Mini-Mus that have an intact *A* gene but have lost the *B* gene and the NEE region do not replicate, as expected. When they are complemented for *B* only, they replicate at a reduced level (Waggoner *et al.*, 1981; Goosen *et al.*, 1982). This suggests that full replication of Mu, in addition to the A and B proteins, requires a function located in the NEE region. In agreement with that hypothesis, Mu mutants that carry an insertion or a deletion inactivating the NEE region make very small plaques and replicate slower than Mu (Goosen *et al.*, 1982; B. T. Waggoner *et al.*, unpublished results). This new function was called *arm* (amplified replication of Mu). It may be directly involved in replication, or it may be a positive regulator of *A* and *B* expression (Waggoner *et al.*, 1981; Goosen *et al.*, 1982). The second hypothesis is usually preferred because it seems unlikely that a function directly involved in DNA replication could be nonessential.

Mini-Mus deleted from *A* and *B* replicate when they are complemented by Mu. The shortest mini-Mu isolated so far that shows that behavior has 1 kb from the *c* end and 800 bp from the *S* end (L. T. Chow *et al.*, unpublished results). As mentioned earlier, MudII301, which has only 116 bp from the *S* end, also replicates (M. Casadaban, unpublished observation). Thus, all the sites necessary for replication are located within the first 1000 bp and the last 116 bp of the Mu genome.

D. Replication–Transposition Intermediates

From the above discussion, it is obvious that transposition, replication, and induced rearrangements are closely associated. After induction, Mu replicates, transposes, and induces rearrangements at a very high frequency, which should make it a very good material for detecting transposition–replication intermediates. However, abnormal DNA configurations are rarely seen by electron microscopic analysis. As mentionned above, circles with and without a tail never account for more than 3% of the total DNA. The infrequent appearance of abnormal structures could result from several facts: (*1*) One replication of Mu is a short (1-min), transient event, and each replication is expected to occur at a different place on the host chromosome. Thus, at any given time after the onset of replication, very few Mus will be in the process of replicating. (*2*) Maturation of the phage DNA, which starts 35 min after the onset of the lytic cycle (Ljungquist and Bukhari, 1977), destroys the structures that have been formed during the replication-transposition process. (*3*) The covalently closed circles may be transient products of rounds of replication involving deletion formation. Upon subsequent replication of their Mu prophage, they should reintegrate by replicon fusion just as efficiently as they were generated which could explain their scarcity.

Using a strain carrying a plasmid harboring a mini-Mu A^-B^- and a complementing Mu prophage on the chromosome, Harshey and Bukhari (1981) observed structures like tailed, open, circular DNA of various sizes, now called *key structures,* and circles the size of the plasmid attached to the chromosomal DNA at a unique site. Key structures were found to be more abundant on deproteinized DNA, suggesting that they might be a by-product, normally present in another form maintained by one or several proteins. Résibois *et al.* (1982) looked at the DNA extracted, as long as 50 min after induction, from a strain lysogenic for a self-replicating mini-Mu unable to mature its DNA and to lyse the host cell. In those preparations, many aberrant DNA structures were found including keys, dumbbells, inverted forks, asymmetrical forks, branched keys, and what were called *ribosome-like structures* (see Fig. 13). In the large majority of the cases, the segments between the different branching points in these structures were shorter than the mini-Mu used, suggesting that they might be partially replicated mini-Mus. A computer analysis of the correlation between a standard partial denaturation map of the mini-Mu and the partially denatured DNA of these structures confirmed that replicating mini-Mus are located at the branching points (Résibois *et al.*, 1982). Evidence that a replicating mini-Mu is located at the junction between the circle and the tail in the key structures has also been obtained by Harshey and Bukhari (1982) using a mini-Mu carrying a cloned *lac* promoter. The mini-Mus were localized with the electron microscope using an immunological reaction between repressor bound to the *Lac* promoter and labeled specific antibodies. Both groups came to the conclusion that, contrary to what was deduced from the analysis of Okazaki fragments (see Section III,C), at late times after induction, Mu can replicate from either the *c* or the *S* end at about the same frequency.

Figure 14 shows how one can explain the generation of these structures by a combination of the two models so far proposed to account for transposition of IS and Tn elements and Mu transposition–replication. As mentioned previously, one type of model predicts the simultaneous binding of both ends of Mu to the target before replication. The other type of model predicts the formation of a nucleoprotein, with binding of the *c* end of Mu to the target before replication, the Mu *S* end being bound after completion of replication. However, as mentioned above, the *S* end of Mu is essential for replication. One must therefore assume that both ends of Mu are part of the DNA–protein complex. One could imagine that in the complex either the *c* end or the *S* end can bind to the target to initiate replication, the other end being able to bind at any time during replication, extruding the new Mu copy from the complex. This would account for the presence of a majority of key structures in deproteinized DNA isolated after Mu induction. It would also explain the ribosome-like and dumbbell structures that, by breakage due either to the use of one of the circular portions of these structures as a target in another transposition event or to the DNA isolation procedure,

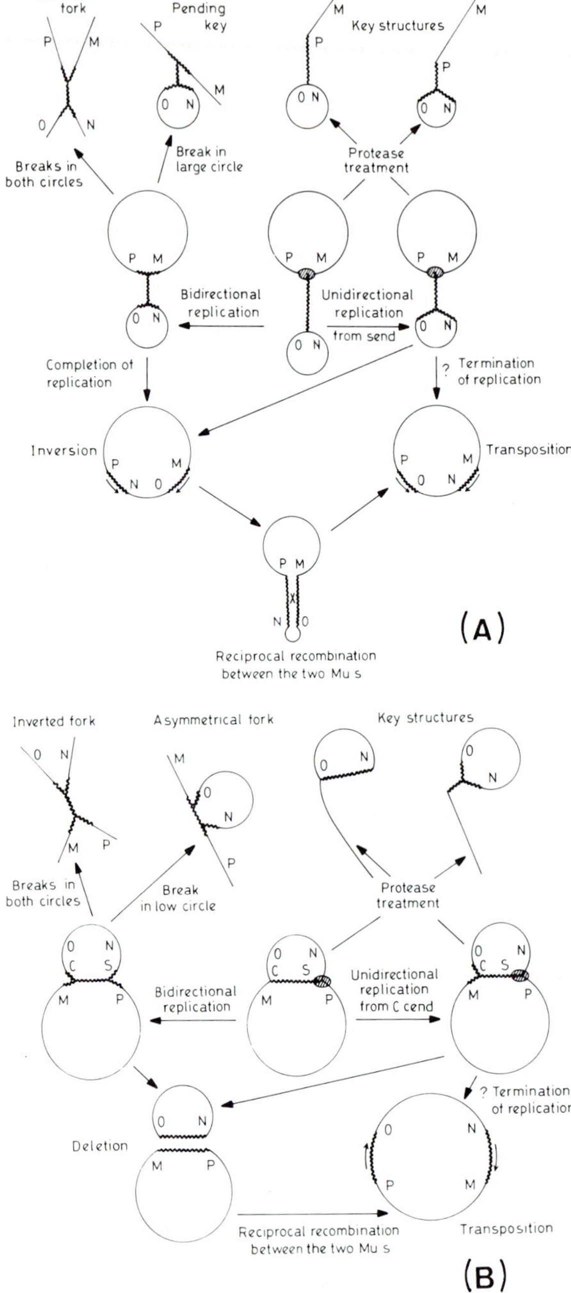

Fig. 14 Transposition intermediates as predicted by different models and seen during mini-Mu replication. The dumbbell and ribosome-like structures shown in the center of (A) and (B) are the result of slight modifications of the models presented in Fig. 1. As described in the text, a complex

generate inverted forks, asymmetrical forks, and pending keys (see the legend of Fig. 14 for a detailed explanation).

IV. D108

Before continuing our discussion of the mechanism of Mu replication, we should briefly summarize what is known about the related mutator phage D108. In 1971, Mise described a new generalized transducing phage that later turned out to be another mutator phage. It induces mutations by integrating its DNA at random locations in host genome (Hull *et al.*, 1978) and by chromosomal rearrangements (Faelen and Toussaint, 1980b). The phage particle has an icosahedral head and a contractile tail with a base plate supporting spikes and fibers (Mise, 1971; Gill *et al.*, 1981). Like Mu DNA, D108 DNA released from the purified phage particles measures about 39 kb, displays a 3-kb-long invertible G segment that separates an α region from a β region, and is flanked by random sequences that, by analogy with Mu, are considered host DNA (Gill *et al.*, 1981). Electron microscopic observations of Mu–D108 heteroduplexes (Gill *et al.*, 1981) reveal several regions of nonhomology between the two genomes (Fig. 15). Three are in the α region, located respectively between 0.1 and 1.5, 4.8 and 5.1, and 7.8 and 8.4 kb from the *c* end, that is, in the segments that in

---

would be formed, including the two ends of Mu, the target site, and the Mu *A* protein. Nicking would occur at one of the phage ends and at the target, and the nicked phage end would be ligated to the target inside the complex. For the sake of simplicity, the complex is represented in a partially opened form, with the ligated end of Mu separated from the rest of the complex. (A) Fate of the dumbbell. The upper part shows the structures generated upon treatment of the central dumbbell form with pronase. Unidirectional replication from the *S* end leads to the appearance of two *S* ends in the small circle. The reverse situation could just as well have occurred, that is, unidirectional replication from the *c* end. As long as both ends have not covalently bound to the target pronase, treatment of the DNA will generate key structures. The dumbbell structure will be pronase resistant once both ends are covalently linked to the target, at which point replication of the Mu could start from the second end. At that stage, a break in one of the circles will generate a pending key, whereas breaks in both circles will generate an inverted fork. Completion of replication of Mu will generate an inversion of the NO DNA segment, which is flanked by inverted copies of Mu. Reciprocal recombination between these two Mus will lead to the return of the NO segment in the original orientation. The transposed Mu copy is in the opposite orientation from the original Mu copy. If the second end of Mu is not linked to the target before the completion of Mu replication, it is still possible to generate a simple transposition by a mechanism similar to the one outlined in Fig 8e. (B) Fate of the ribosome-like structure. The situation is similar to that in (A) except that a break in one circle of the structure will generate an asymmetrical fork, and completion of replication will lead to deletion of the NO segment. After reciprocal recombination between the two Mus, the transposed Mu copy is in the same orientation as the original one. —, bacterial DNA (double-stranded); ﹏, Mu DNA (double-stranded); ○, protein complex. M, N, O, and P are bacterial segments adjacent to the *c* and *S* ends of Mu and flanking the target site.

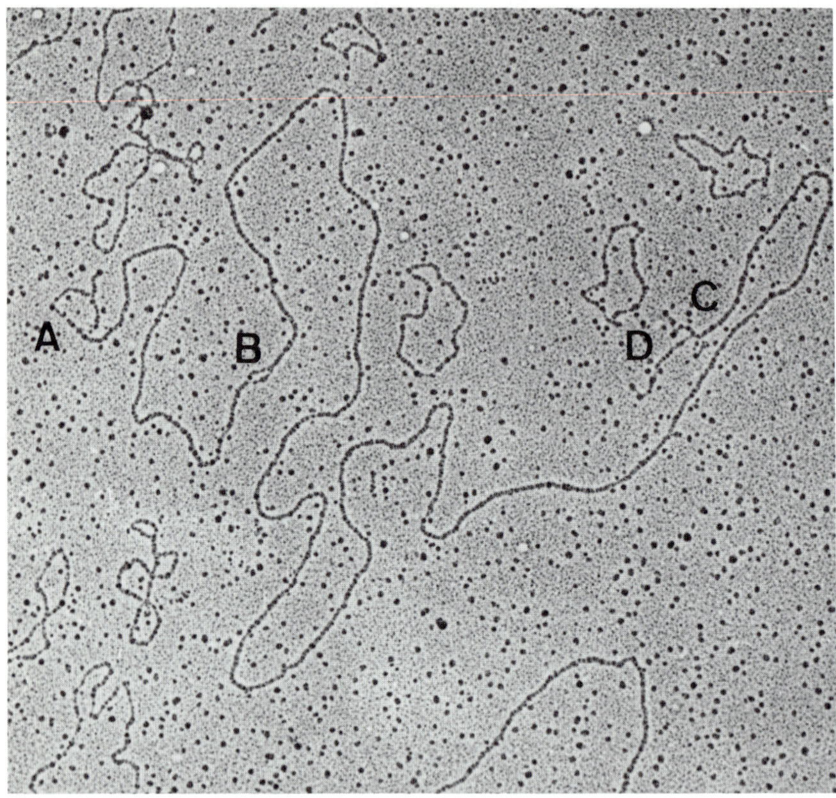

Fig. 15 Electron micrograph of a Mu–D108 heteroduplex DNA. (A) Region of nonhomology in the immunity region. (B) Region of nonhomology at the end of the NEE region. (C) Insertion bubble from D108 at the end of the β region. (D) Asymmetrical SEs resulting from the different lengths of the β regions of the two phages.

Mu encode (*1*) the repressor, *ner,* and the beginning of the *A* gene, (*2*) *gam,* and (*3*) the left end of the late gene region. The G segments, when in the same orientation, are fully reannealed under mild denaturing conditions (40% formamide) but show partially denatured segments in 60% formamide, implying that the G loop sequences in D108 and Mu have diverged. The D108 β region is 0.5 kb longer than the Mu β, and the additional sequence is located 0.1 kb from the *S* end. Consistent with the fact that Mu and D108 genomes are homologous in the region that encodes for the phage coat proteins, the total length of the DNA of both phages is the same and the additional 0.5 kb in the D108 β region is compensated for by an equivalent shortening of the *S* variable end.

In agreement with the region of nonhomology detected on their DNA, Mu and D108 are heteroimmune: Mu grows in bacteria lysogenic for D108, and vice versa. However, both repressors have similarities because neither is inactivated by ultraviolet irradiation or other chemicals that induce lambdoid phages. Thermoinducible mutants of D108 have been isolated (Hull *et al.*, 1978). Mu gene *A* is partially located in the first region of nonhomology. D108 gene *A* is most probably located at the same place because it is inactivated in D108 derivatives carrying an internal deletion extending up to 1.1 kb from the *c* end but not if the deletion ends at 3.5 kb from the *c* end (Résibois *et al.*, 1981). Sequencing the junction between the right end of the first nonhomology region and the adjacent homologous segment, B. Allet has shown that Mu and D108 genes *A* differ only by ~40 amino acids (unpublished). The 70 K protein encoded by D108 (Magazin *et al.*, 1978) is probably the *A* gene product. The host range of D108 is somewhat different from the Mu host range. Bacteria resistant to D108 are resistant to P1, and vice versa; some of them are resistant to Mu, but the reverse is not always true (Hull *et al.*, 1978). Few conditional mutants of D108 have been described so far, but amber mutations in most Mu genes have been recombined into D108 to generate D108–Mu hybrid mutants (M. Faelen, unpublished results). The genetic maps of the two phages can thus be considered extremely similar, if not identical.

V. Mu and Host Factors Required for Transposition–Replication

A. Mu Ends

When Mu DNA integrates in the host chromosome, it duplicates 5 bp of the target site, as do the Tns in the Tn*3* family (Chapter 4) (Rosenberg *et al.*, 1978; Ghosal *et al.*, 1979). These transposable elements differ from Mu in that they possess 40-bp-long, nearly perfect inverted terminal repeats, whereas the first and last 40 bp of Mu show almost no homology. However the first 31 nucleotides of the *c* end are almost all present in the *S* end: The first 14 bp from the *c* end are found at position 79–92 from the *S* end, whereas nucleotides 15–31 from the *c* end are found among the first 20 nucleotides from the *S* end (Kahmann and Kamp, 1979). Whether this kind of inverted repeat is functionally significant remains to be seen. The sequence ACGAAAA is present at a similar position (nucleotides 21–27) in Tn*3*, γδ, IS *101*, and the *c* end of Mu. It is also present in the *S* end of Mu but located at nucleotides 12–18. The significance of this similarity between Mu and other transposable sequences that generate a 5-bp repeat is so far unknown.

If larger parts of the Mu *c* and *S* ends are compared, it appears that the sequence 5' PuCGAAAA is repeated, with minor modifications, five times within the first 62 nucleotides from the *S* end but not the *c* end.

The *c* and *S* ends are not equivalent, as shown by the fact that the structures containing two mini-Mus in opposite orientation flanking a segment of host

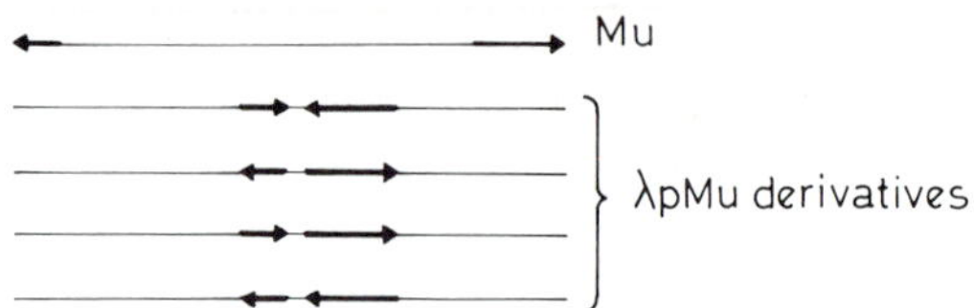

Fig. 16 Structure of different λ phages carrying cloned ends of Mu. The arrows show the ends of Mu and their relative orientations. Only the λpMu, which carries the ends in the same relative orientation as on the Mu genome, can generate mini-Mu-mediated replicon fusion.

DNA, which are seen in large amounts in Mu/mini-Mu mixed lysates by electron microscopy, are not found integrated in the host chromosome after infection with such a mixed lysate (Faelen *et al.*, 1979). Two *c* ends or two *S* ends in opposite orientation are therefore not a substrate for Mu transposition enzyme(s), contrary to what is found with IS*1*. Finally, to be functional, the ends have to be in the proper orientation with respect to each other. The ends of Mu have been cloned in a λ vector in the four possible respective orientations (Fig. 16) (Schumm and Howe 1980, 1981; Howe and Schumm, 1980). Only those phages carrying the ends in the same orientation as Mu can integrate in the host chromosome by Mu-mediated replicon fusion.

B. The *A* Gene Product

The search for integration-deficient mutants of Mu able to make viable phages was unsuccessful (O'Day *et al.*, 1978). This is not surprising in view of what is known about Mu replication. In addition to their inability to replicate, MuA^- mutants neither integrate (O'Day *et al.*, 1978) nor transpose, nor induce rearrangements (Faelen *et al.*, 1978b).

Because both ends of Mu are necessary for integration and transposition (van de Putte and Gruijthuijsen, 1972), it is usually assumed that the *A* protein recognizes the ends of the phage genome. Attempts to clarify the site specificity of *A* have been made by analyzing interactions between D108 and Mu (A. Toussaint *et al.*, unpublished). Although they differ only by about 40 amino acids at their NH_2-end, the *A* proteins of Mu and D108 are not fully interchangeable: Mini-MuA^- are complemented by Mu but only poorly by D108, and mini-D108A^- are complemented by D108 but not by Mu. Two kinds of Mu–D108 hybrids have been constructed. One type has the *c* end, the left end of *A*, and the A specificity of Mu, but the *S* end of D108; the other has the *c* end, the left end of *A*, and the *A* specificity of D108, but the *S* end of Mu. Both types of hybrids yield viable phages, indicating that the *S* ends of Mu and D108 are interchangeable. However, preliminary sequencing suggests that the first 35 bp at the *c* end of the two phages are identical, whereas the right ends are different (Kamp and Kahmann, 1980). These results open the possibility that the *A* protein recognizes a site in the region of nonhomology between Mu and D108 rather than the ends of the genome.

Recent results indicate that a constant synthesis of the *A* protein is required for Mu replication (Pato and Reich, 1982). Whether this reflects stoichiometric use or instability of the *A* gene product remains to be elucidated.

C. The *B* Gene Product

MuB^- mutants, Mu*X* mutants (which are B^- and defective for the NEE region due to the presence and the polarity of an IS sequence in their *B* gene) do integrate, although at a frequency (about 10 times) lower than wild-type Mu (O'Day *et al.*, 1978). As detailed in another section, they induce rearrangements, as do mini-Mus A^+B^-. However, although B^- mutants induce adjacent deletions and inversions at about the same frequency as wild-type Mu, they transpose and mediate the transposition of host sequences at a lower frequency than Mu (about 100 times lower). It is tempting to speculate that in the absence of the *B* protein Mu behaves just like IS and Tn elements, transposing and replicating at a low frequency using the *A* protein only, and that *B* stimulates that process, thereby allowing the phage to use transposition for its lytic replication.

D. Host Replication Functions

Mu replication requires at least the host *dnaC, dnaB, dnaE, dnaZ,* and *dnaG* replication functions (Toussaint and Faelen, 1974; E. Mc Beth, personal communication; A. Toussaint and O. Huisman, unpublished results). The requirements for other host replication factors such as helicase, gyrase, single-stranded DNA, and binding proteins, remain to be studied. There also exist *E. coli* mutants that do not support Mu growth, are defective for Tn*9* transposition, and have been mapped very close to the *dnaB* gene (Illyina *et al.*, 1981; E. Piruzian and R. Upmal, unpublished).

After infection, Mu seems preferentially to integrate near the host replication fork during both lysogenization (Paolozzi *et al.*, 1978, 1979) and lytic growth (Fitts and Taylor, 1980). This could reflect some particularity of the first integration event or more efficient replication of those Mu genomes that happen to transpose into actively replicating DNA. However, it could just as well reflect a very important property of Mu replication that has not been taken into account in the formulation of models and that, once understood, will change our overall view about the mode of Mu replication and transposition.

E. Conclusions and Perspectives

The study of Mu has made an important contribution to our present understanding of transposition in bacteria. Some Mu experiments have been critical in the elaboration of the models that are the basis of our understanding of this process. The fact that Mu uses transposition for its lytic replication has made it possible to devise experimental conditions in which transposition and rearrangements occur at a very high frequency and can even be detected by physical methods, such as electron microscopic analysis of the DNA. This will be very valuable in the future in devising an *in vitro* assay for Mu transposition that should make it possible to dissect the host and phage components involved in the process. Many questions remain to be answered. The function and mode of action of both the *A* protein and the ends of the phage genome require much attention. What is the exact biochemical activity of the *A* protein? With which other proteins (phage or host) does it interract? Where are its recognition and substrate sites? Does the Mu*A* protein interact with the target DNA? If it turns out that the *A* protein does not recognize the ends of the phage genome but rather a site near the left end, and subsequently acts at the ends, it could behave like the type 1 restriction enzymes that track along the phage DNA from their recognition sites to the place of cleavage. This could involve configurational and functional changes, so that an *A* protein that has reached the ends and interacted with them could no longer bind to the recognition site. In that scheme, *A* would be used stoichiometrically, as suggested by Pato and Reich (1982).

Another point that needs to be elucidated is the role of the right end in replication. We know that prophages deleted from the right end do not replicate at a level detectable by biochemical methods because these methods will not detect replication of B^- mutants, even though these were shown to replicate at a low level by genetic methods. Therefore, we cannot rule out the possibility that mutants deleted from the right end could initiate one round of transposition-replication but could not terminate it. In other words, it may be that the right end is necessary for termination of transposition-replication but dispensable for its initiation. However, the right end could just as well be required for replication-initiation, especially if it is true that replication can proceed from either end of the phage genome.

Because of its inducible Tn properties, Mu is a very useful genetical tool, allowing, for instance, the *in vivo* cloning of any gene(s) of bacteria in which it can transpose. Its use thus turns out to be particularly rewarding, both from a fundamental and a practical point of view.

Acknowledgments

We are very grateful to M. Pato, A. L. Taylor and co-workers, and B. Waggoner for their very helpful comments, to the whole "Mu community" for communicating unpublished results, and to M. Faelen, without whom this chapter would not have been what it is. AT is *Chercheur Qualifié* from the Fonds National de la Recherche Scientifique.

References

Abelson, J., Boram, W., Bukhari, A. I., Faelen, M., Howe, M. M., Metlay, M., Taylor, A. L., Toussaint, A., van de Putte, P., Westmaas, G. C., and Wijffelman, C. A. (1973). Summary of the genetic mapping of prophage Mu. *Virology* **54,** 90–92.

Allet, B. (1979). Mu insertion duplicates a 5 base pairs sequence at the host insertion site. *Cell* **16,** 123–129.

Allet, B., and Bukhari, A. I. (1975). Analysis of Mu and λ-Mu hybrid DNAs by specific endonucleases. *J. Mol. Biol.* **92,** 529–540.

Arthur, A., and Sherratt, D. (1979). Dissection of the transposition process: a transposon-encoded site-specific recombination system. *Mol. Gen. Genet.* **175,** 267–274.

Bade, E. G. (1972). Asymmetric transcription of bacteriophage Mu-1. *Virology* **10,** 1205–1207.

Bade, E. G., Howe, M. M., and Rawluk, L. (1978). Preferential generalized transduction by bacteriophage Mu-1. *Mol. Gen. Genet.* **160,** 89–94.

Bialy, H., Waggoner, B. T., and Pato, M. L. (1980). Fate of plasmids containing Mu DNA: chromosome association and mobilization. *Mol. Gen. Genet.* **180,** 377–383.

Boram, W., and Abelson, J. (1973). Bacteriophage Mu integration: on the orientation of the prophage. *Virology* **54,** 102–108.

Bukhari, A. I. (1975). Reversal of mutator phage Mu integration. *J. Mol. Biol.* **96,** 87–99.

Bukhari, A. I. (1976). Bacteriophage Mu as a transposition element. *Annu. Rev. Genet.* **10,** 389–412.

Bukhari, A. I., and Ambrosio, L. (1978). The invertible region of bacteriophage Mu DNA determines the adsorption properties of Mu particles. *Nature (London)* **271,** 575–577.

Bukhari, A. I., and Metlay, M. (1973). Genetic mapping of prophage Mu. *Virology* **54,** 109–116.

Bukhari, A. I., and Taylor, A. L. (1975). Influence of insertions on packaging of host sequences covalently linked to bacteriophage Mu DNA. *Proc. Natl. Acad. Sci. USA* **72,** 4399–4403.

Bukhari, A. I., and Zipser, D. (1972). Random insertion of Mu-1 DNA within a single gene. *Nature (London) New Biol.* **236,** 240–243.

Bukhari, A. I., Froshauer, S., and Botchan, M. (1976). The ends of bacteriophage Mu DNA. *Nature (London)* **264,** 580–583.

Cabezon, T., Faelen, M., Dewilde, M., Bollen, A., and Thomas, R. (1975). Expression of ribosomal protein genes in *E. coli. Mol. Gen. Genet.* **137,** 125–129.

Chaconas, G., Harshey, R. M., Sravetnick, N., and Bukhari, A. I. (1980a). Mechanism of bacteriophage Mu DNA transposition. *Cold Spring Harbor Symp. Quant. Biol.* **45,** 311–318.

Chaconas, G., Harshey, R. M., and Bukhari, A. I. (1980b). Association of Mu-containing plasmids with the *E. coli* chromosome upon prophage induction. *Proc. Natl. Acad. Sci. USA* **77,** 1778–1782.

Chaconas, G., de Bruijn, F. J., Casadaban, M. J., Lupski, J. R., Kwoh, T. J., Harshey, R. M., Du Bow, M. S., and Bukhari, A. I. (1981a). *In vitro* and *in vivo* manipulations of bacteriophage Mu DNA: cloning of Mu ends and construction of mini-Mu's carrying selectable markers. *Gene* **13,** 37–46.

Chaconas, G., Harshey, R. M., Sravetnick, J., and Bukhari, A. I. (1981b). Predominant end-products of prophage Mu DNA transposition during the lytic cycle are replicon fusions. *J. Mol. Biol.* **150,** 341–359.

Chow, L. T., and Bukhari, A. I. (1977). The invertible DNA segments of coliphage Mu and P1 are identical. *Virology* **74,** 242–248.

Chow, L. T., and Bukhari, A. I. (1978). Heteroduplex electron microscopy of phage Mu mutants containing IS*1* insertions and chloramphenicol resistance transposons.. *Gene* **3,** 333–346.

Chow, L. T., Kahmann, R., and Kamp, D. (1977). Electron microscopic characterization of DNA's from non-defective deletion mutants of bacteriophage Mu. *J. Mol. Biol.* **113,** 591–609.

Coelho, A., and Symonds, N. (1981). Analysis of proteins synthesized by plasmids containing cloned fragments of bacteriophage Mu. *FEMS Microbiol. Lett.* **11,** 101–106.

Coelho, A., Leach, D., Maynard-Smith, S., and Symonds, N. (1980). Transposition studies using a *ColE1* derivative carrying phage Mu. *Cold Spring Harbor Symp. Quant. Biol.* **45,** 323–328.

Cohen, S. N., Casadaban, M. J., Chow, J., Tu, C.-P., D. (1978). Studies of the specificity and control of transposition of the Tn*3* element. *Cold Spring Harbor Symp. Quant. Biol.* **43,** 1247–1255.

Coombs, D. H., Robinson, A. J., Bodnar, J. W., Jones, C. J., and Pearson, G. D. (1978). The adenovirus DNA terminal protein complex and HeLa DNA-protein complexes. *Cold Spring Harbor Symp. Quant. Biol.* **43,** 741–753.

Couturier, M. (1976). The integration and excision of the bacteriophage Mu-1. *Cell* **7,** 155–163.

Couturier, M., and van Vliet, F. (1974). Vegetative recombination in bacteriophage Mu-1. *Virology* **60,** 1–8.

Daniell, E., Roberts, R., and Abelson, J. (1972). Mutations in the lactose operon caused by bacteriophage Mu. *J. Mol. Biol.* **69,** 1–8.

Daniell, E., Abelson, J., Kim, J. S., and Davidson, N. (1973a). Heteroduplex structures of bacteriophage Mu DNA. *Virology***51,** 237–239.

Daniell, E., Boram, W., and Abelson, J. (1973b). Genetic mapping of the inversion loop in bacteriophage Mu DNA. *Proc. Natl. Acad. Sci. USA* **70,** 2153–2156.

Daniell, E., Kohne, D. E., and Abelson J. (1975). Characterization of the inhomogeneous DNA in virions of bacteriophage Mu by DNA reannealing kinetics. *J. Virol.* **15,** 739–743.

Dénarié, J., Rosenberg, C., Bergeron, B., Boucher, C., Michel, M., and Barate de Bertalmio, M. (1977). Potential of RP4: Mu plasmids for *in vivo* genetic engeneering of *gram* negative bacteria. *In* "DNA Insertions Elements, Plasmids and Episomes" (A. I. Bukhari, J. A. Shapiro, and S. L. Adhya, eds.), pp. 507–520. Cold Spring Harbor Lab., Cold Spring Harbor, New York.

Emr. D., and Silhavy, T. (1980). Mutations affecting localization of an *E. coli* outer membrane protein, the bacteriophage λ receptor. *J. Mol. Biol.* **141,** 63–90.

Faelen, M., and Toussaint, A. (1973). Isolation of conditional defective mutants of temperate phage Mu-1 and deletion mapping of the Mu-1 prophage. *Virology* **54,** 117–124.

Faelen, M., and Toussaint, A. (1976). Bacteriophage Mu-1, a tool to transpose and to localize bacterial genes. *J. Mol. Biol.* **104,** 525–539.

Faelen, M., and Toussaint, A. (1978). Mu mediated deletions in the chromosome of *E. coli. J. Bacteriol.* **136,** 477–483.

Faelen, M., and Toussaint, A. (1980a). Inversions induced by bacteriophage Mu-1 in the chromosome of *E. coli* K12. *J. Bacteriol.* **142,** 391–399.

Faelen, M., and Toussaint, A. (1980b). Temperate phage D108 induces chromosomal rearrangements. *J. Bacteriol.* **143,** 1029–1030.

Faelen, M., Toussaint, A., and Couturier, M. (1971). Mu-1 promoted integration of a λ-*gal* phage in the chromosome of *E. coli. Mol. Gen. Genet.* **113,** 367–370.

Faelen, M., Toussaint, A., and De Lafonteyne, J. (1975). Model for the enhancement of λ-*gal* integration into partially induced Mu lysogens. *J. Bacteriol.* **121,** 873–882.

Faelen, M., Toussaint, A., Van Montagu, M., Van Den Elsacker, S., Engler, G., and Schell, J. (1977). *In vivo* genetic engineering: Mu mediated transposition of chromosomal DNA onto transmissible plasmids. *In* "DNA Insertion Elements, Plasmids and Episomes" (A. I. Bukhari, J.

A. Shapiro, and S. Adhya, eds.), pp. 521–530. Cold Spring Harbor Lab., Cold Spring Harbor, New York.

Faelen, M., Résibois, A., and Toussaint, A. (1978a). Mini-Mu: an insertion element derived from temperate phage Mu-1. *Cold Spring Harbor Symp. Quant. Biol.* **43,** 1169–1177.

Faelen, M., Huisman, O., and Toussaint, A. (1978b). Involvement of phage Mu-1 early functions in Mu mediated chromosomal rearrangements. *Nature (London)* **271,** 580–582.

Faelen, M., Toussaint, A., and Résibois, A. (1979). Mini-muduction: a new mode of gene transfer mediated by mini-Mu. *Mol. Gen. Genet.* **176,** 191–197.

Faelen, M., Toussaint, A., Lefèbvre, N., Mergeay, M., Braipson-Thiry, J., and Thiry, G. (1981a). Certaines souches de *Erwinia* sont sensibles au bactériophage Mu. *Arch. Int. Physiol. Biochim.* **89,** 55–56.

Faelen, M., Mergeay, M., Gerits, J., Toussaint, A., and Lefèbvre, N. (1981b). Genetic mapping of a mutation conferring sensitivity to bacteriophage Mu in *Salmonella typhimurium* LT2. *J. Bacteriol.* **146,** 914–919.

Fitts, R., and Taylor, A. L. (1980). Integration of bacteriophage Mu at host chromosomal replication forks during lytic development. *Proc. Natl. Acad. Sci. USA* **77,** 2801–2805.

Foster, T. J., Lundblad, V., Hanley-Way, S., Halling, S. M., and Kleckner, N. (1981). Three Tn*10*-associated excision events: relationship to transposition and role of direct and inverted repeats. *Cell* **23,** 215–227.

Franklin, N. C., Dove, W. F., and Yanofsky, C. (1965). The linear insertion of a prophage into the chromosome of *E. coli* shown by deletion mapping. *Biochem. Biophys. Res. Commun.* **18,** 910–923.

Galas, D., and Chandler, M. (1981). On the molecular mechanism of transposition. *Proc. Natl. Acad. Sci. USA* **78,** 4858–4862.

George, M., and Bukhari, A. I. (1981). Heterogenous host DNA attached to the left end of mature bacteriophage Mu DNA. *Nature (London)* **292,** 175–176.

Ghelardini, P., Paolozzi, L., and Liebart, J. C. (1979). Restoration of DNA synthesis at non permissive temperature and of UV resistance induced by bacteriophage Mu-1 in *E. coli lig ts7* strain. *Ann. Microbiol. (Paris)* **130,** 275–285.

Ghelardini, P., Paolozzi, L., and Liebart, J. C. (1980). Restoration of ligase activity in *E. coli* K12 *lig ts7* strain by bacteriophage Mu-1 and cloning of a DNA fragment harbouring the Mu "lig" gene. *Nucleic Acids Res.* **8,** 3157–3173.

Ghosal, D., Sommer, H., and Saedler, H. (1979). Nucleotide sequence of the transposable DNA element IS2. *Nucleic Acids Res.* **6,** 1111–1122.

Gill, G. S., Hull, R. C., and Curtiss, R., III. (1981). Mutator bacteriophage D108 and its DNA; an electron microscopic characterisation. *J. Virol.* **37,** 420–430.

Giphart-Gassler, M., and van de Putte, P. (1979). Thermoinducible expression of cloned early genes of bacteriophage Mu. *Gene* **7,** 33.

Giphart-Gassler, M., Reeve, J., and van de Putte, P. (1981a). Polypeptides encoded by the early region of bacteriophage Mu synthesized in minicells of *E. coli*. *J. Mol. Biol.* **145,** 165–191.

Giphart-Gassler, M., Wijffelman, C., and Reeve, J. (1981b). Structural polypeptides and products of late genes of bacteriophage Mu: characterization and functional aspects. *J. Mol. Biol.* **145,** 139–163.

Giphart-Gassler, M., Plasterk, R. H. A., and van de Putte, P. (1982). G inversion in bacteriophage Mu: a novel way of gene splicing. *Nature* **297,** 339–342.

Goosen, T. (1978). Replication of bacteriophage Mu: direction and possible location of the origin. *In* "DNA Synthesis: Present and Future" (I. Molineux and M. Kohiyama, eds.), pp. 121–126. Plenum, New York.

Goosen, T., Giphart-Gassler, M., and van de Putte, P. (1982). Bacteriophage Mu DNA replication is stimulated by non-essential early functions. *Mol. Gen. Genet.* **186,** 135–139.

Grindley, N., and Sherratt, D. (1978). Sequence analysis at IS*1* insertion sites: models for transposition. *Cold Spring Harbor Symp. Quant. Biol.* **43,** 1257–1261.

Harshey, R. M., and Bukhari, A. I. (1981). A mechanism of DNA transposition. *Proc. Natl. Acad. Sci. USA* **78,** 1090–1094.

Hattman, S. (1979). Unusual modification of bacteriophage Mu DNA. *J. Virol.* **32,** 468–475.

Hattman, S. (1980). Specificity of the bacteriophage Mu *mom*$^+$-controlled DNA modification. *J. Virol.* **34,** 277–279.

Hattman, S. (1982). DNA methylase dependent transcription of the phage Mu *mom* gene. *Proc. Natl. Acad. Sci. USA* (in press).

Howe, M. M. (1972). "Genetic Studies on Bacteriophage Mu." Ph.D. Dissertation, Massachusetts Institute of Technology, Cambridge.

Howe, M. M. (1973a). Prophage deletion mapping of bacteriophage Mu-1. *Virology* **54,** 93–101.

Howe, M. M. (1973b). Transduction by bacteriophage Mu-1. *Virology* **55,** 103–117.

Howe, M. M., and Bade, E. G. (1975). Molecular biolgoy of bacteriophage Mu. *Science (Washington, D.C.)* **190,** 624–632.

Howe, M. M., and Schumm, J. W. (1980). Transposition of bacteriophage Mu. Properties of λ phages containing both ends of Mu. *Cold Spring Harbor Symp. Quant. Biol.* **45,** 337–346.

Howe, M. M., and Zipser, D. (1974). Host deletions caused by the integration of bacteriophage Mu-1. *Am. Soc. Microbiol. Abs.* **5,** 208, 235.

Howe, M. M., O'Day, K. J., and Schultz, D. W. (1979a). Isolation of mutations defining five new cistrons essential for development of bacteriophage Mu. *Virology* **93,** 303–319.

Howe, M. M., Schumm, J. W., and Taylor, A. L. (1979b). The S and U genes of bacteriophage Mu are located in the invertible G segment of Mu DNA. *Virology* **92,** 108–124.

Hsu, M., and Davidson, N. (1972). Structure of inserted bacteriophage Mu-1 DNA and physical mapping of bacterial genes by Mu-1 DNA insertion. *Proc. Natl. Acad. Sci. USA* **69,** 2923–2927.

Hsu, M. T., and Davidson, N. (1974). Electron microscope heteroduplex study of the heterogeneity of Mu phage and prophage DNA. *Virology* **58,** 229–239.

Hull, R. A., Gill, G. S., and Curtis, R., III. (1978). Genetic characterization of Mu-like bacteriophage D108. *J. Virol.* **27,** 513–518.

Ilyina, T. S., Romanova, Y. M., and Smirnov, G. B. (1981). The effect of t*nm* mutations of *E. coli* K12 on transposition of various movable genetic elements. *Mol. Gen. Genet.* **183,** 376–381.

Ito, J., Harding, N. E., and Saigo, K. (1978). Ø29 DNA-protein complex and the structure of replicating DNA molecules. *Cold Spring Harbor Symp. Quant. Biol.* **43,** 525–536.

Jordan, E., Saedler, H., and Starlinger, P. (1968). 0° and strong polar mutations in the *gal* operon are insertions. *Mol. Gen. Genet.* **102,** 353–363.

Kahmann, R., and Kamp, D. (1979). Nucleotide sequences of the attachment sites of bacteriophage Mu DNA. *Nature (London)* **280,** 247–250.

Kahmann, R., Kamp, D., and Zipser, D. (1977). Mapping of restriction sites in Mu DNA. *In* "Insertion Elements, Plasmids and Episomes" (A. I. Bukhari, J. A. Shapiro, and S. Adhya, eds.), pp. 335–339. Cold Spring Harbor Lab., Cold Spring Harbor, New York.

Kahn, S. A., and Novick, R. P. (1980). Terminal nucleotide sequences of Tn*551,* a transposon specifying erythromycin resistance in *Staphylococcus aureus.* Homology with Tn*3*. *Plasmid* **4,** 148–154.

Kamp, D. (1981). Invertible DNA: the G segment of bacteriophage Mu. *In* "Microbiology 1981" (D. Schlessinger, ed.), pp. 73–75. ASM Publications, Washington, D.C.

Kamp, D. and Kahmann, R. (1980). Two pathways in bacteriophage Mu transposition? *Cold Spring Harbor Symp. Quant. Biol.* **45,** 329–335.

Kamp, D., and Kahmann, R. (1981). The relationship of two invertible segments in bacteriophage Mu and *Salmonella typhimurium. Mol. Gen. Genet.* **184,** 564–566.

Kamp, D., Kahmann, R., Zipser, T. R., Broker, L. T., and Chow, L. T. (1978a). Inversion of the G segment of phage Mu controls phage infectivity. *Nature (London)* **271,** 577–580.

Kamp, D., Chow, L. T., Broker, T. R., Kwoh, D., Zipser, D., and Kahmann, R. (1978b). Site specific recombination in phage Mu. *Cold Spring Harbor Symp. Quant. Biol.* **43,** 1159–1168.

Khatoon, H., and Bukhari, A. I. (1978). Bacteriophage Mu-induced modification of DNA is dependent upon a host function. *J. Bacteriol.* **136,** 423–428.

Khatoon, H., and Bukhari, A. I. (1981). DNA rearrangements associated with reversion of bacteriophage Mu induced mutations. *Genetics* **98,** 1–24.

Kleckner, N. (1981). Transposable elements in prokaryotes. *Annu. Rev. Genet.* **15,** 341–404.

Kwoh, D. Y., and Zipser, D. (1979). Specific binding of Mu repressor to DNA. *Nature (London)* **277,** 489–491.

Kwoh, D. Y., and Zipser, D. (1981). Identification of the *gin* protein of bacteriophage Mu. *Virology* **114,** 291–296.

Leach, D., and Symonds, N. (1979). The isolation and characterization of a plaque forming derivative of bacteriophage Mu carrying a fragment of Tn*3* conferring ampicillin resistance. *Mol. Gen. Genet.* **172,** 179–184.

Ljungquist, E., and Bukhari, A. I. (1977). State of prophage Mu DNA upon induction. *Proc. Natl. Acad. Sci. USA* **74,** 3143–3147.

Ljungquist, E., and Bukhari, A. I. (1979). Behavior of bacteriophage Mu DNA upon infection of *E. coli* cells. *J. Mol. Biol.* **133,** 339–357.

Ljungquist, E., Khatoon, H., Dubow, M., Ambrosio, L., De Bruijn, F., and Bukhari, A. I. (1978). Integration of bacteriophage Mu DNA. *Cold Spring Harbor Symp. Quant. Biol.* **43,** 1151–1158.

Magazin, M., and Allet, B. (1978). Synthesis of bacteriophage Mu proteins *in vitro*. *Virology* **85,** 84–97.

Magazin, M., Reeve, J. N., Maynard-Smith, S., and Symonds, N. (1978). Bacteriophage Mu encoded polypeptides synthesized in infected cells. *FEMS Microbiol. Lett.* **4,** 5–9.

Martuscelli, J., Taylor, A. L., Cummings, D. J., Chapman, V. A., De Long, S. S., and Canedo, L. (1971). Electron microscopic evidence for linear insertion of bacteriophage Mu-1 in lysogenic bacteria. *J. Virol.* **8,** 551–563.

Maynard-Smith, S., Leach, D., Coelho, A., Carey, J., and Symonds, N. (1980). The isolation and characteristics of plasmids derived from the insertion of MupAp1 into pML2: their behaviour during transposition. *Plasmid* **4,** 34–50.

Miller, H. I., and Friedman, D. I. (1980). An *E. coli* gene product required for λ site-specific recombination. *Cell* **20,** 711–719.

Mise, K. (1971). Isolation and characterization of a new generalized transducing bacteriophage different from P1 in *E. coli. J. Virol.* **7,** 168–175.

O'Day, K. J., Schultz, D. W., and Howe, M. M. (1978). A search for integration deficient mutants of bacteriophage Mu-1. *In* "Microbiology 1978" (D. Schlessinger, ed.), pp. 48–51. ASM Publications, Washington, D.C.

O'Day, K., Schultz, D., Ericsen, W., Rawluk, L., and Howe, M. (1979). Correction and refinement of the genetic map of bacteriophage Mu. *Virology* **93,** 320–328.

Ohtsubo, H., Ohmori, H., and Ohtsubo, E. (1978). Nucleotide sequence analysis of Tn*3* (Ap): implications for insertion and deletion. *Cold Spring Harbor Symp. Quant. Biol.* **43,** 1269–1277.

Paolozzi, L., Jucker, R., and Calef, E. (1978). Mechanism of phage Mu-1 integration: nalidixic acid treatment causes clustering of Mu-1 induced mutations near replication origin. *Proc. Natl. Acad. Sci. USA* **75,** 4940–4943.

Paolozzi, L., Ghelardini, P., Kepes, A., and Marcovich, H. (1979). The mechanism of integration of phage Mu-1 in the chromosome of *E. coli. Biochem. Biophys. Res. Commun.* **88,** 111–116.

Paolozzi, L., Ghelardini, P., Liebart, J. C., Capozzoni, A., and Marchelli, C. (1980). Two classes of Mu-*lig* mutants: the thermosensitives for integration and replication and the hyperproducers for ligase. *Nucleic Acids Res.* **23,** 5859–5873.

Pato, M. L., and Reich, C. (1982). Instability of transposase activity: evidence from bacteriophage Mu replication. *Cell* **29,** 219–225.

Pato, M. L., and Waggoner, B. T. (1981). Cellular location of Mu DNA replicas. *J. Virol.* **38,** 249–255.

Priess, H., Kamp, D., Kahman, R., Bräuer, B., and Delius, H. (1982). Nucleotide sequence of the immunity region of bacteriophage Mu. *Mol. Gen. Genet.* **186,** 315–321.

Raibaud, O. R. M., Braun-Breton, C., and Schwartz, M. (1979). Structure of the *malB* region in *E. coli* K12. I. Genetic map of the *mal*K-*lam*B operon *Mol. Gen. Genet.* **174,** 241–248.

Razzaki, T., and Bukhari, A. I. (1975). Events following prophage Mu unduction. *J. Bacteriol.* **122,** 437–442.

Reed, R. R., Young, R. A., Steitz, J. A., Grindely, N. D., and Guyer, M. S. (1979). Transposition of the *E. coli* insertion element gamma-delta generates a five base pair repeat. *Proc. Natl. Acad, Sci. USA* **76,** 4882–4886.

Reeve, J. (1977). Bacteriophage infection of mini-cells, a general method for identification of *in vivo* bacteriophage directed polypeptide biosynthesis. *Mol. Gen. Genet.* **158,** 73–79.

Résibois, A., Toussaint, A., Desmet, L., and Lefèbvre, N. (1978). Chromosomal rearrangements by an IS2 insertion in phage Mu-1. *Gene* **4,** 51–68.

Résibois, A., Toussaint, A., Van Gijsegem, F., and Faelen, M. (1981). Physical characterization of mini-Mu and mini-D108 derivatives. *Gene* **14,** 103–113.

Résibois, A., Toussaint, A., and Colet, M. (1982). DNA structures induced by mini-Mu replication. *Virology* **117,** 329–340.

Résibois, A., Colet, M., and Toussaint, A. (1982). Localisation of mini-Mu in its replication intermediates. *EMBO J.* 8, 965–969.

Rosenberg, M., Court, D., Shimatake, H., Brady, C., and Wulff, D. L. (1978). The relationship between function and DNA sequence in an intercistronic regulatory region in phage λ. *Nature* (London) **272,** 414–423.

Salas, M., Mellado, R. P., Vinuela, E. V., and Sogo, J. M. (1978). Characterization of a protein covalently linked to the 5' termini of the DNA of *Bacillus subtilis* phage Ø29. *J. Mol. Biol.* **127,** 411–436.

Schmitt, R., Alterbuchner, J., Wiebauer, K., Arnold, W., Pülher, A., and Schöffl, F. (1980). Basis of transposition and gene amplification by Tn*1721* and related tetracycline-resistance transposons. *Cold Spring Harbor Symp. Quant. Biol.* **45,** 59–65.

Schröder, W., and van de Putte, P. (1974). Genetic study of prophage excision with a temperature inducible mutant of Mu-1. *Mol. Gen. Genet.* **130,** 99–104.

Schröder, W., Bade, E. G., and Deluis, H. (1974). Participation of *E. coli* DNA in the replication of temperate bacteriophage Mu-1. *Virology* **60,** 534–542.

Schumann, W., and Bade, E. G. (1979). *In vitro* constructed plasmids containing both ends of bacteriophage Mu DNA express phage functions. *Mol. Gen. Genet.* **169,** 97–105.

Schumm, J. W., and Howe, M. M. (1980). Analysis of integration and exision of bacteriophage Mu: a new approach. *In* "Microbiology 1980" (D. Schlessinger, ed.), pp. 237–240. ASM Publications, Washington, D.C.

Schumm, J. W., and Howe, M. M. (1981). Mu-specific properties of λ phages containing both ends of Mu depend on the relative orientation of Mu end DNA fragments. *Virology* **114,** 429–450.

Scott, S. H., and Howe, M. M., (1982) Bacteriophage Mu T mutants are defective in synthesis of the major head polypeptide. *Virology* **120,** 264–268.

Shapiro, J. A. (1979). Molecular model for the transposition and replication of bacteriophage Mu and other transposable elements. *Proc. Natl. Acad. Sci. USA* **76,** 1933–1937.

Silhavy, J., Brickman, E., Bassford, P., Jr., Casadaban, M. J., Shuman, H. A., Schwartz, V., Guarente, L., Schwartz, M., and Beckwith, J. R. (1979). Structure of the *mal*B region in *E. coli* K12. II. Genetic map of the *malE, F, G* operon. *Mol. Gen. Genet.* **174,** 249–259.

Southern, E. M. (1975). Detection of specific sequences among DNA fragments. *J. Mol. Biol.* **98,** 503–517.

Symonds, N., and Coelho, A. (1978). Role of the G segment in the growth of phage Mu. *Nature (London)* **271,** 573–574.

Taylor, A. L. (1963). Bacteriophage-induced mutation in *E. coli. Proc. Natl. Acad. Sci. USA* **50,** 1043–1051.

Teifel, J., and Schmieger, H. (1981). The origin of the DNA in transducing particles of bacteriophage Mu; density gradient analysis of DNA. *Mol. Gen. Genet.* **184,** 312–313.

To, C. A., Eisenstark, A., and Toreci, H. (1966). Structure of mutator phage Mu-1 of *E. coli. J. Ultrastruct. Res.* **14,** 441–448.

Toussaint, A. (1976). The DNA modification function of temperate phage Mu-1. *Virology* **70,** 17–27.

Toussaint, A. (1977). The modification function of bacteriophage Mu-1 requires both a bacterial and a phage function. *J. Virol.* **23,** 825–826.

Toussaint, A., and Faelen, M. (1973). Connecting two unrelated DNA sequences with a Mu dimer. *Nature (London) New Biol.* **242,** 1–4.

Toussaint, A., and Faelen, M. (1974). The dependence of temperate phage Mu-1 upon replication functions of *E. coli* K12. *Mol. Gen. Genet.* **131,** 209–214.

Toussaint, A., and Faelen, M. (1981). Formation and resolution of cointegrates upon transposition of mini-Mu. *In* "Microbiology 1981" (D. Schlessinger, ed.), pp. 69–72. ASM Publications, Washington, D.C.

Toussaint, A., Faelen, M., and Bukhari, A. I. (1977). Mu mediated illegitimate recombination as an integral part of the Mu life cycle. *In* "DNA Insertion Elements, Plasmids and Episomes" (A. I. Bukhari, J. A. Shapiro, and S. Adhya, eds.), pp. 275–285. Cold Spring Harbor Lab., Cold Spring Harbor, New York.

Toussaint, A., Lefèbvre, N., Scott, J., Cowan, J. A., De Bruijn, F., and Bukhari, A. I. (1978). Relationships between temperate phages Mu and P1. *Virology* **89,** 146–161.

Toussaint, A., Desmet, L., and Faelen, M. (1980). Mapping of the modification function of temperate phage Mu-1. *Mol. Gen. Genet.* **177,** 351–353.

Toussaint, A., Desmet, L., Van Gijsegem, F., and Faelen, M. (1981). Genetic analysis of Mu or mini-Mu containing F′ *pro lac* episomes after prophage induction. *Mol. Gen. Genet.* **181,** 201–206.

van de Putte, P., and Gruijthuijsen, M. (1972). Chromosome mobilization and integration of F-factors in the chromosome of *recA* strains of *E. coli* under the influence of bacteriophage Mu-1. *Mol. Gen. Genet.* **118,** 173–181.

van de Putte, P., Westmaas, G., Giphart-Gassler, M., and Wijffelman, C. (1977a). On the *kil* gene of bacteriophage Mu. *In* "DNA Insertion Elements, Plasmids and Episomes" (A. I. Bukhari, J. A. Shapiro, S. Adhya, eds.), pp. 287–294. Cold Spring Harbor, New York.

van de Putte, P., Westmaas, G. C., and Wijffelman, C. (1977b). Transfection with Mu-DNA. *Virology* **81,** 152–159.

van de Putte, P., Giphart-Gassler, M., Goosen, T., Van Meeteren, R., and Wijffelman, C. (1978). *In* "Ingegration and Excision of DNA Molecules" (P. Hofschneider, and P. Starlinger, eds.), pp. 33–40. Springer–Verlag, Berlin.

van de Putte, P., Giphart-Gassler, M., Goosens, N., Goosens, T., and Van Leerdam, E. (1980a). Regulation of integration and replication functions of bacteriophage Mu. *Cold Spring Harbor Symp. Quant. Biol.* **45,** 347–353.

van de Putte, P., Cramer, S., and Giphart-Gassler, M. (1980b). Invertible DNA determines host specificity of bacteriophage Mu. *Nature (London)* **286,** 218–222.

van Gijsegem, F., and Toussaint, A. (1982). Chromosome transfer and R-prime formation by an RP4::mini-Mu derivative in *Escherichia coli, Salmonella typhimurium, Klebsiella pneumoniae* and *Proteus mirabilis. Plasmid* **7,** 30–44.

van Leerdam, E., Goosen, T., Plasterk, R. H. A., and van de Putte, P. (1981). Cloning of both ends and the thermo-inducible genes *A* and *B* of bacteriophage Mu on a multicopy plasmid. *Gene* **13,** 111–114.

van Leerdam, E., Karreman, C., and van de Putte, P. (1982). Ner, a cro-like gene of bacteriophage Mu. *Virology* (in press).

van Meeteren, A. (1980). "Transcription of bacteriophage Mu." Ph.D. Dissertation, Rijksuniversiteit, Leiden, The Netherlands.

van Meeteren, R., Giphart-Gassler, M., and van de Putte, P. (1980). Transcription of bacteriophage Mu. II. Transcription of the repressor gene. *Mol. Gen. Genet.* **179,** 185–190.

van Vliet, F., Couturier, M., De Lafonteyne, J., and Jedlicki, E. (1978). Mu-1 directed inhibition of DNA breakdown in *E. coli recA* cells. *Mol. Gen. Genet.* **106,** 109–112.

Waggoner, B. T., and Pato, M. L. (1978). Early events in the replication of Mu prophage DNA. *J. Virol.* **27,** 587–594.

Waggoner, B. T., Gonzales, N. S., and Taylor, A. L. (1974). Isolation of heterogeneous circular DNA from induced lysogens of bacteriophage Mu-1. *Proc. Natl. Acad. Sci. USA* **71,** 1255–1259.

Waggoner, B. T., Pato, M. L., and Taylor, A. L. (1977). Characterization of covelently closed circular DNA molecules isolated after bacteriophage Mu induction. *In* "Insertion Elements, Plasmids and Episomes" (A. I. Bukhari, J. A. Shapiro, S. Adhya, eds.), pp. 263–274. Cold Spring Harbor Lab., Cold Spring Harbor, New York.

Waggoner, B. T., Pato, M. L., Toussaint, A., and Faelen, M. (1981). Replication of mini-Mu. *Virology* **113,** 379–387.

Wijffelman, C., and Lotterman, B. (1977). Kinetics of Mu DNA synthesis. *Mol. Gen. Genet.* **151,** 169–174.

Wijffelman, C., and van de Putte, P. (1977). Asymmetric hybridization of Mu strands with short fragments synthesized during Mu DNA replication. *In* "DNA Insertion Elements, Plasmids and Episomes" (A. I. Bukhari, J. A. Shapiro, S. L. Adhya, eds.), pp. 329–333 Cold Spring Harbor Lab., Cold Spring Harbor, New York.

Wijffelman, C. A., Westmaas, G. C., and van de Putte, P. (1972). Vegetative recombination of bacteriophage Mu-1 in *E. coli. Mol. Gen. Genet.* **116,** 40–46.

Wijffelman, C. A., Westmaas, G. C., and van de Putte, P. (1973). Similarity of vegetative map and prophage map of bacteriophage Mu-1. *Virology* **54,** 125–134.

Wijffelman, C., Gassler, M., Stevens, W. F., and van de Putte, P. (1974). On the control of transcription of bacteriophage Mu. *Mol. Gen. Genet.* **131,** 85–96.

Williams, J. K., and Radding, C. M. (1981). Partial purification and properties of an exonuclease inhibitor induced by bacteriophage Mu-1. *J. Virol.* **39,** 548–558.

Yun, T., and Vapnek, D. (1977). Electron microscopic analysis of bacteriophage P1, P1CM, P7. Determination of genome size, sequence homology and location of antibiotic resistance determinants. *Virology* **77,** 376–385.

CHAPTER 4

Prokaryotic IS Elements

SHIGERU IIDA
JÜRG MEYER
WERNER ARBER

I. Introduction

The concept of chromosomes as invariable entities of genetic information transmitted unchanged from generation to generation has now been abandoned in

Mobile Genetic Elements

ISBN 0-12-638680-3

favor of a more dynamic view. Studies on gene expression and gene organization have revealed DNA rearrangements that have occurred in a number of different ways. Some of these chromosome rearrangements are brought about by mobile genetic elements that are characterized by their ability to insert as discrete DNA segments at other locations in the genome. Prokaryotes contain several classes of such transposable elements as natural constituents of the chromosome or of extrachromosomal plasmids and bacteriophage genomes.

IS elements (insertion sequences) form one class defined as mobile genetic elements containg no detectable genes unrelated to insertion functions (Campbell *et al.*, 1977, 1979). From this arbitrary definition, it is clear that a direct genetic selection for the presence of an IS element is not possible. Tn elements (transposons) form another class that contain additional, detectable gene(s) unrelated to insertion functions. Bacteriophage Mu is representative of a third class of more complex transposable elements. Because IS elements are distinguished from the other transposable elements solely on the basis of the absence of a detectable phenotype, studies with IS elements are more difficult.

IS elements were first detected as a cause of spontaneous mutations in *Escherichia coli* (see the review by Starlinger and Saedler, 1976). Insertion of 0.8–1.5 kb (kilobase pairs) long, discrete DNA segments led to interruption of structural genes or to strong polar effects. Reversion to the original phenotype was observed (e.g., after excision of the elements). Occasionally, insertion of an IS element in front of a gene resulted in an activation of gene expression. Deletion formation was stimulated in the immediate vicinity of IS elements.

Our knowledge of IS elements was expanded by studies on repeated DNA sequences and their involvement in integration and segregation of F and R plasmids (Davidson *et al.*, 1975; Starlinger and Saedler, 1976; Ohtsubo and Ohtsubo, 1977). Such DNA elements were shown to mediate recombination processes leading to gross DNA rearrangements such as replicon fusion and segregation, deletion, and inversion, and were subsequently identified as IS elements. For example, IS*2*, IS*3*, and γδ were found in F plasmids, and IS*1* was found in R plasmids and in bacteriophage P1 DNA.

More recently, a third line of observations has increased our insight into the biology of IS elements: Some transposons are flanked by direct or inverted repeats of IS elements (see the review by Calos and Miller, 1980a), for example, IS*1* as direct repeats in Tn*9*. The detectable (and particularly the selectable) markers contained in these Tn elements have facilitated studies on the activities of the flanking IS element.

The characteristics of bacterial IS elements have been repeatedly reviewed in the last few years (Calos and Miller, 1980a; Cohen and Shapiro, 1980; Kopecko, 1980; Starlinger, 1980; Kleckner, 1981). This review will be structured as follows: First, we will give a somewhat simplified picture of the present knowledge and widely accepted concepts of the structure and function of IS elements. Available experimental data and interpretation will then follow. Next, we will discuss biological functions, particularly gene transmission and expression, in light of the IS functions outlined before. Finally, a synoptical view of these data and concepts will lead to a discussion of consequences, and to questions and speculations.

II. Discussion of Present Knowledge and Concepts

IS elements are discrete genetic entities able to insert into new sites on the same or another replicon without leaving the original location. This translocation occurs rarely and proceeds by mechanisms largely independent of the general, homology-dependent recombination pathways. Transposition involves (*1*) local replication of only the IS element and (*2*) precise DNA cleavage at the ends of the IS element and their ligation to the target DNA, which is cut in a specific way. Little is known about the molecular mechanisms of these replication and break–joining (recombination) steps, and their spatial and sequential relation may be different with different IS elements. Characteristically, the transposition of all IS elements so far studied results in the duplication of a small, discrete number of base pairs (bp) of the target sequence, probably due to specific staggered cutting, ligation, and replication. Sometimes, an additional type of recombination may be required for the resolution of intermediate structures.

IS elements provide sites essential for transposition. Because the ends of IS elements are precisely joined to the target DNA molecules, they have to be recognized specifically. The termini of all IS elements sequenced so far share the structural feature of carrying inverted repeats of about 10–40 bp. Alterations within these inverted repeats often affect transposition activity. Another site in the unique part of a certain IS element is known to be involved in the resolution of transposition intermediates.

Most IS elements also provide one or a few functions. The primary sequences of the IS elements analyzed reveal that one to three polypeptides could be encoded. The activities of a few gene products specified by IS elements have been either detected genetically or identified biochemically. The interaction of these IS functions with host factors and their role in target site selection and in breaking, joining to, and replicating the IS element, in resolving the transposition intermediate, and in regulating transposition activity are largely unknown.

Other IS-mediated DNA rearrangements, such as cointegration of two replicons, and deletion or inversion of sequences adjacent to the IS element are considered related processes and are believed to be initiated with the same mechanisms as transposition. However, IS elements also contribute to DNA rearrangements by providing homologous DNA for general recombination systems. Precise and nearly precise excisions of an IS element are usually less frequent than transposition and are considered to be accomplished by recombination processes unrelated to transposition functions. IS-mediated DNA rearrangements, together with *rec*-dependent recombination between IS elements and excision of the elements, lead to gene duplication and further amplification, the generation and decay of IS-flanked transposons, and other, more complex DNA restructuring.

IS elements usually contain translational and transcriptional stop signals and sometimes start signals. Depending on their location, they may interfere with the regulation of expression of adjacent genes. This regulatory potential, together with the DNA rearrangements associated with IS elements, may help to expand a prokaryotic organism's ability to adapt to new environments and to provide a selective advantage to the population of organisms harboring mobile genetic elements. Furthermore, the IS-mediated association of functional genes with transmissible episomes facilitates the horizontal spreading of bacterial genes even to distantly related organisms. This may reflect the evolutionary significance of IS elements and may explain why they are so widespread, numerous, and diverse. This view has been challenged by the concept that the ability of IS elements to replicate and transpose would suffice to ensure their survival as purely "parasitic" DNA segments.

Different IS elements transpose with different frequencies, and the frequencies may depend on the genotype and the physiological state of the host organism, as well as on the particular structures of the donor and target sequences. Frequencies varying between 10^{-3}, or perhaps more usually 10^{-5}, and 10^{-9} per cell division and per IS element have been reported. Because IS elements cannot be selected for directly, many studies have monitored the occurrence of IS-promoted DNA rearrangements such as cointegration, deletion, and inversion. These and studies on transposons containing IS elements have suggested that some of the properties of IS and Tn elements are either the same or very similar. Many of the notions and concepts about the properties of IS elements represent an extrapolation of results obtained with selectable transposable elements or even of models proposed to interpret such results.

III. Interpretation of Available Data

A. Assignment of IS Elements and Nomenclature

IS elements were originally detected as an acquisition of additional DNA segments when DNA molecules, some of which were known to carry mutations, were physically analyzed (Jordan *et al.*, 1968; Shapiro, 1969; Malamy, 1970; Fiandt *et al.*, 1972; Hirsch *et al.*, 1972).

IS elements were also detected as repetitive DNA segments in electron microscopic studies (Ohtsubo *et al.*, 1974; Davidson *et al.*, 1975; Hu *et al.*, 1975a,b). Identification of such a DNA segment as an IS element can be made structurally or functionally, namely, by (*1*) demonstration of sequence homology to a known IS element (Hu *et al.*, 1975a,b; Saedler *et al.*, 1975) or by (*2*) demonstration of its transposition activity, for example, its translocation to another replicon or its ability to mediate replicon fusion or inversion (Guyer, 1978; Ravetch *et al.*, 1979; Cornelis and Saedler, 1980; Ohtsubo *et al.*, 1980a). Structural identification is usually based on electron microscopic heteroduplex and DNA sequencing studies. The results of restriction analysis and Southern hybridization can occasionally be misleading because of the existence of partially homologous sequences, some of which are unrelated to IS elements (Nisen and Shapiro, 1979).

The terminal inverted repeats of an insertion and the number of base pairs duplicated at the target sequence have been proposed as criteria for the classification of IS elements (Calos and Miller, 1980a). Insertions are also classified as IS elements if they were shown to have duplicated themselves and generated direct repeats at the target sequences (Ravetch *et al.*, 1979; Reed *et al.*, 1979; Ohtsubo *et al.*, 1980a).

If a DNA segment is functionally identified as an IS element and its DNA sequence is shown to be different from that of the known IS elements, it can be called a new IS element. The assignment of a numeral to a new IS element should be given in agreement with the central plasmid registry[1] (Campbell *et al.*, 1979).

The designation for insertion of an IS element in a genome is a double colon, for example, λ :: IS2. The detailed rules for a nomenclature were set up by Campbell *et al.* (1977, 1979).

B. Structural Characteristics

Recent developments in rapid DNA sequencing techniques have allowed the elucidation of the primary structure of several IS elements. However, sequence information is still limited. Table I summarizes the results on those IS elements and their naturally isolated variants and relatives of which all or most of the DNA sequence has been determined. It also includes the less well-characterized elements. However, a number of other IS elements, most of which remain unassigned, have been reported. They include insertions with polar effects in several genes of *E. coli* K12 and insertions found in plasmids and phage genomes such as R and F plasmids and P1 and P1-15 phages (see Szybalski, 1977; Arber *et al.*, 1981).

[1]Addressed to Dr. Esther Lederberg, Department of Medical Microbiology, Stanford University Medical School, Stanford, CA 94305.

Table I List of Some Well-Studied IS Elements

IS elements	Natural occurrence and characterization of IS elements[a]	Length (bp)	Inverted repeats (bp)[b]	Coding capacity[c]	Type of rearrangement observed and length (in bp) of duplication generated at the target sequence as detected after[d]: I	II	III	IV	V	Target specificity[e]	References[f]
IS elements entirely characterized											
IS*1*	*Escherichia coli* chromosome, etc.[a]				9	9/8	9	+	+	AT-rich, preferential cleavage at GC, homology to termini of IS*1*	1–20, 52
	IS*1*b on R100[g]	768	20/23	**91,** 167/125, **(96)**		9	•				16
	*lac*IS58 :: IS*1*[g]	768	20/23	**91,** 167/125, **(96)**	9						9
	IS*1*-D on *Shigella dysenteriae* chromosome[g]	768	18/23	**91,** 167/125, **(96)**							18
υξ[h]	*S. dysenteriae* chromosome	766	25/31	**90, 131, (59)**							18
IS2	*E. coli* chromosome, etc.[a]				5			+			3, 6, 21–25
	*gal*OP-308 :: IS2	1327	32/41	**315, (143)**	5						23
IS*4*	*E. coli* K12 chromosome	1426	16/18	442, **(131)**	11/12/13					$AAACN_{2-3}CAN_{11-19}GTTT$ $C(^{T}_{A})A(^{G}_{A})$	3, 26, 27
IS*5*	*E. coli* K12 chromosome, etc.[a]				4						28–32
	phage Mu :: IS*5*[i]	1195	15/16	**338, (118)**	4						30
	pBR322-*res*(*Hha*II) :: IS*5*[i]	1195	15/16	**338, (118)**	4						32
IS*50*R	Tn*5* on pJR67				9	+	+	+			33–36
	phage fd :: Tn*5*	1534	8/9	**480/476,** (43, 51)			9				33
IS*101*	pSC101					5					37, 38
	phage f1 :: IS*101*[j]	201	31/37			5[j]					38
IS*102*[k]	pSC101					9		+			39, 40
	IS*102* on pSC101	1057	18/18	**307, (114)**							39
IS*903*[k]	Tn*903* on R6					+	9	+			41–46
	IS*903* on Tn*903*	1057	18/18	**307, (114)**							41, 43
IS*10*R[l]	Tn*10* on R100, R6, pSM14				+	+	9	+	+	AT-rich, NGCTNAGCN, homology to termini of IS*10*	44, 47–54
	IS*10* on Tn*10*	1329	17/22	**402**							**49**[l]
IS elements partially characterized[m]											
IS*3*	*E. coli* chromosome[a]	1400			+			+			3, 12, 55, 56
	*gal*OP-308 :: IS2-437 :: IS*3*		32/38		3 or 4						56

IS*30*	*E. coli* chromosome[a]	1250		+			+	57–59
	Tn*2671* on NR1-Basel							58, 59
	Phage P1 :: IS*30*		23/26	2				57, 60
γδ	*E. coli* K12 chromosome[a]	5700		+	+		+	61–66
	pBR322 :: γδ		35/35	5				65
IS*8*[n]	RP4 (= R68) and R68.45	1750			+		+	67, 68
IS*21*[n]	R68 and R68.45	2100		+				69
IS26	Tn*2680* on Rts1	820			+	+	+	70, 71
	pBR322 :: IS*26*		14/14		8			71
ISR*1*	*Rhizobium lupini*	1150		+			+	72

[a] For more detailed information on the natural occurrence of IS*1*, IS2, IS*3*, IS*5*, IS*30*, and γδ, see Table II.

[b] The notation 20/23 indicates that 20 out of 23 bases from each terminus could form normal hydrogen-bonded pairs.

[c] The coding capacity of one strand is given as the number of amino acids of the expected polypeptide(s). IS*1* and IS*50*R could code for two larger overlapping proteins in the same translational frame. These proteins would differ at their N-terminal parts by 42 and 4 amino acids, respectively (Auerswald *et al.*, 1981; Johnsrud, 1979; Ohtsubo *et al.*, 1981a). IS*1* has a second open reading frame on this strand. The coding capacity of the complementary strand is given in parentheses (cf. Fig. 1). The boldface numbers indicate that the coding region fits with the one as determined by the method of J. Shepherd (1981a,b, personal communication).

[d] I, transposition; II, cointegration; III, transposition of a transposon mediated by the indicated IS element; IV, deletion; V, inversion. The arabic numbers given in the respective columns indicate the length in base pairs of the duplication of the target sequence, as measured on the particular rearrangement, whereas + indicates that the rearrangement activity was reported. The number 9 on the first line of IS*1* in column I, for example, indicates that the generation of the 9-bp duplication at the target site of IS*1* integration was observed not only in the case of *lac*IS58 :: IS*1* but also in other instances, whereas the symbol + on the first line of γδ in column I indicates that transposition of γδ was repeatedly observed but that the 5-bp duplication at the target sequences was observed only in pBR322 :: γδ.

[e] The criteria for target site selection are listed. The nucleotide sequences given indicate a consensus sequence found at most integration sites. For IS*5* and IS*10*, it is equal to the target DNA duplication. For IS*4* it is symmetrically located outside, whereby the central CA are found at the beginning of the duplicated sequence.

[f] 1, Calos *et al.* (1978); 2, Cornelis and Saedler (1980); 3, Fiandt *et al.* (1972); 4, Galas *et al.* (1980); 5 Grindley (1978); 6, Hirsch *et al.* (1972); 7, Iida and Arber (1980); 8, Iida *et al.* (1980a, 1981a,d); 9, Johnsrud (1979); 10, Johnsrud *et al.* (1978); 11, Kühn *et al.* (1979); 12, Malamy *et al.* (1972); 13, Meyer *et al.* (1980); 14, Mickel *et al.* (1977); 15, Nyman *et al.* (1981); 16, Ohtsubo and Ohtsubo (1978); 17, Ohtsubo *et al.* (1980b); 18, Ohtsubo *et al.* (1981a); 19, Reif and Saedler (1975, 1977); 20, Saedler *et al.* (1975, 1981); 21, Ahmed *et al.* (1981); 22, Delius *et al.* (1980); 23, Ghosal *et al.* (1979a); 24, Rosenberg *et al.* (1978); 25, Peterson *et al.* (1979); 26, Habermann *et al.* (1979); 27, Klaer *et al.* (1980, 1981a,b); 28, Blattner *et al.* (1974); 29, Chow and Broker (1978); 30, Engler and van Bree (1981); 31, Lusky *et al.* (1981); 32, Schoner and Kahn (1981); 33, Auerswald *et al.* (1981); 34, Berg *et al.* (1975, 1981); 35, Isberg and Syvanen (1981); 35a, Meyer *et al.* (1979); 36, Schaller (1979); 37, Fischhoff *et al.* (1980); 38, Ravetch *et al.* (1979); 39, Bernardi and Bernardi (1981a,b); 40, Ohtsubo *et al.* (1980a); 41, Grindley and Joyce (1980, 1981); 42, Nomura *et al.* (1978); 43, Oka *et al.* (1978, 1981); 44, Sharp *et al.* (1973); 45, Syvanen (1980); 46, Young *et al.* (1980); 47, Botstein and Kleckner (1977); 48, Foster *et al.* (1981a); 49, Kleckner (1979, 1981); 50, Kleckner and Ross (1979); 51, Kleckner *et al.* (1975, 1981); 52, Iida (1982); 53, Watanabe *et al.* (1972); 54, Halling and Kleckner (1982); 55, Deonier *et al.* (1979); 56, Sommer *et al.* (1979b); 57, Arber *et al.* (1979, 1981); 58, Hänni *et al.* (1982); 59, Iida *et al.* (1981b); 60, P. Caspers and W. Arber, unpublished; 61, Davidson *et al.* (1975); 62, Guyer (1978); 63, Guyer *et al.* (1981); 64, Palchaudhuri *et al.* (1976); 65, Reed *et al.* (1979); 66, Reed (1981a); 67, Depicker *et al.* (1980); 68, Leemans *et al.* (1980); 69, Willetts *et al.* (1981); 70, Iida *et al.* (1980b, 1982b); 71, B. Mollet, S. Iida, and W. Arber, unpublished; 72, Priefer *et al.* (1981).

[g] DNA sequences of IS*1* in *lac*IS58 :: IS*1* and IS*1*-D differ from IS*1*b on R100 by 7 and 10 bp, respectively.

[h] Transposition activities of the IS*1*-related element υξ have not yet been determined.

[i] Sequences of two IS*5* elements reported differ by 2 bp.

[j] Phage f1 :: IS*101* was derived from a f1 :: pSC*101* cointegrate.

[k] Sequences of the related elements IS*102* and IS*903* differ by 60 bp.

[l] For the complete DNA sequence of IS*10* determined by Kleckner and collaborators (cited by Kleckner, 1981), see Chapter 6.

[m] Only the approximate size of these elements is known.

[n] IS*8* and IS*21* are related and share common sequences.

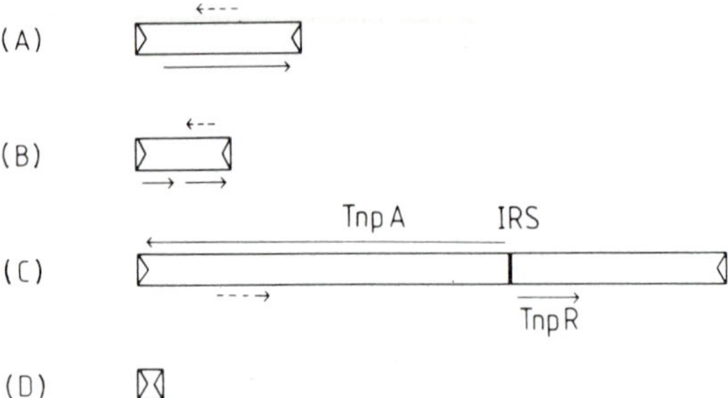

Fig. 1 Simplified common features of IS elements. The open triangles at the ends of elements indicate inverted repeats. Arrows represent the open reading frames (see Table I). (A) shows the structure typical for most of the regular IS elements, such as IS*2*, IS*4*, *IS5*, IS*50*, IS*102*, and IS*903*, whereas (B) shows that of IS*1* and its relatives. (C) gives the features of the $\gamma\delta$ element (and its relative, Tn*3*) where *tnpA* and *tnpR* are the genes for cointegrase and resolvase/repressor, respectively, and IRS identifies an internal resolution site. The IS element shown in (D) carries little besides the terminal inverted repeats, whereas the diffusible transpositional functions are believed to be provided by other elements.

A majority of the IS elements studied range in size between 0.7 and 1.8 kb. Most of them share the following structural features (Fig. 1A,B, Table I).

1. The termini of IS elements carry perfect or nearly perfect inverted repeats of about 10–40 bp. These terminal repeats are believed to serve as recognition sequences for the transposition enzymes (or transposases) in their role of fusing the ends of the IS element with the target DNA.
2. There exist nonsense codons in all three reading frames.
3. One large open reading frame can encode a polypeptide of about 300–400 amino acids.
4. IS elements often carry one smaller open reading frame within the large reading frame, using the same frame but in the opposite direction from the complementary DNA strand.

IS*1* and its relatives differ from this general description with respect to their coding capacity (Table I, Fig. 1B). IS*1* contains two open reading frames in the same direction that could specify shorter peptides of 91 and either 125 or 167 amino acids. Sequencing data on naturally isolated variants of the IS*1* family are already relatively abundant. They include IS*1* from *E. coli* and its relatives isolated from several different microorganisms. Naturally isolated variants of IS*1* are known to have undergone small substitutions, insertions, or deletions of one or a few base pairs. The presence of regions with conserved open reading frames common to all relatives suggests that these sequences are coding for active gene products (Ohtsubo *et al.*, 1981a). It should be pointed out, however, that the transpositional activity of most of the IS*1* relatives sequenced has yet to be demonstrated. Similar conclusions with respect to conserved open reading frames could be drawn by comparing the sequence of the related elements IS*102* and IS*903* (Grindley and Joyce, 1980; Bernardi and Bernardi, 1981a).

A comparison of the primary sequences between IS*2* variants showed small DNA alterations near one end of IS*2* (Musso and Rosenberg, 1977; Ghosal *et al.*, 1979a; C. Sengstag and W. Arber, unpublished results). Several mutant IS*2* elements selected with respect to transcriptional activation of neighboring genes have changes in the terminal sequence. These will be discussed in Section III,D. Data on their transposition functions are scarce.

An interesting variation was reported in IS*50*, the flanking element of Tn*5* (Berg *et al.*, 1981; Isberg and Syvanen, 1981). IS*50*L differs from IS*50*R in only 1 bp (Auerswald *et al.*, 1981; Rothstein and Reznikoff, 1981). This alteration produces a nonsense codon that was shown to affect an IS*50* gene product involved in transposition. At the same time, the mutation generates a better promoter for the expression of the neighboring Km^r (kanamycin-resistance) gene of Tn*5*. Thus, a single-base-pair alteration has resulted here in the loss of the ability of IS*50*L to act as an independent IS element and in the stimulation of the expression of an adjacent gene. Similar functional differences are reported for the two IS*10* elements carried on Tn*10* (Foster *et al.*, 1981a; see Chapter 6).

Another class of naturally isolated mutants of IS elements are those containing an internal deletion or the insertion of an IS element or a transposon. They include IS*1* :: IS*2*,IS*1* :: Tn*2672*, IS*2* :: IS*2*, IS*2* :: IS*3*, and an IS*1* carrying an internal deletion (Sommer *et al.*, 1979b; Ahmed *et al.*, 1981; Saedler *et al.*, 1981; Sommer *et al.*, 1981; Hänni *et al.*, 1982). Mutants of IS elements were also constructed *in vitro* (by DNA manipulation techniques) for the study of transposition functions. These mutants will be discussed later in relation to IS functions (Section III,I).

The structural features so far discussed in this section and illustrated in Figs. 1A and 1B characterize most known IS elements. We will arbitrarily call them *regular* IS elements.

There are two IS elements that have fundamentally different structures. One is the insertion element γδ (Guyer, 1978). It is 5.7 kb long and more complex than regular IS elements, but it still lacks detectable genes unrelated to insertion functions. The γδ element shows a close structural and functional relation to the Ap^r (ampicillin-resistance) transposon Tn*3*. For this reason, the label Tn*1000* was proposed (Reed *et al.*, 1979). Like Tn*3*, γδ carries, in addition to the short terminal inverted repeats, two genes, *tnpA* and *tnpR*, and a site-specific recombination site called *IRS* (Internal Resolution Site) (Fig. 1C) (Chou *et al.*, 1979; Heffron *et al.*, 1979, 1981; Reed *et al.*, 1979; Kostriken *et al.*, 1981; Reed 1981a,b; Reed and Grindley 1981). The *tnpR* gene product of γδ is a polypeptide about 200 amino acids long (Reed, 1981b). It functions as a repressor for both *tnpA* and *tnpR* and as a site-specific recombinase (or resolvase) acting on IRS. The *tnpA* gene product is a long (1015 amino acids long in Tn*3*) polypeptide (Heffron *et al.*, 1979) and functions as a cointegrase, a transposition enzyme that preferentially produces cointegrates. The *tnpR* gene products of γδ and Tn*3* are mutually exchangeable, whereas the *tnpA* gene products are not. Hg^r (mercury-resistance) or Tc^r (tetracycline-resistance) transposons related to γδ have been characterized (Choi *et al.*, 1981; Kitts *et al.*, 1982). Thus, γδ seems to represent the prototype of a different group of transposable elements. For a detailed description and discussion of this Tn*3*/γδ family of movable genetic elements, see Chapter 5.

The other structurally different IS element is the very short IS*101*. IS*101* was originally found as a mediator for cointegrate formation between plasmid pSC101 and bacteriophage fl DNA (Ravetch *et al.*, 1979; Fischhoff *et al.*, 1980). It carries only sites required for transposition, the terminal inverted repeats, and probably an IRS site, whereas transposition functions are likely to be provided in trans by another IS element, γδ (Miller and Cohen, 1980). Whether IS*101* is representative of a whole group of "defective" IS elements providing only sites but no functional genes (Fig. 1D) cannot yet be decided.

C. Sources and Host Range

Some specific IS elements have been isolated from several different bacterial species, in which they were carried either on plasmids or on the chromosome (Table II). Because IS elements could have been transposed into plasmids and phage genomes and become transferred from one species/strain to another, an organism in which a particular IS element was discovered originally need not necessarily be the original host. For instance, the R plasmid R68 and its derivative, R68.45, carrying IS*21* were isolated from *Pseudomonas aeruginosa,* which contains no IS*21* in the chromosome (Holloway and Richmond, 1973; Haas and Holloway, 1976; Willetts *et al.*, 1981). Although only limited data are available concerning the host range of IS elements, we can draw two obvious conclusions:

1. The same kind of IS element, such as IS*1*, can be found on plasmids, phage genomes, and the chromosomes of different bacteria (Table II). Sometimes the isolates from different species represent variants (Cornelis and Saedler, 1980; Iida *et al.*, 1981a; Ohtsubo *et al.*, 1981a).
2. Several IS elements can function efficiently in bacteria that do not contain the same IS element in their chromosome. For example, IS*10* is active in both *E. coli* K12 and *Salmonella typhimurium,* although their chromosomes contain no IS*10* sequence (Ross *et al.*, 1979a). Similarly, IS*50* and IS*26* are active in *E. coli* K12, which carries no such elements in the chromosome (Berg and Drummond, 1978; S. Iida, unpublished). Other IS elements, such as ISR*1*, are active in one microorganism but not in another (Priefer *et al.*, 1981). This difference could be explained either by the degree of expression of the transposition function(s) coded by the IS element or by the host factors involved.

Bacterial DNA has several kinds of repeated segments in the size range of 0.5–5.6 kb (Deonier and Hadley, 1976; Chow, 1977; Ohtsubo and Ohtsubo,

1977). Although some of the large ones may represent the ribosomal RNA genes, it has been speculated that the smaller repeats are IS elements. So far, there are no indications regarding how many different IS elements are carried on a bacterial chromosome, not even for that of the well-studied *E. coli* K12. A systematic search for IS elements in *E. coli* K12 was designed by Arber *et al.* (1979, 1981) using prophage plasmids P1 and its derivative, P1-15, as a target. Besides several known elements, a new insertion sequence, IS*30*, was isolated by this method (Arber *et al.*, 1981; Iida *et al.*, 1981b). Based on the copy number of the known IS elements, one can calculate that at least 1–2% of the *E. coli* K12 chromosome consists of the various IS elements (Table II).

D. Transcriptional Regulation

Several IS elements were discovered because they exert a strong polar effect on the expression of genes located in the same operon distal to the insertion (Jordan *et al.*, 1968; Shapiro, 1969; Malamy, 1970). This polar effect is thought to act, in many cases, on the level of translation due to the presence of nonsense codons found in all three reading frames in the primary sequence of IS elements (Starlinger and Saedler, 1976; Adhya and Gottesman, 1978). Indeed, insertion of IS*1*, Tn*9*, and Tn*10* within a ribosomal RNA operon, the transcripts of which are not translated, showed incomplete polarity, whereas their insertion in other operons is usually polar (Morgan, 1980; Brewster and Morgan, 1981). When it was found that in some cases the polarity can be relieved in *rho*$^-$ bacterial strains, it was concluded that some IS elements carry additional signals for *rho*-dependent termination of transcription (de Crombrugghe *et al.*, 1973; Das *et al.*, 1976, 1977; Besemer and Herpers, 1977; Reyes *et al.*, 1979a). The polar effect of some IS insertions is also suppressed by the antiterminator pN encoded by phage λ (Adhya *et al.*, 1974).

Revertants from polar mutants have been isolated. In some revertants, transcriptional terminators within the IS element were removed by (*1*) precise excision or (*2*) nearly precise excision of the element, (*3*) deletions internal to the element, and (*4*) deletions affecting part of the element together with adjacent DNA (Malamy *et al.*, 1972; Saedler *et al.*, 1972; Ahmed and Scraba, 1978; Ross *et al.*, 1979c; Berg, 1980; Delius *et al.*, 1980; Foster *et al.*, 1981b; Sommer *et al.*, 1981).

Extensive studies on the relief of polar effects caused by IS insertion have been carried out in IS2. Early studies indicated that insertion of IS2 in the *gal* operon in one orientation (called *orientation I*) causes polar effects, whereas the insertion in the other orientation, *orientation II*, leads to a constitutive expression of adjacent genes (Saedler *et al.*, 1974; Starlinger and Saedler, 1976). A similar situation was reported for integrase constitutive mutants of bacteriophage λ (Pilacinski *et al.*, 1977) and in studies on the expression of cloned yeast genes in *E. coli* (Walz *et al.*, 1978; Brennan and Struhl, 1980). However, because this

Table II IS Elements Found in Various Bacterial Strains

IS element	Bacteria	Number of copies: On chromosome[a]	Number of copies: On plasmid	Comments	References[b]
IS*1*	*Escherichia coli* K12	4–10		*lacI* :: IS*1* sequenced	1–4, 36
	E. coli B	14–19			36
	E. coli C	3			4, 36
	E. coli F	≥1			2
	E. coli W	≥1[c]			2
	E. coli Li	—	1 on prophage P1	(variant, G→T at position 757)	5, 6, 18
	E. coli B41	—	2 on EFS0041	Tn*1681*	7, 8
	E. coli 0 : 4	—	2 on R6		9–11
	Shigella dysenteriae	>40		IS*1*-D sequenced	4, 12
	Sh. sonnei	>30			4
	Sh. boydii	2			4
	Sh. flexneri	>40	2 on R100 (NR1)	Tn*2670*, IS*1*b sequenced	4, 13–15
			2 on pSM14 (R14)	Tn*9*	16–18
	Klebsiella aerogenes	1			2, 4
	Serratia marcescens	2[c]			4
	Salmonella typhimurium	≥1[c]			2
	S. paratyphi B	—	3 on R1		11, 19, 20
	S. ordonez	≥1			21
	Yersinia enterocolitica	—	1 on pGC1	Tn*951* (variant, no *Pst*I site)	22
IS2	*E. coli* K12	4–13	1 on F		1, 2, 11, 23, 36
	E. coli B	1			36

	E. coli C	0		36
	E. coli F	≥1		2
	Citrobacter freundii	≥1		2
	Enterobacter aerogenes	≥1		2
	K. aerogenes	≥1		2
	Pseudomonas aeruginosa	≥1		2
	P. putida	≥1		2
IS*3*	*E. coli* K12	5–6	2 on F	23, 24, 36
	E. coli B	4		24, 36
	E. coli C	5		36
IS*4*	*E. coli* K12	1–2		25, 26
IS*5*	*E. coli* K12	10–11		2, 27
	C. freundii	≥1		2
IS*30*	*E. coli* K12	2–8		28, 29
	E. coli C	1		29
γδ	*E. coli* K12	0–3	1 on F	30–32
IS*21*	*P. aeruginosa*	0	1 on R68	33, 35
			2 on R68.45	34, 35

[a] The symbol − indicates that the presence of the respective IS element has not been determined.

[b] 1, Saedler and Heiss (1973); 2, Nisen *et al.* (1979); 3, Johnsrud (1979); 4, Nyman *et al.* (1981); 5, Bertani (1951); 6, Iida *et al.* (1978, 1981a); 7, So *et al.* (1979, 1981); 8, So and McCarthy, 1980; 9, Lebek, 1963; 10, Sharp *et al.* (1973); 11, Hu *et al.*, 1975a; 12, Ohtsubo *et al.* (1981a); 13, Rownd *et al.* (1966); 14, Ohtsubo and Ohtsubo, (1978); 15, Iida *et al.* (1981b); 16, Kondo and Mitsuhashi (1964); 17, MacHattie and Jackowski (1977); 18, Iida (1983); 19, Meynell and Datta (1966); 20, Clerget *et al.* (1980, 1981); 21, Labigne-Roussel *et al.* (1981); 22, Cornelis and Saedler (1980); 23, Deonier *et al.* (1979); 24, Hu *et al.* (1975b); 25, Klaer and Starlinger, (1980); 26, Klaer *et al.* (1980); 27, Schoner and Kahn (1981); 28, Arber *et al.* (1979); 29, P. Caspers and W. Arber, unpublished; 30, Davidson *et al.* (1975); 31, Guyer, (1978); 32, Guyer *et al.* (1981); 33, Holloway and Richmond (1973); 34, Haas and Holloway (1976); 35, Willetts *et al.* (1981); 36, Hu and Deonier (1981).

[c] A strain carrying no IS*1* has been reported (Nisen *et al.*, 1979; Nyman *et al.*, 1981).

transcriptional activation of neighboring genes varied between different IS2 in the bipolar *arg* operon, it has been suspected that the DNA sequence may differ from one IS2 to another (Boyen *et al.*, 1978; Glansdorff *et al.*, 1981). Indeed, small alterations within terminal sequences of two IS2 have been pointed out (Ghosal *et al.*, 1979a). Most of the IS2 mutants were isolated from strains carrying IS2 in orientation I between the promoter and the structural genes in the *gal* operon by strong selection for either relief of transcriptional termination or generation of a promoter activity for the *gal* operon. A majority of the mutants examined had acquired small DNA segments within a specific region near one end of IS2 (Fig. 2). They were thought to be the result of a partial duplication and subsequent small DNA rearrangements associated with DNA replication or repair (Ghosal and Saedler, 1978; Ghosal *et al.*, 1979b; Besemer *et al.*, 1980; Ahmed *et al.*, 1981). Deletions of part of IS2 in the *gal* operon probably removing a transcriptional terminator have also been characterized (Ahmed and Scraba, 1978; Delius *et al.*, 1980), and insertions of IS2 or IS*3* within IS2 that affect the *gal* gene expression have been isolated (Sommer *et al.*, 1979b; Ahmed *et al.*, 1981; see Fig. 2).

Recent studies on IS insertions in the control region of the divergent *argECBH* operons, in which only a relatively weak selection was applied, showed an on–off effect for IS*3* analogous to that originally observed for IS2 (Glansdorff *et al.*, 1981). On the other hand, insertion of IS2 in orientation II did not invariably activate adjacent genes. Based on this observation, Glansdorff *et al.* (1981) formulated the hypothesis that, rather than moving around active promoters (which could potentially be harmful), IS2 and IS*3* could carry a sequence that is inactive by itself but can become part of a promotor when joined to an appropriate sequence.

Integration of IS*1* and IS*5* within an operon usually causes polar effects. However, it has been reported that spontaneous mutations that activate the cryptic *bgl* operon of *E. coli* are due to insertion of IS*1* or IS*5* into a specific region in the operon (Reynolds *et al.*, 1981). In contrast to IS2, which can activate adjacent genes in a constitutive manner, expression of the insertion-activated *bgl* operon is inducible and, in addition, is dependent on cyclic AMP. Two models have been developed to explain this form of cis activation of the *bgl* operon: (*1*) insertion of IS*1* or IS*5* into the specific region could either activate a preexisting, inactive promoter or provide promoter structure, or (2) the operon contains an operator site that is disrupted by the insertion sequence (Reynolds *et al.*, 1981). Transcriptional activity of a yeast gene in *E. coli* by IS*5* insertion has also been reported (Jund and Loison, 1982).

From these observations, one can assume that several different mechanisms operate to activate genes adjacent to a newly integrated IS element: an active promoter can be carried on the IS element; the integration can generate a new promoter at the junction with the IS insertion; a preexisting promoter can be activated; and sequences necessary for negative control can be inactivated. Neighboring sequences may also influence the transcription from these promoters. Thus degree and mode of IS-mediated activation of an adjacent gene depend on the sequence formed by IS integration and on the gene in question. Further point mutations and small DNA rearrangements within or near the IS sequence may act as posttranspositional modulators (Iida *et al.*, 1982a), and they can alter the expression of adjacent genes (see Fig. 2).

The findings discussed in this section illustrate the dynamic nature of genome variations associated with IS elements. They demonstrate that the insertion of an IS element may have a variety of immediate effects on the expression of neighboring genes and that the presence of an IS element can initiate further alterations in gene expression.

E. IS-Mediated DNA Rearrangements

Besides being discovered through their ability to transpose—that is, to insert at new locations on DNA—and to excise subsequently, IS elements were originally revealed as mediators of integration of conjugative F and R plasmids into the bacterial chromosome, as well as of mediators of the fusion between two F plasmid derivatives (Davidson *et al.*, 1975; Palchaudhuri *et al.*, 1976; Ohtsubo and Ohtsubo, 1977). This fusion of two circular replicons is also called *cointegration.* The fused replicon can resolve to yield two replicons, each carrying one copy of the IS element that had mediated the process. IS elements were also observed to mediate deletions during segregation of cointegrates (Davidson *et al.*, 1975; Ohtsubo and Ohtsubo, 1977). The presence of IS*1* in the *gal* operon of *E. coli* K12 leads to an increase in deletion formation in the *gal* region of 30–2000-fold as compared to the wild-type operon (Reif and Saedler, 1975). The IS*1*-mediated deletions start precisely at the end of IS*1* and terminate at various nonrandom sites (Mickel *et al.*, 1977; Reif and Saedler, 1977; Ohtsubo and Ohtsubo, 1978). This is also the case for IS*2*-mediated deletions in the *gal* operon, although these appear less frequently (Peterson *et al.*, 1979). Still another form of DNA rearrangement, inversion of DNA segments adjacent to elements, was also observed (Ross *et al.*, 1979b; Cornelis and Saedler, 1980; Iida *et al.*, 1980a). The inverted segment was found to be flanked by inverted repeats of the same IS element. Figure 3 summarizes elementary DNA rearrangements mediated by IS elements. Combinations of these DNA rearrangements would lead to more complex DNA restructuring (see Section IV).

In early studies, cointegration of F plasmids was interpreted to occur by reciprocal, homologous recombination between preexisting copies of the same IS element carried on both of the participating replicons (Davidson *et al.*, 1975). However, cointegrates that carry directly repeated copies of an IS element at the junction between the two replicons have been isolated even if only one of the two parental replicons contained this IS element (Gill *et al.*, 1978; MacHattie and Shapiro, 1978; Ravetch *et al.*, 1979; Shapiro and MacHattie, 1979; Depicker *et al.*, 1980; Fischhoff *et al.*, 1980; Grindley and Joyce, 1980; Iida and Arber,

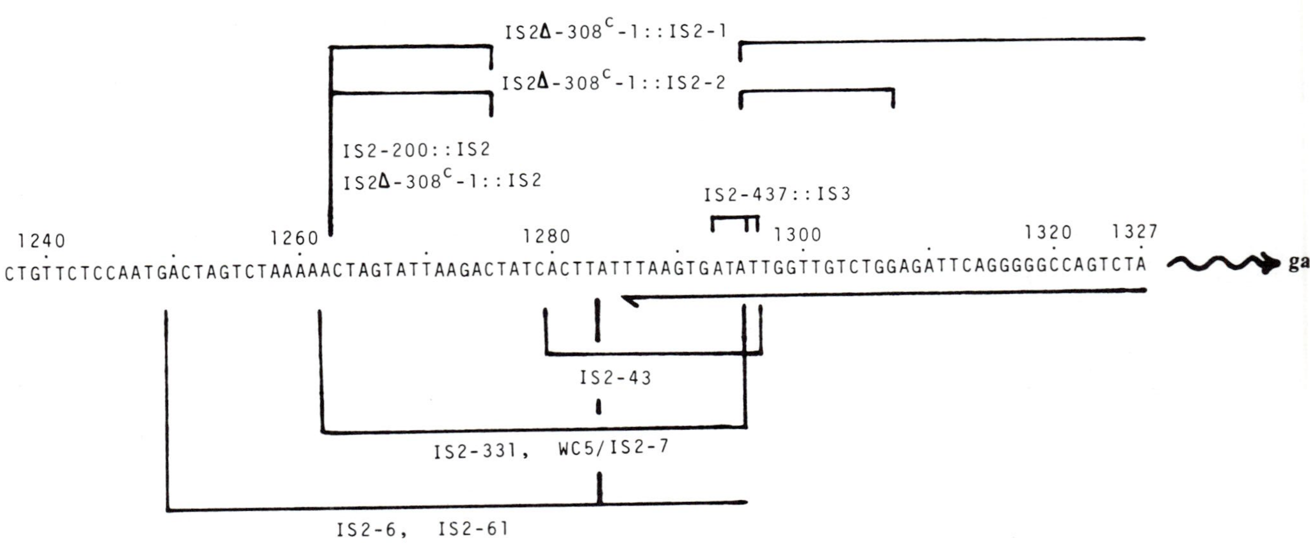
IS2Δ-308c-1::IS2-1
IS2Δ-308c-1::IS2-2
IS2-200::IS2
IS2Δ-308c-1::IS2
IS2-437::IS3
1240
1260
1280
1300
1320
1327
CTGTTCTCCAATGACTAGTCTAAAAACTAGTATTAAGACTATCACTTATTTAAGTGATATTGGTTGTCTGGAGATTCAGGGGGCCAGTCTA
gal
IS2-43
IS2-331, WC5/IS2-7
IS2-6, IS2-61

Fig. 2 DNA sequences of IS2 mutants affecting the expression of adjacent genes. The nucleotide sequence of IS2 between coordinates 1237 and 1327 (end of IS2) is shown as determined on *gal*OP-308 :: IS2 (Ghosal *et al.*, 1979a). In *gal*OP-308 :: IS2, as well as in *gal3* :: IS2, IS2 is inserted in orientation I between the promoter and the structural genes of the *gal* operon, thereby strongly inhibiting the expression of the *gal* genes. Although the two gal^- mutants were isolated independently (Morse *et al.*, 1956; Saedler *et al.*, 1972), the results of DNA sequencing revealed that they are identical (Ghosal and Saedler, 1978; Ahmed *et al.*, 1981). Note that only one other IS2 insertion was found in the *gal* operon (*gal*490 :: IS2; Musso and Rosenberg, 1977), and it was integrated 1 bp to the right of the site of *gal*OP-308 :: IS2. The horizontal arrow under the nucleotide sequence marks the terminal inverted repeat of IS2, and the horizontal wavy arrow indicates the orientation of the *gal* transcription.

Among the derivatives of this gal^- mutant, IS2-6, -61, -7, -331, -WC5, -43, -200 :: IS2, Δ-308^c-1 :: IS2, and Δ-308^c-1 :: IS2-1 are gal^+ revertants. IS2-6 carries a 108-bp insertion either at position 1249/1250 or 1283/1284, as indicated by the vertical lines. This additional 108-bp sequence is derived from the DNA segment between coordinates 1250 and 1295 (as indicated by the horizontal line attached to the vertical lines) by a partial duplication and subsequent small DNA rearrangement (Ghosal and Saedler, 1978). IS2-61 is a derivative of IS2-6, is still gal^+, and carries only a 54-bp insertion (or a 54-bp deletion as compared to IS2-6) either at position 1249/1250 or 1283/1284 (Ghosal and Saedler, 1979). This IS2-61 arose by *recA*-independent recombination between 9-bp direct repeats. Similarly, IS2-331 has a 108-bp insertion either at position 1261/1262 or 1295/1296, and the additional 108-bp sequence also carries a cluster of partially overlapping direct repeats and is derived from the DNA segment between coordinates 1262 and 1295 (Ahmed *et al.*, 1981). IS2-WC5 is a derivative of IS2-331, is still gal^+, and represents IS2 with a 54-bp insertion (or IS2-331 with a 54-bp deletion) at position 1261/1262 or 1295/1296 (Ahmed *et al.*, 1981). IS2-7 is an independently isolated gal^+ revertant of *gal*OP-308 :: IS2 and has proved to be a 54-bp insertion mutant (Ghosal *et al.*, 1979b) identical to IS2-WC5.

In IS2-200 :: IS2 a second IS2 is inserted in orientation II at position 1262/1263 into the resident IS2, resulting in activation of the *gal* operon (Ahmed *et al.*, 1981). A rare deletion derivative that retained only IS2 sequences 121-1 (orientation II) from the second IS2 and 1263-1327 (orientation I) from the resident IS2 was still gal^+ (Ahmed *et al.*, 1981). Incidentally, the same site in IS2 that had served as a transposition target (position 1262/1263 in IS2-200 :: IS2) was observed to represent the endpoint of an IS2-mediated deletion in IS2 Δ-308-1 :: IS2. This deletion fused the end of an IS2 in orientation II to the segment 1263-1327 of the resident IS2 (Delius *et al.*, 1980; Besemer *et al.*, 1980), thus leading to the same fused structure as in IS2-200 :: IS2 and constitutive expression of the *gal* genes. The 65-bp terminus of the resident IS2 was observed to undergo further DNA sequence rearrangements: IS2 Δ-308^c-1 :: IS2-1 and IS2 Δ-308^c-1 :: IS2-2 are derivatives of IS2-Δ-308^c-1 :: IS2, being gal^+ and gal^-, respectively (Besemer *et al.*, 1980). In both the 13-bp sequences between positions 1263 and 1275 of the preexisting IS2 was substituted, namely, by an inverted copy of the 32-bp segment between positions 1296 and 1327 in IS2 Δ-308^c-1 :: IS2-1 and by a 12-bp segment derived from positions 1296 to 1307 in IS2 Δ-308^c-1 :: IS2-2.

A gal^+ mutation IS2-43 carries a 17-bp direct duplication of the sequence between coordinates 1280 and 1296 (Sommer *et al.*, 1979a). This 17-bp duplication is thought to generate a new promoter for the *gal* operon. IS2-437 :: IS3 is a gal^- derivative of IS2-43, and it carries an IS3 inserted at position 1292/1293 to 1295/1296 (or 1296/1297), which is within the duplicated sequence of IS2-43 (Sommer *et al.*, 1979b).

Note that the transposition characteristics of the IS2 derivatives discussed here have not yet been reported, with the exception of IS2-7 (Peterson *et al.*, 1979).

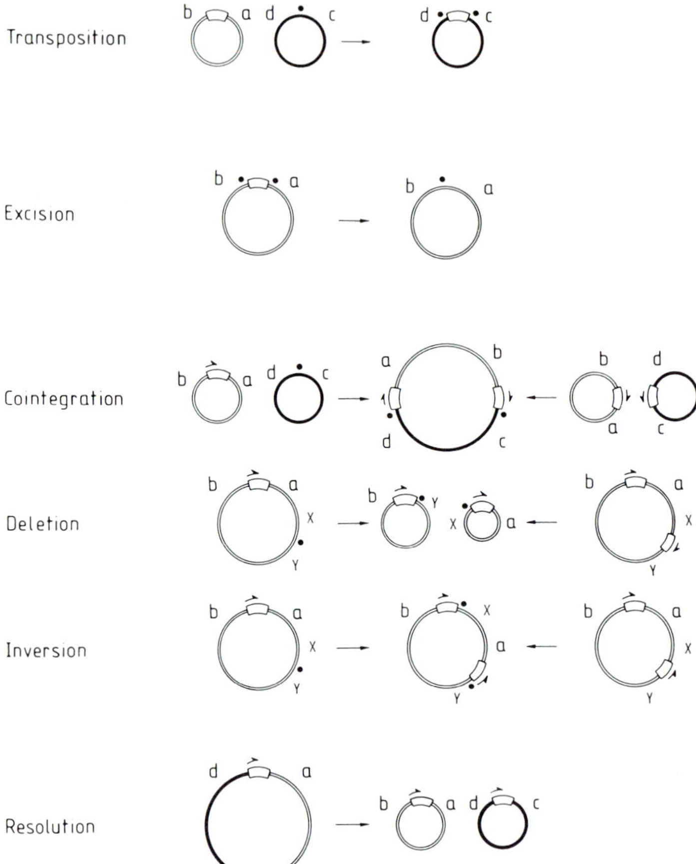

Fig. 3 Schematic representation of DNA rearrangements mediated by IS elements. These simplified schemes represent donor and target DNA molecules before (left and right) and after (center) rearrangement events. Boxes represent IS elements. Small arrows next to the boxes symbolize DNA polarity, where relevant. Small dots above the DNA molecules indicate the position of the target sequences undergoing duplication during the transposition of IS elements. Letters are either genetic or physical markers. Cointegration, deletion, and inversion can be mediated either by transposition of IS elements (as indicated by the arrows pointing right) or by reciprocal recombination between two IS elements (as indicated by the arrows pointing left). Transpositional events generate the duplication of the IS element and of the target sequence, whereas reciprocal recombination does not. Reciprocal recombination resulting in cointegration, deletion, inversion, and resolution may be mediated either by host *rec* enzymes or by site-specific recombinases (or resolvases) encoded by the transposable elements. Combinations of the elementary processes shown may lead to more complex DNA rearrangements, for example, the generation of IS-mediated transposons (Section IV,D).

1980; Ohtsubo *et al.*, 1980a,b; Berg *et al.*, 1981; Foster *et al.*, 1981a; Iida *et al.*, 1981c,d; Isberg and Syvanen, 1981). Because IS elements transpose by themselves, such cointegrate structures could also be formed by reciprocal homologous recombination between an IS element carried by one replicon and another copy of the same IS element newly transposed into the second replicon. Deletion and inversion could be explained by an analogous process, in which an IS element transposed into a replicon that already carried a copy of the same IS element. Depending upon the orientation of the transposed IS element relative to the preexisting copy, reciprocal homologous recombination would result in either deletion or inversion (Fig. 3).

Although the *recA*-dependent pathway discussed above certainly exists, the following findings indicate that it is not the major one. First, it has become evident that a new copy of the IS element (and transposable elements in general) is generated during the transposition process. Second, IS-mediated DNA rearrangements such as cointegration, deletion, and inversion also occur in cells lacking the *recA*-dependent homologous recombination system, whereas reciprocal recombination between two IS elements can be considered to be largely *recA*-dependent.

Indications that the duplication of transposable elements is associated with cointegration were first obtained for bacteriophage Mu and transposon Tn*3* (Toussaint and Faelen, 1973; Gill *et al.*, 1978), and could later be extended to the IS elements IS*1*, IS*8*, IS*10*, IS*26*, IS*50*, IS*101*, IS*102*, and IS*903* (Table I). The crucial experiment was the demonstration that after the mobile element had transposed to a new site, it was still present at the original site. This had been shown for phage Mu and later was also documented for IS elements (Ljungquist and Bukhari, 1977; Klaer *et al.*, 1980; Read *et al.*, 1980). In addition, it was repeatedly observed that locations of IS elements in related *E. coli* K12 strains were conserved (Arber *et al.*, 1979; Deonier *et al.*, 1979; Nyman *et al.*, 1981). Additional IS elements observed were thus thought to be duplicated by transposition during the construction of strains (or isolation of mutants) or upon subsequent propagation.

For several regular IS elements, such as IS*1*, IS2, and IS*10*, reciprocal recombination between two such elements depends mostly on the host *recA* pathway, although it does occur occasionally in $recA^-$ cells (Deonier and Mirels, 1977; Chumley and Roth, 1980; Iida and Arber, 1980; Silver *et al.*, 1980; Iida *et al.*, 1981b, 1982c; Ohtsubo *et al.*, 1981b). Most cointegrates between two replicons, each carrying one copy of the IS*1* element, formed under $recA^-$ conditions contain three copies of the IS*1* element; the two at the junctions between the two replicons must have been derived from only one of the parental IS*1* elements (Iida and Arber, 1980; Ohtsubo *et al.*, 1981b; S. Iida, unpublished). Similarly, phenotypically selected segregants derived under $recA^-$ conditions from a cointegrate carrying two directly repeated IS*1* at the junctions have a variety of physical structures and are formed mainly by IS*1*-mediated deletions rather than by reciprocal recombination between the two IS*1* elements (Iida, *et al.*, 1982c).

In the case of the γδ/Tn*3* family of mobile elements, reciprocal recombination for resolving the cointegrate is not primarily dependent on the *recA* pathway. Stable cointegrates cannot be isolated unless the Tn*3* or γδ contains a $tnpR^-$ mutation, but all transposition products of the mutants in $recA^-$ cells have been found to be cointegrates (Gill *et al.*, 1978). By providing resolvase (*tnpR* gene product) in trans, these cointegrates were resolved to yield a parental replicon and a target replicon with a newly acquired copy of the element (Arthur and Sherratt, 1979; Reed, 1981b). On the other hand, integration of F plasmid into the host chromosome may also be mediated by *recA*-dependent recombination between two γδ elements (Guyer *et al.*, 1981).

A third argument in support of IS duplication during transposition is based on the following observation: Excision of an IS element generally has a lower frequency than its transposition, and the process is independent of functions needed for transposition (Berg, 1977; Egner and Berg, 1981; Foster *et al.*, 1981b). This indicates that excision is not a prerequisite for transposition.

For all of these reasons, it is now widely accepted that duplication of an IS element takes place *during* transposition as well as in transpositional cointegration, inversion, and deletion. However, it should be emphasized here again that cointegration, inversion, as well as deletion can also be a consequence of reciprocal recombination between two identical IS elements. Therefore, these DNA rearrangements may also be formed by recombination of a preexisting element with a second element *after* its transposition and independently of the transposition process. Occurrence of these two pathways for IS*1*-mediated cointegration has been seen in rec^+ cells (Iida *et al.*, 1981a,c).

The precise molecular mechanism of IS duplication during transposition is not known, but two observations may be relevant in this respect. (*1*) The special DNA sequence called *inceptor*, at which phage λ DNA replication is initiated, has also been detected in IS*5* (Lusky *et al.*, 1981). It is not known whether other IS elements carry an inceptor sequence. (*2*) Some $polA^-$ mutations are known to affect the transposition of Tn*5* and Tn*10* (Clements and Syvanen, 1981; Sasakawa *et al.*, 1981). At present, it is not clear whether DNA polymerase I is directly involved in this kind of IS replication.

F. Target Sequences

1. Duplication of Target Sequences

Insertion of an IS element is accompanied by a short duplication of target DNA such that the transposed element is flanked by direct repeats of target DNA. The length of the duplication, but not its nucleotide sequence, is usually specific for the particular IS element (Table I). These 2- to 12-bp direct repeats of target DNA sequences are believed to be formed by the introduction of staggered nicks at the target site and subsequent DNA synthesis associated with the replication of the entire IS element (Grindley and Sherratt, 1979). Some natural isolates lack such a duplicated target DNA sequence, for example, IS*1*b in the R plasmid R100, an IS*1*-D on the *Shigella dysenteriae* chromosome, and γδ in F plasmid (Ohtsubo and Ohtsubo, 1978; Reed *et al.*, 1979; Ohtsubo *et al.*, 1981a). This may indicate that restructuring had occurred subsequent to the initial transposition.

IS*4* is peculiar with respect to the length of the duplication. At the common chromosomal site, it is flanked by an 11-bp duplication, whereas it generates a higher number of duplicated nucleotides at other target sites (Habermann *et al.*, 1979; Klaer *et al.*, 1981a). One IS*1* variant, probably originated from bacteriophage P1, generates an 8-bp instead of a 9-bp duplication during cointegration (Iida *et al.*, 1981a). This IS*1* element carries at least one base pair substitution 12 bp from one end. However, the same alteration was found in one of the two flanking IS*1* elements of Tn*9*, and this Tn*9* generates a 9-bp duplication (Johnsrud *et al.*, 1978; Galas *et al.*, 1980). It is therefore not yet clear whether the 8-bp duplication is dependent solely on this base alteration and whether this IS*1* variant intrinsically generates repeats of variable length at target sequences in a manner similar to IS*4*.

The duplication of the target DNA sequence is also observed in cointegrates (Fig. 3). IS-mediated deletions remove one of the flanking repeats (Ohtsubo and Ohtsubo, 1978; Calos and Miller, 1980b). In IS-mediated inversions, the small repeats should be inverted with the sequence between the IS elements (Fig. 3), although this expectation has not yet found experimental verification. The ability of an IS element to induce transposition, cointegration, and deletion is not dependent on the presence of the small flanking direct duplication (Kleckner, 1979; Reed *et al.*, 1979; Calos and Miller, 1980b; Ohtsubo *et al.*, 1980b). Precise excision, however, is thought to require this small duplication (Foster *et al.*, 1981b; see also Section III,H).

2. Specificity of Target Site Selection

The specificity of insertion site selection seems to be an intrinsic property of an IS element. Some IS elements are found at one or very few sites only, others insert at a larger but still limited number of sites, and a third category seems to transpose more randomly (Table I).

For example, IS*4* resides in one common site in the chromosome of all *E. coli* K12 strains tested, most of which carry only a single copy of IS*4* (Klaer and Starlinger, 1980; Klaer *et al.*, 1981a). But transposition of IS*4* in both possible orientations into a single site within the *galT* gene has been observed repeatedly (Pfeifer *et al.*, 1977; Klaer *et al.*, 1980). A third site of IS*4* is found in an *E. coli* strain that carries two copies of IS*4*, one at the common site and the other at a site different from that in the *galT* gene (Klaer and Starlinger, 1980). A comparison of the DNA sequence around these three IS*4* insertion sites revealed some common structural features (Klaer *et al.*, 1981b), including a hyphenated palindromic sequence outside of the 11- or 12-bp duplicated target sequence.

A much simpler consensus target sequence was determined for IS*5*, and this preferential sequence coincides with the 4-bp duplication sequence (Engler and van Bree, 1981). Whether the presence of the consensus sequence alone is enough to serve as a target for IS*5* integration has yet to be determined.

Despite rather extensive studies on targets for IS*1* insertion [involving not only IS*1* but also the IS*1*-flanked Cm^r (chloramphenicol-resistance) transposons], no apparent consensus sequence was detected. Instead, IS*1* insertion seems to depend upon the following three criteria:

1. Easily denaturable (AT-rich) regions are preferred on a large scale (Galas *et al.*, 1980; Meyer *et al.*, 1980).
2. Within AT-rich regions, the presence of sequences homologous to the termini of IS*1* (the terminal repeats or their adjacent sequences) may stimulate integration (Galas *et al.*, 1980; Saedler *et al.*, 1981).
3. GC pairs at both ends or at least at one end of the 9-bp duplication of the target sequence appeared frequently. Thus, the IS*1*-mediated staggered cleavage appears to be made preferentially at a GC pair (Galas *et al.*, 1980).

A particular 6-bp symmetrical consensus sequence within the 9-bp duplication sequence that is homologous to the ends of the element was also detected in the target region for the IS*10*-flanked Tc^r transposon Tn*10* (Kleckner, 1979; Halling and Kleckner, 1982; see Table I). Tn*10* also seems to integrate preferentially into AT-rich DNA segments, as do several other transposons, including Tn*5* mediated by IS*50* (Miller *et al.*, 1980). IS*2* inserted in the *gal* operon was repeatedly found to integrate into only two sites in the control region (Musso and Rosenberg, 1977; Ghosal and Saedler, 1978; Ahmed *et al.*, 1981; see Fig. 2). On the other hand, IS*2* is inserted at nine different sites on P1 DNA within a 1.7-kb segment that is moderately AT-rich (Arber *et al.*, 1979, 1981; Meyer *et al.*, 1981; C. Sengstag and W. Arber, unpublished).

G. Models for the Transposition Process

Several models describing the transposition process have been proposed. They all take into account the common features of transposable elements: (*1*) the presence of inverted repeats at the termini of the elements; (*2*) generation of short direct repeats at the target sequence; (*3*) duplication of the entire element during transposition; and (*4*) the ability to mediate cointegration, deletion, and inversion (see the reviews by Bukhari, 1981; Kleckner, 1981). A few selected aspects will be discussed here.

The inverted repeats of the elements are regarded as the sites that transposases recognize, cut, and rejoin to the target sequence. The direct duplication of the target sequence is thought to be generated by the introduction of staggered nicks and subsequent DNA synthesis (Grindley and Sherratt, 1979). Deletion and inversion are processes equivalent to cointegration and result from intramolecular transposition, as opposed to intermolecular transposition in cointegration (Arthur and Sherratt, 1979; Shapiro, 1979; Fig. 3).

In one group of models (*replication-first* models), replication of the element takes place before the breakage and rejoining of its ends to the target sequence (Toussaint *et al.*, 1977; Read *et al.*, 1980; Kamp and Kahmann, 1981). In others (*recombination-first* models), the introduction of staggered nicks at the target sequence and ligation with the ends of the element initiate duplication of the entire element (Arthur and Sherratt, 1979; Grindley and Sherratt, 1979; Shapiro, 1979; Galas and Chandler, 1981; Harshey and Bukhari, 1981). The recombination-first models can be further subdivided into *symmetric* and *asymmetric* models. Symmetric models propose that both ends of the element are concomitantly cleaved and ligated to the target DNA molecule (Arthur and Sherratt, 1979; Shapiro, 1979), whereas in asymmetric models transposition is initiated by ligation of one end only (Grindley and Sherratt, 1979; Galas and Chandler, 1981; Harshey and Bukhari, 1981). As a consequence, cointegration is a necessary intermediate in symmetric models, whereas it is a by-product in asymmetric models (Fig. 4). Depending on the model, replication of the element has been proposed to initiate on the donor or immediately after rejoining to the target molecule from one or both ends of the element, and to proceed in a semiconservative or conservative process (Bukhari, 1981).

A replication-first model was originally proposed for bacteriophage Mu, which carries its own replication machinery (Toussaint *et al.*, 1977). For transposition of γδ and its relative, Tn*3*, data are accumulating to support a recombination-first, symmetric model (Gill *et al.*, 1978; Arthur and Sherratt, 1979; Heffron *et al.*, 1981; Kostriken *et al.*, 1981; Muster and Shapiro, 1981; Reed 1981a,b). Cointegrates formed by the activity of the *tnpA* gene product are obligatory intermediates in transposition. They are resolved by site-specific recombination mediated by the *tnpR* gene product, the resolvase. For regular IS elements, most details of the transposition process are still a matter of speculation. Only very few findings can be cited in favor of a replication first model, for example, the presence of the replicational inceptor sequence in IS*5* and an observation supporting the idea of local DNA replication in IS*5* (Lieb, 1980; Lusky *et al.*, 1981). On the other hand, the observation that cointegrates mediated by regular IS elements are stable even in rec^+ cells favors a recombination-first, asymmetric model for transposition of the elements (Grindley and Joyce, 1980; Iida, 1980; Iida and Arber, 1980; Iida *et al.*, 1981c,d, 1982c; Ohtsubo *et al.*, 1980a,b; Galas and Chandler, 1982). At present, we do not know whether all regular IS elements transpose by the same pathway. We cannot even rule out the possibility that the same IS element transposes by different processes depending on the particular conditions.

H. Excision

As discussed above (Section III,D), insertion of an IS element into a structural gene inactivates its function, and insertion into an operon often causes polar effects (Starlinger and Saedler, 1976). Some of these insertion mutations revert due to precise excision of the IS element, thereby restoring the genetic functions

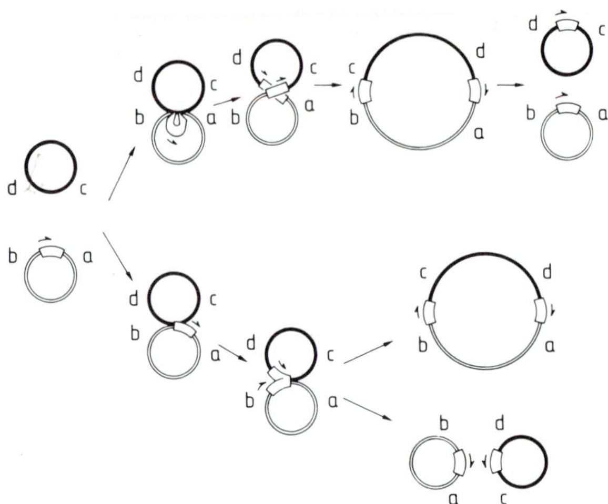

Fig. 4 Simplified version of symmetric and asymmetric models for transposition. In symmetric models (top sequence), ligation of both ends of the IS element with the target sequence initiates the duplication of the IS element to yield a cointegrate as an intermediate (Arthur and Sherratt, 1979; Shapiro, 1979, 1980). The resolution of the cointegrate may occur during the duplication of the IS or subsequently by the action of a resolvase (Shapiro, 1980; Reed, 1981b). In asymmetric models (bottom sequence), ligation of one end of the IS element with the target sequence initiates the duplication of the element (Galas and Chandler, 1981; Harshey and Bukhari, 1981). This does not necessarily mean that transposases recognize only one end of the element. Whether the process leads to cointegration or transposition depends upon the details of strand cleavage and rejoining after the duplication of the IS. The symbols are the same as those in Fig. 3.

of the target DNA. A different kind of excision event was observed occasionally that relieved the polar effects but did not restore the function of the particular gene that had carried the insertion (Ross *et al.*, 1979c; Foster *et al.*, 1981b). It is called *nearly precise* or *imprecise excision.*

The frequency of reversion is generally quite low, varies from element to element, and depends also on the position of the insertion in an operon; but not all mutations due to IS insertion revert at measurable rates (Adhya and Shapiro, 1969; Malamy *et al.*, 1972; Saedler *et al.*, 1972, 1974; Berg, 1977; Arber *et al.*, 1979; Egner and Berg, 1981; Foster *et al.*, 1981b).

Excision proceeds by mechanism(s) that do not lead to transposition. Excision occurs independently of host *recA* functions and of transposition functions encoded by the IS elements (Egner and Berg, 1981; Foster *et al.*, 1981b). Host functions must be involved because several chromosomal mutations affecting precise excision have been identified (see Section III,J). Short, direct repeats of a few base pairs have been observed to be hot spots for ends of spontaneous deletions in the *lacI* gene of *E. coli* (Farabaugh *et al.*, 1978). It has been suggested that the same mechanisms act on the small, direct repeats of target sequences flanking the IS elements and thus mediate precise excision. In addition, inverted repeats within the elements seem to play a role in excision. It has therefore been proposed that the excision of the IS element could be due to slippage during DNA synthesis (Egner and Berg, 1981). Short palindromes within the terminal inverted repeats of IS*10* were involved in imprecise (nearly precise) excision of Tn*10* (Foster *et al.*, 1981b; Chapter 6).

I. Coding Capacity of IS Elements and Gene Products

As mentioned earlier, the nucleotide sequence of most IS elements contains at least one large open reading frame (Fig. 1; Table I), and one may assume that this encodes one component of a putative *transposase,* that is, an enzyme involved in transposition. In order to identify the gene product, polypeptides coded by IS elements have been sought in either mini- or maxi-cells or in *in vitro* protein synthesizing systems. Polypeptides specified by the regular IS elements IS*1*, IS*2*, IS*4*, IS*5*, and IS*50* have been detected (Rothstein *et al.*, 1981; Trinks *et al.*, 1981; Rak *et al.*, 1982; P. Starlinger, personal communication). At present, no active transposase that mediates replication of an element and its recombination with a target sequence has been isolated.

The hypothesis that the open reading frames actually contain biologically significant information can be tested in several ways. First, a comparison of the primary sequence of naturally isolated variants and relatives of an IS element (Table I) will show if the open reading frame is conserved (Bernardi and Bernardi, 1981a; Ohtsubo *et al.*, 1981a). Second, prediction of actual coding frames can be made according to Shepherd's method (Shepherd, 1981a,b). In this analysis, deviation of the actual sequence from a hypothetical primitive codon with only RNY triplets (R = purine, Y = pyrimidine, N = purine or pyrimidine) is determined. As summarized in Table I, the analysis indicates that most of the open reading frames could actually serve as likely coding frames (J. Shepherd, personal communcation). Third, the hypothesis can be examined directly by introducing mutations within the coding region using *in vitro* gene manipulation techniques. Insertions, deletions, and base substitutions thus produced within the element may affect the transposition functions, including cointegration and deletion. Available data from these three lines of study are still limited.

As mentioned earlier, IS*1* may encode two polypeptides of 91 and 125 amino acids (Ohtsubo and Ohtsubo, 1978; Ohtsubo *et al.*, 1981a). Deletions within either one of the two open reading frames reduce the frequency of cointegrate formation. However, introduction of a 5-bp insertion between the two open reading frames also impairs the ability to form cointegrates (Ohtsubo *et al.*, 1981b; Machida *et al.*, 1982). This observation favors the idea that a longer polypeptide consisting of 167 amino acids exerts transposition functions. In its C-terminal part, it is identical to the possible 125-amino-acid-long peptide, whereas its N-terminal part overlaps with the end of the 91-amino-acid-long peptide (Johnsrud, 1979; Ohtsubo *et al.*, 1981a). The insertion of a 4-kb-long

Km^r fragment containing the split IS*26* sequence into one of the two open reading frames was reported to inhibit transposition of the modified IS*1*, although it did not affect cointegration (Reif and Arber, 1981). The apparent differences may stem either from the structure of the modified IS*1* or from the experimental systems. Reduced frequencies of deletion formation mediated by IS*1* mutants containing an internal deletion or insertion have also been reported (Bishop and Davies, 1980; Saedler *et al.*, 1981). Experiments involving IS*1* are complicated by the fact that in the *E. coli* K12 chromosome several additional copies of IS*1* are present and may provide missing functions (Saedler and Heiss, 1973; Nisen *et al.*, 1979; Nyman *et al.*, 1981). Interactions between IS elements will be discussed in Section III,K.

No chromosomal copies of IS*903* and its relative, IS*102*, were detected in *E. coli* K12. The primary sequence of IS*903* indicates that the element could encode a polypeptide of 307 amino acids (Table I). Small insertions or a large deletion in IS*903* were found to reduce cointegration (Grindley and Joyce, 1980). Similarly, an IS*4* derivative containing a Cm^r segment inserted into the *Pst*I site no longer transposed efficiently (Klaer *et al.*, 1981a). But an IS*2* derivative carrying a Km^r segment derived from Tn*5* still transposed (Saint-Girons *et al.*, 1981).

Most of the IS elements also possess a small open reading frame within the large open reading frame but in the opposite orientation (Fig. 1; Table I). It is unknown whether these have any functional importance, but the following observations may be relevant: (*1*) Among the regular IS elements IS*1*, IS*2*, IS*4*, IS*5*, IS*50*, IS*102*, and IS*903*, only IS*50* carries a stop codon in the middle of such a small open reading frame; all the others could encode polypeptides of 96–143 amino acids. (*2*) In all these regular IS elements and in the *tnpA* gene of Tn*3*, the same reading frame would be used for the overlapping, antipolar longer and shorter coding segments. (*3*) Differences in amino acid sequences of putative polypeptides encoded by the related elements IS*102* and IS*903* are much less frequent in the longer open reading frame than in the shorter one. (*4*) Rak *et al.* (1982) found that the small open reading frame of IS*5* can be transcribed and translated. (*5*) IS*10* carries no small open reading frame (Table I).

Little is known about the regulation of expression of transposition function(s) encoded by regular IS elements. Preferential cis action may suggest that transposition functions are expressed scarcely and/or that gene products are rather labile (see Section III,K,1). It has been suggested that IS*50* and IS*10* encode a factor negatively regulating their transposition (Beck *et al.*, 1980; Biek and Roth, 1980, 1981). The DNA sequence of IS*50* indicates a coding capacity for two longer overlapping proteins in the same reading frame (Auerswald *et al.*, 1981). In fact, two polypeptides of the expected sizes were detected in mini-cells (Rothstein *et al.*, 1981). It is not clear yet whether one of these peptides is a repressor or an inhibitor. It is tempting, although probably premature, to speculate that autorepression system(s) similar to that described for the Tn*3*/γδ family of elements (Chou *et al.*, 1979; Heffron *et al.*, 1981; Reed, 1981a; see also Chapter 5) may also exist for regular IS elements. It has been proposed that higher IS*1* activities are due to a high level of expression of IS*1*-encoded proteins from outside promotors (MacHattie and Shapiro, 1978). Expression of transposition functions in different physiological conditions of the host cells will be discussed in Section III,L.

More information is available on the IS element γδ. This IS element carries two transposition genes, *tnpA* and *tnpR* (Fig. 1.). The active *tnpR* gene product, the resolvase/repressor, has been purified, and its mode of action has been examined (Reed, 1981b; Reed and Grindley, 1981; for more details, see Chapter 5).

J. Host Functions

For several reasons, it is generally assumed that host factors must also be involved in the transposition processes. In fact, several *E. coli* chromosomal mutants affecting IS activities have been isolated and characterized by using selective markers carried on transposons mediated by IS elements. These mutants are listed in Table III.

One class of mutants showed either increased or decreased frequencies of transposition, cointegration, and deletion formation. Some of them were identified as *polA* mutants, and in these, exonuclease and polymerase activities were both reduced (Clements and Syvanen, 1981; Sasakawa *et al.*, 1981). This indicated that the *polA* gene product is somehow involved in the replication processes associated with IS transposition.

Mutations in *top,* the structural gene for topoisomerase I, were reported to affect the transposition of some mobile elements but not others (Sternglanz *et al.*, 1981). The effect could be explained either by a direct participation of the topoisomerase I in the cleavage/ligation step or by the requirement of a particular DNA topology for the transposition process. Alternatively, the effect could be indirect, for example, via transcriptional interference.

Although DNA restructuring activities of IS elements were observed in *recA*$^-$ cells, *recA*$^-$ mutants sometimes gave somewhat reduced efficiencies of DNA rearrangements (Young *et al.*, 1980; Grindley and Joyce, 1981; Reif and Arber, 1981; M. Chandler, personal communication; S. Iida, unpublished). Some regular IS elements, such as IS*903*, may even depend on *recA*-mediated processes for cointegration (Young *et al.*, 1980; Grindley and Joyce, 1981). Reciprocal recombination between two homologous sequences of most regular IS elements is mainly dependent on the *recA* pathway (Deonier and Mirels, 1977; Chumley and Roth, 1980; Silver *et al.*, 1980; Ohtsubo *et al.*, 1981b; Iida *et al.*, 1982c).

A particular locus called *ferA* carried on the conjugative F plasmid was shown to reduce precise excision of Tn*5* and Tn*10*. Mutations of *ferB*, located adjacent to *ferA*, enhanced excision in the presence of *ferA* mutations (Hopkins *et al.*, 1980). Interestingly, *ferB* and *ferA* mutations together also stimulate *recA*-mediated reciprocal recombination between two IS*3* elements.

Table III Host Mutants Affecting Functions of IS Elements

Designation	Location on *Escherichia coli* K12 chromosome (min) or on plasmid	Effects on: IS-mediated transposons[a]: Transposition reduced	Effects on: IS-mediated transposons[a]: Precise excision reduced	Effects on: Other mobile elements[b]	Comments[c]	References[d]
gerA	91	Tn*9*, Tn*10*, Tn*903* (*gerA*, *gerB*, *gerC*)			**Ge**nomic **r**earrangements (*gerA*, *gerB*, *gerC*)	1 (*gerA*, *gerB*, *gerC*)
gerB	2					
gerC	44					
himA	38		Tn*5*, Tn*10*	1	**H**ost **i**ntegration **m**ediator	2, 3
hip	19	Tn*9*	Tn*10*	1	**H**ost **i**ntegration **p**rotein	2, 4
polA	86	Tn*5*, Tn*10*			DNA **pol**ymerase	5, 6
recA	58	Tn*903*			General **rec**ombinase I	7
tnd		Tn*5*, Tn*10*			**Tn** transposition-**d**eficient	5
tnm	91	Tn*5*, Tn*9*, Tn*10*, Tn*903*		1, 2	**Tn** **m**igration	1, 8
top	28	Tn*5*, Tn*9*, Tn*10*		3	**Top**oisomerase I (ω protein)	9, 10
ferA	F plasmid		Tn*5*, Tn*10*	4	**F**-factor-mediated **e**xcision and **r**ecombination	11
ferB	F plasmid			4		

[a] Transposition of Tn*5*, Tn*9*, Tn*10*, and Tn*903* (= Tn*601*) is mediated by IS*50*, IS*1*, IS*10*, and IS*903*, respectively.

[b] Key: 1, propagation of phage Mu is affected; 2, transposition of Tn*3* is affected; 3, transposition of Tn*3* is not affected; 4, in *fer*A and *fer*B, double mutants, precise excision of Tn*5* and Tn*10* and recombination between IS*3* are enhanced.

[c] No biochemical functions have been assigned to the products of genes *ger*, *tnd*, *tnm*, and *fer*.

[d] 1, Smirnov *et al*. (1981); 2, Miller *et al*. (1979); 3, Miller and Friedman (1980); 4, A. Kikuchi, personal communication; 5, Clements and Syvanen (1981); 6, Sasakawa *et al*. (1981); 7, Young *et al*. (1980); 8, Ilyina *et al*. (1981a,b); 9, Sternglanz *et al*. (1981); 10, N. Kleckner, cited by Sternglanz *et al*. (1981); 11, Hopkins *et al*. (1980).

It should be pointed out here that some of these host mutations may affect IS functions indirectly, because *top* and *himA* mutations alter the transcriptional characteristics (Miller and Friedman, 1980; Miller *et al.*, 1981; Sternglanz *et al.*, 1981). Still relatively little is known about the involvement of host functions in IS-mediated DNA rearrangement processes, and it is very likely that not all classes of IS elements require the same host functions.

K. Interactions between IS Elements

Functional interaction between IS elements can be expected to occur in cells carrying more than one copy of a particular IS element or more than one different but related IS element. Functional tests may reveal positive (complementation) as well as negative (interference) interactions. In addition, the role played by purely structural interactions between IS elements can be considered.

1. Functional Interactions

Transposons of one class contain two flanking copies of the same IS element. For two of these transposons, Tn*10* and Tn*5*, it was clearly shown that the two copies of the flanking IS elements, IS*10* and IS*50*, respectively, are structurally and functionally different from each other (Rothstein *et al.*, 1980, 1981; Foster *et al.*, 1981a). One of the copies is more active, and transposition is therefore driven by the active IS element. This indicates that an active IS element can complement an inactive one, which may mean that protein(s) produced from the active element can efficiently recognize its interaction sites on both the active and the inactive IS element.

Complementation studies were also performed using IS*1* (Machida *et al.*, 1982). IS*1*-mediated cointegration of plasmids containing one or two directly repeated IS*1* elements was examined in a *recA*$^-$ strain of *E. coli* K12 that also carried 10 copies of IS*1* in the chromosome. Deletions removing one end of an IS*1* completely abolished cointegration mediated by this mutated IS*1*. Deletions or a short insertion internal to IS*1* reduced but did not completely abolish its ability to cointegrate. This reduction was partially but not fully restored by the presence of an additional intact IS*1* on the same plasmid. From these observations, it was concluded that IS*1*-encoded proteins can function in trans, but they act preferentially on the ends of the IS*1* from which they are produced. The same conclusion has been made from experiments with IS*903* (Grindley and Joyce, 1981; N. D. F. Grindley, personal communication). During these studies, it was also found that the end of IS*903* actively transcribed from a promoter outside of the element was less active for cointegration.

Negative interaction (interference) between IS elements was found in studies on Tn*5* transposition. Transposition of Tn*5* is most frequent immediately after entry of Tn*5* into a cell, and it later falls under negative control by a factor encoded within the element itself (Biek and Roth, 1980, 1981). A similar trans-acting repressor or inhibitor function was detected in Tn*10* (Beck *et al.*, 1980). Studies with series of deletions seem to suggest that the repressor or inhibitor in Tn*5* seems to be encoded by IS*50* (Biek and Roth, 1980, 1981). Identification and characterization of these repressor or inhibitor proteins are still lacking. However, it should be pointed out that the mode of this repression is different from the immunity of Tn*3* transposition because the latter is cis-acting (Wallace *et al.*, 1981; see Chapter 5).

2. Structural Interactions

Only a few data are available on the structural interaction between IS elements. Occasionally, IS elements were observed to insert within or next to a preexisting IS element (Chow and Broker, 1978; Sommer *et al.*, 1979b; Ahmed *et al.*, 1981; Saedler *et al.*, 1981; Hänni *et al.*, 1982). It was also reported that recombination between γδ and IS*2* caused a deletion in the F plasmid (Palchaudhuri and Hänni, 1979). However, this deletion could also be explained simply by the activity of IS*2* or γδ alone. As outlined in Section III,E, deletion adjacent to IS*2* and γδ may result from the insertion of one element into the other. Several IS*1* variants have been identified (Ohtsubo *et al.*, 1981a; Section III,B; Table I) that, by homologous recombination between them, could yield new variants. This process may provide insight into the evolution of one particular IS element. Recombination between IS*1* variants has been discussed (Iida *et al.*, 1981a; Iida, 1983).

L. Physiological Conditions

Only limited data are available on the influence of physiological conditions on the transposition of IS elements. It has been indicated, however, that transposition of IS elements does occur or is at least induced in the stationary phase of growth and, in particular, in cultures standing in the refrigerator. Transposition of several kinds of IS elements from the *E. coli* K12 chromosome to the bacteriophage P1 plasmid and its derivative represents the most important source of spontaneous mutants of the prophage genome, at least if lysogenic cultures are kept for a relatively long time in their stationary phase (Arber *et al.*, 1979, 1981). The same lysogenic cells grown exponentially without interruption and for the same number of generations contain less spontaneous prophage mutants (Arber and Iida, 1982; W. Arber, unpublished). From 10% to 25% of spontaneous mutations in the *gal* operon were found to be caused by the insertion of IS elements (Shapiro, 1969; Starlinger and Saedler, 1976). It has also been reported that an unexpectedly high proportion of *E. coli* cells that were selected for IS insertions in one locus of the chromosome also acquired an additional insertion in other unselected locations (Read and Jaskunas, 1980). These scattered observations may suggest that transposition of IS elements is a response of cells under stress conditions.

Reif and Saedler (1975) demonstrated that IS*1*-mediated deletion formation is temperature-sensitive, indicating that a temperature-sensitive protein is involved in transposition processes. Interestingly, temperature sensitivity has also been shown in activities of other transposons and site-specific recombinases (Guarneros and Echols, 1973; Kretschmer and Cohen, 1979). Alternatively, the temperature-sensitive character of transposition may be due to temperature-sensitive expression of transposition enzymes, for example, the presence of cold-sensitive repressor or inhibitor encoded by IS elements (Section III,K).

IV. DNA Rearrangements and Gene Transfer Attributable to IS Elements

In this section, we will briefly compare various mechanisms for homologous and nonhomologous (in the literature, sometimes called *illegitimate*) recombination. We will discuss their relevance to the activities of IS elements and then try to explain situations in which microbial gene transfer and gene expression may sometimes depend on recombinational processes attributable to functional or defective IS elements.

A. Homologous Recombination

General recombination involves alignment and subsequent exchange of extended segments of usually fully homologous sequences of genetic material. The geneticist monitors the process by the use of phenotypically distinct mutants that generally do not affect recombination taking place between the loci of the mutations. Therefore, a recombinant DNA molecule carries the same nucleotide sequences as its two parents except for the genetic markers used. The homologous segments involved may be either an entire genome or a region of it, for example, an IS element. Indeed, general recombination is known to act between two IS elements of the same type carried either on the same DNA molecule or on different replicons (Sections III,E and J).

Intramolecular recombination results in either the inversion of the DNA between the elements or the deletion of this DNA plus one copy of the IS element depending on the orientation in which the two interacting IS elements are carried

(Section III,E; Fig. 3). Intermolecular recombination results in replicon fusion. The efficiency of these recombinational processes depends on the length of the IS element, on the presence of particular sequences within or near the IS elements (Section III,B; Stahl, 1979), and probably on the spatial proximity of the two interacting elements.

The processes so far described here depend on the functions of the host *rec* genes. However, processes of apparently homologous recombination also occur at low rates in *rec*$^-$ mutant bacteria. The molecular bases for these interactions are poorly understood. In all likelihood, more than one mechanism may be involved, and at least some of these might represent site-specific (see the next section) rather than general recombination.

A stimulation of homologous recombination has been observed in regions adjacent to an IS*5* insertion in the *cI* repressor gene of coliphage λ (Lieb, 1980). The stimulatory effect on recombination is observed only in the immediate vicinity (50–100 bp) of the IS*5* insertion. IS*5* or IS*1* insertions into other sites in the *cI* gene of λ appear to show the same local effect. It was speculated to be due to local amplification of the DNA mediated by the inceptor sequence within the IS element. Alternatively, insertion may jam branch migration, thereby accumulating recombinants in which crossing over has taken place adjacent to the IS insertion site (M. Lieb, personal communication).

B. Site-Specific Reciprocal Recombinations

Several site-specific recombination systems have been identified and characterized with respect to both the sites and recombinases involved. They include recombination mediated by *int* and *xis* of phage λ (Miller *et al.*, 1981; Mizuuchi *et al.*, 1981; Nash, 1981), by *cre* of phage P1 (Sternberg *et al.*, 1981), by *tnpR* resolvase of Tn*3* and γδ (Reed, 1981b; Reed and Grindley, 1981), and by the inversion systems *gin*, *cin*, and *hin* of phage Mu, phage P1, and *Salmonella*, respectively (Iino and Kutsukake, 1981; Silverman *et al.*, 1981). Restriction enzymes may also be involved in site-specific recombination (Chang and Cohen, 1977). In all these processes, the specific sequences recognized by the site-specific recombinases and recombined with each other are usually very short (e.g., 15 bp in phage λ integration) and are totally or largely identical. But adjacent, nonhomologous DNA sequences may sometimes play essential roles in the interactions (Landy *et al.*, 1979; Mizuuchi *et al.*, 1981; Nash, 1981.). In the *int* system of phage λ, rare crossing over was also shown to take place in DNA sequences adjacent to the recognition site, probably after branch migration (Enquist *et al.*, 1979). These site-specific recombination systems sometimes mediate DNA rearrangements. For example, the phage λ *int* gene product causes deletions (Weisberg and Adhya, 1977; Landy *et al.*, 1979), and the phage P1 *cin* gene product appears to mediate both cointegrations and deletions (Iida *et al.*, 1981c, 1982d; S. Iida, unpublished). For further details, see Chapters 2 and 12.

C. Replicative Recombinations

For the break and join events characteristic of transposable elements that result in element duplication and target oligonucleotide duplication, the term *duplicative* or *replicative recombination* has been suggested (Campbell, 1981; Muster *et al.*, 1981). Accordingly, *replicombinases* are transposition enzymes that help to mediate this replicative recombination. The interconnection of nonhomologous DNA segments appears always to coincide with their localized replication. In all cases studied and probably as a principle, the IS element itself is always fully preserved in the process, and target sequences joined to the ends of the element are always duplicated. The target sequences are usually not unique, and the variability of the junctions depends on the mechanism of target site selection, which is discussed in Section III,F,2.

Replicative recombinations are thus responsible for transposition of single IS elements as well as for transpositional, IS-mediated cointegration, deletion, and inversion (Section III,E; Fig. 3). The last three types of rearrangements can also be obtained by homologous recombination between copies of the same IS elements (Sections III,E and IV,A). However, the transpositional mechanisms can yield a much greater variety of new structures.

D. Other Kinds of Recombinations

Several other DNA rearrangements, the mechanisms of which are not well understood, have also been reported. Certain deletions and duplications or amplifications were postulated to be consequences of recombination between direct repeats of a few base pairs (Farabaugh *et al.*, 1978; Edlund and Normark, 1981). The possible relevance of these processes for excision of IS elements and its occurrence in *recA*$^-$ bacteria has been discussed in Section III,H. Some restructuring within IS2 (segregation of IS2, IS2-61, and IS2-611 from allele IS2-6; see Section III,D, Fig. 2) can be envisioned to occur by *recA*-independent recombination between short, directly repeated sequences (Ghosal and Saedler, 1979). For other deletions and duplications, it is not known whether small repeated sequences are involved.

A few cointegrates have been isolated, the formation of which was apparently not dependent on genetic homology, transposable elements, or any known site-specific recombination mechanism (Pogue-Geile *et al.*, 1980; Ikeda *et al.*, 1981). Some of them could have been either mediated by DNA gyrase or associated with the initiation of DNA replication (Masters *et al.*, 1979; Kilbane and Malamy, 1980; Ikeda *et al.*, 1981).

RecA-independent, RNA polymerase-dependent homologous recombination was also reported (Ikeda and Kobayashi, 1977; Ikeda and Matsumoto, 1979). This kind of recombination may partly explain rare *recA*-independent homologous recombination between IS elements.

E. Conduction and Transduction

Conduction (or conjugative mobilization) and transduction are potent mechanisms of gene transmission in microorganisms. Accumulating data indicate that IS elements play important roles in gene transmissions mediated by conjugative plasmids and phages.

1. Conduction

Conjugative or self-transmissible plasmids are capable of promoting the conjugative transfer of normally physically unlinked, nonconjugative plasmids and the host chromosome (see the reviews by Clark and Warren, 1979; Holloway, 1979). This process is called *conduction.*

If a segment of either the host chromosome or another replicon is recombined with a conjugative plasmid, this DNA segment can later be transferred from donor to recipient bacteria just like genes of the conjugative plasmid itself. This specialized mode of conjugative transmission is called *F'-duction* or *R'-duction,* depending on the kinds of conjugative plasmids mediating the transfer. After transfer into a recipient cell, the DNA segment may segregate and the segregated fragment may replicate autonomously, provided it is genetically equipped to do so. Alternatively, the segregated DNA may become integrated into another replicon by homologous or nonhomologous recombination. IS elements are involved in such gene transfer by actively or passively mediating integration and segregation of DNA segments (see Section III,E).

Indeed, conjugative F and R plasmids often carry IS elements (Table II). These mediate either transpositional cointegration with DNA segments containing no IS elements or reciprocal recombination with another copy of the same IS element (Iida, 1980; Iida and Arber, 1980; Ohtsubo *et al.*, 1981b). Reciprocal recombination usually proceeds by *recA*-dependent, homologous recombination. It can also proceed by site-specific recombination, provided the IS element is equipped for it, for example, γδ (Section III,B). It has been observed that DNA segments integrated in F′ and R′ plasmids are often flanked by IS elements and that the conduction of pBR322 by F is accompanied by transposition of γδ into pBR322 (Ohtsubo *et al.*, 1974; Davidson *et al.*, 1975; Deonier and Mirels, 1977; Ohtsubo and Ohtsubo, 1977; Guyer, 1978; Deonier and Hadley, 1980; Grindley and Joyce, 1981; Guyer *et al.*, 1981). Obviously, not only IS elements on conjugative plasmids but also IS elements on the chromosome may participate in the formation of cointegrates. Besides IS elements, any other nonhomologous recombination mechanism by which a DNA segment is joined to a conjugative plasmid contributes to conductional gene transfer (Kilbane and Malamy, 1980). An F plasmid derivative deleted for all known IS elements except for a small part of IS*3* can mobilize chromosomal markers with reduced frequencies (Guyer *et al.*, 1981; Willetts and Johnson, 1981).

It has been reported that F-mediated chromosomal transfer is largely *recA*-dependent and can be accounted for only partially by the presence of stable Hfr chromosomes in the F^+ population (Cullum and Broda, 1979). However, the suggested "unstable Hfr chromosomes" or "incomplete integration of F" (Curtiss and Stallions, 1969; Evenchik *et al.*, 1969) may be explained partly by the rapid segregation mediated by the site-specific resolution of $\gamma\delta$ or by other recombinations. Because all *E. coli* K12 strains examined contain several copies of IS*2* and IS*3* and some also contain $\gamma\delta$ (Table II), *recA*-dependent recombination between these elements on F and the chromosome results in a significant contribution to F-mediated chromosome transfer. The loci of IS elements, $\gamma\delta$ in particular, are regarded as sex-factor-affinity (or *sfa*) sites (Guyer *et al.*, 1981).

The IncP-1 plasmid mutant R68.45 showed an increase of up to 10^5-fold in its ability to mobilize the chromosome of *P. aeruginosa* PAO as compared to the parental R68 plasmid (Haas and Holloway, 1976), and it also proved capable of mobilizing efficiently the chromosomes of many other gram-negative bacteria (Holloway, 1979). The mutant plasmid had acquired an extra DNA segment of about 2 kb that is, at least in part, a duplication of a sequence already present nearby on R68 (Leemans *et al.*, 1980; Riess *et al.*, 1980; Willetts *et al.*, 1981). IS activities such as transposition, replicon fusion, and deletion formation were shown to be associated with DNA segments identical to or overlapping with the duplication, and these 2-kb mobile elements were called IS*8* (Depicker *et al.*, 1980; Leemans *et al.*, 1980) and IS*21* (Willetts *et al.*, 1981). respectively. The two IS elements are certainly closely related and may even be identical. Acquisition of these IS elements enabled plasmid R68.45 to faciliate high-frequency conduction of genomes that were shown to contain no copy of this IS element (Willetts *et al.*, 1981).

Analogously, IS elements may also be involved in integrative suppression through the integration of a conjugative plasmid into the chromosome (Nishimura *et al.*, 1971, 1973; Iida, 1977; Chandler *et al.*, 1979c).

2. Transduction

Phage-mediated gene transfer is called *transduction*. In specialized transduction, usually only a limited number of donor genes are transmitted, whereas in generalized transduction, any donor DNA segment can be transmitted by a phage (Ozeki and Ikeda, 1968).

As in conduction mediated by F′ and R′ plasmids, the DNA segment transduced in the specialized mode is usually covalently linked to the phage genome. Integration of phage genomes into the host chromosome or into plasmids can be mediated by IS elements (MacHattie and Shapiro, 1978; Reyes *et al.*, 1979b; Shapiro and MacHattie, 1979; Iida, 1980; Iida and Arber, 1980; Iida *et al.* 1981c,d) as well as by site-specific recombination systems encoded by phages, such as the λint^+, the P1*cre*$^+$, and probably the P1*cin*$^+$ systems (Weisberg and

Adhya, 1977; Chesney *et al.*, 1979; Sternberg *et al.*, 1981; Iida *et al.*, 1981c; S. Iida, unpublished). An IS*1* carried on phage λ was able to substitute the λ integration functions (MacHattie and Shapiro, 1978; Shapiro and MacHattie, 1979), and a similar effect with IS*1* was observed in a DNA segment containing an immunity region for specialized transducing λ phages called λ*alt* (Friedman *et al.*, 1981). Phage P1 carries an IS*1* as a natural constituent (Iida *et al.*, 1978), and this IS*1* is also involved in the formation of many specialized transducers (Iida *et al.*, 1978, 1981b, 1981c, 1981d; Schmitt *et al.*, 1979; Iida, 1980; Iida and Arber, 1980; York and Stodolsky, 1981). Among the first specialized transducing phages isolated, besides λd*gal,* were P1 derivatives, P1*lac* (Luria *et al.*, 1960; Rae and Stodolsky, 1974). Some of these P1*lac* phages were later shown to be generated by the activity of IS*1* alone or in combination with γδ of the host chromosome (S. Iida, J. Elliott, and W. Arber, unpublished).

The formation of generalized transducing particles also appears to be facilitated by the recombination functions of IS elements. Among the generalized transducers, phage Mu itself is a transposable DNA element that is always linked to host sequences (Howe, 1973; Bukhari, 1976; Teifel and Schmieger, 1979; see Chapter 3). For phages P1 and P22, density shift experiments indicated that generalized transducing particles did not contain detectable amounts of phage DNA (Ikeda and Tomizawa, 1965; Ebel-Tsipis *et al.*, 1972). The two phages contain double-stranded linear DNA that is terminally redundant and circularly permuted due to head-full packaging of DNA into phage heads. The first round of packaging starts at a unique site on the phage genome, called the *pac* site, and it is thought to be followed by at least four to five rounds of packaging from concatemeric precursor DNA molecules (Bächi and Arber, 1977; Jackson *et al.*, 1978; Walker *et al.*, 1979; Iida, 1980; Sternberg *et al.*, 1981). Several explanations have been presented for the occurrence of generalized transducing particles: (*1*) The host chromosomal DNA is occasionally and mistakenly encapsulated in place of the intracellular concatemeric phage DNA (Ozeki and Ikeda, 1968; Susskind and Botstein, 1978; Botstein and Potteete, 1979). (*2*) The host chromosome may contain DNA sequences resembling the phage *pac* site, and the chromosomal DNA could be packaged from such pseudo-*pac* sites (Schmieger and Backhaus, 1976). (*3*) If the phage DNA were covalently linked to other replicons, for example, the host chromosome, segments of the cointegrate molecules would be packaged starting from the phage *pac* site. As a result of the second or later rounds of packaging, phage heads must contain only host chromosomal DNA or DNA from another replicon (Iida, 1980; Sternberg *et al.*, 1981; Fig. 5). In the case of P1 transduction, replicon fusion can be brought about by the IS*1* of P1 or by either of two site-specific recombination systems, *cre* and *cin,* carried by P1 (Iida, 1980; Iida and Arber, 1980; Iida *et al.*, 1981c; Sternberg *et al.*, 1981; K. Kennedy and S. Iida, unpublished). Alternatively, IS elements on the chromosome may mediate cointegration with P1 DNA.

For the formation of transducing particles, all three different mechanisms may

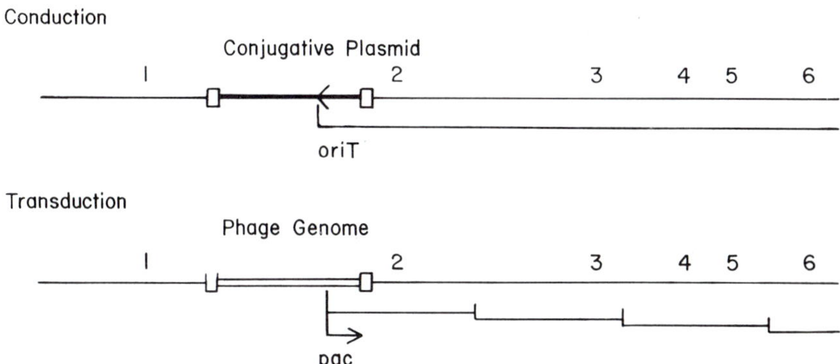

Fig. 5 Possible involvement of IS elements in conduction and transduction. The thin horizontal lines indicate the host chromosomal DNA, and the numbers and boxes flanking either the conjugative plasmid or the phage genome represent chromosomal markers and IS elements, respectively. The arrow at *oriT* (origin of transfer) in the conjugative plasmid marks the starting site and the orientation of conjugative transfer. Phage DNA packaging starts at *pac* in the orientation given by the arrow. The brackets starting at *pac* indicate the segments of DNA (headful amounts) to be packaged into phage particles in consecutive rounds of packaging. Here the integration of the conjugative plasmid, as well as that of the phage genome into the host chromosome between markers 1 and 2, is mediated by an IS element. As discussed in the text, this integration may sometimes be transient, due to rapid resolution of the integrated episome via an element-encoded mechanism, such as *tnpR* of $\gamma\delta$. Alternatively, the integration of episomes could be mediated by an episome-encoded, site-specific recombinase, such as *cre* and *cin* of phage P1. In conjugation, conjugative transfer of the markers 2, 3, 4, 5, and 6 occurs from *oriT* within the plasmid. Similarly, in transduction, the host chromosomal markers are sequentially packaged into the virions and subsequently transferred. Only the chromosomal DNA segment packaged first is covalently linked to phage DNA, whereas most of the transducing phage particles contain only chromosomal DNA.

operate simultaneously with different frequencies. In the case of the R plasmid NR1 or its RTF derivative, both containing IS*1*, it has been shown that under rec^+ conditions at least one-third of the R plasmids transduced by phage P1 are in fact derived from cointegrates brought about by homologous recombination between IS*1* sequences (Iida, 1980; Iida and Arber, 1980; S. Iida and K. Mise, unpublished). Because NR1 and RTF are about 90 and 70 kb, respectively, and the P1 genome has about 90 kb, the cointegrate molecules P1–NR1 and P1–RTF are oversized genomes of which only segments of about 100 kb are accommodated in the mature P1 phage particles. NR1 and RTF sequences in phage heads are thus sometimes flanked by IS*1* and some nonredundant P1 DNA. After injection of these linear DNA molecules, homologous recombination between two IS*1* elements yields circular molecules representing the autonomously replicating plasmids NR1 or RTF without any P1 sequences. Therefore, these "generalized" transductants (Mise and Nakaya, 1977) originate from specialized transduction. The same mechanism was also detected in the transduction of small plasmids, for example, pBR322 and its derivatives (Iida *et al.*, 1981c,d). Chromosomal segments flanked by two IS*1* elements were also shown to form specialized transducing phage derivatives, and the segments were often integrated into the IS*1* of the P1 genome (York and Stodolsky, 1981; S. Iida, unpublished). It is thus conceivable that the cotransduction frequency of markers located adjacent to the chromosomal copies of IS*1* may be affected in P1 transduction.

F. Genesis and Natural History of Transposons Mediated by IS Elements

The demonstration of IS*1* elements as flanking repeats in the Cm^r transposon Tn*9* (MacHattie and Jackowski, 1977) was the first indication that IS elements can also mediate the transposition of genes unrelated to transposition functions. Similar IS*1*-flanked transposons containing various genetic markers were derived from a common precursor, indicating that IS*1*-mediated DNA rearrangements (particularly deletions) occurred and could alter the genetic content of the transposons (Arber *et al.*, 1979; Iida and Arber, 1980). The primary DNA sequences of Tn*9* and its relative, Tn*981* (or TnCm204), revealed that they carry only the structural gene for Cm^r and no putative transposase gene between the flanking IS*1* sequences (Alton and Vapnek, 1979; Marcoli *et al.*, 1980). Another transposon, Tn*1681*, carries only the gene for a heat-stable enterotoxin between two inversely repeated IS*1* elements (So *et al.*, 1979; So and McCarthy, 1980). In addition, a replicon that has acquired an IS*1*-flanked Cm^r transposon now contains two transposons, because both segments flanked by the two IS*1* elements can translocate. Thus, besides the Cm^r transposon, the original replicon with the flanking IS*1* sequences forms a second transposon, a phenomenon called *inverse transposition* (Reif, 1980; Rosner and Guyer, 1980). These various lines of evidence showed that the flanking IS*1* elements alone provided the transposition functions. Analogous findings were derived from studies on Tn*5* (IS*50*), Tn*10* (IS*10*), and Tn*903* (IS*903*) (Chandler *et al.*, 1979b; Grindley and Joyce, 1980; Berg *et al.*, 1981; Foster *et al.*, 1981a; Isberg and Syvanen, 1981; Oka *et al.*, 1981).

The *de novo* genesis of IS*1*-flanked transposons containing a genetic marker was documented in an experimental system (Iida *et al.*, 1981d; Fig. 6). First, an IS*1* element was transposed in the vicinity of the Cm^r determinant on a plasmid. Cm^r genes flanked by direct or inverted repeats of IS*1* were then recovered on the genome of phage λ. The resulting structures were subsequently shown to transpose as a unit. The transposon structures could have been formed by (*1*) IS-mediated cointegration and deletion formation, or (*2*) transposition of a second IS*1* into the plasmid, followed by transposition of this composite structure. The second IS*1* could be a copy of the IS*1* resident either on the same plasmid or on the host chromosome. The genesis of the Tn*5*-flanked *malK*$^+$ transposon Tn*651* has also been reported (Guarente *et al.*, 1980). Many IS-flanked transposons isolated from the natural environment have probably been formed in similar ways (Fig. 6).

The newly formed, IS*1*-flanked Cm^r transposons are quite stable, but occasional decay of the structures was detected (Iida *et al.*, 1981d). This instability was attributed either to (*1*) reciprocal recombination between directly repeated IS*1* elements, (*2*) IS*1*-mediated deletion formation removing the internal part of the transposon, and sometimes extending into neighboring DNA segments, or (*3*) precise or nearly precise excision of the transposon. In *rec*$^+$ cells, the Cm^r transposons flanked by inverted repeats of IS*1* are generally more stable than those flanked by direct repeats (Iida *et al.*, 1981d). The instability of the latter in *rec*$^+$ cells is mainly due to the decay of the transposon by the (1) pathway. This may explain the fact that many naturally isolated transposons carry inverted repeats. Apart from the three pathways for decay of transposons, insertion of another transposon or IS element into a transposon would also alter the structure of the transposon. The restructured transposons may still retain their ability to translocate as units. *In vivo* formation of new composite transposons through such processes was demonstrated (Hänni *et al.*, 1982). Multiple insertions, perhaps followed by IS-mediated DNA rearrangements, are thought to produce the composite transposons found as r-determinants in many R plasmids (Cohen *et al.*, 1978). In addition, spontaneous point mutations or small DNA rearrangements within the newly generated transposon would also alter the nature of the transposon, for example, by reactivation of a gene within a transposon (Iida *et al.*, 1982a).

Transposons flanked by two copies of an IS element are diploid for the IS element. In principle, these copies can be identical, as in Tn*903* (Grindley and Joyce, 1981; Oka *et al.*, 1981). However, as in Tn*5* and Tn*10*, one of the two IS elements can be mutated without loss of the mobility of the transposons (Rothstein *et al.*, 1980, 1981; Foster *et al.*, 1981a). In essence, one complete flanking element provides both the sites and functions required for transposition, whereas the other provides only the site located at the outermost terminus. Therefore, the internal part of the mutated IS element could be deleted without affecting transposition activity (Rothstein *et al.*, 1980; Foster *et al.*, 1981a; Fig. 7D). Although this has been achieved *in vitro*, a transposon that carries such a gene organization with regular IS elements has not yet been isolated from the natural environment.

Figure 7 illustrates the regular IS element derivatives that can act as a new IS element or transposon (depending on whether the DNA segment unrelated to the IS element but contained in the new elements carries a detectable marker or not). If such structures were stable, γδ and its relatives could also form similar new elements. Perhaps the Tc^r transposons Tn*1721*/Tn*1771* (Schmitt *et al.*, 1981; Schöffl *et al.*, 1981) and the Lac transposon Tn*951* (Cornelis *et al.*, 1981) are examples of this kind.

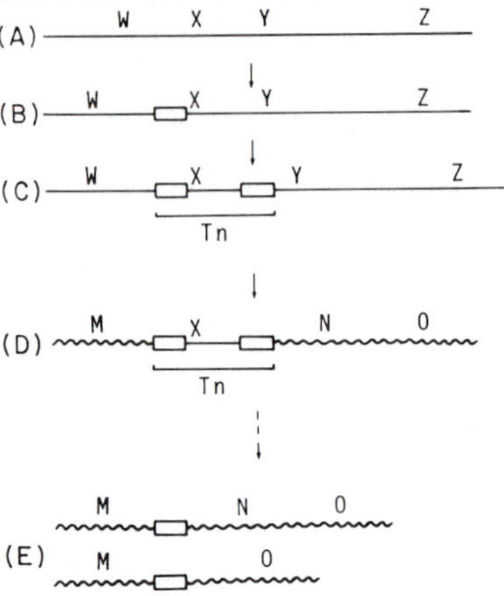

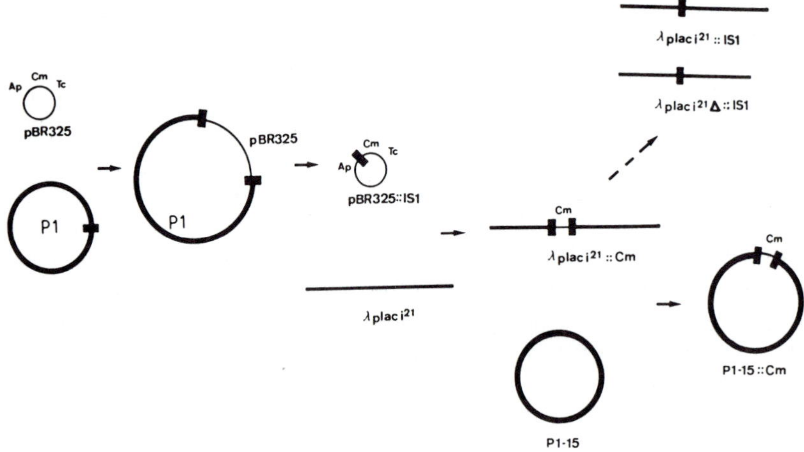

Fig. 6 Genesis of an IS-mediated transposon. The upper drawing represents the general scheme for the genesis and decay of IS-mediated transposons. (A) The horizontal line indicates the DNA carrying the genetic markers W, X, Y, and Z. (B) An IS element (indicated by a box) is transposed into the DNA between markers W and X. (C) Another copy of the same kind of IS element is then transposed into the DNA between X and Y. This can be a copy of the first IS element or one derived from another region of the chromosome. The resulting structure IS–X–IS now is a transposon (Tn). The two flanking IS elements are either in direct or in inverted orientation. (D) The newly generated transposon is transposed into a locus between M and N on the same or another replicon. (E) The

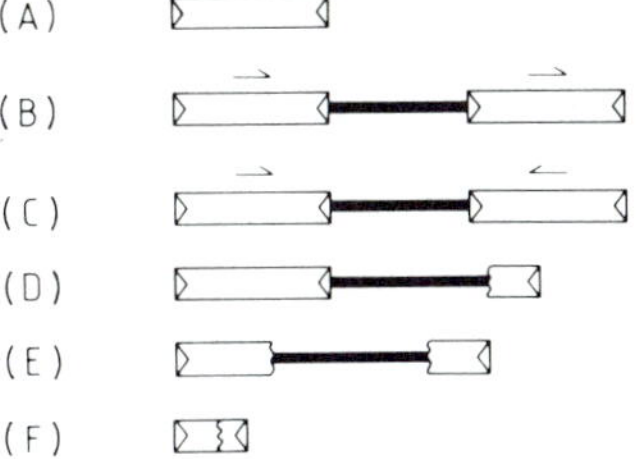

Fig. 7 Comparison of IS elements and their derivatives. An IS element and its terminal inverted repeats are indicated by a box and open triangles within the box, respectively. Black bars indicate a DNA segment unrelated to transposition functions. (A), (B), and (C) represent an IS element, a DNA segment flanked by directly repeated IS elements, and a DNA segment flanked by inversely repeated IS elements, respectively. These structures are known to behave as IS elements or transposons (Berg *et al.*, 1975, 1981; Botstein and Kleckner, 1977; MacHattie and Jackowski, 1977; Arber *et al.*, 1979; Chandler *et al.*, 1979b; So *et al.*, 1979; Iida and Arber, 1980; Iida *et al.*, 1980b, 1981b,d; Reif, 1980; Rosner and Guyer, 1980; Foster *et al.*, 1981a). (D) represents a DNA segment flanked on one side by an intact IS element and on the other side by at least the outer inverted repeat of the same element. This structure sometimes transposes as a unit (Rothstein *et al.*, 1980; Foster *et al.*, 1981a; Schmitt *et al.*, 1981; Schöffl *et al.*, 1981). (E) and (F) show IS elements with an internal insertion (or substitution) and deletion, respectively. Such structures are occasionally active in transposition processes (Meyer *et al.*, 1979; Iida *et al.*, 1981c; Reif and Arber, 1981; Saint-Girons *et al.*, 1981; Machida *et al.*, 1982). The structure shown in (F) is similar to that of element IS*101* (Ravetch *et al.*, 1979; Fischoff *et al.*, 1980). Transposition functions acting on the structure in (F) and some of that in (E) appear to be provided in trans by a related, fully functional element (Miller and Cohen, 1980; Cornelis *et al.*, 1981).

---

transposon has decayed by reciprocal recombination between directly repeated IS elements (shown above) or by transpositional deletion mediated by the IS element at the left side of Tn (shown below). Note that the same decay of the transposon in similar ways may also occur on the replicon where the transposon has been generated.

The lower drawing represents the experimental design to demonstrate the *in vivo* genesis of IS*1*-mediated Cm^r transposons (Iida *et al.*, 1981d). Plasmid pBR325, carrying as markers resistance determinants to ampicillin (Ap), chloramphenicol (Cm), and tetracycline (Tc), and the bacteriophage Pl, carrying IS*1* (indicated by a box), were used for the construction of IS*1*-mediated Cm^r transposons. Plasmids pBR325 :: IS*1* on which IS*1* was integrated between the Ap^r and Cm^r markers were obtained after segregation of pBR325 :: Pl cointegrates. The structure of pBR325 :: IS*1* corresponds to the structure (B) in the upper drawing.

In order to obtain IS*1*-flanked Cm^r transposons *in vivo*, plaque-forming specialized transducing λ phage derivatives carrying only the Cm^r marker derived from pBR325 :: IS*1* were isolated. These λp*laci*^{21} :: Cm contained an IS*1*-flanked Cm^r marker. The transposition capability of the newly generated IS*1*-Cm-IS*1* structure was demonstrated by isolation of P1-15 :: Cm phages, the genome structure of which corresponds to structure (D) in the upper drawing. P1-15 is a Pl phage derivative that lacks IS*1*.

Decay of IS*1*-flanked Cm^r transposons was observed on Cm^s phage derivatives λp*laci*^{21} :: IS*1* and λp*laci*^{21} Δ :: IS*1*.

G. Major DNA Rearrangements Affecting Gene Expression

We have already discussed the multiple effects that IS insertion may have on expression of neighboring sequences (Section III,D). In particular, we have discussed the consequences of small sequence alterations near one end of IS2 (see Fig. 2). However, gene expression may also be affected by DNA rearrangements involving longer segments.

1. Gene Amplification

Tandem duplication and further amplification to partially oligomeric states of genomes have been observed in many systems (Anderson and Roth, 1977). They could have resulted from recombination between small repeats of about 10 bp (Edlund and Normark, 1981) or repeats of two copies of the same IS element (Chandler *et al.*, 1979a; Meyer and Iida, 1979; Shapiro and MacHattie, 1979; Schmitt *et al.*, 1981; Iida *et al.*, 1982b). The IS elements involved need not actively transpose in this process, but they could provide homology for the *rec* pathways or a site-specific recombination system, for example, that of γδ. Therefore, a DNA segment flanked by directly repeated IS elements is ready to amplify under appropriate conditions. In addition, a DNA segment adjacent to the IS element could also be amplified after the formation of flanking direct repeats by the transposition of the element. Such processes have indeed been observed (Meyer and Iida, 1979; S. Iida and J. Meyer, unpublished).

As is true for most mutant selection and enrichment conditions, IS-mediated gene duplication and higher amplification are usually not induced by the selection procedure. Rather, environmental conditions allow cells with a higher number of copies of a particular gene to outgrow the rest of the population. For example, cultures are enriched for cells containing an amplified drug resistance transposon upon growth in increasing concentrations of the antibiotic (Hashimoto and Rownd, 1975; Clewell and Yagi, 1977; Chandler *et al.*, 1979a; Mattes *et al.*, 1979; Meyer and Iida, 1979; Iida *et al.*, 1982b). Interestingly, amplified structures formed by tandem repeats of IS*1*-flanked Cm^r transposons are sometimes relatively stable in *E. coli* K12 rec^+ cells even in the absence of the drug (Meyer and Iida, 1979).

2. Recombinational Switching by Inversion

Recombination between two inverted repeats causes the inversion of the DNA segment between them. Such an inversion can lead to altered expression of genes located on the inverted segment or immediately outside. Inversions may be mediated by a site-specific recombination system, as was discovered to regulate the synthesis of alternative flagellar antigens in *Salmonella* and to determine two alternative host ranges in phage Mu and probably phage P1 (Iino and Kutsukake, 1981; Silverman *et al.*, 1981; Iida *et al.*, 1982d; van de Putte *et al.*, 1980; see Chapter 12). Recombination between inversely repeated IS elements can also act as a genetic switch for gene expression. *Rec*-dependent inversions of DNA segments carrying a promoter have been shown to occur between inverted repeats of IS*10* or between IS*50* and a mutated IS*50* (Berg, 1980; Kleckner and Ross, 1980). Because some IS elements are able to activate an adjacent gene from a promoter within the elements (Section III,D), one may speculate that recombination between the terminal inverted repeats could also produce a switching effect.

3. Operon Fusion

In principle, the strong polar effect exerted by many IS elements could be reverted not only by precise or nearly precise excision but also by deletion of part of the element together with adjacent regions, thereby fusing a new promoter to the operon (Section III,D). This type of deletion was observed among Gal^+ revertants of strains originally carrying IS*1* in either orientation in the regulatory region of the *gal* operon (Sommer *et al.*, 1981). Similarly, deletions removing one IS*10* plus neighboring sequences and fusing Tn*10* genes with *lacZ* were found by Beck (1979). The resulting chimeric proteins had β-galactosidase activity, and their synthesis was under the control of a Tc repressor. Fusion of the *his* operon to Tn*5* operons by deletion of IS*50* has also been reported (Biek and Roth, 1980). In general, IS-mediated deletion followed by the excision of the IS element could result in operon fusion. If IS-mediated deletion extended to the ribosomal RNA operon *rrnX* in which IS insertion causes incomplete polar effects (Morgan, 1980; Brewster and Morgan, 1981), an operon fused to the *rrnX* operon could be activated even in the presence of the IS element.

V. Conclusions, Speculations, and Open Questions

Since the first observation that some spontaneous mutations were caused by inserted DNA sequences, our knowledge of movable genetic elements has increased dramatically within a few years. However, in many respects we have hardly scratched the surface, and many open questions remain to be investigated.

Because we have not yet detected the whole range of possible IS elements in various prokaryotic organisms, their definition and classification remain rather arbitrary. For the time being, we have considered any IS element in the range between 0.7 and 1.8 kb as regular and have classified other elements as exceptions (Section III,B). Hopefully, more elaborate classifications of movable elements will reflect functional and structural rather than size relations.

At present, we do not know if the observed size range of regular IS elements is determined by the functions provided by the IS elements or is related to replication functions. It might perhaps be more than a coincidence that the length of these IS elements corresponds roughly to that of Okasaki fragments. Regular IS elements can code for one or two proteins that are likely to serve in transpositional processes. In this respect, the existence of two types of regular IS elements has been discovered by comparing their DNA sequences (Fig. 1, Table I). Members of the IS*1* family are about 0.8 kb long, and their putative transposases contain fewer than 200 amino acids. The other regular IS elements are more than 1 kb long and can code for peptides of more than 300 amino acids. Most IS elements contain a small open reading frame within a large open reading frame and in opposite orientation to it (Fig. 1). The significance of the former remains to be discovered. The observation that these two overlapping open reading frames often obey the criteria for derivation from an ancient coding pattern is striking and may point to a functional relevance rather than to a random coincidence (Table I). Demonstration that the small open reading frame of IS*5* can be transcribed and translated (Rak *et al.*, 1982) supports this view.

The γδ/Tn*3*-related elements carry two nonoverlapping genes determining a replicombinase (see Section IV,C), here also called *cointegrase,* and a bifunctional protein acting as a resolvase and a repressor. In addition, γδ/Tn*3* elements carry an internal resolution site. Their transposition is likely to proceed in a recombination-first, symmetric manner, and the cointegrate is an obligate intermediate (Section III,G).

If transposases can act in trans, as is suggested for γδ, short, IS*101*-like elements carrying only sites for the action of transposases, that is, pairs of terminal inverted repeats can serve as substrates and thus behave as IS elements by complementation, although perhaps at low frequency (Figs. 1, 7).

Little is known about how the expression of transposition functions is regulated. Because transposition of IS elements is usually an infrequent event, these functions are thought to be repressed under normal conditions (Section III,K). The difficulty of measuring precise transposition frequencies experimentally (Arber and Iida, 1982) is an obstacle in estimating the degree of repression. Frequency considerations have suggested that simultaneous multiple transpositions of IS elements could be related (Read and Jaskunas, 1980; Section III,L). Also, actual IS transposition may be more frequent than observed, if a fraction of cells does not survive the transposition process, for example, as a consequence of insertion into or deletion of essential genes. We also know little about the factors influencing both the regulation and the processes of transposition. Some factors could be encoded by the host chromosome or plasmids (Section III,J). Others may be physiological, such as temperature, growth conditions, and the state of cells. Some of these factors may also influence the host range, copy number, and specific activities of IS elements.

In some cases, transposition enzymes can act in trans, but in others they preferentially act in cis (Section III,K). Therefore, one may speculate that transposases are often unstable and/or stoichiometric rather than catalytic enzymes. Alternatively, the preferential cis action may be related to a possible transcriptional activation of sites for protein–nucleic acid interations.

Some IS elements carry perfect inverted repeats at their ends, and others carry nearly perfect repeats with several mismatched base pairs (Table I). The significance of this mismatching and of the adjacent sequences inside the elements is not known. In some regular IS elements, such as IS*1*, IS2, IS4, and IS*10* (Ohtsubo and Ohtsubo, 1978; Ghosal *et al.*, 1979a; Klaer *et al.*, 1981b; Kleckner, 1981), the large open reading frame terminates within one of the inverted repeats. It is thus likely that such an end sequence is transcribed and partly translated. These processes at the end may be related to the preferential cis effects in transposition, for example, by affecting the ability of this particular end sequence to allow for the initiation of the transpositional processes. The effect that transcription of DNA segments outside an IS element may have on its transposition is unclear. Similarly, it is unknown to what extent transcription of an operon or gene affects the probability that this DNA sequence will become a target for IS transposition.

Although we now know the entire DNA sequences of some IS elements (Table I), we wonder if the nucleotide sequences—particularly of regular IS elements—still hide unknown functional features. If these sequences were used as much as is potentially possible, one might expect overlapping genes and multifunctional sequences to exist. If so, IS elements may generally contain a great deal of information.

Besides insertion into new sites, IS elements have been shown to mediate several kinds of DNA rearrangements, such as cointegration, deletion, inversion, and excision (Section III,E and H). Because IS elements themselves are not phenotypically detectable, their activities are often experimentally monitored as cointegration or deletion rather than as transposition itself. Few data are available on the quantitative relations between these processes and on the parameters influencing the consequences of IS activities, for example, that determine if the outcome will be transposition or cointegration.

Our conceptions of the pathways of IS-mediated DNA rearrangements are still rather speculative. Isolation and characterization of additional host mutants would certainly help to clarify the processes involving IS elements. It is widely accepted that IS elements are duplicated during the transposition processes. However, it is not clear whether the IS element on the donor molecule is *always* recovered in the transposition process, or whether there exists a strict relationship between the duplication and the recombination during transposition of IS elements (Section III,G). So far, most of the investigations that have been carried out are structural analyses of DNA molecules before and after transposition and DNA rearrangements. Detailed elucidation of transposition reactions must await reconstitution of IS-mediated processes in the test tube. Such *in vitro* systems could allow us to examine the proteins, factors, and substrates required for the transposition processes. They would also be expected to yield information regarding intermediate steps, the mode of DNA replication, and the possible role of superhelicity of DNA or other structural particularities, and finally, to reveal the mechanism(s) regulating the transposition reactions.

Still unclear are the parameters that govern the choice of a target sequence and the mechanisms involved in the subsequent introduction of staggered nicks (Section III,F). It remains unknown if a specific nucleotide sequence on the target DNA molecule stringently controls or facultatively modulates these processes. An analogy with the activation at recognition sites and the different modes of DNA cleavage by the various types of restriction endonucleases (Yuan, 1981; Bickle, 1982) can be speculated upon: Some transposition enzymes may introduce staggered cleavage right at a recognition sequence (e.g., probably IS*5* and IS*10*), others at a short, measured distance from a recognition site (e.g., probably IS*4*), and still others at a long and perhaps random distance from a specific sequence serving for the activation of the enzyme. Another possibility is that some transposition systems may not depend on interaction with a specific sequence on a target DNA molecule. If a specific DNA sequence is involved in target site selection, it may still be present in the DNA molecule near the IS element inserted and may also influence transposition events in which this DNA molecule serves as donor (Iida *et al.*, 1981a). However, normal transposition has been observed from donor DNA molecules carrying a deletion on either side of the insertion, indicating that the flanking sequences in the donor are not necessary for transposition per se (Kleckner, 1979; Calos and Miller, 1980b).

Several variants and relatives of a number of IS elements were found in either the same or different microorganisms. The members of such families of IS elements usually differ from each other by small DNA sequence alterations (Tables I and II). However, IS*1* and $\nu\xi$, which also belongs to the IS*1* family, share only 55% sequence homology (Ohtsubo *et al.*, 1981a). In an analogy to the situation found with related phage genomes, for example, the lambdoid family, recombination within the homologous segments may result in viable hybrid IS elements. IS*102* and IS*903* are also related. Nucleotide differences between these two elements are found only in their left halves (Bernardi and Bernardi, 1981a). Perhaps one of these elements represents a parental line and the other a recently formed recombinational hybrid. Such a hybrid has been found in yeast Ty elements (see Chapter 7).

Hybrid IS elements could also evolve under natural conditions through nonhomologous recombination. Some of these newly formed hybrid sequences could have acquired new characteristics, for example, a different target specificity. Therefore, these processes appear to be important for the evolution and development of IS elements, although the evolutional origin of IS elements remains uncertain.

The existence of IS elements and their significance in evolution are controversial issues. On the one hand, their ability to duplicate and to transpose has been postulated to be sufficient to ensure the preservation of IS elements under nonselective conditions (Doolittle and Sapienza, 1980; Orgel and Crick, 1980; Sapienza and Doolittle, 1981). On the other hand, it is important to point out the multiple roles played by IS elements in genetic alterations. We have seen that IS elements in prokaryotes may allow for the generation of new regulatory signals (Section III,D). What is probably more significant is that they generate new gene constellations by a variety of mechanistically different pathways (Section IV). IS elements can mediate gene fusion, gene amplification, operon fusion, generation and decay of transposons, replicon fusion, and replicon segregation. Point mutations and small DNA rearrangements then could act as posttranspositional modulators and could repair or alter a function affected by such IS-mediated DNA rearrangements (Iida *et al.*, 1982a). Through these processes, IS elements also develop the long-term ability possibly to mediate the mobilization of any gene for horizontal gene transfer and to establish stable inheritance of the transferred genetic material in the recipient cells. Microbial geneticists have long taken advantage of these natural characters of IS elements for various genetic manipulations, including isolation of mutants, *in vivo* construction of strains, and genetic mapping through transduction and conjugation (see Bachmann and Low, 1980). There is no doubt that all these processes, daily observed under experimental conditions, also occur in the natural environment. Microorganisms indeed seem to take advantage of these elements for adaptation to environmental conditions. Transposition of IS elements may occur more readily in resting cells (and probably under stress conditions) than in exponentially growing cells (Section III,L). Because growth conditions in the natural environment undergo frequent alterations, IS-mediated DNA rearrangements may be found more frequently in nature than under controlled laboratory conditions.

Evolution affects an individual microbial cell rarely, if at all. It acts at the level of the microbial population. Similarly, IS elements are genetically significant in microbial populations rather than in single cells. Their action is rare enough not to affect equilibrated growth drastically but frequent enough to allow both adaptative switches and evolutionary drifts to take place. Conceptually, it may be interesting to consider to what degree evolution driven by IS elements is random, either in space or in time. Because transposition, and therefore other genetic rearrangements as well, often obey rules in target selection, the genesis of new gene constellations should not be considered to be fully random in space. However, it is an entirely open question if such rules represent an advantage for a given evolving microbial population. Even less is known about the randomness of transposition in time. It seems likely that some internal physiological and some external environmental factors may influence the rate of transpositional events, which would imply nonrandomness in time of IS-driven DNA rearrangements.

In conclusion, we consider IS elements as basic agents of primary importance for genetic variability at the level of populations, although they could survive in nature by their replicative ability alone. Through DNA rearrangement, they act in vertical as well as in horizontal gene exchange, and they may represent one of the principal devices for evolution.

Acknowledgments

We are indebted to John C. W. Shepherd, who provided the computer analysis of DNA sequences for Table I. We thank our many colleagues for helpful discussions and for providing materials prior to publication. The continued support by the Swiss National Science Foundation is gratefully acknowledged.

References

Adhya, S., and Gottesman, M. (1978). Control of transcription termination. *Annu. Rev. Biochem.* **47,** 967–996.

Adhya, S., and Shapiro, J. (1969). The galactose operon of *E. coli* K12. I. Structural and pleiotropic mutations. *Genetics* **62,** 231–247.

Adhya, S., Gottesman, M., and de Crombrugghe, B. (1974). Release of polarity in *Escherichia coli* by gene N of phage λ: termination and antitermination of transcription. *Proc. Natl. Acad. Sci. USA* **71,** 2534–2538.

Ahmed, A., and Scraba, D. (1978). Nature of deletions formed in response to IS*2* in a revertant of the *gal3* insertion of *E. coli. Mol. Gen. Genet.* **163,** 189–196.

Ahmed, A., Bidwell, K., and Musso, R. (1981). Internal rearrangements of IS*2* in *Escherichia coli*. *Cold Spring Harbor Symp. Quant. Biol.* **45,** 141–151.

Alton, N. K., and Vapnek, D. (1979). Nucleotide sequence analysis of the chloramphenicol resistance transposon Tn*9*. *Nature (London)* **282,** 864–869.

Anderson, R. P., and Roth, J. R. (1977). Tandem genetic duplications in phage and bacteria. *Annu. Rev. Microbiol.* **31,** 473–505.

Arber, W., and Iida, S. (1982). The involvement of IS elements of *E. coli* in the genesis of transposons and in spontaneous mutagenesis. *In* "Drug Resistance in Bacteria" (S. Mitsuhashi, ed.), pp. 3–13. Japan Sci. Soc. Press, Tokyo and Thieme-Stratton, New York.

Arber, W., Iida, S., Jütte, H., Caspers, P., Meyer, J., and Hänni, C. (1979). Rearrangements of genetic material in *Escherichia coli* as observed on the bacteriophage P1 plasmid. *Cold Spring Harbor Symp. Quant. Biol.* **43,** 1197–1208.

Arber, W., Hümbelin, M., Caspers, P., Reif, H. J., Iida, S., and Meyer, J. (1981). Spontaneous mutations in the *Escherichia coli* prophage P1 and IS-mediated processes. *Cold Spring Harbor Symp. Quant. Biol.* **45,** 38–40.

Arthur, A., and Sherratt, D. (1979). Dissection of the transposition process: a transposon-encoded site-specific recombination system. *Mol. Gen. Genet.* **175,** 267–274.

Auerswald, E. A., Ludwig, G., and Schaller, H. (1981). Structural analysis of Tn*5*. *Cold Spring Harbor Symp. Quant. Biol.* **45,** 107–113.

Bächi, B., and Arber, W. (1977). Physical mapping of *Bgl*II, *Bam*HI, *Eco*RI, *Hin*dIII, and *Pst*I restriction fragments of bacteriophage P1 DNA. *Mol. Gen. Genet.* **153,** 311–324.

Bachmann, B. J., and Low, K. B. (1980). Linkage map of *Escherichia coli* K12, Edition 6. *Microbiol. Rev.* **44,** 1–56.

Beck, C. F. (1979). A genetic approach to analysis of transposons. *Proc. Natl. Acad. Sci. USA* **76,** 2376–2380.

Beck, C. F., Moyed, H., Ingraham, J. L. (1980). The tetracycline resistance transposon Tn*10* inhibits translocation of Tn*10*. *Mol. Gen. Genet.* **179,** 453–455.

Berg, D. E. (1977). Insertion and excision of the transposable kanamycin resistance determinant Tn*5*. *In* "DNA Insertion Elements, Plasmids and Episomes" (A. I. Bukhari, J. A. Shapiro, S. L. Adhya, eds.), pp. 205–212, Cold Spring Harbor Lab., Cold Spring Harbor, New York.

Berg, D. E. (1980). Control of gene expression by a mobile recombinational switch. *Proc. Natl. Acad. Sci. USA* **77,** 4880–4884.

Berg, D. E., and Drummond, M. (1978). Absence of DNA sequences homologous to transposable element Tn*5* (Km) in the chromosome of *Escherichia coli* K-12. *J. Bacteriol.* **136,** 419–422.

Berg, D. E., Davis, J., Allet, B., and Rochaix, J.-D. (1975). Transposition of R factor genes to bacteriophage λ. *Proc. Natl. Acad. Sci. USA* **72,** 3628–3632.

Berg, D. E., Egner, C., Hirschel, B. J., Howard, J., Johnsrud, L., Jorgensen R. A., and Tlsty, T. D. (1981). Insertion, excision and inversion of Tn*5*. *Cold Spring Harbor Symp. Quant. Biol.* **45,** 115–123.

Bernardi, A., and Bernardi, F. (1981a). Complete sequence of an IS element present in pSC101. *Nucleic Acids Res.* **9,** 2905–2911.

Bernardi, A., and Bernardi, F. (1981b). Site-specific deletions in the recombinant plasmid pSC101 containing the redB-ori region of phage λ. *Gene* **13,** 103–109.

Bertani, G. (1951). Studies on lysogenesis. I. The mode of phage liberation by lysogenic *Escherichia coli*. *J. Bacteriol.* **62,** 293–300.

Besemer, J., and Herpers, M. (1977). Suppression of polarity of insertion mutations within the *gal* operon of *E. coli*. *Mol. Gen. Genet.* **151,** 295–304.

Besemer, J., Görtz, G., and Charlier, D. (1980). Deletions and DNA rearrangements within the transposable DNA element IS*2*. A model for the creation of palindromic DNA by DNA repair synthesis. *Nucleic Acids Res.* **8,** 5825–5833.

Bickle, T. (1982). The ATP dependent types I and III restriction endonucleases. *In* "The Nucleases" (R. J. Roberts and S. Linn, eds.), Cold Spring Harbor Lab. Cold Spring Harbor, New York. (in press)

Biek, D., and Roth, J. R. (1980). Regulation of Tn*5* transposition in *Salmonella typhimurium*. *Proc. Natl. Acad. Sci. USA* **77,** 6047–6051.

Biek, D., and Roth, J. R. (1981). Regulation of Tn*5* transposition. *Cold Spring Harbor Symp. Quant. Biol.* **45,** 189–191.

Bishop, J. O., and Davies, J. A. (1980). Plasmid cloning vectors that can be nicked at a unique site. *Mol. Gen. Genet.* **179,** 573–580.

Blattner, F. R., Fiandt, M., Haas, K. K., Twose, P. A., and Szybalski, W. (1974). Deletions and insertions in the immunity region of coliphage lambda: revised mesurement of the promoter–startpoint distance. *Virology* **62,** 458–471.

Botstein, D., and Kleckner, N. (1977). Translocation and illegitimate recombination by the tetracycline resistance element Tn*10*. *In* "DNA Insertion Elements, Plasmids, and Episomes (A. I. Bukhari, J. A. Shapiro, S. L. Adhya, eds.), pp. 185–203. Cold Spring Harbor Lab., Cold Spring Harbor, New York.

Botstein, D., and Potteete, A. R. (1979). Formation of P22 generalized transducing particles in vitro. *Virology* **95,** 574–576.

Boyen, A., Charlier, D., Crabeel, M., Cunin, R., Palchaudhuri, S., and Glansdorff, N. (1978). Studies on the control region of the bipolar *argECBH* operon of *Escherichia coli*. I. Effect of regulatory mutations and IS2 insertions. *Mol. Gen. Genet.* **161,** 185–196.

Brennan, M. B., and Struhl, K. (1980). Mechanisms of increasing expression of a yeast gene in *Escherichia coli*. *J. Mol. Biol.* **136,** 333–338.

Brewster, J. M., and Morgan, E. A. (1981). Tn9 and IS*1* inserts in a ribosomal ribonucleic acid operon of *Escherichia coli* are incompletely polar. *J. Bacteriol.* **148,** 897–903.

Bukhari, A. I. (1976). Bacteriophage Mu as a transposition element. *Annu. Rev. Genet.* **10,** 389–412.

Bukhari, A. I. (1981). Models of DNA transposition. *Trends Biochem. Sci.* **6,** 56–60.

Calos, M. P., and Miller, J. H. (1980a). Transposable elements. *Cell* **20,** 579–595.

Calos, M. P., and Miller, J. H. (1980b). Molecular consequences of deletion formation mediated by the transposon Tn9. *Nature (London)* **285,** 38–41.

Calos, M. P., Johnsrud, L., and Miller, J. H. (1978). DNA sequence at the integration sites of the insertion element IS*1*. *Cell* **13,** 411–418.

Campbell, A. (1981). Some general questions about movable elements and their implications. *Cold Spring Harbor Symp. Quant. Biol.* **45,** 1–9.

Campbell, A., Berg, D., Botstein, D., Lederberg, E., Novick, R., Starlinger, P., and Szybalski, W. (1977). Nomenclature of transposable elements in prokaryotes. *In* "DNA Insertion Elements, Plasmids and Episomes" (A. I. Bukhari, J. A. Shapiro, and S. L. Adhya, eds.), pp. 15–22. Cold Spring Harbor Lab., Cold Spring Harbor, New York.

Campbell, A., Berg, D. E., Botstein, D., Lederberg, E. M., Novick, R. P., Starlinger, P., and Szybalski, W. (1979). Nomenclature of transposable elements in prokaryotes. *Plasmid* **2,** 466–473, and *Gene* **5,** 197–206.

Chandler, M., Boy de la Tour, E., Willems, D., and Caro, L. (1979a). Some properties of the chloramphenicol resistance transposon Tn9. *Mol. Gen. Genet.* **176,** 221–231.

Chandler, M., Roulet, E., Silver, L., Boy de la Tour, E., and Caro, L. (1979b). Tn*10*-mediated integration of the plasmid R100-1 into the bacterial chromosome: inverse transposition. *Mol. Gen. Genet.* **173,** 23–30.

Chandler, M., Silver, L., Lane, D., and Caro, L. (1979c). Properties of an autonomous r-determinant from R100.1. *Cold Spring Harbor Symp. Quant. Biol.* **43,** 1223–1231.

Chang, S., and Cohen, S. N. (1977). *In vivo* site-specific genetic recombination promoted by the *Eco*RI restriction endonuclease. *Proc. Natl. Acad. Sci. USA* **74,** 4811–4815.

Chesney, R. H., Scott, J. R., and Vapnek, D. (1979). Integration of the plasmid prophages P1 and P7 into the chromosome of *Escherichia coli*. *J. Mol. Biol.* **130,** 161–173.

Choi, C. L., Grinsted, J., Altenbuchner, J., Schmitt, R., and Richmond, M. H. (1981). Transposons Tn*501* and Tn*1721* are closely related. *Cold Spring Harbor Symp. Quant. Biol.* **45,** 64–65.

Chou, J., Casadaban, M. J., Lemaux, P. G., and Cohen, S. N. (1979) Identification and characterization of a self-regulated repressor of translocation of the Tn*3* element. *Proc. Natl. Acad. Sci. USA* **76,** 4020–4024.

Chow, L. T. (1977). Sequence arrangements of the *Escherichia coli* chromosome and of putative insertion sequences, as revealed by electron microscopic heteroduplex studies. *J. Mol. Biol.* **113,** 611–621.

Chow, L. T., and Broker, T. R. (1978). Adjancent insertion sequences IS2 and IS5 in bacteriophage Mu mutants and an IS5 in a lambda d*arg* bacteriophage. *J. Bacteriol.* **133,** 1427–1436.

Chumley, F. G., and Roth, J. R. (1980). Rearrangement of the bacterial chromosome using Tn*10* as a region of homology. *Genetics* **94,** 1–14.

Clark, A. J., and Warren, G. J. (1979). Conjugal transmission of plasmids. *Annu. Rev. Genet.* **13,** 99–125.

Clements, M. B., and Syvanen, M. (1981). Isolation of a *polA* mutation that affects transposition of insertion sequences and transposons. *Cold Spring Harbor Symp. Quant. Biol.* **45,** 201–204.

Clerget, M., Chandler, M., and Caro, L. (1980). Isolation of an IS*1*-flanked kanamycin resistance transposon from R1drd19. *Mol. Gen. Genet.* **180,** 123–127.

Clerget, M., Chandler, M., and Caro, L. (1981). The structure of R1drd19: a revised physical map of the plasmid. *Mol. Gen. Genet.* **181,** 183–191.

Clewell, D. B., and Yagi, Y. (1977). Amplification of the tetracycline resistance determinant on plasmid pAMα1 in *Streptococcus faecalis*. *In* "DNA Insertion Elements, Plasmids, and Episomes" (A. I. Bukhari, J. A. Shapiro, and S. L. Adhya, eds.), pp. 235–246. Cold Spring Harbor Lab., Cold Spring Harbor, New York.

Cohen, S. N., and Shapiro, J. A. (1980). Transposable genetic elements. *Sci. Am.* **242** (2), 36–45.

Cohen, S. N., Brevet, J., Cabello, F., Chang, A. C. Y., Chou, J., Kopecko, D. J., Kretschmer, P. J., Nisen, P., and Timmis, K. (1978). Macro- and microevolution of bacterial plasmids. *In* "Microbiology—1978" (D. Schlessinger, ed.), pp. 217–220. Amer. Soc. Microbiol., Washington, D.C.

Cornelis, G., and Saedler, H. (1980). Deletions and inversion induced by a resident IS*1* of the lactose transposon Tn*951*. *Mol. Gen. Genet.* **178,** 367–374.

Cornelis, G., Sommer, H., and Saedler, H. (1981). Transposon Tn*951* (TnLac) is defective and related to Tn*3*. *Mol. Gen. Genet.* **184,** 241–248.

Cullum, J., and Broda, P. (1979). Chromosome transfer and Hfr formation by F in rec^+ and *recA* strains of *Escherichia coli* K-12. *Plasmid* **2,** 358–365.

Curtiss, R., III, and Stallions, D. R. (1969). Probability of F integration and frequency of stable Hfr donors in F^+ populations of *Escherichia coli* K-12. *Genetics* **63,** 27–38.

Das, A., Court, D., and Adhya, S. (1976). Isolation and characterization of conditional lethal mutants of *Escherichia coli* defective in transcription termination factor rho. *Proc. Natl. Acad. Sci. USA* **73,** 1959–1963.

Das, A., Court, D., Gottesman, M., and Adhya, S. (1977). Polarity of insertion mutations is caused by rho-mediated termination of transcription. In "DNA Insertion Elements, Plasmids, and Episomes" (A. I. Bukhari, J. A. Shapiro, and S. L. Adhya, eds.), pp. 93–97. Cold Spring Harbor Lab., Cold Spring Harbor, New York.

Davidson, N., Deonier, R. C., Hu, S., and Ohtsubo, E. (1975). Electron microscope heteroduplex studies of sequence relations among plasmids of *Escherichia coli*. X. Deoxyribonucleic acid sequence organization of F and of F-primes, and the sequences involved in Hfr formation. *In* "Microbiology 1974" (D. Schlessinger, ed.), pp. 56–65. Amer. Soc. Microbiol., Washington, D.C.

de Crombrugghe, B., Adhya, S., Gottesman, M., and Pastan, I. (1973). Effect of rho on transcription of bacterial operons. *Nature (London) New Biol.* **241,** 260–264.

Delius, H., Charlier, D., and Besemer, J. (1980). The structure of unstable constitutive revertants of mutant *gal*OP-308 :: IS*2*-I. *Mol. Gen. Genet.* **179,** 391–397.

Deonier, R. C., and Hadley, R. G. (1976). Distribution of inverted IS-length sequences in the *E. coli* K-12 genome. *Nature (London)* **264,** 191–193.

Deonier, R. C., and Hadley, R. G. (1980). IS*2*-IS*2* and IS*3*-IS*3* relative recombination frequencies in F integration. *Plasmid* **3,** 48–64.

Deonier, R. C. and Mirels, L. (1977). Excision of F plasmid sequences by recombination at directly repeated insertion sequence 2 elements: involvement of *recA*. *Proc. Natl. Acad. Sci. USA* **74,** 3965–3969.

Deonier, R. C., Hadley, R. G., and Hu, M. (1979). Enumeration and identification of IS*3* elements in *Escherichia coli* strains. *J. Bacteriol.* **137,** 1421–1424.

Depicker, A., de Block, M., Inzé, D., van Montagu, M., and Schell, J. (1980). IS-like element IS*8* in RP4 plasmid and its involvement in cointegration. *Gene* **10,** 329–338.

Doolittle, W. F., and Sapienza, C. (1980). Selfish genes, the phenotype paradigm and genome evolution. *Nature (London)* **284,** 601–603.

Ebel-Tsipis, J., Botstein, D., and Fox, M. S. (1972). Generalized transduction by phage P22 in *Salmonella typhimurium*. I. Molecular origin of transducing DNA. *J. Mol. Biol.* **71,** 433–448.

Edlund, T., and Normark, S. (1981). Recombination between short DNA homologies causes tandem duplication. *Nature (London)* **292,** 269–271.

Egner, C., and Berg, D. E. (1981). Excision of transposon Tn*5* is dependent on the inverted repeats but not on the transposase function of Tn*5*. *Proc. Natl. Acad. Sci. USA* **78,** 459–463.

Engler, J. A., and van Bree, M. P. (1981). The nucleotide sequence and protein-coding capability of the transposable element IS*5*. *Gene* **14,** 155–163.

Enquist, L. W., Nash, H., and Weisberg, R. A. (1979). Strand exchange in site-specific recombination. *Proc. Natl. Acad. Sci. USA* **76,** 1363–1367.

Evenchik, Z., Stacey, K. A., and Hayes, W. (1969). Ultraviolet induction of chromosome transfer by autonomous sex factors in *Escherichia coli*. *J. Gen. Microbiol.* **56,** 1–14.

Farabaugh, P. J., Schmeissner, U., Hofer, M., and Miller, J. H. (1978). Genetic studies of the *lac* repressor. VII. On the molecular nature of spontaneous hotspots in the *lacI* gene of *Escherichia coli*. *J. Mol. Biol.* **126,** 847–863.

Fiandt, M., Szybalski, W., and Malamy, M. H. (1972). Polar mutations in *lac, gal* and phage λ consist of a few IS-DNA sequences inserted with either orientation. *Mol. Gen. Genet.* **119,** 223–231.

Fischhoff, D. A., Vovis, G. F., and Zinder, N. D. (1980). Organization of chimeras between filamentous bacteriophage f1 and plasmid pSC101. *J. Mol. Biol.* **144,** 247–265.

Foster, T. J., Davis, M. A., Roberts, D. E., Takeshita, K., and Kleckner, N. (1981a). Genetic organization of transposon Tn*10*. *Cell* **23,** 201–213.

Foster, T. J., Lundblad, V., Hanley-Way, S., Halling. S. M., and Kleckner, N. (1981b). Three Tn*10*-associated excision events: relationship to transposition and role of direct and inverted repeats. *Cell* **23,** 215–227.

Friedman, D., Tomich, P., Parsons, C., Olson, E., Deans, R., and Flamm, E. (1981). λ*alt*SF: a phage variant that acquired the ability to substitute specific sets of genes at high frequency. *Proc. Natl. Acad. Sci. USA* **78,** 410–414.

Galas, D. J., and Chandler, M. (1981). On the molecular mechanisms of transposition. *Proc. Natl. Acad. Sci. USA* **78,** 4858–4862.

Galas, D. J., and Chandler, M. (1982). Structure and stability of Tn*9*-mediated cointegrates. Evidence for two pathways of transposition. *J. Mol. Biol.* **154,** 245–272.

Galas, D. J., Calos, M. P., and Miller, J. H. (1980). Sequence analysis of Tn*9* insertions in the *lacZ* gene. *J. Mol. Biol.* **144,** 19–41.

Ghosal, D., and Saedler, H. (1978). DNA sequence of the mini-insertion IS*2-6* and its relation to the sequence of IS*2*. *Nature (London)* **275,** 611–617.

Ghosal, D., and Saedler, H. (1979). IS*2-61* and IS*2-611* arise by illegitimate recombination from IS*2-6*. *Mol. Gen. Genet.* **176,** 233–238.

Ghosal, D., Sommer, H., and Saedler, H. (1979a). Nucleotide sequence of the transposable DNA-element IS*2*. *Nucleic Acids Res.* **6,** 1111–1122.

Ghosal, D., Gross, J., and Saedler, H. (1979b). DNA sequence of IS*2-7* and generation of mini-insertions by replication of IS*2* sequences. *Cold Spring Harbor Symp. Quant. Biol.* **43,** 1193–1196.

Gill, R., Heffron, F., Dougan, G., and Falkow, S. (1978). Analysis of sequences transposed by complementation of two classes of transposition deficient mutants of Tn*3*. *J. Bacteriol.* **136,** 742–756.

Glansdorff, N., Charlier, D., and Zafarullah, M. (1981). Activation of gene expression by IS*2* and IS*3*. *Cold Spring Harbor Symp. Quant. Biol.* **45,** 153–156.

Grindley, N. D. F. (1978). IS*1* insertion generates duplication of a nine base pair sequence at its target site. *Cell* **13,** 419–426.

Grindley, N. D. F., and Joyce, C. M. (1980) Genetic and DNA sequence analysis of the kanamycin resistance transposon Tn*903*. *Proc. Natl. Acad. Sci. USA* **77,** 7176–7180.

Grindley, N. D. F., and Joyce, C. M. (1981). Analysis of the structure and function of the kanamycin-resistance transposon Tn*903*. *Cold Spring Harbor Symp. Quant. Biol.* **45,** 125–133.

Grindley, N. D. F., and Sherratt, D. J. (1979). Sequence analysis at IS*1* insertion sites: models for transposition. *Cold Spring Harbor Symp. Quant. Biol.* **43,** 1257–1261.

Guarente, L. P., Isberg, R. R., Syvanen, M., and Silhavy, T. J. (1980). Conferral of transposable properties to a chromosomal gene in *Escherichia coli*. *J. Mol. Biol.* **141,** 235–248.

Guarneros, G., and Echols, H. (1973). Thermal asymmetry of site-specific recombination by bacteriophage λ. *Virology* **52,** 30–38.

Guyer, M. S. (1978). The γδ sequence of F is an insertion sequence. *J. Mol. Biol.* **126,** 347–365.

Guyer, M. S., Reed, R. R., Steitz, J. A., and Low, K. B. (1981). Identification of a sex-factor-affinity site in *E. coli* as γδ. *Cold Spring Harbor Symp. Quant. Biol.* **45,** 135–140.

Haas, D., and Holloway, B. W. (1976). R factor variants with enhanced sex factor activity in *Pseudomonas aeruginosa*. *Mol. Gen. Genet.* **144,** 243–251.

Habermann, P., Klaer, R., Kühn, S., and Starlinger, P. (1979). IS*4* is found between eleven or twelve base pair duplications. *Mol. Gen. Genet.* **175,** 369–373.

Halling, S. M., and Kleckner, N. (1982). A symmetrical six-base-pair target site sequence determines Tn*10* insertion specificity. *Cell* **28,** 155–163.

Hänni, C., Meyer, J., Iida, S., and Arber, W. (1982). Occurrence and properties of the composite transposon Tn*2672:* evolution of multiple drug resistance transposons. *J. Bacteriol.* **150:** 1266–1273.

Harshey, R. M., and Bukhari, A. I. (1981). A mechanism of DNA transposition. *Proc. Natl. Acad. Sci. USA* **78,** 1090–1094.

Hashimoto, H., and Rownd, R. H. (1975). Transition of the R factor NR1 in *Proteus mirabilis:* level of drug resistance of nontransitioned and transitioned cells. *J. Bacteriol.* **123,** 56–68.

Heffron, F., McCarthy, B. J., Ohtsubo, H., and Ohtsubo, E. (1979). DNA sequence analysis of the transposon Tn*3:* three genes and three sites involved in transposition of Tn*3*. *Cell* **18,** 1153–1163.

Heffron, F., Kostriken, R., Morita, C., and Parker, R. (1981). Tn*3* encodes a site-specific recombination system: identification of essential sequences, genes, and the actual site of recombination. *Cold Spring Harbor Symp. Quant. Biol.* **45,** 259–268.

Hirsch, H.-J., Starlinger, P., and Brachet, P. (1972). Two kinds of insertions in bacterial genes. *Mol. Gen. Genet.* **119,** 191–206.

Holloway, B. W. (1979). Plasmids that mobilize bacterial chromosome. *Plasmid* **2,** 1–19.

Holloway, B. W., and Richmond, M. H. (1973). R-factors used for genetic studies in strains of *Pseudonomas aeruginosa* and their origin. *Genet. Res.* **21,** 103–105.

Hopkins, J. D., Clements, M. B., Liang, T. Y., Isberg, R. R., and Syvanen, M. (1980). Recombination genes of the *Escherichia coli* sex factor specific for transposable elements. *Proc. Natl. Acad. Sci. USA* **77,** 2814–2818.

Howe, M. M. (1973). Transduction by bacteriophage Mu-1. *Virology* **55,** 103–117.

Hu, M., and Deonier, R. (1981). Comparison of IS*1*, IS*2* and IS*3* copy number in *Escherichia coli* strains K12, B and C. *Gene* **16,** 161–170.

Hu, S., Ohtsubo, E., Davidson, N., and Saedler, H. (1975a). Electron microscope heteroduplex studies of sequence relations among bacterial plasmids: identification and mapping of the insertion sequences IS*1* and IS*2* in F and R plasmids. *J. Bacteriol.* **122,** 764–775.

Hu, S., Ptashne, K., Cohen, S. N., and Davidson, N. (1975b). αβ sequence of F is IS*3*. *J. Bacteriol.* **123,** 687–692.

Iida, S. (1977). Directed integration of an F′ plasmid by integrative suppression. Isolation of plaque forming lambda transducing phage for the *dnaC* gene. *Mol. Gen. Genet.* **155,** 153–162.

Iida, S. (1980). A cointegrate of the bacteriophage P1 genome and the conjugative R plasmid R100. *Plasmid* **3,** 278–290.

Iida, S. (1983). On the origin of the chloramphenicol resistance transposon Tn*9*. *J. Gen. Microbiol.* (in press).

Iida, S., and Arber, W. (1980). On the role of IS*1* in the formation of hybrids between the bacteriophage P1 and the R plasmid NR1. *Mol. Gen. Genet.* **177,** 261–270.

Iida, S., Meyer, J., and Arber, W. (1978). The insertion element IS*1* is a natural constituent of coliphage P1 DNA. *Plasmid* **1,** 357–365.

Iida, S., Meyer, J., and Arber, W. (1980a). IS*1*-mediated inversion observed in phage P1CmTc1 DNA. *Experientia* **36,** 748.

Iida, S., Meyer, J., Linder, P., and Reif, H. J. (1980b). The kanamycin transposon derived from the R plasmid Rtsl and carried by phage P1Km has flanking direct repeats of 0.8 kb. *Experientia* **36,** 1451.

Iida, S., Marcoli, R., and Bickle, T. A. (1981a). Variant insertion element IS*1* generates 8-base pair duplications of the target sequence. *Nature (London)* **294,** 374–376.

Iida, S., Hänni, C., Echarti, C., and Arber, W. (1981b). Is the IS*1*-flanked r-determinant of the R plasmid NR1 a transposon? *J. Gen. Microbiol.* **126,** 413–425.

Iida, S., Meyer, J., and Arber, W. (1981c). Cointegrates between bacteriophage P1 DNA and plasmid pBR322 derivatives suggest molecular mechanisms for P1-mediated transduction of small plasmids. *Mol. Gen. Genet.* **184,** 1–10.

Iida, S., Meyer, J., and Arber, W. (1981d). Genesis and natural history of IS-mediated transposons. *Cold Spring Harbor Symp. Quant. Biol.* **45,** 27–43.

Iida, S., Marcoli, R., and Bickle, T. A. (1982a). Phenotypic reversion of an IS*1* mediated deletion mutation: a combined role for point mutations and deletions in transposon evolution. *EMBO J.* **1,** 755–759.

Iida, S., Meyer, J., Linder, P., Goto, N., Nakaya, R., Reif, H.-J., and Arber, W. (1982b). The Kanamycin resistance transposon Tn*2680* derived from the R plasmid Rts1 and carried by phage P1Km has flanking 0.8 kb long direct repeats. *Plasmid* **8,** 187–198.

Iida, S., Schrickel, S., and Arber, W. (1982c) On the segregation of IS*1*-mediated cointegrates between bacteriophage P1 DNA and plasmid pBR322 derivatives. *FEMS Microbiology Letters* (in press).

Iida, S., Meyer, J., Kennedy, K. E., and Arber, W. (1982d). A site-specific, conservative recombination system carried by bacteriophage P1. Mapping the recombinase gene *cin* and the cross-over site *cix* for the inversion of the C segment. *EMBO J.* (in press).

Iino, T., and Kutsukake, K. (1981). *Trans*-acting genes of bacteriophages P1 and Mu mediate inversion of a specific DNA segment involved in flagellar phage variation of *Salmonella*. *Cold Spring Harbor Symp. Quant. Biol.* **45,** 11–16.

Ikeda, H., and Kobayashi, I. (1977). Involvement of DNA-dependent RNA polymerase in a *recA*-independent pathway of genetic recombination in *Escherichia coli*. *Proc. Natl. Acad. Sci. USA* **74,** 3932–3936.

Ikeda, H., and Matsumoto, T. (1979). Transcription promotes *recA*-independent recombination mediated by DNA-dependent RNA polymerase in *Escherichia coli*. *Proc. Natl. Acad. Sci. USA* **76,** 4571–4575.

Ikeda, H., and Tomizawa, J. (1965). Transducing fragments in generalized transduction by phage P1. I. Molecular origin of the fragments. *J. Mol. Biol.* **14,** 85–109.

Ikeda, H., Moriya, K., and Matsumoto, T. (1981). *In vitro* study of illegitimate recombination: involvement of DNA gyrase. *Cold Spring Harbor Symp. Quant. Biol.* **45,** 399–408.

Ilyina, T. S., Nechaeva, E. V., Romanova, Y. M., and Smirnov, G. B. (1981a). Isolation and mapping of *Escherichia coli* K-12 mutants defective in Tn*9* transposition. *Mol. Gen. Genet.* **181,** 384–389.

Ilyina, T. S., Romanova, Y. M., and Smirnov, G. B. (1981b). The effect of *tnm* mutations of *Escherichia coli* K-12 on transposition of various movable genetic elements. *Mol. Gen. Genet.* **183,** 376–379.

Isberg, R. R., and Syvanen, M. (1981). Replicon fusions promoted by the inverted repeats of Tn*5*. The right repeat is an insertion sequence. *J. Mol. Biol.* **150,** 15–32.

Isberg, R. R., and Syvanen, M. (1982). DNA gyrase is a host factor required for transposition of Tn*5*. *Cell* **30,** 9–18.

Jackson, E. N., Jackson, D. A., and Deans, R. J. (1978). *Eco*RI analysis of bacteriophage P22 DNA packaging. *J. Mol. Biol.* **118,** 365–388.

Johnsrud, L. (1979). DNA sequence of the transposable element IS*1*. *Mol. Gen. Genet.* **169,** 213–218.

Johnsrud, L., Calos, M. P., and Miller, J. H. (1978). The transposon Tn*9* generates a 9 bp repeated sequence during integration. *Cell* **15,** 1209–1219.

Jordan, E., Saedler, H., and Starlinger, P. (1968). 0^0 and strong-polar mutations in the *gal* operon are insertions. *Mol. Gen. Genet.* **102,** 353–363.

Jund, R., and Loison, G. (1982). Activation of transcription of a yeast gene in *E. coli* by an IS*5* element. *Nature (London)* **296,** 680–681.

Kamp, D., and Kahmann, R. (1981). Two pathways in bacteriophage Mu transposition? *Cold Spring Harbor Symp. Quant. Biol.* **45,** 329–336.

Kilbane, J. J., and Malamy, M. H. (1980). F factor mobilization of non-conjugative chimeric plasmids in *Escherichia coli:* general mechanisms and a role for site-specific *recA*-independent recombination at *oriV1*. *J. Mol. Biol.* **143,** 73–93.

Kitts, P., Symington, L., Burke, M., Reed, R., and Sherratt, D. (1982). Transposon-specified site-specific recombination. *Proc. Natl. Acad. Sci. USA* **79,** 46–50.

Klaer, R., and Starlinger, P. (1980). IS*4* at its chromosomal site in *E. coli* K12. *Mol. Gen. Genet.* **178,** 285–291.

Klaer, R., Pfeifer, D., and Starlinger, P. (1980). IS*4* is still found at its chromosomal site after transposition to *galT*. *Mol. Gen. Genet.* **178,** 281–284.

Klaer, R., Kühn, S., Fritz, H. J., Tillmann, E., Saint-Girons, I., Habermann, P., Pfeifer, D., and Starlinger, P. (1981a). Studies on transposition mechanisms and specificity of IS*4*. *Cold Spring Harbor Symp. Quant. Biol.* **45,** 215–224.

Klaer, R., Kühn, S., Tillmann, E., Fritz, H. J., and Starlinger, P. (1981b). The sequence of IS*4*. *Mol. Gen. Genet.* **181,** 169–175.

Kleckner, N. (1979). DNA sequence analysis of Tn*10* insertions: origin and role of 9 bp flanking repetitions during Tn*10* translocation. *Cell* **16,** 711–720.

Kleckner, N. (1981). Transposable elements in prokaryotes. *Annu. Rev. Genet.* **15,** 341–404.

Kleckner, N., and Ross, D. G. (1979). Translocation and other recombination events involving the tetracycline-resistance element Tn*10*. *Cold Spring Harbor Symp. Quant. Biol.* **43,** 1233–1245.

Kleckner, N., and Ross, D. G. (1980). *recA*-dependent genetic switch generated by transposon Tn*10*. *J. Mol. Biol.* **144,** 215–221.

Kleckner, N., Chan, R. K., Tye, B. K., and Botstein, D. (1975). Mutagenesis by insertion of a drug-resistance element carrying an inverted repetition. *J. Mol. Biol.* **97,** 561–575.

Kleckner, N., Foster, T. J., Davis, M. A., Hanley-Way, S., Halling, S. M., Lundblad, V., and Takeshita, K. (1981). Genetic organization of Tn*10* and analysis of Tn*10*-associated excision events. *Cold Spring Harbor Symp. Quant. Biol.* **45,** 225–238.

Kondo, E., and Mitsuhashi, S. (1964). Drug resistance of enteric bacteria. IV. Active transducing bacteriophage P1CM produced by the combination of R factor with bacteriophage P1. *J. Bacteriol.* **88,** 1266–1276.

Kopecko, D. J. (1980). Specialized genetic recombination systems in bacteria: their involvement in gene expression and evolution. *Prog. Mol. Subcell. Biol.* **7,** 135–234.

Kostriken, R., Morita, C., and Heffron, F. (1981). Transposon Tn*3* encodes a site-specific recombination system: identification of essential sequences, genes, and actual site of recombination. *Proc. Natl. Acad. Sci. USA* **78,** 4041–4045.

Kretschmer, P. J., and Cohen, S. N. (1979) Effect of temperature on translocation frequency of the Tn*3* element. *J. Bacteriol.* **139,** 515–519.

Kretschmer, P. J., and Cohen, S. N. (1979) Effect of temperature on translocation frequency of the Tn*3* element. *J. Bacteriol.* **139,** 515–519.

Kühn, S., Fritz, H. J., and Starlinger, P. (1979). Close vicinity of IS*1* integration sites in the leader sequence of the *gal* operon of *E. coli*. *Mol. Gen. Genet.* **167,** 235–241.

Labigne-Roussel, A., Gerbaud, G., and Courvalin, P. (1981). Translocation of sequences encoding antibiotic resistance from the chromosome to a receptor plasmid in *Salmonella ordonez*. *Mol. Gen. Genet.* **182,** 390–408.

Landy, A., Hoess, R. H., Bidwell, K., and Ross, W. (1979). Site-specific recombination in bacteriophage λ-structural features of recombining sites. *Cold Spring Harbor Symp. Quant. Biol.* **43,** 1089–1097.

Lebek, G. (1963). Ueber die Entstehung mehrfachresistenter Salmonellen. Ein experimenteller Beitrag. *Zentralbl. Bakteriol. Parasitenk. Infektionskr. Hyg. Abt. 1: Orig.* **188,** 494–505.

Leemans, J., Villarroel, R., Silva, B., van Montagu, M., and Schell, J. (1980). Direct repetition of a 1.2 Md DNA sequence is involved in site-specific recombination by the P1 plasmid R68. *Gene* **10,** 319–328.

Lieb, M. (1980). IS*5* increases recombination in adjacent regions as shown for the repressor gene of coliphage lambda. *Gene* **12,** 277–280.

Ljungquist, E., and Bukhari, A. I. (1977). State of prophage Mu DNA upon induction. *Proc. Natl. Acad. Sci. USA* **74,** 3143–3147.

Luria, E. S., Adams, J. N., and Ting, R. C. (1960). Transduction of lactose-utilizing ability among strains of *E. coli* and *S. dysenteriae* and the properties of the transducing phage particles. *Virology* **12,** 348–390.

Lusky, M., Kröger, M., and Hobom, G. (1981). Detection of replicational inceptor signals in IS*5*. *Cold Spring Harbor Symp. Quant. Biol.* **45,** 173–176.

MacHattie, L. A., and Jackowski, J. B. (1977). Physical structure and deletion effects of the chloramphenicol resistance element Tn*9* in phage lambda. *In* "DNA Insertion Elements, Plasmids, and Episomes" (A. I. Bukhari, J. A. Shapiro, and S. L. Adhya, eds.), pp. 219–228. Cold Spring Harbor Lab., Cold Spring Harbor, New York.

MacHattie, L. A., and Shapiro, J. A. (1978). Chromosomal integration of phage λ by means of a DNA insertion element. *Proc. Natl. Acad. Sci. USA* **75,** 1490–1494.

Machida, Y., Machida, C., Ohtsubo, H., and Ohtsubo, E. (1982a). Factors determining frequency of plasmid cointegration mediated by insertion sequence IS*1*. *Proc. Natl. Acad. Sci. USA* **79,** 277–281.

Machida, Y., Machida, C., and Ohtsubo, E. (1982b). A novel type of transposon generated by insertion element IS*102* present in a pSC101 derivative. *Cell* **30,** 29–36.

Malamy, M. H. (1970). Some properties of insertion mutations in the *lac* operon. *In* "The Lactose Operon" (J. R. Beckwith and D. Zipser, eds.), pp. 359–373. Cold Spring Harbor Lab., Cold Spring Harbor, New York.

Malamy, M. H., Fiandt, M., and Szybalski, W. (1972). Electron microscopy of polar insertions in the *lac* operon of *Escherichia coli*. *Mol. Gen. Genet.* **119,** 207–227.

Marcoli, R., Iida, S., and Bickle, T. A. (1980). The DNA sequence of an IS*1*-flanked transposon coding for resistance to chloramphenicol and fusidic acid. *FEBS Lett.* **110,** 11–14.

Masters, M., Andresdottir, V., and Wolf-Watz, H. (1979). Plasmids carrying *oriC* can integrate at or near the chromosome origin of *Escherichia coli* in the absence of a functional *recA* product. *Cold Spring Harbor Symp. Quant. Biol.* **43,** 1069–1072.

Mattes, R., Burkhardt, H. J., and Schmitt, R. (1979). Repetition of tetracycline resistance determinant genes on R plasmid pRSD1 in *Escherichia coli*. *Mol. Gen. Genet.* **168,** 173–184.

Meyer, J., and Iida, S. (1979). Amplification of chloramphenicol resistance transposons carried by phage P1Cm in *Escherichia coli*. *Mol. Gen. Genet.* **176,** 209–219.

Meyer, R., Boch, G., and Shapiro, J. (1979). Transposition of DNA inserted into deletions of the Tn*5* kanamycin resistance element. *Mol. Gen. Genet.* **171,** 7–13.

Meyer, J., Iida, S., and Arber, W. (1980). Does the insertion element IS*1* transpose preferentially into A+T-rich DNA segments? *Mol. Gen. Genet.* **178,** 471–473.

Meyer, J., Ståhlhammar-Carlemalm, M., and Iida, S. (1981). Denaturation map of bacteriophage P1 DNA. *Virology* **110,** 167–175.

Meynell, E., and Datta, N. (1966). The relation of resistance transfer factors of the F-factor (sex-factor) of *Escherichia coli*. *Genet. Res.* **7,** 134–140.

Mickel, S., Ohtsubo, E., and Bauer, W. (1977). Heteroduplex mapping of small plasmids derived from R-factor R12: *in vivo* recombination occurs at IS*1* insertion sequences. *Gene* **2,** 193–210.

Miller, C. A., and Cohen, S. N. (1980). F plasmid provides a function that promotes *recA*-independent site-specific fusions of pSC101 replicon. *Nature (London)* **285,** 577–579.

Miller, H. I., and Friedman, D. I. (1980). An *E. coli* gene product required for λ site-specific recombination. *Cell* **20,** 711–719.

Miller, H. I., Kikuchi, A., Nash, H. A., Weisberg, R. A., and Friedman, D. I. (1979). Site-specific recombination of bacteriophage lambda: the role of host gene products. *Cold Spring Harbor Symp. Quant. Biol.* **43,** 1121–1126.

Miller, J. H., Calos, M. P., Galas, D., Hofer, M., Büchel, D. E., and Müller-Hill, B. (1980). Genetic analysis of transpositions in the *lac* region of *Escherichia coli*. *J. Mol. Biol.* **144,** 1–18.

Miller, H. I., Abraham, J., Benedik, M., Campbell, A., Court, D., Echols, H., Fischer, R., Galindo, J. M., Guarneros, G., Hernandez, T., Mascarenhas, D., Montanez, C., Schindler, D., Schmeissner, U., and Sosa, L. (1981). Regulation of the integration–excision reaction by bacteriophage λ. *Cold Spring Harbor Symp. Quant. Biol.* **45,** 439–445.

Mise, K., and Nakaya, R. (1977). Transduction of R plasmids by bacteriophages P1 and P22. Distinction between generalized and specialized transduction. *Mol. Gen. Genet.* **157,** 131–138.

Mizuuchi, K., Weisberg, R., Enquist, L., Mizuuchi, M., Buraczynska, M., Foeller, C., Hsu, P. L., Ross, W., and Landy, A. (1981). Structure and function of the phage λ *att* site: size, Int-binding sites, and location of the crossover point. *Cold Spring Harbor Symp. Quant. Biol.* **45,** 429–437.

Morgan, E. A. (1980). Insertions of Tn*10* into an *E. coli* ribosomal RNA operon are incompletely polar. *Cell* **21,** 257–265.

Morse, M. L., Lederberg, E. M., and Lederberg, J. (1956). Transductional heterogenotes in *Escherichia coli*. *Genetics* **41,** 758–779.

Musso, R. E., and Rosenberg, M. (1977). Nucleotide sequences at two sites for IS2 DNA insertion. *In* "DNA Insertion Elements, Plasmids, and Episomes" (A. I. Bukhari, J. A. Shapiro, and S. L. Adhya, eds.), pp. 597–600. Cold Spring Harbor Lab., Cold Spring Harbor, New York.

Muster, C. J., and Shapiro, J. A. (1981). Recombination involving transposable elements: on replicon fusion. *Cold Spring Harbor Symp. Quant. Biol.* **45,** 239–242.

Muster, C. J., MacHattie, L. A., and Shapiro, J. A. (1981). Transposition and rearrangements in plasmid evolution. *In* "Molecular Biology, Pathogenicity, and Ecology of Bacterial Plasmids" (S. B. Levy, R. C. Clowes, and E. L. Koenig, eds.), pp. 349–358. Plenum, New York.

Nash, H. A. (1981). Integration and excision of bacteriophage λ: the mechanism of conservative site specific recombination. *Annu. Rev. Genet.* **15,** 143–167.

Nisen, P., and Shapiro, L. (1979). *E. coli* ribosomal RNA contains sequences homologous to insertion sequences IS*1* and IS*2*. *Nature* (*London*) **282,** 872–874.

Nisen, P., Purucker, M., and Shapiro, L. (1979). Deoxyribonucleic acid sequence homologies among bacterial insertion sequence elements and genomes of various organisms. *J. Bacteriol.* **140,** 588–596.

Nishimura, Y., Caro, L., Berg, C. M., and Hirota, Y. (1971). Chromosomal replication in *Escherichia coli*. IV. Control of chromosomal replication and cell division by an integrated episome. *J. Mol. Biol.* **55,** 441–456.

Nishimura, A., Nishimura, Y., and Caro, L. (1973). Isolation of Hfr strains from R^+ and ColV2$^+$ strains of *Escherichia coli* and derivation of an R'*lac* factor by transduction. *J. Bacteriol.* **116,** 1107–1112.

Nomura, N., Yamagishi, H., and Oka, A. (1978). Isolation and characterization of transducing coliphage fd carrying a kanamycin resistance gene. *Gene* **3,** 39–51.

Nyman, K., Nakamura, K., Ohtsubo, H., and Ohtsubo, E. (1981). Distribution of the insertion sequence IS*1* in Gram-negative bacteria. Nature (*London*) **289,** 609–612.

Ohtsubo, H., and Ohtsubo, E. (1977). Repeated DNA sequences in plasmids, phages and bacterial chromosomes, *In:* "DNA Insertion Elements, Plasmids, and Episomes" (A. I. Bukhari, J. A. Shapiro, and S. L. Adhya, eds.), pp. 49–63. Cold Spring Harbor Lab., Cold Spring Harbor, New York.

Ohtsubo, H., and Ohtsubo, E. (1978). Nucleotide sequence of an insertion element, IS*1*. *Proc. Natl. Acad. Sci. USA* **75,** 615–619.

Ohtsubo, E., Deonier, R. C., Lee, H. J., and Davidson, N. (1974). Electron microscope heteroduplex studies of sequence relations among plasmids of *Escherichia coli*. IV. The F sequences in F14. *J. Mol. Biol.* **89,** 565–584.

Ohtsubo, H., Zenilman, M., and Ohtsubo, E. (1980a). Insertion element IS*102* resides in plasmid pSC*101*. *J. Bacteriol.* **144,** 131–140.

Ohtsubo, E., Zenilman, M., and Ohtsubo, H. (1980b). Plasmids containing insertion elements are potential transposons. *Proc. Natl. Acad. Sci. USA* **77,** 750–754.

Ohtsubo, H., Nyman, K., Doroszkiewicz, W., and Ohtsubo, E. (1981a). Multiple copies of iso-insertion sequences of IS*1* in *Shigella dysenteriae* chromosome. *Nature* (*London*) **292,** 640–643.

Ohtsubo, E., Zenilman, M., Ohtsubo, H., McCormick, M., Machida, C., and Machida, Y. (1981b). Mechanism of insertion and cointegration mediated by IS*1* and Tn*3*. *Cold Spring Harbor Symp. Quant. Biol.* **45,** 283–295.

Oka, A., Nomura, N., Sugimoto, K., Sugisaki, H., and Takanami, M. (1978). Nucleotide sequence at the insertion sites of a kanamycin transposon. *Nature* (*London*) **276,** 845–847.

Oka, A., Sugisaki, H., and Takanami, M. (1981). Nucleotide sequence of the kanamycin resistance transposon Tn*903*. *J. Mol. Biol.* **147,** 217–226.

Orgel, L. E., and Crick, F. H. C. (1980). Selfish DNA: the ultimate parasite. *Nature* (*London*) **284,** 604–607.

Ozeki, H., and Ikeda, H. (1968). Transduction mechanisms. *Annu. Rev. Genet.* **2,** 245–278.

Palchaudhuri, S., and Hänni, C. (1979). Deletion mutants of F, FΔ(8.5–17.6) and the mechanism of their formation. *Plasmid* **3,** 598–604.

Palchaudhuri, S., Maas, W. K., and Ohtsubo, E. (1976). Fusion of two F-prime factors in *Escherichia coli* studied by electron microscope heteroduplex analysis. *Mol. Gen. Genet.* **146,** 215–231.

Peterson, P. A., Ghosal, D., Sommer, H., and Saedler, H. (1979). Development of a system useful for studying the formation of unstable alleles of IS*2*. *Mol. Gen. Genet.* **173,** 15–21.

Pfeifer, D., Habermann, P., and Kubai-Maroni, D. (1977). Specific sites for integration of IS elements within the transferase gene of the *gal* operon of *E. coli* K12. *In* "DNA Insertion Elements, Plasmids, and Episomes" (A. I. Bukhari, J. A. Shapiro, and S. L. Adhya, eds.), pp. 31–36. Cold Spring Harbor Lab., Cold Spring Harbor, New York.

Pilacinski, W., Mosharrafa, E., Edmundson, R., Zissler, J., Fiandt, M., and Szybalski, W. (1977). Insertion sequence IS*2* associated with *int*-constitutive mutants of bacteriophage lambda. *Gene* **2,** 61–74.

Pogue-Geile, K. L., Dassarma, S., King, S. R., and Jaskunas, S. R. (1980). Recombination between bacteriophage lambda and plasmid pBR322 in *Escherichia coli*. *J. Bacteriol.* **142,** 992–1003.

Priefer, U. B., Burkhardt, H. J., Klipp, W., and Pühler, A. (1981). IS*R1:* an insertion element isolated from the soil bacterium *Rhizobium lupini*. *Cold Spring Harbor Symp. Quant. Biol.* **45,** 87–91.

Rae, M. E., and Stodolsky, M. (1974). Chromosome breakage, fusion and reconstruction during P1*dl* transduction. *Virology* **58,** 32–54.

Rak, B., Luskey, M., and Hable, M. (1982). Expression of two proteins from overlapping and oppositely oriented genes on transposable DNA insertion element IS*5*. *Nature (London)* **297,** 124–128.

Ravetch, J. V., Ohsumi, M., Model, P., Vovis, G. F., Fischhoff, D., and Zinder, N. D. (1979). Organization of a hybrid between phage fl and plasmid pSC101. *Proc. Natl. Acad. Sci. USA* **76,** 2195–2198.

Read, H. A., and Jaskunas, S. R. (1980). Isolation of *E. coli* mutants containing multiple transpositions of IS sequences. *Mol. Gen. Genet.* **180,** 157–164.

Read, H. A., Sarma, S. D., and Jaskunas, S. R. (1980). Fate of donor insertion sequence IS*1* during transposition. *Proc. Natl. Acad. Sci. USA* **77,** 2514–2518.

Reed, R. R. (1981a). Resolution of cointegrates between transposons γδ and Tn*3* defines the recombination site. *Proc. Natl. Acad. Sci. USA* **78,** 3428–3432.

Reed, R. R. (1981b). Transposon-mediated site-specific recombination:a defined *in vitro* system. *Cell* **25,** 713–719.

Reed, R. R., and Grindley, N. D. F. (1981). Transposon-mediated site-specific recombination *in vitro:* DNA cleavage and protein–DNA linkage at the recombination site. *Cell* **25,** 721–728.

Reed, R. R., Young, R. A., Steitz, J. A., Grindley, N. D. F., and Guyer, M. S. (1979). Transposition of the *Escherichia coli* insertion element γδ generates a five-base-pair repeat. *Proc. Natl. Acad. Sci. USA* **76,** 4882–4886.

Reif, H. J. (1980). Genetic evidence for absence of transposition functions from the internal part of Tn*981*, a relative of Tn*9*. *Mol. Gen. Genet.* **177,** 667–674.

Reif, H. J., and Arber, W. (1981). Analysis of transposition of IS*1-kan* and its relatives. *Cold Spring Harbor Symp. Quant. Biol.* **45,** 40–43.

Reif, H. J., and Saedler, H. (1975). IS*1* is involved in deletion formation in the *gal* region of *E. coli* K12. *Mol. Gen. Genet.* **137.** 17–28.

Reif, H. J., and Saedler, H. (1977). Chromosomal rearrangements in the *gal* region of *E. coli* K12 after integration of IS*1*. *In* "DNA Insertion Elements, Plasmids, and Episomes" (A. I. Bukhari, J. A. Shapiro, and S. L. Adhya, eds.), pp. 81–91. Cold Spring Harbor Lab., Cold Spring Harbor, New York.

Reyes, O., Gottesman, M., and Adhya, S. (1979a). Suppression of polarity of insertion mutations in the *gal* operon and N mutations in bacteriophage lambda. *J. Bacteriol.* **126,** 1108–1112.

Reyes, O., Gottesman, M., and Adhya, S. (1979b). Formation of lambda lysogens by IS2 recombination: *gal* operon-lambda p_R promoter fusions. *Virology* **94,** 400–408.

Reynolds, A. E., Felton, J., and Wright, A. (1981). Insertion of DNA activates the cryptic *bgl* operon in *E. coli* K12. *Nature (London)* **293,** 625–629.

Riess, G., Holloway, B. W., and Pühler, A. (1980). R68.45, a plasmid with chromosome mobilizing ability (Cma) carries a tandem duplication. *Genet. Res.* **36,** 99–109.

Rosenberg, M., Court, D., Shimatake, H., Brady, C., and Wulff, D. L. (1978). The relationship between function and DNA sequence in an intercistronic regulatory region in phage λ. *Nature (London)* **272,** 414–422.

Rosner, J. L., and Guyer, M. S. (1980). Transposition of IS*1*-λBIO- IS*1* from a bacteriophage λ derivative carrying the IS*1-cat*-IS*1* transposon (Tn*9*). *Mol. Gen. Genet.* **178,** 111–120.

Ross, D. G., Grisafi, P., Kleckner, N., and Botstein, D. (1979a). The ends of Tn*10* are not IS*3*. *J. Bacteriol.* **139,** 1097–1101.

Ross, D. G., Swan, J., and Kleckner, N. (1979b). Physical structures of Tn*10*-promoted deletions and inversions: role of 1400 bp inverted repetitions. *Cell* **16,** 721–731.

Ross, D. G., Swan, J., and Kleckner, N. (1979c). Nearly precise excision: a new type of DNA alteration associated with the translocatable element Tn*10*. *Cell* **16,** 733–738.

Rothstein, S. J., and Reznikoff, W. S. (1981). The functional differences in the inverted repeats of Tn*5* are caused by a single base pair nonhomology. *Cell* **23,** 191–199.

Rothstein, S. J., Jorgensen, R. A., Postle, K., and Reznikoff, W. S. (1980). The inverted repeats of Tn*5* are functionally different. *Cell* **19,** 795–805.

Rothstein, S. J., Jorgensen, R. A., Yin, J. C. P., Yong-di, Z., Johnson, R. C., and Reznikoff, W. S. (1981). Genetic organization of Tn*5*. *Cold Spring Harbor Symp. Quant. Biol.* **45,** 99–105.

Rownd, R., Nakaya, R., and Nakamura, A. (1966). Molecular nature of the drug-resistance factors of the enterobacteriaceae. *J. Mol. Biol.* **17,** 376–393.

Saedler, H., and Heiss, B. (1973). Multiple copies of the insertion-DNA sequences IS*1* and IS*2* in the chromosome of *E. coli* K12. *Mol. Gen. Genet.* **122,** 267–277.

Saedler, H., Besemer, J., Kemper, B., Rosenwirth, B., and Starlinger, P. (1972). Insertion mutations in the control region of the *gal* operon of *E. coli*. *Mol. Gen. Genet.* **115,** 258–265.

Saedler, H., Reif, H. J., Hu, S., and Davidson, N. (1974). IS*2*, a genetic element for turn-off and turn-on of gene activity in *E. coli*. *Mol. Gen. Genet.* **132,** 265–289.

Saedler, H., Kuboi, D. F., Nomura, M., and Jaskunas, S. R. (1975). IS*1* and IS*2* mutations in the ribosomal protein genes. *Mol. Gen. Genet.* **141,** 85–89.

Saedler, H., Cornelis, G., Cullum, J., Schumacher, B., and Sommer, H. (1981). IS*1*-mediated DNA rearrangements. *Cold Spring Harbor Symp. Quant. Biol.* **45,** 93–98.

Saint-Girons, I., Fritz, H. J., Shaw, C., Tillmann, E., and Starlinger, P. (1981). Integration specificity of an artificial kanamycin transposon constructed by the *in vitro* insertion of an internal Tn*5* fragment into IS*2*. *Mol. Gen. Genet.* **183,** 45–50.

Sapienza, C., and Doolittle, W. F. (1981). Genes are things you have whether you want them or not. *Cold Spring Harbor Symp. Quant. Biol.* **45,** 177–182.

Sasakawa, C., Uno, Y., and Yoshikawa, M. (1981). The requirement for both DNA polymerase and 5′ to 3′ exonuclease activities of DNA polymerase I during Tn*5* transposition. *Mol. Gen. Genet.* **182,** 19–24.

Schaller, H. (1979). The intergenic region and the origins for filamentous phage DNA replication. *Cold Spring Harbor Symp. Quant. Biol.* **43,** 401–408.

Schmieger, H., and Backhaus, H. (1976). Altered cotransduction frequencies exhibited by HT-mutants of *Salmonella*-phage P22. *Mol. Gen. Genet.* **143,** 307–309.

Schmitt, R., Mattes, R., Schmid, K., and Altenbuchner, J. (1979) Raf plasmids in strains of *Escherichia coli* and their possible role in enteropatogeny. *In* "Plasmids of Medical, Environmental and Commercial Importance: Development in Genetics" (K. N. Timmis and A. Pühler, eds.), Vol. 1, pp. 199–210. Elsevier, Amsterdam.

Schmitt, R., Altenbuchner, J., Wiebauer, K., Arnold, W., Pühler, A., and Schöffl, F., (1981). Basis of transposition and gene amplification by Tn*1721* and related tetracycline-resistance transposons. *Cold Spring Harbor Symp. Quant. Biol.* **45,** 59–65.

Schöffl, F., Arnold, W., Pühler, A., Altenbuchner, J., and Schmitt, R. (1981). The tetracycline resistance transposons Tn*1721* and Tn*1771* have three 38-base-pair repeats and generate five-base-pair direct repeats. *Mol. Gen. Genet.* **181,** 87–94.

Schoner, B., and Kahn, M. (1981). The nucleotide sequence of IS*5* from *Escherichia coli. Gene* **14,** 165–174.

Shapiro, J. A. (1969). Mutations caused by the insertion of genetic material into the galactose operon of *Escherichia coli. J. Mol. Biol.* **40,** 93–105.

Shapiro, J. A. (1979). Molecular model for the transposition and replication of bacteriophage Mu and other transposable elements. *Proc. Natl. Acad. Sci. USA* **76,** 1933–1937.

Shapiro, J. A. (1980). A model for the genetic activity of transposable elements involving DNA replication. *In* "Plasmids and Transposons—Environmental Effects and Maintenance Mechanisms" (C. Stuttard and K. R. Rozee, eds.), pp. 229–244. Academic Press, New York.

Shapiro, J. A., and MacHattie, L. A. (1979). Integration and excision of prophage λ mediated by the IS*1* element. *Cold Spring Harbor Symp. Quant. Biol.* **43,** 1135–1142.

Sharp, P. A., Cohen, S. N., and Davidson, N. (1973). Electron microscope heteroduplex studies of sequence relations among plasmids of *Escherichia coli*. II. Structure of drug resistance (R) factors and F factors. *J. Mol. Biol.* **75,** 235–255.

Shepherd, J. C. W. (1981a). Method to determine the reading frame of a protein from the purine/pyrimidine genome sequence and its possible evolutionary justification. *Proc. Natl. Acad. Sci. USA* **78,** 1596–1600.

Shepherd, J. C. W. (1981b). Periodic correlations in DNA sequences and evidence suggesting their evolutionary origin in a commaless genetic code. *J. Mol. Evol.* **17,** 94–102.

Silver, L., Chandler, M., Lane, H. E. D., and Caro, L. (1980). Production of extrachromosomal r-determinant circles from integrated R100.1: involvement of the *E. coli* recombination system. *Mol. Gen. Genet.* **179,** 565–571.

Silverman, M., Zieg, J., Mandel, G., and Simon, M. (1981). Analysis of the functional components of the phase variation system. *Cold Spring Harbor Symp. Quant. Biol.* **45,** 17–26.

Smirnov, G. B., Ilyina, T. S., Romanova, Y. M., Markov, A. P., and Nechaeva, E. V. (1981). Mutants of *Escherichia coli* affected in the processes of transposition and genomic rearrangements. *Cold Spring Harbor Symp. Quant. Biol.* **45,** 193–200.

So, M., and McCarthy, B. J. (1980). Nucleotide sequence of the bacterial transposon Tn*1681* encoding a heat-stable (ST) toxin and its identification in enterotoxigenic *Escherichia coli* strains. *Proc. Natl. Acad. Sci. USA* **77,** 4011–4015.

So, M., Heffron, F., and McCarthy, B. J. (1979). The *E. coli* gene encoding heat stable toxin is a bacterial transposon flanked by inverted repeats of IS*1*. *Nature (London)* **277,** 453–456.

So, M., Atchinson, R., Falkow, S., Moseley, S., and McCarthy, B. J. (1981). A study of the dissemination of Tn*1681:* a bacterial transposon encoding a heat-stable toxin among enterotoxigenic *Escherichia coli* isolates. *Cold Spring Harbor Symp. Quant. Biol.* **45,** 53–58.

Sommer, H., Cullum, J., and Saedler, H. (1979a). IS*2-43* and IS*2-44:* new alleles of the insertion sequence IS*2* which have promoter activity. *Mol. Gen. Genet.* **175,** 53–56.

Sommer, H., Cullum, J., and Saedler, H. (1979b). Integration of IS*3* into IS*2* generates a short sequence duplication. *Mol. Gen. Genet.* **177,** 85–89.

Sommer, H., Schumacher, B., and Saedler, H. (1981). A new type of IS*1*-mediated deletion. *Mol. Gen. Genet.* **184,** 300–307.

Stahl, F. W. (1979). Special sites in general recombination. Annu. Rev. Genet. **13,** 7–24.

Starlinger, P. (1980). IS elements and transposons. *Plasmid* **3,** 241–259.

Starlinger, P., and Saedler, H. (1976). IS elements in micro-organisms. *Curr. Top. Microbiol. Immunol.* **75,** 111–153.

Sternberg, N., Hamilton, D., Austin, S., Yarmolinsky, M., and Hoess, R. (1981). Site-specific recombination and its role in the life cycle of bacteriophage P1. *Cold Spring Harbor Symp. Quant. Biol.* **45,** 297–309.

Sternglanz, R., DiNardo, S., Voelkel, K. A., Nishimura, Y., Hirota, Y., Becherer, K., Zumstein, L., and Wang, J. C. (1981). Mutations in the gene coding for *Escherichia coli* DNA topoisomerase I affect transcription and transposition. *Proc. Natl. Acad. Sci. USA* **78,** 2747–2751.

Susskind, M., and Botstein, D. (1978). Molecular genetics of bacteriophage P22. *Microbiol. Rev.* **42,** 385–413.

Syvanen, M. (1980). Tn*903* induces inverted duplications in the chromosome of bacteriophage lambda. *J. Mol. Biol.* **139,** 1–17.

Szybalski, W. (1977). IS elements in *Escherichia coli,* plasmids and bacteriophages. *In* "DNA Insertion Elements, Plasmids, and Episomes (A. I. Bukhari, J. A. Shapiro, and S. L. Adhya, eds.), pp. 583–590. Cold Spring Harbor Lab., Cold Spring Harbor, New York.

Teifel, J., and Schmieger, H. (1979). The origin of the DNA in transducing particles of bacteriophage Mu. Density gradient analyses of intact phages. *Mol. Gen. Genet.* **176,** 293–295.

Toussaint, A., and Faelen, M. (1973). Connecting two unrelated DNA sequences with a Mu dimer. *Nature (London) New Biol.* **242,** 1–4.

Toussaint, A., Faelen, M., and Bukhari, A. I. (1977). Mu-mediated illegitimate recombination as an integral part of the Mu life cycle. *In* "DNA Insertion Elements, Plasmids, and Episomes" (A. I. Bukhari, J. A. Shapiro, and S. L. Adhya, eds.), pp. 275–285. Cold Spring Harbor Lab., Cold Spring Harbor, New York.

Trinks, K., Habermann, P., Beyreuther, K., Starlinger, P., and Ehring, R. (1981). An IS*4*-encoded protein is synthesized in minicells. *Mol. Gen. Genet.* **182,** 183–188.

van de Putte, P., Cramer, S., and Giphart-Gassler, M. (1980). Invertible DNA determines host specificity of bacteriophage Mu. *Nature (London)* **286,** 218–222.

Walker, J. T., Iida, S., and Walker, D. H. (1979). Permutation of the DNA in small-headed virions of coliphage P1. *Mol. Gen. Genet.* **167,** 341–344.

Wallace, L. J., Ward, J. M., Bennett, P. M., Robinson, M. K., and Richmond, M. H. (1981). Transposition immunity. *Cold Spring Harbor Symp. Quant. Biol.* **45,** 183–188.

Walz, A., Ratzkin, B., and Carbon, J. (1978). Control of expression of a cloned yeast (*Saccharomyces cerevisiae*) gene (*trp5*) by a bacterial insertion element (IS2). *Proc. Natl. Acad. Sci. USA* **75,** 6172–6176.

Watanabe, T., Ogata, Y., Chan, R. K., and Botstein, D. (1972). Specialized transduction of tetracycline resistance by phage P22 in *Salmonella typhimurium.* I. Transduction of R factor 222 by phage P22. *Virology* **50,** 874–882.

Weisberg, R. A., and Adhya, S. (1977). Illegimitate recombination in bacteria and bacteriophage. *Annu. Rev. Genet.* **11,** 451–473.

Willetts, N., and Johnson, D. (1981). pED100, a conjugative F plasmid derivative without insertion sequences. *Mol. Gen. Genet.* **182,** 520–522.

Willetts, N. S., Crowther, C., and Holloway, B. W. (1981). The insertion sequence IS*21* of R68.45 and the molecular basis for mobilization of the bacterial chromosome. *Plasmid* **6,** 30–52.

York, M. K., and Stodolsky, M. (1981). Characterization of P1*argF* derivatives from *Escherichia coli* K12 transduction. I. IS*1* elements flank the *argF* gene segment. *Mol. Gen. Genet.* **181,** 230–240.

Young, R., Grillo, D. S., Isberg, R., Way, J., and Syvanen, M. (1980). Transposition of the kanamycin-resistance transposon Tn*903*. *Mol. Gen. Genet.* **178,** 681–689.

Yuan, R. (1981). Structure and mechanism of multifunctional restriction endonucleases. *Annu. Rev. Biochem.* **50,** 285–315.

Note Added in Proof

Isberg and Syvanen (1982) demonstrated that supercoiled DNA molecules are preferred targets for Tn*5* transposition. A DNA gyrase mutant affected in supercoiling of DNA reduces Tn*5* transposition. This reduction is suppressed by introduction of an additional *top* mutation, which causes an increase in the superhelical density of DNA (see Section III,J). These findings suggest that single-stranded, that is, preferentially AT-rich, regions in supertwisted DNA may represent the targets for transposition (Section III,F).

Machida *et al.* (1982b) reported the generation of new transposons flanked by one copy of IS*102* at one end, whereas the other end is formed by a very short sequence that is partially homologous to the end of IS*102* and is carried in inverted orientation (see Section IV,F; Fig. 7D). Apparently, IS*102* transposition enzyme can recognize not only the inverted repeat of IS*102*, but also a sequence with homology to the inverted repeat and that is carried on the donor replicon (Section III,B,1). The finding may imply that IS-mediated DNA rearrangements can sometimes also initiate at sequences only partially homologous to the inverted repeat of the IS element, thereby generating a wider variety of new gene constellations (see Sections III,E; IV,C and F; V).

CHAPTER 5

Tn*3* and Its Relatives

FRED HEFFRON

I. Introduction

Tn*3* is the best-studied member of a class of related transposons that are medically important because they confer antibiotic resistance to many pathogenic bacteria. About 20 transposons belonging to this group have been isolated from at least 50 different genera (see Table I). Most of these genera are gram-negative bacteria, but Tn*551* (encoding erythromycin resistance) was isolated from *Staphylococcus aureus* (Kahn and Novick, 1980). Other members of the Tn*3* family encode tetracycline resistance, several encode multiple antibiotic re-

Mobile Genetic Elements

ISBN 0-12-638680-3

Table I Transposons from Various Sources

Element	Associated determinants	Length (kb)	Terminal	Source and comments	References[a]
Tn*3*	Ampicillin-resistant (Amp^r)	4957	38/38	R1drd19	1–6
Tn*1*, Tn2, Tn*801*, Tn*802*, Tn*901*, Tn*902*, Tn*401*, Tn*1701*, Tn*2601*, Tn*2602*, Tn*2660*	Amp^r	5	—	Plasmids in *Pseudomonas*, *Salmonella*, *Homophilus influenzae*, *Neisseria gonorrhoeae*, and *Proteus*	7–12
γδ	?	5.8	36/37	F episome	13–16
IS*101*	None	0.209	35/38	*Salmonella panama*, IS101 probably contains only sites, no functions	17, 18
Tn*501*	Mercury-resistant (Hg^r)	8.2	35/38	*Pseudomonas* plasmid, pUS1	19–24
Tn*1721*, Tn*1771*	Tetracycline-resistant (Tc^r)	11.4	35/38	pSRD1, pFS202	25–28
Tn*2603*	Oxacillin-resistant (Ox^r), streptomycin-resistant (Sm^r), sulfonamide-resistant (Su^r), Hg^r	22	—	RGN238; *Escherichia coli* only known transposon specifying resistance to oxacillin	29

Tn*21*	Streptomycin-resistant (Sm^r), Su^r, Hg^r, Amp^r, Sm^r, Su^r	19.6	35/38	R100, *Shigella flexneri*	30
Tn*4*	Amp^r, Sm^r, Su^r	23.5	—	R65, *E. coli*	31–35
Tn*551*	Erythromycin-resistant (Ery^r)	5.3	35/35	*Staphylococcus aureus* plasmid	36–38
Tn*917*	Ery^r	5.3	—	pAD2 from *Streptococcus faecalis*.Tn*917* is probably identical to Tn*551*. Tn*917* has been shown to be inducible for transposition functions in the presence of small amounts of erythromycin. It appears to be similar to Tn*501* in that regard.	

[a] 1, Kopecko and Cohen (1975); 2, Heffron *et al.* (1977b,1979); 3, Gill *et al.* (1978, 1979); 4, Kostriken *et al.* (1981); 5, Chow *et al.* (1979a,b); 6, Casadaban *et al.* (1982); 7, Hedges and Jacob (1974), 8, Bennett and Richmond (1976); 9, Chiang and Clowes (1980); 10, Bennett *et al.* (1977); 11, Heffron *et al.* (1975a,b, 1976); 12, Wang *et al.* (1980); 13, Guyer (1978); 14, Reed (1981a,b); 15, Reed *et al.* (1979); Reed and Grindley (1981); 16, Kitts *et al.* (1982); 17, Ravetch *et al.* (1979); 18, Fischoff *et al.* (1981); 19, Stanisich *et al.* (1977); 20, Bennett *et al.* (1978); 21, Schmitt *et al.* (1981); 22, Brown *et al.* (1980); 23, Choi *et al.* (1981); 24, Altenbucher *et al.* (1981); 25, Choi *et al.* (1981); 26, Schmitt *et al.* (1979, 1981); 27, Schoffl *et al.* (1981); 28, Altenbucher *et al.* (1981); 29, Yamamoto *et al.* (1981a,b); 30, de la Cruz and Grinsted (1982); 31, Kopecko and Cohen (1975); 32, Kopecko *et al.* (1976); 33, Nisen *et al.* (1977); 34, Sharp *et al.* (1973); 35, Clerget *et al.* (1981); 36, Kahn and Novick (1980); 37, Novick *et al.* (1979, 1981); 38, Tomich *et al.* (1980).

sistance, and one carries the genes for lactose metabolism. Tn*3* and its immediate relatives, Tn*1* and Tn*2*, carry the most common penicillinase gene found in gram-negative bacteria. (These transposons, and several like them, are so closely related that they probably differ by only a few base changes. For clarity, they will all be referred to as *Tn3*.) This penicillinase is of medical importance because it has been identified in disease-causing bacteria such as *Haemophilus influenzae, H. ducreyi,* and *Neisseria gonorrhoeae* (Brunton *et al.*, 1981; Fayet *et al.*, 1982; Elwell *et al.*, 1975). Pencillin resistance in these organisms has necessitated the use of new antibiotics to combat them. The rise of so many new antibiotic-resistant bacteria in only the past few years must be due in part to the transposability of such antibiotic resistance genes, as well as to other factors, such as indiscriminate use of antibiotics in Third World countries and the addition of antibiotics to animal feed.

The Tn*3* family, as well as other transposons, are usually found on plasmids from antibiotic-resistant bacteria, but they may transpose to bacteriophages and to the chromosome of *Escherichia coli* and many other bacteria. Infectious antibiotic resistance (encoded on plasmids) usually results from production of an enzyme that inactivates the antibiotic. In the case of Tn*3*, the enzyme is a β-lactamase that cleaves the lactam ring of penicillin, thus rendering it harmless to the bacteria. The origin of such resistance genes, and the resistance plasmids on which these genes are found, have long been a mystery. Surprisingly, antibiotic resistance predates the antibiotics era (Smith, 1967), although undoubtedly its epidemic spread in the past 20 years has resulted from the large-scale use and misuse of antibiotics.

Before the discovery of antibiotic-resistance transposons in 1974, there were several observations suggesting that plasmid antibiotic resistance was unusual. For example, it had been realized that enteric penicillin resistance was able to move from one plasmid to another R factor (Datta *et al.*, 1971) and to the *E. coli* chromosome (Richmond and Sykes, 1972). These observations were not understood until Hedges and Jacob (1974) showed that plasmids that acquired penicillin resistance always showed the same increase in molecular weight. Thus, the same sequence carrying a gene for penicillin resistance must be inserted each time. Those workers first realized that this jumping sequence must be a close kin of the IS sequences identified earlier (Jordan *et al.*, 1968; Shapiro, 1969), but that unlike the IS sequences, it carried an antibiotic-resistance gene. This provided a plausible explanation for the reason so many unrelated plasmids could carry the same antibiotic-resistance genes, that is, by transposition. This fact was first demonstrated by Heffron *et al.* (1975b), who showed that unrelated plasmids conferring penicillin resistance, isolated from many different bacteria, all contained the same transposable sequence, Tn*3*.

Tn*3*-like elements have a similar structure and are transposed by a common mechanism. Compared with the composite elements (see Chapter 4), the Tn*3* family can be thought of as analogous to single IS elements. Every member of this family has a short inverted repeated sequence (35–40 bp) at either end. The short repeats of different members of the family are all related. In addition, they

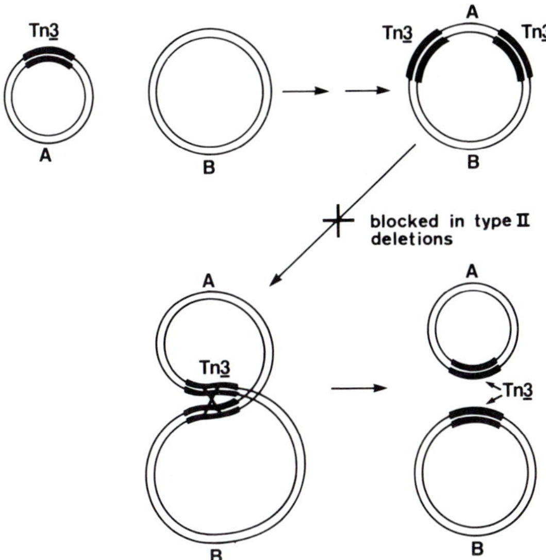

Fig. 1 Transposition of Tn*3* takes place via a cointegrate intermediate. Cointegrates in Tn*3* transposition are observed for some internal deletions in the transposon. The correct explanation for this observation was first proposed by Gill *et al*. (1978), who hypothesized that the cointegrates were intermediates in Tn*3* transposition and that their resolution was blocked because the internal deletion removed an internal recombination site (IRS).

all encode a related high molecular weight protein that is essential for transposition. Another common feature is that they are flanked by a 5-bp duplication of target DNA. In addition to these structural similarities, we now know that the Tn*3*-like transposons are all transposed by a two-step mechanism (see Fig. 1). In the first step, the donor and recipient DNA molecules become fused via a duplication of the transposon (a cointegrate). In the second step, the duplicated copies undergo recombination to yield a precise transposition.

Analysis of Tn*3* transposition mutants first suggested that Tn*3* might encode a site-specific recombination system. A subset of internal deletions in Tn*3* (type II deletions, as seen in Fig. 3) formed fusions with F rather than a precise transposition event (Heffron *et al.*, 1977). Heteroduplexes demonstrated that the fusions had the cointegrate structure shown in Figs. 1 and 2 (Gill *et al.*, 1978). It was proposed that cointegrates might be intermediates in Tn*3* transposition and that the function deleted was part of a site-specific recombination system (see Fig. 1; Gill *et al.*, 1978; Shapiro, 1979). Furthermore, because the phenotype of type II deletions is dominant, type II deletions would have removed an internal recombination site (Gill *et al.*, 1978). In view of the small amount of data available to substantiate this idea at the time, I find it surprising that it turned out to be completely correct. Cointegrates can be observed only in Tn*3* mutants blocked at the recombination step. For other transposons, the recombination step is not this

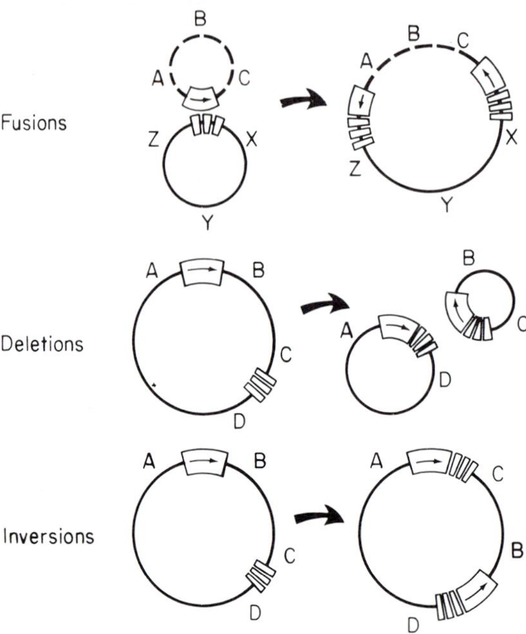

Fig. 2 Fusions, deletions, and inversions mediated by a transposon. This figure shows three of the most common events in which Tn*3* participates. Tn*3* is depicted as a box containing an arrow. The locations of the 5-bp target sequence and the duplicated copies are shown as narrow boxes. During cointegration, a 5-bp sequence is duplicated at either end of the two transposons (McCormick *et al.*, 1981). The predicted location of the 5-bp duplication is shown for the other two events, adjacent deletion and replicative inversion. During replicative inversion, the DNA intervening between the two copies of the transposon is also inverted (Chiang *et al.*, 1982a,b).

efficient or takes place only under certain conditions (e.g., induction with mercury for Tn*501*; see Section II,A).

Tn*3*-like transposons share several other properties. Their specificity of insertion is complex. Most insertions are clustered in narrow regions of the target DNA, whereas insertion seldom takes place in other regions. Finally, Tn*3*, along with its relatives, is unique among transposons in showing immunity in cis to subsequent insertions of Tn*3* from another replican (Robinson *et al.*, 1977). That is, a resident copy of Tn*3* is somehow able to prevent further insertions of Tn*3* into the same molecule. Thus, the Tn*3* family shows many differences from composite transposons. Composite transposons are made up of two long repeated sequences (the two ISs), which are able to transpose on their own. The short repeats at the ends of Tn*3* are not capable of independent transposition. Composite transposons generally duplicate a 9-bp sequence of target DNA on insertion; Tn*3* duplicates a 5-bp sequence. Furthermore, in composite transposons, either precise transposition or cointegration can be the end product of transposition, and cointegrates when formed, appear to be relatively stable. Thus, composite elements are not transposed by a discrete two-step mechanism, as are Tn*3*-like transposons.

II. Element-Specific Recombination Events

A. Fusions

Like members of the composite group, Tn*3* transposons participate in a number of nonhomologous recombination events aside from transposition. As has been mentioned, they appear to transpose by a mechanism in which cointegrates are intermediates. Thus, these transposons can fuse together separate plasmids, as shown in Figs. 1 and 2. For Tn*3*, however, the cointegrate is unstable except in Tn*3* mutants blocked at the recombination step. In Tn*501*, which confers resistance to mercury, cointegrate resolution is slow or inefficient, except when the cells are exposed to low levels of mercury (Sherratt *et al.*, 1981; Arthur *et al.*, 1981). The mercury appears to function as an inducer of Tn*501* transposition functions. Many of the larger transposons in this family, for example, Tn*21*, form cointegrates that are only slowly broken down, perhaps because of their size (de la Cruz and Grinsted, 1982). The recombination sites may be so far apart in the cointegrate that recombination is limited by the low probability that two sites will come into contact.

B. Adjacent Deletion Formation

DNA located adjacent to the end of a transposon suffers frequent deletions and inversions catalyzed by the transposon (see Fig. 2). Both adjacent deletions and inversions are a prediction of the molecular models for transposition, as will be discussed later, and are identical to the first step of an intramolecular transposition event. Deletions adjacent to Tn*3* always extend precisely from the end of the transposon to an outside sequence. Adjacent deletions that started at one end of a Tn*3* insertion in phage P22 all had breakpoints in P22 at only a few sites, suggesting that these deletions terminated at specific sequences (Weinstock *et al.*, 1979). This high specificity for deletion endpoints contrasted to the regional specificity for insertion of Tn*3*, as described in the next paragraph. The frequency of adjacent deletion from Tn*3* and its specificity seem to depend on the plasmid or bacteriophage in which it is located. For example, when Tn*3* was transposed to F, over 95% of the insertions examined contained such deletions (Miller *et al.*, 1980). Deletion endpoints were not clustered in this case. It is possible that the very tight clustering of deletion endpoints in P22 arose from some unknown barriers in P22 necessary for phage viability. Deletions in several other plasmids occurred less frequently; these included R388, RSF1010, pSC101, and ColE1 (Nisen *et al.*, 1977; So *et al.*, 1975; Heffron *et al.*, 1975a; R. Gill and F. Heffron, unpublished). We have no current explanation for why adjacent deletion frequency should be high in some plasmids but low in others. Tn*3* may also stimulate another type of deletion in which the deletion actually starts within the transposon (and not at the inverted repeat) and terminates somewhere within the plasmid (McCormick *et al.*, 1981; Ohtsubo *et al.*, 1981).

C. Inversions

During replicative inversion, a second inverted copy of the transposon is inserted elsewhere on a plasmid. In the process, the DNA that intervenes between the old and new copies of the transposon is inverted (see Fig. 2). Inversion formation occurs with the same frequency as adjacent deletion and within an order of magnitude of the frequency of transposition (Chiang and Clowes, 1980; Chiang *et al.*, 1981). Adjacent deletions and inversion formation require the transposase but not the resolvase (see Section III,C functions; McCormick *et al.*, 1981; Wong and F. Heffron, unpublished observation).

D. Precise Excision

Tn*3* does not appear to excise precisely. For example, insertion of Tn*3* was used to generate auxotrophic mutations in *E. coli* by transposition to the chromosome (Harayama *et al.*, 1981; Wang *et al.*, 1980). Only a few of the auxotrophic mutations could be reverted to prototrophy. Even in these cases, revertants were ampicillin-resistant and the ampicillin resistance was at or near its original location. Thus, these revertants might have arisen in some way other than precise excision, such as by rearrangement of Tn*3* or by a mutation within the transposon. In what is probably the best example of a nonreverting Tn*3*-family insertion, a sequenced IS*101* insertion (known not to contain an adjacent deletion) failed to revert at a frequency of one in 10^{11} (Ravetch *et al.*, 1979).

E. Duplication of Target DNA

Insertion of Tn*3* generates a duplication of a 5-bp sequence originally present once in the target DNA at either end of the transposon (Ohtsubo *et al.*, 1979; Tu and Cohen, 1980). In Tn*3* mutants blocked after the cointegration step, the 5-bp duplication has been identified as flanking the sites where the two copies of the transposon join the recipient DNA (McCormick *et al.*, 1981). Most of the sequenced 5-bp duplications are high in A+T, reflecting the fact that insertion of Tn*3* occurs preferentially into A+T-rich DNA (Tu and Cohen, 1980). The 5-bp duplication following transposition apparently arises as part of the mechanism of integration and is unrelated to the 5-bp duplication on the donor transposon. The duplication is lost by adjacent deletion and replicative inversion (Chiang *et al.*, 1982b; Chiang and Clowes, 1980; Ohtsubo *et al.*, 1979; McCormick *et al.*, 1981). Copies of Tn*3* missing the 5-bp duplication can still transpose (Ohtsubo *et al.*, 1979). IS*101*, Tn*1721*, Tn*501*, Tn*551*, γδ, Tn*951*, and Tn*21* all generate 5-bp duplications on insertion (see Table I for references).

F. Target Specificity

Transposable elements show surprising differences in their specificity of insertion. The bacteriophage Mu inserts promiscuously into *lacZ*. In fact, out of 75 insertions into *lacZ*, every one could be separated from the next by recombination (Bukhari and Zipser, 1972; Daniell *et al.*, 1971). Insertion of IS*5*, in contrast, is highly specific. It inserts at only a few sites, and these must contain the sequence C T/A A G/A (Engler and van Bree, 1981). Study of a large collection of Tn*3* insertions has shown that insertions are clustered in some parts of the recipient molecule but occur infrequently in others (Weinstock *et al.*, 1979; Grinsted *et al.*, 1978; Kretschmer and Cohen, 1977; Heffron *et al.*, 1975a). Even within a cluster, their distribution is nonrandom (Heffron *et al.*, 1975a). This type of target specificity has been called *regional specificity*. Independent insertions can take place at the same nucleotide (Tu and Cohen, 1980). A detailed analysis of hot spots for Tn*3* insertion has suggested that many insertions take place preferentially into A+T-rich DNA and near a sequence that is homologous with the end of Tn*3* (Tu and Cohen, 1980). A similar homology has been observed between a sequence in IS*1* and hot spots for Tn*9* insertion in *lacZ* (Miller *et al.*, 1980). Tn*3*'s preference for A+T-rich sequences may suggest that local denaturing of the DNA must take place before insertion. The frequent insertion near a sequence related to the end of Tn*3* may reflect a preference of the Tn*3* transposase for sequences related to its normal substrate (i.e., the ends of Tn*3*).

A particularly thorough study of Tn*3* and Tn*501* insertions has demonstrated how complicated this phenomenon is. In that study, the plasmid RP1, together with several derivatives of RP1 containing either Tn*501* or Tn*3*, were used as targets (Grinsted *et al.*, 1978). The authors found that Tn*501* was a hot spot for Tn*3* insertion in one location and a cold spot for Tn*3* insertion in another. Insertions took place near the ends of Tn*501*, suggesting a preference for a sequence related to the inverted repeats (IRs), as later observed by Tu and Cohen (1980). What determines this complicated transposition specificity is not known. Perhaps there is something about the physical structure of the recipient DNA that is responsible. It may be that particular regions of a plasmid are more likely to be exposed, in contrast to other regions of the plasmid, which are always inside the molecule and inaccessible to an incoming transposon.

III. The Genetic Organization of Tn3

More is known about the structure of Tn*3* than about other members of this group; its sequence is known, there is a large collection of mutations within Tn*3*, and fusion proteins with β-galactosidase (*lacZ*) have been constructed to *tnpA* and *tnpR* (Ohtsubo *et al.*, 1979; Heffron *et al.*, 1977b, 1978, 1979; Gill *et al.*, 1979; Chou *et al.*, 1979a,b). This thorough analysis has resulted in a relatively simple map, shown in Fig. 3. We can divide Tn*3* sites and functions into those required for replicative recombination and those required for resolution of cointegrates. The transposase (product of *tnpA*) and the two IRs are both required for replicative recombination. The resolvase (product of *tnpR*) and the internal resolution site (IRS) are required for resolution of cointegrates. Regulation of *tnpA* and *tnpR* expression is complex and will be dealt with separately. In addition, Tn*3* displays transposition immunity (Robinson *et al.*, 1977). That is, the frequency of Tn*3* insertion into a plasmid already containing a copy of Tn*3* is reduced by 10^{-3} to 10^{-6}. The frequency of transposition to other plasmids in the same cell is unaffected. The mechanism of transposition immunity is unknown, and we do not yet know what parts of Tn*3* are required for immunity.

The genetic map of Tn*3* was initially formulated from the analysis of two sets of Tn*3* mutants, both constructed by the techniques of *in vitro* mutagenesis. Mutants of the first set contained deletions of various sizes having one or both endpoints within the transposon (Heffron *et al.*, 1977). The deletions were constructed by treating DNA of RSF1050 (a small plasmid containing Tn*3*) with a combination of nonspecific nucleases so as first to linearize the DNA and then to remove a short double-stranded sequence. The linear DNA was transformed directly into *E. coli* where the bacteria recircularized the DNA. The second collection was constructed by random *in vitro* insertion into Tn*3* of an oligonucleotide containing an *Eco*RI restriction site (a linker), as shown in Fig. 4 (Heffron *et al.*, 1978). This method has the advantage that the site of the mutation can be determined merely by cleavage with the cognate restriction enzyme (*Eco*RI in this case). Furthermore, the new restriction sites can be used in sequencing the transposon and to make second-generation mutations such as deletions between any two linker mutations. The construction technique is shown in Fig. 4 and described in the legend.

Both kinds of mutations were screened for their effect on transposition by means of a "transfer-out assay" (see Fig. 5). This assay makes use of the fact that RSF1050 depends on DNA polymerase I for its replication and maintenance in *E. coli*. When donor cells containing the self-transmissible plasmid F and RSF1050 are mated to a *polA* recipient cell, only F is able to transfer, and the frequency of transposition onto F can be determined by the number of recipient cells that are resistant to penicillin. If the recipient is not *polA*, recipient cells resistant to penicillin can arise by mobilization of RSF1050 by F. Wild-type Tn*3* yields about one penicillin-resistant F per 10^3 penicillin-sensitive Fs, whereas the

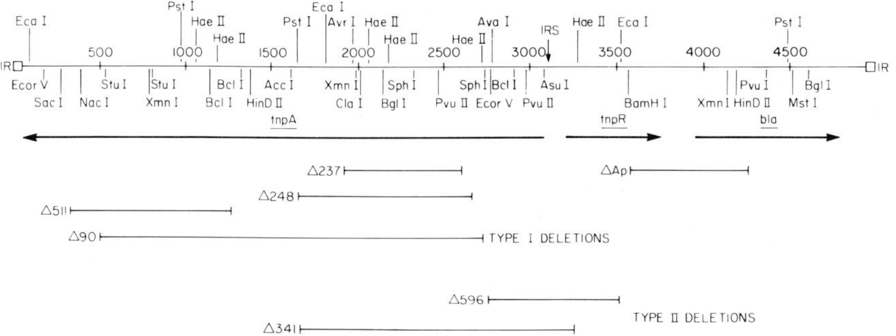

Fig. 3 The structure of Tn*3*. Tn*3* encodes three genes and three sites. The transposon is 4957 bp long and is flanked at both ends by perfect 38-bp IRs. Both IRs are required for transposition (Heffron *et al.*, 1977; Gill *et al.*, 1979). The transposase is encoded at *tnpA* and negatively regulated by the product of *tnpR* (which also negatively regulates itself; Heffron *et al.*, 1979; Gill *et al.*, 1979; Chou *et al.*, 1979a,b). The identity and location of the IRS were first postulated on the basis of the two classes of deletions shown here. Type I deletions give a perfect transposition event when complemented to transpose; type II deletions give only cointegrates (Heffron *et al.*, 1977; Gill *et al.*, 1978). This suggested that type II deletions lack a site necessary for resolution of the cointegrate. The phenotype of the deletion ΔAp supplied the first evidence that *tnpR* was required for resolution (Heffron *et al.*, 1977). Transposition with ΔAp also gave cointegrates, but unlike the type II deletions, this phenotype was recessive to a wild-type transposon (Arthur and Sherratt, 1979). The phenotype of mutations in the site-specific recombination system suggests that cointegrates are intermediates in Tn*3* transposition.

frequency is at least 100-fold lower for transposition-defective mutants of Tn*3*. Complementation was carried out by the construction of strains containing a second copy of Tn*3* in order to supply missing transposition functions. By complementation, transposition-negative mutations in Tn*3* could be divided into two types: those that could be complemented to transpose by a second copy of Tn*3* (trans-recessive) and those that could not (cis-dominant). Cis-dominant mutations defined the sites essential for Tn*3* transposition; trans-recessive mutations defined coding sequences for transposition functions. The ability of the Tn*3* transposase to be supplied in trans is another difference between Tn*3* and many IS elements and composite transposons. For Tn*9*, Tn*10*, and Tn*903*, essential transposition functions do not appear to work well in trans (Ohtsubo *et al.*, 1981; Grindley and Joyce, 1981; Chapter 6).

A. Replicative Recombination Functions

1. The Terminal Repeated Sequences

Tn*3* contains identical 38-bp inverted repeated sequences at each end (for the sequence, see Fig. 6; Ohtsubo *et al.*, 1979; Cohen *et al.*, 1979; Kopecko and Cohen, 1975; Heffron *et al.*, 1975b). Mutations that delete the terminal inverted

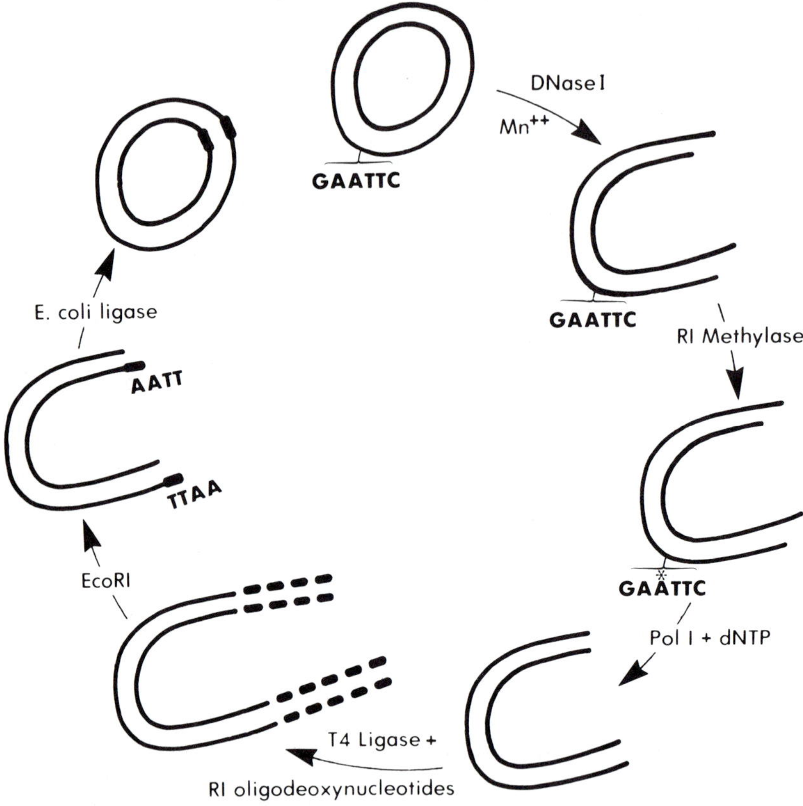

Fig. 4 Construction of mutant plasmids containing new *Eco*RI restriction sites (Heffron *et al.*, 1978). Supercoiled DNA is first digested with DNase I in the presence of Mn^{2+} (Carbon *et al.*, 1975). To prevent internal cleavage of the plasmid at a later step, the *Eco*RI site present is blocked by methylation. The ends of the linearized plasmid are made duplex by treatment with the Klenow fragment of *E. coli* DNA polymerase I, and the *Eco*RI octamer $\frac{\text{GGAATTCC}}{\text{CCTTAAGG}}$ is added to the ends by blunt-end ligation with T4 DNA ligase. Cleavage with *Eco*RI is used to remove excess octamers and make the ends cohesive. Finally, the DNA is recircularized with T4 DNA ligase. The last enzymatic reaction is carried out at a lower concentration of T4 ligase than the linker ligation to avoid a high background of transformants that do not contain an added restriction site. The result of this procedure is an insertion mutation that can be mapped by cleavage with *Eco*RI. About 70% of the mutations contain small deletions (<40 bp) or duplications (<8 bp) adjacent to the inserted *Eco*RI linker. The remaining 30% contain deletions longer than 40 bp.

repeat (IR) of Tn*3*, or a deletion that removes an internal part of one IR, cannot be complemented to transpose (Heffron *et al.*, 1977b, 1979; Gill *et al.*, 1979). Therefore, the IRs must contain sequences that play a structural role in transposition. These sequences must function as recognition sites to define the ends of the transposon. Thus, several Tn*3* derivatives have been constructed containing additional DNA sequences, and these transpose normally, although at a lower frequency (Manis *et al.*, 1980; Goebel *et al.*, 1977). The IRs probably function

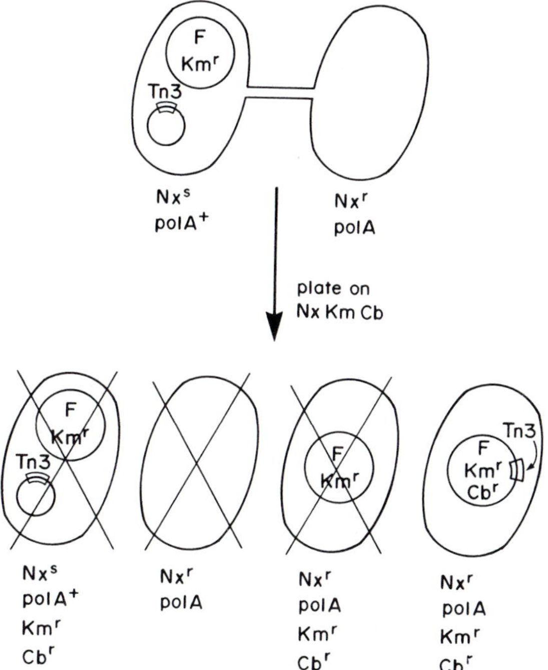

Fig. 5 Detection of transposition. This assay to detect transposition (Heffron *et al.*, 1977b) uses a kanamycin resistant (Km^r) derivative of the sex factor F as a recipient for transposition. The assay takes advantage of the strict dependence of *ColE*1 on polymerase I for its replication and maintainence in *E. coli*. By using a *polA* recipient, mobilization of the donor plasmid by F is prevented. The recipient cell is both resistant to nalidixic acid (Nx^R) and *polA*. As shown in the figure, the only cells surviving the selection contain F derivatives to which Tn*3* has transposed. The transposition frequency is determined as the ratio of the number of carbenicillin-resistant (Cb^r) transconjugant colonies to kanamycin-resistant transconjugant colonies from the mating depicted.

as sites at which the transposase can bind and carry out symmetrical enzymatic reactions with both ends of the transposon. The terminal repeated sequences are highly conserved in the Tn*3* family (see Fig. 6). γδ is a member of this family and one of Tn*3*'s closest relatives. Tn*3* and γδ share extensive DNA homology (R. A. Young and R. R. Reed, unpublished observations). The terminal IRs of γδ share 27/38-bp homology with the ends of Tn*3* (see Fig. 6; Reed *et al.*, 1979). In spite of the close similarity, these two transposons will not complement each other for transposition (Heffron *et al.*, 1977b; M. McCormick and E. Ohtsubo, unpublished). Comparison of the sequences of the ends of γδ and Tn*3* shows that specific recognition of the IRs by the homologous transposase must lie in only six noncontiguous bases of the IRs that differ between the two transposons. It will be interesting to see which of the base changes between the two IRs are essential for their specific recognition.

Within the terminal repeated sequences of the Tn*3*-like transposons γδ, Tn*951*

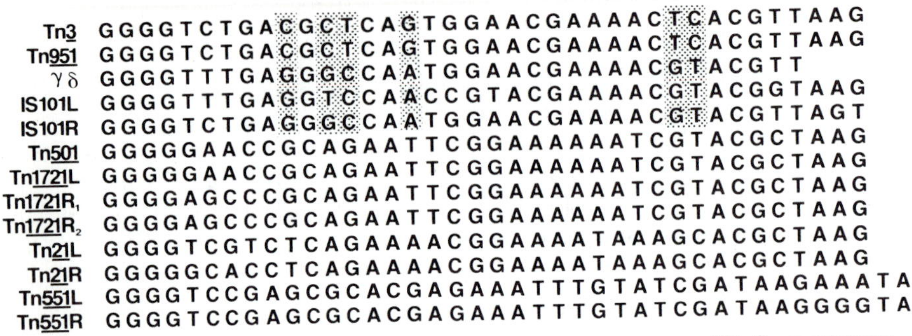

Tn3	G G G G T C T G A C G C T C A G T G G A A C G A A A A C T C A C G T T A A G
Tn951	G G G G T C T G A C G C T C A G T G G A A C G A A A A C T C A C G T T A A G
γδ	G G G G T T T G A G G G C C A A T G G A A C G A A A A C G T A C G T T
IS101L	G G G G T T T G A G G T C C A A C C G T A C G A A A A C G T A C G G T A A G
IS101R	G G G G T C T G A G G G C C A A T G G A A C G A A A A C G T A C G T T A G T
Tn501	G G G G G A A C C G C A G A A T T C G G A A A A A A T C G T A C G C T A A G
Tn1721L	G G G G G A A C C G C A G A A T T C G G A A A A A A T C G T A C G C T A A G
Tn1721R1	G G G G A G C C C G C A G A A T T C G G A A A A A A T C G T A C G C T A A G
Tn1721R2	G G G G A G C C C G C A G A A T T C G G A A A A A A T C G T A C G C T A A G
Tn21L	G G G G T C G T C T C A G A A A A C G G A A A A T A A A G C A C G C T A A G
Tn21R	G G G G G C A C C T C A G A A A A C G G A A A A T A A A G C A C G C T A A G
Tn551L	G G G G T C C G A G C G C A C G A G A A A T T T G T A T C G A T A A G A A A T A
Tn551R	G G G G T C C G A G C G C A C G A G A A A T T T G T A T C G A T A A G G G G T A

Fig. 6 Comparison of the ends of members of the Tn*3* family of transposons. This figure compares the sequences of the terminal IRs in Tn*3*-like transposons. The ends are obviously related. The transposons are grouped according to whether the transposase genes can complement each other for transposition. Tn*3* transposase can complement Tn*951* but not γδ. This is not surprising, because Tn*951* and Tn*3* have identical ends. Tn*551* was identified in a gram-positive bacterium (Kahn and Novick, 1980), and it is not known whether it can complement other transposons. Comparing the sequences of the ends, it can be seen that many bases are conserved in all transposons of this family. For example, the terminal four Gs are conserved. γδ transposase will complement IS*101* but not Tn*501* or Tn*3*. Therefore, the five nucleotide differences between the ends of IS*101* and γδ cannot play a role in recognition by γδ transposase. Ruling out any of these same base differences between Tn*3* and γδ eliminates all but six noncontiguous bases that must account for specific recognition. These bases are highlighted in the figure. References for these sequences are given in Table I.

and IS*101*, and the bacteriophage Mu, there is a specific heptanucleotide sequence, ACGAAAA, that is conserved. All of these transposable elements generate a 5-bp duplication of target DNA on insertion. This may suggest a common progenitor for the Tn*3*-like transposons and Mu (Reed *et al.*, 1979). It will be interesting to see if Mu's transposition functions are related to Tn*3*'s once they have been sequenced.

2. THE TRANSPOSASE

All deletions and insertion mutations within a contiguous 3-kb region from the left-hand side of Tn*3* (see Fig. 3) were transposition-negative but could be complemented to transpose. Several unsuccessful attempts were made to complement these transposition-negative mutants with one another before the DNA sequence of Tn*3* was determined (Gill *et al.*, 1979). The DNA sequence confirmed that there must be only a single complementation group because there is only a single long open reading frame in this region of Tn*3*. The protein encoded by the *tnpA* gene came to be known as the *transposase,* because it was essential for Tn*3* transposition. We now know that transposition is a two-step process and that the transposase is required only for the first step. Instead of the term *transposase,* which implies that this protein can do everything, *tnpA activity* or *replicative recombinase* may be more appropriate. Most of *tnpA* appears to be essential for transposition. For example, a frameshift mutation only 45 bp from the 3′ end of *tnpA* is transposition-negative, although it can be complemented to transpose (Heffron *et al.*, 1979). Sequences homologous to the transposase have been identified in Tn*1721*, Tn*501*, Tn*21*, and γδ (Schmitt *et al.*, 1981; R. A. Young and R. R. Reed, unpublished).

Transcription of *tnpA* begins at nucleotide 3095 (see Fig. 7; a G) as deduced from sequencing the 5′-terminus of transcripts initiated *in vitro* (W. Wishart, C. Machida, and E. Ohtsubo, unpublished observation). RNA polymerase transcribes the entire 3 kb of the gene, and translation is terminated at an ochre codon at nucleotide 36 within the terminal IR. Transcription does not appear to be terminated within the IR and can transcribe a gene adjacent to a Tn*3* insertion. The product of *tnpR* was shown to regulate streptomycin resistance in such a transcriptional fusion between *tnpA* and the gene for streptomycin phosphotransferase (see Section III,C). Because transcription from the *tnpA* promotor reads out through the end of the transposon, Tn*3* can be used as a flying promotor.

Translation of the transposase transcript may begin from either of two initiation codons (see Fig. 7). This is suggested by the transposition phenotype for mutations at the amino terminus. The amino acid sequence of the amino terminus of purified transposase demonstrates that translation is initiated predominantly at the first methionine in *tnpA,* as indicated in Fig. 7 (Ditto *et al.*, 1982). On the other hand, linker mutation 25 (indicated in the figure) contains a deletion that removes this same methionine, and the mutant shows only about a sevenfold reduction in transposition frequency (Heffron, *et al.*, 1978; McCormick *et al.*, 1981). Thus, translation must be initiated at another codon. Mutant 25 does not shift the reading frame, and so it is possible that translation originated upstream at a second start site. To test this, we constructed the deletion shown in Fig. 7, which both shifts the reading frame and removes this methionine (Heffron *et al.*, 1981; Kostriken *et al.*, 1981). This deletion transposes 100-fold less frequently than the wild type, suggesting that translation is initiated at a second upstream site (Heffron *et al.*, 1981, unpublished), most likely the valine (GTG) indicated.

The Tn*3* transposase was first identified as a protein band (Gill *et al.*, 1979; Chou *et al.*, 1979b). Transposase from wild-type Tn*3* is made in such minute amounts that it can never be identified, but mutants in *tnpR* are derepressed for transposition and make more transposase. This finding suggests that the amount of transposase made from a wild-type Tn*3* (estimated to be <0.01% of the total cell protein) limits the frequency of transposition. The Tn*3* transposase is a very large protein (120,000 daltons). The protein, as deduced from the DNA sequence, has an isoelectric point near neutrality but contains several basic regions (Heffron *et al.*, 1979). On the basis of secondary structure predictions, the protein has two separate, generally helical domains separated by a long random coil (Heffron *et al.*, 1979). Transposase has been purified to homogeneity from an overproducing strain (Fennewald *et al.*, 1981; the overproducer is described in the next paragraph). Without divalent cations, the protein is insoluble. Purified transposase binds tightly but nonspecifically to single-stranded DNA. In

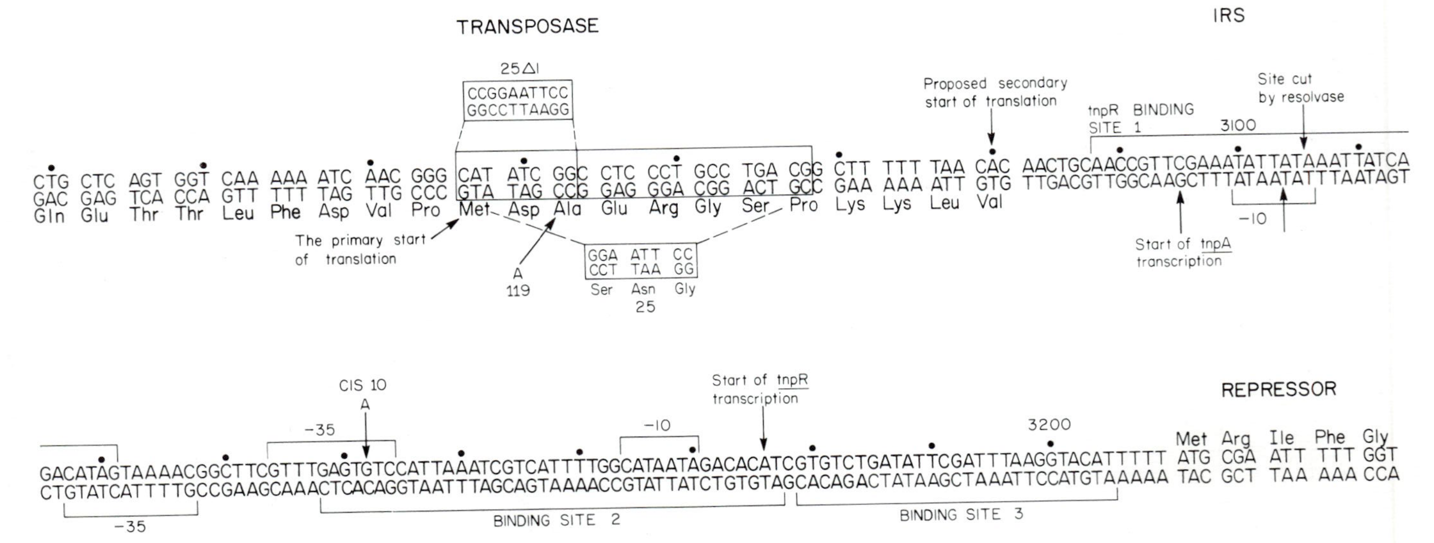
TRANSPOSASE
IRS
25Δ1
CCGGAATTCC
GGCCTTAAGG
Proposed secondary start of translation
tnpR BINDING SITE 1
3100
Site cut by resolvase
CTG CTC AGT GGT CAA AAA ATC AAC GGG CAT ATC GGC CTC CCT GCC TGA CGG CTT TTT TAA CAC AACTGCAACCGTTCGAAATATTATAAATTATCA
GAC GAG TCA CCA GTT TTT TAG TTG CCC GTA TAG CCG GAG GGA CGG ACT GCC GAA AAA ATT GTG TTGACGTTGGCAAGCTTTATAATATTTAATAGT
Gln Glu Thr Thr Leu Phe Asp Val Pro Met Asp Ala Glu Arg Gly Ser Pro Lys Lys Leu Val
−10
The primary start of translation
A
119
GGA ATT CC
CCT TAA GG
Ser Asn Gly
25
Start of tnpA transcription
CIS 10
A
Start of tnpR transcription
REPRESSOR
−35
−10
3200
Met Arg Ile Phe Gly
GACATAGTAAAACGGCTTCGTTTGAGTGTCCATTAAATCGTCATTTTGGCATAATAGACACATCGTGTCTGATATTCGATTTAAGGTACATTTTT ATG CGA ATT TTT GGT
CTGTATCATTTTGCCGAAGCAAACTCACAGGTAATTTAGCAGTAAAACCGTATTATCTGTGTAGCACAGACTATAAGCTAAATTCCATGTAAAAA TAC GCT TAA AAA CCA
−35
BINDING SITE 2
BINDING SITE 3

fact, single-stranded binding was used as an assay in its purification. Transposase does not appear to bind more tightly to the IRs, although the genetic results would suggest that it must. The fact that only single-stranded binding activity has been identified may suggest that the enzyme is only partially active as isolated. There are likely to be tight enzymatic controls to prevent excessive activity of an enzyme that we assume can join nonhomologous DNAs.

Several overproducers of the transposase have been constructed. One set of overproducers was selected as mutations from a transposase-β-galactosidase fusion protein showing increased translation (Casadaban *et al.*, 1982). All of these mutations except one are located in a probable ribosome binding site before the methionine used as the translational start. The one exception was a G to A transition in the valine codon, described previously as a possible second start. One of the best overproducers is indicated in the sequence shown in Fig. 7 and was used as a source of transposase in its purification. The transposase has also been overproduced by construction of a translational fusion with the first eight amino acids of *lacZ*. The construction used one of the linker mutations that fortuitously lay 9 bp in front of the methionine start. From either strain, overproduction of the transposase (1–10% of total cell protein) is detrimental to the cell, and the transposition frequency is as high as or higher than the frequency of transposition for a *tnpR*- (Casadaban *et al.*, 1982; M. Wong and F. Heffron, unpublished observation). This effect does not require a complete transposon:

Fig. 7 The nucleotide sequence between the amino terminus of the transposase and the repressor. Several features that are of interest are identified in the sequence. Transcription of *tnpA* begins at a G (nucleotide 3095) according to the sequence of transcripts initiated *in vitro* (W. Wishart, C. Machida, and E. Ohtsubo, unpublished observation). The −10 homology region corresponds with the site cut *in vitro* by the resolvase (Reed, 1981a,b; Reed and Grindley, 1981). Binding of resolvase–repressor in this region very probably has a twofold role in regulating expression of the transposase and in mediating resolution of cointegrates. How these two activities are controlled is not yet known. The methionine indicated is the primary start of translation, as determined from the protein sequence of the amino terminus of purified transposase (Ditto *et al.*, 1981). A secondary start is suggested by the phenotypes of the two substitution mutations shown in the figure. Mutation 25, which is in frame but deletes this methionine, causes at most a sevenfold decrease in transposition frequency (Heffron *et al.*, 1979; McCormick *et al.*, 1981). Mutation 25 Δ1, which both shifts the reading frame and deletes the methionine, produces more than a 100-fold decrease in transposition frequency (Heffron *et al.*, 1981). One possible alternative start is the valine indicated in the figure, although there is no evidence that this valine is used. The C to A transversion, mutation 119 (Casadaban *et al.*, 1982), appears to be an up translational mutation. Cells containing Tn*3* derivatives with this mutation makes transposase at 1% total cell protein. These cells were used as a source in purifying the transposase (Fennewald *et al.*, 1981). The mutation appears to improve a ribosome binding site for the transposase, making its translation more efficient. Three binding sites for γδ resolvase have been identified in footprinting experiments on γδ and Tn*3* DNA (Grindley, 1982).

The start of *tnpR* transcription, indicated in the figure, was also determined from the sequence of *in vitro*-initiated transcripts (W. Wishart, C. Machida, and E. Ohtsubo, unpublished observation). This transcriptional start places the operator mutation *cis*10 in the most likely −35 homology region and suggests that the promoter and operator for *tnpR* must overlap (Chou *et al.*, 1979a).

The overproduction is detrimental even in cells containing an overproducer that is not part of a complete transposon. But overproducing cells that overproduce an inactive transposase (due to an inframe insertion) grow normally (M. Wong and F. Heffron, unpublished). This finding shows that an activity of the transposase itself is detrimental. Tn*3* transposition is temperature-sensitive (Kretschmer and Cohen, 1979; Weinstock *et al.* 1979). The frequency of transposition is at least 100-fold lower at 37°C than it is at 30°C. The low frequency of transposition at 37°C may reflect the instability of the transposase itself because the detrimental effect of the transposase overproducers is observed only at temperatures of 30°C or lower.

B. Tn*3*'s Site-Specific Recombination System

1. IDENTIFICATION OF THE INTERNAL RESOLUTION SITE (THE IRS)

Resolution of cointegrates is normally highly efficient. Within the cell, the equilibrium between resolved and unresolved cointegrates is more than 10^5 in favor of the resolved products (C. J. Muster and J. Shapiro, unpublished). However, a subset of internal deletions in Tn*3* (type II deletions in Fig. 3) were blocked at the replicative recombination step in transposition and formed cointegrates when they were complemented to transpose. The fact that cointegrates formed by type II deletions were stable when complemented suggested that the deletion had removed a sequence essential for resolution of cointegrates (Gill *et al.*, 1978). This essential sequence is the IRS. More extensive deletion mapping has demonstrated that at least 130 bp appear to be essential for resolution (see Fig. 7; Grindley *et al.*, 1982; R. G. Wells and N. D. F. Grindley, unpublished observation).

Recombination between Tn*3* and γδ *in vivo* was carried out to define the actual recombination site (Kostriken *et al.*, 1981; Reed, 1981a). γδ is a close relative of Tn*3*, which shares extensive DNA homology. The actual site of recombination within Tn*3* could be defined if recombination could take place between these two transposons. A hybrid recombination site would be generated by recombination between γδ and Tn*3* in cis. As there are a number of base pair changes in this region of the two transposons, sequencing across the junction fragment would define the recombination site. The sequence of the junction fragment demonstrates that recombination takes place within a 19-bp, A+T-rich sequence located between nucleotides 3095 and 3113 (see Fig. 7). On the evidence of deletion mapping, additional sequences away from the recombination site are also required, as mentioned in the previous paragraph. *In vitro* experiments, described later, have confirmed that recombination takes place within this sequence. Tn*3*'s recombination system bears a resemblance to the lambda integration system. In both systems, recombination takes place at a very A+T-rich sequence but requires sequences outside the recombination site. There do not appear to be DNA homologies between these two systems, however, nor does resolution of cointegrates depend on the same bacterial functions that are required for λ integration (Heffron *et al.*, 1981).

2. The Resolvase

The first hint that the *tnpR* product could be the resolvase necessary for resolution of cointegrates came from the phenotype of ΔAp, a deletion that removed part of *tnpR* and *bla* (see Fig. 3; Heffron *et al.*, 1977). This deletion does not overlap the type II deletions, but it also transposed to form cointegrates. Unlike type II deletions, cointegrates generated by transposition of ΔAp resolved when transposition functions were supplied in trans (Arthur and Sherratt, 1979). Subsequent experiments confirmed that *tnpR* is the function deleted in ΔAp that is required for resolution of cointegrates (Heffron *et al.*, 1981; Kostriken *et al.*, 1981; Kitts *et al.*, 1982). One of these experiments is illustrated in Fig. 8. A small restriction fragment from Tn*3* containing the IRS was used to construct a tester plasmid containing a direct repeat of the IRS. By analogy with resolution of cointegrates, recombination between the two copies of the IRS would delete the intervening DNA that contains an *Eco*RI restriction site. Thus, complementation for the resolvase can be assessed by purifying plasmid DNA from strains containing this tester plasmid and digesting the DNA with *Eco*RI. Recombination invariably took place when the tester plasmid was complemented with another plasmid containing an intact copy of *tnpR*. It never took place when the tester plasmid was complemented with a plasmid containing a mutant copy of *tnpR*. Thus, the *tnpR* product must be required for resolution. Figure 5 also illustrates that the *tnpR* analogue of γδ can complement Tn*3* cointegrates for resolution. In summary, resolution of Tn*3* cointegrates takes place by recombination between the repeated copies Tn*3* at an internal site and requires the *tnpR* product. Construction of plasmids similar to the tester plasmid described here provided a substrate for an *in vitro* assay and opened the door to purification of the resolvase.

The resolvase from γδ was purified to over 95% homogeneity from an overproducing strain (Reed, 1981b; Reed and Grindley, 1981). A plasmid containing two copies of the *res* site was used as a substrate (*res* site: *resolution* site; the abbreviation for the IRS in γδ). No accessory proteins are required for resolution. The protein requires a supercoiled substrate and Mg^{2+} for activity but not ATP. Omitting Mg^{2+} apparently blocks an intermediate step in recombination. Without Mg^{2+}, linear molecules are recovered that still have the enzyme attached to the 5′-ends and contain a 2-bp 3′ extension (see Fig. 7). This observation permitted determination of the site of the endonucleolytic break. The site of the cut is between the second T and A in a palindromic sequence $\substack{\text{TTATAA}\\\text{AATATT}}$ at nucleotide 3105 (see Figs. 3 and 7). *In vitro* cleavage thus occurs within the

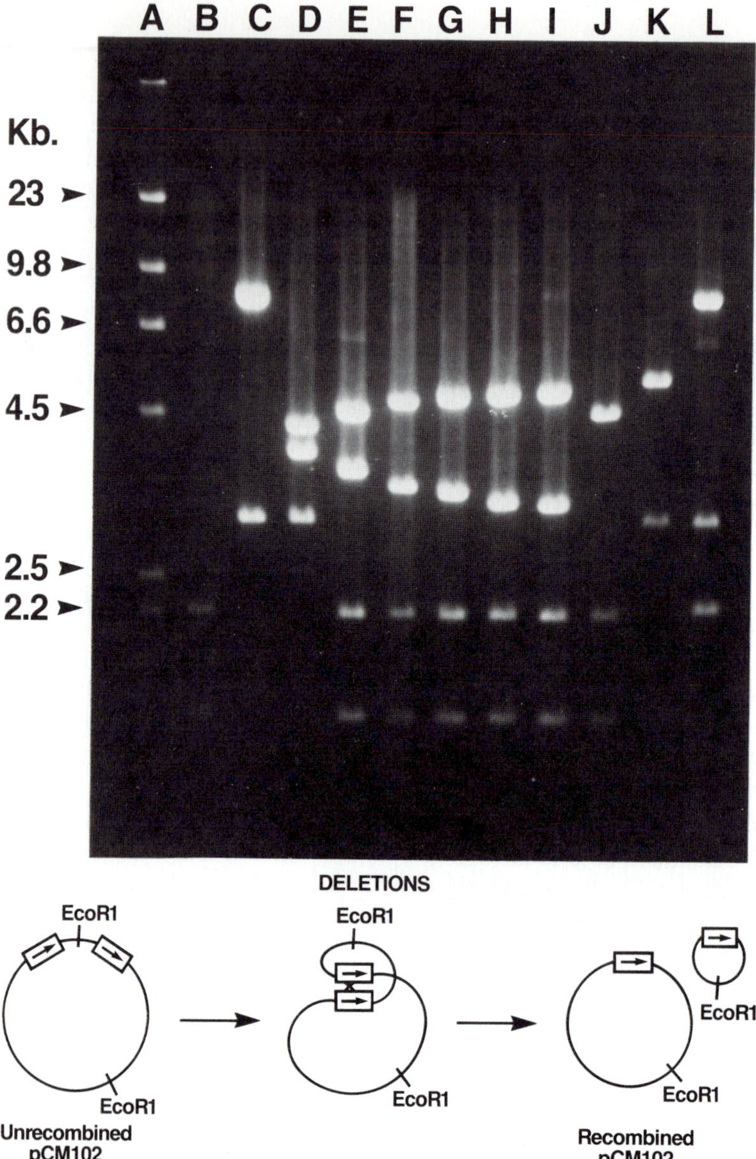

Fig. 8 The product of *tnpR* is required for resolution of cointegrates (Kostriken *et al.*, 1981). pCM102 contains two directly repeated copies of the internal resolution site and provides a convenient way to examine resolution *in vivo*. The figure shows an agarose gel of *Eco*RI-cleaved plasmid DNA from cells containing pCM102 and a second plasmid containing complete or mutant copies of Tn*3*. Lane A contains molecular weight markers, and lane B shows the two *Eco*RI restriction fragments from pCM102 (no recombination has taken place). Lane C shows the pattern from cells

sequence predicted from the genetic experiment described in the last paragraph. Linearized DNA can be detected when a substrate is used which contains only a single *res* site (at high enzyme concentration and without Mg^{2+}). However, two *res* sites in the same substrate molecule greatly stimulate cutting. This result may indicate that interaction between two *res*–resolvase complexes is required before cleavage and joining can take place. The two *res* sites must be oriented as direct repeats in the substrate for recombination to be detected. This result appears to parallel the genetic results. *In vivo,* recombination between IRs of Tn*3* or γδ does not appear to occur efficiently (Chiang and Clowes, 1982; Reed and Grindley, 1981a; Kostriken *et al.,* 1981), although recombination does take place.

In vitro binding studies with purified resolvase have identified three binding sites in the intercistronic region between *tnpA* and *tnpR* (see Fig. 7; Grindley *et al.,* 1982). Binding site I corresponds to the *res* site or IRS and presumably the operator for *tnpA*. Binding site II coincides with the operator mutation for *tnpR* described by Chou *et al.* (1979a) and also indicated on Fig. 7. If this binding site corresponds to the operator for *tnpR,* the function of the third site is still unknown. Deletion mapping has shown that all three binding sites are required for resolution (Grindley *et al.,* 1982). Thus these binding sites must have a twofold role in recombination and in regulation of *tnpA* and *tnpR* transcription. Three binding sites for λ repressor have been identified in the region between pR and pL. The function of these three operators is to regulate transcription from the two promoters. It is likely that a similar finely tuned regulatory system is present in Tn*3*. No *in vitro* binding by resolvase was detected to the left of the IRS in spite of deletion results that suggested that there are essential sequences here for resolution (Kostriken *et al.,* 1981). Perhaps resolvase does not bind to a sequence that is nevertheless required for resolution of cointegrates.

There appear to be many site-specific recombination systems present in nature similar to the one found in Tn*3*. The Tn*501*/Tn*1721* subgroup encodes a related resolvase, although this resolvase is transcribed in the same direction as the transposase. There is 36% DNA homology between *tnpR* and *hin,* a gene required for *Salmonella* phase variation (Simon *et al.,* 1980). In *Salmonella,* an invertible sequence controls phase variation by expression of different flagellar antigens (see Chapter 12). Inversion of this sequence in *Salmonella* requires the

---

that contain pCM102 and a second plasmid with a complete copy of Tn*3* (RSF1050; Heffron *et al.,* 1977). A function from Tn*3* has complemented pCM102 for resolution. Lanes D–I show the results when pCM102 is placed in the cell with mutant copies of Tn*3*. In D, pCM102 is complemented with a transposase mutant (recombination still takes place), whereas recombination is blocked in lanes E–I, which are from cells containing mutations in *tnpR*. Lanes J and K demonstrate that only *tnpR* is required for resolution. Lane J shows that pBR322 (the upper band) does not complement for resolution, but when *tnpR* is cloned into pBR322 (lane K), resolution takes place. The closely related transposon γδ can also complement resolution. This result is shown in lane L, where plasmid DNA was prepared from cells containing pCM102 and a pBR322 derivative with γδ (Guyer, 1978).

hin gene product. *hin* is closely related to an invertible system found in Mu and to another found in the bacteriophage P1. It is very likely that *tnpR* and *hin* both arose from a common ancestral gene. How *tnpR* was picked up by Tn*3* is an interesting puzzle.

C. Regulation of Tn*3* Functions

The product of *tnpR* regulates its own expression and acts as a negative regulator of transposase transcription (Heffron *et al.*, 1978, 1979; Gill *et al.*, 1979; Chou *et al.*, 1979a,b). Regulation of transposase expression probably occurs at the level of transcription, translation, and enzyme stability. There appears to be a fail-safe mechanism built into Tn*3* to prevent runaway transposition. These safeguards include transposition immunity (described later), low-frequency transposition to the chromosome, and higher expression of transposase in stationary phase cultures than in log phase cultures (Fennewald *et al.*, 1981).

1. Regulation of *tnpR*

Expression of *tnpR* is negatively regulated by the product of *tnpR*. This self-regulation has been clearly demonstrated by construction of fusions between *tnpR* and *lacZ* in which expression of β-galactosidase is under *tnpR* control. These fusions show a 50-fold decrease in β-galactosidase expression when the product of *tnpR* is supplied in trans (Chou *et al.*, 1979a,b). Transcription of *tnpR* *in vitro* is initiated at nucleotide 3175 (W. Wishart, C. Machida, and E. Ohtsubo, unpublished observation). The site of transcription suggests that initiation sequences essential for initiation of transcription overlap the operator for *tnpR*. An operator mutation for *tnpR* (*cis*10 in Fig. 7) has been identified that lies within the −35 homology region of the *tnpR* promoter. The fact that *tnpR* is autogenously regulated could have important biological consequences. For example, there appears to be little zygotic induction of Tn*3* on entering a virgin cell, presumably because the repressor is itself autogenously regulated and thus high-level transcription of the transposase is rapidly shut down.

2. Regulation of *tnpA*

Mutations in *tnpR* show an increase in both transposition frequency and the amount of transposase produced. Presumably, the small amount of transposase found in wild-type cells normally limits the frequency of transposition, and the 10–100-fold increase in transposition frequency observed in *tnpR* mutants results solely from the increased synthesis of transposase. This regulation appears to be at the level of transcription because mutants of *tnpR* show a much higher level of expression of genes downstream from the transposase that are read from the transposase promoter (Heffron *et al.*, 1979). Transcriptional regulation of *tnpA* by *tnpR* was demonstrated by the construction of a transcriptional fusion between *tnpA* and a second gene. Tn*3* was integrated into the first gene in a plasmid (RSF1010) operon that encodes resistance to streptomycin and sulfonamides (Heffron *et al.*, 1977a). Insertion into the first gene in the operon, for sulfonamide resistance, simultaneously eliminates expression of the second gene, presumably by interrupting its transcription. A revertant was selected (ΔAp in Fig. 3) in which expression of streptomycin resistance was normal. This revertant resulted from a deletion of *tnpR*. It would thus appear that transcription initiated at the *tnpA* promoter was reading out through the end of the transposon and transcribing the gene for streptomycin resistance. Transcription must be turned on in ΔAp because the *tnpR* product normally negatively regulates *tnpA* transcription. This proved to be the case, for when the repressor is supplied from another plasmid, the level of streptomycin resistance drops at least 10-fold.

The operator at which the repressor binds to regulate transcription of *tnpA* most likely corresponds to binding site 1 in Fig. 7. As was the case for the *tnpR* operator, this operator appears to overlap sequences required for initiation of transcription. This sequence also overlaps the internal resolution site, as described previously. Expression of transposase is also limited by inefficient translation of the *tnpA* transcript. Fusions between β-galactosidase and the transposase show low amounts of β-galactosidase activity. Mutations selected for increased expression of β-galactosidase all improved a ribosome binding site in front of the methionine used for translational initiation of the transposase. Finally, pulse-chase experiments suggest that the transposase itself turns over quite rapidly (R. Gill, unpublished). Because turnover is more rapid at 37°C than at 30°C, the fact that transposition occurs less frequently at 37°C may result from the increased turnover of transposase. All these factors probably help keep the amount of active transposase in a cell to a minimum.

IV. Mechanism of Transposition

Because Tn*3* transposition is a two-step process, it should be discussed in terms of what is known about the mechanism of each step. This section will also include a short discussion of the host functions that affect both steps in Tn*3* transposition.

A. Replicative Recombination Step

There are many pieces of evidence that suggest that transposition involves replication. The most compelling evidence is that Tn*3* mutants defective in the site-specific recombination system invariably transpose to form cointegrates (Kostriken *et al.*, 1981; McCormick *et al.*, 1981; Muster and Shapiro, 1981). When formed in cells that contain two distinguishable copies of Tn*3*, cointegrates always contain identical copies. A number of other observations also suggest that transposition must be replicative, but because these concern composite transposons or Mu, they will not be described here (Read *et al.*, 1980; Ljungquist and Bukhari, 1977; Berg, 1977; Botstein and Kleckner, 1977; Fennewald and Shapiro, 1979). One piece of evidence suggesting that a copy of Tn*3* is transposed comes from experiments in which Tn*3* is transposing from plasmid to plasmid. Within these cells, 10% of the recipient plasmid copies can contain a transposed copy of Tn*3*. However, the donor plasmid missing Tn*3* is not recovered at an appreciable frequency (Bennett *et al.*, 1977).

Transposition models can be divided into two types. The first type pictures transposition as asymmetric, that is, the two ends of the transposon are not treated identically. For example, one end of Mu becomes attached to recipient DNA. Starting from this end, a copy is made and inserted into the recipient DNA molecule, and then the other end of Mu is joined to the recipient DNA (Harshey and Bukhari, 1981; Galas and Chandler, 1981; Grindley and Sherratt, 1979). This type of model is particularly favored for Mu because its ends are not identical. At first blush, in an asymmetric model, there appears to be no reason why cointegrates should result. However, as inspection of Fig. 9 will show, whether cointegrates can result from products of transposition depends on what happens at the Mu terminus. The second type of model pictures transposition as a two-ended, symmetric event. That is, both ends of the transposon are treated identically, although the enzymatic reactions that occur in the cell need not occur simultaneously at both ends of the transposon. The transposon becomes joined to the recipient DNA at both ends and is then duplicated (or the order could be reversed). An example of this model is also shown in Fig. 9 (Shapiro, 1979). In this type of model, a cointegrate will be an obligate intermediate in transposition, as fits most closely with Tn*3*. Both models explain the high frequency of adjacent deletion and replicative inversion, as well as the structure of such inversions.

Both adjacent deletions and inversions can be understood to result from intramolecular transposition. This outcome can be seen in Fig. 9 by connecting the ends of the intermediate shown in the bottom of the figure such that the molecule has resulted from transposition into the same plasmid (see the legend of Fig. 7 as well). As predicted by the model, subsequent findings have shown that these two events do not require the site-specific recombination system (McCormick *et al.*, 1981; M. Wong and F. Heffron, unpublished). Another verified prediction of the model is the structure of inversions. According to the models, the DNA that intervenes between the two copies of the transposon should itself be inverted. Precisely this inversion has been found (Chiang and Clowes, 1980; Chiang *et al.*, 1981). (The DNA that intervenes between the two copies could itself be inverted by recombination between the two copies of Tn*3* at the IRS. This recombination

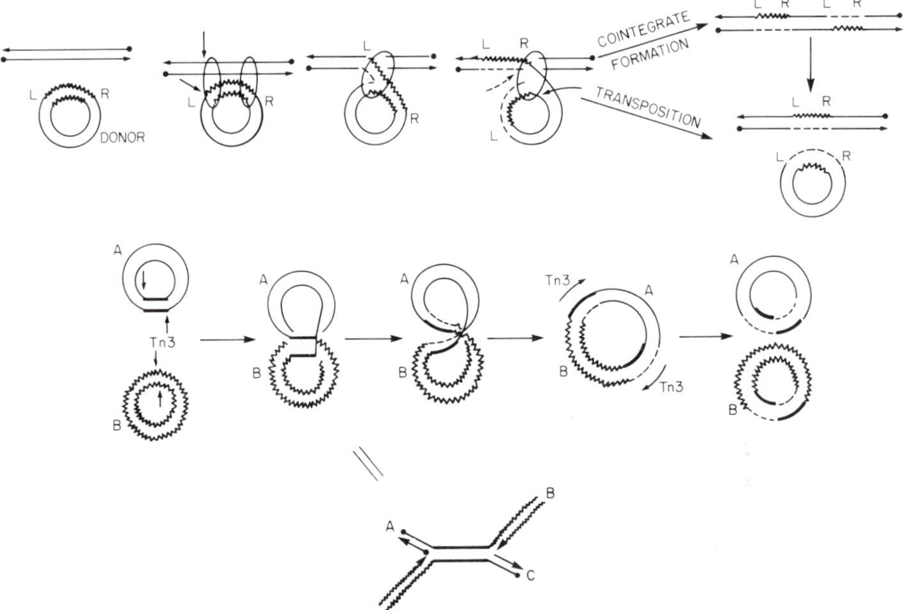

Fig. 9 Molecular mechanisms of transposition. Models for the mechanism of transposition can be divided into two types, as shown here. In the models, the transposon can be asymmetrically (Grindley and Sherratt, 1978; Harshey and Bukhari, 1981; Galas and Chandler, 1981), or symmetrically (Shapiro, 1979). transposed. In an asymmetric model, such as the one illustrated at the top (Harshey and Bukhari, 1981), a copy of the transposon is inserted via a rolling circle type of replication. Cointegrates can arise either by failure of the transposon to terminate replication at its end for one more round of replication (Grindley and Sherratt, 1978) or by a strand exchange in which the free end of the transposon is ligated to the recipient, as shown here (Harshey and Bukhari, 1981). Thus, in these models, cointegrates are not intermediates but end products in transposition. In the symmetric model illustrated (Shapiro, 1979), the entire transposon is first joined to the recipient DNA. The structure of this hypothetical intermediate is illustrated at the bottom and looks identical to a terminus of replication. Replication through the transposon results in a cointegrate as an obligate intermediate, which must then be resolved to give a precise transposition event. In the structure shown at the bottom, if the bonds that join the transposon to the donor plasmid were broken, the result would be a precise transposition, although the donor molecule would be lost (Ohtsubo *et al.*, 1981). In the bottom figure, connection of A with B and C with D would produce an adjacent deletion, and connection of A with D and B with C would lead to replicative inversion, both outcomes of intramolecular transposition.

is inefficient, and thus the inversion of the target DNA is maintained.) One disturbing prediction of these models for replicative recombination is that intramolecular transposition into the same transposon (and not just the same plasmid) results in a structure that will break the plasmid DNA. Perhaps transposable elements are prohibited from insertion into themselves, as will be described in more detail in Section 7.

B. Site-Specific Recombination Step

The elegant biochemistry of γδ's resolvase suggests at least an outline of how resolution must occur, although there remain many biochemical conundrums (see Fig. 10). Presumably, at least two monomers of resolvase bind at the *res* site. The enzyme must normally have little endonucleolytic activity unless it comes in contact with another *res*–resolvase complex. Interaction of the two complexes might cause a conformational change, and the enzyme then cleaves the DNA and exchanges cut ends with the other *res*–resolvase complex.

In vitro, recombination takes place only when a molecule contains directly repeated copies of the *res* site. Inversion is not observed to occur between inverted copies of the *res* site (Reed, 1981b; Reed and Grindley, 1981). This surprising result does not appear to jibe completely with *in vivo* results in which inversions can be observed. Most likely, the rate of inversion is simply so slow *in vivo* that it is not detected. Why inversions do not occur is still not understood. One likely explanation is that once resolvase is bound to a *res* site it can undergo linear diffusion along the DNA molecule in search of a second *res* site. Thus the interaction with the second *res* site would be orientation dependent.

It is not yet clear what supplies the energy for recombination. *In vivo,* resolution is so efficient that the frequency of unrecombined molecules is less than one in 10^5 (C. J. Muster and J. Shapiro, unpublished). Perhaps the energy to drive recombination this far toward the resolved products comes from superhelicity or from the enzyme itself. The latter possibility would explain why the enzyme is used up during the reaction, that is, if the energy for the reaction came from breaking a peptide bond in the resolvase. The products might contain less supercoils than the substrate; this, too, would favor the formation of products. Almost as many questions about the mechanism of resolution have been raised by the biochemical results as have been answered.

C. Host Functions Required for Transposition

A number of mutations in host functions have been isolated that affect transposition. The normal function of the products of these host genes would obviously bear on our thinking about the mechanism. A mutation in *polA* was one of the first host mutants identified (Syvanen *et al.*, 1982). The mutation reduces the transposition frequency 50-fold, cotransduces with *polA*, shows sensitivity to methyl methanesulfonic acid (MMS), and shows a decrease in both polymerase I and 5′-exonuclease activity. As expected, most revertants selected for resistance to MMS had simultaneously regained their transposition phenotype. Surprisingly, a small fraction of the revertants selected in this way still showed the same low frequency of transposition. This may suggest that an activity of *polA* other than its polymerase or 5′-exonuclease is essential for transposition. That *polA* is required for transposition has also been reported by another group (Sasakawa *et al.*, 1981).

The requirement for supercoiling, and the effect of genes that control super-

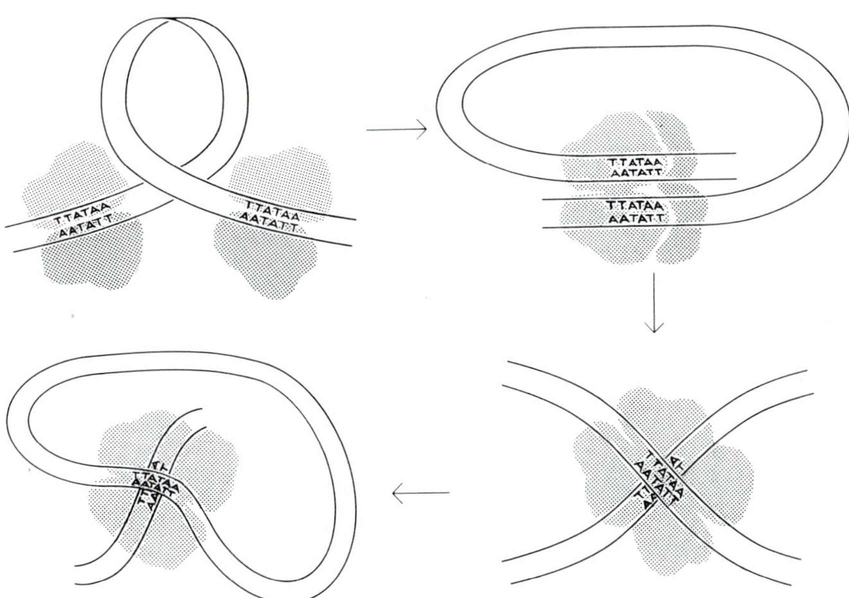

Fig. 10 Resolution of cointegrates by the resolvase. The resolvase binds at the IRS. Two monomers of resolvase are shown here to explain the symmetric cutting on both strands at the IRS. When the two resolvase–IRS complexes meet, cutting takes place, followed by a conformational change to exchange strands. The strands become ligated to complete resolution. The actual mechanism will be more complicated and must take into account such facts as multiple binding sites for the resolvase, the necessity for a superhelical substrate, and the fact that the enzyme is used up during the reaction (Reed, 1981b; Reed and Grindley, 1981).

coiling, appear to be more complex. The level of supercoiling in a bacterial cell is apparently controlled by two antagonistic enzymes, DNA gyrase and topoisomerase I. Gyrase puts in negative supercoils, whereas topoisomerase I takes them out. Thus, mutations in *gyrA* or *gyrB* (the genes for DNA gyrase) reduce supercoiling, and mutations in *top* (the gene for topoisomerase I) increase it. Both kinds of mutation apparently reduce transposition frequency. *Him*B is a mutation in *gyrB*, a subunit of DNA gyrase, isolated as a mutation that decreases the frequency of λ integration. This mutation decreases supercoiling and shows a concomitant decrease in transposition frequency of Tn*3* and Tn*5*, as do drugs that specifically inhibit DNA gyrase (Isberg and Syvanen, 1982; R. Parker, H. Miller, and F. Heffron, unpublished observations). The effects of mutations in the *top* gene are complicated by the fact that they are lethal by themselves and can be isolated only in a background containing a compensatory mutation in *gyrA* or *GyrB* or in other unidentified genes (R. Sternglanz, unpublished observations). The effect of mutations in *top*, unlike those in DNA gyrase, appears to depend on the recipient molecule. Thus, transposition from λ to the chromosome was reduced 50-fold in the top^-/gyr^- double mutant (but was normal in the top^+/gyr^- transductant), whereas in the same experiment the frequency of transposition to F was unchanged. This result may be considered in light of other observations, such as the fact that Tn*3* is not transposed to the chromosome as frequently as it is to a plasmid in the same cell. Perhaps there are still unknown structural differences between the chromosome and some (or all) plasmids. The mutations in *top*, *gyrA*, *gyrB*, and *polA* are the best-characterized mutations affecting transposition; however, several unlinked mutations that affect transposition have been reported (Ilyina *et al.*, 1981).

V. *IS*101 *and Tn*951*: The Degenerate Transposons*

Two members of this family provide the only known examples of naturally occurring parasitic transposons that depend on other transposons for transposition functions. IS*101* was mentioned earlier. It is only 209 bp long and does not appear to encode any genes, only the sequences necessary for both transposition and resolution (Ravetch *et al.*, 1979; Fischoff *et al.*, 1981). It is dependent on γδ for both its transposase and its resolvase (Miller and Cohen, 1980; Fischoff *et al.*, 1981). The sequences of its IRs are closely related to those of γδ, although they are not identical (see Fig. 6). Cointegrates formed by IS*101* are resolved even in a *recA*− host, although the resolution site has not yet been defined. It does not have the sequence TTATAA, at which resolution actually takes place within γδ, but it does have another sequence found in Tn*3*'s and γδ's IRS (nucleotides 3095–3113).

Tn*951* encodes *lacZ* and *lacY*, the genes for lactose metabolism, but not genes for its transposition, apparently (Cornelius *et al.*, 1978, 1979, 1981). The *lac* genes show DNA homology to the analogous genes in *E. coli* and show similar internal restriction sites. The terminal 40 bp of Tn*951* are a perfect IR. The outside 38 bp are identical to the ends of Tn*3*. Tn*951* contains 100 bp corresponding to the *c* terminus of the transposase gene of Tn*3* with two mismatches, but the 6.5 kb of DNA between this homologous stretch and the end of the *lacY* gene show no DNA homology either to Tn*3* or to the *E. coli* chromosome. Tn*951* does not appear to encode a repressor/resolvase gene either. This is suggested by the fact that its frequency of transposition is higher when complemented by *tnpR*− derivatives of Tn*3* than when complemented by wild-type copies of Tn*3*.

VI. *The Tn*501/*Tn*1721 *Subgroup*

Tn*501*, Tn*1721*, Tn*4*, Tn*21*, and Tn*2603* form a subgroup of transposons more closely related to each other than to Tn*3*. They are closely related by their transposition functions, but they also specify different antibiotic-resistance determinants. Tn*501*, Tn*21*, and Tn*1721* share a 4-kb segment that specifies *tnpA*,

TnpR, and the IRS (Schmitt *et al.*, 1982). Their transposition functions appear to be completely interchangeable (Schmitt *et al.*, 1982; Grinsted and de la Cruz, 1981). These findings are based on complementation experiments carried out similarly to the experiments for Tn*3*. The terminal repeats of Tn*501* and Tn*1721* are nearly identical but show a number of mismatches with those of Tn*21* (see Fig. 6). The structures of these transposons are compared in Fig. 11. They are interesting from several standpoints. Unlike Tn*3*, this subgroup encodes a *tnpA* and a *tnpR* gene that are most probably transcribed in the same direction, although the *A* gene has its own promoter (Schmitt *et al.*, 1981). In fact, for Tn*501*, both the transposition frequency and the efficiency of cointegrate resolution are substantially increased in the presence of low amounts of mercury, suggesting that *tnpA* and *tnpR* can be transcribed from the same promoter as an inducible gene for mercury resistance (Arthur *et al.*, 1981). I would like to speculate that Tn*917*, a transposon isolated from *Streptococcus faecilis*, has a sequence similar to that of Tn*501*. Like Tn*501*'s induction with Hg^2, Tn*917* is inducible for its transposition functions with erythromycin (Tomich *et al.*, 1980). Tn*917* is probably identical to Tn*551*, a member of the Tn*3* family isolated from another gram-positive bacterium. Both are 5.3 kb in length, and both encode erythromycin resistance. It will be interesting to see whether Tn*917* is closely related to Tn*501* in its transposition functions.

Tn*1721* encodes an amplifiable gene for tetracycline resistance. In the presence of high amounts of tetracycline, the resistance gene can be duplicated many times to increase the level of antibiotic resistance of the bacteria. Amplification probably occurs by homologous unequal crossing over between long, directly repeated sequences present in the transposon (Schmitt *et al.*, 1979). As shown in Fig. 11, a complete copy of the transposase is present in the right-hand side of the transposon, and a second, truncated copy in the left end of Tn*1721*. The ability to amplify the tetracycline-resistance (*tet*) gene and increase bacterial resistance to tetracycline is probably an advantage to the bacteria if they are exposed to high levels of tetracycline.

In fact, the right end of Tn*1721* not only contains a complete set of transposition genes; it is capable of independent transposition (Schmitt *et al.*, 1979). This minor transposon uses a third internal repeat that is identical to the two terminal repeated sequences for its transposition. The third repeat is located between the A gene and the *tet* gene. The defective copy of the transposase is thought to be vestigial, and it may provide insight into the evolution of Tn*1721*. At one time, Tn*1721* probably contained two copies of the minor transposon flanking *tet*, and its structure was thus identical to that of the composite transposons. An internal deletion in one copy of the minor transposon removed one IR, *tnpR*, and part of *tnpA*, resulting in the structure observed today.

Tn*4*, Tn*2603*, and Tn*21* are large, closely related transposons that encode multiple antibiotic resistances (Yamamoto *et al.*, 1981a,b; Nisen *et al.*, 1977; Sharp *et al.*, 1973; Clerget *et al.*, 1981; Kopecko and Cohen, 1975; Kopecko *et al.*, 1976). Tn*4* and Tn*2603* can be thought of as arising from Tn*21* by insertion of additional DNA sequences (see Fig. 11). Such complex transposons were

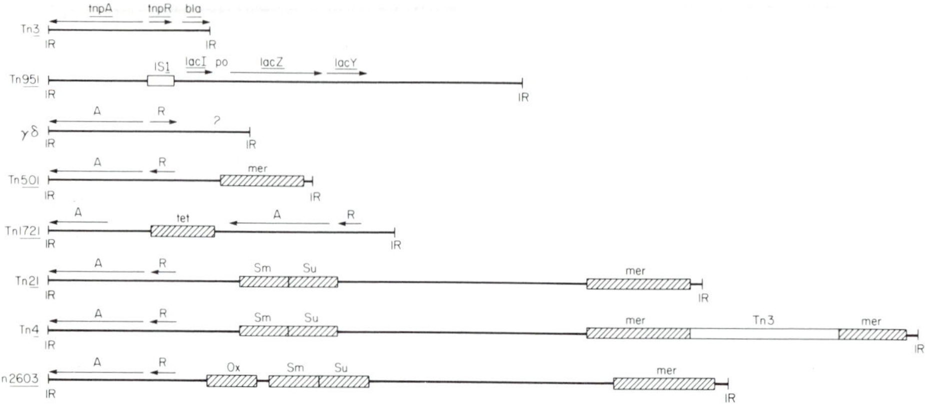

Fig. 11 Structure of Tn*3*-like transposons. The Tn*3* family must have evolved from a common progenitor. All members have related ends, as shown by the sequences in Fig. 6. They contain transposase genes (A) that are related by heteroduplexing and a related repressor–resolvase function (R) (Schmitt *et al.*, 1981; R. Young and R. Reed, unpublished observation; Heffron *et al.*, 1979; Gill *et al.*, 1979; Chou *et al.*, 1979a,b; Grinsted and de la Cruz, 1981). The resolvase of Tn*501*, Tn*1721*, and Tn*21* appears to be encoded in the same direction as the transposase rather than divergently transcribed. This conclusion is based on the facts that in Tn*501* the resolvase can be induced with low levels of mercury (Sherratt *et al.*, 1981; Arthur *et al.*, 1981) and that heteroduplexes form a substitution loop between Tn*3* and Tn*501* in the R region of the two transposons (Schmitt *et al.*, 1981). Tn*951* is defective and depends on Tn*3* for transposition functions. A, *tnpA*; R, *tnpR; bla-*, β-lactamase gene; IR, inverted repeat; mer, mercury-resistance determinant; Sm, the gene that confers streptomycin resistance; Su, sulfonamide resistance; Ox, a β-lactamase that will hydrolyze oxacillin types of penicillin derivatives.

probably built up by insertion of new antibiotic determinants, which are themselves transposable. These three transposons may have evolved from Tn*501*, but such a rearrangement appears to be more complicated than a simple insertion.

The way in which other members of the Tn*3* family have evolved is an open question and may be compared to the evolution of transforming retroviruses. Closely related members of both types of elements can carry along very different genes, either antibiotic-resistance genes for transposons or transforming genes for retroviruses. Both types of transposable elements are compactly organized, and for both types of elements, the mechanism for picking up such genes has not been defined. Tn*3* may have picked up the *bla* gene by a mechanism analogous to that by which Tn*1721* is thought to have incorporated tetracycline resistance. If this is the mechanism, then it is surprising that no vestigial part of the second copy of the transposase remains.

VII. Transposition Immunity

Among transposable elements, the Tn*3* family is unique in showing a phenomenon that limits multiple insertions of the same transposon into a plasmid

(Robinson *et al.*, 1977). The surprising part of this finding is that the transposition frequency to other plasmids in the same cell is unaffected (Robinson *et al.*, 1977). *Transposition immunity* can be defined as a transposon-encoded mechanism that acts in cis to limit transposition of a second copy of the transposon into a plasmid. This limitation may not apply to intramolecular transposition because both replicative inversion and adjacent deletion occur at approximately the same frequency as transposition to other plasmids. Transposition immunity has been observed for several plasmids to which Tn*3* has been transposed. These plasmids include R388, RP1, and R1, pHS1, and pMB8 (Robinson *et al.*, 1977; C. Lee, M. McCormick, E. Ohtsubo, and F. Heffron; unpublished). In all cases, transposition frequency was reduced between 10^3 and 10^6-fold by a resident copy of Tn*3*. There is a delay between the time at which a transposon integrates into a plasmid and the establishment of immunity (Robinson *et al.*, 1977). Presumably, this lag reflects the time it takes for a plasmid to be "altered" by a new resident copy of a transposon so that it becomes immune.

Transposition immunity is highly specific. Tn*3* confers immunity to insertion of another Tn*3* but not to insertion of Tn*501*. In fact, many times Tn*501* is a hot spot for Tn*3* insertion. Conversely, insertion of the mercury transposon Tn*501* occurs readily into a plasmid with Tn*3* and is frequently accompanied by deletions (Grinsted *et al.*, 1978; Stanisich *et al.*, 1977). On the other hand, the two closely related transposons Tn*501* and Tn*1721* show immunity to each other (Schmitt *et al.*, 1982). For immunity to be this specific, models such as changes in plasmid supercoiling due to a resident transposon must be ruled out. Several naturally occurring plasmids appear to be immune to Tn*3* insertions. R391, a member of the J incompatibility group, and R100-1 are both immune (Bennett and Richmond, 1976). It is not clear whether their natural immunity results from the same phenomenon as transposition immunity. The most obvious explanation is that they contain sequences that can confer transposition immunity. Finally, the *E. coli* chromosome itself is resistant to Tn*3* insertion (Kretschmer and Cohen, 1977).

Only part of Tn*3* need be present on a plasmid to confer immunity, but which part is essential for immunity appears to depend on the plasmid in which Tn*3* is resident (Wallace *et al.*, 1981a,b; C. Lee, M. McCormick, E. Ohtsubo, and F. Heffron; unpublished). Deletions of the right-hand end of Tn*3*, including those that remove just the right inverted repeat, are nonimmune, although deletions of the left-hand end are still immune. But either inverted repeat can confer immunity when cloned into another plasmid. For example, pBR322 is immune to Tn*3* transposition. Most surprising of all is the observation that internal deletions that remove the region between A and R are nonimmune in one plasmid but immune when the same deletion is present in a Tn*3* in another plasmid. This suggests that immunity is at least a two-factor system; perhaps transcription is required through the inverted repeat, for example.

There is still no consensus as to which functions in Tn*3*, if any, are necessary for immunity. We find that *tnpR* is not required for immunity, in contradiction to earlier work (Wallace *et al.*, 1981a,b,c). The frequency of cointegration (transposition) is reduced by more than four orders of magnitude when both plasmids contain a copy of Tn*3*, even though both copies are *tnpR*− and there is no other source in the cell (C. Lee, M. McCormick, E. Ohtsubo, and F. Heffron, unpublished). One clue as to how this mechanism might be working comes from all the insertion specificity studies. Some Tn*501* insertions can be strong hot spots for integration of Tn*3*. The actual frequency of Tn*3* transposition into the same plasmid with a new hot spot does not increase (Grinsted *et al.*, 1978). We are left with the conclusion that Tn*3* can identify a particular sequence within a plasmid. It is almost as if Tn*3* first travels around a plasmid to determine if particular sequences are present. Thus it is no accident that Tn*501*, which contains many sequences closely related to those in Tn*3*, should be such a hot spot for insertion of Tn*3*.

Transposition immunity is an example of one of the many aspects of Tn*3* recombination that are still not understood. The inability to map an immunity site may suggest that it is a more complex structure of the DNA and not one specific sequence that is recognized during transposition. It is possible that solution of this problem will change our ideas about the structure of DNA in the bacterial cell

Acknowledgments

I would like to thank Agnes Fisher for her careful reading of this chapter and her many helpful suggestions, Louisa Dalessandro for typing the manuscript, and Nancy Kleckner for her review.

References

Altenbuchner, J., Choi, C.-L., Grinsted, J., Schmitt, R., and Richmond, M. H. (1981). The transposons Tn501 (Hg) and Tn1721 (Tc) are related. *Genet. Res.* **37,** 285–289.

Arthur, A., and Sherratt, D. J. (1979). Dissection of the transposition process: a transposon-encoded site-specific recombination system. *Mol. Gen. Genet.* **175,** 267–274.

Arthur, A., Burke, M., and Sherratt, D. J. (1981). Inter-replicon transposition of the Hgr transposon Tn501 is inducible by Hg^{2+} ions and proceeds through cointegrate intermediates. *Soc. Gen. Microbiol. Q.*, **6,** 96–103.

Bennett, P. M., and Richmond, M. H. (1976). The translocation of a discrete piece of DNA carrying an amp gene between Replicons in *Escherichia coli*. *J. Bacteriol.* **126,** 1–6.

Bennett, P. M., Grinsted, J., and Richmond, M. H. (1977). Transposition of TnA does not generate deletions. *Mol. Gen. Genet.* **154,** 205–212.

Bennett, P. M., Grinsted, J., Choi, C. L., and Richmond, M. H. (1978). Characterization of Tn*501*, a transposon determining resistance to mercuric ions. *Mol. Gen. Genet.* **159,** 101–106.

Berg, D. F. (1977). Insertions and excisions of the transposable kanamycin resistance determinant Tn5. *In* "DNA Insertion Elements, Plasmids, and Episomes" (A. I. Bukhari, J. A. Shapiro, and S. Adhya, eds.), pp 205–212. Cold Spring Harbor Lab., Cold Spring Harbor, New York.

Botstein, D., and Kleckner, N. (1977). Translocations and illegitimate recombination by the tetracycline resistance element Tn10. *In* "DNA Insertion Elements, Plasmids and Episomes" (A. I. Bukhari, J. A. Shapiro, S. Adhya, eds.), pp 185–204. Cold Spring Harbor Lab., Cold Spring Harbor, New York.

Brown, N. L., Choi, C. L., Grinsted, J., Richmond, M. H., and Whitehead, P. R. (1980). Nucleotide sequences at the ends of the mercury transposon, Tn501. *Nucleic Acids Res.* **8,** 1933–1945.

Brunton, J., Bennett, P., and Grinsted, J. (1981). Molecular nature of a plasmid specifying beta-lactamase production in *Haemophilus ducreyi. J. Bacteriol.* **148,** 788–795.

Bukhari, A. I., and Zipser, D. (1972). Random insertion of Mu-1 DNA within a single gene. *Nature (London) New Biol.* **236,** 240–243.

Carbon, J., Shenk, J. E., and Berg, P. (1975). Biochemical procedure for production of small deletions in simian Virus 40 DNA. *Proc. Natl. Acad. Sci. USA* **71,** 1392–1396.

Casadaban, M. J., Chou, J., Lemaux, P., Tu, C.-P.F., and Cohen, S. N. (1981). Tn3 transposition and control. *Cold Spring Harbor Symp. Quant. Biol.* **45,** 269–274.

Casadaban, M. J., Chou, J., and Cohen, S. N. (1982). Overproduction of the Tn3 transposition protein and its role in DNA transposition. *Cell* **18,** 345–354.

Chiang, S. J., and Clowes, R. C. (1980). Intramolecular transposition and inversion in plasmid R6K. *J. Bacteriol.* **142,** 668–682.

Chiang, S. J., and Clowes, R. C. (1982a). Recombination between two TnA transposon sequences oriented as inverse repeats is found less frequently than between direct repeats. *Mol. Gen. Genet.* **185,** 169–175.

Chiang, S. J., Jordan, E., and Clowes, R. C. (1982b). Inter- and intramolecular transposition and transposition immunity in *Tn3* and Tn*2660*. *Mol. Gen. Genet.* (in press).

Choi, C.-L., Grinsted, J., Altenbuchner, J., Schmitt, R., and Richmond, M. S. (1981). Transposons Tn501 and Tn1721 are closely related. *Cold Spring Harbor Symp. Quant. Biol.* **45,** 64–66.

Chou, J., Casadaban, M. J., Lemaux, P. G., and Cohen, S. N. (1979a). Identification and characterization of a self-regulated repressor of translocation of the *Tn3* element. *Proc. Natl. Acad. Sci. USA* **76,** 4020–4024.

Chou, J., Lemaux, P. G., Casadaban, M. J., and Cohen, S. N. (1979b). Transposition protein of *Tn3* identification and characterization of an essential repressor controlled gene product. *Nature (London)* **282,**801–806.

Clements, M. B., and Syvanen, M. (1981). Isolation of a *pol*A mutation that affects transposition of insertion sequences and transposons. *Cold Spring Harbor Symp. Quant. Biol.* **45,** 201–204.

Clerget, M., Chandler, M., and Caro, L. (1981). The structure of R1 drd 19: a revised physical map of the plasmid. *Mol. Gen. Genet.* **181,** 183–191.

Clowes, R. C., Holmans, P. L., and Chiang, S. J. (1981). Intramolecular transposition of a Beta-lactamase sequence and related genetic rearrangements. *Cold Spring Harbor Symp. Quant. Biol.* **45,** 167–172.

Cohen, S. N., Casabadan, M. J., Chou, J., and Tu, C. P. D. (1979). Studies of the specificity and control of transposition of the Tn*3* element. *Cold Spring Harbor Symp. Quant. Biol.* **43,** 1247–1255.

Cornelis, G. (1980). Transposition of Tn951 (Tn*lac*) and cointegrate formation are thermosensitive processes. *J. Gen. Microbiol.* **117,** 243–247.

Cornelis, G., Ghosal, D., and Saedler, H. (1978). Tn951: a new transposon carrying a lactose operon. *Mol. Gen. Genet.* **160,** 215–224.

Cornelis, G., Ghosal, D., and Saedler, H. (1979). Multiple integration sites for the lactose transposon Tn951 on plasmic RP1 and establishment of a coordinate system for Tn951. *Mol. Gen. Genet.* **168,** 61–67.

Cornelis, G., Sommer, H., and Saedler, H. (1981). Transposon Tn951 (TnLAC) is defective and related to Tn*3* *Mol. Gen. Genet.* **184,** 241–248.

Daniell, E., Roberts, R., and Abelson, J. (1971). Mutations in the lactose operon caused by bacteriophage Mu. *J. Mol. Biol.* **69,** 1–8.

Datta, N., Hedges, R. W., Shaw, E. J., Sykes, R., and Richmond, M. H. (1971). Properties of an R factor from *Pseudomonas aeroginosa*. *J. Bacteriol.* **108,** 1244–1249.

de la Cruz, F., and Grinsted, J. (1981). Genetic and molecular characterization of Tn21, a multiple resistance transposon from R100.1. *J. Bacteriol.* **151,** 222–228.

Ditto, M. D., Chou, J., Hunkapillar, M. W., Fennewald, M. A., Gerrard, S. P., Cozzarelli, N. R., Hood, L. E., Cohen, S. N., and Casadaban, M. J. (1982). The amino terminal sequence of the *Tn3* transposase protein. *J. Bacteriol.* **149,** 407–410.

Elwell, L. P., De Graaff, J., Seibert, D., and Falkow, S. (1975). Plasmid-linked ampicillin resistance in *Haemophilus influenzae* type b. *Infect. Immun.* **12,** 404–410.

Engler, J. A., and van Bree, M. P. (1981). The nucleotide sequence and protein-coding capability of the transposable element IS5. *Gene*.

Fayet, O., Froment, Y., and Piffaretti, J. C. (1982). Beta-lactamase-specifying plasmids isolated from *Neisseria gonorrhoeae* have retained an intact right part of a *Tn3*-like transposon. *J. Bacteriol.* **149,** 136–144.

Fennewald, M. A., and Shapiro, J. A. (1979). Transposition of Tn7 in *Pseudomonas aeruginosa* and isolation of Alk-Tn7 mutations. *J. Bacteriol.* **139,** 264–269.

Fennewald, M. A., Gerrard, S. P., Chou, J., Casadaban, M., and Cozzarelli, N. R. (1981). Purification of the *Tn3* transposase and analysis of its binding to DNA. *J. Biol. Chem.* **256,** 4687–4690.

Fischoff, D. A., Vovis, G. F., and Zinder, N. D. (1981). Organization of chimeras between filamentous bacteriophage F1 and plasmid pSC101. *J. Mol. Biol.* **144,** 247–265.

Galas, D. J., and Chandler, M. (1981). On the molecular mechanisms of transposition. *Proc. Natl. Acad. Sci. USA* **78,** 4858–4862.

Gill, R., Heffron, F., Dougan, G., and Falkow, S. (1978). Analysis of sequences transposed by complementtion of 2 classes of transposition deficient mutants of transposition element *Tn3*. *J. Bacteriol.* **136,** 742–756.

Gill, R., Heffron, F., and Falkow, S. (1979). Identification of the protein encoded by the transposable element *Tn3* which is required for its transposition. *Nature (London)* **282,** 797–801.

Gill, R. E., Falkow, S., Ohtsubo, E., Ohtsubo, H., So, M., and Heffron, F. (1980). A genetic analysis of the transposon *Tn3*: evidence for cointegrates as intermediates in transposition. *In* "Mobilization and Reassembly of Genetic Information" (W. A. Scott, R. Werner, D. R. Joseph, and J. Schultz, eds.), pp. 47–64. Academic Press, New York.

Goebel, W., Lindenmaier, W., Pfeifer, F., Schrempf, I., and Schelle, B. (1977). Transposition and insertion of intact, deleted, and ampicillin transposon *Tn3* from mini-R1 (RSC) plasmid into transfer factor. *Mol. Gen. Genet.* **157,** 119–129.

Grindley, N. D. F., and Joyce, C. M. (1981). Analysis of the structure and function of the kanamycin-resistant transposon Tn903. *Cold Spring Harbor Symp. Quant. Biol.* **45,** 125–133.

Grindley, N. D., and Sherratt, D. J. (1978). Sequence analysis of IS1 insertion sites: models for transposition. *Cold Spring Harbor Symp. Quant. Biol.* **63,** 1257–1261.

Grindley, N. D. F., Lauth, M. R., Wells, R. G., Wityk, R. J., Salvo, J. J., and Reed, R. R. (1982). Transposon-mediated recombination: identification of three binding sites for resolvase at the *res* sites of γδ and Tn*3*. *Cell* **30,** 19–27.

Grinsted, J., and de la Cruz, F. (1981). Complementation of transposition of *tnpA* mutants of Tn1721 and Tn501. *Plasmid* (in press).

Grinsted, J., Bennett, P. M. Higginson, S., and Richmond, M. H. (1978). Regional preference of insertion of Tn501 and Tn801 into Rp1 and its derivatives. *Mol. Gen. Genet.* **166,** 313–320.

Guyer, M. S. (1978). The Gamma Delta sequence of F is an insertion sequence. *J. Mol. Biol.* **126,** 347–365.

Harayama, S., Tsuda, M., and Iino, T. (1981). Tn1 insertion mutagenesis in *Escherichia coli* K-12 using a temperature sensitive mutant of plasmid RP4. *Mol. Gen. Genet.* **184,** 52–55.

Harshey, R. M., and Bukhari, A. I. (1981). A mechanism of DNA transposition. *Proc. Natl. Acad. Sci. USA* **78,** 1090–1094.

Hedges, R. W., and Jacob, A. E. (1974). Transposition of ampicillin resistance from RP4 to other replicons. *Mol. Gen. Genet.* **132,** 31–40.

Hedges, R. W., Matthew, M., Smith, D. I., Cresswell, J. M., and Jacob, A. E. (1977). Properties of a transposon conferring resistance to penicillins and streptomycin. *Gene* **1,** 241–253.

Heffron, F., Rubens, C., and Falkow, S. (1975a). The translocation of a plasmid DNA sequence which mediates ampicillin resistance: molecular nature and specificity of insertion. *Proc. Natl. Acad. Sci. USA* **72,** 3623–3627.

Heffron, F., Sublett, R., Hedges, R. W., Jacob, A., and Falkow, W. (1975b). Origin of the TEM Beta-lactamase gene found on plasmids. *J. Bacteriol.* **122,** 250–256.

Heffron, F., Rubens, C., and Falkow, S. (1976). Transposition of a plasmid DNA sequence which mediates ampicillin resistance: identity of laboratory constructed plasmids and clinical isolates. *J. Bacteriol.* **129,** 530–533.

Heffron, F., Rubens, C., and Falkow, S. (1977a). Transposition of a plasmid DNA sequence which mediates ampicillin resistance: general description and epidemiologic considerations. *In* "DNA Insertions, Elements, Plasmids, and Episomes" (A. I. Bukhari, J. A. Shapiro, and S. L. Adhya, eds.), pp. 151–160. Cold Spring Harbor Lab., Cold Spring Harbor, New York.

Heffron, F., Bedinger, P., Champoux, J. J., and Falkow, S. (1977b). Deletions affecting the transposition of an antibiotic resistance gene. *Proc. Natl. Acad. Sci. USA* **74,** 702–706.

Heffron, F., So, M., and McCarthy, B. J. (1978). In vitro mutagenesis of a circular DNA molecule by using synthetic deoxyoligonucleotides. *Proc. Natl. Acad. Sci. USA* **75,** 6012–6016.

Heffron, F., McCarthy, B. J., Ohtsubo, H., and Ohtsubo, E. (1979). DNA sequence analysis of the transposon *Tn3* three genes and three sites involved in transposition of *Tn3*. *Cell* **18,** 1153–1164.

Heffron, F., Kostriken, R., Morita, C., and Parker. R. (1981). *Tn3* encodes a site-specific recombination system: identification of essential sequences, genes and the actual site of recombination. *Cold Spring Harbor Symp. Quant. Biol.* **45,** 259–268.

Holmans, P. L., and Clowes, R. C. (1979). Transposition of a duplicate antibiotic resistance gene and generation of deletions in plasmic R6K. *J. Bacteriol.* **137,** 977–989.

Ilyina, T. S., Mechaera, E. V., Ramonida, Y. A., and Smirnoj, G. B. (1981). Isolation and mapping of *Escherichia coli* Tn2 mutants defective in Tn9 transposition. *Mol. Gen. Genet.* **181,** 384–389.

Isberg, R. R., and Syvanen, M. (1982). DNA gyrase is a host factor required to transposition of Tn*5*. *Cell* **30,** 9–18.

Jordon, E., Saedler, H., and Starlinger, P. (1968). 00- and strong-polar mutations in the gal operon are insertions. *Mol. Gen. Genet.* **102,** 353–365.

Kahn, S. A., and Novick, R. P. (1980). Terminal nucleotide sequences of Tn551, a transposon specifying eruthromycin resistance in staphylococcus aureus: homology with Tn*3*. *Plasmid* **4,** 148–154.

Kitts, P., Reed, R., Symington, L., Burke, M., and Sherratt, D. (1982). Transposon-specified site-specific recombination. *Proc. Natl. Acad. Sci. USA* **79,** 46–50.

Klaer, R., Pfiefer, D., and Starlinger, P. (1980). IS4 insertion sequence is still found at its chromosomal site after transposition to GalT. *Mol. Gen. Genet.* **178,** 281–284.

Kleckner, N. (1981). Transposable elements in prokaryotes. *Annu. Rev. Genet.* **15,** 341–404.

Kopecko, D. J. (1980a). Specialized genetic recombination systems in bacteria: their involvement in gene expression and evolution. *In* "Progress in Molecular and Subcellular Biology" (F. Hahn, ed.), pp. 135–234. Springer-Verlag, New York and Berlin.

Kopecko, D. J. (1980b). Involvement of specialized recombination in the evolution and expression of bacterial genomes. *In* "Plasmids and Transposons" (C. Stuttard and K. R. Rozee, eds.), pp. 165–205. Academic Press, New York.

Kopecko, D. J., and Cohen, S. N. (1975). Site-specific *rec*A-independent recombination between bacterial plasmids: involvement of palindromes at the recombinational loci. *Proc. Natl. Acad. Sci. USA* **72,** 1373–1377.

Kopecko, D. J., Brevet, J., and Cohen, S. N. (1976). Involvement of multiple translocating DNA segments and recombinational hot spots in the structural evolution of bacterial plasmids. *J. Mol. Biol.* **108,** 333–360.

Kostriken, R., Morita, C., and Heffron, F. (1981). The transposon *Tn3* encodes a site-specific recombination system: identification of essential sequences, genes and the actual site of recombination. *Proc. Natl. Acad. Sci. USA* **78,** 4041–4045.

Kretschmer, P. J., and Cohen, S. N. (1977). Selected translocation of plasmic genes: frequency and regional specificity of translocation of the Tn*3* element. *J. Bacteriol.* **130,** 888–899.

Kretschmer, P. J., and Cohen, S. N. (1979). Effect of temperature on translocation frequency of the *Tn3* element. *J. Bacteriol.* **139,** 515–519.

Ljungquist, E., and Bukhari, A. I. (1977). State of prophage Mu DNA upon induction. *Proc. Natl. Acad. Sci. USA.* **74,** 3143–3147.

Manis, J., Kopecko, D., and Kline, B. (1980). Cloning of a Lac+ BamHI fragment into transposon *Tn3* and transposition of the *Tn3* (Lac) element. *Plasmid* **4,** 170–174.

McCormick, M., Wishart, W., Ohtsubo, H., Heffron, F., and Ohtsubo, E. (1981). The structure of recombinant plasmids mediated by the transposable DNA element *Tn3* and *Tn3* mutants. *Gene* **15,** 103–118.

Miller, C. A., and Cohen, S. N. (1980). F plasmid provides a function that promotes RecA-independent site-specific fusions of pSC101 replicon. *Nature (London)* **285,** 577–579.

Miller, J. H., Calos, M. P., Galas, D., Hofer, M., Buchel, D. E., and Muller-Hill, B. (1980). Genetic analysis of transposition in the *lac* region of *E. coli*. *J. Mol. Biol.* **144,** 1–18.

Muster, C. J., and Shapiro, J. A. (1981). Recombination involving transposable elements on replicon fusions. *Cold Spring Harbor Symp. Quant. Biol.* **45,** 239–242.

Nisen, P. O., Kopecko, D. J., Chow, J., and Cohen, S. N. (1977). Site specific DNA deletions occurring adjacent to the terminal of a transposable ampicillin resistant element (*Tn3*) *J. Mol. Biol.* **117,** 975–998.

Novick, R. P., Edelman, I., Schwesinger, M. D., Gruss, A. D., Swanson, E. C., and Pattee, P. A. (1979). Genetic translocation in *Staphylococcus aureus*. *Proc. Natl. Acad. Sci. USA.* **76,** 400–404.

Novick, R. P., Khan, S. A., Murphy, W., Iordaneslo, S., Edelman, I., Kroleewski, J., and Ruch, M. (1981). Hitchhiking transposons and other mobile genetic element and site-specific recombination systems in *Staphylococcus aureus*. *Cold Spring Harbor Symp. Quant. Biol.* **45,** 67–76.

Ohtsubo, E., Rosenbloom, M., Schrempf, H., Goebel, W., and Rosen, J. (1978). Site-specific recombination involved in the generation of small plasmids. *Mol. Gen. Genet.* **159,** 131–141.

Ohtsubo, S., Ohmori, H., and Ohtsubo, E. (1979). Nucleotide sequence analysis of *Tn3* (Ap): implications for insertion and deletion. *Cold Spring Harbor Symp. Quant. Biol.* **45,** 1269–1278.

Ohtsubo, E., Zenilman, M., Ohtsubo, H., McCormick, M., Machida, C., and Machida, Y. (1981). Mechanism of insertion and cointegration mediated by IS1 and *Tn3* *Cold Spring Harbor Symp. Quant. Biol.* **45,** 283–295.

Ravetch, J. V., Ohsumi, M., Model, P., Vovis, G. F., Fischoff, D., and Zinder, N. D. (1979). Organization of a hybrid between Phage F1 and plasmid pSC101. *Proc. Natl. Acad. Sci. USA* **76,** 2195–2198.

Read, H. A., Das, S. S., and Jaskunas, S. R. (1980). Fate of donor insertion sequence IS1 during transposition. *Proc. Natl. Acad. Sci. USA* **77,** 2514–2518.

Reed, R. R. (1981a). Resolution of cointegrates between transposons Gamma-delta and *Tn3* defines the recombination site. *Proc. Natl. Acad. Sci. USA* **78,** 3428–3432.

Reed, R. R. (1981b). Transposon-mediated site-specific recombination: a defined in vitro system. *Cell* **25,** 713–719.

Reed, R. R., and Grindley, N. D. F. (1981). Transposon-mediated site-specific recombination in vitro: DNA cleavage and protein-DNA linkage at the recombination site. *Cell* **25,** 721–728.

Reed, R. R., Young, R. A., Steitz, J. A., Grindley, N. D., and Guyer, M. S. (1979). Transposition of the *E. coli* insertion element Gamma-delta generates a five base pair repeat. *Proc. Natl. Acad. Sci. USA* **76,** 4882–4886.

Richmond, M. H., and Sykes, R. B. (1972). The chromosomal integration of a β-lactamase gene derived from the P-type R factor RP1 in *Escherichia coli. Genet. Res.* **20,** 231–237.

Robinson, M. K., Bennett, P. M., and Richmond, M. H. (1977). Inhibition of TnA translocation by TnA. *J. Bacteriol.* **129,** 407–414.

Robinson, M. K., Bennett, P. M., Grinsted, J., and Richmond, M. H. (1978). The stable carriage of two TnA units on a single replicon. *Mol. Gen. Genet.* **160,** 339–346.

Sasakawa, G., Uno, Y., and Yoshikawa, M. (1981). The requirement for both DNA polymerase and 5′ to 3′ exonuclease activities of DNA polymerase during Tn5 transposition. *Mol. Gen. Genet.* **182,** 19–29.

Schmitt, R., Bernhard, E., and Mattes, R. (1979). Characterisation of Tn1721, a new transposon containing Tetracycline resistance genes capable of amplification. *Mol. Gen. Genet.* **172,** 53–56.

Schmitt, R., Altenbuchner, J., Wiebauer, K., Arnold, W., Puhler, A., and Schoffl, F. (1981). Basis of transposition and gene amplification by Tn1721 and related Tetracycline-resistance transposons. *Cold Spring Harbor Symp. Quant. Biol.* **45,** 59–65.

Schmitt, R., Altenbuchner, J., and Grinsted, J. (1982). Complementation of transposition function encodes by transposons Tn501 (Hg^R), and Tn1721 (Tet^R),. *In* "Molecular Biology, Pathogenecity and Ecology of Bacterial Plasmids" (S. B. Levy, R. C. Clowes, and E. L. Koenigh, eds.), pp. 359–370. Plenum Press, New York and London.

Schoffl, F., Arnold, W., Puhler, A., Altenbuchner, J., and Schmitt, R. (1981). The tetracycline resistance transposons Tn1721 and Tn1771 have three 38 base-pair repeats and generate five base-pair direct repeats. *Mol. Gen. Genet.* **181,** 87–94.

Shapiro, J. A. (1969). Mutations caused by insertion of genetic material into the galactose operon of *Escherichia coli. J. Mol. Biol.* **40,** 93–105.

Shapiro, J. A. (1979). Molecular model for the transposition and replication of bacteriophage Mu and other transposable elements. *Proc. Natl. Acad. Sci. USA* **76,** 1933–1937.

Sharp, P., Cohen, S. N., and Davison, N. (1973). Electron microscope heteroduplex studies of sequence relations among plasmids of *Escherichia coli*. II. Structure of drug resistance (R) factors and F factors. *J. Mol. Biol.* **75,** 235–255.

Shenk, T., Carbon, J., and Berg, P. (1976). Construction and analysis of viable deletion mutants of Simian Virus 40. *J. Virol.* **18,** 664.

Sherratt, D., Arthur, A., and Burke, D. J. (1981). Transposon-specified, site-specific recombination systems. *Cold Spring Harbor Symp. Quant. Biol.* **45,** 275–282.

Simon, M., Zieg, J., Silverman, M., Mandel, G., and Doolittle, R. (1980). Genes whose mission is to jump. *Science (Washington, D.C.)* **209,** 1370–1374.

Smith, D. H. (1967). R-factor infection of *Escherichia coli* lyophilized in 1946. *J. Bacteriol.* **94,** 2071.

So, M., Gill, R., and Falkow, S. (1975). The generation of a ColEl-Ap^r cloning vehicle which allows detection of inserted DNA. *Mol. Gen. Genet.* **142,** 239–249.

Stanishich, V. A., Bennett, P. M., and Richmond, M. H. (1977). Characterisation of a translocation unit encoding resistance to mercuric ions that occurs on a nonconjugative plasmid in *Pseudomonas cerosinosea. J. Bacteriol.* **129,** 1221–1223.

Sternglanz, R., Dinardo, S., Voelkel, K., Nislimura, Y., Hirota, Y., Becherer, K., Zumstein, L., and Wang, J. (1981). Mutations in the gene coding for *E. coli* DNA topoisomerase I after transcription and transposition. *Proc. Nat. Acad. Sci. USA* **78,** 2747–2751.

Symington, L., and Sherratt, D. J. (1981). Functional relationships between transposon Tn1/3 and the transposable element Gamma-delta. *Symp. Soc. Gen. Microbiol.* **31,** 95–103.

Syvanen, M., Hopkins, J. D., and Clements, M. (1982). A new class of mutants in DNA polymerase I that affects gene transposition. *J. Mol. Biol.* **158,** 203–212.

Tomich, P. K., An, F. Y., and Clewell, D. B. (1980). Properties of Erythromycin-inducible transposon Tn917 in *Streptococcus faecalis. J. Bacteriol.* **141,** 1366–1374.

Tu, C. P., and Cohen, S. N. (1980). Translocation specificity of the *Tn3* element: characterization of sites of multiple insertions. *Cell* **19,** 151–160.

Wallace, L. J., Ward, J. M., Bennet, P. M., Robinson, M. K., and Richmond, M. H. (1981a). Transposition immunity. *Cold Spring Harbor Symp. Quant. Biol.* **45,** 183–188.

Wallace, L. J., Ward, J. R., and Richmond, M. H. (1981b). The location of sequences of TnA required for the establishment of transposition immunity. *Mol. Gen. Genet.* **184,** 80–86.

Wallace, L. J., Ward, J. M., and Richmond, M. H. (1981c). The *tnp*R product of TnA is required for transposition immunity. *Mol. Gen. Genet.* **184,** 87–91.

Wang, A., Dia, X., and Lu, D. (1980). The transposition properties of Tn2 in *E. coli. Cell* **21,** 251–256.

Weinstock, G. M., Susskind, M. M., and Botstein, D. (1979). Regional specificity of illegitimate recombination by the translocable ampicillin-resistance element Tn1 in the genome of phage P22. *Genetics* **92,** 685–710.

Wiebauer, K., Schraml, S., Shales, S. W., and Schmitt, R. (1981). Tetracycline resistance transposon Tn1721:recA-Dependent gene amplification and expression of tetracycline resistance. *J. Bacteriol.* **147,** 851–859.

Yamamoto, T., Tanaka, M., Baba, R., and Yamagishi, S. (1981a). Physical and functional mapping of Tn*2603,* a transposon encoding ampicillin streptomycin, sulfonamide, and mercury resistance. *Mol. Gen. Genet.* **181,** 464–469.

Yamamoto, T., Tanaka, M., Nohara, C., Fukunaga, Y., and Yamagishi, S. (1981b). Transposition of oxacillin-hydrolysing penicillinase gene. *J. Bacteriol.* **145,** 808–813.

Zheng, Z. X., Chandler, M., Hipskind, R., Clerget, M., and Caro, L. (1982). Dissection of the r-determinant of the plasmid R100.1: the sequence at the extremities of Tn21. *Nucleic Acids Res.* **9,** 6265–6278.

CHAPTER 6

Transposon Tn*10*

Nancy Kleckner

I. Introduction

Transposon Tn*10* was originally isolated from nature on the conjugative drug-resistance plasmid R100 (also called R222) (Watanabe and Lyang, 1962;

Mobile Genetic Elements

ISBN 0-12-638680-3

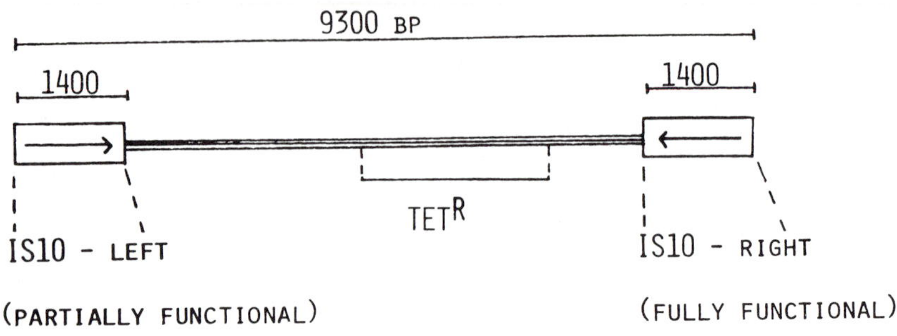

Fig. 1 Transposon Tn*10*.

Watanabe *et al.*, 1972; Sharp *et al.*, 1973). The intact transposon is 9300 bp in length and has 1400-bp inverted repeats at its ends. The 6500 bp of nonrepeated material include the 2500-bp tetracycline-resistance determinant and (probably) other determinants the functions of which are not known (Kleckner *et al.*, 1975; Jorgensen and Reznikoff, 1979). The 1400-bp repeat sequences are closely related but nonidentical IS elements, IS*10*-Right and IS*10*-Left, that encode the sites and functions responsible for Tn*10* transposition and that cooperate with one another to mediate transposition of the intervening *tet*R material (Foster *et al.*, 1981a; Fig. 1; see below).

Tn*10* elements account for an important fraction of the plasmid-encoded tetracycline-resistance determinants isolated from nature. The *tet*R genes of Tn*10* are readily distinguishable in both DNA sequence and phenotype from other types of tetracycline-resistance genes, including those found on the commonly used laboratory plasmid pSC101 (and its derivatives pMB9 and pBR322) or the broad host range plasmids RP1 and RP4 (Mendez *et al.*, 1980; McMurry *et al.*, 1980).

IS*10* has thus far been identified only in association with Tn*10*. IS*10* does not occur in the chromosomes of *Escherichia coli* K12 or *Salmonella typhimurium* LT2 and has not been identified on plasmids other than those containing a full Tn*10* element (Davidson *et al.*, 1974; Ross *et al.*, 1979a).

*II. Transposition of Tn*10 *as a Discrete Unit*

Tn*10* can undergo many successive cycles of transposition as a genetically discrete and functionally intact unit (Kleckner *et al.*, 1975, 1978). Transposition of the element can be detected genetically by the appearance of the *tet*R genes at a new position in the chromosome or in an entirely new replicon. When Tn*10* inserts into an identifiable structural or regulatory region, appearance of *tet*R

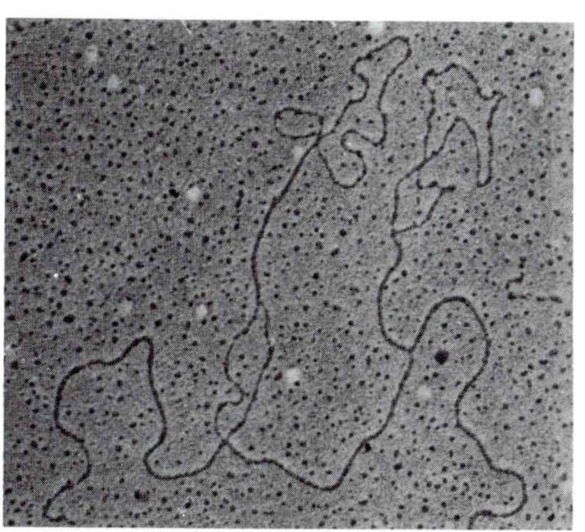

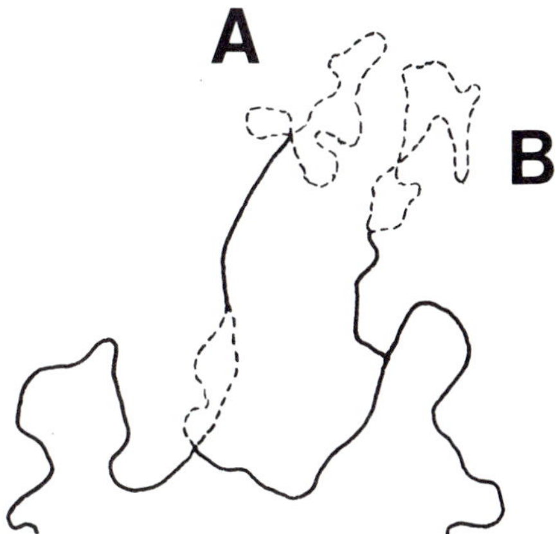

Fig. 2 Transposon stem-and-loop structures. DNA heteroduplex between two λ phage genomes containing transposons: *c*I :: Tn*10* *nin5*/*c*I+ *nin* :: TnHACIO. The stem-and-loop structure of (B) is *c*I :: Tn*10;* the stem-and-loop structure of (A) is an insertion of another IS*10*-flanked transposon derived from Tn*10* and having a similar overall structure (Foster *et al.*, 1981a). The TnHACIO insertion occurs in a region of nonhomology between the two phage genomes, the *nin5* deletion.

genes at a new site is correlated with acquisition of a new mutation in that region. Tn*10* insertions can also be visualized directly in artificial DNA–DNA heteroduplex molecules composed of one strand with and one strand without the insertion. In such molecules Tn*10* DNA, with its long inverted repeats, forms a stem-and-loop structure of characteristic dimensions (Fig. 2). Heteroduplex analysis has shown that the original Tn*10* element on R222 is a simple insertion of Tn*10* into a region of the plasmid that is otherwise homologous with the conjugative plasmid F (Sharp *et al.*, 1973).

Transposition of drug-resistance elements can be assayed quantitatively. When Tn*10* is introduced into a tetracycline-sensitive bacterial strain on a nonreplicating, nonintegrating, nonkilling bacteriophage genome, each of the tetracycline-resistant organisms arising from the infection contains an insertion of the element into the bacterial chromosome or associated episome (Kleckner *et al.*, 1975, 1978). Drug-resistance elements that transpose at higher frequencies, such as Tn*5* (*kan*R), can be introduced in an analogous way on phage-generated generalized transducing fragments (Biek and Roth, 1980). The activity of a Tn*10* element that is already resident in the bacterial chromosome can be determined by monitoring its transposition into a coresident conjugative plasmid. In such a donor population, the proportion of cells containing an episomal Tn*10* insertion can be found by mating with a suitable recipient population and determining the proportion of episome-receiving exconjugants that have also acquired the transposon's *tet*R marker (Foster *et al.*, 1981a). For Tn*5*, integration of a stably resident transposon into the genome of a superinfecting bacteriophage can also be used as an assay (Isberg and Syvanen, 1981).

Wild-type Tn*10* transposes at a very low frequency: approximately 10^{-7} transpositions per element per generation, as measured by conjugation assays (see Section IV), and approximately 10^{-5}–10^{-6} transpositions per input phage genome in abortive phage infection assays (Foster *et al.*, 1981a; N. Kleckner, unpublished).

*III. Genetic Organization of Tn*10

A. Tn*10* as a Composite Transposon

Each of the 1400-bp inverted repeats is a structurally intact IS sequence that is (or once was) capable of independent transposition as a discrete unit. The structural integrity of the two IS*10* elements is revealed by the fact that both ends of both elements are active sites for Tn*10*-promoted rearrangements (Ross *et al.*, 1979c; Foster *et al.*, 1981a; Section VII,C).

IS*10*-Right and IS*10*-Left cooperate to mediate transposition of the intervening *tet*R segment. The IS*10* sequences clearly provide important sites at the termini of Tn*10* In fact, all of the Tn*10*-encoded sites and Tn*10*-encoded functions required for transposition lie within these two IS*10* sequences.

The ability of IS*10*-Left and IS*10*-Right to mobilize intervening material is illustrated by the experiment in Fig. 3. A circular plasmid genome containing Tn*10* can give rise to two types of transposition products: normal Tn*10* transposition involving the "outside" ends of the IS*10* elements, and a second, symmetrically related transposition event involving the "inside" ends of the two elements and resulting in movement of the plasmid replicon flanked by inverted repeats of IS*10*-Right and IS*10*-Left. This newly transposed segment, with IS*10* elements at its ends, is itself a new transposon that is capable of independent transposition as a discrete unit at frequencies comparable to those of Tn*10* (Foster *et al.*, 1981a).

B. Functional Analysis of the IS*10* Elements

Although IS*10*-Right and IS*10*-Left are both structurally intact, they are not both functionally intact. Complementation analysis shows that IS*10*-Right encodes at least one diffusible function that is essential for Tn*10* transposition and that acts at the ends of the element. The nature and role of this function are discussed below (Section IV). IS*10*-Right can promote normal levels of Tn*10* transposition even when IS*10*-Left is specifically inactivated, and can provide

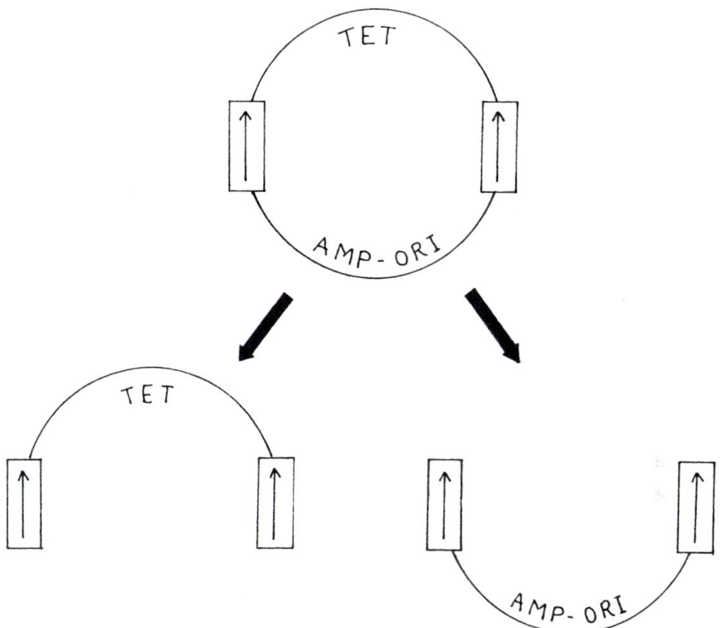

Fig. 3 Symmetrically related IS*10*-flanked transposons. A circular *amp*R-*ori* plasmid genome containing Tn*10* gives rise to two different types of transposition products: insertions of the standard 9300-bp Tn*10* element and insertions of the symmetrically related IS*10*-*amp*R-*ori*-IS*10* segment (Foster *et al.*, 1981a).

the same level of complementing functions in trans to a mutant element as does wild-type Tn*10*. IS*10*-Left, however, is functionally defective and can provide only 1–10% of the transposition activity of IS*10*-Right (Foster *et al.*, 1981a; N. Kleckner and M. A. Davis, unpublished).

The functional difference between the two IS*10* elements is reflected at the nucleotide sequence level. The two sequences differ at 15 of the 665 positions compared thus far. Some of these differences are within a single long coding region that extends from one end to the other of IS*10*-Right and specifies a protein essential for transposition; other differences are located upstream of the coding region in a presumptive regulatory region (Halling *et al.*, 1982). Although it is not yet clear how each of the observed nucleotide sequence differences contributes to the functional differences between the two elements, the behavior of laboratory-constructed IS*10* elements that are hybrids between -Left and -Right suggests that changes in the coding region are primarily responsible for the nonfunctionality of IS*10*-Left (R. W. Simons and N. Kleckner, unpublished).

C. Essential Sites near the Ends

All sites required for Tn*10* transposition lie very near (perhaps within 13 bp of) the ends of the element. Artificially constructed deletion mutants of Tn*10* that retain only the outer 27 bp of each IS*10* element plus the *tet*R determinant are still capable of efficient transposition if diffusible functions are provided in trans from another element. In addition, four cis-dominant point mutations that abolish Tn*10* transposition even in the presence of added complementing functions all lie within 13 bp of the termini of the element. (Foster *et al.*, 1981a; J. C. Way, unpublished).

D. Evolution of Tn*10*

It seems most likely that IS*10*-Left and IS*10*-Right were originally identical in sequence and subsequently diverged from one another. Simple mechanisms by which a composite transposon having Tn*10*s inverted repeat structure can arise by replicative transposition from a single IS sequence have been suggested (Shapiro, 1979), and events of this kind have been directly observed among the array of IS*10*-promoted rearrangements (Kleckner and Ross, 1980; Section VII,C). There is no need to postulate formation of Tn*10* by independent insertion of related but distinct sequences to positions flanking the *tet*R determinant. IS*10*-Left mediates transposition at such a low frequency that it is unlikely to have transposed in its present form. IS*10*-Right, on the other hand, mediates DNA rearrangements at about the same frequency as IS elements identified as independent entities in bacterial genomes. We favor the view that IS*10*-Right is similar to the original IS*10* element that gave rise to Tn*10*. IS*10*-Left may have degenerated to its present nonfunctional state because only one set of IS*10*-encoded functions is needed for efficient transposition of *tet*R genes (Halling *et al.*, 1982). An alternative possibility is suggested by the observations that p-IN of IS*10*-Left is more active than p-IN of IS*10*-Right, and that transfer of two particular base changes from IS*10*-Left to IS*10*-Right results in increased expression of transposition functions and a "high-hopper" phenotype (R. W. Simons, B. C. Hoopes, W. R. McClure, M. A. Davis, and N. Kleckner, unpublished). It is possible that transposable elements with high transposition rates would not be stably maintained in a bacterial population. If this were true, mutations eliminating the activity of IS*10*-Left could have arisen as a secondary response to initial mutation(s) that increased transposition activity above the acceptable level. It is also possible that IS*10*-Left may have some unidentified regulatory role or that it is important for transposition in organisms other than *E. coli* and *Salmonella*.

IS*10*-Right and IS*10*-Left differ in sequence to a much greater extent than do the flanking IS sequences of two kanamycin-resistance transposons, Tn*5* and

Tn*903*. These two elements also have a composite inverted-repeat structure, but the flanking IS elements differ by 1 and 0 bp, respectively. The greater divergence may mean that Tn*10* is older than the other elements and has had more time to accumulate changes. In this regard, it may be relevant that tetracycline resistance was observed in human and animal populations in 1950, whereas kanamycin resistance was not observed until 1964 (Falkow, 1974; references therein).

The inverted repeats of Tn*10* are separated by some 6500 bp of DNA, of which only 2500 are involved in expression of tetracycline resistance. Several Tn*10*-like *tet*R elements have been isolated from nature, and all have the same structure as Tn*10* (Kaulfers *et al.*, 1978; Bennett *et al.*, 1980; Jahn *et al.*, 1979; Jorgensen *et al.*, 1980). This suggests that Tn*10*'s "mystery" material may confer some selective advantage in nature that is either irrelevant in the laboratory or has not been tested.

IV. Genetic Analysis of Tn10 Transposition

A. Length Dependence in Conjugation Assays

Transposition frequencies of wild-type Tn*10* have been compared with those of several Tn*10* deletion variants that are still fully capable of promoting their own transposition but that are shorter than the wild-type element. Each variant retains intact all of IS*10*-Right, the *tet*R genes, and the outer end of IS*10*-Left (which provides the left terminus of the element) but lacks some portion of the nonrepeated material outside the *tet*R region and/or the inner portion of IS*10*-Left. Figure 4 shows that transposition of such elements is length-dependent. The longer the transposing segment, the lower the observed frequency of transposition. Approximately 30–40% of all transposition events are lost for every kilobase of transposon length. Extrapolation of the curve in Fig. 4 to zero length yields a 100-fold increase in transposition frequency. Thus, Tn*10* must initiate a minimum of 100 transposition events for every one that yields a complete product.

The length dependence of Tn*10* transposition is similar to that first reported for a series of IS1-flanked transposons (Chandler *et al.*, 1982). In both instances, the degree of dependence is too great to be accounted for by a requirement that both ends of the transposing segment be in the same place at the same time. Length dependence could reflect disintegration of a replication complex as it moves across the element, the exposure of nuclease-sensitive, single-stranded regions during transposition, or any other process that results in a finite probability of decay or disruption per unit length. It is not yet clear whether Tn*10* transposition is length-dependent in other types of assays.

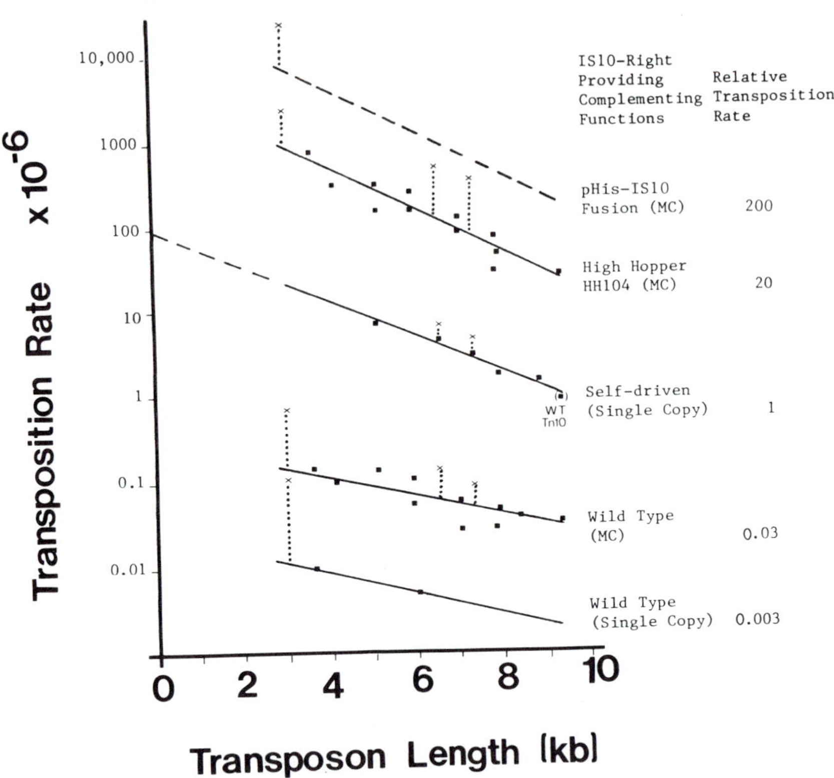

Fig. 4 Tn*10* transposition varies with transposon length and level of transposition functions (N. Kleckner and H.-J. Kim, unpublished). Deletion derivatives of Tn*10*, each of which retains the *tet*R determinant and the termini of the element but lacks other internal segment(s), were assayed for their transposition from the bacterial chromosome to an F′*lacpro* episome by a conjugation "mating-out" assay (described by Foster *et al.*, 1981a). Transposition of those elements that retain all of IS*10*-Right intact and thus can still provide their own transposition functions is shown in the self-driven curve. Other deletion elements that have internal lesions in both IS*10*-Right and IS*10*-Left were assayed for transposition in the presence of various levels of complementing functions. These functions were provided by a single copy of IS*10*-Right present at a separate site in the chromosome, or by pBR322 derivatives (MC) carrying either wild-type IS*10*-Right, the high-hopper IS*10*-Right mutant HH104, or an artificial fusion between IS*10*'s transposase gene and the *Salmonella* histidine operon promoter. Some Tn*10* derivatives always fall above the line defined by other elements at a given level of complementing functions; these derivatives are indicated by x. The transposition rate is the rate per transposon per bacterial generarion.

B. Inefficient Complementation of Defective Tn*10* Elements: IS*10* Functions Act in cis

When a functionally defective Tn*10* element is present in a single copy in the bacterial chromosome and a single copy of IS*10*-Right is present elsewhere in the genome, the defective element can be complemented to transpose, but at less than 1% of the normal "self-driven" wild-type level. Even when functions are present in larger amounts, provided by IS*10*-Right on a multicopy plasmid, transposition reaches less than 5% of the self-driven level. Complementation data for defective Tn*10* elements of varying lengths are shown in Fig. 4. Noncomplementability may be a general property of IS elements (Machida *et al.*, 1982; Isberg and Syvanen, 1981; see Kleckner, 1981 for review).

We have shown directly that this low level of complementation is due to the fact that IS*10* transposition functions act preferentially in cis. In artificial constructions that place the IS*10* transposase gene (expressed from a heterologous promoter) at different alternative positions relative to suitably marked function-defective Tn*10* elements, it can be seen that transposition functions work 150 times better on transposon ends that are 1 kb away than they do on ends that are in an entirely different part of the bacterial chromosome, and that functions work 10 times better on ends that are 50 kb away than on a very distant element (Fig. 5; D. Morisato, J. C. Way, H.-J. Kim, and N. Kleckner, unpublished).

An alternative and not mutually exclusive explanation for inefficient complementation is that actions of IS*10* functions on the ends of an IS*10* or Tn*10* element are mechanistically linked to expression of those functions. Such coupling would, by definition, not be mimicked in any situation in which transposition functions from one element are provided in trans to the ends of a second element. The experiment in Fig. 5 does not directly address this possibility. However, the magnitude of the effect of cis action is sufficient to account for most and perhaps all of the observed complementation inefficiency.

Other DNA-binding proteins (φX174 *cis*A, P2 gene A, and λ Q) appear *in vivo* to act preferentially in cis (Echols *et al.*, 1976; Francke and Ray, 1972; Lindahl, 1970; Lindqvist and Sinsheimer, 1967; Tessman, 1966; Burt and Brammer, 1982). One suggested explanation for this phenomenon is that certain DNA-binding proteins might move by one-dimensional diffusion along DNA rather than being freely diffusible throughout the cell (Berg *et al.*, 1981; Adam and Delbruck, 1968; Echols *et al.*, 1976; Richter and Eigen, 1974). Preferential cis action is only one of several ways in which the action of transposition functions may resemble action of φX174 *cis*A protein (Shapiro, 1979; Harshey and Bukhari, 1981; K. Mizuuchi, unpublished; Read *et al.*, 1980; Ohtsubo *et al.*, 1981; Machida *et al.*, 1982)

C. Importance and Role of IS*10* Functions

The apparent cis action of IS*10* proteins is not absolute. Defective Tn*10* elements can be complemented to transpose at very high frequencies when high levels of IS*10* functions are present. In fact, transposition of both defective and nondefective Tn*10* elements (present as single-copy insertions in the chromosome) can reach 10–100 times the wild-type level in the presence of multicopy plasmids containing either high-hopper IS*10*-Right mutants that produce in-

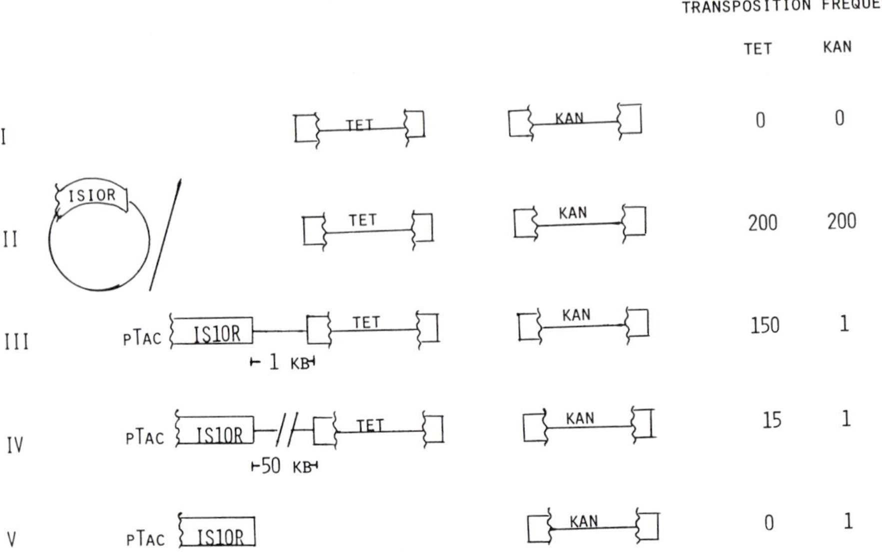

Fig. 5 Transposition of two differentially marked (*tet*R and *kan*R) function-defective Tn*10* variants was measured in the absence of any IS*10* functions (strain I) or with IS*10* functions under control of a heterologous (pTac) promoter and placed in different configurations with respect to the two variants. Both elements transpose at comparable frequencies when the transposase fusion is present on a multicopy plasmid (strain II). When the transposase gene is placed near the *tet*R element, transposition of that nearby element is much more frequent than transposition of the *kan*R element, which is located in a distant region of the chromosome (strain III). Even transposition of a *tet*R element located 50 kb away from the transposase gene is complemented more efficiently than the distant element (strain IV). The failure of the distant element in strain II to be efficiently complemented is not due to active interference by the nearby *tet*R element (see strain V). The *kan*R element was introduced by complemented transposition into the donor chromosome. The *tet*R element and chromosomally located transposase fusions were carried within integrated λ prophage genomes.

creased levels of transposition functions or artificial constructions in which IS*10*-Right's transposase gene is fused to a heterologous promoter (Fig. 4).

These data show that the frequency of Tn*10* transposition varies over a remarkable four orders of magnitude as complementing functions vary from wild-type single-copy to multicopy overproducer levels. Several lines of evidence suggest that transposition activity increases linearly in direct proportion to the amount of transposition functions (D. Morisato, J. C. Way, H.-J. Kim, and N. K., unpublished). We conclude that the level of IS*10* functions (probably of the major transposition protein) is the single most important component that normally limits the frequency of Tn*10* transposition *in vivo*. We also expect that the level of IS*10* protein should be the most important target of IS*10*- or host-encoded regulation of transposition (Section V).

Figure 4 shows that length dependence and function dependence of Tn*10* transposition reflect two independent aspects of the transposition process. The slope of the length-dependence curve is the same at all levels of transposition functions, and the effect of increasing transposition functions is the same regardless of the transposon length. Conceptually, transposition can also be separated into two parts: the cut-and-paste part, in which transposon and chromosomal sequences are broken and/or joined, and the replication part, in which transposon sequences are duplicated. The function-dependent and length-dependent aspects of transposition may correspond to the cut-and-paste and replication steps, respectively.

The association of length dependence with transposon duplication has been discussed above. The idea that IS*10* functions are intimately involved in the cut-and-paste steps in transposition is supported by four other observations:

1. IS*10* functions are known to act at the ends of the element (Foster *et al.*, 1981a; Section III). During transposition, these ends are precisely and specifically separated from adjacent donor DNA sequences and joined to target DNA (Kleckner, 1979; Section VII).

2. Preferred target DNA sites are selected by Tn*10* during insertion (Kleckner *et al.*, 1979b; Halling and Kleckner, 1982; Section VII). Because the sites chosen by Tn*10* are different from the sites chosen by other elements, IS*10* functions must participate in recognition of the target DNA.

3. The target site sequence that determines Tn*10* insertion specificity is itself symmetrical and is symmetrically positioned between the positions, 9 bp apart, where target DNA is cleaved during transposon insertion (Halling and Kleckner, 1982; Section VII). This juxtaposition of symmetrical recognition and cleavage sites would be economically explained if a single protein, the IS*10* transposase, were responsible for both recognition and cleavage of the target DNA during insertion.

4. If transposase interacts with the target DNA site, and if transposase is preferentially cis-acting, then one might anticipate that Tn*10* would preferentially choose target sites that are near the moving element itself. This prediction is upheld by the observation that 90% of all Tn*10*-promoted inversions are less than 30 kb in length (Kleckner *et al.*, 1979a, unpublished). This degree of preference corresponds well with the fact that transposition functions work about 10% as well on transposon ends 50 kb away as on ends only 1 kb away (Fig. 5 and preceding discussion).

D. Other Factors

A few Tn*10* derivatives fall reproducibly above the length dependence line defined by the majority of the elements regardless of the level of transposition functions involved (Fig. 4). Thus, additional factors must influence transposition in ways yet to be defined. From the structures of derivatives tested thus far, we can unambiguously exclude one possibility: The frequency of transposition is not influenced by the presence or absence of the inside IS*10* ends.

E. Do Tn*10*-Promoted Rearrangements Occur in Bursts?

It has been suggested that transposon-promoted rearrangements might occur in bursts. That is, a cell that has undergone one such rearrangement might have a higher than average probability of having undergone a second rearrangement. Evidence for such a situation has been presented for IS*1* (Read and Jaskunas, 1981). We have investigated this possibility in Tn*10*. Many small, independent, exponentially growing cultures of a strain carrying Tn*10*-derived elements at two widely separated loci were examined for the frequency of Tn*10*-promoted rearrangements at either or both positions. The frequency of "doubly rearranged" bacteria was exactly that expected on the basis of the individual frequencies of the two single rearrangements. We conclude that Tn*10* elements at distant loci on the bacterial chromosome usually behave independently of one another. These data do not rule out the possibility that under different conditions, transposition of elements in all cells might be specifically activated by an appropriate change in cellular physiology or by an appropriate environmental signal.

V. *Genetic Organization of IS*10*-Right*

A. Sites, Symmetries, Coding Regions, and Promoters

IS*10*-Right is the functional transposition module in Tn*10*. It has been observed to undergo replicative transposition as a discrete unit in its own right (Kleckner and Ross, 1980; E. A. Raleigh and N. Kleckner., unpublished; Section VII,C). Genetic, DNA sequence, and *in vitro* transcription experiments have provided the overall picture of IS*10*-Right, as shown in detail in Fig. 6 (Halling *et al.*, 1982; R. W. Simons, B. C. Hoopes, W. R. McClure, and N. Kleckner, unpublished).

The ends of IS*10*-Right are short, 23-bp, nearly perfect inverted repeats. These repeats can be subdivided into two parts. The outer 13 bp are a sequence that occurs only twice in IS*10*, once at each terminus of the element. The four cis-dominant mutations that alter sites at the ends of Tn*10* lie in this 13-bp sequence (J. C. Way, unpublished). The inner 10 bp are a sequence that occurs five times near the ends of IS*10*-Right: four times as two pairs of inverted repeats near the outside end and a fifth time near the inside end. We argue elsewhere that this sequence is likely to be a protein recognition site (Halling *et al.*, 1982).

The short inverted repeats at the ends of IS*10* probably reflect an underlying functional equivalence between those two ends. The inside-out transposon (Fig. 3) is capable of independent transposition, and we know that the normal inside ends of the IS*10* elements can serve efficiently as the outside termini of a transposing segment (Foster *et al.*, 1981a). This observation places constraints on transposition mechanisms. For example, if transposition involves DNA replication originating at the ends of the element (Section VII,C), such replication must be able to proceed from either (or both) ends of IS*10*. Functional symmetry between the ends of the element is also suggested by the symmetrical way in which the target DNA is cleaved and transposon sequences are inserted during both normal and inside-out transposition events (Section VII,A,B).

IS*10*-Right encodes a single long open reading frame that extends from one end of the element (bp 108) to the other (bp 1313), ending within the inverted repeat at the inside end of the element. This coding region is expressed *in vivo*: In-frame fusions of this region to the structural gene for β-galactosidase result in significant expression of that enzyme (R. W. Simons, unpublished). More importantly, this coding region specifies a function essential for transposition. Deletions and point mutations that specifically alter this coding region abolish the ability of IS*10*-Right to provide complementing transposition functions in trans, and fusion of this coding region to a heterologous promoter places expression of transposition functions under control of that promoter. The polypeptide encoded by this gene should be 402 amino acids in length, with a molecular weight of 47,000. IS*10*-Right contains only one other (short) coding region that might plausibly be important *in vivo;* its possible significance has not yet been assessed.

In vitro and *in vivo* analysis of IS*10*-Right has revealed the presence of three different promoters, all of which lie near the ends of the element (Fig. 6). p-IN is located just upstream of the major coding region and is responsible for expression of that gene *in vivo*. Mutations that increase the expression of transposition functions *in vivo* also increase the activity of this promoter *in vitro*. The two other promoters, p-OUT and p-III, are symmetrically positioned near the two ends of IS*10*. Each one directs transcription outward towards its respective end and the adjacent chromosomal material. p-OUT lies just inside the start of the long coding region, and its transcript overlaps and opposes that of p-IN. The region of overlap includes the ATG start codon for the major structural gene. p-IN is a weak promoter both *in vivo* and *in vitro;* p-OUT is a relatively strong promoter under both conditions. A great deal of circumstantial evidence, summarized below, suggests that other IS elements will turn out to have outward-directed promoters analogous to p-OUT and/or p-III.

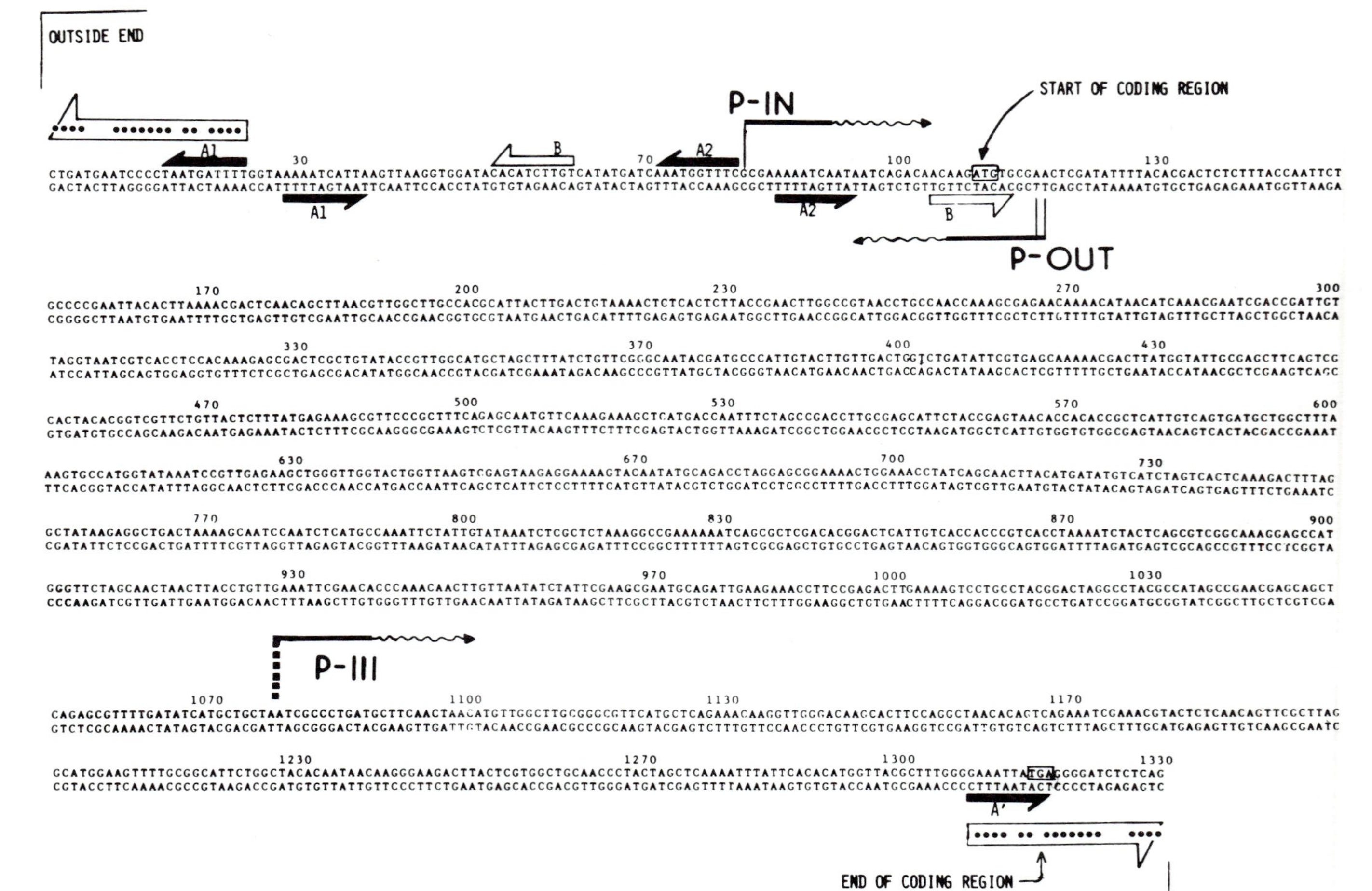
OUTSIDE END
START OF CODING REGION
P-IN
A1
B
A2
P-OUT
P-III
A'
END OF CODING REGION
INSIDE END

B. Regulation of IS*10* Transposition

IS*10* directly regulates expression of its own transposition functions. This regulation is manifested in a phenomenon called *multicopy inhibition*. Transposition of a single-copy Tn*10* element resident in the bacterial chromosome is reduced three- to tenfold in the presence of a multicopy plasmid carrying IS*10*-Right (Foster *et al.*, 1981a). Genetic analysis has established three important features of multicopy inhibition:

1. The region of IS*10*-Right that is both necessary and sufficient for multicopy inhibition has been defined by deletion, insertion, and point mutations (Fig. 7). The essential region corresponds to p-OUT and the template for its transcript.
2. Multicopy inhibition is due to a decrease in the expression of transposition functions from the single-copy Tn*10* element. Multicopy inhibition is not observed when transposition functions for the resident Tn*10* element are expressed under the control of a heterologous plac promoter instead of under IS*10* control.
3. Multicopy inhibition acts to depress translation but not transcription of the transposase gene. Multicopy inhibiting plasmids depress β-galactosidase expression in artificially constructed fusions where the *lac*Z gene is translationally fused to the transposase gene in IS*10*-Right, but they do not affect expression in constructions where the intact *lac*Z gene, with its normal translation start region, is transcriptionally fused to p-IN.

The transcript from p-OUT overlaps that from p-IN for a region of 33 bp that includes the start of the transposase coding region. On the basis of our genetic observations, we propose that the transcript specified by p-OUT inhibits translation of that coding region by pairing directly with the message from p-IN, thereby sequestering the ATG start codon for the transposase gene. Other genetic experiments support this possibility (R. W. Simons and N. Kleckner, unpublished). Direct pairing between complementary RNA molecules encoded by opposite strands of a single region has been shown to regulate replication, copy number, and incompatibility in plasmid ColE1 (Tomizawa and Itoh, 1981a,b).

Fig. 6 Structure of IS*10*-Right at the nucleotide sequence level. DNA sequence of IS*10*-Right, 1329 bp (Halling *et al.*, 1982). Basepair 1 is the outside end of IS*10*-Right and thus is the right terminus of Tn*10*, as shown in Fig. 1. The sequence of the top strand is written in a 5′ to 3′ direction from left to right. Several features of the sequence are emphasized: (a) 23-bp nearly perfect homology between the inside and outside ends of the element is indicated by large open arrows, with homologous basepairs denoted by filled circles; (b) two related internal symmetries (A1 and A2) and a related 10-bp sequence at the inner terminus are indicted by filled arrows; (c) two widely separated halves of a third symmetry, (B), are indicated by open arrows; and (d) the start and stop codons of the long open coding region at bp 108 and bp 1314, respectively. Also shown are the exact transcription start points for two promoters, p-IN and p-OUT, located near the outside end of IS*10*-Right and the approximate startpoint (5 bp) for a third promoter, p-III, located near the inside end (R. W. Simons, B. C. Hoopes, W. R. McClure, and N. Kleckner, unpublished).

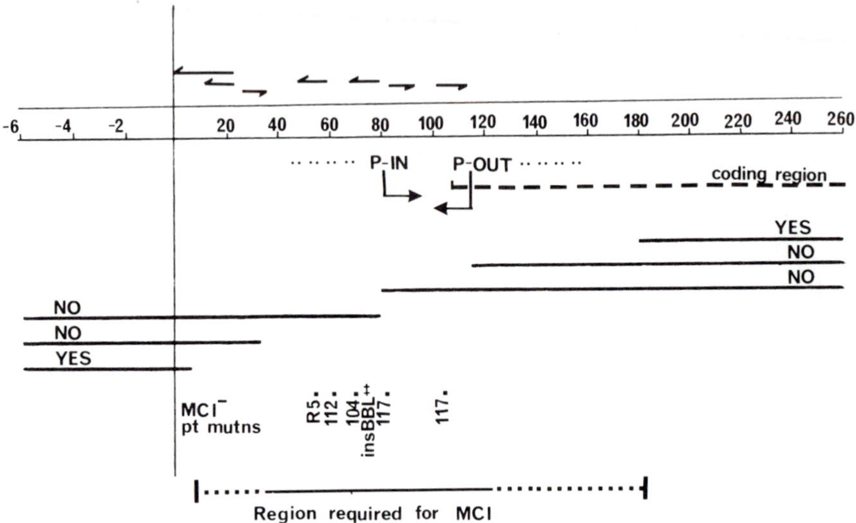

Fig. 7 The IS*10*-Right region required for multicopy inhibition N. Kleckner, M. A. Davis, T. J. Foster and R. W. Simons, unpublished). Deletion mutants of IS*10*-Right that have retained (YES) or lost (NO) the ability to exert multicopy inhibition demonstrate that information between bp 7 and bp 37 on the left and between bp 119 and bp 180 on the right is essential for this effect. The latter interval contains the Pribnow box and the -35 region of p-OUT. The mutation BBL, which is a 14-bp insertion of an artificial linker fragment at a convenient restriction site, shows that the region by bp 70 must also be intact for proper inhibition. Finally, four different point mutations that abolish multicopy inhibition span the region defined by the deletions.

Because the frequency of Tn*10* transposition is directly related to the level of IS*10* functions, it is reasonable that IS*10* should regulate its own activity by regulating expression of those functions. Self-regulation may only serve to keep the frequency of IS*10* transposition very low, or it may provide a mechanism by which transposition can be made sensitive to some as yet undefined host or environmental signal.

Some transposable elements (Mu, Tn*5*, and elements of the Tn*3*/γδ family) are regulated by trans-acting repressors. When these elements enter a naive host, transposition functions are expressed at a high level until the repressed state is established. Regulation of the type proposed for IS*10* could have very different consequences. Because p-OUT is strong relative to p-IN, the p-OUT transcript would be present to inhibit expression of transposition functions before any such functions could be made. Consistent with this idea, Tn*10* transposition frequencies are about the same during stable and established residence in a host cell as upon introduction into a new host.

The trans-acting effects of multicopy inhibition are probably a weak manifestation of a regulatory process that normally works in cis. Point mutations that abolish multicopy inhibition in trans confer a much stronger high-hopper phenotype in cis. Furthermore, transposition of Tn*10* from a newly introduced λ genome is not affected by the presence of an established single-copy IS*10* or Tn*10* element in the recipient host (M. A. Davis, S. H. Way and N. Kleckner., unpublished).

No role for IS*10*'s third promoter, pIII, has yet been assigned. It is not excluded that p-III and/or p-OUT may have either direct mechanistic roles in the transposition process or additional undiscovered regulatory roles. The possible participation of the IS*10* transposase protein in regulation of IS*10* transposition is also being investigated.

C. Protection from External Promoters

Inspection of the nucleotide sequence of IS*10* reveals a possible mechanism by which the element may protect itself from fortuitous activation by external promoters (Halling *et al.*, 1982). The outside end of IS*10* contains numerous internal symmetries. One of these, symmetry B in Fig. 6, includes the start codon for the long coding region. The nature of these symmetrics is such that a transcript entering IS*10* from the outside end can form a stable secondary structure in which the ATG start codon is sequestered within a double-stranded region. Such sequestering could prevent efficient translation of inappropriate read-through messages initiated outside of the element. When transcription is initiated at the normal promoter, p-IN, there is no such opportunity for formation of a secondary structure or sequestering of the start codon. An IS element may be at special risk from external activation because every cycle of transposition places the element in a new chromosomal context.

D. Tn*10*-Mediated Turn-Off and Turn-On of Adjacent Chromosomal Genes

Insertion sequences and transposable elements usually exert strong polar effects on expression of promoter-distal genes when they are inserted into polycistronic transcription units. IS elements were originally identified as unusual strongly polar spontaneous mutations in the *gal* and *lac* operons of *E. coli* (Fiandt *et al.*, 1972; Saedler *et al.*, 1974). Transposon Tn*10*, in particular, exerts strong polarity in the *Salmonella* histidine operon, and the polarity pattern of *his* :: Tn*10* insertions revealed the presence of low-level secondary promoters at internal sites in the operon (Kleckner *et al.*, 1975). Somewhat paradoxically, insertion sequences have more recently been associated with turn-on of adjacent downstream genes in some situations and have been isolated from genetic selections that demand gene turn-on or independence from normal negative regulatory controls (Blattner *et al.*, 1974; Boyen *et al.*, 1978; Brennan and Struhl, 1980; Pilacinski *et al.*, 1977; Saedler *et al.*, 1974; Reynolds *et al.*, 1981; Glansdorff *et al.*, 1981).

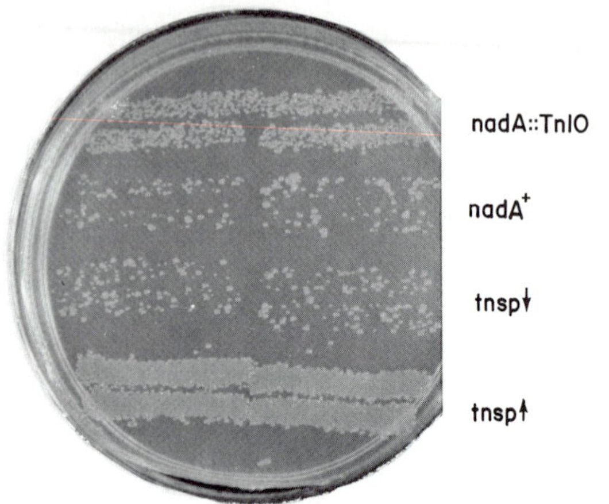

Fig. 8 Screen for bacterial host mutations that alter Tn*10* transposition. Galactose-sensitive (GalE⁻) strains of *E. coli* fail to grow on nutrient plates containing galactose. Galactose-resistant derivatives of such strains arise spontaneously at a frequency of about 10^{-6}; these derivatives contain additional lesions in the galactose operon. The rare galactose-resistant mutants can be detected in single colonies or small patches of cells that are first grown on a permissive medium and then replica-plated onto a medium containing galactose. Only the galactose-resistant cells in a colony or patch form colonies on the selective plates. The frequency of galactose-resistant mutants increases when the GalE⁻ strain also carries a Tn*10* insertion near the *gal* operon (in the *nad*A gene). This increase is detectable by replica plating. A comparison between the nad⁺ parent and its *nad*A :: Tn*10* derivative in such a test is shown. Mutants of the *nad*A :: Tn*10* strain that give increased or decreased frequencies of Tn*10*-promoted galactose resistance can be identified by replica plating of individual colonies from a mutagenized population. Replica patches of two mutants isolated in this way are shown (D. E. Roberts and N. Kleckner, unpublished).

These apparently contradictory effects of transposon insertion can now be understood in the case of Tn*10*. Ciampi *et al.* (1982) have shown that Tn*10* insertions turn on distal genes in the histidine operon if, and only if, the host cell is deficient for transcription termination factor rho (Rho⁻) or if all rho-dependent termination sites between the Tn*10* insertion and the distal gene are deleted. We suspect that the transcription responsible for this Tn*10*-dependent activation of adjacent genes probably originates at p-OUT. The observed level of distal gene turn-on is compatible with the activity of p-OUT *in vivo* as assessed from artificial gene fusions. Analogous effects of Tn*10* in the *ilv*GEDA operon have also been observed (Blazey and Burns, 1982).

The Tn*10* picture is probably directly applicable to many other IS/Tn elements as well. The effects of Tn*5* insertions on distal genes in the histidine operon are similar to those of Tn*10* (J. R. Roth *et al.*, unpublished). *In vitro* experiments are suggestive of outward-directed promoters in other systems. Finally, a large number of apparently contradictory or confusing observations and phenomena can all be explained by the combination of outward promoters and position-dependent termination of the resulting read-through transcription (see Ciampi *et al.*, 1982 for a complete discussion).

VI. Role of the Host in Tn10 Transposition

The role of host-encoded functions in Tn*10* transposition is not known. Because the IS*10* protein is probably too small to have DNA polymerase activity, the host may at least be responsible for the replication associated with transposition. We have developed a genetic screen that identifies mutations of *E. coli* that alter the frequency of Tn*10*-promoted DNA rearrangements (D. E. Roberts and N. Kleckner, unpublished). GalE$^-$ strains of *E. coli* fail to grow in the presence of galactose unless they contain further appropriate mutations in the galactose operon (Adhya and Shapiro, 1969). The presence of a Tn*10* insertion near the gal operon increases the frequency with which a GalE$^-$ strain mutates to galactose resistance, and host mutations that alter the activity of the Tn*10* element can be identified by their effects of this enhancement. The actual assay used for isolation of such mutations is shown in Fig. 8. Analysis of host mutations that increase or decrease Tn*10* activity has thus far revealed only that each mutation confers its own spectrum of changes on the different DNA rearrangements promoted by Tn*10* and by other transposable elements.

VII. Mechanism of Tn10 Transposition

A. Precise Cleavage and Joining at Transposon Ends and Target Sites

Integration of Tn*10* into a target genome always results in insertion of the same discrete set of transposon sequences, and the inserted sequences always occur sandwiched between a short direct repeat of a 9-bp sequence that occurs only once at the original target site prior to insertion (Kleckner 1979; Fig. 9). Although all Tn*10* insertions are flanked by 9-bp direct repeats, the exact 9-bp sequence repeated is different for insertions at different target sites (but see below).

The integration of a specific set of transposon sequences implies that each time the element transposes to a new position, exactly the same nucleotide at each end of the element is joined to target DNA. The presence of the short (9-bp) direct repeat probably means that the two strands of the target DNA molecule are broken at slightly staggered positions during the insertion process. If breaks were made in this fashion, replication across the resulting single-stranded regions would result in duplication of the intervening 9-bp region. Sandwiching of transposon sequences between the duplicated regions would then give the observed final product. Regardless of exactly how or in what order these steps may occur

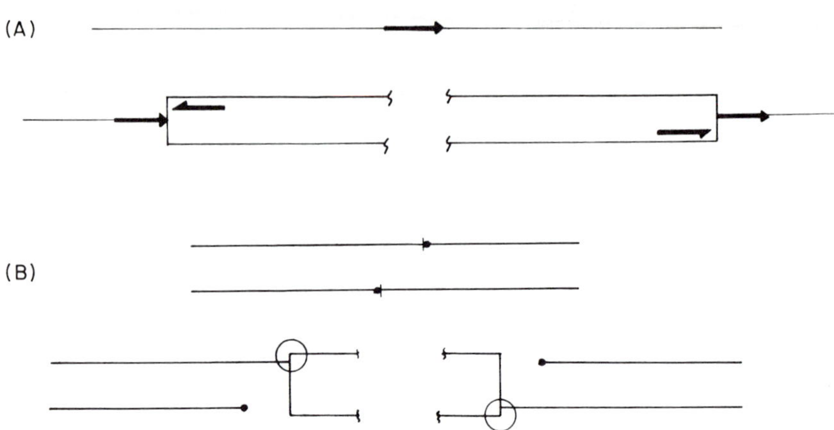

Fig. 9 Joining of transposon sequences to target DNA. (A) Transposon insertion is accompanied by duplication of a short target DNA sequence at the insertion site. (B) Target DNA is broken by a pair of staggered nicks; transposon sequences are joined by symmetrical single-stranded connections to resulting overhanging target DNA ends. (From Kleckner, 1981, with permission of Annual Reviews, Inc.)

(see below), the critical event that determines the structure of the product is the single-strand connection at each new junction between the specific nucleotide that defines the end of the transposon and the "overhanging" end of the cleaved target DNA (Fig. 9).

Short, flanking direct repeats could arise by an alternative mechanism in which the incoming transposon carries with it one copy of the 9-bp sequence and searches for a matching one in the target DNA during the insertion process. For Tn*10*, this possibility is made less likely by the observation that a transposon can move from a given starting position to several different alternative positions where different 9-bp sequences are repeated, and conversely, that transposons at two different positions can both select the same target DNA sites (Kleckner, 1979). These observations are also supported by additional information about Tn*10* insertion specificity (Section VII,B). Experiments with Tn*9* also argue against this alternative possibility (Johnsrud *et al.*, 1978).

The presence of an exact 9-bp direct repeat is not important for subsequent transposition of Tn*10* to a new position: Artificially constructed elements lacking this repeat still transpose at normal frequencies (Kleckner, 1979). This observation argues against any interaction between the repeats that depends upon exact DNA sequence homology. It does not exclude the possibility that the sequences that flank the element contribute in other ways to the transpositions process.

B. Preferential Insertion at Specific Target Sites

Tn*10* inserts at many different sites in a large target genome, such as a bacterial or bacteriophage chromosome. However, fine-structure genetic map-

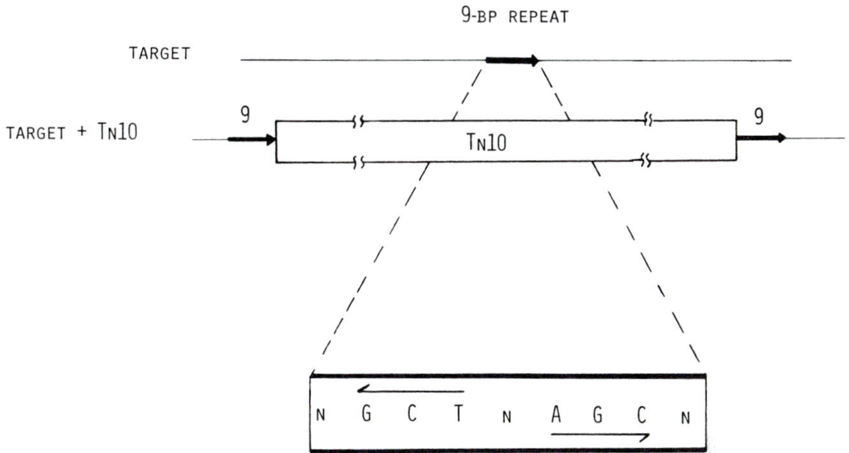

Fig. 10 Tn*10* insertion hot spots share a symmetrical 6-bp consensus sequence. This specificity-determining sequence is symmetrically located within the 9-bp target DNA sequence that is duplicated during Tn*10* insertion.

ping of large numbers of insertions into smaller targets has revealed the existence of preferred insertion hot spots (Kleckner *et al.*, 1979b; Foster, 1977). The distribution of Tn*10* insertions in the *Salmonella* histidine operon and elsewhere suggests that Tn*10* inserts at high efficiency into certain especially favored sites that occur approximately once in every 1000 bp of target DNA, and that Tn*10* also recognizes a larger number of sites less specifically and at a much lower efficiency (Kleckner *et al.*, 1979b; Halling and Kleckner, 1982; Miller *et al.*, 1980).

DNA sequence analysis has shown directly that all of the Tn*10* insertions at a genetically defined hot spot occur at identical nucleotide positions, that is, at the same 9-bp target DNA site. Comparisons among many different target site sequences have identified the target DNA signal responsible for this insertion specificity: a particular symmetrical 6-bp sequence (GCTNAGC) that is located within the duplicated 9-bp target site sequence (Fig. 10; Halling and Kleckner, 1982). The sequences at strong hot spots conform more closely to the consensus sequence than do sequences at other sites, and the consensus sequence and closely related sequences are generally absent from potential target regions where Tn*10* is known not to insert. Other aspects of the target DNA, yet to be defined, influence the efficiency with which a particular target site sequence is used.

The 6-bp specificity sequence is symmetrically located within the 9-bp target DNA sequence where cleavage and duplication occur during Tn*10* insertion (Fig. 10). This juxtaposition of recognition and cleavage sites, plus the intrinsic symmetry of the perfect consensus sequence and its symmetrical location within the 9-bp target site, all suggest that target DNA may be both recognized and cleaved by the symmetrically disposed subunits of a single protein. An attractive candidate for such a protein is the IS*10*-encoded transposase. Each transposable element exhibits a unique pattern of insertion specificity (see Kleckner, 1981 for a review), and Tn*10* must therefore contribute to its own choice of insertion sites. Because it is known that the IS*10* transposase acts at the ends of the transposon and that those ends are joined to the broken target DNA molecule in a symmetrical way, it would be simple if that single protein were responsible for both symmetrical recognition and cleavage of the target and symmetrical cleavage and joining of transposon ends to that target.

The basic properties that characterize Tn*10* insertion almost certainly characterize insertion of IS*10* as well. Transpositions of single IS*10* elements have not been examined directly. However, Tn*10*-promoted deletions and deletion/inversions occur at the same favored target DNA sites as do Tn*10* insertions, and insertions of the inside-out IS*10* transposon are flanked by 9-bp target DNA repeats the sequences of which include an appropriate specificity determinant (Noël and Ames, 1978; Halling and Kleckner, 1982).

C. Tn*10*-Promoted Rearrangements Other Than Transposition

Tn*10* has been shown to promote five different DNA rearrangements other than transposition. Each of these rearrangements can be understood as the product of an event promoted either by a single IS*10* element or by the two IS*10* elements acting in concert. Each of the five rearrangements is shown in Fig. 11.

1. Simple Deletions

Among tetracycline-sensitive derivatives of a Tn*10*-containing genome, an important fraction have suffered deletions that begin at the inside end of one IS*10* element, extend across Tn*10* into or beyond the tetracycline-resistance genes, and sometimes extend into adjacent chromosomal material (Kleckner *et al.*, 1979a; Ross *et al.*, 1979c; Fig. 11A). These are classic IS-promoted deletions, examples of which have been observed in many other cases (Ohtsubo *et al.*, 1978; Mickel *et al.*, 1977; Reif and Saedler, 1977; Davidson *et al.*, 1974). This structure makes it clear why any given Tn*10*-promoted deletion that affects adjacent chromosomal material always removes material on only one side of the original Tn*10* element (Kleckner *et al.*, 1979a).

2. IS*10* Replicative Inversions

Genomes containing a single Tn*10* element have been observed to acquire a third copy of IS*10*. Appearance of a third IS*10* element adjacent to a Tn*10* insertion in λ was accompanied by inversion of the intervening chromosomal

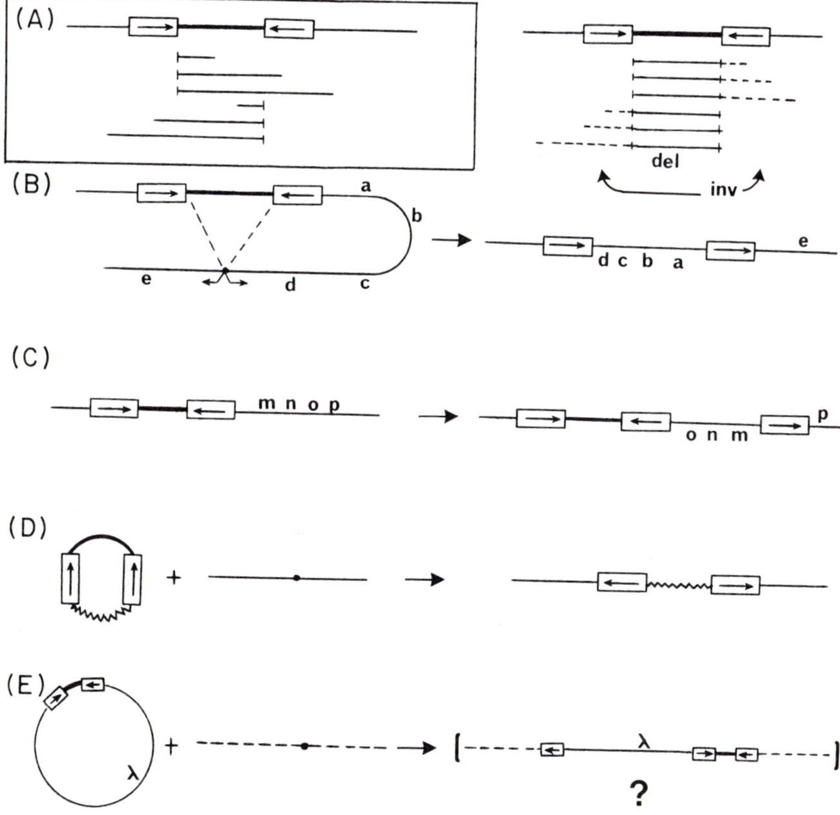

Fig. 11 Tn*10*-promoted rearrangements (A) Tetracycline-sensitive deletions. Each deletion extends from the inside end of IS*10*-Right or IS*10*-Left across the element, ending within Tn*10* or adjacent chromosomal material (Kleckner *et al.*, 1979a; Ross *et al.*, 1979c). These deletions can be generated by one of the individual IS*10* elements acting alone (see Fig. 12) or by the combined action of both elements in a manner analogous to inversion/deletions. (B) Tetracycline-sensitive inversion/deletions. Deletion of the specific DNA segment between the IS*10* elements (solid line) plus inversion of one adjacent segment extending outward from the inside end of one of the IS*10* sequences (dashed line). These structures can be accounted for by the interaction of both inside IS*10* ends with a target DNA site located either within an IS*10* element or in the adjacent chromosomal non-Tn*10* material. The latter possibility is diagrammed in the bottom portion of this part of the figure. (Kleckner *et al.*, 1979a; Ross *et al.*, 1979c). (C) Intramolecular replicative transposition of IS*10* with concomitant inversion of material between the moving element and the target site (Kleckner and Ross, 1980). This event is accounted for in Fig. 12. (D) Inside-out transpositions, as described in Fig. 3. (E) Replicon-fusion cointegrates. Cointegration of both Tn*10* and λ genome sequences from a λ :: Tn*10* phage into the bacterial chromosome has been observed genetically (N. Kleckner, unpublished). The physical structure suggested here is that predicted for a cointegration product promoted by one individual IS*10* element according to the views depicted in Figs. 12 and 13.

material (Kleckner and Ross, 1980; Fig. 11C). The coincident IS*10* insertion and chromosomal inversion suggest that the transposition event involved was intramolecular and, thus, that the acquisition of a third IS*10* element must necessarily have involved replication of one of the two IS*10*s present in the original Tn*10* element. This type of replicative inversion is neatly accounted for by most transposition models (see below). In two other instances, IS*10* transpositions have been identified, in which a third IS*10* element has inserted within Tn*10*'s own *tet*R genes, causing a mutation to tetracycline sensitivity. Inversion of the adjacent segment was not examined in these cases (Sharp *et al.*, 1973; Bennett *et al.*, 1980). Additional examples of replicative transposition of a single IS*10* element have also been observed in *E. coli* strains carrying chromosomal Tn*10* insertions (E. A. Raleigh and N. Kleckner, unpublished).

3. Inside-Out Transpositions

As discussed in Section III, a circular genome containing Tn*10* can give rise to inside-out transpositions in which the inside ends of the IS*10* elements interact with a target site in a manner exactly analogous to the interaction of the outside IS*10* ends with target sites during normal Tn*10* insertion (Foster *et al.*, 1981a; Chandler *et al.*, 1979; Figs. 3 and 11D).

4. Tetracycline-Sensitive Inversion/Deletions

A second important class of spontaneous tetracycline-sensitive derivatives isolated from Tn*10*-containing genomes have undergone the specific deletion of all Tn*10* material between the IS*10* elements plus inversion of a contiguous DNA segment beginning at one IS*10* inside end and ending within that IS element or the adjacent chromosomal material. These events are the intramolecular equivalent of inside-out transpositions; they involve an interaction between the two inside IS*10* ends and a new target site that lies outside the element in adjacent chromosomal material (Kleckner *et al.*, 1979a; Ross *et al.*, 1979b; Chaing and Clowes, 1980; Kehoe and Foster, 1977; Fig. 11B).

5. Cointegrate Replicon Fusions

When Tn*10* is introduced into a new host on a nonreplicating, nonkilling bacteriophage vehicle (a λ derivative that is defective for *N, P, int, att,* and *c*I functions), approximately 10% of all tetracycline-resistant bacteria arising from the infection contain the λ genome as well as Tn*10* (N. Kleckner, unpublished). The structures of these cointegrate molecules have not been examined in detail, but they are probably replicon-fusion cointegrates analogous to those promoted by other elements. The length dependence of Tn*10* transposition (see above) suggest that the element duplicated during these cointegration events is likely to be IS*10* rather than an entire Tn*10* (Fig. 11E).

All of the Tn*10*-promoted rearrangements described above involve the precise joining of one or more IS*10* ends to a new target DNA site. This feature is absolutely characteristic of IS-promoted recombination events (Kleckner and Ross, 1979). Simple excision of Tn*10* does occur, as do partial excisions of Tn*10* sequences. However, all of these excision events involve rejoining of two sequences, neither of which is the terminus of an IS*10* element, and none is promoted by IS*10* functions (Section VIII).

D. Replication: Cointegrations and Simple Insertions

It is generally agreed than many transposon-promoted rearrangements involve specific replication of transposon sequences (see Kleckner, 1981 for a review). The evidence for this view comes primarily from the observation of cointegrate structures in which precise joining of transposon sequences to a new target molecule is accompanied by specific replication of the moving element but not of flanking donor information or the target genome. The observation of intramolecular replicative inversions (Fig. 11C) provides direct evidence for transposition-associated replication in the case of IS*10*.

A number of models have been proposed for the formation of cointegrate structures. Mechanisms that account for the formation of cointegrates will automatically account for two other IS*10*-promoted events, simple deletions and replicative inversions. When the chosen target site is on a different molecule from the transposon, the outcome of an appropriate series of steps will be a cointegrate structure. If the target site is on the same molecule as the element, the analogous series of events will lead to a simple deletion or to a replicative inversion, depending on the orientation of that site relative to the ends of the transposon (Shapiro, 1979; Faelen *et al.*, 1974; Fig. 12).

This unified explanation greatly reduces the apparent diversity of Tn*10*-promoted rearrangements. However, mechanisms that account economically for cointegrate formation do not directly account for intermolecular insertion of a transposon (Tn*10* or the inside-out element) into a second target genome, nor do they necessarily account for Tn*10*'s *tet*S inversion/deletions.

Different transposable elements probably use different mechanisms for generating simple insertions. The Tn*3*/γδ family of transposons separates the integration process into two discrete steps: A stable, covalently closed cointegrate structure is generated by a replicative mechanism, and the cointegrate is then reduced by site-specific, nonreplicative recombination into a genetically unaltered donor molecule and a simple insertion product molecule (Kitts *et al.*, 1982; Ohtsubo *et al.*, 1981; Kostriken *et al.*, 1981; Reed, 1981; Chapter 5; Fig. 13).

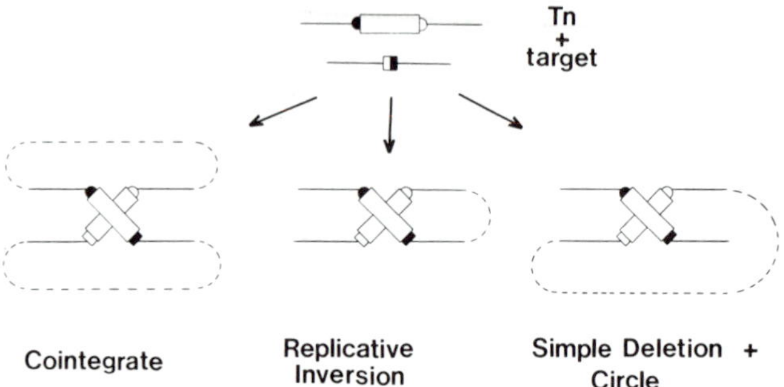

Fig. 12 Three possible products from a single type of transposon–target DNA interaction. Interaction between a transposable element on one circle and a target DNA site on another circle can generate a cointegrate structure in which the transposon is duplicated and the two circles are joined; one copy of the transposon occurs at each new junction (see also Fig. 13). In most transposition models, each transposon in the cointegrate structure keeps one end attached to the donor circle (semicircular symbols) and the other end attached to the target DNA site (rectangular symbols) as diagrammed at the left. Exactly identical transposon–target DNA connections (indicated by the solid lines and symbols) will yield replicative inversions or simple deletions if the target site is on the same molecule as the transposon. The orientation of the target site relative to the transposon determines which of these two intramolecular products will be formed. These two alternatives are depicted in the center and right-hand panels. The three different products are drawn above in such a way as to emphasize that they involve fundamentally identical transposon–target DNA interactions. Conversion of the three structures to their untangled forms will make clear the exact nature of each product.

Observations in many other systems suggest that the two other major classes of transposable elements, IS sequences and Mu-like transposing bacteriophages, probably use other mechanisms to generate simple insertions (see Harshey and Bukhari, 1981; Grindley and Joyce, 1981; Galas and Chandler, 1982; Ohtsubo *et al.*, 1981; Machida *et al.*, 1982; and Kleckner, 1981 for a summary). In the case of IS*10*, both the genetic structure of the transposon and the behavior of transposition products are very different from Tn*3* or γδ. IS*10* encodes only a single essential transposition protein of any significant size; Tn*3* encodes two large proteins, each of which is responsible for one of the two steps in transposition. All of the sites required for normal Tn*10* transposition lie within 70 bp of the ends of the element; reduction of Tn*3* cointegrates occurs at a site in the middle of the element. Tn*3* cointegrates are very unstable in the presence of the reduction function; IS*10*/Tn*10* cointegrates do not reduce at a detectable rate ($<10^{-4}$) (N. Kleckner, unpublished). Tn*3* cointegrates can be recovered as a very large fraction of all transposition products; IS*10*/Tn*10* cointegrates are recovered much less frequently (see above). Although none of the properties of IS*10* necessarily precludes a two-step pathway analogous to that of Tn*3*, the extent of the differences between the two elements is suggestive of important differences in mechanism.

Several types of alternatives to the Tn*3*/γδ mechanism have been suggested

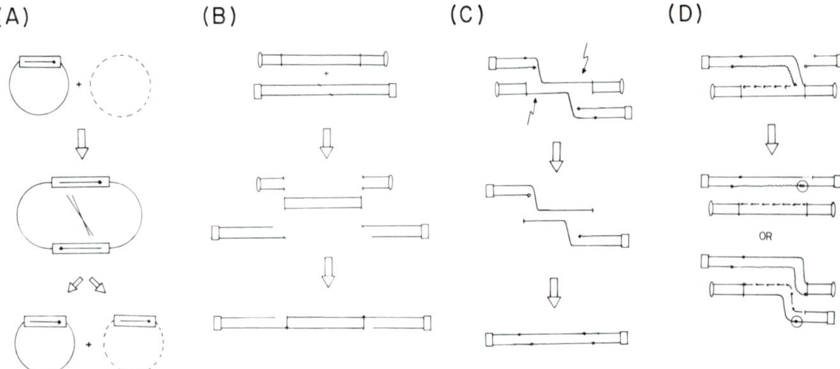

Fig. 13 Alternative mechanisms for obtaining simple insertions of a transposon. (A) Two-step mechanism: formation of a stable, covalently closed cointegrate structure by a replicative process, followed by reciprocal site-specific, nonreplicative recombination between the duplicate transposon copies. This is the mechanism used by elements in the Tn*3*/γδ family (see text and Kleckner, 1981 for references). (B) Cut-and-paste insertion without replication: Target DNA is broken by a pair of staggered nicks, the transposon is separated from adjacent donor molecule sequences by double-stranded breaks at each end of the element, and transposon sequences are joined by symmetrical, single-stranded connections to overhanging ends of the broken target molecule (see also Fig. 9). (C) Interrupted of the symmetrical replicative transposition event. Symmetrical single strands at each end of the transposon are disconnected from the donor molecule (vertical ellipses at the ends) and reconnected to the broken target molecule (rectangles at the ends). DNA replication begins at each new junction on the target DNA molecule and proceeds symmetrically toward the center of the element. Interruption of the remaining single-stranded connections between transposon and donor molecule sequences will result in incorporation of partially replicated transposon sequences at the target site and (presumably) loss of the disrupted donor molecule (see Kleckner, 1981; Ohtsubo *et al.*, 1981 for details). (D) Asymmetrical transposition. Transposition is initiated at only one end of the element. A single-stranded transposon end is separated from adjacent donor sequences and connected to a target molecule. Replication proceeds across the element. In this example, synthesis is primed on the target; it could equally well be primed on the donor molecule. When the replication fork reaches the other end of the transposon, a simple insertion will result if the newly replicated "bottom" strand is connected to the target; but a cointegrate will result if the old original bottom strand is disconnected from adjacent donor sequences and reconnected to the target molecule while the newly replicated bottom strand is connected to donor sequences (see text and Kleckner, 1981 for references and details).

(see Fig. 13B–D; Kleckner, 1981). There is no evidence about which, if any, of these possibilities is correct.

1. Simple insertions occur by a nonreplicative cut-and-paste pathway (Fig. 13B). Replicative transposition models involve single-stranded breaks at each end of the transposon on the donor molecule. If events may sometimes involve double-stranded breaks at each end of the moving element, the excised molecule could be integrated at the new target position by single-stranded connections and the remaining target DNA gaps sealed by repair processes to yield a simple cut-

and-paste transposition event with concomitant destruction of the DNA molecule that donated the transposon. All genetic experiments are compatible with such a donor-suicide process (see Kleckner, 1981 for a discussion).

2. Transposition and simple insertion events begin in the same way and are initiated symmetrically at both ends of the element. However, as replication across the element proceeds, nonspecific nucleases sometimes interrupt the intermediate structure in such a way as to separate transposon sequences from adjacent donor DNA (Fig. 13C; Ohtsubo *et al.*, 1981). The consequence of such a process will be integration at the new site of duplex transposon DNA, some of which will have been semiconservatively replicated and some of which will not have been replicated at all. This mechanism is also suicidal for the donor molecule.

3. Both transposition and simple insertion events are initiated in the same way at one end of the moving element. Replication proceeds from that end across to the other end of the element, where the process can be resolved in two alternative ways, one of which gives a cointegrate product and the other a simple insertion (Fig. 13D; Harshey and Bukhari, 1981; Grindley and Sherratt, 1978; Galas and Chandler, 1982; Kamp and Kahmann, 1981; K. Mizuuchi, unpublished). In such models, however, the molecular details of the two alternative events must be different, and the event that gives a simple insertion product must in fact involve a double-stranded break to separate the transposon from adjacent donor chromosomal sequences (Kleckner, 1981).

VIII. Tn*10*-Associated Excision

Analysis of spontaneously occurring, tetracycline-sensitive derivatives of Tn*10* insertions in bacteriophage λ and the *Salmonella* histidine operon has identified three different excision events involving specific transposon sequences (Fig. 14; Botstein and Kleckner, 1977; Kleckner *et al.*, 1979a; Ross *et al.*, 1979b; Foster *et al.*, 1981b).

1. Precise excision. Insertion of Tn*10* into a new target site results in duplication of a 9-bp target DNA sequence and integration of the transposon between the resulting 9-bp direct repeats. Precise excision of Tn*10* is a deletion between these 9-bp repeats that exactly reconstructs the target site to its original wild-type sequence. Precise excision is detected genetically as reversion of a Tn*10* insertion mutation. The frequency of precise excision varies from one insertion site to another but is usually less than 10^{-8} per element per generation.

2. Nearly precise excision. Nearly precise excision is a deletion event similar in structure to precise excision but involving short repeat sequences that occur

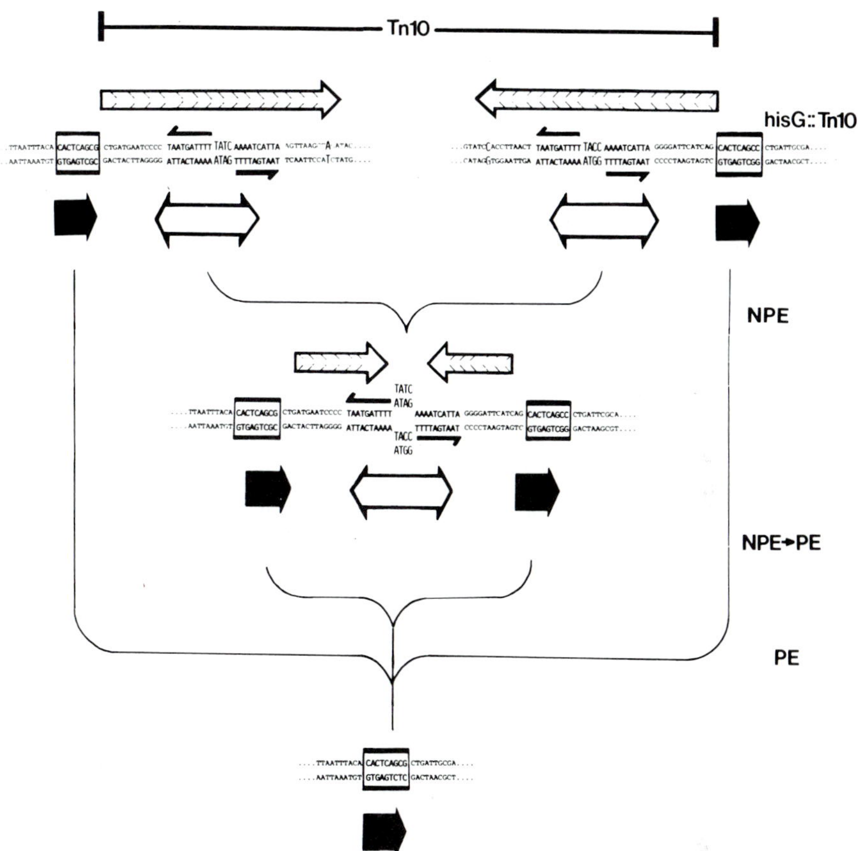

Fig. 14 Three Tn*10*-associated excision events (Foster *et al.*, 1981b). A Tn*10* insertion in the *Salmonella* histidine operon can undergo nearly precise excision (NPE) or precise excision PE). The 50-bp NPE remnant can itself undergo precise excision (NPE to PE).

within the Tn*10* element, one near each end (these repeats are symmetry A1 in Fig. 6). Deletion of material between these repeats, analogous to excision between the 9-bp repeats in precise excision, results in excision of all but 50 bp of Tn*10*. Nearly precise excisions are detected genetically as a distinctive class of polarity-relief revertants from *his* :: Tn*10* insertions or as derivatives of λ :: Tn*10* phages having an unusual plaque morphology. The frequency of nearly precise excision is about 10^{-6} per element per generation.

3. Precise excision of nearly precise excision remnant. The 50-bp sequence remaining after nearly precise excision retains sequences from each end of Tn*10* plus the flanking 9-bp direct repeats. This material can be further excised to give complete restoration of the wild-type target DNA sequence. These excision events are identified genetically as full prototrophic revertants of *his* :: Tn*10*

nearly precise excision derivatives. The frequency with which a nearly precise excision remnant gives rise to a precise excision product depends upon the site and is about 100-fold higher than the frequency of Tn*10* precise excision at that site.

Each of the three excision events can occur in the absence of the host homologous recombination function, *rec*A, and each can occur even when the excising Tn*10* element is defective for transposition functions. Thus, these three events appear to be promoted, at least primarily, by host-encoded DNA handling functions. Mutations of *E. coli* that increase Tn*10* excision have been isolated and shown to affect several such functions (Foster *et al.*, 1981b; Lundblad and Kleckner, 1981; also, see below).

The three excision events are structurally similar. Each involves excision between a pair of short direct repeats, and in each case those direct repeats lie on either side of an inverted repeat. For precise and nearly precise excision, the inverted repeats are the 1400-bp IS*10* elements that are 6500 bp apart. For precise excision of the nearly precise excision remnant, the inverted repeats are only 23 bp long and are separated by only 4 bp.

Precise and nearly precise excision appear to occur by closely related or identical pathways. All host mutations that increase precise excision also increase nearly precise excision (Foster *et al.*, 1981b). Both types of excision appear to involve an interaction between the IS*10* inverted repeats. Deletion variants of Tn*10* having shorter inverted repeats exhibit lower frequencies of precise and nearly precise excision. Interaction between the inverted repeats may help to bring the flanking direct repeats closer together in space or into a structure that favors subsequent excision and rejoining across the repeats. A favored possibility is that excision may be stimulated by the formation of intramolecular snapbacks between the inverted repeats (Foster *et al.*, 1981b).

The properties of *E. coli* mutations that increase precise excision suggest that host DNA-handling functions are indeed involved in precise and nearly precise excision, but they do not yet define specific mechanisms by which these events could occur. One major class of *E. coli* mutants exhibiting increased frequencies of Tn*10* precise excision (called TexA$^-$) appear to be unusual alleles of the RecB and RecC genes in which both precise and nearly precise excision are altered without affecting the frequency of recombination in Hfr$^-$ and P1-mediated crosses. (Lundblad and Kleckner, 1981; V. Lundblad, A. Taylor, G. Smith and N. Kleckner, unpublished). Classic alleles (including amber mutations) in RecB and RecC have the opposite phenotype: They reduce general recombination without affecting Tn*10* excision. Partial dominance of TexA$^-$ mutations and other properties of TexA$^-$ strains suggest that TexA$^-$ mutations qualitatively alter RecBC protein.

A second group of Tex$^-$ mutations are indistinguishable from previously isolated mutations in the *E. coli* pathway for correction of basepair mismatches. Mutations in the *mut*S, *mut*H, *mut*L, *dam,* and *uvr*D genes have all been identified among Tex$^-$ strains, and previously isolated alleles in each of these genes have been shown to have a Tex$^-$ phenotype. A third group of Tex$^-$ mutations do not map in *rec*BC or mismatch correction genes but have not yet been assigned to any other loci (Lundblad and Kleckner, 1981; V. Lundblad, unpublished).

Precise excision of the 50-bp nearly precise excision remnant involves a much smaller region of DNA than do precise and nearly precise excision and probably occurs by a different pathway. This third event is unaffected by any Tex− mutations isolated thus far, and other mutations in other genes do affect this event without altering precise and nearly precise excision. Our working hypothesis is that precise excision of the nearly precise excision remnant may occur during DNA replication or repair. At such times, small, single-stranded regions will certainly occur in DNA, and aberrant replication across the resulting inverted repeat snapback may then generate an appropriate excision product.

IX. Tn10 as a Substrate for Homologous Recombination

A. Tn*10*/IS*10* as a Portable Region of Homology

Transposons play an important role *in vivo* as portable regions of homology. RecA- dependent gene amplification and Hfr and F′ formation frequently involve homologous recombination between two copies of an IS sequence or transposon located at two different positions (see Kleckner, 1981 for summary and references).

An example of this phenomenon involving IS*10* sequences has been described in bacteriophage λ (Kleckner and Ross, 1980). One of the IS*10* sequences in a λ :: Tn*10* phage has undergone what appears to be a replicative inversion event (Figs. 11, 12) that places IS*10* sequences as inverted repeats on either side of an important phage promoter, pL. The same event simultaneously inverts the intervening promoter-bearing segment with respect to flanking markers on the λ genome, creating a characteristic defective phenotype. Subsequent homologous recombination between the IS*10* inverted repeats then reversibly inverts the promoter-bearing segment back and forth between the defective and original normal configurations. (Most of these recombination events are probably intramolecular, although intermolecular recombination can be observed in certain circumstances; D. Ennis and G. Smith, unpublished.) As a consequence of such recombination, adjacent genes outside of the inverted region are switched on or off depending on the orientation of the promoter-bearing segment. In this case, in response to a laboratory-imposed selective pressure, transposition of IS*10* has given rise to a genetic switch, but operation of that switch depends upon homologous recombination between two IS*10* elements serving as regions of homology.

Recombination between homologous Tn*10* elements on different genomes or at different positions in the same genome has also been exploited for genetic manipulation of bacteria in the laboratory. Deletions and duplications with predetermined endpoints, directed transpositions, and directed Hfr formation can all be constructed genetically using Tn*10* as a region of homology (Kleckner *et al.*, 1977; Chumley *et al.*, 1979; Chumley and Roth, 1980; Schmid and Roth, 1980).

B. Intramolecular Recombination between Two IS*10* Elements

Several instances of recombination between two IS*10* elements on the same genome have been described. The λ promoter inversion event described above is one example. In this particular case, recombination between IS*10* elements, which are present as inverted repeats separated by about 3.5 kb, can occur in the absence of λ recombination functions. Such recombination is almost exclusively dependent upon bacterial RecA function and is highest if RecBC function is present, but can occur at a reduced level even in its absence (Kleckner and Ross, 1980). This recombination process has been quantitated in single-cycle growth experiments (D. Ennis and G. Smith, unpublished). The proportion of λ genomes that have undergone recombination between the two 1400-bp IS*10* elements during a single round of infection is about 1×10^{-3} (if the genomes do not contain any Chi sites; see below). This frequency corresponds well to the frequency of recombination observed between standard genetic markers in λ, approximately 4×10^{-3} recombinants per 1000 bp of DNA (Campbell, 1971). As with many RecBC-dependent process, recombination between IS*10* sequences is stimulated by the genetic element Chi. In fact, Tn*10* itself contains at least one Chi site within the 6500 bp of non-IS*10* material (D. Ennis, E. Yagil, F. W. Stahl, G. Smith and N. Kleckner, unpublished). In the presence of Tn*10* Chi site(s), the frequency of the promoter inversion event increases to 6×10^{-3}.

Recombination between the IS*10* elements of a Tn*10* insertion in the *Salmonella* chromosome occurs at a similarly low frequency. An upper limit can be placed on this frequency by experiments in which Hfr strains are generated by recombination between one Tn*10* element in the chromosome and a second one on a conjugative plasmid. The direction of transfer to the resulting Hfr depends upon the relative orientations of the two Tn*10* elements. Intramolecular recombination between the IS*10* sequences on either Tn*10* element will invert the intervening nonrepeated segment, and such events can be detected by the occasional occurrence of an Hfr with an opposite direction of transfer. Such Hfrs constitute less than 1% of all Hfr strains constructed by this method; hence, fewer than 1% of the chromosomal or plasmid-borne Tn*10* elements in the experimental populations have undergone intramolecular recombination (F.

Chumley and J. Roth, unpublished). This low frequency is consistent with the frequency of homologous recombination between standard markers in the bacterial chromosome.

Thus, for a first approximation, homologous recombination between IS*10* sequences appears to occur at the frequency expected for DNA segments of about 1400 bp and exhibits the usual dependence on bacterial homologous recombination systems.

Intramolecular recombination between Tn*10*'s IS*10* elements does occur at a high frequency during induction of an excision-defective P22 :: Tn*10* prophage (S. M. Halling and N. Kleckner, unpublished). At least 20% of the phage genomes replicated and packaged in this situation have undergone exchange between IS*10*-Right and IS*10*-Left. Such exchanges can be detected by Southern blot hybridization analysis of DNA from phage particles produced in such an induction. Because the two IS*10* elements differ in sequence and in the occurrence of certain restriction enzyme cleavage sites, exchanges between the two elements generate specific new restriction fragments the appearance of which is diagnostic of recombination.

The consequences of these exchange events have been examined directly at the nucleotide sequence level. Transpositions of Tn*10* from phage genomes in the induction lysate into the *Salmonella* histidine operon were isolated; eight individual independent insertions into a particular *his*G hot spot were examined by DNA sequence analysis. Four of these insertions appear not to have undergone exchanges. They are identical to wild-type Tn*10*, with characteristic differences between IS*10*-Right and IS*10*-Left. The other four have undergone a process that can be loosely referred to as *gene conversion,* although the fates of all markers in the recombination event have not been followed: Some or all of the sequences of IS*10*-Left have been "converted" to those of IS*10*-Right. The two IS*10* elements in these transposons are identical in sequence to one another and to wild-type IS*10*-Right for at least 350 bp, beginning at the very outside end of each sequence where it is joined to adjacent histidine operon sequences. Because the outermost difference between IS*10*-Right and IS*10*-Left is at bp 24, the track of converted sequences must extend essentially all the way to the outer terminus of IS*10*-Left in all four cases. In one case, the track of gene-converted sequence extends inward as far as bp 1200 of IS*10*; in the other cases, conversion has extended less far inward. In all four cases, conversion between the IS*10* elements has occurred without concomitant inversion of the intervening *tet*R segment. The failure to recover Tn*10* elements in which IS*10*-Right sequences have been converted to those of IS*10*-Left is accounted for by the fact that IS*10*-Left is functionally defective, and such Tn*10* elements would never transpose out of the P22 genome. We presume that the high frequency of gene conversion observed in this situation reflects either the action of P22 phage recombination functions, which are known to be exceedingly active during phage induction, and/or the unusual arrangement of the phage genome while it is being amplified by replication of a "locked-in" prophage. The significance of this unusual pattern of gene conversion and the special importance of the Tn*10* termini in this process remain to be understood.

Other examples of recombination between Tn*10*'s IS*10* elements have been encountered fortuitously in the laboratory. One instance of recombination between the elements that inverted the intervening *tet*R segment without gene conversion of sequences near the ends of the element has been observed in a λ :: Tn*10* phage derivative (Kleckner, 1979), and one Tn*10* mutant isolated as a high-hopper transposon that transposes more frequently than the wild type (HH117; see Fig. 7) has two changes near the outside end of IS*10*-Right, each of which is a change from the IS*10*-Right sequence to that of IS*10*-Left. Clearly, some information has moved from IS*10*-Left to IS*10*-Right; however, sequences on both sides of these changes are exclusively those of IS*10*-Right (M. A. Davis and N. Kleckner, unpublished).

Acknowledgments

I gratefully acknowledge David Botstein and Russell Chan, who originally discovered Tn*10* and recognized its importance. I am equally grateful for the contributions of students and colleagues in my laboratory whose thoughts and experiments have made possible our current understanding of Tn*10*: Michael Davis, Timothy Foster, Shirley Halling, Hyun-Joo Kim, Victoria Lundblad, Donald Morisato, Elizabeth Raleigh, Denise Roberts, Robert Simons, Judith Swan, Kenichi Takeshita, and Jeffrey Way. I am especially grateful to John Roth for stimulating discussions and for exchange of information prior to publication. I also thank Anna Ferri for her important role in preparation of this manuscript. My research is supported by grants from the National Science Foundation, the National Institutes of Health, and the Camille and Henry Dreyfus Foundation.

References

Adam, G., and Delbruck, M. (1968). Reduction of dimensionality in biological diffusion processes. *In* "Structural Chemistry and Molecular Biology" (A. Rich and N. Davidson, eds.), pp. 198–215. Freeman, San Francisco.

Adhya, S. L., and Shapiro, J. A. (1969). The galactose operon of *E. coli* K12. I. Structural and pleiotropic mutations of the operon. *Genetics* **62,** 231–247.

Bennett, P. M., Richmond, M. H., and Petrocheilou, V. (1980). The inactivation of genes on a plasmid by the duplication of one inverted repeat of a transposon-like structure which itself mediates tetracycline-resistance. *Plasmid* **3,** 135–149.

Berg, O. G., Winter, R. B., and von Hippel, P. H. (1981). Diffusion-driven mechanisms of protein translocation on nucleic acids. I. Models and theory. *Biochemistry* **20,** 6929–6948.

Biek, D., and Roth, J. R. (1980). Regulation of Tn5 transposition in *Salmonella typhimurium. Proc. Natl. Acad. Sci. USA* **77,** 6047–6051.

Blattner, F. R., Fiandt, M., Hass, K. K., Twose, P. A., and Szybalski, W. (1974). Deletions and insertions in the immunity region of coliphage lambda: revised measurement of the promoter–startpoint distance. *Virology* **62,** 458–471.

Blazey, D. L., and Burns, R. O. (1982). Transcriptional activity of the transposable element Tn10 in the *Salmonella ilv*GEDA operon. *Proc. Natl. Acad. Sci. USA* **79,** 5011–5015.

Botstein, D., and Kleckner, N. (1977). Translocation and illegitimate recombination by the tetracycline resistance element Tn10. *In* "DNA Insertion Elements, Plasmids and Episomes" (S. Adhya, ed.), pp. 185–204. Cold Spring Harbor Lab., Cold Spring Harbor, New York.

Boyen, A., Charlier, D., Crabeel, M., Cunin, R., Palchaudhuri, S., and Glansdorff, N. (1978). Studies on the control region of the bipolar ArgECBHJ operon of *Escherichia coli.* I. Effect of regulatory mutations and insertion sequence 2 insertions. *Mol. Gen. Genet.* **161,** 185–196.

Brennan, M. B., and Struhl, K. (1980). Mechanisms of increasing expression of a yeast gent in *E. coli. J. Mol. Biol.* **136,** 333–338.

Burt, D. W., and Brammar, W. J. (1982). The *cis* specificity of the Q gene product of bacteriophage lambda. *Mol. Gen. Genet.* **185,** 486–472.

Campbell, A. (1971). Genetic structure. *In* "The Bacteriophage Lambda" (A. D. Hershey, ed.), pp. 13–44. Cold Spring Harbor Lab., Cold Spring Harbor, New York.

Chaing, S. J., and Clowes, R. C. (1980). Intramolecular transposition and inversion in plasmid R6K. *J. Bacteriol.* **142,** 668–682.

Chandler, M., Roulet, E., Silver, L., Boy de la Tour, E., and Caro, L. (1979). Tn10 mediated integration of the plasmid R100.1 into the chromosome inverse transposition. *Mol. Gen. Genet.* **173,** 23–30.

Chandler, M., Clerget, M., and Caro, L. (1982). The transposition frequency of IS1 flanked transposons is a function of their size. *J. Mol. Biol.* **154,** 229–243.

Chumley, F. G., and Roth, J. R. (1980). Rearrangement of the bacterial chromosome using Tn10 as a region of homology. *Genetics* **94,** 1–14.

Chumley, F., Menzel, R., and Roth, J. R. (1979). Hfr formation directed by Tn10. *Genetics* **91,** 639–655.

Ciampi, M. S., Schmid, M. B., and Roth, J. R. (1982). The transposon Tn10 provides a promoter for transcription of adjacent sequences. *Proc. Natl. Acad. Sci. USA* **79,** 5016–5020.

Davidson, N., Deonier, R. C., Hu, S., and Ohtsubo, E. (1974). *In* "Microbiology 1974" D. Schlessinger, ed.), pp. 56–65. *Am. Soc. Microbiol.,* Washington, D.C.

Echols, H., Court, D., and Green, L. (1976). On the nature of cis-acting regulatory proteins and genetic organization in bacteriophage: the example of gene Q of bacteriophage lambda. *Genetics* **83,** 5–10.

Faelen, M., Toussaint, A., and de la Fonteyne, J. (1974). Model for the enhancement of lambda-gal integration into partially induced Mu-1 lysogens. *J. Bacteriol.* **121,** 873–882.

Falkow, S. (1974). "Infectious Multiple Drug Resistance." Arrowsmith, Bristol, England.

Fiandt, M., Szybalski, W., and Malamy M. H. (1972). Polar mutations in *lac, gal* and phage lambda consist of a few IS-DNA sequences inserted in either orientation. *Mol. Gen. Genet.* **119,** 223–231.

Foster, T. J. (1977). Insertion of the tetracycline resistance translocation unit Tn10 in the lac operon of *Escherichia-coli* K12. *Mol. Gen. Genet.* **154,** 305–310.

Foster, T. J., Davis, M. A., Roberts, D. E., Takeshita, K., and Kleckner, N. (1981a) Genetic organization of transposon Tn10. *Cell* **23,** 201–213.

Foster, T. J., Lundblad, V., Hanley-Way, S., Halling, S. M., and Kleckner N. (1981b). Three Tn10-associated excision events: relationship to transposition and role of direct and inverted repeats. *Cell* **23,** 215–227.

Francke, B., and Ray, D. S. (1972). Cis-limited action of the gene-A product of bacteriophage OX174 and the essential bacterial site. *Proc. Natl. Acad. Sci. USA* **69,** 475–479.

Galas, D. J., and Chandler, M. (1982). Structure and stability of Tn9-mediated cointegrates. *J. Mol. Biol.* **154,** 245–272.

Gill, R. E., Falkow, S., Ohtsubo, E., Ohtsubo, H., So, M., and Heffron, F. (1980). A genetic analysis of the transposon Tn3: evidence for cointegrates as intermediates in transposition. *In* "Mobilization and Reassembly of Genetic Information" (W. A. Scott, R. Werner, D. R. Joseph, and J. Schultz, eds.), pp. 47–64. Academic Press, New York.

Glansdorff, N., Charlier, D., and Zafarullah, M. (1981). Activation of gene expression by IS2 and IS3. *Cold Spring Harbor Symp. Quant. Biol.* **45,** 153–156.

Grindley, N. D., and Joyce, C. M. (1981). Analysis of the structure and function of the kanamycin-resistance transposon Tn903. *Cold Spring Harbor Symp. Quant. Biol.* **45,** 125–134.

Grindley, N. D., and Sherratt, D. J., (1978). Sequence analysis at IS1 insertion sites: models for transposition. *Cold Spring Harbor Symp. Quant. Biol.* **43,** 1257–1261.

Halling, S. M., and Kleckner, N. (1982). A symmetrical six-basepair target site sequence determines Tn10 insertion specificity. *Cell* **28,** 155–163.

Halling, S. M., Simons, R. W., Way, J. C., Walsh, R. B., and Kleckner, N. (1982). DNA sequence organization of Tn10's IS10-Right and comparison with IS10-Left. *Proc. Natl. Acad. Sci. USA* **79,** 2608–2612.

Harshey, R. M., and Bukhari, A. I. (1981). A mechanism of DNA transposition. *Proc. Natl. Acad. Sci. USA* **78,** 1090–1094.

Isberg, R. R., and Syvanen, M. (1981). Replicon fusions promoted by the inverted repeats of Tn5. *J. Mol. Biol.* **150,** 15–32.

Jahn, G., Laufs, R., Kaulfers, P.-M., and Kolenda, H. (1979). Molecular nature of two haemopilus-influenzae R factors containing resistances and the multiple integration of drug resistance transposons. *J. Bacteriol.* **138,** 584–597.

Johnsrud, L., Calos, M. P., and Miller, J. H. (1978). The transposon Tn9 generates a 9 basepair repeated sequence during integration. *Cell* **15,** 1209–1219.

Jorgenson, R. A., and Reznikoff, S. W. (1979). Organization of structural and regulatory genes that mediate tetracycline resistance of transposon Tn10. *J. Bacteriol.* **138,** 705–714.

Jorgensen, S. T., Oliva, B., Grinsted, J., and Bennett, P. M. (1980). New translocation sequence mediating tetracycline resistance found in *E. coli* pathogenic for piglets. *Antimicrob. Agents Chemother.* **18,** 200–205.

Kamp, D., and Kahmann, R. (1981). Two pathways in bacteriophage Mu transposition? *Cold Spring Harbor Symp. Quant. Biol.* **45,** 329–336.

Kaulfers, P. M., Laufs, R., and Jahn, G. (1978). Molecular properties of transmissible R factors of *Haemophilus influenzae* determining tetracycline resistance. *J. Gen. Microbiol.* **105,** 243–252.

Kehoe, M. A., and Foster, T. J. (1977). Genetic analysis of mutations in the transfer genes of PDU-202 Tra Tn10 plasmids caused by the excision of Tn10. *Mol. Gen. Genet.* **157,** 109–118.

Kitts, P., Reed, R., Symington, L., Burke, M., and Sherratt, D. (1982). Transposon-specified site-specified recombination. *Proc. Natl. Acad. Sci. USA* **79,** 46–50.

Kleckner, N. (1979). DNA sequence analysis of Tn10 insertions: origin and role of 9bp flanking repetitions during Tn10 translocation. *Cell* **16,** 711–720.

Kleckner, N. (1981). Transposable elements in prokarynotes. *Annu. Rev. Genet.* **15,** 341–404.

Kleckner, N., and Ross, D. G. (1979). Translocation and other recombination events involving the tetracycline-resistant transposon Tn10. *Cold Spring Harbor Symp. Quant. Biol.* **43,** 1233–1246.

Kleckner, N., and Ross, D. G. (1980). A *rec*A-dependent genetic switch generted by transposon Tn10. *J. Mol. Biol.* **144,** 215–221.

Kleckner, N., Chan, R. K., Tye, B.-K., and Botstein, D. (1975). Mutagenesis by insertion of a drug-resistance element carrying an inverted repetition. *J. Mol. Biol.* **97,** 561–575.

Kleckner, N., Roth, J. R., and Botstein, D. (1977). Genetic engineering *in vivo* using translocatable drug resistance elements. *J. Mol. Biol.* **116,** 126–159.

Kleckner, N., Barker, D. E., Ross, D. G., Botstein, D., Swan, J. A., and Zabeau, M. (1978). Properties of the translocatable tetracycline resistance element Tn10 in *E. coli* and bacteriophage lambda. *Genetics* **90,** 427–450.

Kleckner, N., Reichardt, K., and Botstein, D. (1979a). Inversions and deletions of the *Salmonella* chromosome generated by the translocatable tetracycline-resistance element Tn10. *J. Mol. Biol.* **127,** 89–115.

Kleckner, N., Steele, D. A., Reichardt, K., and Botstein, D. (1979b). Specificity of insertion by the translocatable tetracycline-resistance element Tn10. *Genetics* **92,** 1023–1040.

Kostriken, R., Morita, C., and Heffron, F. (1981). The transposon Tn3 encodes a site-specific recombination system: identification of essential sequences, genes and the actual site of recombination. *Proc. Natl. Acad. Sci. USA* **78,** 4041–4045.

Lindahl, G. (1970). Bacteriophage P2: replication of the chromosome requires a protein which acts only on the genome that coded for it. *Virology* **42,** 522–533.

Lindqvist, B. H., and Sinsheimer, R. L. (1967). The process of infection with bacteriophage OX174. Bacteriophage DNA synthesis in abortive infections with a set of conditional lethal mutants. *J. Mol. Biol.* **30,** 69–80.

Lundblad, V., and Kleckner, N. (1981). Mutants of *E. coli* K12 which affect excision of transposon Tn10. *In* "Molecular and Cellular Mechanism of Mutagenesis" (W. M. Generoso, ed.), pp. 245–258. Plenum, New York.

Machida, Y., Machida, C., Ohtsubo, H., and Ohtsubo, E. (1982). Factors determining frequency of plasmid cointegration mediated by insertion sequence IS1. *Proc. Natl. Acad. Sci. USA* **79,** 277–281.

McMurry, L., Petrucci, R. E., and Levy, S. B. (1980). Active efflux of tetracycline encoded by four genetically different tetracycline resistance determinants in *E. coli*. *Proc. Natl. Acad. Sci. USA* **77,** 3974–3977.

Mendez, B., Tachibana, C., and Levy, S. (1980). Heterogeneity of tetracycline resistance determinants. *Plasmid* **3,** 99–108.

Mickel, S., Ohtsubo, E., and Bauer, W. (1977). Heteroduplex mapping of small plasmids derived from R factor R-12 *in vivo* recombination occurs at IS1 insertion sequences. *Gene* **2,** 193–210.

Miller, J. H., Calos, M. P., Galas, D., Hofer, M., Buchel, D. E., and Muller-Hill, B. (1980). Genetic analysis of transposition in the *lac* region of *E. coli*. *J. Mol. Biol.* **144,** 1–18.

Noel, K. D., and Ames, G. F. (1978). Evidence for a common mechanism for the insertion of the Tn10 transposon and for the generation of Tn10-stimulated deletions. *Mol. Gen. Genet.* **166,** 217–223.

Ohtsubo, E., Rosenbloom, M., Schrempf, H., Goebel, W., and Rosen, J. (1978). Site-specific recombination involved in the generation of small plasmids. *Mol. Gen. Genet.* **159,** 131–141.

Ohtsubo, E., Zenilman, M., Ohtsubo, H., McCormick, M., and Machida, Y. (1981). Mechanism of insertion and cointegration mediated by IS1 and Tn3. *Cold Spring Harbor Symp. Quant. Biol.* **45,** 283–296.

Pilacinski, W., Mosharrafa, E., Edmundson, R., Zissler, J., Fiandt, M., and Szybalski, W. (1977). Insertion sequence IS2 associated with Int-constitutive mutants of bacteriophage lambda. *Gene* **2,** 61–74.

Read, H. A., and Jaskunas, J. S. (1981). Isolation of *E. coli* containing multiple transpositions of IS sequences. *Mol. Gen. Genet* **180,** 157–64.

Read, H. A., Das, S., and Jaskunas, J. S. (1980). Fate of donor insertion sequence IS1 during transposition. *Proc. Natl. Acad. Sci. USA* **77,** 2514–2518.

Reed, R. R. (1981). Resolution of cointegrates between transposons gamma-delta and Tn3 defines the recombination site. *Proc. Natl. Acad. Sci. USA* **78,** 3428–3431.

Reif, H. J., and Saedler, H. (1977). Chromosomal rearrangements in the gal region of *E. coli* after integration of IS1. *In* "DNA Insertion Elements, Plasmids and Episomes" (S. Adhya, ed.), pp. 81–91. Cold Spring Harbor Lab. Cold Spring Harbor, New York.

Reynolds, A. E., Felton, J., and Wright, A. (1981). Activation of the cryptic *bgl* operon in *E. coli* K12 is due to insertion of DNA. *Nature (London)* **293,** 625–629.

Richter, P. H., and Eigen, M. (1974). Diffusion-controlled reaction rates in spheroidal geometry. Application to repressor-operator association and membrane-bound enzymes. *Biophys. Chem.* **2,** 255.

Ross, D., Grisafi, P., Kleckner, N., and Botstein, D. (1979a). The ends of Tn10 are not IS3. *J. Bacteriol.* **129,** 1097–1101.

Ross, D. G., Swan, J., and Kleckner, N..(1979b). Nearly precise excision: a new type of DNA alteration associated with the translocatable element Tn10. *Cell* **16,** 733–738.

Ross, D. G., Swan, J., and Kleckner, N. (1979c). Physical structures of Tn10 promoted deletions and inversions role of 1400 basepairs inverted repetitions. *Cell* **16,** 721–732.

Saedler, H., Reif, H. J., Hu, S., and Davidson, N. (1974). IS2, a genetic element for turn-off and turn-on of gene activity in *E. coli. Mol. Gen. Genet.* **132,** 265–289.

Schmid, M., and Roth, J. R. (1980). Circularization of transduced fragments: a mechanism for adding segments to the bacterial chromosome. *Genetics* **94,** 15–29.

Shapiro, J. A. (1979). Molecular model for the transposition and replication of bacteriophage Mu and other transposable elements. *Proc. Natl. Acad. Sci. USA* **76,** 1933–1937.

Sharp, P. A., Cohen, S. N., and Davidson, N. (1973). Electron microscope heteroduplex studies of sequence relations among plasmids of *E. coli.* II. Structure of drug resistance (R) factors and F. factors. *J. Mol. Biol.* **75,** 235–255.

Tessman, E. (1966). Mutants of bacteriophage M13 blocked in infectious DNA synthesis. *J. Mol. Biol.* **16,** 218–236.

Tomizawa, J., and Itoh, T. (1981a). Plasmid ColE1 incompatibility determined by interaction of RNAI with primer transcript. *Proc. Natl. Acad. Sci. USA* **78,** 6096–6100.

Tomizawa, J., and Itoh, T. (1981b). Inhibition of ColE1 RNA primer formation by a plasmid-specified small RNA. *Proc. Natl. Acad. Sci. USA* **78,** 1421–1425.

Watanabe, T., and Lyang, K. W. (1962). Episome-mediated transfer of drug resistance in Enterobacteriaceae. *J. Bacteriol.* **84,** 422–430.

Watanabe, T., Ogata, Y., Chan, R., and Botstein, D. (1972). Specialized transduction of tetracycline resistance by phage P22 in *Salmonella typhimurium. Virology* **50,** 874–882.

CHAPTER 7

Transposable Elements in Yeast

G. SHIRLEEN ROEDER
GERALD R. FINK

Mobile Genetic Elements

ISBN 0-12-638680-3

I. Introduction

The Ty (*t*ransposon *y*east) elements of *Saccharomyces cerevisiae* are a family of dispersed, repetitive DNA sequences. As diagrammed in Fig. 1, each Ty element consists of a central region of about 5.6 kb pairs of DNA flanked by direct repeats of an approximately 330-bp sequence called δ. There are 30–35 Ty elements per haploid yeast genome. In addition to the δ sequences associated with Ty elements, the yeast genome contains at least 100 solo δ sequences.

Ty elements were originally described by Cameron *et al.* (1979), who characterized a Ty sequence present at the yeast *SUP4* locus. The demonstration that this element occurred at the *SUP4* locus of some yeast strains but not others suggested that the sequence was transposable. The presence of terminally repeated sequences indicated structural similarities to the transposable elements of prokaryotes.

II. Transposition of Ty Elements

The demonstration that Tys were transposable came from the analysis of mutations at a number of genetic loci. Two mutations at the *HIS4* locus (Roeder and Fink, 1980; Roeder *et al.*, 1980), five mutations at the *ADR3* locus (Williamson *et al.*, 1981), and one mutation at the *CYC7* locus (Errede *et al.*, 1980a) have been shown to result from the insertion of Ty elements. The following sections describe the selective systems employed to isolate these mutants.

A. Selective Systems

Spontaneous *his4* mutants were isolated by inositol starvation, as described by Henry *et al.* (1975). This technique selectively kills growing cells and leads to an enrichment of rare auxotrophic mutants within a cell population. Two out of 20 *his4* mutants isolated were shown to result from Ty insertions (Chaleff and Fink, 1980; Roeder and Fink, 1980; Roeder *et al.*, 1980). These mutations, called *his4–912* and *his4–917*, result from the insertion of Ty elements, known as Ty912 and Ty917, respectively, into the 5′-noncoding region of the *HIS4* gene.

The total complement of cytochrome *c* in yeast normally consists of 95% iso-1-cytochrome *c*, coded by the *CYC1* gene, and 5% iso-2-cytochrome *c*, coded by the *CYC7* gene. The amount of iso-2-cytochrome *c* produced by the wild-type *CYC7* gene is insufficient to permit the growth of *cyc1*$^-$ mutant cells

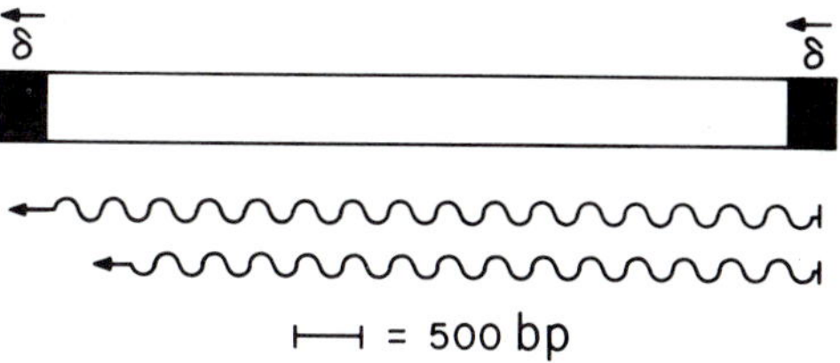

Fig. 1 Ty structure and transcription. The closed bars represent the 330-bp δ sequences found as direct repeats at the ends of the element. The open bar represents the 5600-bp internal region of the Ty. The wavy lines indicate the two major Ty transcripts. The direction of transcription is from right to left, as indicated by the arrowheads.

on media containing lactate as the sole carbon source. Selection for growth on lactate leads to the isolation of cells in which iso-2-cytochrome *c* is overproduced due to mutation at the *CYC7* locus. Of three mutations isolated in this way, one of them, known as *CYC7-H2*, has been shown to result from the insertion of a Ty element into the 5′-noncoding region of the *CYC7* gene (Errede *et al.*, 1980a,b).

Yeast produces three kinds of alcohol dehydrogenase—ADHI, coded by the *ADCI* gene; m-ADH, coded by the *ADM* gene; and ADHII, coded by the *ADR2* gene (Ciriacy, 1975a,b). Because expression of the *ADR2* gene is repressed by glucose, mutant cells that lack ADHI and m-ADH cannot grow on glucose-containing medium under anaerobic conditions. Selection for growth under these conditions leads to the isolation of mutants in which ADHII synthesis is no longer repressible by glucose (Ciriacy, 1976, 1979). Five of seven such mutants were shown to result from Ty insertions into the 5′-noncoding region of the ADR2 gene (Williamson *et al.*, 1981). These mutants are designated $ADR3{-}2^{c}$, -3^{c}, -6^{c}, -7^{c}, and -8^{c}.

B. Homology Independence

The wild-type *HIS4*, *CYC7*, and *ADR2* genes have been cloned and subjected to extensive restriction mapping and DNA sequence analysis. In addition, the mutant genes containing Ty elements have been cloned and analyzed. Comparison of the mutant genes with the wild-type genes indicates that all the mutations result from the insertion of an intact Ty element into the 5′-noncoding region of the affected gene. There is no DNA sequence homology between the Ty element and the DNA into which the element inserts (Errede *et al.*, 1980b; Farabaugh and Fink, 1980; Roeder *et al.*, 1980; Young *et al.*, 1982).

C. Generation of 5-bp Repeats

Ty transposition results in a duplication of DNA present at the target site. A 5-bp sequence that occurs only once in the sequence of the wild-type gene is found at both junctions between the inserted Ty and the wild-type gene sequence (Gafner and Philippsen, 1980; Farabaugh and Fink, 1980; Roeder *et al.*, 1980; Young *et al.*, 1982). The duplication of target DNA at the insertion site is a feature of transposition for bacterial transposons (summarized by Grindley and Sherrat, 1978), *Drosophila* mobile elements (Dunsmuir *et al.*, 1980), and retroviruses (Shoemaker *et al.*, 1980). It is thought to reflect a basic aspect of the mechanism of transposition.

The 5-bp repeats generated by Ty transposition are shown in Table II. Because these sequences show no obvious similarities, it is unlikely that the pentanucleotide is a recognition site for Ty insertion. In all Ty insertion mutations, the region on either side of the Ty is high in AT base pairs.

D. Frequency

It is not yet possible to determine accurately the frequency of Ty transposition. Estimates are based upon the frequency of insertion mutations at particular genetic loci. Ty insertions leading to loss of *HIS4* function occur spontaneously at frequency of approximately 2×10^{-8} cells. Ty insertion mutations occur at a frequency of 10^{-9} among His^{+} revertants of a promoterless *HIS3* gene (Scherer *et al.*, 1982). Analysis of spontaneous mutations within the structural gene at the *ADR2* locus failed to demonstrate any Ty insertions in 10^{8} cells. However, Ty insertions in the *ADR2* regulatory region, designated *ADR3*, occur at a frequency of approximately 10^{-8} cells after ultraviolet (UV) irradiation (Ciriacy, 1976). Because it was not demonstrated that UV irradiation increased the frequency of insertion, it is possible that the regulatory mutants at *ADR3* are of spontaneous origin.

These experiments underestimate the frequency of Ty transposition for two reasons. First, only transposition into a limited segment of the genome is detected. Second, only Ty transpositions leading to the selected phenotype are recovered. The data suggest a minimum frequency of 10^{-8} cells. It is unknown whether all Ty elements transpose at equal frequency or whether some are more active than others.

E. Specificity of Insertion Sites

All of the Ty insertion mutations characterized thus far occurred in the regulatory regions adjacent to genes and not in the structural genes themselves. This result has an obvious explanation in those cases in which the Ty insertions appeared in mutant hunts designed to select regulatory mutants. However, the absence of Ty insertions in the structural genes for *HIS4* and *ADR2* could be an indication of sequence or structural specificity during Ty insertion. More muta-

Table I Properties of Ty Elements[a]

Element	Mutation	Phenotype	Genetic location	Class	Reference[b]
Ty1			Multiple sites	Ty1	1
Ty1-B10			Near *SUP4*	Ty1	2
Ty1-D15			Unknown	Ty1	2
Ty1-17			Near *LEU2*	Ty917	3
Ty1-161			Near *PGK*	Ty1	3
	DEL1	Promotes deletions of the *CYC1-OSM1-RAD7* gene cluster	Two Tys flanking *CYC1-OSM1-RAD7*	Ty1	4
	CYC7-H2	Constitutive overproduction of iso-2-cytochrome *c*	*CYC7*	Ty1	5
	ADR3-2[c]	Constitutive overproduction of alcohol dehydrogenase II	*ADR3*	Ty1	6
	ADR3-3[c]	Constitutive overproduction of alcohol dehydrogenase II	*ADR3*	Ty917	6
	ADR3-6[c]	Constitutive overproduction of alcohol dehydrogenase II	*ADR3*	Ty1	6
	ADR3-7[c]	Constitutive overproduction of alcohol dehydrogenase II	*ADR3*	Ty1	6
	ADR3-8[c]	Constitutive overproduction of alcohol dehydrogenase II	*ADR3*	Ty1	6
Ty912	*his4-912*	His$^-$	*HIS4*	Ty1	7
Ty917	*his4-917*	His$^-$	*HIS4*	Ty917	8

[a] The table summarizes the properties of 15 different Ty elements. The first column contains the name of each Ty. The second column shows the mutations, if any, associated with each element. The third column describes the phenotype of the mutation resulting from Ty insertion. The fourth column indicates the genetic location of each Ty. The fifth column indicates whether the Ty restriction map is similar to the map of Ty1 or similar to the map of Ty917. Sources of information are given in the last column.

[b] 1, Cameron *et al.* (1979); 2, Gafner and Philippsen (1980); 3, Kingsman *et al.* (1981); 4, Liebman *et al.* (1981); 5, Errede *et al.* (1980a,b); 6, Williamson *et al.* (1981); 7, Roeder and Fink (1980); 8, Roeder *et al.* (1980).

tions will have to be examined in order to determine whether the absence of Ty insertions into structural gene sequences is statistically significant.

III. Catalog of Ty Elements

The properties of the Ty elements that have been studied most extensively are summarized in Table I. This table includes the Ty elements responsible for the *his4-912*, *his4-917*, *CYC7-H2*, and *ADR3-2*[c], *-3*[c], *-6*[c], *-7*[c], and *-8*[c] mutations. In addition, the table includes several Tys that are not associated with insertion mutations. These elements occur naturally at the same position in many different yeast strains and have no known effects on the expression of adjacent genes. They include Ty1-17, which is near the *LEU2* gene on chromosome III, and Ty1-161, which is near the *PGK* locus on chromosome III (Kingsman *et al.*,

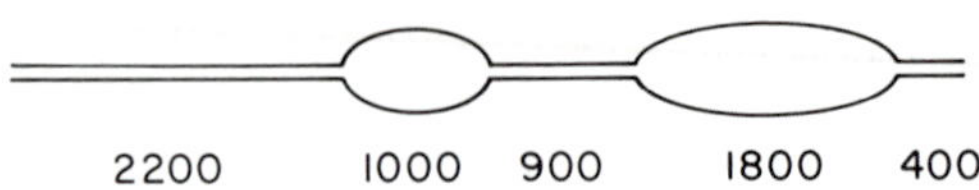

Fig. 2 Diagrammatic representation of a heteroduplex between Ty1 and Ty1-17. The parallel lines indicate double-stranded DNA formed in regions of sequence homology. The bubbles indicate single-stranded DNA that is unable to hybridize due to DNA sequence heterology. The numbers indicate the length in bases of the double- and single-stranded regions. The orientation of the elements is the same as that shown in Fig. 1, in which Ty transcription proceeds from right to left. The information is taken from Kingsman *et al.* (1981).

1981). Also included are Ty1-B10, near the *SUP4* locus on chromosome X, and Ty1-D15, of unknown genetic location (Gafner and Philippsen, 1980). Table I includes two Ty elements found on chromosome X in certain yeast strains known as *DEL1* (Liebman *et al.*, 1979; Shalit *et al.*, 1981). These Tys were detected because of their ability to promote deletions of the *CYC1-OSM1-RAD7* gene cluster (Liebman *et al.*, 1979, 1981). The Tys in Table I will be described in more detail in the remaining sections of this chapter.

In addition to the Ty elements described in Table I, we shall refer to two Ty elements known as Ty912(ura) and Ty917(ura). These elements are derivatives of Ty912 and Ty917 that have been genetically marked through the insertion of a 1200-bp restriction fragment containing the yeast *URA3* gene into the single *Hind*III site carried by these transposons (Roeder and Fink, 1982).

IV. Heterogeneity of Ty Elements

Yeast Ty elements display considerable sequence divergence. They differ from each other by simple base pair changes and also by insertion, deletion, and substitution mutations. In general, Ty elements can be divided into two main classes—those that are similar to Ty912 and Ty1 and those that are similar to Ty917 and Ty1-17. Members of the Ty912 and Ty1 class exist in about 25–30 copies per cell. Members of the Ty917 and Ty1-17 class are present in about six copies per cell (Roeder *et al.*, 1980; Kingsman *et al.*, 1981).

Southern hybridization analysis and electron microscopic heteroduplex analysis indicate that members of the Ty912 and Ty917 classes differ from each other by two large substitution mutations (Kingsman *et al.*, 1981). Figure 2 shows a diagrammatic representation of a heteroduplex between Ty1 and Ty1-17. The elements are homologous for about 2.2 kb at the left end. They then diverge such that the right ends are nonhomologous except for a 0.9-kb-pair region near the middle of the Ty and a 0.4-kbp kilobase pair region at the extreme right end.

Figure 3 shows detailed restriction maps of Ty912 and Ty917. The figure also indicates restriction sites that are polymorphic within the Ty912 class. Kingsman

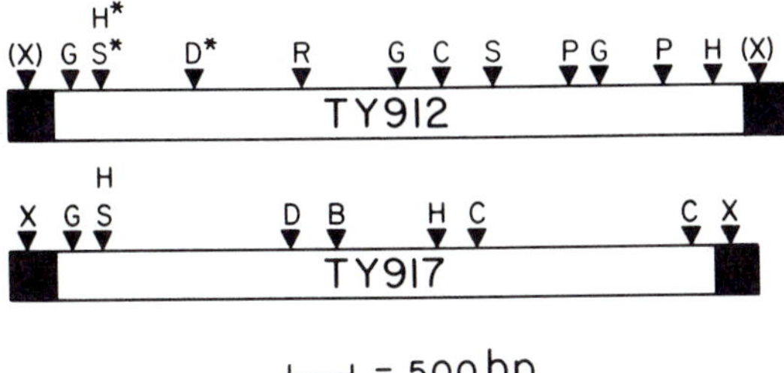

Fig. 3 Restriction maps of Ty912 and Ty917. As in Fig. 1 and in the remaining figures in this chapter, the closed bars represent the δ sequences and the open bars represent the internal regions of the elements. The arrows indicate restriction endonuclease cleavage sites for *Xho*1 (X), *Bgl*II (G), *Sal*1 (S), *Hpa*1 (H), *Hind*III (D), *Eco*R1 (R), *Cla*1 (C), *Pst*1 (P), and *Bam*H1 (B). The asterisks indicate restriction sites that are present in Ty912 but are not found in some other members of the Ty912 class of elements. The brackets indicate restriction sites that are not found in Ty912 but are found in most members of the Ty912 class of elements. The orientation of the elements is the same as that shown in Fig. 1, in which Ty transcription proceeds from right to left.

et al. (1981) have reported that Ty1-161 differs from other members of the Ty912 class by a 1200-bp insertion near the right end of the element.

V. Heterogeneity of δ Sequences

The δ sequences associated with Ty elements show considerable sequence heterogeneity. The sequences of 18 different δs are presented in Fig. 4 and compared in Table II. The sequences contain long stretches that have been completely conserved interspersed with regions that show considerable variation. The 3′-terminal decanucleotide TATTCCAACA is found in all Ty-associated δ sequences. The δ sequences found in Ty elements fall into two general classes—those like the Ty912 δs and those like the TyB10 δs. The two types show considerable sequence divergence differing by as many as 52 out of 333 bp.

The two δs flanking a single Ty element usually have identical sequences. However, three out of nine elements analyzed by DNA sequencing have nonidentical δs. In the most extreme case, there are six differences out of 332 nucleotides. These differences are usually base substitutions, but the δs of Ty917 differ by 1 bp in length. Similar δ sequences can flank Ty elements with very dissimilar internal regions. For example, the δs of Ty912 and Ty917 differ by only 15 bp changes, but they flank internal regions that differ by substitution mutations covering two-thirds of the element.

Solo δs show considerable divergence from the δs associated with Ty elements. Sequence analysis of two solo δs near *ARG4* has revealed several dramat-

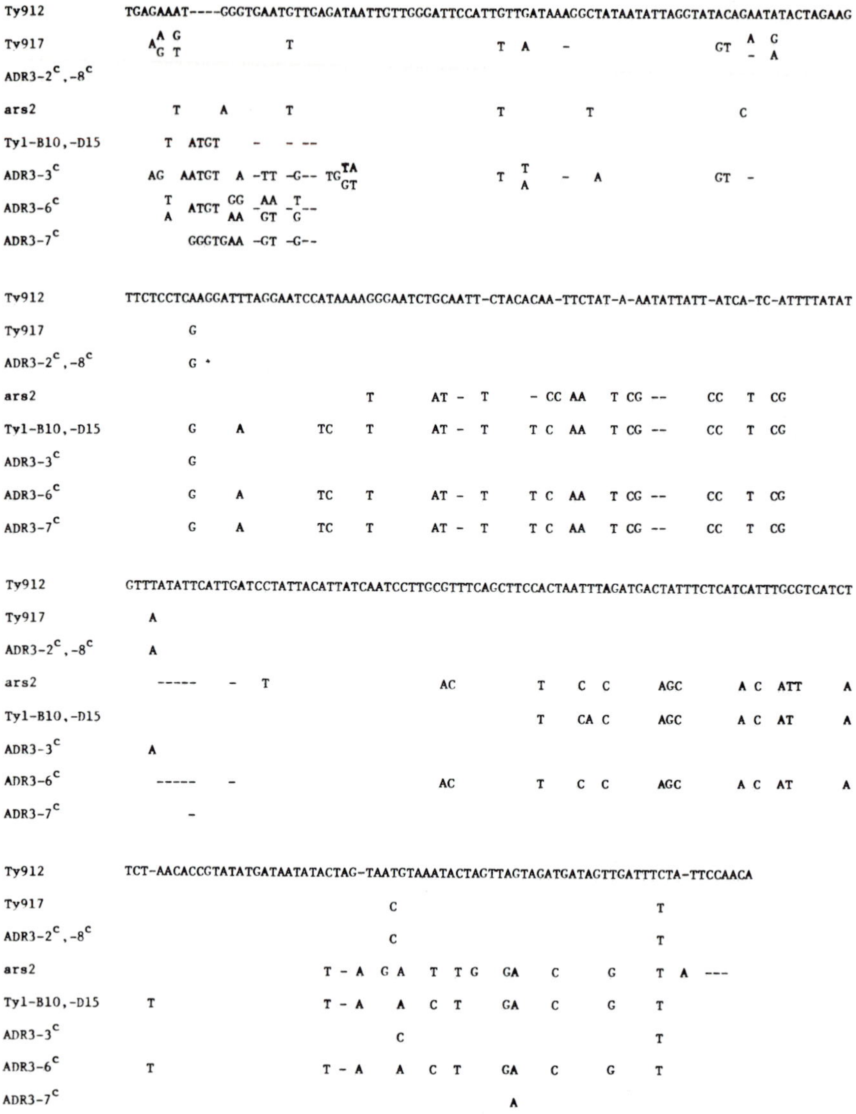

```
Ty912          TGAGAAAT----GGGTGAATGTTGAGATAATTGTTGGGATTCCATTGTTGATAAAGGCTATAATATTAGGTATACAGAATATACTAGAAG

                A G                                                                        A G
Ty917          A           T                          T A   -                       GT
                G T                                                                        - A
ADR3-2c,-8c

ars2             T   A     T                          T       T                        C

Ty1-B10,-D15    T ATGT   -  - --

                                 TA                    T
ADR3-3c        AG AATGT A -TT -G-- TG                 T     -  A                    GT -
                                 GT                    A
                T      GG  AA  T
ADR3-6c           ATGT    -   ---
                A      AA  GT  G
ADR3-7c            GGGTGAA -GT -G--


Ty912          TTCTCCTCAAGGATTTAGGAATCCATAAAAGGGAATCTGCAATT-CTACACAA-TTCTAT-A-AATATTATT-ATCA-TC-ATTTTATAT

Ty917              G

ADR3-2c,-8c        G *

ars2                           T     AT - T    - CC AA   T CG --    CC  T CG

Ty1-B10,-D15       G   A    TC   T     AT - T    T C  AA   T CG --    CC  T CG

ADR3-3c            G

ADR3-6c            G   A    TC   T     AT - T    T C  AA   T CG --    CC  T CG

ADR3-7c            G   A    TC   T     AT - T    T C  AA   T CG --    CC  T CG


Ty912          GTTTATATTCATTGATCCTATTACATTATCAATCCTTGCGTTTCAGCTTCCACTAATTTAGATGACTATTTCTCATCATTTGCGTCATCT

Ty917           A

ADR3-2c,-8c     A

ars2             -----  -  T                 AC      T  C C    AGC    A C ATT   A

Ty1-B10,-D15                                         T  CA C   AGC    A C AT    A

ADR3-3c         A

ADR3-6c          -----  -                    AC      T  C C    AGC    A C AT    A

ADR3-7c             -


Ty912          TCT-AACACCGTATATGATAATATACTAG-TAATGTAAATACTAGTTAGTAGATGATAGTTGATTTCTA-TTCCAACA

Ty917                                    C                    T

ADR3-2c,-8c                              C                    T

ars2                               T - A G A  T T G  GA  C  G  T A ---

Ty1-B10,-D15    T                  T - A   A  C T    GA  C  G  T

ADR3-3c                                  C                    T

ADR3-6c         T                  T - A   A  C T    GA  C  G  T

ADR3-7c                                          A
```

Fig. 4 Comparison of the δ sequence of Ty912 with the δ sequences of eight Ty elements and one solo δ sequence. The source of the δ appears to the left of each sequence. The δ sequence of Ty912 is shown in its entirety on the top line as a single DNA strand (5′–3′). The differences between the Ty912 δ's and the other δ sequences are shown below the Ty912 sequence. A dash indicates a 1-bp deletion of the sequence present in another δ sequence. Where the two δ's within a Ty element differ, the entry contains two base designations; the top entry is the sequence of the right δ (in which Ty transcription initiates), and the bottom entry is the sequence of the left δ (in which Ty transcription terminates). The sequence begins at the left end of the δ and ends at the right end of the δ (δs oriented as in Fig. 1). The sequences of the *ADR3*-2^c, -3^c, -6^c, -7^c, and -8^c are from Williamson *et al.* (1982). The *ars2* δ sequence is from Tschumper and Carbon (1982). The Ty912 and Ty917 δ sequences are from Farabaugh and Fink (1980) and Roeder *et al.* (1980), with several corrections.

Table II Properties of δ Sequences[a]

Source	5-bp repeat	Number of nucleotides	Class	Differences
Ty912	T-A-A-G-A	334	Ty912	0
Ty917	C-A-T-A-A	333/332	Ty912	4
Ty1-B10	G-A-A-A-C	338	Ty1–B10	0
Ty1-D15	A-T-T-T-T	338	Ty1–B10	0
ADR3-2[c]	C-A-T-A-G	334	Ty912	0
ADR3-3[c]	G-G-T-T-C	332	Ty1–B10	3
ADR3-6[c]	C-T-A-T-C	332	Ty1–B10	6
ADR3-7[c]	T-A-T-A-A	337	Ty1–B10	0
ADR3-8[c]	G-G-A-A-A	334	Ty912	0

[a] The table contains a comparison of the δ sequences associated with nine different Ty elements. The name of each element or the mutation associated with each element appears in the first column. The sequence of the 5-bp repeat of the target DNA flanking each Ty is shown in the second column. The number of nucleotides in each δ sequence is shown in the third column. The fourth column indicates whether the δ sequence is similar to the Ty912 δ sequence or similar to the Ty1–B10 δ sequence. The last column indicates the number of nucleotide differences between the left and right δ sequences within a Ty element. See Fig. 4 for complete sequence information.

ic alterations in structure (Tschumper and Carbon, 1982). One of the δs, which abuts the *ars2* origin of DNA replication, is deleted for half of the δ sequence and contains many mutational alterations in the remaining 165 bp. The other δ, which is contained within the *ars2* segment, is intact but has undergone changes in the decanucleotide at the 3′-end, which is conserved in all Ty elements.

The Ty-associated δ sequences seem to be more highly conserved than the solo δ sequences. This suggests that the δs associated with Ty elements have some functional significance, whereas the solo δs do not. It is known that the δs contain the signals required for Ty transcription, and it is likely that they are also required for Ty transposition. The striking similarity between δs within the same Ty element could be explained by gene conversion between δs at opposite ends of the same transposon.

VI. Heterogeneity in the Distribution of Ty Elements

Strains of yeast from different laboratories show variation in the distribution of Ty elements. This variation can be visualized when total genomic DNA from different strains is analyzed by Southern hybridization using Ty DNA as a probe. Most laboratory stocks have roughly the same number of restriction fragments that hybridize to the probe; however, the sizes of the fragments and the intensity of hybridization can be very different. Cloning and characterization of DNA segments from the same chromosomal location in different strains demonstrate that these differences are not simply the result of differences in restriction sites between Ty elements in the same positions. For example, the *Eco*R1 fragment containing *SUP4* in *S. cerevisiae* B596 contains a Ty element, whereas the same fragment in *S. cerevisiae* S288C does not (Cameron *et al.*, 1979).

Natural isolates of yeast strains show considerable variation in the number and position of Ty elements. In one study, 21 natural isolates of *cerevisiae* strains from all over the world were analyzed by Southern hybridization using Ty DNA as a probe (Eibel *et al.*, 1980). One of these strains appeared to have only seven elements, and another had only five elements. Another study indicated that *S. norbensis* has sequences that hybridize weakly with a δ sequence probe but does not have any sequences capable of hybridization with the internal regions of the Ty element (Fink *et al.*, 1980). However, other observations indicate that *S. norbenis* and *S. cerevisiae* are very closely related to each other. Southern hybridization analysis using single-copy genes such as *URA3* and *HIS4* demonstrates extensive cross-hybridization. Crosses between *S. cerevisiae* and *S. norbensis* yield fertile diploids that produce four viable meiotic products. Moreover, meiotic recombination in these hybrids occurs between homologous chromosomes, indicating that the genes in *S. norbensis* have the same relative positions as those in *S. cerevisiae*. The similarities between single-copy genes and the variation among Ty elements indicate that Ty sequences undergo a greater rate of evolutionary change than does the rest of the genome.

VII. Recombination between Ty and δ Sequences

Recombination between Ty elements at dispersed sites in the yeast genome results in a variety of DNA rearrangements. Most of these recombination events are apparently unrelated to the ability of the elements to transpose. The Ty simply acts as a portable region of homology, making it possible for the normal recombination system to promote rearrangements.

Recombination events involving Ty elements fall into two main classes—those that lead to a recombinant configuration of markers outside the Tys and those that leave the flanking markers in the parental configuration. In both kinds of recombination, the first step is presumed to be the pairing or alignment of homologous Tys from different sites in the genome (Fig. 5a). Figure 5b shows the result of a single reciprocal crossover. This event can be most easily thought of as the breakage of the Tys at homologous sites, followed by their rejoining in a recombinant configuration. This exchange leads to recombination between markers within the Tys and also to recombination between genetic markers that lie outside the Ty elements. Depending on the position and orientation of the Ty elements involved, such a reciprocal crossover can result in deletion, transloca-

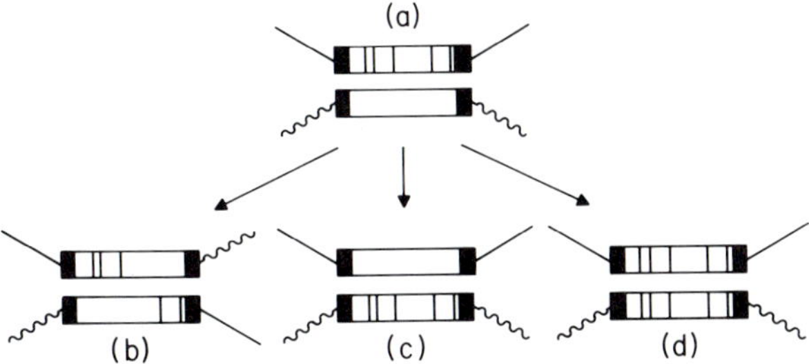

Fig. 5 Recombination events involving Ty elements. The solid and wavy lines represent unique sequence yeast DNA flanking the Ty elements. The vertical lines indicate sequence differences between the two recombining Tys. These differences could be either single base pair changes or insertion, deletion, or substitution mutations. (a) Two Ty elements from different sites in the genome pair with each other as the first step in recombination. (b) Products of a single reciprocal crossover between the Tys shown in (a). (c) Products of two reciprocal crossovers between the Tys shown in (a). (d) Products of gene conversion between the Tys shown in (a).

tion, inversion, duplication, or transposition. Because there are more than 30 Ty elements distributed throughout 17 chromosomes, the potential for DNA rearrangements due to Ty–Ty recombination is very great. When the solo δ sequences are also considered, this potential is even more striking.

Figure 5c,d shows Ty–Ty recombination events that do not lead to recombination of flanking markers. Figure 5c represents a recombination event in which the two Tys engage in two reciprocal crossovers—one near each end of the Ty. This results in a reciprocal exchange of markers within the Tys but leaves the flanking markers in the parental configuration.

Figure 5d shows a nonreciprocal recombination event. In nonreciprocal recombination or gene conversion, the two Tys interact with each other in such a way that part or all of the sequence of one Ty becomes identical to the sequence of the other Ty. The sequence of one of the Tys and the sequences outside the Tys are unaffected by this process.

Either gene conversion or double reciprocal crossing-over can result in the replacement of one Ty sequence by another. In gene conversion, the sequence of one Ty is lost and the sequence of the other Ty is duplicated. In contrast, reciprocal recombination recombines sequences without destroying or duplicating them. Neither gene conversion nor double reciprocal crossing-over alters the configuration of markers outside the Tys.

Recombination between Ty and δ sequences has been detected in several different experimental systems. Many revertants of Ty insertion mutations, such as *his4-912*, *his4-917*, and *ADR3*^{c}, result from recombination between Tys or δs. *Ura3*$^{-}$ derivatives of Ty912(ura) and Ty917(ura) can also result from Ty–Ty recombination. In addition, spontaneous mutations at the *SUP4* and *CYC1* loci sometimes result from recombination between Ty or δ sequences in these chromosomal regions. Except where specified, all of the recombination events that we will describe occurred in mitotically growing cells.

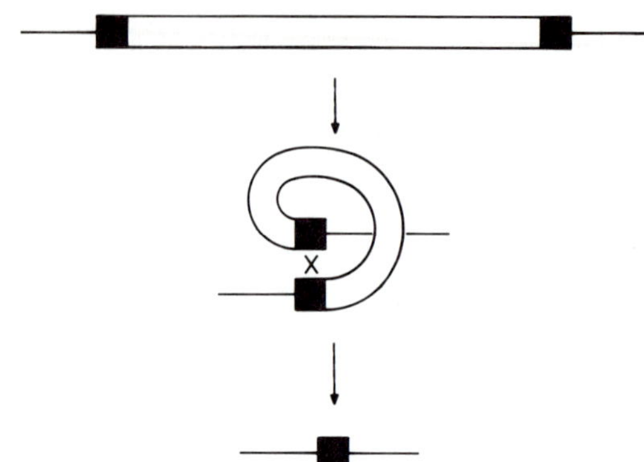

Fig. 6 Ty excision. The solid lines indicate unique sequence yeast DNA flanking the Ty. The X indicates the site of the crossover. Reciprocal recombination between homologous δ sequences at the ends of the Ty results in excision of the internal region of the Ty and leaves behind a single δ sequence.

A. Excision

One of the most frequent kinds of recombination events involving Tys is reciprocal crossing-over between the homologous δ sequences at opposite ends of a single Ty element. This recombination results in deletion of the internal region of the Ty and leaves behind a single copy of the δ sequence (Fig. 6). The majority of revertants isolated from the *his4-912* (Roeder and Fink, 1980; Farabaugh and Fink, 1980) and *ADR3*^{c} (Ciriacy and Williamson, 1981) insertion mutations result from Ty excision due to δ–δ recombination.

If a Ty element contains nonidentical δs, then excision of the Ty should sometimes leave behind the left δ, sometimes the right δ, and somtimes a δ that is a hybrid of the original two. Three independent strains resulting from excision of Ty917 have been isolated (G. S. Roeder and G. R. Fink, unpublished). Although they all contain a single δ sequence in the *HIS4* regulatory region, they have different phenotypes with respect to *HIS4* expression. It seems likely that the differences in phenotype are due to differences in the sequence of the δ remaining after Ty917 excision. The sequence of one of these δs has been determined and is a hybrid between the left and right δ sequences of Ty917 (P. J. Farabaugh and G. R. Fink, unpublished).

The haploid genome of *S. cerevisiae* contains approximately 100 solo δ sequences that are unassociated with Ty elements (Cameron *et al.*, 1979; Eibel *et al.*, 1980). It seems likely that most of these solo δs are the results of Ty excision of the sort just described because transposition of solo δs has not been observed.

Excision of a δ sequence, or of an entire Ty element, to restore the original wild-type sequence has never been detected. Excisions of this type are theoretically possible because Ty transposition generates a 5-bp duplication of target DNA at the ends of the element (Farabaugh and Fink, 1980; Gafner and Philip-

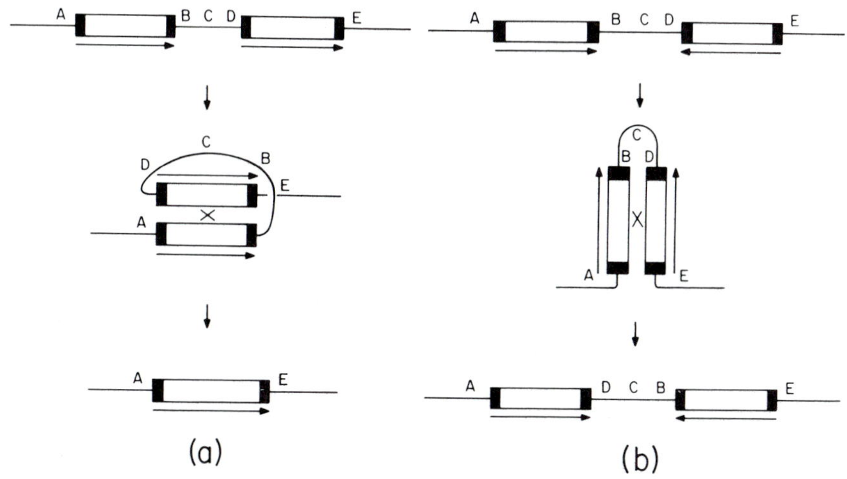

Fig. 7 Deletion and inversion due to δ–δ recombination. A, B, C, D, and E indicate genetic markers in the unique sequence DNA flanking the Ty elements. The arrows indicate the orientations of the Tys. The Xs indicate the sites of crossovers. (a) Reciprocal crossing-over between Tys that lie in the same orientation on the same chromosome results in the deletion of one of the Ty elements and the DNA lying between the two Tys. (b) Reciprocal crossing-over between two Tys that lie in opposite orientations on the same chromosome results in inversion of the DNA lying between the two elements.

psen, 1980). Reciprocal crossing-over between these 5-bp repeats could excise the entire Ty, including both δ sequences. However, events of this sort have never been seen among several hundred revertants derived from the *his4-912*, *his4-917*, and *ADR3*^{c} mutations. Apparently, the yeast recombination system acts only on homologous DNA segments greater than 5 bp in length.

B. Deletions and Inversions

Recombination between δs or Ty elements that are in the same orientation on the same chromosome results in deletion of the DNA between the recombining δ or Ty sequences (Fig. 7a). Recombination between δs that are separated by non-Ty DNA results in deletion of unique sequence chromosomal DNA. The *SUP4* region of chromosome X contains six solo δs arranged such that the tRNA gene is flanked by directly repeated δ sequences (Cameron *et al.*, 1979). Deletion of the *SUP4* gene resulting from δ–δ recombination occurs spontaneously at a frequency of 3×10^{-7} cells (Rothstein, 1979).

A segment of unique sequence DNA that is flanked by intact Ty elements should be deleted even more frequently than a segment flanked by δs. In this case, deletion can result from crossing-over anywhere within a 6000-bp region. Scherer and Davis (1980a,b) have used yeast transformation to construct strains carrying nontandem Ty duplications. In these strains, the duplicated Tys are separated by plasmid sequences that include the *URA3* gene. Reciprocal recombination between the Ty elements results in loss of the plasmid and renders the cell *ura3*$^-$. After 10 generations of growth under nonselective conditions, three out of 1000 cells examined had lost the plasmid sequences.

Tandem duplications of Ty elements occur at a number of sites in the yeast genome, as demonstrated by the isolation of yeast DNA fragments containing circular permutations of Ty elements (Cameron *et al.*, 1979). One might expect these structures to recombine at a rate similar to that for the nontandem elements just described.

Deletions promoted by Ty–Ty recombination occur at high frequency in certain yeast strains known as *DEL1*. In *DEL1* strains, the *CYC1* region of chromosome X contains two Ty elements that surround a 13-kb-pair DNA segment containing the *CYC1, OSM1*, and *RAD7* genes (Liebman *et al.*, 1981). Ty–Ty recombination results in deletion of all three genes and occurs at a spontaneous rate of 10^{-5}–10^{-6} per cell division. In *del1* strains, there are no Ty elements in the *CYC1* region, and the deletions do not occur. In most of the deletions in *DEL1* strains, crossing-over occurs within the internal regions of the Tys, resulting in excision of the 13-kb-pair unique segment and one of the Ty elements. However, recombination can also occur between the δs at the internal ends of the Tys, resulting in excision of the unique sequence DNA and a single δ sequence. In these instances, a tandem duplication of Tys remains at the original site (Liebman *et al.*, 1981).

Reciprocal recombinations between Ty elements in opposite orientations on the same chromosome results in inversion of the DNA between the two Tys (Fig. 7b). One revertant of the *his4-912* mutant carries an inversion of most of the left arm of chromosome III (Chaleff and Fink, 1980). There are intact Ty elements at both inversion breakpoints (Roeder and Fink, 1980), but the restriction maps of these Tys are slightly different from that of Ty912 (P. J. Farabaugh and G. R. Fink, unpublished). These data suggest that the inversion occurred by reciprocal recombination between Ty912 and a Ty element near the left end of chromosome III.

C. Translocations

Recombination between two Ty elements on different chromosomes results either in the generation of a dicentric chromosome and a chromosome fragment or in a reciprocal chromosome translocation. Events of the first sort are lethal. Chromosome translocations have been detected in His4$^+$ revertants of the *his4-912* mutant.

Two Ty912-promoted translocations have been described, one between chromosomes I and III and the other between chromosomes III and XII (Chaleff and Fink, 1980). In both cases, the chromosome III translocation breakpoint occurs within Ty912 (Roeder and Fink, 1980). Cloning and characterization of DNA

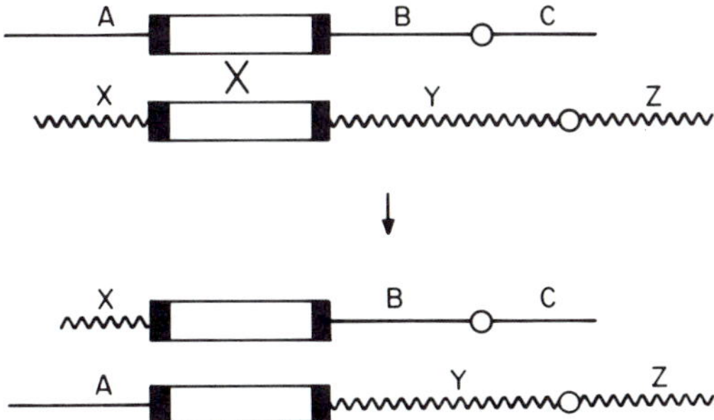

Fig. 8 Reciprocal chromosome translocation due to Ty–Ty recombination. A, B, and C indicate genetic markers on the chromosome represented by the solid line. X, Y, and Z indicate genetic markers on the chromosome represented by the wavy line. The open circles represent centromeres. The X indicates the site of crossing-over. A reciprocal crossover between two Tys on different chromosomes results in the generation of two new genetic linkage groups.

segments containing the translocation breakpoints indicate that intact Ty elements are present at both of the junctions between chromosome III DNA and chromosome I (or XII) DNA (P. J. Farabaugh and G. R. Fink, unpublished). These results suggest that both translocations are the result of reciprocal recombination between homologous Tys on different chromosomes (Fig. 8).

D. Unequal Crossing-Over

Recombination between Ty elements at different sites on homologous chromosomes results in a deletion on one chromosome and a duplication on the other (Fig. 9). Unequal crossover events of this sort have been observed between Ty elements on chromosome III (G. S. Roeder and G. R. Fink, unpublished). In strains that carry the *his4-917* mutation, there are at least two Ty elements on the left arm of chromosome III—Ty917 at the *HIS4* locus and Ty1-17 at the *LEU2* locus. These elements share extensive homology and lie in the same orientation on the chromosome. In diploid cells that are undergoing meiosis, reciprocal recombination between the Ty917 element on one chromosome III and the Ty1–17 element on the homologous chromosome occurs in about 1% of all meiotic events. This unequal crossover generates two aberrant chromosomes, one of which carries a duplication of all of the DNA between *HIS4* and *LEU2* and one of which lacks this information. Surprisingly, haploid cells that are missing the 40-kb pairs of DNA between *HIS4* and *LEU2* are still viable.

E. Transposition

In addition to being able to promote their own transposition, Ty elements can sometimes mediate the transposition of unique sequence DNA. As described

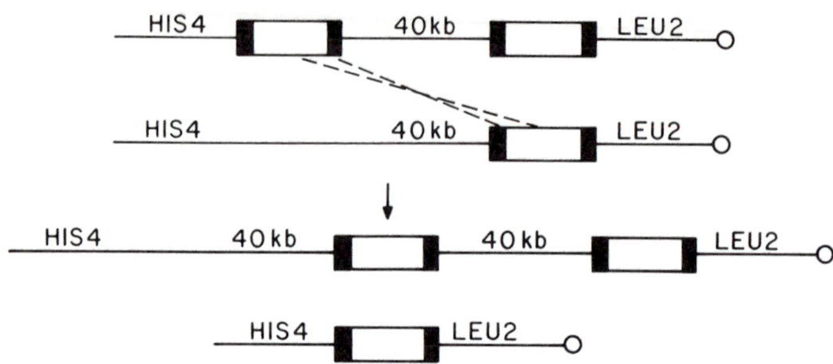

Fig. 9 Unequal crossing-over due to recombination between Ty elements. The figure shows recombination between two homologous chromosomes in a diploid yeast strain. Only the left arm of chromosome III is shown. The open circles represent the chromosome III centromeres. Both of the recombining chromosomes contain a Ty element near the *LEU2* gene. Only one of the chromosomes contains a Ty element at the *HIS4* locus. A reciprocal crossover between the Ty element at *HIS4* and the Ty element near *LEU2* on the other homologue results in the generation of two aberrant chromosomes. One chromosome carries a duplication of the 40 kb of DNA between *HIS4* and *LEU2*, and the other chromosome is missing this information.

previously, some yeast strains display a high frequency of deletion of the *CYC1-OSM1-RAD7* region of chromosome X. Stiles *et al.* (1980, 1981) have characterized a yeast strain in which this same 13-kb-pair DNA segment has undergone transposition to chromosome VII. This transposition is thought to have involved a circular DNA molecule resulting from recombination between the Tys flanking the *CYC1* region. Such a circle would contain the 13-kb pair unique segment and a single Ty element. Transposition would result when the circular chromosomal fragment reintegrated into the genome, possibly by recombination with another Ty element (Fig. 10). It has been suggested that the *CYC1-OSM1-RAD7* gene cluster on chromosome X and the very similar *CYC7-RAD23-ANP1* gene cluster on chromosome V are related by an ancient transpositional event of this type (McKnight *et al.*, 1981).

In two of the His$^+$ revertants derived from the *his4-912* mutant, the *HIS4* gene is transposed to a new chromosomal location (Chaleff and Fink, 1980). However, the molecular events leading to these transpositions are not yet clearly understood.

F. Excision and Reintegration

As described previously, Ty elements can become excised from the yeast genome by reciprocal recombination between the δ sequences at the ends of the element. This leaves behind a single copy of the δ sequence on the chromosome. The other product of such a crossover should be a circular Ty element containing a single δ sequence. Because this circular element is unable to replicate, it is most often lost from the cell. However, the excised element can sometimes

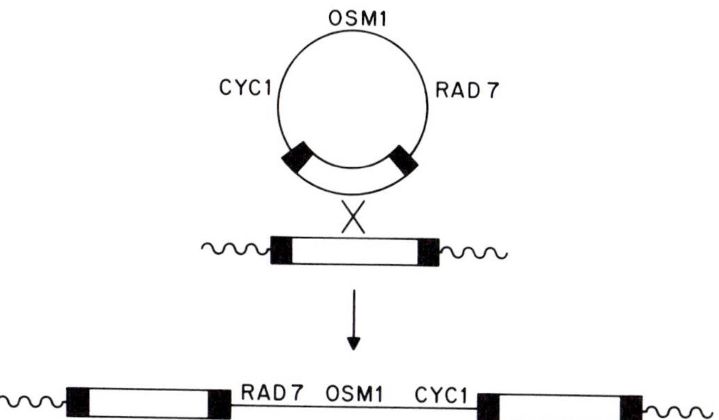

Fig. 10 Possible mechanism of transposition of the *CYC1-OSM1-RAD7* gene cluster. Recombination between the Tys flanking the *CYC1-OSM1-RAD7* gene cluster has generated a circle containing one Ty and the *CYC1*, *OSM1*, and *RAD7* genes. When the Ty contained within the circle engages in a reciprocal crossover with a Ty in a yeast chromosome, the *CYC1-OSM1-RAD7* gene cluster is reintegrated into the yeast genome.

become reintegrated into the genome. This reintegration can be detected if the excised element contains a marker such as the *URA3* gene in Ty912(ura).

Analysis of strains containing Ty912(ura) indicates that approximately 1% of the excised Ty elements are reintegrated. The reintegrated Ty912(ura) maps near *LEU2* in 90% of the cases and near *PGK* in the remaining 10% of the cells. Because both *LEU2* and *PGK* are the sites of preexisting Ty elements (Kingsman *et al.*, 1981), these observations suggest that the excised Ty reintegrates by reciprocal crossing-over with a homologous Ty elsewhere (Fig. 11). Because *HIS4*, *LEU2*, and *PGK* are all on chromosome III, the data also indicate that the excised Ty preferentially reintegrates into its chromosome of origin. This preference for movement to new sites on the same chromosome is reminiscent of the behavior of maize controlling elements. *Ac* transposes most frequently short distances along the same chromosome, less frequently to more distant sites, and least frequently to other chromosomes (Fincham and Shastry, 1974; Van Schaik and Brink, 1959).

G. Replacement of Ty Sequences

All of the DNA rearrangements just described result in recombination between genetic markers that lie outside the Ty elements. However, Ty elements can also engage in recombination events that do not alter the arrangement of flanking markers (Fig. 5c,d). Several *ura3*$^-$ derivatives of Ty912(ura) and Ty917(ura) retain an intact Ty element at the *HIS4* locus. The Ty elements are in the same position and in the same orientation as the original elements; however, they differ from the parental Tys by the loss of *URA3* DNA and by changes in the

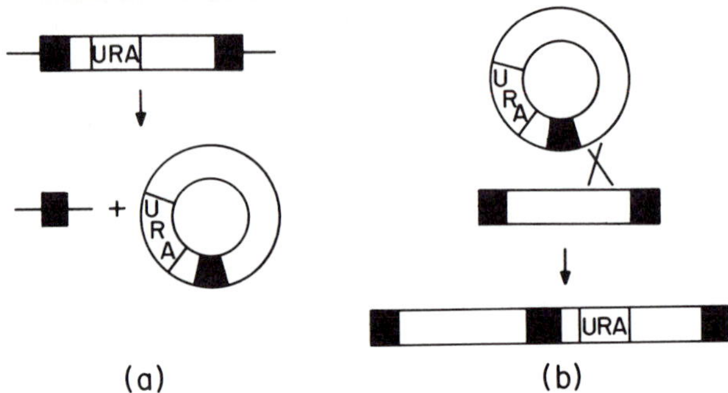

Fig. 11 Excision and reintegration of a Ty element. (a) A Ty element containing the *URA3* gene is excised from the yeast genome by δ–δ recombination. This generates a circle containing the internal region of the Ty and a single δ sequence. A solo δ sequence is left behind in the chromosome. (b) When the circular Ty engages in reciprocal recombination with a Ty in a yeast chromosome, the Ty containing *URA3* is reintegrated into the yeast genome and a tandem duplication of Tys is generated.

restriction endonuclease cleavage patterns (Roeder and Fink, 1982). The new restriction maps are characteristic of Ty elements known to exist elsewhere in the genome. The simplest interpretation of these observations is that the *HIS4* Ty has engaged in recombination with a Ty elsewhere in the yeast genome. This recombination is different from the events described in Figs. 5b, 7, 8, and 9 because it does not lead to chromosome aberrations or to alterations in the arrangement of markers flanking the Ty. The two recombination mechanisms that can explain this class of events are gene conversion and double crossing-over, as diagrammed in Fig. 5c and 5d.

Several observations indicate that these replacement events result from gene conversion rather than double crossovers. First of all, mitotic gene conversion of non-Ty sequences occurs more frequently than reciprocal crossing-over. Furthermore, if the Ty replacements result from double crossovers, the frequency of these events should be the product of the probabilities of two single crossovers. However, the frequency of chromosome aberrations (10^{-8}) due to single crossovers and the frequency of Ty replacement (10^{-7}) are very similar. Therefore, the double crossover hypothesis requires an unusual coupling between single crossover events. In addition, the double crossover hypothesis does not explain the loss of the *URA3* marker from cells in which the *HIS4* Ty has been replaced. Therefore, the double crossover hypothesis requires that the recombination occur at the two chromatid stage of chromosome replication and that the *URA3*-containing chromatids be lost by a subsequent segregation event.

When Ty917(ura) or Ty912(ura) are corrected (converted) using another Ty element as template, they lose the *URA3* DNA and acquire the restriction map displayed by the second Ty. The conversion occurs between Ty elements that share extensive homology and also between Ty elements that differ by large substitution mutations. It is surprising that Ty elements that exist at dispersed sites in the genome and that differ by extensive regions of nonhomology can still pair with each other and engage in recombination. However, conversion between

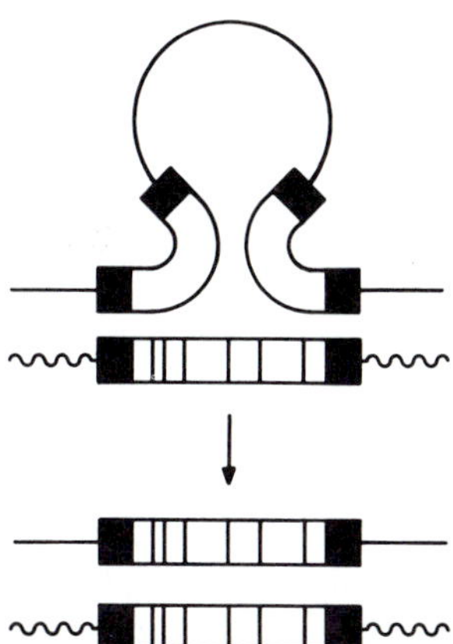

Fig. 12 Recombination between a Ty element and a DNA segment containing two Tys. The solid and wavy lines represent unique sequence DNA flanking the Tys. The vertical lines indicate sequence differences between the Tys. A single Ty (in the lower chromosome in the diagram) interacts with a DNA segment (in the upper chromosome in the diagram) containing two Tys that are separated by unique sequence DNA. The single Ty pairs with the left end of the leftmost Ty and the right end of the rightmost Ty in the upper chromosome. This generates a loop containing Ty sequences and unique sequence DNA. Conversion of the sequence of the upper chromosome to the sequence of the Ty in the lower chromosome results in deletion of the DNA within the loop and leaves behind a single Ty element. The sequence of this Ty is identical to that of the Ty used as template during conversion.

dispersed repeated sequences is not unique to Ty elements. It occurs between *HIS3* genes inserted at different sites in the genome by transformation (Scherer and Davis, 1980a), between the nonallelic *CYC1* and *CYC7* genes (Ernst *et al.*, 1981), and between dispersed tRNA genes (Munz and Leupold, 1981). Furthermore, high-frequency gene conversion in a unique direction is responsible for mating-type switching in homothallic yeast strains (Chapter 13). In higher eukaryotic cells, gene conversion is thought to be involved in the correction of globin genes (Slightom *et al.*, 1980) and in the evolution of immunoglobulin genes (Bothwell *et al.*, 1981; Schreier *et al.*, 1981).

An unusual case of recombination involves the interaction between a single Ty element and a DNA segment that contains two Ty elements with unique sequence DNA in the middle. Conversion in one direction will result in transposition of the unique sequence DNA to the site of the single Ty element. An event of this sort may be responsible for the *CYC1* transposition described by Stiles *et al.* (1981). Conversion in the other direction will result in deletion of the unique sequence DNA (Fig. 12). In the place of the original two Ty elements and the unique sequence DNA will be a single Ty element that differs from either of the two Tys originally present at the site. Events of this sort occur at the *CYC1* region in *DEL1* strains (Liebman *et al.*, 1981) and in the *HIS4–LEU2* region of chromosome III (Roeder and Fink, 1982).

The conversion events just described involve recombination between two Ty elements. However, it is conceivable that conversion can also occur between a Ty element and a solo δ. In this case, the result will be either transposition of the

Ty to the site of the δ or excision of the Ty element. Events of this kind have not yet been reported.

Gene conversion is generally considered a mechanism for maintaining sequence homogeneity. Because one sequence is corrected to become identical to another, conversion will eventually force identity on any family of homologous sequences. However, gene conversion may also be an important mechanism for generating diversity. Consider a gene conversion event between two different Tys in which the conversion covers only part of the element. In this case, conversion can generate a new Ty by combining parts of previous ones. It has been suggested that conversion may be an important mechanism in the generation of immunoglobulin diversity (Baltimore, 1981).

As described in Section IX, different Ty elements can have very different effects on the expression of adjacent genes. Therefore, conversion offers a way of effecting changes in the expression of genes that abut Ty elements.

H. Recombination between Nonhomologous DNA Sequences

All of the events described thus far result from recombination between homologous Ty or δ sequences at different sites in the genome. However, reciprocal crossing-over between a Ty element and a nonholomogous DNA sequence has also been observed. These events were detected in derivatives of the *his4-912* mutant that had recovered only one or two of the three enzymatic activities specified by the *HIS4* gene (Chaleff and Fink, 1980). These revertants all resulted from deletions that have one endpoint in the *HIS4* structural gene and one endpoint in the internal region of Ty912 (Roeder and Fink, 1980). DNA sequence analysis of some of these deletions indicates that they result in a fusion of the *HIS4* coding region with a coding region in the Ty element (P. J. Farabaugh and G. R. Fink, unpublished). Two observations indicate that these *HIS4* deletions are causally related to the presence of Ty912. First, the frequency of deletions involving the *HIS4* locus is 100-fold greater in *his4-912* mutant strains than in wild-type strains or other mutant strains. Second, all of the deletions have one endpoint within the Ty even though this was not required by the selective conditions employed. Thus, Ty elements may be highly recombinogenic sequences that engage in a high frequency of both homology-dependent and homology-independent recombination events.

I. Concerted Events

More than one Ty–Ty recombination event can occur within a single cell. The frequency of cells sustaining multiple events is much greater than would be expected if these events were independent. The data suggest that a cell that is undergoing one recombination event has an increased probability of undergoing another related rearrangement.

A significant fraction of the *SUP4* deletion mutants resulting from δ–δ recombination also carry inversions of DNA segments adjacent to the *SUP4* locus (R. Rothstein, unpublished). These inversions result from recombination between δs that are in opposite orientations and that lie outside the deleted segment. Similar complex events are reported to occur at *DEL1* (Liebman *et al.*, 1981).

Many revertants of the *his4-912* mutant contain mutational alterations that are unrelated to the reversion event. Some of the revertants isolated from diploid cells sustained DNA rearrangements at the *HIS4* loci on both chromosomes. Others sustained one event at *HIS4* and another alteration elsewhere in the genome. It has been suggested that the *his4–912* revertants represent cells in which Ty elements at several chromosomal locations have been activated (Chaleff and Fink, 1980).

VIII. Ty Transcription

A. Structure of Ty-RNA

Yeast cells contain two abundant polyadenylated RNAs that are homologous to Ty DNA (R. T. Elder *et al.*, 1980, unpublished). The major species of RNA is 5.7 kb long, and the less abundant species is 5.0 kb long (Fig. 1). Hybrids between yeast RNA and a recombinant λ phage carrying a Ty element are continuous duplexes without intervening sequences.

The 5′- and 3′-ends of Ty-RNA have been determined by sequence analysis of cDNA copies (R. T. Elder *et al.*, unpublished). The 5′-end of the 5.7-kb RNA is 236–240 bases from the right end of the element. Transcription proceeds through the entire internal region of the Ty and terminates 33–46 bases from the left end of the element. Thus, the transcript contains δ sequences at both the 5′- and 3′-termini, and approximately 45 bases of the δ sequence are repeated at both ends of the transcript. The less abundant 5.0-kb transcript has the same 5′-end as the major transcript, but it terminates in the internal region of the Ty.

B. Control of Ty Transcription

The 5′- and 3′-endpoints of the 5.7-kb RNA raise a number of questions about Ty transcription signals. Because the δs in a Ty are identical or almost identical to each other, transcription initiation and termination signals must be present at both ends of the element. Yet, only the rightmost promoter and the leftmost terminator appear to be functional. Transcripts that initiate in the rightmost δ stop at a termination site present in the left δ even though transcription proceeded through the identical sequence present in the right δ. A transcript initiating in the left δ could either terminate at the end of the δ or proceed out of the δ into the flanking DNA. Studies on transcription of DNA adjacent to Ty912, Ty917, and

Ty elements at the *ADR3* locus (Young *et al.*, 1982) failed to find a transcript emerging from the Ty element into the adjacent DNA. On the other hand, analysis of RNA in strains containing a solo Ty912–δ at *HIS4* revealed a *HIS4* transcript longer than that found in wild-type strains (S. Silverman and G. R. Fink, unpublished). Sequence analysis shows that this transcript initiates within the δ at the site identified by Elder as the initiation sequence for the 5.7-kb Ty-RNA. No *HIS4* transcript is initiated at this site when the complete Ty912 is present. These results indicate that transcription initiating at one end of the Ty affects the activity of the promoter present in the δ at the other end of the element. This effect could result from transcription termination at the end of the δ or from a block in initiation.

The amount of the 5.7-kb Ty transcript present in a yeast cell is related to the mating type of the cell (Elder *et al.*, 1980). In cells with an a, α, a/a, or α/α mating type, 5–10% of the total polyadenylated RNA is Ty-RNA. In cells with the a/α *MAT* configuration, Ty-RNA is about one-fifth this amount in log phase cells and only one-twentieth this amount in stationary phase cells. The amount of the 5.0-kb RNA is reduced in stationary phase cells irrespective of the mating type of the cell.

C. Function of Ty-RNA

It is unknown whether Ty-RNA is translated into protein. The considerable sequence divergence among Ty elements makes it unlikely that all elements have retained the capacity to encode functional gene products. One explanation for the high level of Ty-RNA is that these transcripts are intermediates in the transposition process. This possibility is supported by the structural similarities between Ty-RNA and the RNA genomes of retroviruses. Both kinds of RNA are flanked by directly repeated sequences of approximately 50 bases. Furthermore, both retrovirus RNA and Ty-RNA contain sequences homologous to tRNA at the 5′-end (Eibel *et al.*, 1980; Taylor and Illmensee, 1975). In the case of the retrovirus RNA, both the terminal repeats and the tRNA homology are known to be important in the synthesis of a cDNA copy that is then stably integrated into the genome (Shoemaker *et al.*, 1980; Temin, 1981).

IX. Ty Effects on Gene Expression

A. ROAM Mutations

Ty insertions at the *ADR3* (Young *et al.*, 1982) *CYC7* (Rothstein and Sherman, 1979), *DUR1*, and *CARGA* (D. Jauniaux *et al.*, unpublished) loci cause regulatory mutations leading to constitutive overproduction of the products of the adjacent genes. Remarkably, the overproduction caused by these insertions depends upon the constitution of the mating-type locus (Deschamps and Wiame, 1979; Errede *et al.*, 1980a,b; Lemoine *et al.*, 1978; Rothstein and Sherman, 1979). The overproduction that occurs in haploids is greatly diminished in diploids homozygous for the Ty insertion mutation but heterozygous at the *MAT* locus (*MATa*/*MAT*α). This effect is not a result of diploidy because diploids homozygous at the *MAT* locus have the elevated enzyme levels characteristic of haploids. Mutations showing this behavior have been termed ROAM (regulated overproducing alleles responding to mating signals) (Errede *et al.*, 1980a).

Strains heterozygous at the *MAT* locus are distinguished from haploids and from diploids homozybous at *MAT* by their inability to mate and by their ability to sporulate. In order to determine whether the ability to mate per se controlled the ROAM phenotype, sterile mutations that prevent cells from mating were combined with ROAM mutations (Errede *et al.*, 1980a,b). The *ste7* mutation prevented both conjugation and overproduction of ROAM gene products, whereas the *ste5* mutation prevented conjugation without affecting the level of ROAM gene products. This result suggests that the signal causing the reduction in ROAM gene expression is not simply the inability to mate but some more specific regulatory signal. In order to study the effect of sporulation competence on ROAM mutations, diploids capable of both mating and sporulating were constructed using mutations at the *MAT* and *SAD* loci. These cells displayed the same levels of ROAM gene products as haploid strains, suggesting that the overproduction in ROAM mutants is not simply the result of sporulation ability.

The unusual behavior of ROAM mutations could be explained very simply if a transcript subject to mating-type control started in the Ty and proceeded through the adjacent structural gene. However, two observations make this explanation untenable. First, in several different ROAM mutants, the direction of Ty transcription is opposite to that of the adjacent gene. Second, the transcript of the affected gene is known to start at the normal initiation site in some ROAM mutants (Young *et al.*, 1982).

B. Differential Effects on Gene Expression Produced by Various Tys

Analysis of Ty insertion mutations at the *HIS4* locus indicates that different Ty elements can have very different effects on gene expression. Ty912 and Ty917 have inserted in opposite orientations into the *HIS4* regulatory region. Ty912 transcription proceeds in the same direction as *HIS4* transcription, and Ty917 transcription proceeds away from *HIS4*. Both the *his4-912* and *his4-917* mutations prevent transcription of the *HIS4* gene.

As described previously (Section VII,G), the sequence of the Ty element at *HIS4* can be replaced by the sequence of a different Ty element. Thus, it has been possible to isolate a series of derivatives of the *his4-912* and *his4-917* mutants that differ only by the precise sequence of the Ty. Although these strains contain Ty elements at the same site and in the same orientation as the starting strains, they display considerable variation with respect to *HIS4* expression. Derivatives of the *his4-917* mutant include those that are His$^-$, those that are weakly His$^+$, those that are strongly His$^+$, those that display a cold-sensitive His$^+$ phenotype, and those in which expression of the *HIS4* gene is under mating-type control. Derivatives of the *his4-912* mutant include those that are His4$^-$ and those that are His$^+$, but they do not yet include any that are under mating-type control. These results suggest that the effect that a Ty element has on the expression of the adjacent gene is determined largely by the DNA sequence of the element itself and that different Ty elements can have radically different effects on gene expression.

X. Control of Tys

As described previously, the expression of yeast Ty elements is affected by the mating-type configuration and by the metabolic state of the cell. However, many questions concerning Ty regulation remain unanswered. For example, do Ty elements encode proteins that promote their own transcription, transposition, or recombination? Is Ty recombination promoted by the same enzymes that catalyze general mitotic and meiotic recombination in yeast, or are there special enzymes that act only on Ty sequences?

The presence of more than 30 Tys in the yeast genome renders mutational analysis of these elements difficult. A mutation in one element that destroys a functioning gene is unlikely to have a phenotypic effect because the remaining Ty elements will still be functional. The study of Ty function is further complicated by the lack of suitable assay systems. The transposition of Ty elements is difficult to measure because these events occur at very low frequencies and because the elements lack suitable genetic markers. In spite of these difficulties, some genes affecting Ty expression and recombination have been identified. Three of these genes, defined by the *spm1*$^-$, *spm2*$^-$, and *spm3*$^-$ mutations, have both suppressor and mutator effects on Ty912 and Ty917 (Roeder *et al.*, 1980).

A. Suppressor-Mutators

The *spm*$^-$ mutations were originally detected in His$^+$ revertants of the *his4-912* and *his-917* mutants. The Spms, which are unlinked to *HIS4*, lead to a phenotypic suppression of Ty insertion mutations at the *HIS4* locus. The *spm1*$^-$ mutation suppresses the His$^-$ phenotype of both the *his4-912* and *his4-917* mutations. Thus, strains that carry the *spm1*$^-$ mutation and either the *his4-912* or *his4-917* mutation are phenotypically His$^+$. The *spm2*$^-$ and *spm3*$^-$ mutations suppress the *his4-917* mutation but not the *his4-912* mutation. Strains resulting from Ty912 excision carry a single Ty912 δ in the *HIS4* regulatory region and display a cold-sensitive His$^+$ phenotype. All three Spms suppress this cold sensitivity, leading to growth in the absence of histidine at all temperatures.

The *spm1*$^-$, *spm2*$^-$, and *spm3*$^-$ mutations all affect the frequency with which Ty912 undergoes excision by δ–δ recombination. The *spm1*$^-$ and *spm3*$^-$ mutations lead to a 100-fold increase in the frequency of excision, whereas the *spm2*$^-$ mutation leads to a 1000-fold increase. The effects of the *spm*$^-$ mutations on Ty917 excision or on other kinds of Ty-promoted recombination events have not yet been determined. The Spms do not affect the frequency of *SUP4* deletions due to δ–δ recombination, nor do they affect general meiotic and mitotic recombination frequencies.

The *spm*$^-$ mutations affect both Ty912 excision and transcription of the adjacent *HIS4* gene. One explanation for these diverse functions of the Spms is that the *spm*$^-$ mutations affect more than one gene. They may be regulatory mutations that affect the production of one gene product involved in suppression and another gene product involved in Ty excision. Mutational analysis of the *SPM2* locus and extensive subcloning of the cloned *SPM2* gene have failed to separate suppressor and mutator activities. These observations suggest that a single protein is responsible for all Spm activities. We present a model showing how a single protein might have such diverse effects.

This model proposes that Spms affect the superhelicity of DNA in and around a Ty element. This proposal is based on two observations. First, transcription in many different systems has been shown to be enhanced by supercoiling (Gellert, 1981). Second, supercoiling has been shown to affect several kinds of recombination events, including transposon-mediated, site-specific recombination (Reed, 1981), bacteriophage integration (Nash *et al.*, 1980), mating-type switching (Nasmyth, 1982), and some types of general recombination (Hays and Boehmer, 1978; Raina and Ravin, 1979). The Spms may directly affect supercoiling of DNA, thus indirectly affecting both transcription and recombination of these sequences.

B. RAD52

The *RAD52* gene product is required for mitotic and meiotic gene conversion, for homothallic switching of mating type, and for repair of radiation-induced DNA damage (Game and Mortimer, 1974; Game *et al.*, 1980; Prakash *et al.*, 1980; Malone and Esposito, 1980; Jackson and Fink, 1981). The *rad52* mutation has no effect on the frequency of Ty912 excision in *SPM*$^+$ strains. In the *spm2*$^-$ strain, in which Ty912 excision occurs in 10^{-3} cells, the *rad52* mutation reduces the frequency of recombination 100-fold (D. T. Chaleff and G. R. Fink, unpublished). These observations indicate that there are two distinct pathways of Ty912 excision, one of which occurs in *spm2*$^-$ cells and is *RAD52*-dependent and one of which occurs in *Spm*$^+$ cells and is *RAD52*-independent. An additional difference between these pathways is that the *RAD52*-dependent pathway acts on both Ty912 and Ty912(ura), whereas the *RAD52*-independent pathway acts on Ty912 but not Ty912(ura).

The *rad52* mutation has also been examined for its effects on *CYC1-OSM1-RAD7* deletions due to Ty–Ty recombination. The *rad52* mutation does not alter the frequency of these deletions (Liebman and Downs, 1980).

C. Other Genes Controlling Ty Expression

Several ADHII$^-$ revertants of the *ADR*^{c} insertion mutations carry mutations that are unlinked to the *ADR3* gene. These mutations fall into four complementation groups, known as *tye1–tye4* (Ciriacy and Williamson, 1981). Although *tye* mutations mainly interfere with ADHII synthesis in the insertion mutants, they exert significant effects on ADHII levels in wild-type strains. The *tye1* and *tye2* mutations display a strongly delayed mating-type reaction and are thus similar to *ste* mutations.

Scherer and Davis (1980b) have constructed yeast strains in which a *HIS3* gene lacking its own promoter sits adjacent to a Ty element. These strains are His$^-$ but revert to His$^+$ at high frequency. The mutations leading to His$^+$ are unlinked to *HIS3* and fall into three main classes. Class I mutants display a recessive, cold-sensitive His$^+$ phenotype. Class II mutations are dominant and lead to poor growth under all conditions tested. Class III mutants are temperature-sensitive and lethal.

XI. Summary and Conclusions

The yeast cell has several mechanisms for moving Ty elements to new sites in the genome. One of these is gene conversion, during which the DNA sequence of one Ty element is replaced by the DNA sequence of another Ty element. Another mechanism of movement is excision and reintegration. This involves excision of a Ty from one site in the genome by δ–δ recombination, followed by its reintegration due to recombination with a Ty element elsewhere. An additional means of moving Ty sequences is by the classic transposition process. During transposition, a Ty inserts into a segment of DNA that did not previously contain any Ty or δ sequences. The first two mechanisms of movement involve recombination between homologous DNA sequences. Transposition is a homology-independent process. Ty movement by homology-dependent recombination events occurs approximately 100-fold more frequently than transposition into nonhomologous DNA segments.

The mechanism of Ty transposition remains to be elucidated. Transposition may involve replication of a Ty, as demonstrated for prokaryotic transposons, or it may proceed through an RNA intermediate, as demonstrated for retroviruses. The genes that control Ty transposition and the gene products encoded by the Tys themselves remain to be identified. The study of the mechanism of transposition should be facilitated by the use of genetically marked transposons such as Ty912(ura) and Ty917(ura). These markers should make it possible to follow a Ty through repeated cycles of transposition. However, the absence of a simple assay for transposition and the low frequency of transposition will make these experiments difficult.

One of the most surprising aspects of Ty elements is their heterogeneity. Ty elements differ by many base pair changes and by insertion and substitution mutations. Some of this heterogeneity is the result of gene conversion. A conversion event between Tys that differ at more than one position can produce a Ty that is different from both parental Tys. Although gene conversion can generate new combinations of mutations, it cannot generate new mutations. In fact, repeated cycles of conversion will tend to eliminate mutations and make all of the Tys identical to each other. In order to maintain heterogeneity in spite of continuing cycles of conversion, the Tys must have some means of acquiring new mutations. One possibility is that mutations are generated during the transposition process.

Reciprocal recombination between Tys at different sites in the genome can generate a variety of chromosome aberrations. These include deletions, translocations, inversions, and duplications. Because the yeast genome contains more than 30 Ty elements, the potential for genome rearrangement due to Ty–Ty recombination is very great. Recombination between Ty elements leads not only to DNA rearrangements but also to alterations in the expression of genes near the recombining Tys. Different Ty elements have very different effects on the expression of adjacent genes. When a Ty becomes changed by recombination with a Ty elsewhere, the expression of nearby genes may also be changed. Thus, recombination between Tys can be significant in the evolution of the yeast cell, not only as a means of deleting and duplicating genes but also as a means of altering gene expression.

Alterations in the expression of neighboring genes is the most puzzling aspect of Ty behavior. Clearly, the Tys do not exert these effects by means of transcripts that initiate in the Ty and proceed through the adjacent gene. Molecules that control Ty transcription may effect changes in chromatin structure at the sites of Tys, and these changes may indirectly alter the expression of neighboring genes. The nature of these alterations is far from understood.

Acknowledgments

We thank Janet Bernardo for secretarial assistance. Some of the studies reported in this review were supported by U.S. Public Health Research Grants GM15408 and GM28904-02 from the National Institutes of Health. G. R. Fink is an American Cancer Society Professor of Biochemistry.

References

Baltimore, D. (1981). Gene conversion: some implications for immunoglobulin genes. *Cell* **24,** 592–594.

Bothwell, A. L. M., Paskind, M., Reth, M., Imaniski-Kari, T., Rajewsky, K., and Baltimore, D. (1981). Heavy chain variable region contribution to the NP^b family of antibodies: somatic mutation evident in a γ2a variable region. *Cell* **24,** 625–637.

Cameron, J. R., Loh, E. Y., and Davis, R. W. (1979). Evidence for transposition of dispersed repetitive DNA families in yeast. *Cell* **16,** 739–751.

Chaleff, D. T., and Fink, G. R. (1980). Genetic events associated with an insertion mutation in yeast. *Cell* **21,** 227–237.

Ciriacy, M. (1975a). Genetics of alcohol dehydrogenase in *Saccharomyces cerevisiae*. I. Isolation and genetic analysis of *adh* mutants. *Mutat. Res.* **29,** 315–326.

Ciriacy, M. (1975b). Genetics of alcohol dehydrogenase in *Saccharomyces cerevisiae*. II. Two loci controlling synthesis of the glucose repressible *ADHII*. *Mol. Gen. Genet.* **138,** 157–164.

Ciriacy, M. (1976). Cis-dominant regulatory mutations affecting the formation of glucose-repressible alcohol dehydrogenase (ADHII) in *Saccharomyces cerevisiae*. *Mol. Gen. Genet.* **145,** 327–333.

Ciriacy, M. (1979). Isolation and characterization of further cis- and trans-acting regulatory elements involved in the synthesis of glucose-repressible alcohol dehydrogenase (ADHII) in *Saccharomyces cerevisiae*. *Mol. Gen. Genet.* **176,** 427–431.

Ciriacy, M., and Williamson, V. W. (1981). Analysis of mutations affecting Ty-mediated gene expression in *Saccharomyces cerevisiae*. *Mol. Gen. Genet.* **182,** 159–163.

Deschamps, J., and Wiame, J. (1979). Mating type influence on cis-mutations leading to constitutivity of ornithine transaminase in diploids of *Saccharomyces cerevisiae*. *Genetics* **92,** 749–759.

Dunsmuir, P., Brorein, W. J., Jr., Simon, M. A., and Rubin, G. M. (1980). Insertion of the Drosophila transposable element *copia* generates a 5 base pair duplication. *Cell* **21,** 575–579.

Eibel, H., Gafner, J., Stotz, A., and Philippsen, P. (1980). Characterization of the yeast mobile element Ty1. *Cold Spring Harbor Symp. Quant. Biol.* **45,** 609–617.

Elder, R. T., St. John, T. P., Stinchcomb, D. T., and Davis, R. W. (1980). Studies on the transposable element Ty1 of yeast. I. RNA homologous to Ty1. *Cold Spring Harbor Symp. Quant. Biol.* **45,** 581–584.

Ernst, J. F., Stewart, J. W., and Sherman, F. (1981). The *cycl-11* mutation in yeast reverts by recombination with a nonallelic gene: composite genes determining the iso-cytochromes *c*. *Proc. Natl. Acad. Sci. USA* **78,** 6334–6338.

Errede, B., Cardillo, T. S., Sherman, F., Dubois, E., Deschamps, J., and Wiame, J. M. (1980a). Mating signals control expression of mutations resulting from insertion of transposable repetitive element adjacent to diverse yeast genes. *Cell* **25,** 427–436.

Errede, B., Cardillo, T. S., Wever, G., and Sherman, F. (1980b). Studies on transposable elements in yeast. I. ROAM mutations causing increased expression of yeast genes: their activation by signals directed toward conjugation functions and their formation by insertion of TY1 repetitive elements. *Cold Spring Harbor Symp. Quant. Biol.* **45,** 593–602.

Farabaugh, P. J., and Fink, G. R. (1980). Insertion of the eukaryotic transposable element Ty1 creates a 5-base pair duplication. *Nature (London)* **286,** 352–356.

Fincham, J. R. S., and Shastry, G. R. K. (1974). Controlling elements in maize. *Annu. Rev. Genet.* **8,** 15–50.

Fink, G. R., Farabaugh, P. J., Roeder, G. S., and Chaleff, D. (1980). Transposable elements (Ty) in yeast. *Cold Spring Harbor Symp. Quant. Biol.* **45,** 575–580.

Gafner, J., and Philippsen, P. (1980). The yeast transposon Ty1 generates duplications of target DNA upon insertion. *Nature (London)* **286,** 414–418.

Game, J., and Mortimer, R. (1974). A genetic study of X-ray sensitive mutants in yeast. *Mutat. Res.* **24,** 281–292.

Game, J. C., Zamb, T. J., Braun, R. J., Resnick, M., and Roth, R. M. (1980). The role of radiation (*rad*) genes in meiotic recombination in yeast. *Genetics* **94,** 51–68.

Gellert, M. (1981). DNA topoisomerases. *Annu. Rev. Biochem.* **50,** 879–910.

Grindley, N. D. F., and Sherrat, D. (1978). Sequence analysis at IS1 insertion sites: models for transposition. *Cold Spring Harbor Symp. Quant. Biol.* **43,** 1257–1261.

Hays, H. B., and Boehmer, S. (1978). Antagonists of DNA gyrase inhibit repair and recombination of UV-irradiated phage. λ. *Proc. Natl. Acad. Sci. USA.* **75,** 4125–4129.

Henry, S., Donahue, T., and Culbertson, M. (1975). Selection of spontaneous mutants by inositol starvation in yeast. *Mol. Gen. Genet.* **143,** 5–11.

Jackson, J., and Fink, G. R. (1981). Gene conversion between duplicated genetic elements in yeast. *Nature (London)* **292,** 306–311.

Kingsman, A. J., Gimlich, R., Clarke, L., Chinault, C., and Carbon, J. (1981). Sequence variation in dispersed repetitive sequences in *Saccharomyces cerevisiae*. *J. Mol. Biol.* **145,** 619–632.

Lemoine, Y., Dubois, E., and Wiame, J. M. (1978). The regulation of urea amidolyase of *Saccharomyces cerevisiae*. *Mol. Gen. Genet.* **166,** 251–258.

Liebman, S. W., and Downs, K. M. (1980). The *RAD52* gene is not required for the function of the *DEL1* mutator gene in *Saccharomyces cerevisiae*. *Mol. Gen. Genet.* **179,** 703–705.

Liebman, S. W., Singh, A., and Sherman, F. (1979). A mutator affecting the region of the iso-1-cytochrome *c* gene in yeast. *Genetics* **92,** 783–802.

Liebman, S. W., Shalit, P., and Picologlous, S. (1981). Ty elements are involved in the formation of deletions in *DEL1* strains of *Saccharomyces cerevisiae*. *Cell* **26,** 401–409.

Malone, R., and Esposito, R. (1980). The *RAD52* gene is required for homothallic interconversion of mating types and spontaneous mitotic recombination in yeast. *Proc. Natl. Acad. Sci. USA.* **78,** 503–507.

McKnight, G. L., Cardillo, T. S., and Sherman, F. (1981). An extensive deletion causing overproduction of yeast iso-2-cytochrome *c*. *Cell* **25,** 409–419.

Munz, P., and Leupold, U. (1981). Heterologous recombination between redundant tRNA genes in *Schizosaccharomyces pombe*. *Alfred Benzon Symp.* **16,** 264–275.

Nash, H., Mizuuchi, K., Enquist, L. W., and Weisberg, R. A. (1980). Strand exchange in λ integrative recombination: genetics, biochemistry, and models. *Cold Spring Harbor Symp. Quant. Biol.* **45,** 417–428.

Nasmyth, K. A. (1982). The regulation of yeast mating-type chromatin structure by *SIR:* an action at a distance affecting both transcription and transposition. *Cell* **30,** 567–578.

Prakash, S., Prakash, L., Burke, W., and Montelone, B. A. (1980). Effects of the *RAD52* gene on recombination in *Saccharomyces cerevisiae*. *Genetics* **94,** 31–50.

Raina, J. L., and Ravin, A. W. (1979). Superhelical DNA in *Streptococcus sanguis:* role in recombination *in vivo*. *Mol. Gen. Genet.* **176,** 171–181.

Reed, R. (1981). Transposon-mediated site-specific recombination: a defined *in vitro* system. *Cell* **25,** 713–719.

Roeder, G. S., and Fink, G. R. (1980). DNA rearrangements associated with a transposable element in yeast. *Cell* **21,** 239–249.

Roeder, G. S., and Fink, G. R. (1982). Movement of yeast transposable elements by gene conversion. *Proc. Natl. Acad. Sci. USA* **79,** 5621–5625.

Roeder, G. S., Farabaugh, P. J., Chaleff, D. T., and Fink, G. R. (1980). The origins of gene instability in yeast. *Science (Washington, D.C.)* **209,** 1375–1380.

Rothstein, R. (1979). Deletions of a tyrosine tRNA gene in *S. cerevisiae*. *Cell* **17,** 185–190.

Rothstein, R., and Sherman, F. (1979). Dependence on mating type for the overproduction of iso-2-cytochrome *c* in the yeast mutant *CYC7-H2*. *Genetics* **94,** 891–898.

Scherer, S., and Davis, R. W. (1980a). Recombination of dispersed repeated DNA sequences in yeast. *Science (Washington, D.C.)* **209,** 1380–1384.

Scherer, S., and Davis, R. W. (1980b). Studies on the transposable element Ty1 of yeast. II. Recombination and expression of Ty1 and adjacent sequences. *Cold Spring Harbor Symp. Quant. Biol.* **45,** 584–591.

Scherer, S., Mann, C., and Davis, R. W. (1982). Reversion of a promoter deletion in yeast. *Nature* **298,** 815–819.

Schreier, P. H., Bothwell, A. L. M., Mueller-Hill, B., and Baltimore, D. (1981). Multiple differences between the nucleic acid sequences of the IgG2a^a and IgG2a^b alleles of the mouse. *Proc. Natl. Acad. Sci. USA*. **78,** 4495–4499.

Shalit, P., Loughney, K., Olsen, M., and Hall, B. D. (1981). Physical analysis of the *CYC1-sup4* interval in *Saccharomyces cerevisiae*. *Mol. Cell. Biochem.* **1,** 228–236.

Shoemaker, D., Goff, S., Gilboa, E., Paskind, M., Mitra, S. W., and Baltimore, D. (1980). Structure of a cloned circular Moloney murine leukemia virus DNA molecule containing an inverted segment: implications for retrovirus integration. *Proc. Natl. Acad. Sci. USA*. **77,** 3932–3936.

Slightom, J. L., Blechl, A. L., and Smithies, O. (1980). Human fetal $^G\gamma$- and $^A\gamma$-globin genes: complete nucleotide sequences suggest that DNA can be exchanged between these duplicated genes. *Cell* **21,** 627–638.

Stiles, J. I., Friedman, L. R., and Sherman, F. (1980). Studies on transposable elements in yeast. II. Deletions, duplications, and transpositions of the COR segment that encompasses the structural gene of yeast iso-1-cytochrome *c*. *Cold Spring Harbor Symp. Quant. Biol.* **45,** 602–607.

Stiles, J. I., Friedman, L. R., Helms, C., Consaul, S., and Sherman, F. (1981). Transposition of the gene cluster *CYC1-OSM1-RAD7* in yeast. *J. Mol. Biol.* **148,** 331–336.

Taylor, J. M., and Illmensee, R. (1975). Site on the RNA of an avian sarcoma virus at which primer is bound. *J. Virol.* **16,** 553–558.

Temin, H. (1981). Structure, variation and synthesis of retrovirus long terminal repeat. *Cell* **27,** 1–3.

Tschumper, G., and Carbon, J. (1982). Delta sequences and double symmetry in a yeast chromosomal replicator region. *J. Mol. Biol.* **156,** 293–307.

Van Schaik, N. W., and Brink, R. A. (1959). Transpositions of modulator, a component of the variegated pericarp allele in maize. *Genetics* **44,** 725–738.

Williamson, V. M., Young, E. T., Ciriacy, M. (1981). Transposable elements associated with constitutive expression of yeast alcohol dehydrogenase II. *Cell* **23,** 605–614.

Young, T., Williamson, V., Taguchi, A., Smith, M., Sledziewski, A., Russell, D., Osterman, J., Denis, C., Cox, D., and Beier, D. (1982). The alcohol dehydrogenase genes of the yeast, *Saccharomyces cerevisiae:* isolation, structure, and regulation. *In* ''Genetic Engineering of Microorganisms for Chemicals'' (A. Hollaender, R. D. DeMoss, S. Kaplan, J. Konisky, D. Savage, and R. S. Wolfe, eds.), pp. 335–361. Plenum, New York.

CHAPTER 8

Dispersed Repetitive DNAs in *Drosophila*

GERALD M. RUBIN

I. Introduction

The first evidence for mobile DNA elements in *Drosophila* was provided by the behavior of a small number of unstable mutations. These mutations display genetic properties that are analogous in many respects to those of mutations caused by transposable element insertions in maize and prokaryotes (reviewed in Green, 1976, 1977a, 1980). Such unstable mutations revert to a wild-type phenotype at unusually high rates and generate deletions and other chromosomal rearrangements having one endpoint at the site of the mutation. As discussed in this chapter, recent molecular studies have confirmed that many *Drosophila*

Mobile Genetic Elements

ISBN 0-12-638680-3

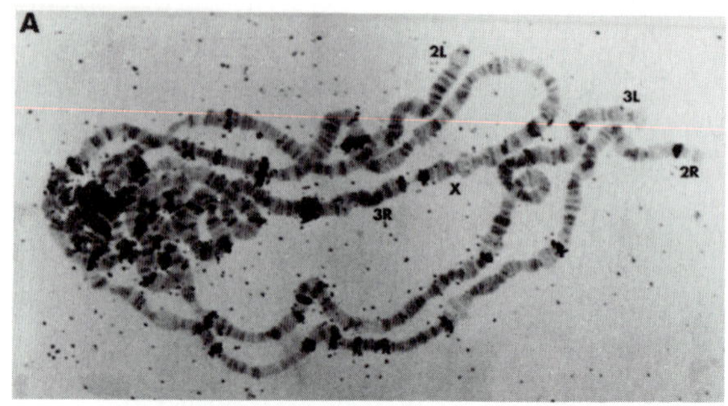

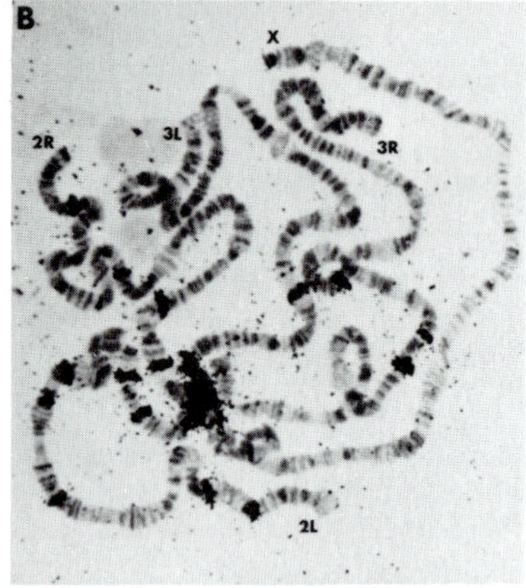

Fig. 1 An example of the use of the *in situ* hybridization technique to determine the genomic location(s) of a cloned DNA sequence. *In situ* hybridization of ^{3}H-labeled *copia* sequences to salivary gland polytene chromosomes of the Seto (A) and Canton S (B) strains of *D. melanogaster* is shown (Strobel *et al.*, 1979). There are 43 sites labeled on the chromosome arms of Seto and only 19 in Canton S. The chromocenter (a region that results from the association of the centric heterochromatin of all chromosomes) in both strains also contains sequences homologous to *copia*. The five major chromosome arms are indicated.

mutations are indeed the result of DNA insertions. Some, but not all, of these insertion mutations are highly unstable. At least three distinct structural classes of transposable elements have been identified in the *Drosophila* genome, and a strong correlation is emerging between the genetic behavior of an insertion mutation and the structure of the responsible transposable element.

Drosophila has proved to be an excellent experimental system in which to study transposable elements. Not only is it amenable to classic genetic approaches, but a method exists for determining the genomic location(s) of any isolated DNA sequence. This technique, called *in situ hybridization,* takes advantage of the presence of giant polytene chromosomes in certain specialized cells of *Drosophila* larvae (see Fig. 1). Using this method, the number and locations of copies of a particular transposable element can be determined rapidly regardless of whether these elements have a recognizable phenotypic effect on the organism.

The first biochemical evidence for mobile *Drosophila* DNA sequences came from observations of polymorphisms in the number and positions of several dispersed, moderately repetitive DNA sequences between different stocks of *D. melanogaster* (Ilyin *et al.*, 1978; Strobel *et al.*, 1979), between cell lines and their parent fly stocks (Potter *et al.*, 1979), and between *D. melanogaster* and its sibling species *D. simulans* (Livak *et al.*, 1978). These observations have been extended to include a large number of other moderately repetitive DNA sequences (Young, 1979; Meselson *et al.*, 1980; Young and Schwartz, 1981; Pierce and Lucchesi, 1981; Dawid *et al.*, 1981). Several of the repetitive DNA sequences were shown to move as discrete, well-conserved DNA elements (Potter *et al.*, 1979). The conclusion from these studies is that the positions occupied by each dispersed, moderately repetitive sequence are highly variable between stocks or species. This is in marked contrast to nonrepetitive sequences the positions of which are relatively quite constant. A similar conclusion is drawn from comparisons of cloned DNA from homologous chromosomal regions of different *D. melanogaster* stocks (W. Bender, P. Spierer, and D. Hogness, personal communication).

In this chapter, dispersed, moderately repetitive sequences will be grouped into classes based primarily on their internal structures. The biochemical and genetic properties of each class will be considered in turn. For a general review of *Drosophila* genome organization, the reader is referred to Spradling and Rubin (1981).

II. copia-Like Elements

One clearly distinct structural class of *Drosophila* middle repetitive sequences is the *copia*-like elements (see Fig. 2). Seven such repeated-sequence families with element sizes ranging from 5 to 8.5 kb have been studied in detail, and their properties are summarized in Table I. Although nonhomologous in nucleotide sequence, they share several properties. Elements from each family carry long direct terminal repeats and occur at approximately 10–100 widely scattered and highly polymorphic locations in the chromosome arms of *D. melanogaster,* as well as in the centric heterochromatin. There is no evidence for tandem arrangement; a single element probably occurs at each site. The sequences of these elements are closely conserved and nonpermuted at each genomic location, and it is in these respects that they differ markedly from other *Drosophila* dispersed, moderately repetitive DNA sequences (see below). Many sequence families in addition to the seven listed in Table I have been described that, although sharing many of the properties of *copia*-like elements, have not been characterized as fully (Rubin *et al.*, 1981; Young, 1979; Pierce and Lucchesi, 1981; Meyerowitz and Hogness, 1981). In total, there may be over 30 different families of *copia*-

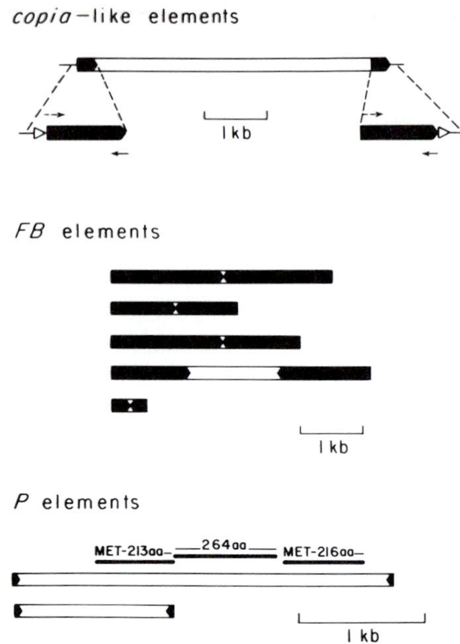

Fig. 2 Summary of the structures of three classes of *Drosophila* transposable elements. The *copia*-like elements carry long direct terminal repeats. Each repeat makes up about 5% of the length of the element. These repeats are shown on an expanded scale below the element to illustrate the presence of short, imperfect inverted repeats at the ends of each long direct repeat (-→ ←) and the presence of a few base pairs of duplicate target sequence (▷) flanking the element that were present in one copy before insertion. The different genomic copies of the elements of one family are very similar in structure to one another.

The *FB* elements comprise a family of heterogeneous but cross-homologous sequences ranging in size from a few hundred base pairs to several kilobases. Each *FB* element carries long terminal inverted repeats. In some cases, the entire element consists of these inverted repeats. In other cases, a central sequence is located between the inverted repeats. The inverted repeat sequences themselves are internally repetitious, having a substructure made up primarily of 31-bp tandem repeats. The number of these 31-bp tandem repeats can differ not only between *FB* elements but also between the termini of a single *FB* element.

The *P* elements have a structure very different from that of both the *copia*-like and *FB* elements. *P* elements carry perfect terminal inverted repeats of 31 bp. A fraction of the *P* elements (about one-third in the one strain examined) are very similar in sequence to one another and are 2.9 kb in length. The remainder of the *P* elements are more heterogeneous, but all appear to have structures that are consistent with their having been derived from the 2.9-kb element by one or more internal deletions. DNA sequence analysis of the 2.9-kb element revealed three long open translational reading frames, which are indicated.

like elements accounting for about half of the moderately repetitive DNA in the *Drosophila* genome.

This class of *Drosophila* elements is strikingly similar in structure to the yeast Ty1 element (see Chapter 7) and to the integrated proviruses of RNA tumor viruses (see Chapter 10).

Table I *copia*-Like Elements

Element	Length (kb)	Length of direct repeat	Copies/genome in *D. melanogaster*	Number of base pairs duplicated upon insertion	Terminal inverted repeat	Copies in cell culture genome/ copies in embryo genome
copia	5.0[a,b]	276 bp[c]	20–60[b]	5[d]	13/17	2, 3[b]
412	7.0[a]	481/571 bp[c]	~40[b]	4[e]	8/10[e]	1, 3[b]
297	6.5[b,f]	412 bp[f]	~30[b]	4[f]	3[f]	3, 5[b]
mdg1[k]	7.2[g]	442/444 bp[g]	20–30[g]	4[g]	13/16[g]	8[g]
mdg3	5.5[h,j]	269 bp[h,j]	15[h,j]	3 or 5[h,j,m]	2 or 15/18[h,j,m]	13[h,j]
B104	8.5[i]	429 bp[i]	100[i]	5[i]	3[i]	N.D.
gypsy	7.3[l]	0.5 kb[l]	~10[l]	N.D.	N.D.	N.D.

[a] Finnegan *et al.* (1978).
[b] Potter *et al.* (1979).
[c] Levis *et al.* (1980).
[d] Dunsmuir *et al.* (1980).
[e] Will *et al.* (1981).
[f] Young and Rubin (manuscript in preparation).
[g] Kulguskin *et al.* (1981).
[h] Bayev *et al.* (1980).
[i] Scherer, Tschudi, and Pirrotta (manuscript in preparation).
[j] Ilyin *et al.* (1980b).
[k] The fragments described in earlier publications as *Dm225* and *Dm234* are parts of the complete *mdg1* element.
[l] W. Bender (personal communication).
[m] Ilyin *et al.* (1980b) interpret their data to indicate that *mdg3* makes a 5-bp duplication upon insertion; however, because the corresponding "empty target site" was not isolated, the data are ambiguous. This author favors the interpretation that the element makes a 3-bp duplication and that the ends of the element correspond to the ends of the direct repeat, which carries 2-bp inverted repeats.

A. Structure

In general, the 10–100 copies of the element of one family are quite close to one another in sequence, as judged by restriction site mapping and electron microscopic heteroduplex analysis (see references in Table I), although variant elements do exist. The prevalence of variant elements ranges from less than 5% for *412* (Rubin *et al.*, 1976; Finnegan *et al.*, 1978) to greater than 20% for *297* (Potter *et al.*, 1979; E. Young and G. M. Rubin, unpublished). Most variant elements that have been analyzed contain predominantly small deletions within the element (Finnegan *et al.*, 1978; Saigo *et al.*, 1981; E. Young and G. M. Rubin, unpublished), but larger deletions have also been detected (E. Young and G. M. Rubin, unpublished).

The elements of each of the seven families listed in Table I have direct terminal repeats of 276–571 bp. The terminal repeats on a single element have been shown to be identical for two elements each of the *copia, mdg1,* and *412* families and for a single element of *mdg3* (see references in Table I). Differences were observed, however, between the terminal repeat sequences of different elements within a family. The repeats on the two *copia* elements examined differ by two base substitutions, whereas the differences between repeats on different *mdg1* and *412* elements are more extensive, involving changes in length as well as base substitutions. The fact that terminal repeats on a single element are identical to each other, but differ between elements, indicates that a mechanism exists for preserving the identity of repeats on a single element. The possible nature of such a mechanism(s) is discussed further below.

The first and last few nucleotides of the direct repeats are imperfect inverted repeats of each other. The length and fidelity of these terminal inverted repeats differ between families (see Table I). The direct repeats of *copia, 412, mdg1,* and *B104,* but not *297* or *mdg3,* carry the terminal dinucleotides 5′TG . . . CA3′. The similarity of the *copia*-like elements to the yeast Ty1 element and to certain provirus genomes can include the nucleotide sequences of their extreme termini. For example, the last 10 bp of *copia* and the yeast Ty1 element are identical, and the terminal repeats of *copia* and the chicken retroviruses, SNV, share the sequences 5′TGT . . . TACAACA3′ (see Levis, *et al.*, 1980). Those proviruses of RNA tumor viruses that have been studied at the DNA sequence level share a number of features. These include (*1*) DNA sequences contained in their direct terminal repeats (LTRs) that are thought to signal the start of RNA transcription and polyadenylation, (*2*) an 18-bp tRNA binding sequence starting 2 bp to the right of the left-hand LTR, and (*3*) an 11-base, purine-rich sequence adjacent to and including the first two bases of the right-hand LTR. Sequences analogous to these shared proviral sequences are found to a varying extent in the *Drosophila copia*-like elements (see Table I for references and Chapter 10 for a detailed list and discussion of these sequence homologies). The biological significance of the above sequence homologies is unclear; they may be relics of a common evolutionary ancestor or convergent evolution to utilize a conserved cellular enzyme system. These alternatives have been discussed more fully by Temin (1980).

The terminal direct repeats of *copia* (Levis *et al.*, 1980), *412* (Will *et al.*, 1981), and *mdg1* (Kulguskin *et al.*, 1981) are rarely, if ever, found separate from the main body of the element. A small number of "free" terminal repeats for the *gypsy* element have been observed (W. Bender, personal communication). It has not been determined whether free terminal repeats derived from other families exist in the genome. The paucity of free repeats observed for *copia*-like elements is somewhat surprising given the fact that such repeats could be generated by a homologous recombination, either between the direct repeats of a single element or between two elements by unequal crossover. Such homologous recombinations occur between directly repeated IS elements in *recA*$^+$ cells of *Escherichia coli* (see Chapter 4) and between the terminal repeats of the yeast element Ty1 (see Chapter 7). In light of these examples, it is unclear why single copies of the direct repeat sequence of *copia* do not accumulate in the genome.

The structural analogy between *copia*-like elements and retroviruses is strengthened by the finding of circular forms of the *copia* element (Flavell and Ish-Horowicz, 1981). A heterogeneous population of small circular DNA molecules is found at low levels (3–40 molecules per cell) in *Drosophila* embryos and tissue culture cells (Stanfield and Helinski, 1976). Stanfield and Lengyel (1979) found that these molecules consist of a major component with a sequence complexity of 18 kb as well as a minor high-complexity component. Preparations of small circular DNA exhibited complementarity to middle repetitive chromosomal sequences, and to nuclear and poly(A)-containing cytoplasmic RNA (Stanfield and Lengyel, 1980). Covalently closed circles of the *copia* element containing either one or two copies of the terminal direct repeat have been found in this DNA fraction at levels corresponding to one copy per 10–50 cells (Flavell and Ish-Horowicz, 1981). Circular forms of the *412* element are also found in this fraction (A. J. Flavell, personal communication); the existence of circular forms of other elements has not been determined. A. J. Flavell (personal communication) has examined the structure of two independently cloned *copia* circles that contain two terminal direct repeats. In both cases, the repeats were separated by a single base pair of non-*copia* DNA. This base pair was different in the two circles. The possible origins of circular forms of *copia*-like elements and their potential role in the mechanism(s) of transposition of these elements are discussed below.

B. Transposition

The genomic positions that can be occupied by the members of a family of *copia*-like elements do not appear to be highly specified. Dispersed repetitive sequences have been found within highly repetitive satellite DNA (Carlson and

Brutlag, 1978), within other moderately repetitive DNA sequences (Finnegan *et al.*, 1978; Tchurikov *et al.*, 1980) including tandemly repeated gene families (Potter *et al.*, 1979; Goldberg, 1979; Saigo *et al.*, 1981), and within nonrepetitive DNA (Potter *et al.*, 1979). Moreover, the DNA sequences of the insertion sites of three *copia* elements have been determined, and they appear to share no common sequences that might function as a recognition signal for insertion (Dunsmuir *et al.*, 1980), although very weak sequence homology has been detected in hybridization experiments between *copia* and the area surrounding some of its insertion sites (Potter *et al.*, 1979; Dunsmuir *et al.*, 1980). In the case of *copia,* it also appears that there is no site that is invariably occupied, as determined by *in situ* hybridization. A comparison of four stocks showed that of the 19 or more euchromatic sites where *copia* sequences are found in each strain, only one site was common to all four strains (Strobel *et al.*, 1979), and this site is not occupied by a *copia* element in another strain (Young and Schwartz, 1981). A tendency to find elements preferentially in certain chromosomal locations has been noted, however (Ananiev *et al.*, 1978; Tchurikov *et al.*, 1980; Young and Schwartz, 1981; Pierce and Lucchesi, 1981; E. Strobel, J. Merriam, J. Lim, and G. M. Rubin, unpublished). There is also evidence for site preference at the level of DNA sequence in the case of the *297* element; it has a strong preference for insertion at the nucleotides ATAT (E. Young and G. M. Rubin, unpublished).

It is not possible to derive a rate of transposition from data on the degree of polymorphism observed in the position of various *copia*-like elements between strains. Moreover, there may not be a single rate for a given element because the stability of these sequences is almost certainly influenced by both genetic background and environmental factors. In cases in which transposition has been looked for in inbred stocks grown under controlled laboratory conditions, the rate of sequence rearrangement is very low or undetectable; however, the laborious nature of the *in situ* hybridization assay technique precludes the examination of large enough numbers of individuals to detect low rates. For example, Pierce and Lucchesi (1981) detected no changes in the position of the dispersed repetitive element Dm25 in several inbred lines during a 1-year period. One case of a gain or loss of a *copia* element within a laboratory stock has been observed (Young and Schwartz, 1981); however, this change could have occurred at any time during the past 50 years. A recent transposition of a *copia* element into the *white* locus was observed (Rubin *et al.*, 1982) following a cross between two strains known to produce hybrid dysgenesis (see Fig. 3 and Chapter 9). It is possible that hybrid dysgenesis, in addition to the dramatic increase it produces in the transposition rate of *P* elements, also increases the rate of transposition of *copia*-like elements. Such an increase in overall transposition rates might be caused by a dysgenesis-induced "genomic shock" similar to the genomic shock induced by the breakage–fusion–bridge cycle in maize (see Chapter 1).

An example of altered transposition rates is provided by the behavior of *copia*-like elements in cell culture lines. Elements of several different families undergo transposition and increase in number severalfold when cells are grown in culture

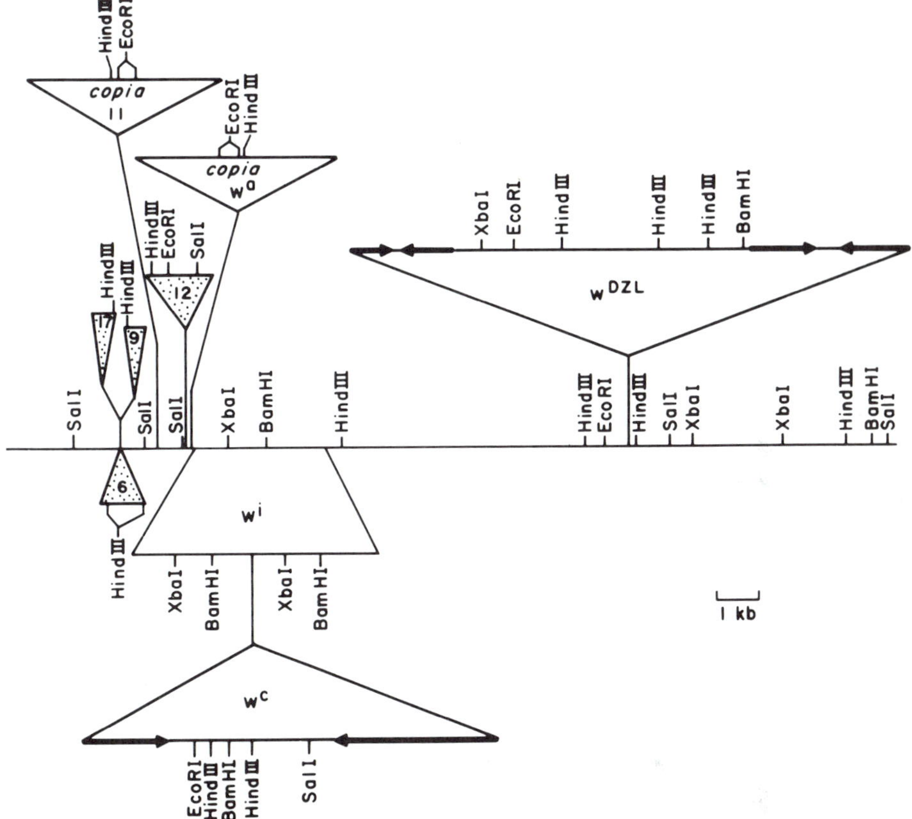

Fig. 3 Summary of the structures of several mutant alleles at the *white* locus (see text for references). The long horizontal line depicts a restriction site map of the wild-type *white* locus. The *white-ivory* (w^i) allele results from a 3-kb tandem duplication of *white* locus sequences. The *white-crimson* (w^c) allele arose by the insertion of a 10-kb *FB* element into the w^i duplication. The *white-dominant zeste-like* (w^{DZL}) allele results from a 13-kb insertion consisting of two *FB* elements flanking a segment of nonrepetitive DNA. The locations of *FB* terminal repeat sequences in the w^c and w^{DZL} insertions are indicated by the heavy arrows. The *white-apricot* (w^a) allele was apparently caused by the insertion of a *copia* element, as was the mutation $w^{hd81b11}$ (11). The insertions causing mutations $w^{\#6}$ (6), $w^{\#12}$ (12), $w^{hd80k17}$ (17), and w^{h381b9} (9) are members of the *P*-element sequence family.

(see Table I and references therein). The rates of transposition of these elements in cell culture, if uniform, would be 10^{-3}–10^{-4} per element per generation (Potter *et al.*, 1979). The increase in copy number of certain elements in cell cultures is in marked contrast to the relatively constant number of these same elements (Strobel *et al.*, 1979) and a wide variety of other dispersed, repeated sequences (Young, 1979) in different fly stocks. These results suggest that the control of transposition rates and element number is altered in cells during adaptation to or growth in cell culture.

Little is known about the mechanism(s) of transposition used by *copia*-like elements. Insertion of a *copia*-like element is accompanied by the duplication of a few base pairs of DNA present once in the target DNA at the site of insertion and found immediately adjacent to the ends of the element after insertion (see Table I). The generation of such a duplication appears to be a universal property of transposable elements in prokaryotes and eukaryotes, and probably results from a staggered cut in the initial cleavage of the target DNA. Two other structural features also yield clues about possible transposition mechanisms. First, the direct repeats of *copia*-like elements are conserved within an element but have diverged between elements. Second, circular forms of the *copia*-like elements exist.

A mechanism must exist to account for the absolute conservation of the sequences of the direct repeats bounding a single element, juxtaposed against a background of divergence between those of different elements of the same family. It is possible that uniformity is maintained by rectification independent of transposition, such as gene conversion. Alternatively, the information of only one of the direct repeats might be used during transposition in the generation of an element with two direct repeats. Two such transposition pathways are suggested by analogy to the known properties of bacterial transposable elements and retroviruses. The transposable element-mediated fusion of two plasmids, one of which contains a single copy of IS*1* or phage Mu, can result in a stable cointegrate in which the transposable element is present at both junctions of the fusion product (see Chapter 4). A circular episome containing the main body of a *copia*-like element with one copy of the direct repeat sequence could be generated, for example, by a homologous recombination between the direct repeats of a chromosomal element. Such circular forms of *copia,* containing one copy of the terminal repeat, have been observed (Flavell and Ish-Horowicz, 1981). The fusion of such a circle at the direct repeat sequence with a new chromosomal target could produce a cointegrate structure with the body of the element flanked by direct repeat sequences, both of which would contain the information from only one of the direct repeats of the parent element. A second mechanism by which rectification of the direct repeat sequences might occur during transposition involves an RNA intermediate in transposition and is based on the reverse transcription of vertebrate retrovirus RNA. The RNA genome of retroviruses contains only a small terminal direct repeat. The DNA provirus that is reverse-transcribed from this template (see Chapter 10) includes much longer terminal direct repeats containing sequences present only once in the RNA. Among the products of reverse transcription are DNA circles containing either one or two copies of the long terminal repeat. As pointed out by Flavell and Ish-Horowicz (1981) circles with two terminal repeats might also be generated by a homologous recombination between the short duplication (5 bp in the case of *copia*) of flanking host DNA. If *copia* circles with two direct repeats are generated by such a homologous recombination, one copy of the flanking 5-bp repeat of host DNA should be present between the direct repeats. Because this 5-bp sequence differs among genomic elements, it would be expected to differ among the circles. However, only a single base pair is found between the two direct repeats in such circles (A. J. Flavell, personal communication). It should also be pointed out that it is not known whether any of the transcripts of *copia*-like elements contain all the sequence information of the element, as would be necessary if reverse transcription were used in transposition. Of course, it is also possible that none of the circular forms play a role in transposition.

C. Gene Products

The RNA transcripts of the *copia*-like elements are unusual in several respects. These transcripts are often highly abundant; *copia* transcripts, for example, can constitute as much as 3% of the polyadenylated RNA in certain cell culture lines (Finnegan *et al.*, 1978). Six of the seven characterized *copia*-like elements listed in Table I are known to be transcribed; the transcription of *gypsy* has not been studied. Both full-length and shorter transcripts have been observed for *copia* (Carlson and Brutlag, 1978; Young and Schwartz, 1981), *mdg1* and *mdg3* (Ilyin *et al.*, 1980a), *412* (Finnegan *et al.*, 1978; Young and Schwartz, 1981), and *B104* (G. Scherer, C. Tschudi, and V. Pirrotta, unpublished). RNA complementary to the full length of *297* has been detected (G. M. Rubin, unpublished), but the number and size(s) of *297* transcripts have not been determined. The transcription of *mdg1* and *mdg3* is not fully asymmetric; about 95% of the transcripts are complementary to one strand and 5% to the other (Georgiev *et al.*, 1981). Georgiev *et al.* (1981) suggest that transcription of both DNA strands may be a common property of all *copia*-like elements. Equivalent measurements for *copia, 412, 297,* and *B104* have not been reported. It is not known whether the low-level transcription of the second strand is due to the presence of a weak promoter within the element or to read-through from a promoter in sequences flanking the element at one or more of its many genomic locations.

In contrast to the transcripts of single-copy genes, only 10–20% of the cytoplasmic *mdg1* and *mdg3* RNAs cosediment with polysomes in a sucrose gradient (Ilyin *et al.*, 1980a). The vast majority of cytoplasmic *mdg1* and *mdg3* RNAs are in smaller complexes with protein that have sedimentation velocities of 80 S. Likewise, most *copia* (Falkenthal and Lengyel, 1980; Young and Schwartz, 1981) and *412* (Young and Schwartz, 1981) transcripts are found in small RNP particles. The nature of these particles is unknown. Moreover, the majority of *412* and *copia* RNA is found associated with nuclei (Young and Schwartz, 1981).

The structure of *copia* transcripts has been studied extensively in several laboratories. The major transcripts are 5 and 2 kb in length (Carlson and Brutlag, 1978), although smaller transcripts have also been detected (Young and Schwartz, 1981; Flavell *et al.*, 1981). The 5- and 2-kb RNAs have structures typical of eukaryotic mRNAs; they are polyadenylated (Falkenthal and Lengyel,

1980; Young and Schwartz, 1981; Flavell *et al.*, 1981) and have modified 5′-termini (Flavell *et al.*, 1981). Both 5- and 2-kb RNAs have heterogeneous, but identical, sets of 5′-termini, and fine scale mapping places these 5′-termini at the same place in the middle of the terminal repeat (Flavell *et al.*, 1981). This is consistent with less precise mapping carried out previously (Flavell *et al.*, 1980; Young and Schwartz, 1981).

Total *copia* RNA can be translated *in vitro* to produce a 51,000-dalton and several smaller polypeptides (Flavell *et al.*, 1980; Falkenthal and Lengyel, 1980). The mRNA activity cosediments with the 2-kb RNA species (Flavell *et al.*, 1980).

The amounts of RNAs encoded by different *copia*-like elements appear to be modulated independently during development. In one study, the steady-state level of total cell *copia*-specific RNA changed severalfold during development, being low in 6- to 8-hr embryos, abundant in larvae, and low in adults (Flavell *et al.*, 1980). The times of onset of accumulation of cytoplasmic polyadenylated RNA homologous to *copia, 412, B104,* and *297* during the 22 hr of embryonic development were determined by Scherer *et al.* (1981). *Copia*-specific RNA begins to accumulate at 10–15 hr, *412* at 15 hr, and *B104* at 2–5 hr. Transcripts of the *297* element are present at low levels in RNA isolated from 0–90-min embryos but disappear at later times, suggesting that they are present in the unfertilized egg but are not transcribed in the embryo.

D. Genetic Effects

Elements such as *copia, 412, 297, mdg1, mdg3, B104,* and *gypsy,* which change their location in the genome, will affect the expression of other genes as long as their movement is not limited to genetically inert sites. There is no indication that the movement of these elements is limited in this way. Moreover, there is now direct evidence that several *Drosophila* mutations result from the insertion of *copia*-like elements. For example, Gehring and Paro (1980) reported suggestive evidence that the *white-apricot* (w^a) mutation, which has a phenotype intermediate between wild type and the bleached-white of *white* null alleles, was caused by the insertion of a *copia* element into the *white* locus. This suggestion was greatly strengthened by two observations. First, fine scale genetic mapping experiments showed that the w^a phenotype and sequence homology to the *copia* element were recombinationally inseparable (Bingham and Judd, 1981). Second, comparison of DNA segments cloned from the *white* locus of wild-type and w^a flies revealed the presence of an inserted complete *copia* element within the *white* locus of the w^a strain (Bingham *et al.*, 1981; Levis *et al.*, 1982a; Goldberg *et al.*, 1982). A recently isolated *white* mutation with a null phenotype, w^{hd81b6}, has also been shown to be the result of a *copia* insertion. The *copia* element inserted in w^{hd81b6} is about 1 kb from the *copia* insertion site in w^a and in the opposite orientation (Rubin *et al.*, 1982; see Fig. 3). Several bithorax mutations, including bx^3 and bxd^1, have been shown to result from *gypsy* insertions (W. Bender, E. Lewis, and D. S. Hogness, unpublished). In the two phenotypic revertants of bxd^1 and one phenotypic revertant of bx^3 examined, a single copy of the *gypsy* terminal repeat is left behind. This is analogous to the reversion of the Ty1-induced *his4-912* mutation in yeast (see Chapter 7). No information exists concerning the structure of revertants of w^a, and no revertants of w^{hd81b6} have been isolated.

The phenotypes of some *copia*-like element insertion mutations are affected by unlinked genes. For example, the w^a allele is suppressed by *suppressor of* w^a, $su(w^a)$ and enhanced by *suppressor of forked, su(f)* (Green, 1959). The effect, if any, of $su(w^a)$ or *su(f)* on w^{hd81b6} has not been determined. The bxd^1 and bx^3 mutations are suppressed by *su(Hw)*, an allele-specific suppressor that is known to suppress mutations in several loci (Lindsley and Grell, 1968). J. Modolell, W. Bender, and M. Meselson (unpublished) have tested the hypothesis that all the mutations suppressible by *su(Hw)* may be due to *gypsy* insertions by carrying out a series of *in situ* hybridizations to chromosomes bearing suppressible mutant loci using *gypsy* sequences as probe. In 11 of 11 cases, the predicted correlation is observed, strongly suggesting that all mutations suppressed by *su(Hw)* are *gypsy* insertions. It is not known, however, whether all *gypsy* insertions are suppressible. Whether or not a mutation is suppressible may well depend on the position of the insertion within the affected locus.

Determination of how these *copia* and *gypsy* insertions produce their phenotypes will require a greater understanding of the expression of the wild-type *white* and *bithorax* loci. However, at least three mechanisms by which *copia*-like elements could exert an effect on gene expression can easily be envisioned. First, the element could act as a simple insertional mutator, disrupting the integrity of a structural gene or transcription unit. Any transposable element that does not exhibit a high degree of site specificity for integration would be expected to act as a mutator in this way. Second, the element could alter the transcription pattern of a gene by the introduction of a strong promoter or RNA processing signal. The structure of the *copia*-like elements suggests that they may be capable of promoting the transcription of downstream genes. The major transcripts of the element begin in one of the two identical direct terminal repeats, raising the possibility that transcription of adjacent sequences may be initiated in the other terminal repeat. There is evidence that transcription of cellular sequences adjacent to integrated retroviruses may occur in this way (see Chapter 10). The observation that the abundance of transcripts from several families of *copia*-like elements changes during development suggests that the insertion of an element may, in some cases, place an adjacent gene under a new developmental program. For example, Ty1 insertions in yeast sometimes place adjacent genes under mating-type control (see Chapter 7). The third mechanism, a type of position effect, may operate when a heavily transcribed, transposable element integrates into an inactive chromosome region. Changes in chromosome structure are known to have an effect on gene expression. For example, the activity of euchromatic loci is often suppressed when they are translocated adjacent to heterochromatic regions of the genome; conversely, certain loci normally located within or very near heterochromatin can be induced to variegate when repositioned in euchromatin or within certain other heterochromatic regions (Spofford, 1976). Transcription of a gene affects the local chromatin structure at that site, and this change could affect either the availability for transcription of the promoter for an adjacent gene or the normal processing of RNA transcripts.

The mechanism(s) by which the *su*(w^a) and *su*(*Hw*) may suppress the effects of at least certain insertions of *copia* and *gypsy* are more difficult to envision. M. Meselson (personal communication) has pointed out that these suppressors are almost completely recessive. The suppressor appears functionally equivalent to a deletion of the wild-type suppressor locus. This suggests that the wild-type suppressor gene produces a diffusible substance that interacts with the transposable element to turn off genes in which, or perhaps adjacent to which, the transposable element resides. It should also be pointed out that the suppressor loci do not appear to be modified elements; DNA sequences homologous to *gypsy* are not present in at least one allele of *su*(*Hw*) (W. Bender, personal communication).

III. Foldback Elements

A. Structure

Potter and co-workers (Potter *et al.*, 1980; Truett *et al.*, 1981) have described a family of transposable elements called *foldback*, or *FB*, elements with a structure very different from that of the *copia*-like elements (see Fig. 2). The members of this dispersed repetitive family have terminal inverted repeats. In some cases, these repeats flank a central region of nonrepeat sequences. The inverted repeats of all family members are homologous; however, both the length of the inverted repeat and the central region are highly variable. The inverted repeats themselves have an unusual structure, being composed primarily of tandem copies of simple sequence DNA. Near the outer ends of each inverted repeat, there are multiple copies of a 10-bp sequence; however, these 10-bp repeats are generally separated by large stretches of more diverse DNA. The precise arrangements may be important because this terminal portion of the inverted repeat appears to be very similar in different elements of the family. At about 300 bp in from the end of the element, the 10-bp repeat expands to 20 bp, and the stretches of DNA between repeats are shorter and less diverse. At 500 bp into the element, the repeat unit again expands to 31 bp, and these repeats are now contiguous and may extend for hundreds of base pairs. The two halves of the inverted repeat in the single element examined are not identical; they differ by both single base changes and insertions/deletions.

Of the nine *FB* elements examined (Potter *et al.*, 1980; Truett *et al.*, 1981), four carried no DNA between the inverted repeats. In four of the five elements with internal DNA, these sequences were homologous to those of the inverted repeats themselves; it is as if one-half of the inverted repeat was larger than the other. The longest internal DNA segment observed was about 1.5 kb.

The *FB* elements exhibit a degree of polymorphism in their genomic positions similar to that of the *copia*-like elements, and again, no transposition rate can be calculated. Most of the potential transposition mechanisms discussed above for the *copia*-like elements are unlikely to apply to the *FB* elements, given the structural differences between these two classes of elements. In the one case analyzed in detail, insertion of an *FB* element created a 9-bp duplication of target DNA sequences (Truett *et al.*, 1981).

It is not known whether there are other families of *Drosophila* transposable elements that carry large inverted repeat sequences distinct from those of the *FB* family. Schmidt *et al.* (1975) estimated that there are 2000–4000 inverted repeat pairs in the *Drosophila* genome accounting for about 3% of the genome. These sequences are slightly enriched in moderately repetitive DNA, but both highly repetitive and nonrepetitive DNA are also found in inverted repeats. Thus, the one characterized *FB* family of sequences could account for about 10% of the 500–1000 inverted repeat pairs that contain moderately repetitive DNA.

B. Genetic Properties

The genetic properties of mutations caused by the insertion of *FB* elements are revealed by the behavior of two such mutations—*white-crimson* (w^c) and *white-Dominant zeste-like* (w^{DZL}). Both are highly unstable mutations at the *white* locus. The *white* locus is a complex genetic locus with alleles affecting both the amount and pattern of pigmentation of the eye. Flies homozygous for deletions of this locus have bleached-white eyes but are fully viable and fertile. A wide variety of perturbations in the normal expression of *white* are thus easily recognized phenotypically. I will first describe the mutation *white-ivory* (w^i) because of its interesting properties and because the highly unstable w^c allele was recovered as a partial phenotypic revertant of w^i. Next, I will discuss w^c and w^{DZL}. Finally, I will review the properties of a large transposable segment of the X chromosome, called *TE*, which usually carries *white* and an adjacent gene. *FB* elements flanking this chromosomal segment appear to be responsible for its movement.

1. White-ivory (w^i)

The *ivory* allele of the *white* locus (w^i) was isolated by Muller (1920) as a spontaneous mutant. Its genetic properties are briefly summarized here (see Bowman and Green, 1966; see Bowman, 1965 for more details). The *ivory* allele results in a faintly tinted eye, somewhat darker in homozygous females than in males, as it is not dosage compensated. The phenotype includes a reduced frequency of recombination between w^i and flanking alleles of *white*. *White-ivory* reverts spontaneously, restoring both the original eye color and the normal recombination frequency within the locus. The frequency of reversion is approximately 5 per 10^5 X chromosomes in homozygous females and about one-tenth this rate in males and deletion heterozygote females. X irradiation results in nearly equal reversion rates in both males and females. Exchange of outside markers does not accompany the reversion of w^i to wild type, and conditions known to augment the frequency of interchromosomal recombinational events fail to increase the frequency of w^i reversion, suggesting that other mechanisms are involved.

Molecular analysis (Karess and Rubin, 1982) revealed that w^i was not caused by an insertion of foreign DNA, but rather was a small tandem duplication of approximately 3 kb within the *white* locus (Fig. 3). The genomic structure of four complete phenotypic revertants of w^i and one partial revertant was examined by Karess and Rubin (1982). In three of the four complete phenotypic revertants the structure of the *white* locus also reverts to wild type, confirming that the 3-kb duplication is responsible for the *ivory* phenotype. The reversion events are more complex in the two other cases, however. The fourth phenotypic reversion to wild type was accompanied by the insertion of approximately 2 kb of moderately repetitive DNA into the *ivory* duplication. Moreover, in addition to this insertion, a portion of the original duplicated 3-kb sequence was removed, presumably in the same event. Likewise, the one partial phenotypic revertant examined also contains an insertion of moderately repetitive DNA into, and a partial deletion of, the duplicated 3-kb sequence. It may be that the 3-kb duplication causes the w^i phenotype by interfering with production of a functional gene transcript and that the DNA insertions found in two of the revertants carry promoters. Evidence supporting this hypothesis is provided by the properties of a second partial phenotypic revertant of w^i, w^c.

2. WHITE-CRIMSON (w^c)

The w^c allele was derived from w^i in a screen for X-ray induced revertants and confers a light reddish-orange eye color as a recessive phenotype (for details of the genetic properties of w^c, see Green, 1967, 1969a,b). w^c mutates at a frequency of about 0.1% to wild-type, w^i, and white-eyed derivatives. The wild-type revertants are mutationally stable and phenotypically indistinguishable from wild type. The w^i revertants are phenotypically identical to the original *white-ivory* and revert to wild type at the same frequency as the original *white-ivory* allele. The *white* derivatives fall into two classes, stable and unstable. The stable derivatives include deletions that have one endpoint at the *white* locus and extend to the left or right of the *white* locus. The unstable *white* derivatives mutate at frequencies similar to those of w^c, generating wild-type, w^i, and w^c derivatives. The w^c derivatives of mutable *white* alleles retain the mutability of the original

w^c. The w^c allele is also capable of transposition from the X chromosome to autosomes. In these new autosomal positions, w^c retains its property of mutability and mutates to *white* and w^{di}, a phenotype lighter than w^c, but never to wild type. Genetic data suggest that only a portion of the *white* locus has been transposed. It may be that only the structural gene but not its normal promoter is transposed. Thus, the transposed segment could never revert to wild type, because when the w^c insertion excises, the w^c promoter is lost.

The production of these w^c derivatives does not appear to be mediated by normal homologous recombination events. Reversions and other derivatives occur in clusters, suggesting a premeiotic origin, and are not accompanied by recombination of flanking markers. Furthermore, mutations occur at similar frequencies in males and females, whereas meiotic recombination is virtually absent in male *Drosophila*. Mutations are almost entirely limited to the germ line, as somatic events are only rarely detected. w^c has maintained the same high level of mutability for 15 years. Collins and Rubin (1982) obtained mutation rates for w^c that are indistinguishable from those reported by Green in 1967. This maintenance of a high mutation rate within a stock distinguishes the w^c mutation from mutations in *Drosophila* that are induced by hybrid dysgenesis. These mutations are unstable only when certain genetic criteria are met (see below).

Comparison of the structure of the w^c allele with its parent chromosome, w^i, indicates that the w^c phenotype results from a 10-kb DNA insertion into the center of the w^i duplication, as shown in Fig. 3 (Collins and Rubin, 1982). One of the more interesting consequences of the w^c insertion is an increase in expression of the *white* locus, as judged by the phenotype. The w^i duplication reduces *white* locus function dramatically, resulting in a very lightly pigmented eye. The insertion of 10 kb into this duplication results in an increase in eye pigmentation, indicating that *white* locus function has been partially restored. It is likely that the w^c insertion contains a promoter that restores partial function by allowing read-through transcription of *white* locus sequences. An alternate explanation is that the w^c insertion may function simply by disrupting the w^i duplication. It is possible that the close juxtaposition of the two copies of the w^i duplication is incompatible with *white* locus function. Preliminary analysis of a *white*-eyed derivative of w^c suggests that disruption of the w^i duplication is not sufficient to enhance gene expression. This allele has apparently been derived from w^c by an element internal rearrangement that leaves the flanking w^i duplication sequences intact (M. Collins and G. M. Rubin, unpublished).

Excision of the w^c insertion occurs frequently, resulting in a phenotypic reversion of w^c to w^i or wild type. In five independent cases of reversion to a w^i phenotype, the insertion has been precisely excised within the experimental limits of about 50 bp (Collins and Rubin, 1982). This precise excision event occurs with the extremely high frequency of 1 per 10^3 X chromosomes (Green, 1967). Precise excision has also been observed for mutations caused by transposable elements in prokaryotes. These mutations are detected as reversion events, and nucleotide sequencing has indicated, for at least some of these revertants, that the excision is truly precise (Kleckner, 1977; Foster *et al.*, 1981). Precise excisions of transposable elements in prokaryotes are quite rare, occurring with a frequency ranging from 1 per 10^6–10^{10} cells (Kleckner, 1977; Calos and Miller, 1980). Mutations caused by the insertion of the transposable element Ty1 into his *his4* locus in yeast revert at a frequency of 1 per 10^5 (see Chapter 7). The reversions are not precise excision events but occur by recombination between the directly repeated δ sequences at the ends of the element, leaving one copy of the δ element behind (see Chapter 7). Truly precise excision of Ty1 from the *his4* mutations has not been detected.

The structure of the w^c element must account for the high frequency of these apparently precise excisions. The ends of the w^c insertion could be recognized as sites for recombination. Alternately, chromosome breaks introduced at the ends of the insertion could remove the insertion; rejoining of the broken ends would restore the w^i duplication. Further structural analysis of the w^c insertion indicates that the ends of the w^c insertion sequence are large inverted repeats composed of tandemly repeated subunits of about 150 bp in length (see below). Excision of the element may involve pairing of these inverted repeats. Nucleotide sequencing of the ends of the w^c insertion and of the w^i duplication before and after excision of the w^c insertion will be needed to elucidate the mechanism of excision.

Excision of the w^c insertion also occurs in reversions of w^c to wild type. An examination of six independent revertants of w^c to wild type indicates that all six occurred by excision of both the insertion sequence and one copy of the w^i duplication, restoring the gene structure to wild type (Collins and Rubin, 1982). Reversion of w^c to wild type could occur by intrachromatid recombination within the w^i duplication and a looping out of the w^c insertion sequence. Reversion of w^i to wild type has been proposed to occur by a similar mechanism (Bowman, 1965; Karess and Rubin, 1982). w^c reverts to wild type nearly two orders of magnitude more frequently than does w^i. If both mutations revert by recombination within the w^i duplication accompanied by a looping out of internal sequences, an explanation must be found for the difference in reversion rates. One could argue that recombination between two copies of small duplication occurs more frequently if the copies are no longer contiguous. Insertion of 10 kb of DNA between the two copies could relieve a topological constraint and allow pairing and recombination to occur more frequently. Alternately, reversion of w^c to wild type could occur by a pathway similar to that of reversion of w^c to w^i. Chromosome breaks introduced at the ends of the w^c insertion could be repaired by recombination within the w^i duplication. Further analysis of wild-type and w^i revertants and other derivatives of w^c should provide more information about mechanisms involved in w^c mutability.

Levis *et al.* (1982b) demonstrated that the w^c insertion is a member of the *FB* element family (see Fig. 3). The w^c insertion is much larger than the *FB* elements studied by Potter and co-workers. As described above, *FB* differ among themselves in the presence of a segment between their inverted repeats and in the sequences of this segment in cases in which one is present. The 4-kb central segment between the inverted repeats of the w^c insertion is present at only about four locations in the w^c genome, as opposed to the inverted repeats, which are present at more than 30 sites. It is not known whether these other copies of the w^c central region are also associated with *FB* inverted repeats.

3. WHITE-DOMINANT ZESTE-LIKE (w^{DZL})

White-dominant zeste-like w^{DZL} was isolated as a spontaneous female mutant having a yellow eye color among a population of red-eyed flies of the Oregon-R wild-type strain (Bingham, 1980). This mutation partially represses the function of the *white* locus in cis and in trans, but apparently only when two copies of the *white* locus are present on chromosome homologues that are able to pair synaptically in the *white* locus region. Females of the genotype w^{DZL}/w^{DZL} or $w^{DZL}/+$ have yellow eyes, whereas w^{DZL}/Y males have reddish-brown eyes.

The recessive z^1 allele of the *zeste* locus interacts with the *white* locus to impart a similar yellow eye color, and similarly requires pairing of two copies of the *white* locus to exert its repression (Jack and Judd, 1979). However, recombination mapping places w^{DZL} at or just beyond the proximal end of the eye color *white* locus on the X chromosome (Bingham, 1980). It is proposed to be an allele of *white* on the basis of its phenotype and proximity, although its dominance precludes tests of complementation.

About 0.1% of the progeny of w^{DZL} flies differ from their parents in phenotype (Bingham, 1981). These include derivatives that are phenotypically wild type and others that are white-eyed. Although the original mutation, as well as many of its derivatives, has no chromosomal abnormalities detectable by examination of polytene chromosomes, some of the derivatives contain deletions or inversions that in all cases have one breakpoint in the region of the *white* locus. The generation of chromosomal rearrangements by w^{DZL} and its instability suggested that this allele results from a DNA insertion at the *white* locus. That w^{DZL} is in fact caused by a DNA insertion has been demonstrated (Levis and Rubin, 1982; P. M. Bingham, personal communication). The only difference observed between the structure of the *white* locus in w^{DZL} and the parent Oregon-R chromosomes is the presence in w^{DZL} of a 13-kb insertion at the position indicated in Fig. 3 (Levis and Rubin, 1982).

The w^{DZL} insertion is composed of two nonidentical *FB* elements, each of which has a pair of inverted repeats, flanking a 6.5-kb central segment (Levis *et al.*, 1982b). In the wild-type strain in which the w^{DZL} mutation arose, this central segment is present as a unique sequence on the second chromosome. In w^{DZL} stocks it is present in two sites—in its original second chromosome site and at *white*. The central segments of w^{DZL} and w^c are nonhomologous (Levis *et al.*, 1982b).

In 12 independent cases examined, Levis and Rubin (1982) found that reversion to a wild-type eye-color phenotype is correlated with the excision of a portion the 13-kb w^{DZL} insertion, thus indicating that the insertion is the cause of the mutation. The portion of the insertion that is excised in these eye-color revertants is heterogeneous in size but probably includes the central 6 kb of the insertion in all cases. Levis *et al.* (1982b) proposed that recombination between the homologous sequences of the terminal *FB* elements of the w^{DZL} insertion is involved in the generation of this class of simple eye-color revertants of w^{DZL}. Recombination between long, directly repeated sequences has been observed in both yeast (see Chapter 7) and bacteria (Chumley and Roth, 1980; Berg *et al.*, 1981). It is the major pathway in yeast for the excision of the *his4-912* and *his4-917* Ty insertion mutations (see Chapter 7) and in *E. coli* for the excision of Tn*9* (MacHattie and Jackowski, 1977). Consistent with this proposal, the w^{DZL} simple eye color revertants appear to have deleted the sequences between the terminal repeats of the insertion (Levis and Rubin, 1982).

Imprecise excisions of the bacterial element Tn*9* and the yeast elements Ty912 and Ty917 leave one copy of the terminal repeat as a remnant at the site of the original insertion. Each of the terminal repeats of the w^{DZL} insertion is itself internally repetitious. Therefore, deletions between different subrepeats within the terminal repeats will give rise to insertion remnants of different lengths. Based on the structure of the terminal repetitions of w^{DZL}, Levis *et al.* (1982b) showed that this model can account for the generation of residual insertions of about 1.9–4.1 kb. Most of the residual insertions in the 12 revertants they examined do fall within this range. The larger residual insertions seen in some revertants could have arisen by expansion of an originally smaller residual insertion by unequal exchanges between the remaining tandem repeats. Whatever mechanism(s) operate to cause the imprecise excision of the w^{DZL} insertion accompanying eye color reversion, genetic evidence indicates that they must be independent of meiotic recombination and occur preferentially in germ cell precursors (Bingham, 1981).

Many revertants of w^{DZL} for eye color maintain their mutability; new eye color alleles appear among their progeny (Levis and Rubin, 1982). The ability of the eye-color reversion events to separate the eye-color and instability phenotypes of w^{DZL} may be a reflection of the bipartite structure of the w^{DZL} insertion; different regions of the w^{DZL} insertion appear to confer its eye-color and instability phenotypes.

The dominance of the w^{DZL} phenotype in both cis and trans is not expected *a priori* for an insertion mutation. Bingham (1980) has elaborated on two basic types of mechanisms to account for the dominance of the w^{DZL} allele. Levis and Rubin (1982) have restated these hypotheses in light of their molecular findings as follows: The first hypothesis is that the simple presence of an insertion of this size at this position may somehow lead to a perturbation of the *white*-locus region that is propagated to the other homologue lacking the insertion. Alternately, the insertion may play an active role in the repression of *white*-locus expression, that is, gene products of the insertion may be involved in this repression. Flies carrying a 13-kb insertion show the mutant phenotype, whereas those carrying a 1.9–6.2-kb subset of this insertion at the same position are wild type in phenotype. This suggests that particular DNA sequences from the central region of the insertion, which are present in the original mutation but absent in eye-color revertants, play some role in the repression of *white*-locus expression.

The position of the insertion associated with the w^{DZL} mutation, 10 kb proximal to the insertion associated with w^{a}, places it at the very edge of the 14-kb region shown to be sufficient to produce a w^{+} phenotype (Levis *et al.*, 1982a). It is quite possible, in fact, that the sequences into which this insertion occurred are not part of the region required for a w^{+} phenotype, but that insertion into this region can nevertheless alter *white*-locus expression by a position effect. The transcribed and protein coding regions of *white* have not yet been localized. For reasons that are not understood, transposable element insertions in yeast often occur outside of structural gene sequences. Their insertion adjacent to genes can nevertheless alter the expression of the gene, increasing it, decreasing it, or altering its regulation (see Chapter 7).

The properties of the w^{c} and w^{DZL} mutations suggest that *FB* elements can excise from the genome and promote chromosomal rearrangements at a high frequency. The w^{c} and w^{DZL} insertions also have a number of different properties that can probably be attributed to differences in the structure of these two insertions and to their different positions within the *white* locus. These points are discussed more fully in Levis *et al.* (1982b).

4. *TE*

The large transposable chromosomal segment, *TE,* is over 100 kb long and usually carries both *white* and the adjacent gene *roughest* (Ising and Ramel, 1976; Ising and Block, 1981). Approximately 150 locations for *TE* have been mapped. Although these sites are widely scattered, they are not randomly distributed, indicating some site preference (Ising and Block, 1981). Moreover, the distribution of these preferred sites depends on the chromosomal position of the donor element. Ising and Block (1981) interpret this result as evidence for a relation between somatic chromosome pairing and the transposition event. The pairing of homologous chromosomes during the prophase and metaphase of somatic divisions in *Drosophila* is well known and probably also occurs during interphase (see Jack and Judd, 1979). At least in some cases, Ising and Block (1981) found that *TE* may transpose into a new position and still remain in the donor position. The transposition of the *TE* element can be followed genetically and is on the order of 10^{-4} transpositions per generation (Ising and Block, 1981).

The mobility of *TE* is most likely dependent on the presence of specific DNA sequences at its two ends. Contrary to the original hypothesis of Gehring and Paro (1980) that the *copia* transposable element is present at the ends of *TE,* Goldberg *et al.* (1982) have reported that *FB* elements are found at or near the ends of *TE* at a number of its locations. This suggests that *FB* elements are responsible for the movement of *TE*. Thus, *TE* would be analogous in structure to the w^{DZL} insertion. Both contain a central DNA segment, which occurs in a unique chromosomal location in most strains, flanked by *FB* elements. The *FB* elements at the two ends of the w^{DZL} insertion differ (Levis *et al.*, 1982b). Those at the ends of *TE* have not been characterized as fully; however, one appears to be very similar to the w^{c} insertion (Levis *et al.*, 1982b). Mobilization of chromosomal sequences by flanking transposable elements, like that described above for *FB* elements, is well established in prokaryotes (see Chapter 4). It is thus likely that any chromosomal segment between two *FB* elements can be mobilized. Given the large number of *FB* elements in the genome and their polymorphic distributions, this hypothesis would predict that, given the proper detection scheme, a *TE* for any chromosomal region could be isolated. This hypothesis could also provide an explanation for the observation that in many cases genes located adjacent to a donor *TE* transpose together with the *TE* to a new site (Ising and Block, 1981). Such transposition events may represent cases in which a nearby *FB* element is used instead of one of the *FB* elements flanking the original *TE*.

IV. P *Elements*

A family of elements, called *P elements,* with structures clearly distinct from those of the *copia*-like or *FB* elements, play a central role in hybrid dysgenesis. Hybrid dysgenesis is a ''syndrome of correlated genetic traits that is spontaneously induced in hybrids between certain mutually interacting strains, usually in one direction only'' (Kidwell *et al.*, 1977). These traits can include sterility, male recombination, mutation, and chromosomal aberration. *D. melanogaster* exhibits at least two independent systems of interacting strains: *I–R*, and *P–M*. Traits induced by *MR* chromosomes (Slatko and Hiraizumi, 1975; Green, 1977b) appear to be closely related or identical to those induced in the *P–M* system. These systems of hybrid dysgenesis are reviewed in detail elsewhere in this volume (Chapter 9). In the *P–M* system, genetic evidence indicates that the genome of certain strains, called *P strains,* contain multiple copies of an apparently mobile element known as the *P factor*. A cross between a male from such a *P* strain to a female from a strain that lacks *P* factors, an *M* strain, results in hybrid dysgenesis. The *P* factors do not produce dysgenesis within the *P* strains, but only when placed in the cytoplasmic background of an *M* strain.

Many of the mutations arising in dysgenic crosses are unstable, suggesting that they are due to DNA insertions (Green, 1977b; Golubovsky *et al.*, 1977; Engels, 1979; Simmons and Lim, 1980). Moreover, the stability of those mutations tested appears to be under the same complex control system as all other manifestations of hybrid dysgenesis (Engels, 1979). They do not revert when maintained in a *P* strain (*P* cytotype), but they may revert at high frequencies when placed in the cytoplasmic background of an *M* strain (*M* cytotype).

These and other observations (reviewed in Chapter 9) led to the proposal that *P–M* hybrid dysgenesis results from the action of a family of transposable elements, the *P* factors. In its simplest form, this hypothesis states that *P* factors are present in *P* strains, where their transposition is repressed, and are absent from *M* strains. When chromosomes carrying *P* factors are placed in the *M* cytotype, it is proposed that these elements become derepressed and transpose at high rates. Among other effects, *P* factors would then induce mutations by inserting into and disrupting genetic loci. Such dysgenesis-induced mutations would be expected to be stable in the *P* cytotype, where *P*-factor transposition is repressed, but unstable in the *M* cytotype, where they could revert by excision of the *P* factor.

Biochemical evidence in support of this hypothesis has recently been obtained; several hybrid dysgenesis-induced mutations in the *white* locus are due to DNA insertions (Rubin *et al.*, 1982). The inserted DNAs are members of a family of DNA sequences, the *P* elements, absent in *M* strains but present in about 50 dispersed copies in *P* strains (Bingham *et al.*, 1982). The transposition rate of these elements is greatly increased under conditions of hybrid dysgenesis (Bingham *et al.*, 1982).

Four *P*-element insertions into the *white* locus have been analyzed in detail (Rubin *et al.*, 1982; see Fig. 3). These four *P* elements, although homologous to each other in sequence, are quite heterogeneous in size, ranging from 0.5 to 1.4 kb in length. These *P* elements, however, all appear to be capable of responding to signals produced during dysgenesis by dramatically increasing the rates at which they both insert into and excise from the *white* locus (Rubin *et al.*, 1982). Perhaps these heterogeneous *P* elements arose by various internal deletions of an ancestral *P* element, which itself might encode the *P*-element-specific transposase. Thus, the transposition of the small elements would require the presence elsewhere in the genome of an intact *P* element. Such a transposable element system would then be formally analogous to the complementation documented in prokaryotes of an element carrying a defective transposase gene by a wild-type copy of the same element (see Chapter 5) and to two element systems, such as *Ac/Ds*, in maize (see Chapter 1).

The above hypothesis predicts not only the existence of a larger conserved *P* element but also that the small, heterogeneous *P* elements retain its ends. Such a conserved *P* element has been isolated and is approximately 2.9 kb in length (G. M. Rubin and K. O'Hare, unpublished; see Fig. 2). These 2.9-kb elements comprise about one-third of the *P* elements in the genome of the one *P* strain examined. The three copies of this element studied show a high level of sequence homogeneity similar to that observed for the *copia*-like elements. Sequence studies on one of these *P* elements demonstrate the presence of 31-bp terminal inverted repeats (K. O'Hare and G. M. Rubin, unpublished). Moreover, the one small *P* element the sequence of which was determined (No. 6 in Fig. 3) retains these 31-bp terminal repeats and differs from the 2.9-kb element by a simple 1.7-kb internal deletion (K. O'Hare and G. M. Rubin, unpublished). Examination of the sequence of the 2.9-kb *P* element reveals the presence of three long open reading frames (see Fig. 2). These may encode the proposed transposase and perhaps also the protein(s) that regulate its synthesis. Moreover, when plasmid DNA containing a cloned copy of this 2.9-kb P element is microinjected into an *M* embryo, transposition of the *P* element into the chromosomes of the embryo can occur (A. C. Spradling and G. M. Rubin, unpublished results).

Three of the *P*-element insertions, those causing mutations $w^{\#6}$, $w^{hd80k17}$, and w^{hd81b9}, occur at tightly clustered or identical sites within the *white* locus, suggesting some insertional site specificity. Site preference for insertion of *P* elements could account in large part for the observed differences in the frequencies at which mutations arise at various genetic loci in dysgenic individuals (see Chapter 9).

Examination of the structures of wild-type phenotypic revertants of four different *P*-element insertion mutations in *white* (Rubin *et al.*, 1982) indicated that phenotypic reversion was accompanied by excision of the *P* element. The excisions appeared to be precise within the limits of resolution of the analysis, which was estimated to be less than ±50 bp. Direct DNA sequence analysis will, of course, be required to determine if the excisions are precise to the nucleotide. The fact that full wild-type phenotype is restored, however, suggests that these excisions were indeed precise. It is important to note, however, that because the *P*-element insertion alleles display an extreme mutant phenotype (bleached-white eye color), any imprecise excisions would have probably escaped detection.

The rates at which *P*-element excisions occur in the *M* cytotype are much higher than those at which precise excision of prokaryotic transposable elements are observed (see Kleckner, 1981). Moreover, *P*-element excision apparently depends on the *M* cytotype, suggesting that, unlike most prokaryotic transposable elements, *P* elements require action of the *P*-element-specific functions to excise. These differences suggest that *P*-element excision and the precise excision of prokaryotic transposable elements occur by different pathways.

V. Other Dispersed Repetitive Families

About 12% of the *Drosophila* genome consists of moderately repetitive DNA, most of which is dispersed throughout the genome (see Spradling and Rubin, 1981 for review). Not all of these dispersed, repetitive sequences fall into one of the three sequence types described above. Moreover, many of these remaining sequence families have at least some properties consistent with their being mobile elements. The structure and mobility of these sequences are reviewed in this section.

A. Insertions Found in Ribosomal Genes

About half of the ribosomal genes in *D. melanogaster* contain a sequence insertion that disrupts the continuity of the 28 S gene (Glover and Hogness, 1977; White and Hogness, 1977; Wellauer and Dawid, 1977; Pellegrini *et al.*, 1977). The insertions have been classified into two types (I and II) depending upon whether or not they contain cleavage sites for the restriction endonuclease *Eco*RI (Wellauer *et al.*, 1978; Dawid *et al.*, 1978; Wellauer and Dawid, 1978). The type I insertions, which do not contain *Eco*RI sites, are found in about 60% of the rDNA units of the X chromosome but are rare or absent on the Y chromosome (Tartoff and Dawid, 1976; Wellauer *et al.*, 1978; Peacock *et al.*, 1981). The major type I insertion is 5 kb in length, although shorter type I insertions of 1.0 and 0.5 kb have been characterized; all of these share homology at their right-hand ends (Wellauer and Dawid, 1978). Type II insertions, which are found in about 15% of the rDNA units of both the X and Y chromosomes (Wellauer *et al.*, 1978; Glover, 1977; Dawid *et al.*, 1978; Roiha and Glover, 1980), vary in size from 1.5 to 4.0 kb, with a predominant class of 3.2 kb (Wellauer *et al.*, 1978; Roiha and Glover, 1980). The type I and type II insertions are nonhomologous in sequence (Roiha and Glover, 1980).

The type I and type II insertions occur within the 28 S rDNA gene at sites 50–80 nucleotides apart (Roiha *et al.*, 1981; I. B. Dawid and M. Rebbert, personal communication). 28S rRNA genes containing 5-kb type I insertions also contain a deletion of 9 bp of an rDNA coding region adjacent to the left-hand side of the insertion (Roiha *et al.*, 1981; P. M. M. Rae, personal communication). In 28 S rRNA genes with 1.0- or 0.5-kb type I insertions, there is a duplication of 11 or 14 bp of an rDNA coding sequence at the ends of the insertion (I. B. Dawid and M. Rebbert, personal communication). In contrast, there is no duplication or deletion of rDNA sequences adjacent to type II insertions (Roiha *et al.*, 1981; I. B. Dawid and M. Rebbert, personal communication). The deletion of sequences adjacent to 5-kb type I insertions would apparently prevent rDNA repeat units containing these type I insertions from contributing to the production of mature rRNA.

Type II insertion sequences appear to be limited to the nucleolus organizer (Roiha and Glover, 1980; Long *et al.*, 1981). In contrast, sequences homologous to the type I insertion also occur in the chromocenter of salivary gland chromosomes and in one band of the fourth chromosome, as revealed by *in situ* hybridization (Dawid and Botchan, 1977; Dawid and Wellauer, 1978; Dawid *et al.*, 1980; Kidd and Glover, 1980; Peacock *et al.*, 1981). The type I insertions in the chromocenter have diverged in sequence from the 5-kb type I insertion found in the 28 S RNA gene. The chromocentral type I sequences are usually interspersed with or interrupted by other middle repetitive sequences, but they sometimes occur in tandem arrays (Kidd and Glover, 1980; Dawid *et al.*, 1981) containing adjacent 28 S rDNA sequences as part of the repeating unit (Roiha *et al.*, 1981). This suggests the possibility that these sequences could insert into intact 28 S genes by homologous recombination.

In *D. virilis*, many rDNA units are interrupted by a DNA sequence inserted at the same site within the 28 S RNA gene, as are the type I insertions of *D. melanogaster* (Barnett and Rae, 1979; P. M. M. Rae, personal communication). The inserted DNA is either 5 kb or 10 kb in length; the 10-kb insertions contain two 5-kb units in tandem. In contrast to the type I insertions of *melanogaster*, there is no loss of 28 S rRNA coding sequences. Rather, the insertion is flanked on both sides by 14 bp of coding sequence present once in the uninterrupted 28 S gene (Rae *et al.*, 1980). The two 5-kb units found in the 10-kb insertion are themselves separated by one copy of this 14-bp sequence, suggesting that the 10-kb insertions may arise from unequal crossing-over between two rDNA repeats with 5-kb insertions (P. M. M. Rae, personal communication). The reciprocal product of this exchange would be an intact 28 S gene, suggesting that genes with insertions are capable of entering the pool of intact genes. As for *D. melanogaster* type I insertions, the *virilis* insertions are not effectively transcribed, and DNA sequences homologous to the insertions are found elsewhere in the chromocenter (P. M. M. Rae, personal communication).

B. Scrambled, Clustered Arrays

Scrambled, clustered arrays of moderately repetitive DNA sequences in which the individual sequence elements are only 300–1000 bp in length have been described by Wensink *et al*. (1979). Each cluster is several kilobases in length and contains many different such sequences; individual sequences are often repeated within a cluster, sometimes in inverted orientation. Different clusters share some, but not all, sequences; however, the arrangement of shared sequences is not conserved between clusters. Furthermore, the results of Wensink *et al*. (1979) suggest that sequences from within a single cluster are found in several hundred clusters located both in the centric heterochromatin and on the chromosome arms. A detailed analysis of four clusters revealed that together they contain representatives of a total of at least 17 different repeated-sequence families. Perhaps as much as one-quarter of the mass of moderately repetitive sequences are found in such clusters. DNA sequences that occur in the "scrambled–clustered" arrangement described by Wensink *et al*. (1979) do not appear to increase in number in cell culture, as do the *copia*-like elements, and on average show less strain polymorphism in their chromosomal locations than do the *copia*-like elements (P. Dunsmuir and G. M. Rubin, unpublished).

Dawid *et al*. (1981) and Pardue and Dawid (1981) have described the properties of several dispersed, moderately repetitive sequences that were originally selected for study because they were interspersed with the chromocentral copies of the type I ribosomal insertions. Like the scrambled, clustered arrays, these sequences are heterogeneous in structure; the homologous sequence elements making up one family have different restriction maps. These sequences are homologous to rare nuclear, heterogeneous poly(A)$^-$ RNAs and are polymorphic in their genomic locations when different *D. melanogaster* strains are compared. Dawid *et al.* (1981) have characterized in detail one such sequence element, 101F, which appears to have inserted into a chromocentral tandem array of type I ribosomal insertions. Although this sequence element is flanked by an apparent 13-bp duplication of target DNA sequences, the element itself contains neither direct nor inverted terminal repeats.

VI. Evolution of Drosophila *Transposable Elements and Their Relationship to the Rest of the Genome*

Most dispersed, repetitive sequences isolated from *D. melanogaster* are found in fewer copies in the genome of the sibling species *D. simulans* (Meselson *et al.*, 1980; Young and Schwartz, 1981; Pierce and Lucchesi, 1981; Martin and Schedl, 1982) and are often totally absent from more distantly related species (Pierce and Lucchesi, 1981; Martin and Schedl, 1982). Martin and Schedl (1982) have studied the distribution of five *D. melanogaster* middle repetitive sequences throughout the genus *Drosophila*. *copia*- and *412*-Homologous sequences were apparently present prior to the several major successive radiations of the *Drosophila* genus that occurred about 40×10^6 years ago, although neither is found in more distantly related dipterans. There are dramatic differences in the copy number of sequences homologous to these elements in the genomes of different fly species. Moreover, the copy number does not necessarily reflect the evolutionary relationship between the different species. For example, the sibling species *D. simulans* has only two to four copies of *copia*-homologous sequences, whereas *D. pseudoobscura* has 30. Comparison of restriction fragments homologous to *copia* found in *D. melanogaster* and other species shows that the *copia* sequences are well preserved only within the immediate sibling species. Despite differences from the *melanogaster* pattern, the restriction fragments hybridizing to *copia* or *412* within a given species tend to be relatively homogeneous. Sequences for *copia* and 412 are sometimes lost entirely. For example, neither is found in *D. hydei*. In contrast to the widespread occurrence of *copia*- and *412*-homologous sequences, sequences homologous to *297* appear only within the *melanogaster* group. The two other dispersed, repetitive sequence analyzed, *T1P56* and the αβ sequence, are not *copia*-like elements. Their distribution is even more restricted than that of *297;* they are found only in the *melanogaster* sibling species, suggesting a recent origin.

It is likely that moderately repetitive sequences in *Drosophila* are constantly being reshuffled. Functional transposable elements may be destroyed by deletion or by the insertion of other elements. The scrambled, clustered repeats found on the chromosome arms and in the centric heterochromatin may be the graveyards of such dead elements. Out of these may emerge new favorable combinations of sequences that can multiply and disperse themselves within the genome by duplicative transposition mechanisms.

The molecular analysis of mutations in several *Drosophila* loci suggests that the majority of spontaneous mutations may be due to insertion of moderately repetitive DNA sequences. The mutations analyzed, however, are not a random sample and were selected by virtue of their unusual phenotypes. It may be that many of these are mutations in control regions rather than structural genes. Studies in yeast (see Chapter 7) suggest that this would strongly bias the sample in favor of insertion mutations. Despite this caveat, it seems likely that a large fraction of spontaneous mutations and chromosomal rearrangements in *Drosophila* are caused by transposable elements. This is perhaps not surprising, given that about 10% of the genome is composed of such sequences.

Drosophila transposable elements appear to be preferentially active in the germ line. Mutations such as w^{c}, w^{DZL}, and many of the hybrid dysgenesis-induced mutations revert at high enough frequencies so that one would expect, at least occasionally, to see somatic mosaics. The absence of such mosaics indicates that their rates of reversion in somatic tissues are much lower than in the germ line. This is not simply due to a difference between mitotic and meiotic divisions; many of the germ-line events occur in clusters, indicating a premeiotic origin.

The properties of the transposable elements described above suggest that the mutations they cause may display a wide range of complex phenotypes. For example, some elements are subject to regulation by unlinked genes, and the expression of many appears to be developmentally modulated. The existence of such regulation indicates that many elements have evolved an interactive, perhaps symbiotic relationship with the rest of the genome.

In addition to their role in increasing genetic variability by mutations and chromosomal rearrangement, some elements, such as those involved in hybrid dysgenesis, also decrease the reassortment of genetic variations by preventing interbreeding between certain populations. More experimental evidence is needed before the importance of these various effects to the population biology and evolution of the elements themselves, as well as the flies and genomes that harbor them, can be assessed.

References

Ananiev, E. V., Gvozdev, V. A., Ilyin, Y. V., Tchurikov, N. A., Georgiev, G. P. (1978). Reiterated genes with varying location in intercalary heterochromatin of *Drosophila melanogaster* polytene chromosomes. *Chromosoma* **70,** 1–17.

Barnett, T., and Rae, P. M. M. (1979). A 9.6 kb intervening sequence in *D. virilis* rDNA, and sequence homology in rDNA interruptions of diverse species of Drosophila and other diptera. *Cell* **16,** 763–775.

Bayev, A. A., Krayev, A. S., Lyubomirskaya, N. V., Ilyin, Y. V., Skryabin, K. G., and Georgiev, G. P. (1980). The transposable element Mdg3 in *Drosophila melanogaster* is flanked with the perfect and mismatched inverted repeats. *Nucleic Acids Res.* **8,** 3263–3273.

Berg, D. E., Egner, C. Hirschel, B. J., Howard, J., Johnsrud, L., Jorgensen, R. A., and Testry, T. D. (1981). Insertion, excision and inversion of Tn5. *Cold Spring Harbor Symp. Quant. Biol.* **45,** 115–123.

Bingham, P. M. (1980). The regulation of white locus expression: a dominant mutant allele at the white locus of *Drosophila melanogaster. Genetics* **95,** 341–353.

Bingham, P. M. (1981). A novel dominant mutant allele at the white locus of *Drosophila melanogaster* is mutable. *Cold Spring Harbor Symp. Quant. Biol.* **45,** 519–525.

Bingham, P. M., and Judd, B. H. (1981). A copy of the *copia* transposable element is very tightly linked to the w^a allele at the white locus of *D. melanogaster. Cell* **25,** 705–711.

Bingham, P. M., Levis, R., and Rubin, G. M. (1981). The cloning of DNA sequences from the *white* locus of *Drosophila melanogaster* using a novel and general method. *Cell* **25,** 693–704.

Bingham, P. M., Kidwell, M. G., and Rubin, G. M. (1982). The molecular basis of *P–M* hybrid dysgenesis: the role of the *P* element, a *P* strain-specific transposon family. *Cell* **29,** 995–1004.

Bowman, J. T. (1965). Spontaneous reversion of the *white-ivory* mutant of *Drosophila melanogaster. Genetics* **52,** 1067–1079.

Bowman, J. T., and Green, M. M. (1966). X-ray induced reversion of the *white-ivory* mutant of *Drosophila melanogaster. Genetics* **37,** 7.

Calos, M. P., and Miller, J. H. (1980). Transposable elements. *Cell* **20,** 579–595.

Carlson, M., and Brutlag, D. L. (1978). One of the copia genes is adjacent to satellite DNA in *Drosophila melanogaster. Cell* **15,** 733–742.

Chumley, F. G., and Roth, J. R. (1980). Rearrangement of the bacterial chromosome using Tn10 as a region of homology. *Genetics* **94,** 1–14.

Collins, M., and Rubin, G. M. (1982). Structure of the Drosophila mutable allele, *white-crimson,* and its *white-ivory* and wild type derivatives. *Cell* **30,** 71–79.

Dawid, I. B., and Botchan, P. (1977). Sequences homologous to the ribosomal insertion occur in the *Drosophila* genome outside the nucleolus organizer. *Proc. Natl. Acad. Sci. USA* **74,** 4233–4237.

Dawid, I. B., and Wellauer, P. K. (1978). Ribosomal DNA and related sequences in *Drosophila melanogaster. Cold Spring Harbor Symp. Quant. Biol.* **42,** 1185–1194.

Dawid, I. B., Wellauer, P. K., and Long, E. O. (1978). Ribosomal DNA in *Drosophila melanogaster.* I. Isolation and characterization of cloned fragments. *J. Mol. Biol.* **126,** 749–768.

Dawid, I. B., Lauth, M., and Wellauer, P. K. (1980). Repetitive DNA elements related to the ribosomal insertion of *Drosophila melanogaster. Miami Winter Symp.* **17,** 217–234.

Dawid, I. B., Long, E. O., Di Nocera, P. P., Pardue, M. L. (1981). Ribosomal insertion-like elements in *Drosophila melanogaster* are interspersed with mobile sequences. *Cell* **25,** 399–408.

Dunsmuir, P., Brorein, W. J., Jr., Simon, M. A., and Rubin, G. M. (1980). Insertion of the Drosophila transposable element *copia* generates a 5 base pair duplication. *Cell* **21,** 575–579.

Engels, W. (1979). Extrachromosomal control of mutability in *Drosophila melanogaster. Proc. Natl. Acad. Sci. USA* **76,** 4011–4015.

Falkenthal, S., and Lengyel, J. A. (1980). Structure, translation, and metabolism of the cytoplasmic copia ribonucleic acid of *Drosophila melanogaster. Biochemistry* **19,** 5842–5850.

Finnegan, D. J., Rubin, G. M., Young, M. W., and Hogness, D. S. (1978). Repeated gene families in *Drosophila melanogaster. Cold Spring Harbor Symp. Quant. Biol.* **42,** 1053–1063.

Flavell, A. J., and Ish-Horowicz, D. (1981). Extrachromosomal circular copies of the eukaryotic transposable element *copia* in cultured Drosophila cells. *Nature (London)* **292,** 591–595.

Flavell, A. J., Ruby, S. W., Toole, J. J., Roberts, B. E., and Rubin, G. M. (1980). Translation and developmental regulation of RNA encoded by the eukaryotic transposable element *copia. Proc. Natl. Acad. Sci. USA* **77,** 7107–7111.

Flavell, A. J., Levis, R., Simon, M., and Rubin, G. M. (1981). The 5′ termini of RNAs encoded by the transposable element copia. *Nucleic Acids Res.* **9,** 6279–6291.

Foster, T. J., Lunblad, V., Hanley-Way, S., Halling, S. M., and Kleckner, N. (1981). Three Tn10-associated excision events: relationship to transposition and role of direct and inverted repeats. *Cell* **23,** 215–227.

Gehring, W. J., and Paro, R. (1980). Isolation of a hybrid plasmid with homologous sequences to a transposing element of *Drosophila melanogaster. Cell* **19,** 897–904.

Georgiev, G. P., Ilyin, Y. V., Chmeliavskaite, V. G., Ryskov, A. P., Kramerov, D. A., Skryabin, K. G., Krayev, A. S., Lukanidin, E. M., and Grigoryan, M. S. (1981). Mobile dispersed genetic elements and other middle repetitive DNA sequences in the genome of *Drosophila* and mouse: transcription and biological significance. *Cold Spring Harbor Symp. Quant. Biol.* **45,** 641–654.

Glover, D. M. (1977). Cloned segment of *Drosophila melanogaster* rDNA containing new types of sequence insertion. *Proc. Natl. Acad. Sci. USA* **7,** 4932–4936.

Glover, D. M., and Hogness, D. S. (1977). A novel arrangement of the 18S and 28S sequences in a repeating unit of *Drosophila melanogaster* rDNA *Cell* **10,** 167–176.

Goldberg, M. L. (1979). Ph.D. Dissertation. Stanford Univ., Stanford, California.

Goldberg, M. L., Paro, R., and Gehring, W. J. (1982). Molecular cloning of the white locus region of Drosophila melanogaster using a large transposable element. *EMBO J.* **1,** 93–98.

Golubovsky, M. D. Ivanouv, Yu. N., and Green, M. M. (1977). Genetic instability in Drosophila melanogaster: putative multiple insertion mutants at the singed bristle locus. *Proc. Natl. Acad. Sci. USA* **74,** 2973–2975.

Green, M. M. (1959). Spatial and functional properties of pseudoalleles at the white locus in Drosophila melanogaster. *Heredity* **13,** 302–315.

Green, M. M. (1967). The genetics of a mutable gene at the white locus of *Drosophila melanogaster. Genetics* **56,** 467–482.

Green, M. M. (1969a). Mapping a *Drosophila melanogaster* "controlling element" by intrallelic crossing over. *Genetics* **61,** 423–428.

Green, M. M. (1969b). Controlling element mediated transpositions of the *white* gene in *Drosophila melanogaster. Genetics* **61,** 429–441.

Green, M. M. (1976). Mutable and mutator loci. *In* "The Genetics and Biology of *Drosophila*" (M. Ashburner and E. Novitski, eds.), Vol. Ib, pp. 929–946. Academic Press, New York and London.

Green, M. M. (1977a). The case for DNA insertion mutations in *Drosophila. In* "DNA Insertion Elements, Plasmids and Episomes" (A. I. Bukhari, J. A. Shapiro, and Adhya, S. L., eds.), pp. 437–445. Cold Spring Harbor Lab., Cold Spring Harbor, New York.

Green, M. M. (1977b). Genetic instability in *Drosophila melanogaster: de novo* induction of putative insertion mutants. *Proc. Natl. Acad. Sci. USA* **74,** 3490–3493.

Green, M. M. (1980). Transposable elements in *Drosophila* and other diptera. *Annu. Rev. Genet.* **14,** 109–120.

Ilyin, Y. V., Tchurikov, N. A., Ananiev, E. V. Ryskov, A. P., Yenikopolov, G. N., Limborska, S. A., Maleeva, N. E., Gvozdev. V. A., and Georgiev, G. P. (1978). Studies on the DNA fragments of mammals and *Drosophila* containing structural genes and adjacent sequences. *Cold Spring Harbor Symp. Quant. Biol.* **42,** 959–969.

Ilyin, Y. V., Chmeliauskaite, V. G., Kulguskin, V. V., and Georgiev, G. P. (1980a). Mobile dispersed genetic element MDG1 of *Drosophila melanogaster:* transcription pattern. *Nucleic Acids Res.* **8,** 5333–5346.

Ilyin, Y. V., Chmeliauskaite, V. G., Ananiev, E. V., and Georgiev, G. P. (1980b). Isolation and characterization of a new family of mobile dispersed genetic elements, *mdg3,* in *Drosophila melanogaster. Chromosoma* **81,** 27–53.

Ising, G., and Block, K. (1981). Derivation-dependent distribution of insertion sites for a *Drosophila* transposon. *Cold Spring Harbor Symp. Quant. Biol.* **45,** 527–544.

Ising, G., and Ramel, C. (1976). The behaviour of a transposing element in *Drosophila melanogaster. In* "The Genetics and Biology of Drosophila" (M. Ashburner and E. Novitski, eds.), Vol. Ib, pp. 947–954. Academic Press, New York and London.

Jack, J. W., and Judd, B. H. (1979). Allelic pairing and gene regulation: a model for the *zeste-white* interaction in *Drosophila melanogaster. Proc. Natl. Acad. Sci. USA* **76,** 1368–1372.

Karess, R., and Rubin, G. M. (1982). A small tandem duplication is responsible for the unstable *white-ivory* mutation in *Drosophila. Cell* **30,** 63–69.

Kidd, S. J., and Glover, D. M. (1980). A DNA segment from *D. melanogaster* which contains five tandemly repeating units homologous to the major rDNA insertion. *Cell* **19,** 103–229.

Kidwell, M. G., Kidwell, J. F., and Sved, J. A. (1977). Hybrid dysgenesis in *Drosophila melanogaster.* A syndrome of aberrant traits including mutation, sterility and male recombination. *Genetics* **86,** 813–833.

Kleckner, N. (1977). Translocatable elements in prokaryotes. *Cell* **11,** 11–23.

Kleckner, N. (1981). Transposable elements in prokaryotes. *Annu Rev. Genet.* **15,** 341–404.

Kulguskin, V. V. Ilyin, Y. V., and Georgiev, G. P. (1981). Mobile dispersed genetic element MDG1 of *Drosophila melanogaster:* nucleotide sequence of long terminal repeats. *Nucleic Acids Res.* **9,** 3451–3464.

Levis, R., and Rubin, G. M. (1982). The unstable w^{DZL} mutation of *Drosophila* is caused by a 13 kilobase insertion that is imprecisely excised in phenotypic revertants. *Cell* **30,** 543–550.

Levis, R., Dunsmuir, P., and Rubin, G. M. (1980). Terminal repeats of the *Drosophila* transposable element *copia:* nucleotide sequence and genomic organization. *Cell* **21,** 581–588.

Levis, R., Bingham, P. M., and Rubin, G. M. (1982a). Physical map of the white locus of *Drosophila melanogaster. Proc. Natl. Acad. Sci. USA* **79,** 564–568.

Levis, R., Collins, M., and Rubin, G. M. (1982b). FB elements are the common basis for the instability of the w^{DZL} and w^{c} *Drosophila* mutations. *Cell* **30,** 551–565.

Lindsley, D. L., and Grell, E. H. (1968). Genetic variation of *Drosophila melanogaster.* Carnegie Inst. Washington, Washington, D.C.

Livak, K. J., Freund, R., Schweber, M., Wensink, P. C., and Meselson, M. (1978). Sequence organization and transcription of two heat shock loci in *Drosophila. Proc. Natl. Acad. Sci. USA* **75,** 5613–5617.

Long, E. O., Rebbert, M. L., and Dawid, I. B. (1981). Structure and expression of ribosomal RNA genes of *Drosophila melanogaster* interrupted by type-2 insertions. *Cold Spring Harbor Symp. Quant. Biol.* **45,** 667–672.

MacHattie, L. A., and Jackowski, J. B. (1977). Physical structure and deletion effects of the chloramphenicol resistance elements Tn9 in phage lambda. *In* "DNA Insertion Elements, Plasmids and Episomes" (A. I. Bukhari, J. A. Shapiro, and S. L. Adhya, eds.), pp. 219–234. Cold Spring Harbor Lab., Cold Spring Harbor, New York.

Martin, G., and Schedl, P. (1982). (manuscript in preparation).

Meselson, M., Dunsmuir, P., Schweber, M., and Bingham, P. (1980). Unstable DNA elements in the chromosomes of *Drosophila. In* "Genes, Cells, and Behavior: a View of Biology Fifty Years

Later'' (H. Horowitz and E. Hutchings, eds.), pp. 88–92. Calif. Inst. Tech., Pasadena. (50th Anniv. Symp.).

Meyerowitz, E. M., and Hogness, D. S. (1982). DNA sequence organization at the 68C *Drosophila* polytene chromosome puff. *Cell* **28,** 165–176.

Muller, H. (1920). Further changes in the white-eye series of *Drosophila* and their bearing on the manner of occurrence of mutation. *J. Exp. Zool.* **31,** 443–473.

Pardue, M. L., and Dawid, I. B. (1981). Chromosomal locations of two DNA segments that flank ribosomal insertion-like sequences in *Drosophila:* flanking sequences are mobile elements. *Chromosoma* **83,** 29–43.

Peacock, W. J., Appels, R., Endow, S., Glover, D. M. (1981). Chromosomal distribution of the major insert in *D. melanogaster* 28 S ribosomal RNA genes. *Genet. Res.* **37,** 209–214.

Pellegrini, M., Manning, J., and Davidson, N. (1977). Sequence arrangement of the rDNA of *Drosophila melanogaster. Cell* **10,** 213–224.

Pierce, D. A., and Lucchesi, J. C. (1981). Analysis of a dispersed repetitive DNA sequence in isogeneic lines of *Drosophila. Chromosoma* **82,** 471–492.

Potter, S. S., Brorein, W. J., Dunsmuir, P., and Rubin, G. M. (1979). Transposition of elements of the *412, copia,* and *297* dispersed repeated gene families in *Drosophila. Cell* **17,** 415–427.

Potter, S. S., Truett, M., Phillips, M., and Maher, A. (1980). Eukaryotic transposable genetic elements with inverted terminal repeats. *Cell* **20,** 639–647.

Rae, P. M. M., Kohorn, B. D., and Wade, R. P. (1980). The 10 kb *Drosophila virilis* 28 S rDNA intervening sequence is flanked by a direct repeat of 14 base pairs of coding sequence. *Nucleic Acids Res.* **8,** 3491–3504.

Roiha, H., and Glover, D. M. (1980). Characterization of complete type II insertions in cloned segments of ribosomal DNA from *Drosophila melanogaster. J. Mol. Biol.* **140,** 341–355.

Roiha, H., Miller, J. R., Woods, L. C., and Glover, D. M. (1981). Arrangements and rearrangements of the sequences flanking the two types of rDNA insertions in *D. melanogaster. Nature (London)* **290,** 749–753.

Rubin, G. M., Finnegan, D. J., and Hogness, D. S. (1976). The chromosomal arrangement of coding sequences in a family of repeated genes. *Prog. Nucleic Acid. Res. Mol. Biol.* **19,** 221–226.

Rubin, G. M., Brorein, W. J., Jr., Dunsmuir, P., Flavel, A. J., Levis, R., Strobel, E., Toole, J. J., and Young, E. (1981). *Copia*-like transposable elements in the *Drosophila* genome *Cold Spring Harbor Symp. Quant. Biol.* **45,** 619–628.

Rubin, G. M., Kidwell, M. G., and Bingham, P. M. (1982). The molecular basis of *P–M* hybrid dysgenesis: the nature of induced mutations. *Cell* **29,** 987–994.

Saigo, K., Millstein, L., and Thomas, C. A., Jr. (1981). The organization of *Drosophila melanogaster* histone genes. *Cold Spring Harbor Symp. Quant. Biol.* **45,** 815–827.

Scherer, G., Telford, J., Baldari, C., and Pirrotta, V. (1981). Isolation of cloned genes differentially expressed at early and late stage of *Drosophila* embryonic development. *Dev. Biol.* **86,** 438–447.

Schmidt, C. W., Manning, J. E., and Davidson, N. (1975). Inverted repeat sequences in the *Drosophila* genome. *Cell* **5,** 159–172.

Simmons, M. T., and Lim, J. K. (1980). Site specificity of mutations arising in dysgenic hybrids of *Drosophila melanogaster. Proc. Natl. Acad. Sci. USA* **77,** 6042–6046.

Slatko, B., and Hiraizumi, Y. (1975). Genetic elements causing male crossing over in *Drosophila melanogaster. Genetics* **81,** 313–324.

Spofford, J. (1976). Position effect variegation in *Drosophila. In* ''The Genetics and Biology of *Drosophila*'' (M. Ashburner and E. Novitski, eds.), pp. 955–1018. Academic Press, New York.

Spradling, A. C., and Rubin, G. M. (1981). *Drosophila* genome organization: conserved and dynamic aspects. *Annu. Rev. Genet.* **15,** 219–264.

Stanfield, S., and Helinski, D. R. (1976). Small circular DNA in *Drosophila melanogaster*. *Cell* **9,** 333–345.

Stanfield, S., and Lengyel, J. A. (179). Small Circular DNA of *Drosophila melanogaster:* chromosomal homology and kinetic complexity. *Proc. Natl. Acad. Sci. USA* **76,** 6142–6146.

Stanfield, S., and Lengyel, J. A. (1980). Small circular deoxyribonucleic acid of *Drosophila melanogaster:* homologous transcripts in the nucleus and cytoplasm. *Biochemistry* **19,** 3873–3877.

Strobel, E., Dunsmuir, P., Rubin, G. M. (1979). Polymorphism in the chromosomal locations of elements of the *412, copia* and *297* dispersed repeated gene families in *Drosophila*. *Cell* **17,** 429–439.

Tartoff, K. D., and Dawid, I. B. (1976). Similarities and differences in the structure of X and Y chromosome rRNA genes in *Drosophila*. *Nature (London)* **263,** 27–30.

Tchurikov, N. A., Zelentosova, E. S., and Georgiev, G. P. (1980). Clusters containing different mobile dispersed genes in the genome of *Drosophila melanogaster*. *Nucleic Acids Res.* **8,** 1243–1258.

Temin, H. (1980). The origin of retroviruses from cellular moveable genetic elements. *Cell* **21,** 599–600.

Truett, M. A., Jones, R. S., and Potter, S. S. (1981). Unusual structure of the FB family of transposable elements in *Drosophila Cell* **24,** 753–763.

Wellauer, P. K., and Dawid, I. B. (1977). The structural organization of ribosomal DNA in *Drosophila melanogaster*. *Cell* **10,** 193–212.

Wellauer, P. K., and Dawid, I. B. (1978). Ribosomal DNA in *Drosophila melanogaster*. II. Heteroduplex mapping of cloned and uncloned rDNA. *J. Mol. Biol.* **126,** 769–782.

Wellauer, P. K., Dawid, I. B., and Tartoff, K. D. (1978). X and Y chromosomal ribosomal DNA of *Drosophila:* comparison of spacers and insertions. *Cell* **14,** 269–278.

Wensink, P. C., Tabata, S., and Pachl, C. (1979). The clustered and scrambled arrangement of moderately repetitive elements in *Drosophila* DNA *Cell* **18,** 1231–1246.

White, R. L., and Hogness, D. S. (1977). R-loop mapping of the 18S and 28S sequences in the long and short repeating units of *Drosophila melanogaster* rDNA. *Cell* **10,** 177–192.

Will, B. M., Bayev, A. A., and Finnegan, D. J. (1981). Nucleotide sequence of terminal repeats of 412 transposable elements of *Drosophila melanogaster*. A similarity to proviral long terminal repeats and its implications for the mechanism of transposition. *J. Mol. Biol.* **153,** 897–915.

Young, M. W. (1979). Middle repetitive DNA: a fluid component of the *Drosophila* genome. *Proc. Natl. Acad. Sci. USA* **76,** 6274–6278.

Young, M. W., and Schwartz, H. E. (1981). Nomadic gene families in *Drosophila*. *Cold Srping Harbor Symp. Quant. Biol.* **45,** 629–640.

CHAPTER 9

Hybrid Dysgenesis Determinants

JEAN-CLAUDE BREGLIANO
MARGARET G. KIDWELL

I. Introduction

Hybrid dysgenesis in *Drosophila melanogaster* has been defined as "a syndrome of correlated genetic traits that is spontaneously induced in hybrids between

Mobile Genetic Elements

ISBN 0-12-638680-3

certain mutually interacting strains, usually in one direction only'' (Sved, 1976; Kidwell and Kidwell, 1976). During the past decade, several groups of workers were surprised to observe one or more of a variety of unusual genetic phenomena when males from natural populations were crossed with females from long-established laboratory strains. These phenomena include high frequencies of partial or complete sterility, male recombination, visible and lethal mutation, reversion of mutations, further mutation of unstable alleles, chromosomal rearrangement, transmission ratio distortion (including sex ratio distortion), chromosomal nondisjunction, and modification of female recombination frequencies. These traits are almost always restricted to the germ line; they are rarely observed in somatic cells.

Two independent systems of interacting strains have been identified (Kidwell, 1979), denoted *I–R* (for *inducer* and *reactive*) and *P–M* (for paternal and maternal). Their independence notwithstanding, these two systems exhibit many striking similarities. Both involve complex interstrain interactions with mobile genetic elements as key components. These elements are able to transpose at very high frequencies under specific conditions and appear to be regulated by genetic mechanisms with an unusual and complex mode of inheritance. In both systems, environmental factors play a major role in the manifestation of dysgenic traits. The most striking difference between the two systems is that dysgenic traits are restricted to females in the *I–R* system but appear in both sexes in the *P–M* system. For detailed reviews of hybrid dysgenesis and associated traits, see Sved (1979), Bregliano *et al.* (1980), Woodruff *et al.* (1982), and Kidwell (1982a).

Hybrid dysgenesis systems have a number of interesting implications for molecular, cellular, and developmental genetics and for population and evolutionary biology. In particular, the study of the regulation of the transposable elements involved in hybrid dysgenesis is relatively advanced compared to that of many other elements in eukaryotic organisms. Hybrid dysgenesis systems, therefore, offer great promise for increased understanding of the regulation of mobile elements in higher organisms. Further, the observed unusual distribution of hybrid dysgenesis determinants in natural and laboratory populations provides an opportunity to gain important insights into the population dynamics and the evolution of mobile element systems.

II. Phenomenology of Hybrid Dysgenesis

A. Two Distinct Systems

Dysgenic traits usually appear in only one of the two possible reciprocal crosses between interacting strains; thus, there are both maternally and paternally contributing categories. The two mutually interacting categories of strains are

Table I Expected Outcomes of Various Interstrain Matings in the I–R System

Male parent	Female parent	
	I (Inducer)	*R (Reactive)*
I	Male and female progeny normal	Male hybrids normal; female hybrids (*SF* females) show dysgenic traits[a]
R	Female hybrids (*RSF* females) and male hybrids normal	Male and female progeny normal

[a] Dysgenic traits include female sterility (*SF*), high mutation frequency, mutational instability, and nondisjunction of the X chromosome.

called *inducer* (*I*) and *reactive* (*R*) in the *I–R* system, and *paternal* (*P*) and *maternal* (*M*) in the *P–M* system, as shown in Tables I and II. There are also neutral strains called *N* and *Q* in the *I–R* and *P–M* systems, respectively. Later, we shall see that these strains do not constitute true third categories; they are merely special cases of one of the two main categories.

The main differences between the two systems may be summarized briefly: (*1*) Dysgenic traits occur in hybrids of both sexes in the *P–M* system, whereas they occur only in females in the *I–R* system. (*2*) The spectrum of dysgenic traits differs in the two systems (Tables I and II). In particular, the sterilities that are characteristic of each system are of different types. In the *I–R* system, although hybrid females lay normal numbers of eggs, a certain proportion of them do not hatch. This kind of sterility has been named *SF* sterility (for *sterilité femelle*). In the *P–M* system, sterility is usually the result of the failure of the gonads to develop. This kind of sterility is called *GD* sterility (for *gonadal dysgenesis*).

The first evidence that the categories of the two systems are independently distributed was provided by Kidwell (1979). She demonstrated that inducer strains may be either *P*, *M*, or *Q*. Subsequently, it was learned that the variation existing within each category of the *I–R* system is also independent of the classification in the *P–M* system. *M* strains may be either strong or weak inducer, strong or weak reactive, or even neutral (M. Kidwell and J. Bregliano, unpublished data). All *P* strains so far characterized are also *I* strains; none is *R*. We do not know whether this restriction reflects a biological constraint or whether it can be explained in terms of sequential evolution of *I* and *P* types in natural populations. As will be shown in Section VII, the distribution of different categories of strains in populations from many parts of the world is related to the dates when these strains were collected in the wild and, to a lesser extent, to their geographical origins. As proposed by Kidwell (1979), a dual terminology has been used to describe *Drosophila* strains: *IP*, *IM*, *IQ*, *RM*, *NM*. The missing type is *RP*. Biochemical evidence for the independence of the two systems has re-

Table II Expected Outcomes of Various Interstrain Matings in the P–M System

	Female parent	
Male parent	*P (Paternal)*	*M (Maternal)*
P	Normal	Male and female hybrids show dysgenic traits[a]
M	Normal[b]	Normal

[a] Dysgenic traits include (*GD*) sterility, high mutation frequency, mutational instability, chromosomal rearrangements, nondisjunction, male recombination, female recombination increase, transmission ratio distortion, and F_2 embryo lethality.
[b] Occasionally, dysgenic traits may be observed in this cross at low but nontrivial frequencies.

cently been obtained (Bingham *et al.*, 1982). Details will be presented in Section III,B.

B. Characterization of Strains

Several hundred strains of various geographical origins have been tested and classified in the *I–R* system on the basis of the *SF* sterility test. As indicated in Table I, reduced fertility appears only in females, called *SF* females, produced by crosses between reactive females and inducer males. The reciprocal cross leads to normally fertile daughters (named *RSF* females). Crosses involving neutral strains and crosses between strains in the same class are also normally fertile (Picard and L'Héritier, 1971; Picard *et al.*, 1972, 1976).

Within the two main classes of strains, there is considerable quantitative variation. On the basis of the hatchability of the eggs laid by young *SF* females, strong and weak stocks in both categories can be distinguished, giving low and high hatchabilities, respectively, when crossed with a standard stock of the appropriate interacting category. The quantitative variation observed within each class is independent of the choice of the standard interacting strain (Bucheton *et al.*, 1976). Several lines of evidence (e.g., see Section V,B) show that neutral (*N*) strains represent one extreme limit of the reactive condition because their level of reactivity is too low to produce any detectable reduction in fertility (Picard, 1978a).

In the *P–M* system, *GD* sterility and other dysgenic traits (Table II) are induced in the germ line of hybrid progeny resulting from matings between *M* strain females and *P* strain males (Kidwell *et al.*, 1977b). Dysgenic traits may sometimes also be observed in the progeny of the reciprocal cross, but only at low frequencies. In nonhybrids, frequencies are normally at a trivial level, although a few exceptions have been observed (e.g., Periquet, 1978). Matings between *Q* and *M* strains do not produce hybrid *GD* sterility in either direction. However, *Q* strain males mated with *M* strain females produce nonreciprocal male recombination, lethal mutations, and other traits characteristic of dysgenesis in the *P–M* system (Simmons *et al.*, 1980; Engels and Preston, 1981a; Kidwell *et al.*, 1981). Based on this and other evidence, the hypothesis has been advanced that *Q* strains represent a special subset of *P* strains possessing most, but not all, *P* strain functions.

In addition to the qualitative distinction between *P* and *M* strains, there is a quantitative continuum of variation within each category. For example, *GD* sterility may vary from essentially 0 to 100% in matings of individual *P* or *M* strains with standard reference strains (Kidwell, 1979).

C. Description of Dysgenic Traits

1. The *I–R* System

Dysgenic events associated with the *I–R* system take place only in the germ line of hybrid females, never in males (Picard *et al.*, 1978; Proust and Prudhommeau, 1982). In addition to *SF* sterility, dysgenic traits include X chromosome nondisjunction and the production of visible mutations (Table I). In the *I–R* system, mutational events appear to occur at meiosis because they generally lead to isolated mutants in the progeny of *SF* females. In some cases, however, small clusters of mutants are found, which apparently result from premeiotic events.

2. The *P–M* System

In addition to *GD* sterility, several other traits have been intensively studied in *P–M* hybrids of both sexes (Fig. 2). Although recombination is not normally observed in the *D. melanogaster* male, low frequencies of male crossing-over have been independently observed many times (e.g., Hiraizumi, 1971). Male recombination is often associated with *GD* sterility (Kidwell and Kidwell, 1975; Yannopoulos, 1978b; Engels and Preston, 1980; Matthews and Gerstenberg, 1979). This appears to be one of the most consistently observed attributes of the *P–M* system. Male recombination may occur concurrently in both the second and third chromosomes (Hiraizumi *et al.*, 1973; Kidwell and Kidwell, 1976; Sved, 1976; Woodruff and Thompson, 1977) and seems to be largely premeiotic (Sved, 1978; Hiraizumi, 1979). The distribution of male recombination breakpoints shows marked differences from the distribution of meiotic breakpoints in female recombination in that a larger proportion of breaks occur in and near the centric heterochromatin (Slatko and Hiraizumi, 1975; Kidwell and Kidwell, 1976). In addition, male recombination does not appear to be affected by factors known to influence normal meiotic crossing-over in females (Woodruff *et al.*, 1978). It is more difficult to determine whether increases in female recombination associated with *P–M* hybrid dysgenesis (Kidwell, 1977; Slatko, 1978a; Sinclair and Green, 1979) are the result of premeiotic or meiotic events.

Several other unusual phenomena have been commonly reported in crosses producing male recombination, notably chromosome breakage, distortion of transmission ratios, and high frequencies of mutation. Frequent chromosome breakage has been observed in spermatocytes of putative *P–M* dysgenic hybrids with male parents from strains from Oklahoma (Henderson *et al.*, 1978) and southern Greece (Yannopoulos, 1978b; Yannopoulos and Zacharopoulov, 1980). It has been argued that chromosome breakage is the basic event underlying most, if not all, dysgenic traits (Thompson and Woodruff, 1978).

Transmission distortion was first observed by Hiraizumi (1971) as a negative correlation between male recombination frequency and the transmission ratio of chromosomes derived from wild and laboratory strains. Similar results were reported by Kidwell and Kidwell (1976) and Kidwell *et al.* (1977). They concluded that in *P–M* dysgenic hybrids, there is a tendency for *P* strain-derived chromosomes to be transmitted less frequently than *M* strain-derived chromosomes. Subsequent studies of chromosomes from *T-007*, a Texas population (Hiraizumi, 1977), have provided further insights into this phenomenon. Distorted transmission in the hybrid male was mainly due to "elimination" of *T-007* (putative *P*) chromosomes among the progeny. No such elimination of the *M* homologues occurred. The negative correlation between male recombination frequency and distortion frequency was due to an increased number of recombinants produced per male and to a decreased number of progeny receiving the *T-007* chromosome. Electron microscopic analyses suggest that several different defects in spermatogenesis may be involved in *P* chromosome loss (Matthews, 1981).

High frequencies of mutation have been observed in a number of experiments. Mutations may produce visible phenotypes (Green, 1977; Golubovsky *et al.*, 1977; Engels, 1979c, 1981b; Simmons and Lim, 1980) or may be lethal (Kidwell *et al.*, 1977a; Simmons *et al.*, 1980; Raymond and Simmons, 1981). Both types tend to be mutationally unstable, reverting frequently to wild-type or mutating further to other unstable alleles. Site specificity has often been observed, details of which will be described later.

3. PHYSIOLOGY OF STERILITY

Because most of the genetic data have been obtained on the basis of the two types of sterility (*SF* and *GD*), it is necessary to describe the main physiological characteristics of these dysgenic traits.

a. SF Sterility. *SF* sterility, resulting from *I–R* interactions, occurs only in hybrid females; their brothers are normally fertile. *SF* females lay a normal number of eggs, but a certain percentage of these fail to hatch. This percentage varies from 100% down to nearly normal values, depending on the age of the egg-laying females and the strength of the interacting strains used as parents. The genotype of the mates of *SF* females has no influence on hatching percentage. Although the unhatched eggs are fertilized and initiate mitosis, their development is arrested between the fifth and eighth cleavage divisions (Picard *et al.*, 1977). This precise timing is a characteristic of *SF* sterility (see Table III). No other female sterility known in *D. melanogaster* exhibits such a uniform timing of embryo death at late cleavage (Gans *et al.*, 1975).

One aspect of the *I–R* interaction, currently being investigated, is the mode of biochemical determination of developmental arrest. A detailed cytological investigation performed by Lavige and Lecher (1982) has shown that the early cleavage divisions of *SF* female eggs appear to be normal, having neither visible chromosomal aberrations nor spindle abnormalities. However, there seems to be a slowing down of the rate of mitosis. The first detectable cytological defect, appearing at the time of embryo death, is the fragmentation of chromosomes into small pieces. Spindle anomalies appear simultaneously.

Until the twelfth cleavage division, normal embryos are in a syncitial state and the nuclei are dividing very rapidly and synchronously. Throughout the cleavage stage nuclear genes are not transcribed (Zalokar, 1976), and development is supported by messengers and proteins produced by the mother, notably by nurse cells of the egg chambers during late oogenesis. Therefore, the developmental arrest of eggs of *SF* females probably results from a maternal effect due to some precise biochemical deficiency in the oocyte pool. The fragmentation of chromosomes is the first observable abnormality, suggesting that this deficiency may involve DNA metabolism.

b. GD Sterility. In contrast to *SF* sterility, hybrids of both sexes may be affected by *GD* sterility resulting from *P–M* interactions (Yannopoulos, 1978b; Engels and Preston, 1979; Schaefer *et al.*, 1979). However, there is a tendency for female sterility frequencies to be somewhat higher than those in males (Engels and Preston, 1979; Kidwell and Novy, 1979). The mating ability of sterile hybrids shows no obvious abnormalities, but the gonads of both sexes typically do not develop beyond an early rudimentary stage (see Table III). In addition, a minority of affected hybrids may possess one rudimentary and one normal gonad. These unilaterally affected individuals may be either sterile or fertile (Schaefer *et al.*, 1979). As in the *I–R* system, the frequency of sterility depends both on the strain cross and on the early developmental temperature.

Very recently, a second type of partial sterility has been observed in the *P–M* system (M. G. Kidwell, unpublished results). This sterility, called *EL* sterility (for *embryo lethality*), results from embryo death in the progeny of both male and female F_1 dysgenic hybrids that have escaped *GD* sterility at an earlier developmental stage. Although *EL* sterility and *SF* sterility are alike, because arrest of development occurs in F_2 embryos, preliminary experiments suggest that the two types may not occur at precisely the same developmental stage and that they differ in a number of other respects, including their sensitivity to environmental factors.

Table III Sequence of Dysgenic Events in the Two Systems of Hybrid Dysgenesis in Relation to Developmental Stage and Germ-Line Activity

Approximate time (hr)[a]	Developmental stage	Germ-line activity[b]		Hybrid dysgenic events and temperature-sensitive periods		
				I–R system	*P–M* system	
0 (F_1)	Fertilization	Fusion of sperm and ovum in egg containing maternally derived cytoplasm		R/I hybrid zygote in R cytoplasm	M/P hybrid zygote in M cytoplasm	
0–1½		8 cleavage divisions				
1½	Egg (F_1)	Pole-cell formation				
6–8		Pole-cell proliferation and migration				
16–21		1–2 germ-cell mitoses				
		Females	Males			
21–49	First-instar larva	8–12 oogonia in primordial ovary	36–38 spermatogonia	♀ heat sensitive for *SF* sterility throughout this period	♀&♂ heat sensitive for *GD* sterility	Induction of *GD* sterility, premeiotic recombination, mutation, etc.
50–71	Second-instar larva					
72–74	Third-instar larva		First spermatocytes appear Spermatocytes begin meiosis			Chromosome fragmentation and other anomalies at ♂ meiosis

120	Puparium formation	Ovariole differentiation has begun		↓	
154		First oocytes appear			
		Oocytes begin meiosis		Mutation at ♀ meiosis	Mutation and recombination at ♀ meiosis
216	Adult (F_1)	Ovaries immature	Testes contain mature spermatocytes	↑	
		Vitellogenesis begins			
264	Egg laying begins	Meiosis completed in laid uninseminated eggs		♀ cold sensitive for *SF* sterility ↓	Reduced nos. of eggs and sperm (*GD* sterility)
272 = 0 (F_2)	Fertilization ↑				Mating behavior normal
0–2½		12–13 cleavage divisions		Embryo death	
	Egg (F_2)				
2½–3		Blastoderm formation		(*SF* sterility) prior to blastoderm formation	Embryo death (*EL* sterility); (specific stage not yet defined)

[a] At 25°C.
[b] Following Bodenstein (1950).

D. Environmental Effects

Two environmental factors have been found to have a great influence on the production of dysgenic traits, especially on sterility. They are aging and temperature.

1. AGING EFFECT

We have seen above that the precise timing of embryo death is one characteristic of *SF* sterility. Another striking observation is that the hatching frequency of eggs increases with the age of egg-laying hybrid females and can eventually reach control values, that is, values obtained with *RSF* females (see Table I) coming from the reciprocal cross (Picard *et al.*, 1977). This unusual maternal aging effect provides a convenient method for the unambiguous identification of *SF* sterility (Fig. 1). Similar aging effects on nondisjunction and lethal mutations have been demonstrated (Picard *et al.*, 1978; Proust and Prudhommeau, 1982). There is also a partly heritable effect of aging on the reactivity level of females that is cumulative over several generations (Bucheton, 1979b). This will be described in detail in Section VI.

Although there is no strong effect of aging on *GD* sterility, some data seem to indicate that there are some similarities between the two systems. A weak tendency has been observed for *GD* sterility to decrease with increasing age of *M* mothers (M. G. Kidwell, unpublished results), as has been described for the *I–R* system (see Section VI). Also, the frequency of male recombination tends to decrease with increasing male age (Kidwell *et al.*, 1977b).

1. TEMPERATURE EFFECTS

Developmental temperature is important in the manifestation of *SF* sterility, *GD* sterility, and some of the other dysgenic traits associated with the *P–M* system.

The effects of temperature on *SF* sterility may be summarized as follows (see also Table III). Various developmental stages of *SF* females have been subjected to heat treatments at 29°C. Picard *et al*. (1977) and Bucheton (1978) reported the effect of heat treatments of various lengths during imaginal life. They observed that treatments of 48 hr or more have important effects on the fertility of *SF* females. The hatching percentage rises sharply after a delay of 24-hr (Fig. 1). When the temperature is returned to 20°C, after about 24 hr hatchability undergoes a sharp decline. These observations reveal the existence of a period during oogenesis, 1 or 2 days before oviposition, during which high temperature has a curative effect. This sensitive period may coincide with vitellogenesis. In contrast, Bucheton (1979a,b) observed that the effect of extended heat treatments earlier in development (larval and pupal stages) consistently reduced the hatching

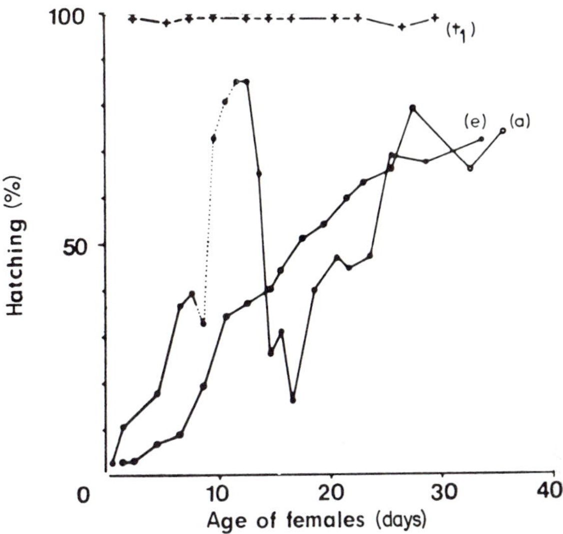

Fig. 1 Change of fertility of *SF* or *RSF* females with aging and thermic treatments. (t_1), *RSF* females maintained at 20°C during their whole life; (a), *SF* females maintained at 20°C during their whole life; (e), *SF* females grown at 30°C between day 8 and day 12 of imaginal life. (From Picard *et al.*, 1977.)

percentage. As in the case of aging, these effects were partly heritable and always reversible.

GD sterility induction is usually highly sensitive to developmental temperature (Engels and Preston, 1979; Kidwell and Novy, 1979). At temperatures below 25°C, the fertility of dysgenic hybrids is not usually affected to an observable degree, but in one or two exceptional cases *GD* sterility occurs at 20°C (Stamatis *et al.*, 1981; M. G. Kidwell, unpublished results). There is typically a narrow critical developmental temperature range between 24° and 29°C in which sterility rises sharply from a low level to a high maximum (Fig. 2). The temperature-sensitive stage for female sterility occurs during late embryonic and early larval development (see Table III). The frequency of *EL* sterility appears to be positively correlated with temperature, but unlike *SF* sterility, there is no curative effect of high temperature immediately prior to egg laying (M. G. Kidwell, unpublished results).

Male recombination and transmission ratio distortion are also affected by temperature. Maximum frequencies of these dysgenic traits were observed at intermediate temperatures (25°C) by Engels (1979a) and Kidwell and Novy (1979), but Yannopoulos and Pelecanos (1977) observed maximum male recombination at high temperatures.

A comparison of the effect of F_1 developmental temperature on the frequency of dysgenic traits in the two systems reveals some striking similarities. In both

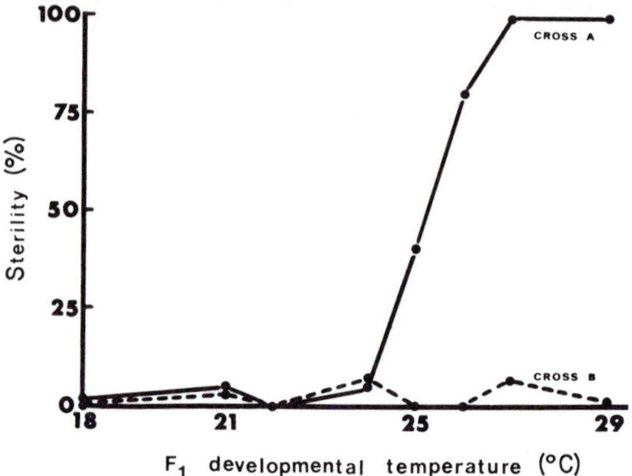

Fig. 2 Percentage of female sterility in the progeny of reciprocal crosses between Canton-S and Harwich strains raised at a range of developmental temperatures. Cross A is the dysgenic cross: Cross B hybrid females have chromosomal complements identical to those of cross A but differ in their maternally derived cytoplasm. (From Kidwell and Novy, 1979.)

systems high temperature during early development tends to increase dysgenesis, particularly the two types of sterility. The mediating effect of high temperature on *SF* sterility during late oogenesis cannot be considered to be a difference between *SF* and *GD* sterility because complete *GD* sterility prevents the existence of this stage in *P–M* hybrids.

III. Inheritance of the Chromosomal Components

The genetic determination of hybrid dysgenesis appears to be surprisingly similar in the two systems, with only minor differences. In both systems, the strains responsible for the paternal contribution (*I* and *P* categories) bear chromosomal determinants that exhibit many characteristics typical of transposable elements.

A. *I* Factors

The genetic determinants of the inducer character are borne by chromosomes and are called *I factors*. Unlike classic markers, they may be linked to any of the four chromosomes (Picard, 1976; Pelisson, 1977). *I* factors act as independent entities. A single *I* factor can induce the maximum frequency of dysgenic traits possible for the level of reactivity of the females used as mates. No cumulative effect has been observed among inducer chromosomes (Pelisson, 1979), in contrast to that observed among *P* factors (Engels, 1979b).

In heterozygous males bearing both inducer and reactive chromosomes, inheritance of *I* factors follows a Mendelian pattern. In heterozygous females, however, the situation is completely different. An interesting phenomenon called *chromosomal contamination* occurs in their germ line (Picard, 1976). In such females, all chromosomes of reactive origin may acquire *I* factors, with a probability that can reach 100%. The results of this process are shown diagrammatically in Fig. 3. A single inducer chromosome is sufficient to contaminate heterologous as well as homologous chromosomes. Chromosomal contamination appears to be irreversible. After being contaminated, a chromosome of reactive origin exhibits all known characteristics of chromosomes of inducer origin, including the ability to contaminate other reactive chromosomes. The evidence suggests that during the contamination process the donor chromosome remains unchanged (Picard, 1979; Pelisson, 1981).

1. Evidence for the Transposability of *I* Factors

There are two possible explanations for chromosomal contamination. One is that contamination results from the derepression of *I* factors carried by all chromosomes but repressed in *R* strains. The other is that *I* factors are mobile sequences the transposition of which is normally repressed in *I* strains but greatly enhanced by reactivity. Several lines of evidence indicate that the second explanation is the correct one.

The first evidence was provided by Pelisson (1978). He used inducer chromosomes differing in their ability to induce *SF* sterility in order to contaminate identical reactive chromosomes. He observed a correlation between the strength of the contaminating chromosomes and the strength of the contaminated ones. This result is most easily explained if contamination involves the transposition of genetic elements.

Other evidence has been provided by the study of mutational events resulting from the *I–R* interaction. (For some of these events, however, the *P–M* interaction cannot be completely excluded.) Of 15 visible sex-linked mutations recovered in females by Picard *et al.* (1978), 14 were associated with recessive lethality. This may mean that they were not point mutations. Other mutations were recovered in male progeny and were therefore not associated with recessive lethals; however, some of these were unstable in reactive females, with reversion frequencies ranging from 0.001 to 0.01. This instability provides indirect circumstantial evidence for the presence of insertions. Proust and Prudhommeau (1982) have analyzed a number of recessive lethal mutations that they obtained from *I* × *R* strain crosses. Several of them are associated with deletions or inversions or tandem duplications visible on salivary gland chromosomes. Mapping experiments show that mutations and chromosomal rearrangements are not

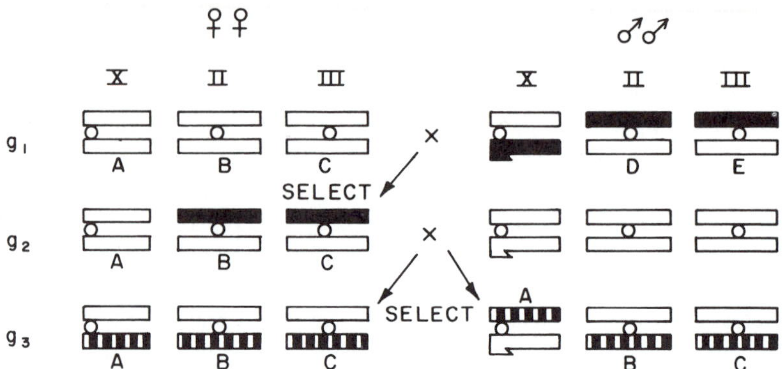

Fig. 3 Schematic representation of chromosomal contamination in the *I–R* system. The g_1 hybrid males are the offspring of matings between reactive (*R*) females and inducer (*I*) males. A, B, C, D, and E denote balancer chromosomes that suppress crossing-over; each carries a dominant marker gene for easy identification. Both homologous and heterologous contamination are possible in g_2 females that have reactive cytoplasm transmitted from their mothers. They do not occur in males. Chromosomal contamination in the *P–M* system could be illustrated in a similar way if maternal (*M*) and paternal (*P*) chromosomes were substituted for *R* and *I* chromosomes in this mating scheme. However, in that case, chromosomal contamination would be possible in both male and female g_2 progeny having *M* cytotype. ■, *I* chromosome; □, *R* chromosome; ▮▮▮, potentially contaminated chromosome of reactive origin.

random, but that there are several hot spots on the X chromosome, notably near the *yellow* and *cut* loci.

The most direct evidence for transposition is provided by a mutant of the *white* locus found by Pelisson (1981). This mutant, denoted w^{IR1}, exhibits a brownish eye color at 20°C. It was recovered in the progeny of an *SF* female on a reactive parental chromosome. The most interesting feature is that it is tightly linked to an *I* factor, detected by the inducer property; it has not been possible to separate the mutation from the *I* factor by recombination. Therefore, it appears very likely that the mutation is due to the insertion of the *I* factor itself. Further evidence is provided by the instability of this mutant when present in *SF* females. In the germ line of such females, it gives rise to less colored and colorless derivatives with a frequency of about 0.001. Moreover, the instability decreases as the *SF* females grow older and, therefore, behaves just like other traits of the *I–R* system in depending on the strength of the *I–R* interaction. Preliminary molecular data show that the w^{IR1} mutant contains a 5-kb insert (A. Bucheton, H. Sang, and D. J. Finnegan, personal communication).

One important question is whether or not the transposition of *I* factors involves their excision from the donor-contaminating chromosomes. Two lines of circumstantial evidence suggest that a copy of the *I* factor remains at the contaminating site after transposition has occurred. First, Picard (1979) showed that a chromosome retained its inducer properties after it had acted as a contaminating element. Second, the mutation frequency of the w^{IR1} mutant is at least 100 times lower than that of the contamination frequency associated with the *I* factor closely linked to this allele (Pelisson, 1981). This implies that most of the time, transposition does not change the w^{IR1} DNA sequence.

Another interesting problem is raised by the association between mutation and chromosomal contamination. The w^{IR1} mutation was the only one of nine *I–R*-produced mutations tested by Pelisson (1981) that was found to be associated with the inducer property. At least two hypotheses may account for this result. All these mutations may result from *I* factor transposition, but the process leading to mutation may also lead to deletion of part of the element, resulting in the loss of ability to induce dysgenic traits. Another possibility is that most of the mutations are produced by movable elements other than *I* factors the transposition of which is also enhanced by the *I–R* interaction.

The results of several experiments indicate that not all *I* factors are functionally equivalent. Picard (1976) and Picard and Pelisson (1979) found in some *I* strains a polymorphism for functional inducer chromosomes (denoted i^+) and other chromosomes that show no detectable inducer activity (denoted i^0). These i^0 chromosomes are stable in their strain of origin but can be contaminated in dysgenic females by i^+ chromosomes. There are two kinds of i^0 chromosomes (Pelisson and Bregliano, 1981). The first type may be considered as true i^0; it does not exhibit any feature of inducer chromosomes, and thus *I* factors may be either absent or functionally defective. The second type is able to change reactive chromosomes into strong i^+ chromosomes by the contamination process. After detailed studies, one chromosome of this category was shown to have a slight inducer potential. For these reasons, this kind of chromosome is now assumed to bear a functionally different *I* factor called I^ϵ. One explanation is that a changed mode of insertion of this *I* factor allows only a low level of expression of inducer properties and that it may recover its complete efficiency after transposition and reinsertion in the normal mode into another chromosome.

Two kinds of data support the idea that varying efficiencies of *I* factors may be due sometimes to transient differences in expression and sometimes to differences that are genetically stable. Pelisson and Bregliano (1981) crossed males from a synthetic I^ϵ stock with females from an *R* strain. They showed that a relatively high frequency of chromosomal contamination took place in the hybrid female progeny, leading to chromosomes with strong inducer properties. This means that the *I* factors borne by i^ϵ chromosomes are not completely genetically defective. This conclusion is consistent with an observation made in another contamination experiment in which some reactive genomes, originating from *SF* females, were found to have escaped contamination (J.-C. Bregliano, unpublished results). However, these genomes spontaneously became strong inducers in succeeding generations. This indicates that $I \rightarrow I^\epsilon$ and $I^\epsilon \rightarrow I$ changes may occur in successive rounds of transposition. The varying efficiencies of these *I* factors cannot be explained by stable genetic differences. They might be explained by either varying insertion patterns or by sequence rearrangements easily modifiable during the transposition process.

Conversely, some other observations support the idea that stable genetic differences between *I* factors might exist. We have already mentioned that in some cases a tendency to maintain the inducer strength of *I* factors through the contamination process has been observed (Pelisson, 1978).

The i^0 and i^ϵ chromosomes are of interest for additional reasons; for example, they allow the mapping of *I* factors by means of females heterozygous for i^+/i^0 chromosomes. The high frequency of contamination precludes such mapping in females heterozygous for inducer and reactive chromosomes, even in *RSF* females. The results obtained by Pelisson and Picard (1979) and Pelisson (1981) show that i^+ chromosomes originating from *I* strains bear only a few functional *I* factors, but their locations vary in different strains. For example, in one strain, a single *I* factor was mapped approximately in the middle of the X chromosome and, in another strain, at least two factors were mapped to the X, one at each end.

Finally, in considering the mode of inheritance of *I* factors, it is noteworthy that every attempt to detect any infectious property (by injection or ovary transplantation) has given negative results (J.-C. Bregliano, J. M. Lavige, N. Plus and J. Proust, unpublished data).

B. *P* Factors

The chromosomal component of the *P–M* interaction consists of one or more chromosomally integrated *P* factors. Evidence has accumulated indicating that *P* factors belong to a somewhat heterogeneous sequence family of elements, dispersed throughout the genome of *P* strains, and that they are mobile under certain conditions. Earlier observations that suggested male recombination could be induced by injection of extracts from male recombination strains (Sochacka and Woodruff, 1976) could not be reproduced (Sved *et al.*, 1978). Together with the negative evidence from the *I–R* system, there is little reason to support the idea of an infectious agent as the basis of hybrid dysgenesis.

Genetic mapping studies have shown that the potential for *GD* sterility may be associated with all the major chromosomes of *P* strains (Engels, 1979b; Engels and Preston, 1980; M. G. Kidwell, unpublished results). However, the genomic location of *P* factors appears to vary widely among different *P* strains. *P*-bearing chromosomes appear to act independently with respect to sterility production, suggesting that they are qualitatively identical irrespective of their genomic location (Engels, 1979b, 1981a).

The pattern of distribution of *MR* (*male recombination*) factors and their inheritance has been shown to be essentially similar to that of *P* factors, as described above. Slatko and Hiraizumi (1975) mapped a major *MR* element to the centromeric region of the second chromosome. Secondary elements with relatively minor effects were also suggested by their results. Later, Slatko (1978b) localized newly induced male recombination activity to the centromeric region of the third chromosome.

Recently, observations at the molecular level have confirmed that *P* elements[1] are dispersed, repeated DNA segments having substantial DNA sequence homology with each other. By *in situ* hybridization, Bingham *et al.* (1982) have found that *P* and *Q* strains carry 30–50 sites of homology to cloned copies of the *P* element, distributed among all the major chromosomes, and that the precise locations of these sites are specific to each *P* or *Q* strain examined. Further, these authors found no sites of homology to the *P* element in most *M* strains, which could be either inducer or reactive in the *I–R* system. These results provide further evidence for the independence of the two systems.

1. EVIDENCE FOR THE TRANSPOSABILITY OF *P* FACTORS

Both direct and indirect evidence that transposable genetic elements are involved in the *P–M* system of hybrid dysgenesis has come from a number of sources over the last 5 years.

On the basis of their mutational properties, including inordinately high spontaneous mutation and reversion frequencies, Green (1977) and Golubovsky *et al.* (1977) claimed that a number of X-linked recessive visible mutants were insertion mutants. The production of these putative insertion mutants was associated with the presence of so-called male recombination chromosomes derived from natural populations. The *sn, ras,* and *y* loci were subject to increased mutation rates, but the *w* and *ct* loci were not.

Slatko (1978b) reported that new male recombination activity occurred when chromosomes of a marker strain were associated both homologously and nonhomologously with *T-007,* a second chromosome exhibiting high male recombination potential. This could not be accounted for by regular meiotic recombination. Matthews *et al.* (1978) demonstrated the presence of new *MR* elements in certain chromosomal regions after homologous association with the same region known to carry major *MR* elements. The new *MR* elements were apparently stable and not capable of further transposition. Furthermore, the new *MR* factors appeared to differ from the originals in that little transmission distortion was associated with them. Yannopoulos (1979) provided evidence for successive transposition of *MR* factors from one autosome to another. The data suggested that the sites of integration were not completely random.

In a screen for mutations in the *zeste-white* region of the X chromosome of *P–M* dysgenic hybrids, Simmons and Lim (1980) identified insertions in two

[1]We wish here to clarify the distinction made in our usage of the terms *P factor* and *P element*. The general term *P element* is used to include all elements with sufficient structural similarities so that DNA sequence homology can be demonstrated among them. It implies nothing concerning functional potential. The term *P factor* is used in relation to *P* elements that do have a demonstrated functional potential, such as sterility or *MR* production. In this sense, all *P* factors are *P* elements, but some *P* elements may not be *P* factors.

different mutant stocks using cytological methods. Many mutations were associated with other structural changes, including inversions and deficiencies. The *zwl* locus was extremely sensitive to mutator activity, in contrast to the apparently insensitive *zw2* locus, suggesting preferential insertion or preferential excision of mobile elements at the *zwl* locus.

Berg *et al.* (1980) observed X chromosome rearrangements in up to 10% of the gametes produced by *P/M* dysgenic hybrids. Aberrations included both simple inversions and complex multibreak rearrangements. Most of the breakpoints occurred at a few precise hot spots, possible points of insertion of mobile *P* factors. Subsequently, Engels and Preston (1981b) presented evidence that the chromosomal positions of at least some *P* factors are indeed chromosome breakage hot spots. These hot spots, in *P*-derived X chromosomes, vanished when their part of the chromosome was replaced by the homologous part from an *M* strain. The hot spots in the nonsubstituted parts of the chromosome remained functional, indicating autonomous action. A new breakage site was observed, coinciding with the appearance of an unstable mutation at the singed locus. There was also evidence of rapid scrambling of hot spot positions when the genome was associated with *M* cytotype for several generations.

Additional indirect evidence for transposition of *P* factors was provided by Raymond and Simmons (1981). They found a statistical coincidence between secondary mutation of an unstable allele at the *singed* locus and X-linked lethal mutation induction in *P–M* dysgenic hybrids. This observation may be interpreted by postulating excision of a *P* factor from the *singed* locus and reinsertion elsewhere in the chromosome, resulting in lethality. However, other interpretations of these results cannot be excluded.

Kidwell (1982b) demonstrated the occurrence of chromosomal contamination in the *P–M* system in a way analogous to Picard's (1976) demonstration in the *I–R* system (see Fig. 3). Under certain circumstances, the transfer of *GD* sterility potential from *P* autosomes to *M* X chromosomes was demonstrated to take place at quite high frequencies. The main results and conclusions of these experiments may be summarized as follows: (*1*) Analogous to the control of *I* factor contamination by reactivity, *M* cytotype was necessary for significant frequencies of *P* factor contamination. In *P* cytotype, contamination frequencies were low or undetectable. (*2*) In contrast to the *I–R* system, there was evidence of chromosomal contamination in males, but the frequency observed was considerably lower than that in females. (*3*) Individual *P/M* hybrid females from the same cross varied widely in their capacity for contamination. (*4*) There was evidence of a negative correlation between chromosomal contamination frequency and certain fitness traits, such as the number of adults produced. (*5*) There was some indication that a moderate developmental temperature (23–25°C) was optimum for the recovery of contaminated chromosomes, but we cannot rule out the possibility that temperature and intrinsic contamination rate are positively related over the entire range of temperature because of the known high incidence of hybrid sterility at high temperatures.

As with chromosomal contamination in the *I–R* system, *P* factor contamination is most readily interpreted in terms of transposition of members of a mobile element family from one chromosomal location to another in the presence of a destabilizing cytoplasmic state (*M* cytotype). This interpretation is considerably strengthened by recent findings on the biochemical nature of a series of *white* locus mutants induced by *P–M* hybrid dysgenesis and not by *I–R* dysgenesis (Rubin *et al.*, 1982). Two *white* mutants generated by Simmons and Lim (1980) in *P* strain chromosomes and *five* white mutants generated in *M* strain chromosomes by a scheme analogous to that used to demonstrate chromosomal contamination (Kidwell, 1982b) were subjected to detailed biochemical analysis. The results of this analysis are described elsewhere (Rubin *et al.*, 1982; see also Chapter 8), and only a summary is included here.

In every case, the occurrence of a mutation was accompanied by an insertion of non-*white* locus DNA into the *white* locus at approximately the same location in the left-hand side of the locus. Five of the insertions responsible for the *white* mutants were shown to have sequence homology with one another. There appears to be considerable variation in the size of these inserted *P* elements (see Chapter 8). The three smallest inserts (approximately 0.5 kb) differed in their internal restriction sites, despite their similar length and strong common homology. A high frequency of reversion to wild type was observed in all mutations identified as *P* insertions. Reversion is apparently accompanied by complete excision of the element.

Two independent mutants isolated in the same screening procedure were observed to have the *copia* element inserted into the same site in the *white* locus (Rubin *et al.*, 1982). In contrast to the high reversion rates of *P* element insertions in *M* cytotype but not *P* cytotype, the *copia* insertions in the *white* locus were never observed to revert in either cytotype (for a description of cytotype see Section IV,B).

One of the experiments that produced the *white* mutants, described above, also generated a large number of visible mutants at other X-chromosome loci (M. G. Kidwell, unpublished results). The majority of mutations occurred at mutational hot spots already recognized to be associated with male recombination chromosomes and the *P–M* system (Green, 1977; Engels, 1979c). The most active site was the *singed* locus, where the frequency of newly induced putative insertion mutants was approximately 0.0006. In addition to *white*, other loci with moderately high frequencies of mutation were *lozenge* and *scalloped*. Similar to the results of Engels (1979c), high rates of reversion and mutation to other alleles were common in *M* cytotype but not in *P* cytotype. Note that the chromosomes in which the mutations were induced were of *M* strain origin and therefore were free of *P* elements at the start of the experiment. The high degree of instability of many of these newly induced mutations in *M* cytotype, therefore, strongly suggested that they were also produced by insertion of *P* elements that had transposed from *P* strain autosomes. Direct evidence for this conclusion comes from *in situ* hybridization experiments in which cloned *P* elements were demonstrated to have homology with DNA sites at the newly mutated loci (P. M. Bingham, personal communication).

A direct chromosomal contamination experiment using biochemical methods (Bingham *et al.*, 1982) further supports the genetic evidence for transposition in the *P–M* system and provides an estimate of the rate at which this occurs in a strongly interacting strain combination. X chromosomes originating from the Canton-S *M* strain were observed to acquire, on average, approximately one copy of a *P* element per fly per generation in F_1 *P/M* dysgenic hybrid males. The distribution of numbers of copies in individual males appeared to follow the Poisson distribution, suggesting independent acquisition of multiple copies. The results of this experiment, compared with those of Kidwell (1982b), strongly suggest that only a small fraction of *P* elements, which have become newly integrated into an X chromosome of *M* origin, have the sterility-inducing potential of *P* factors. The rate of transposition of *P* elements in hybrid dysgenic females has not yet been estimated in a similar way, but other experiments (Kidwell, 1982b) suggest that it may be considerably higher than the rate in males.

There are a number of questions concerning the nature and control of *P* factors, some of which may be answered by the biochemical analyses now underway. One set of questions concerns the size and nucleotide sequence of *P* elements and the extent of homology among elements in the same strains and among different strains. There is growing evidence that not all *P* elements may be functionally the same. For example, although some sequence homology between *P* elements from *P* and *Q* strains can be inferred from *in situ* hybridization experiments (Bingham *et al.*, 1982), these elements are functionally different. In *Q* strains, *P* elements appear to lack the sterility potential that is possessed by *P* factors in *P* strains while retaining male recombination and other functions.

Another question concerns the nature and molecular properties of hot spots. It has been suggested that a distinction be made between mutational hot spots and those for chromosome breakage (Engels and Preston, 1981b). Berg *et al.* (1980) observed that most breakpoints of dysgenesis-induced chromosome rearrangements occurred at a few highly specific chromosomal locations. Engels and Preston (1981b) suggest that this type of hot spot, exemplified by the X-linked *heldup* locus, marks the site of integration of a *P* element and is therefore found only in paternal chromosomes. However, the location of such sites varies in chromosomes from different strains. A second type of hot spot, exemplified by the *singed* locus and found in both paternal and maternal chromosomes, is hypothesized to be a preferential landing site for transposing *P* elements. Such a mutational hot spot is not expected to be a site of chromosome breakage except when a *P* element is already integrated within the locus.

Finally, the molecular basis of sterility production is not at all understood. We do not know whether the transposition of *P* factors is necessary for the production of sterility. We do know, however, that transposition of *P* factors can take place when the developmental temperature is held at 23°C (Rubin *et al.*, 1982).

At this temperature, *GD* sterility is not normally observed in *P–M* dysgenic hybrids (Engels and Preston, 1979; Kidwell and Novy, 1979). Therefore, transposition can take place without detectable sterility production.

IV. Inheritance of the Maternal Regulatory Components

The maternal component of hybrid dysgenesis apparently controls the stability and other properties of the chromosomal *I* and *P* elements. This regulatory component is genetically determined in both systems by complex mechanisms involving both cytoplasmic and chromosomal inheritance.

A. Reactivity in the *I–R* System

The activity of *I* factors is controlled by a cellular state called *reactivity*. Although the reactive state is passed from a mother to her daughters through the maternally transmitted cytoplasm or nucleoplasm, it is completely determined by the genotype over a period of several generations.

It will be recalled from an earlier section (see Table I) that *SF* sterility results from crosses between *R* strain females and *I* strain males.The reciprocal cross produces normally fertile progeny. This result, taken alone, may be explained either by true extrachromosomal inheritance or by the action of classic genes with a maternal effect. Further investigations have shown that the situation is more complex. With unselected reactive stocks, there may be a wide variation in the level of reactivity among individual flies. However, homogeneous strains, either strong or weak, that remain stable over several years may be obtained by selection (Picard and L'Héritier, 1971; Picard *et al.*, 1972). In experiments using genetically marked chromosomes from strong and weak *R* strains, it has been possible to elucidate the rules of inheritance of the reactive state without any interference from inducer factors (Bucheton, 1973; Bucheton and Picard, 1978). The results of these experiments are summarized as follows and in Tables IV and V.

1. Crosses were performed between strong and weak *R* strains in both directions. The level of reactivity of the F_1 females was measured by crossing them with males of a standard inducer stock. Although this level tended to be similar to that of the mothers, there also appeared to be a small effect, in a single generation, of reactive chromosomes contributed by the *P* strain (see Table IV). As in classic quantitative inheritance, detailed analysis showed that each of the three major paternally derived chromosomes had a minor influence on reactivity

Table IV Levels of Reactivity of F_1 Females from Matings Involving Strong and Weak Reactive Strains

Male parent (source of paternally derived chromosomes)	Female parent (source of maternally derived cytoplasm and chromosomes)	
	R (Strong)	*N* (Weak)
R	Strong	Moderately weak
N	Moderately strong	Weak

level and that their action tended to be additive. Nevertheless, the different chromosomes were not equivalent in their influence on reactivity.

2. A simple experiment gave particularly informative results concerning the mutual influence of maternal inheritance and chromosomal control upon reactivity. With the use of a weak reactive (*N*) stock and a strong reactive one bearing marked chromosomes, several lines were synthesized in two generations. These originated in their maternal cytoplasmic lineage from the strong strain and in their chromosomes from the weak strain. The females of the first generation of these lines exhibited a strong mean reactive level, and the variance among lines was high. Over several generations, the reactivity of these lines decreased regularly and finally reached the low level corresponding to their genotype (see Table V). The surprising result was that more than 10 generations were sometimes required to reach this new equilibrium.

3. A further experiment provided evidence that reactive chromosomes alone are able to determine the reactive state (Bucheton, 1979c). Starting with a cross between inducer females bearing marked chromosomes and males of a reactive stock, Bucheton repeatedly backcrossed the male progeny to females of the maternal inducer stock. After several generations of backcrossing, the chromosomes of the males were used to reconstitute a homozygous reactive genotype in an inducer maternal lineage; particular care was necessary in order to ensure that chromosomes contaminated by *I* factors were discarded. The lines thus obtained exhibited a low level of reactivity in the first generation. This level increased progressively over several following generations.

Although this result cannot easily be explained by the hypothesis that transmission of reactivity by male gametes is dependent on an extrachromosomal element [as is the case for CO_2 sensitivity induced by σ virus (L'Héritier, 1970)], it does not completely rule out this possibility in the determination of the reactive condition. Such elements might exist in all stocks, whatever their category, with inducer stocks being characterized by only a very low, undetectable frequency. According to such a hypothesis, the role of reactive chromosomes would be to favor the multiplication of these elements, which might be ubiquitous symbionts or mitochondrial DNA variants. We have tried to test this possibility in several

Table V Levels of Reactivity, at Equilibrium,[a] in Females of Synthesized Lines Having Different Combinations of Genome and Maternal Cytoplasmic Lineage, Derived from Originally Strong and Weak Reactive Strains

Original strain source of all chromosomes	Original strain source of maternal cytoplasmic lineage	
	R (Strong)	*N* (Weak)
R	Strong	Strong
N	Weak	Weak

[a] The number of generations required to reach equilibrium after line synthesis may vary from about 4 to more than 10 (Bucheton and Picard, 1978).

kinds of experiments, but all gave negative results (e.g., Bucheton, 1979c; Picard and Wolstenholme, 1980).

The overall results suggest that reactivity is a quantitative cellular character with limited cytoplasmic transmission controlled by a polygenic chromosomal system. Its main unusual characteristic is the long-delayed chromosomal control.

B. Cytotype in the *P–M* System

The strain characteristic mediating resistance or susceptibility to the action of *P* factors has been termed *cytotype* (Engels, 1979b). Strains possessing *P* cytotype (*P* and *Q* strains) are essentially immune to *P–M* dysgenesis, and those having *M* cytotype (*M* strains) are susceptible. The transmission of cytotype has mainly been studied in relation to *GD* sterility, but the stability of putative *P* insertion mutants (Engels, 1979c) has also been shown to be similarly controlled.

Like that of the *I–R* system, the maternal or regulatory component of the *P–M* system is complex and exhibits some unusual genetic characteristics. The first clue that two components were involved in hybrid dysgenesis was provided by the observation of large reciprocal differences in the frequency of dysgenic traits (Kidwell and Kidwell, 1975). F_1 hybrid females produced by reciprocal matings between *P* and *M* strains have identical chromosome complements. Yet, hybrids with *M* mothers may exhibit complete or partial sterility and other dysgenic traits, whereas those with *P* mothers are usually normal. As in the *I–R* system, an extrachromosomal property of the egg is critical in determining whether or not *P–M* dysgenesis will occur. This property, conferring susceptibility or immunity to the action of *P* factors, was called *cytotype* (Engels, 1979b) to indicate that it is a property of the whole cell, as opposed to a trait controlled by strictly cytoplasmic or Mendelian inheritance.

The fact that cytotype may be transmitted unchanged through the female line for two or more generations, irrespective of the genomic constitution (Engels,

1979b; Kidwell, 1981), rules out a simple maternal effect and suggests that cytotype is a self-replicating property of the germ line. However, when the chromosomes of a *P* strain are replaced by those of an *M* strain, or vice versa, by backcrossing or whole chromosome substitution (Engels, 1979b; Yannopoulos, 1978a; Kidwell, 1981), the cytotype will sooner or later switch to that specified by the chromosomes. The switching of cytotype, by genome substitution, can occur with equal efficiency in either the *P* to *M* or the *M* to *P* direction. Moreover, with a few exceptions, the effect of switching tends to be of the all-or-nothing variety (Engels, 1979b). The number of generations required for chromosomal substitutions to manifest fully their cytotype conversion effect varies from at least one up to as many as 10 or more, depending on the strain and chromosome involved (Kidwell, 1981). Thus, cytotype is characterized by short-term cytoplasmic transmission but long-term chromosomal dependence. As with reactivity in the *I–R* system, simple models of cytoplasmic and maternal inheritance do not apply.

Although most *M* strains (including laboratory marker stocks) are very strongly *M* (M. G. Kidwell, unpublished results), limited data suggest that there is strain variability in both the chromosomal locations and the relative magnitude of the effects of genomic cytotype determinants. In *P*/*M* strain hybrids, Engels (1979b) found that all three major chromosomes were associated with the ability to determine cytotype and that the chromosomal distribution of putative *P* cytotype determinants was closely correlated with that of *P* factors themselves in the same strain. Similar to its effect on *GD* sterility, each chromosome tended to act approximately independently to influence cytotype. Working with *Q*/*M* strain hybrids, Kidwell (1981) found that X chromosomes had a major effect on cytotype determination. The factors involved were localized to the distal half of the chromosome.

V. Interactions between the Two Components

Two general types of hybrid dysgenesis determinants have been described in the last two sections: first, the chromosomally associated, dispersed families of elements, called *I* and *P factors,* and second, the cell properties called *reactivity* and *cytotype,* which regulate the stability of their respective element families.

It should again be emphasized that, in addition to the qualitative distinction between the two types of determinants within each system (i.e., between *I* and *R* and between *P* and *M*), a wide range of quantitative variability exists in the level of activity or ''strength'' of each determinant in different strains of the same qualitative type. If two mutually interacting types of strains are brought together

Table VI Summary of Interaction Effects in the Two Systems

Effect of interaction	Type of interaction[a]	
	I–R	*P–M*
Induction of dysgenic traits	Female sterility (*SF*), mutation, etc., positively correlated in dysgenic hybrid females	Gonadal sterility (*GD*), mutation, etc., positively correlated in dysgenic hybrids of both sexes
Chromosomal contamination	Reactive (*R*) chromosomes changed to inducer (*I*) in reactive females	Maternal (*M*) chromosomes changed to paternal (*P*) in *M* cytotype in both sexes
Short-term changes in maternal component	Level of reactivity largely follows the maternal cytoplasmic lineage	Cytotype largely follows the maternal cytoplasmic lineage
Long-term changes in maternal component	Level of reactivity completely determined by the inducer (*I*) or reactive (*R*) chromosomal constitution	Cytotype completely determined by the paternal (*P*) or maternal (*M*) chromosomal constitution

[a] For the specific strain combinations that lead to interactions, see Tables I and II.

in the appropriate combination, the degree of interaction appears to depend on the strength of both types of determinant. When the results of many dysgenic crosses are compared, this leads to a broad range of frequencies of manifested dysgenic traits.

The various effects of interaction in the two systems are summarized in Table VI. The most conspicuous effects of interaction are various dysgenic traits the frequencies of which are usually positively correlated. In addition, we now have clear evidence that chromosomal contamination is a direct result of interaction. We must also consider the short- and long-term intrastrain interactions that occur when two incompatible types of chromosomes are brought together in the same strain.

A. Interaction and Dysgenic Traits

A positive correlation between hatching percentage of eggs from *SF* females and the frequency of nondisjunction, producing X0 sons, was pointed out by Picard *et al*. (1978). This correlation has been observed in two cases: (*1*) between young *SF* females obtained from mothers of different reactive stocks and (*2*) within broods of individual *SF* females at various stages of their life; the percentage of X0 sons decreased sharply as the mothers became older. In the same experiments, visible mutations were also scored, but a similar quantitative study was not possible because of their low frequency. More complete data have been obtained by Proust and Prudhommeau (1982) with X-chromosome recessive lethal mutants induced by *I–R* interactions. They observed a close relation between the increase in hatchability of eggs of *SF* females during the aging process and the decrease in lethal frequency detected in their progeny. The latter values varied from about 6 ± 2.5% for the more sterile young *SF* females to about 1% for the same females one month later. Control values given by *RSF* females were about 0.2%.

Another interesting result was obtained in crosses between a strong reactive strain and a very weak inducer stock called *BT*. As described earlier, this stock bears only i^0 and i^ϵ chromosomes. Its inducer level is very weak at the extreme limit of detectability by hatching percentage (Perlisson and Bregliano, 1981). When this *BT* stock was crossed with a strong *R* strain, the frequency of recessive X chromosome lethals from *SF* females was relatively low (0.78%) and the frequency of X0 males was near control values. This means that production of these dysgenic traits is correlated not only with the level of reactivity but also with the inducer strength.

Many published studies have provided evidence that, as with the *I–R* system, the frequencies of different dysgenic traits resulting from *P–M* interaction are correlated. For example, Engels and Preston (1980) reported statistically significant correlations between sterility and male recombination. In a more extensive study, the frequencies of various traits were measured in *P–M* dysgenic hybrids produced independently by approximately 200 natural population lines. There were significant correlations ($P < 0.001$) for all pairwise combinations between *GD* sterility, male recombination frequency, and transmission distortion of the third chromosome (M. G. Kidwell, unpublished).

There is one notable exception to this generalization about the behavior of *Q* strains. These exhibit male recombination, mutation and other dysgenic traits in hybrids with *M* strains but no readily observable *GD* levels (Engels and Preston, 1981a). Other differences in the interaction potential of one *P* and one *Q* strain were observed by Simmons *et al.* (1980) in a detailed comparative study of recessive lethal frequencies in *P* × *M* and *Q* × *M* hybrids. Although the general pattern of hybrid dysgenesis was observed, differences appeared when frequencies of lethals were compared among hybrids from the two types of reciprocal crosses and nonhybrid controls. Male progeny from both the dysgenic *P* male × *M* female and *Q* male × *M* female crosses had mutation rates significantly higher than those of *P* strain nonhybrid males. However, whereas the *M* male × *P* female reciprocal cross resulted in a mutation rate three to four times higher than that of nonhybrids (compared to 15 times higher in the *P* × *M* dysgenic cross), the mutation rate of the *M* male × *Q* female cross showed no increase over that of nonhybrids. Also, there was a larger fraction of unstable lethals from the *P* × *M* dysgenic cross than from the *Q* × *M* dysgenic cross. Further studies involving other *P* and *Q* strains are required before the generality of these results can be assessed.

In summary, it appears that the interactions that produce various dysgenic traits can, in general, be interpreted in terms of destabilization of *I* and *P* factors in the appropriate interacting maternal lineage. This interpretation is not difficult to accept with respect to some dysgenic traits, such as mutations and chromosomal aberrations, but with others the mechanisms involved are not at all clear. For example, the stability of *P* insertion mutations appears to be clearly under the control of *M* cytotype, as demonstrated by Engels (1981b) with the hypermutable allele sn^{w}. It is far more difficult to understand the mechanism by which the destabilization of *I* and *P* factors produces some other dysgenic traits. For example, we do not know how *I* and *P* factors produce arrest of development of embryos and *GD* sterility. This is one of the most interesting unsolved problems relating to hybrid dysgenesis.

B. Interaction and Chromosomal Contamination

In the *I–R* system, the dependence of chromosomal contamination upon the *I–R* interaction emerges from the following lines of evidence.

1. Contamination does not occur within inducer stocks; it requires the presence of the reactive state. This is demonstrated by the existence of i^0 chromosomes, described earlier, which do not exhibit any inducer potential and remain stable over time within inducer stocks.
2. The frequency of chromosomal contamination in hybrids is correlated with the level of reactivity of the *M* strain. This was shown by Picard (1978a) using six different maternal strains ranging in strength from low to high reactivity. A seventh maternal stock was used: the so-called neutral *Paris* stock; it produced a contamination frequency nearly identical to that obtained with low reactive stocks. This means that this neutral strain actually belongs to the reactive category, but its reactivity is too weak to produce any detectable sterility in hybrids.
3. Additional information has been obtained using two reactive lines having the same genotype but differing transiently in their level of reactivity because of their different maternal origins. The frequency of chromosomal contamination was significantly higher in *SF* females produced from the strongest reactive line. It may be concluded that the efficiency of contamination is not directly controlled by the chromosomes themselves but by the extrachromosomal state of the reactive cells (Picard, 1978a,b).

Taken together, these results indicate that of all the phenomena produced by *I–R* interactions, chromosomal contamination is the most sensitive. It occurs at a relatively high frequency in crosses involving inducer or reactive stocks too weak to produce any detectable sterility or nondisjunction (Picard, 1978a; Pelisson and Bregliano, 1981). Thus, contamination may be used as the most efficient criterion for detecting low levels of either inducer or reactive potential.

Evidence for the dependence of chromosomal contamination on *P–M* interaction is not as extensive as that for the *I–R* interaction. Kidwell (1982b) demonstrated the dependence of high contamination frequencies on *M* cytotype. However, more detailed studies are needed before we can completely rule out the hypothesis that a low level of contamination occurs in *P* cytotype individuals. Also, it has not always been possible to obtain consistent results on *P* factor contamination, even in the presence of a strong *P–M* interaction. The reason for these difficulties is not at present understood. Certainly, the situation seems to differ from that in the *I–R* system, in which chromosomal contamination is the most sensitive indicator of interaction.

C. Incompatibilities between Interacting Strains

As described above, in the *I–R* system a change in genotype induces a corresponding change in the level of reactivity, but with a long delay. However, this is true only for experimental schemes involving reactive or neutral genotypes. Crosses involving an inducer genotype lead to a different result. Using stocks with marked chromosomes, Picard (1978c) synthesized lines in two generations bearing the chromosomes of an inducer stock and the cytoplasm of a reactive stock. In such lines, any detectable reactivity disappears abruptly even if the reactive stock is strong. More important is the fact that the same result is obtained when contaminated chromosomes of reactive origin are used instead of chromosomes of inducer origin. However, in this case the disappearance of any detectable reactivity requires four or five generations (Picard, 1978c; J.-C. Bregliano, unpublished results).

It therefore follows that it is easy to transform a reactive stock into an inducer one without any change in its genotype other than chromosomal contamination by *I* factors. This kind of transformation has been performed quite easily using several reactive stocks, either strong or weak, and is irreversible.

The problem raised by this incompatibility is that it is not clear whether it acts at the cellular level or at the population level. The rapidity of the change from the reactive to the inducer condition favors the hypothesis that there is some kind of biochemical incompatibility, possibly mediated by regulatory mechanisms, between *I* factors and the expression of putative chromosomal elements responsible for the control of reactivity. (Note, however, that in the *P–M* system there is no evidence that excludes the possibility that *M* cytotype represents the absence of function.) In the *I–R* system, we cannot rule out the alternative hypothesis that the incompatibility lies at the population level. The appearance of chromosomes bearing *I* factors within a reactive cytoplasm triggers *SF* sterility, which acts as a powerful screen and allows the hatching only of eggs that bear genetic combinations unable to maintain the reactive state. In this view, the disappearance of reactivity in lines bearing contaminated chromosomes of reactive origin would be due only to selective pressure.

Kidwell *et al.* (1981) showed that in mixed populations of *P* and *M* flies, *M* cytotype rapidly disappeared within four or five generations irrespective of the initial proportions of *P* and *M* types. In this case, the disappearance of *M* cytotype was most likely due to incompatibility at the cellular level because there was no *GD* sterility at the temperature used for the experiments. However, the relationship between chromosomal cytotype determinants and *P* factors is not understood. One possibility is that *P* factor activity and *P* cytotype determination represent two distinct functions of the *P* family of transposable elements (Rubin *et al.*, 1982; Bingham *et al.*, 1982). Another possibility is that a cytoplasmic or nucleoplasmic intermediate is involved in the determination of cytotype.

VI. Regulation of *I* and *P* Elements

A. The Nature of Reactivity and Cytotype

Although the resolution of the biochemical basis of the maternal component of hybrid dysgenesis promises to be a difficult and complex problem, it is a very important one. An understanding of the biochemical determination of reactivity and cytotype may throw light on areas of wider general interest. It would not only allow us to understand the molecular basis of inheritance but may also provide some understanding of the more general question of how the transposition of mobile elements is controlled in eukaryotic systems.

Our present incomplete knowledge allows us only a limited understanding of the degree of specificity of *M* cytotype and reactivity in the regulation of their respective transposable elements. Although we know that reactivity is not necessary for *P* factor transposition, because chromosomal contamination takes place in the progeny of *IP* male × *IM* female crosses (Kidwell, 1982b), we cannot determine directly whether reactivity is completely lacking an effect on *P* factor mobility so long as we have no *RP* strains. There is abundant evidence that *M* cytotype is not sufficient for *I* factor transposition, because many *IM* strains exist in which *I* factors do not transpose at a detectable frequency. However, because all known *R* strains have *M* cytotype, we do not know whether, in addition to reactivity, the *M* cytotype is also necessary for *I* factor destabilization. It is also not known whether the mobility of other transposable elements, such as *copia*, is affected by the regulatory components of either system of hybrid dysgenesis.

The example of recovery of a *copia* insertion into the *white* locus in a *P–M* dysgenic cross, described in Section III, does not of itself constitute strong evidence for the increased frequency of mobilization of this element in *M* cytotype. The absence or low frequency of reversion of this insertion mutant (Rubin *et al.*, 1982) and the observed stability of all *copia* elements in many *M* strains (Young and Schwartz, 1981) argue strongly against a general destabilization of *copia* in *M* cytotype. However, a general increase in transposition rates of elements other than *P* factors, including *copia,* may be induced by hybrid dysgenesis due to a phenomenon similar to "genomic shock" (see Chapters 1 and 8).

Among a number of possible explanations for the unusual mode of inheritance of reactivity and *M* cytotype, two hypotheses have been proposed (Bucheton and Picard, 1978; Engels, 1981a). One hypothesis supposes that the direct genetic determinants of reactivity and cytotype are extrachromosomal, self-replicative elements. These elements form an intracellular population, tending to be transmitted through cell division but liable to undergo quantitative or qualitative variation under the control of nuclear genes. According to this hypothesis, the delayed effect of a genotypic substitution is not surprising. One alternative hypothesis proposes that the genetic determinants of reactivity and cytotype are strictly chromosomal. In order to explain the delayed effects observed, it is postulated that the products of certain genes provide a complex feedback regulation of the expression of the genes themselves. Note that these hypotheses represent just two extreme possibilities; we may easily imagine intermediate models and also totally different models.

At present, some explanations for the unusual mode of inheritance of reactivity and cytotype appear to be more plausible than others. Several results argue against the symbiont hypothesis (viruses, bacteria, or mycoplasma). These are the failure to remove reactivity or *M* cytotype irreversibly by thermic treatments or antibiotics and the extensive negative results of horizontal transmission experiments. Reactivity and *M* cytotype are therefore probably different from entities such as the *killer* particles in *Saccharomyces cerevisiae* (Wickner, 1974, 1978), the σ virus in *D. melanogaster* (L'Héritier, 1970), or the hybrid male sterility in *D. paulistorum* (Ehrman and Kernaghan, 1971). Although we cannot completely rule out the hypothesis of a mitochondrial DNA variant, at this time it is not supported by any experimental evidence.

The more plausible hypothesis is that chromosomal genetic information is sufficient to produce the reactive biochemical state and *M* cytotype. Several distinct mechanisms may be imagined in order to explain the sometimes long-delayed chromosomal effects. Reactivity and *M* cytotype may be comparable to some known phenomena of structural inheritance (Beisson *et al.,* 1976). In this context, Sved's (1976) model of hybrid dysgenesis in terms of spatial organization of chromosomes may be applicable. He suggested that chromosome–membrane associations in the nucleus may be involved in the determination of the female component of hybrid dysgenesis. Alternatively, reactivity and *M* cytotype may be comparable to the regulation of surface antigen expression in *Paramecium aurelia* (Sonneborn and Lesuer, 1948) and *P. primaurelia* (Capdeville *et al.,* 1978). We may even imagine as a unifying hypothesis that reactive and cytotype-determining "genes" are some kind of mobile genetic elements that may produce free copies. Like that of *I* and *P* factors, the DNA responsible for reactivity and *M* cytotype may belong in the category of middle repetitive sequences.

An understanding of the genetic mechanisms responsible for reactivity and cytotype may throw some light on several other phenomena that have similar patterns of inheritance and the molecular determination of which is unknown. For example, in *D. melanogaster,* the extrachromosomal δ element studied by Minamori (1972), like reactivity and cytotype, exhibits a delayed effect of genotype and undergoes reversible changes with heat treatment. The mutant denoted *459* (Mariol, 1981) also exhibits a delayed effect. Moreover, it produces a phenotype similar to that of *GD* sterility at the same permissive temperature. Examples in other organisms include the nuclear mutants in *Paramecium,* described by Beisson *et al.* (1974) and Sainsart-Chanet (1976), and a mutant resistant to basic fuchsin in *S. cerevisiae* (Barrere and Mounolou, 1976). We may also recall the numerous phenomena named *Dauermodification* that were observed for many years in various organisms (see the review by Jinks, 1964), which exhibit heritable effects of environmental factors, particularly thermic treatments.

B. Environmental Factors in Regulation

As described in Section II,C, environmental factors may have an important effect on the occurrence and level of expression of some dysgenic traits, particularly sterility, in both systems. In addition, some of these effects may be cumulative over several generations. Another interrelated aspect is that the stage in germ line development at which environmental changes take place (Table III) may actually determine the specific dysgenic trait that is eventually manifested. Of particular interest, when considering the regulation of *I* and *P* factors, are the striking effects of cell type, developmental stage, and external environment (notably temperature) on destabilization. When a dysgenic hybrid zygote is formed in a cell with a reactive or *M* maternal lineage, the nuclei of somatic cells show no signs of instability, as inferred from the normal appearance of the soma of hybrid individuals and particularly from the absence of somatic mosaics (e.g., Thompson *et al.,* 1978; Engels, 1979c). However, in the germ line of dysgenic hybrids the outcome is completely different, even though the genetic and maternal cytoplasmic origins of the cells are identical to those of the soma. Thus, a question of fundamental importance for understanding regulation concerns the reason for the marked difference in stability of the genome in somatic and germ line cells in the same dysgenic individuals.

A related problem concerns the identification of the stages in the developmental cycle (see Table III) when interaction takes place. Most of the chromosomal dysgenic traits associated with *I–R* interaction appear as isolated events, which suggests that they are produced at the meiotic stage. However, a few of them appear in small clusters, suggesting that premeiotic events may occur.

It is possible to identify more precisely the stage of induction of *SF* sterility. Several lines of evidence (already discussed in Section II), particularly the timing of the curative effect of heat treatment at late oogenesis, indicate that the critical step is the synthesis of the oocyte pool. The presence of both *I* factors and a significant level of reactivity during this period appears to lead to anomalies in the biochemical constitution of the ooplasm. This prevents the completion of the first step of embryonic development normally supported by RNA and proteins of the oocyte pool.

In the *P–M* system, there appears to be considerable variation in the stage at which dysgenic events are induced. The very nature of *GD* sterility implies an arrest of development in the germ line at a very early stage (Engels and Preston, 1979; Kidwell and Novy, 1979). Frequent small clusters of identical mutations and products of male crossing-over strongly suggest a gonial origin closer to the time of meiosis than events leading to *GD* sterility.

In the *P–M* system, a strong case may be made for the idea that the stage at which a particular disruption of the germ line occurs may, to a large extent, determine the type of dysgenic trait that results. For example, if complete hybrid sterility results from *GD* sterility due to an early arrest in development of the germ line, it automatically precludes the observation of any other dysgenic traits. Conversely, if *GD* sterility is partial or nonexistent, there is an opportunity for destabilization later in development. This may result in mutation, recombination, and other dysgenic traits in the same generation and even partial sterility in the following generation.

The differential effect of temperature on different dysgenic traits may be explained by postulating a general tendency for high temperature to increase genome instability, with variable sensitivity during different stages of development. For example, in the *P–M* system, *GD* sterility may result from the temperature sensitivity of developmental arrest at a very early stage. The observation, described earlier, that an intermediate temperature gives maximum effect on other dysgenic traits, such as male recombination, may be an artifact imposed by the exclusion of sterile individuals from breeding. High temperatures, by sterilizing the individuals most sensitive to dysgenesis early in development, may effect a truncation of the high end of the distribution of a later-developing trait, giving the appearance of an intermediate optimum.

It appears that in the *I–R* system, the factors of aging and temperature exert their effect through modification of the expression of reactivity. Their cumulative effect on the level of reactivity over several generations is of particular interest. For example, Bucheton (1979b) was able to shift an *R* strain from strong toward weak reactivity by breeding each generation from old egg-laying females (Fig. 4). However, this change was always reversible; when young egg-laying females were used again, reactivity gradually returned to its original level. It was verified that the chromosomes of those lines that were temporarily modified by the repetitive action of aging exhibited the same properties as chromosomes from the original *R* strain (Bucheton, 1979c). As expected, aging acted only on the extrachromosomal state.

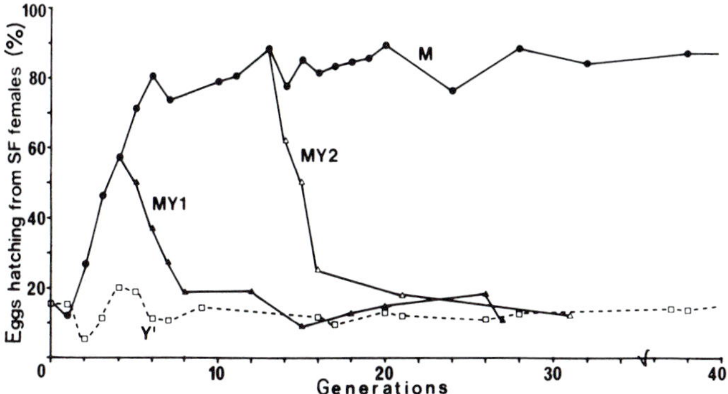

Fig. 4 Cumulative and reversible effects induced by aging at the level of reactivity of females of a reactive strain. Y′, In this line the flies of each generation were collected from eggs laid by females 2–4 days old. M, Each generation of this line was obtained from egg-laying females 11–12 days old. MY_1 and MY_2, These lines were established from generations G_3 and G_{13}, respectively, of the M line and maintained in the same way as the Y′ line.

C. Other Aspects of Regulation

Although little detailed information is yet available, it is becoming increasingly clear that the ability to perform specific functions is partly inherent in the *I* and *P* elements themselves. This may be determined by structural properties, by the chromosomal location of sites of integration, or by interrelationships with other members of the element family in the same genome. An example of this aspect of regulation is the I^{ϵ} element, described earlier, which may have only weak inducer strength under some conditions but may acquire a strong inducer function after passage through a reactive female. Another example is the wide variability of reversion frequencies of putative *P* insertion mutants. The hypermutable allele sn^{w} (Engels, 1979c, 1981b) mutates to other allelic forms at frequencies between 40 and 60% per generation, a rate considerably higher than that of other alleles at this locus and of insertion mutants at other loci (Green, 1976). Also, a preliminary survey of insertion mutants induced by *P–M* hybrid dysgenesis (M. G. Kidwell, unpublished results) revealed a wide variability of reversion rates among different loci in the same *M* cytotype. For example, *sn* and *lz* mutants appeared to be particularly unstable compared to those at *w* (Rubin *et al.*, 1982) and other X chromosome loci.

A different aspect of regulation is addressed by the experiments of Eeken and Sobels (1981). Their results suggested that the observed level of at least one manifestation of *P* factor activity may be modified by the disruption of enzymatic pathways involved in DNA repair. In hybrid dysgenic males carrying an *MR* second chromosome and deficient for both excision and postreplication repair (because they also carried the $mei\text{-}9^{a}$ and $mei\text{-}41^{D5}$ meiotic mutants), the frequency of mutations at the *sn* and *ras* X-chromosome loci was significantly enhanced. One of several possible interpretations is that *MR*-induced potential primary lesions are susceptible to repair.

VII. Hybrid Dysgenesis and Evolution

A. Strain Distribution in Laboratory and Natural Populations

In Section II, brief reference was made to striking trends that had been observed in the distribution of natural and laboratory populations with respect to their potential for hybrid dysgenesis. Out of a large number of strains so far tested, all of those derived from recently caught wild flies seem to belong to the inducer category in the *I–R* system. This is true whatever the geographical origin of the flies; some of them were caught in the forests of Africa or Russia, far from any inhabited area. The reactive category includes only those strains kept in laboratories for at least 20 years. Until recently, a similar dichotomy was also thought to be true for the *P–M* system, with only a slight difference being noted; strains kept in the laboratory for as short a period as 10 years were found to belong to the *M* category. However, recent results strongly suggest that a few *M* wild populations do exist today in some parts of the world. Further details of this potentially important observation are given later in this section.

Table VII gives a summary of preliminary information on the classification of a large number of strains relative to their dates of collection in the wild. More details will be available with the complete results of an extensive worldwide survey of strains (M. G. Kidwell, unpublished). Striking shifts in strain distribution are observed to have occurred, from *R* to *I* and from *M* to *P*, with decrease in laboratory age. However, there is some suggestion that in American strains, this shift occurred somewhat earlier than that in the other continents represented (mainly Europe). This is particularly true of the *M* category and may also hold for the reactive one. It is noteworthy that many of the stocks included in Table VII were kept continuously in the same laboratory after collection from the wild. Therefore, temporal and geographical differences cannot, in general, be attributed to differences in subsequent breeding conditions.

The observed distribution of strains has been explained in two different ways. The first hypothesis was suggested by Bucheton *et al.* (1976) with respect to the *I–R* system. It assumes that the inducer state has always been the general condition for all wild populations and that the reactive state, resulting from loss of *I* factors, is restricted to small isolates, mainly those kept under laboratory conditions. This hypothesis has been extended to the *P–M* system. Engels (1981a) has proposed a mechanism to account for the types of strain changes implied in both

Table VII Percentage Distribution of Strains Collected from Nature at Various Times According to Their Characteristics in Both Systems of Hybrid Dysgenesis

System	Decade of collection in the wild					
	1920s	1930s	1940s	1950s	1960s	1970s
I–R system						
Americas						
R or *N*	67 (2)[a]	50 (4)	—	13 (2)	3 (1)	0 (0)
I	33 (1)	50 (4)	—	87 (13)	97 (37)	100 (304)
Other continents						
R or *N*	100 (1)	100 (5)	75 (3)	27 (7)	9 (3)	0 (0)
I	0 (0)	0 (0)	25 (1)	73 (19)	91 (30)	100 (34)
P–M system						
Americas						
M	100 (3)	100 (8)	—	87 (13)	58 (22)	3 (9)
P or *Q*	0 (0)	0 (0)	—	13 (2)	42 (16)	97 (295)
Other continents						
M	100 (1)	100 (5)	100 (4)	92 (22)	97 (31)	32 (11)
P or *Q*	0 (0)	0 (0)	0 (0)	8 (2)	3 (1)	68 (23)

[a] Figures in parentheses indicate the number of strains characterized.

systems in his stochastic loss hypothesis. Note that the hypothesis regarding the change in laboratory culture implies two corollaries:

1. A stock that has lost *I* or *P* factors must quickly acquire reactivity or *M* cytotype, respectively, because true neutral strains have not yet been found in either system and must therefore be assumed to be exceptional or nonexistent.
2. The change from *I* to *R* and from *P* to *M* must be relatively easy under a wide variety of laboratory conditions because *R* and *M* strains are ubiquitous in laboratories scattered over different continents. Many of them are wild-type strains that certainly have no recent common origin. This is especially evident for wild-type *M* strains, which are very numerous in laboratory collections.

A quite different explanation, the recent-invasion hypothesis, has been proposed by Kidwell (1979, 1982a) and states that the essential changes have occurred in natural populations and have been in the reverse direction, that is, from *R* to *I* and *M* to *P*. This implies the rapid spread of *I* and *P* elements in formerly *R* and *M* natural populations. It is suggested that such a change may be achieved even in the absence of any kind of selective pressure because of the efficiency of transposition of *I* and *P* elements in hybrids. Although at first this idea may seem unlikely because of the implied extremely rapid spread of *I* and *P* factors, given the tendency for dysgenic traits to lower drastically the reproductive fitness of hybrids, new data from natural populations enhance the attractiveness of this hypothesis. The lines of evidence that lead us to this conclusion come from both *I–R* and *P–M* studies.

1. EVIDENCE FROM THE *I–R* SYSTEM

1. At least thirty *I* strains have been studied and regularly tested for more than 10 years in the Clermont laboratory, and no changes from *I* to *R* have been observed.

2. The *BT* stock composed of i^0 and i^ϵ chromosomes, described earlier, exhibits only a very weak inducer property at the limits of detectability of the sterility test. It may be considered a good candidate for a change toward the *R* type, constituting a necessary intermediate step because it lacks any functionally efficient *I* factors. However, this stock was synthesized in 1976, and since that time no trace of reactivity has been detected. Therefore, even when we try to promote transformation from *I* toward *R* in the laboratory, it does not succeed.

One may argue that the *BT* strain is not the most favorable intermediate because it still possesses *I* factors. Although the I^ϵ element has lost almost all its potential for producing *SF* sterility, it may have retained its ability to suppress reactivity. Nevertheless, the above observations indicate that the spontaneous occurrence of a strain having no *I* elements is an unlikely event, and therefore, the simultaneous occurrence of similar such strains in many different laboratories is highly improbable.

At this point, it should be remembered that the reverse transformation, from *R* to *I*, takes place easily (see Section III). A similar transformation (from *M* to *P*) occurred rapidly in a number of mixed populations without artificial selection (Kidwell *et al.*, 1981).

2. EVIDENCE FROM THE *P–M* SYSTEM

As already mentioned, recent findings concerning the geographical distribution of strains in the *P–M* system have important implications.

1. Strains derived from flies caught in four locations in the USSR in 1980 and 1981 and three locations in Australia in 1979 and 1980 appear to be of the *M* type (M. G. Kidwell and N. Plus, unpublished results). This means that the *M* condition is not exclusively a laboratory oddity but actually exists in natural populations of some countries. Therefore, it may constitute strong evidence in favor of the hypothesis that changes from one category to another actually take place in natural populations.

2. In the United States, the majority of approximately 200 lines collected from 20 dispersed wild populations in 1977 and 1978 were of the *P* type (M. G. Kidwell, unpublished results). In France, of 10 samples from wild populations tested recently, eight are *Q*, one is strong *P*, and one is weak *P*. Although this

result must be confirmed by more extensive studies, it may mean that significant geographical differences in the frequency of *P* chromosomes exist between the American and European continents. Such geographical differences are consistent with the recent invasion hypothesis. *Q* elements may be defective *P* elements that are unable to produce a complete syndrome of hybrid dysgenic traits. The relative frequencies of the two types may be expected to vary in different subpopulations due to chance or other factors.

3. Taken together, the data of Table V may be easily explained by assuming that the distribution patterns reflect the dates of appearance, or the rapidity of spread, of *P* and *Q* factors in natural populations of various geographical areas. Conversely, these distributions are difficult to explain in terms of the hypothesis of laboratory changes from *P* to *M* unless it is postulated that major geographical differences existed in the original numbers of *P* factors carried by these strains.

3. Conclusions

The relationship between hybrid dysgenesis and evolution is puzzling. However, it seems that we may now have enough concordant pieces of evidence to suggest an evolutionary process of recent origin. This view, although surprising at first, increasingly appears to be the more plausible explanation. At the very least, it may be considered as a useful working hypothesis amenable to verification. We may summarize it as follows.

Before 1920 or 1930, it is postulated that most, if not all, populations of the world were of the *RM* type, that is to say, reactive and with *M* cytotype and lacking both *I* and *P* factors. In the period 1930–1960, *I* factors appeared in natural populations (perhaps first in America) and spread throughout the populations of the world within a period of 20–30 years. The great transposition efficiency of *I* factors, with the possible addition of adaptive pressures (see below), would have allowed them to invade all populations, except perhaps some geographical isolates. *R* strains kept in laboratories would be considered fossil strains originating from flies caught in the wild during a period preceding the spread of *I* factors.

Beginning about 30 years ago, it is suggested that *P* factors evolved and radiated in a similar way. This putative second wave may still be continuing in some countries, such as the USSR. It is tempting to suggest that these successive waves are responsible for some mutational bursts directly observed in natural populations by some authors (e.g., Golubovsky, 1974, 1980; Berg, 1979).

The recent invasion hypothesis is amenable to verification, at least for the *P–M* system. If the spread is actually continuing, in some countries we may expect to observe a rapid transformation of *M* populations toward the *Q* or *P* types similar to that demonstrated in laboratory populations by Kidwell *et al.* (1981). The USSR and Australia are promising countries for observing this transformation because *M* populations apparently still exist there. Moreover, there is other evidence that *P* factors are also present in these countries. Indeed, M. D. Golubovsky (personal communication) has found a mutant at the *singed* locus, from a wild population, that is hypermutable in the presence of *M* cytotype. This mutant appears to be similar to the *singed* mutant obtained by *P–M* interaction in the laboratory by Engels (1979c).

The hypothesis of rapid evolution in the wild raises many questions. For example, what was the mode of origin of *P* and *I* factors in natural populations? Did both of them really appear first in America, and how? One of the most important questions concerns the causes of the hypothesized rapid spread of *I* and *P* elements to diverse parts of the world. Could this spread be explained wholly by the great efficiency of transposition of these mobile elements, or was it also favored by selective pressures? In other words, are *I* and *P* factors examples of so-called selfish DNA, or do they confer selective value to their carriers at the phenotypic level?

B. Selfish or Adaptive DNA?

The question of selfish DNA versus a possible adaptive role of *I* and *P* factors is relevant irrespective of which proposed interpretation is accepted for the observed distribution of strains in both systems. This is because of the observation that present-day wild populations are of the inducer type and the majority of them are also of the *P* type.

At present, there are no clear answers; only speculative hypotheses can be put forward. Two hypotheses may be of interest for directing future studies. The first hypothesis assumes that *I* and *P* factors are selfish DNA, following the views of Doolittle and Sapienza (1980) and Orgel and Crick (1980). These DNA elements may have spread and then been maintained in wild *Drosophila* populations in the absence of any kind of positive natural selection, either recently or in the remote past. Transposability is postulated as a mechanism sufficient to ensure their propagation. Supporting this hypothesis is the fact that the change from *R* to *I* or from *M* to *P* may be easily reproduced in the laboratory, as described earlier. In this view, *I* and *P* factors would be parasite sequences and may be compared to the σ virus, another parasite known in *Drosophila* and responsible for CO_2 sensitivity. This rhabdovirus is not at all contagious. It is perpetuated only by nonchromosomal hereditary transmission, more efficiently than is a Mendelian gene (L'Héritier, 1970). In spite of a slight disadvantage incurred by its host, it is maintained in many populations of the world at an observable frequency. In population cages it rapidly invades all populations, whatever the input frequency of infected flies (Fleuriet, 1976).

However, note that with respect to mobile elements, such as the *I* and *P* factors, the term *selfish* DNA may be misleading. Even if the absence of a short-term selective advantage of these elements for their carriers could be demonstrated, the possibility of their long-term advantage for the species would not be precluded (see Section VII,D).

The second hypothesis must be viewed in the context of a recent evolutionary change in natural populations and assumes that the rapidity of the change has been partly due to a selective pressure in favor of the transmission of *I* and *P* factors.

If Table VII reflects the approximate data on the appearance of *I* and *P* factors in wild populations, this timing roughly coincides with the appearance of strong new selective pressures in populations of insects. We refer to the intensive use of insecticides for agricultural use. Present-day natural populations of *D. melanogaster* are considerably more resistant to insecticides than old laboratory stocks; the difference is as much as 100 times for DDT (N. Plus, unpublished). The rapidity of acquisition of chemical resistance by insects was unexpected (Crow, 1957); therefore, the involvement of transposing sequences in this process may be a highly plausible explanation. The coincident timing of general insecticide use with the data presented in Table VII leads us to question whether *I* or *P* factors have been involved either directly or indirectly in the control of insecticide resistance. This suggestion was first proposed by N. Plus. Since then, she has performed many experiments, in collaboration with the authors of this chapter, in order to test this hypothesis.

The general conclusions are that although some results have suggested that correlations exist, at present no data decisively support a link, either direct or indirect, between *I* and *P* factors and the spread of resistance to insecticides. However, it is worthwhile to consider this hypothesis because it is a very attractive one, in the sense that it appears to fit well with many data. First, it may explain the curious double coincident timing between the dates of appearance of *I* strains (Table VII) and intensive use of DDT, on the one hand, and the dates of appearance of *P* strains and the first use (in 1950) of organophosphorous compounds, on the other. Second, it may explain the surprisingly rapid spread of genetic resistance to insecticides in wild populations of flies. Last, it leads to the attractive idea that *I* and *P* factors may have similarities to bacterial transposons that are also responsible for the rapid spread of many types of antibiotic resistance in bacteria.

C. Relationships to Other *Drosophila* Systems

The question has been raised of whether *MR* elements (Hiraizumi, 1971; Slatko and Hiraizumi, 1975; Green, 1980) should be classified as forming a third system of hybrid dysgenesis rather than as an integral part of the *P–M* system. The available evidence overwhelmingly supports the interpretation that *MR* factors may be indistinguishable from *P* factors, although some may form a subset of *P* elements. Two X-linked recessive lethals induced in the germ line of hybrid dysgenic males by MRh^{12} second chromosomes have in fact been shown to be *P* factor insertions in the C4 locus (R. Voelker, A. Greenleaf, and P. Bingham, personal communication). In addition to male recombination, the majority of *MR* strains have the potential for *GD* sterility (Engels and Preston, 1980; M. G.

Kidwell, unpublished); those that have no *GD* sterility potential (*Q* strains) possess *P* cytotype, which has the capacity to suppress both *GD* sterility and male recombination (Engels and Preston, 1981a; Kidwell *et al.*, 1981). Furthermore, the statistically significant correlations referred to earlier, between male recombination frequency and *GD* sterility frequency in hybrids of natural population lines with laboratory *M* stocks provides strong evidence that *MR* factors and *P* factors are not independent systems. It is possible that essentially the same *P* factors are responsible for the different dysgenic traits observed and that the stage in development at which the germ line aberration occurs to a large extent determines the actual trait manifested. Both internal and external factors may affect this critical timing.

Several unstable mutants and X chromosomes in *D. melanogaster* have some of the properties of hybrid dysgenesis-induced unstable mutants. For example, Lim (1979) described an X chromosome with site-specific instability. High frequencies of recessive lethal mutations have often occurred in clusters and have showed inordinately high reversion frequencies, often associated with chromosomal rearrangements. For a review of other unstable mutants at the *white* locus, see Green (1980).

There are a number of other mutator systems in *melanogaster* and other *Drosophila* species (reviewed by Woodruff *et al.*, 1982) that have similarities to hybrid dysgenesis systems but the etiology of which is unknown. In most cases, transposable elements have not yet been specifically implicated, but they cannot yet be ruled out. In addition, two systems in *D. ananassae* and *D. simulans* have striking similarities to hybrid dysgenesis systems in *D. melanogaster*.

In *D. ananassae,* Hinton (1979, 1981) described a two-component system consisting of a clastogenic mutator of sperm chromosomes and a suppressor that functions in the oocyte soon after fertilization. The mutator component showed some similarities to hybrid dysgenesis *P* factors. The inheritance of the suppressor component appears to be analogous to that of reactivity and *M* cytotype in that there are alternative cytoplasmic states that interact with paternally originating chromosomes. These states are determined by the genotype and, parallel with hybrid dysgenesis in *D. melanogaster,* this determination shows delayed expression.

In *D. simulans,* Periquet (1981) reported two types of interacting strains, analogous to *P* and *M* strains of *D. melanogaster,* that also produce a similar type of associated *GD* sterility. Woodruff and Bortolozzi (1976) had earlier observed low levels of male recombination in *D. simulans* strain crosses.

D. Generation of Diversity and Reproductive Isolation

In a general context, the possible significance of hybrid dysgenesis determinants for the evolution of the species may be considered in two ways. A number of experiments have demonstrated that the potential for generation of genetic diversity by mutation can be inordinately high when *I* and *P* factors are in a destabilized state. This diversity may be generated at different levels, for example, changes at the level of DNA sequence or structural changes in chromosomes. As with all new mutations, it is expected that a large fraction will be eliminated by natural selection because of negative effects on the fitness of the organism. Although we still have a great deal to learn about the type and magnitude of mutation effects induced by dysgenesis determinants, we cannot rule out the possibility that some mutations may be selectively advantageous.

The question of regulation is of paramount concern when we consider the actual impact that transposable elements may have on genetic diversity in natural populations. If, as suggested by some of our studies, a population has the ability for rapid evolution of suppressors of transposition, then over long periods of time, transposable elements may exist largely in a stable state. The production of new variability through mobilization of elements may be limited to brief periods when suppression is temporarily relaxed by hybridization.

A second way that hybrid dysgenesis determinants may have implications for evolution is by acting as postmating reproductive isolating mechanisms. Crosses between long-established laboratory strains and natural population strains tend to produce hybrids with lower relative fitness than the parental strains. Such hybrids may be similar to those produced in nature when population boundaries meet again after a period of geographical isolation. Studies with hybrid dysgenesis suggest that populations may evolve rapidly as carriers of transposable elements, and these may act, separately or in conjunction with other isolating mechanisms, to promote reproductive isolation and eventual speciation. (see Bingham *et al.*, 1982, for a specific model of reproductive isolation involving transposable elements.)

It is interesting to note that there are many examples of intersubspecific and interspecific crosses in the genus *Drosophila* that have parallels with hybrid dysgenic crosses (see Kidwell, 1982a for a review). For example, the sterility of hybrids produced by crosses between the sibling species *D. melanogaster* and *D. simulans* is associated with rudimentary gonads similar to those observed in *GD* sterility associated with the *P–M* system of hybrid dysgenesis. The tendency of sterility to be unequal in the progeny of hybrids produced by reciprocal crosses is also a quite common observation in *Drosophila* interspecific hybrids. As with hybrid dysgenesis, interspecific hybrid sterility is not necessarily accompanied by hybrid inviability. There are many examples of interspecific crosses similar to hybrid dysgenesis in that incompatibility is restricted to the germ line, with the soma being viable and often vigorous.

In conclusion, we suggest that three aspects of hybrid dysgenesis systems are of particular interest for a general understanding of mobile elements in eukaryotes. These are the existence of complex genetic mechanisms that regulate the stability of *I* and *P* factors, the strong influence of environmental factors during various stages of development, and the complete absence of *P* factors (and probably *I* factors) in some strains of the species, which appears to be related to their history and geographic origin. Further significant advances are promised by the developing synthesis of results from studies at the molecular, chromosomal, and population levels, in conjunction with those from more classic genetic approaches.

Acknowledgments

We thank James Shapiro, Nadine Plus, Michael Simmons, William Engels, Paul Bingham, and all workers of the Clermont Group of Genetics for helpful comments on the first draft.

Preparation of the manuscript was supported in part by the Centre National de la Recherche Scientifique (ERA 692), the University of Clermont-Ferrand II, the Fondation pour la Recherche Medicale, the National Science Foundation (DEB76-82630), and the National Institutes of Health (GM-25399).

References

Barrere, G., and Mounolou, J. C. (1976). Yeast mutants resistant to basic fuchsin: a genetic approach to the integration of nuclear and mitochondrial information. *Mutat. Res.* **35,** 39–52.

Beisson, J., Sainsard, A., Adoutte, A., Beale, G. H., Knowles, J., and Tait, A. (1974). Genetic control of mitochondria in *Paramecium. Genetics* **78,** 403–413.

Beisson, J., Lefort-Tran, M., Pouphile, M., Rossignol, M., and Satir, B. (1976). Genetic analysis of membrane differentiation in *Paramecium.* Freeze-fracture study of the trichocyst cycle in wild-type and mutant strains. *J. Cell Biol.* **69,** 126–143.

Berg, R. L. (1979). Global patterns of mutability in natural populations of *Drosophila melanogaster. Genetics* **91,** s8–9.

Berg, R. L., Engels, W. R., and Kreber, R. A. (1980). Site-specific X-chromosome rearrangements from hybrid dysgenesis in *Drosophila melanogaster. Science* (Washinton, D.C.) **210,** 427–429.

Bingham, P. M., Kidwell, M. G., and Rubin, G. M. (1982). The molecular basis of *P–M* hybrid dysgenesis: the role of the *P* element, a *P*-strain-specific transposon family. *Cell* **29,** 995–1004.

Bodenstein, D. (1950). The postembryonic development of *Drosophila. In* "Biology of Drosophila" (M. Demerec, ed.), pp. 275–367. Wiley, New York.

Bregliano, J. J., Picard, G., Bucheton, A., Pelisson, A., Lavige, J. M., and L'Héritier, P. (1980). Hybrid dysgenesis in *Drosophila* melanogaster. *Science* (Washington, D.C.) **207,** 606–611.

Bucheton, A. (1973). Contribution à l'étude de la stérilité femelle non mendelienne chez *Drosophila melanogaster*. Transmission héréditaire des degrés d'efficacité du facteur réacteur. *C. R. Hebd. Seances Acad. Sci. Ser. D* **276,** 641–644.

Bucheton, A. (1978). Non-Mendelian female sterility in *Drosophila melanogaster:* influence of aging and thermic treatments. I. Evidence for a partly inheritable effect of these two factors. *Heredity* **41,** 357–369.

Bucheton, A. (1979a). Non-Mendelian female sterility in *Drosophila melanogaster:* influence of aging and thermic treatments. II. Action of thermic treatments on the sterility of *SF* females and on the reactivity of reactive females. *Biol. Cell.* **34,** 43–49.

Bucheton, A. (1979b). Non-Mendelian female sterility in *Drosophila melanogaster:* influence of aging and thermic treatments. III. Cumulative effects induced by these factors. *Genetics* **93,** 131–142.

Bucheton, A. (1979c). Contribution à l'étude d'un cas d'interaction nucleocytoplasmique: le système inducteur–reacteur de *Drosophila melanogaster.* Controle génétique de la reactivité. Ph.D. Dissertation, Univ. Clermont-Ferrand II, Aubière, France.

Bucheton, A., and Picard, G. (1978). Non-Mendelian female sterility in *Drosophila melanogaster:* hereditary transmission of reactivity levels. *Heredity* **40,** 207–223.

Bucheton, A., Lavige, J. M., Picard, G., and L'Héritier, P. (1976). Non-Mendelian female sterility in *Drosophila melanogaster:* quantitative variations in the efficiency of inducer and reactive strains. *Heredity* **36,** 305–314.

Capdeville, Y., Vierny, C., and Keller, A. M. (1978). Regulation of surface antigens expression in *Paremecium primaurelia.* Genetic and physiological factors involved in allelic exclusion. *Mol. Gen. Genet.* **161,** 23–29.

Crow, J. F. (1957). Genetics of insect resistance to chemicals. *Annu. Rev. Entomol.* **2,** 227–246.

Doolittle, W. F., and Sapienza, C. (1980). Selfish genes, the phenotype paradigm and genome evolution. *Nature* (London) **284,** 601–603.

Eeken, J. C. J., and Sobels, F. H. (1981). Modification of MR mutator activity in repair-deficient strains of *Drosophila melanogaster. Mutat. Res.* **83,** 191–200.

Ehrman, L., and Kernaghan, R. P. (1971). Microorganismal basis of infectious hybrid male sterility in *Drosophila paulistorum. J. Hered.* **62,** 67–71.

Engels, W .R. (1979a). Germ line aberrations associated with a case of hybrid dysgenesis in *Drosophila melanogaster* males. *Genet. Res.* **33,** 137–146.

Engels, W. R. (1979b). Hybrid dysgenesis in *Drosophila melanogaster:* rules of inheritance of female sterility. *Genet. Res.* **33,** 219–136.

Engels, W. R. (1979c). Extrachromosomal control of mutability in *Drosophila melanogaster. Proc. Natl. Acad. Sci.* **76,** 4011–4015.

Engels, W. R. (1981a). Hybrid dysgenesis in *Drosophila* and the stochastic loss hypothesis. *Cold Spring Harbor Symp. Quant. Biol.* **45,** 561–565.

Engels, W.R. (1981b). Germline hypermutability in *Drosophila* and its relation to hybrid dysgenesis and cytotype. *Genetics* **98,** 565–587.

Engels, W. R., and Preston, C. R. (1979). Hybrid dysgenesis in *Drosophila melanogaster:* the biology of female and male sterility. *Genetics* **92,** 161–174.

Engels, W. R., and Preston, C. R. (1980). Components of hybrid dysgenesis in a wild population of *Drosophila melanogaster. Genetics* **95,** 111–128.

Engels, W. R., and Preston, C. R. (1981a). Characteristics of a neutral strain in the *P–M* system of hybrid dysgenesis. *Drosophila Inf. Serv.* **56,** 35–37.

Engels, W. R., and Preston, C. R. (1981b). Identifying *P* factors in *Drosophila* by means of chromosome breakage hotspots. *Cell* **26,** 421–428.

Fleuriet, A. (1976). Presence of the hereditary rhabdovirus *Sigma* and polymorphism for a gene for resistance to this virus in natural populations of *Drosophila melanogaster. Evolution* (*Lawrence Kans.*) **30,** 735–739.

Gans, M., Audit, C., and Masson, M. (1975). Isolation and characterization of sex-linked female-sterile mutants in *Drosophila melanogaster. Genetics* **81,** 683–704.

Golubovsky, M. D. (1974). Investigation of synchronous and similar changes of the gene pool in geographically separated natural populations of *Drosophila melanogaster*. *Genetika (Moscow)* **4,** 72–83. [Sov. Genet. (Engl. transl.)].

Golubovsky, M. D. (1980). Mutational process and microevolution. *Genetica (The Hague)* **52–53,** 139–149.

Golubovsky, M. D., Ivanov, Yu. N., and Green, M. M. (1977). Genetic instability in *Drosophila melanogaster:* putative multiple insertion mutants at the singed bristly locus. *Proc. Natl. Acad. Sci. USA*. **74,** 2973–2975.

Green, M. M. (1976). Mutable and mutator loci. *In* "The Genetics and Biology of *Drosophila*" (M. Ashburner and E. Novitski, eds.), Vol. 1B, pp. Academic Press, New York/London.

Green, M. M. (1977). Genetic instability in *Drosophila melanogaster*. *De novo* induction of putative insertion mutants. *Proc. Natl. Acad. Sci. USA*. **74,** 3490–3493.

Green, M. M. (1980). Transposable elements in *Drosophila* and other Diptera. *Annu. Rev. Genet.* **14,** 109–120.

Henderson, S. A., Woodruff, R. C., and Thompson, J. N., Jr. (1978). Spontaneous chromosome breakage at male meiosis associated with male recombination in *Drosophila melanogaster*. *Genetics* **88,** 93–107.

Hinton, C. W. (1979). Two mutators and their suppressors in *Drosophila ananassae*. *Genetics* **92,** 1153–1171.

Hinton, C. W. (1981). Nucleocytoplasmic relations in a mutator–suppressor system of *Drosophila ananassae*. *Genetics* **98,** 77–90.

Hiraizumi, Y. (1971). Spontaneous recombination in *Drosophila melanogaster* males. *Proc. Natl. Acad. Sci. USA*. **68,** 268–270.

Hiraizumi, Y. (1977). The relationship among transmission frequency, male recombination and progeny production in *Drosophila melanogaster*. *Genetics* **87,** 83–93.

Hiraizumi, Y. (1979). A new method to distinguish between meiotic and premeiotic recombinational events in *Drosophila melanogaster*. *Genetics* **92,** 543–554.

Hiraizumi, Y., Slatko, B., Langley, C., and Nill, A. (1973). Recombination in *Drosophila melanogaster* male. *Genetics* **73,** 439–444.

Jinks, J. L. (1964). "Extrachromosomal inheritance." Prentice-Hall, Englewood Cliffs, New Jersey.

Kidwell, M. G. (1977). Reciprocal differences in female recombination associated with hybrid dysgenesis in *Drosophila melanogaster*. *Genet. Res.* **30,** 77–88.

Kidwell, M. G. (1979). Hybrid dysgenesis in *Drosophila melanogaster:* the relationship between the *P–M* and *I–R* interaction systems. *Genet. Res.* **33,** 205–217.

Kidwell, M. G. (1981). Hybrid dysgenesis in *Drosophila melanogaster:* the genetics of cytotype determination in a neutral strain. *Genetics* **98,** 275–290.

Kidwell, M. G. (1982a). Intraspecific hybrid sterility. *In* "The Genetics and Biology of *Drosophila*" (M. Ashburner, H. L. Carson, and J. N. Thompson, Jr., eds.), Vol. 3C. Academic Press, London and New York.

Kidwell, M. G. (1982b). Hybrid dysgenesis in *Drosophila melanogaster:* evidence for transposability of *P* factors. *Genetics*

Kidwell, M. G., and Kidwell, J. F. (1975). Cytoplasm–chromosome interactions in *Drosophila melanogaster*. *Nature (London)* **253,** 755–756.

Kidwell, M. G., and Kidwell, J. F. (1976). Selection for male recombination in *Drosophila melanogaster*. *Genetics* **84,** 333–351.

Kidwell, M. G., and Novy, J. B. (1979). Hybrid dysgenesis in *Drosophila melanogaster:* sterility resulting from gonadal dysgenesis in the *P–M* system. *Genetics* **92,** 1127–1140.

Kidwell, M. G., Kidwell, J. F., and Ives, P. T. (1977a). Spontaneous non-reciprocal mutation and sterility in strain crosses of *Drosophila melanogaster*. *Mutat. Res* **42,** 89–98.

Kidwell, M. G., Kidwell, J. F., and Sved, J. A. (1977b). Hybrid dysgenesis in *Drosophila melanogaster:* a syndrome of aberrant traits including mutation, sterility and male recombination. *Genetics* **86,** 813–833.

Kidwell, M. G., Novy, J. B., and Feeley, S. M. (1981). Rapid unidirectional change of hybrid dysgenesis potential in *Drosophila*. *J. Hered.* **72,** 32–38.

Lavige, J. M., and Lecher, P. (1982). Mitoses anormales dans les embryons à developpement bloqué dans le système *I–R* de dysgénésie hybride chez *Drosophila melanogaster*. *Biol. Cell.* **44,** 9–14.

L'Héritier, Ph. (1970). *Drosophila* viruses and their role as evolutionary factors. *Evol. Biol.* **4,** 185–209.

Lim, J. K. (1979). Site-specific instability in *Drosophila melanogaster:* the origin of the mutation and cytogenetic evidence for site specificity. *Genetics* **93,** 681–701.

Mariol, M. C. (1981). Genetic and developmental studies of a new *grandchildless* mutant of *Drosophila melanogaster*. *Mol. Gen. Genet.* **181,** 505–511.

Matthews, K. A. (1981). Developmental stages of genome elimination resulting in transmission ratio distortion of the *T-007* male recombination (MR) chromosome of *Drosophila melanogaster*. *Genetics* **97,** 95–111.

Matthews, K. A., and Gerstenberg, M. V. (1979). Non-reciprocal female female sterility associated with male recombination chromosomes from Texas populations of *Drosophila melanogaster*. *Genetics* **93,** s77.

Matthews, K. A., Slatko, B. E., Martin, D. W., and Hiraizumi, Y. (1978). A consideration of the negative correlation between transmission ratio and recombination frequency in a male recombination system of *Drosophila melanogaster*. *Jpn. J. Genet.* **53,** 13–25.

Minamori, S. (1972). Extrachromosomal element delta in *Drosophila melanogaster*. VIII. Inseparable association with sensitive second chromosome. *Genetics* **70,** 557–566.

Orgel, L. E., and Crick, F. H. C. (1980). Selfish DNA: the ultimate parasite. *Nature (London)* **284,** 604–607.

Pelisson, A. (1977). Contribution à l'étude d'une sterilité femelle non mendelienne chez *Drosophila melanogaster:* mise en evidence d'un chromosome 4 inducteur. *C.R. Hebd. Seances Acad. Sci. Ser. D* **284,** 2399–2402.

Pelisson, A. (1978). Non-Mendelian female sterility in *Drosophila melanogaster:* variations of chromosomal contamination when caused by chromosomes of various inducer efficiencies. *Genet. Res.* **32,** 113–122.

Pelisson, A. (1979). The *I–R* system of hybrid dysgenesis in *Drosophila melanogaster:* influence on *SF* females sterility of their inducer and reactive paternal chromosomes. *Heredity* **43,** 423–428.

Pelisson, A. (1981). The *I–R* system of hybrid dysgenesis in *Drosophila melanogaster:* are *I* factor insertions responsible for the mutator effect of the *I–R* interaction? *Mol. Gen. Genet.* **183,** 123–129.

Pelisson, A., and Bregliano, J. C. (1981). The *I–R* system of hybrid dysgenesis in *Drosophila melanogaster:* construction and characterization of a non-inducer stock. *Biol. Cell.* **40,** 159–164.

Pelisson, A., and Picard, G. (1979). Non-Mendelian female sterility in *Drosophila melanogaster:* I-factor mapping on inducer chromosomes. *Genetica (The Hague)* **50,** 141–148.

Periquet, G. (1978). Recherche sur le déterminisme génétique de caractère atrophie gonadique chez *Drosophila melanogaster*. *Biol. Cell.* **33,** 33–38.

Periquet, G. (1981). Hybrid sterility in *Drosophila simulans:* relationships with the hybrid dysgenesis syndrome in *Drosophila melanogaster*. *Heredity* **46,** 255–261.

Picard, G. (1976). Non-Mendelian female sterility in *Drosophila melanogaster:* hereditary transmission of *I* factor. *Genetics* **83,** 107–123.

Picard, G. (1978a). Non-Mendelian female sterility in *Drosophila melanogaster:* further data on chromosomal contamination. *Mol. Gen. Genet.* **164,** 235–247.

Picard, G. (1978b). Non-Mendelian female sterility in *Drosophila melanogaster:* sterility in the daughter progeny of *SF* and *RSF* females. *Biol. Cell.* **31,** 235–244.

Picard, G. (1978c). Non-Mendelian female sterility in *Drosophila melanogaster:* sterility in stocks derived from the genotypically inducer or reactive offspring of *SF* and *RSF* females. *Biol. Cell.* **31,** 245–254.

Picard, G. (1979). Non-Mendelian female sterility in *Drosophila melanogaster:* principal characteristics of chromosomes from inducer and reactive origin after chromosomal contamination. *Genetics* **91,** 455–471.

Picard, G., and L'Héritier, P. (1971). A maternally inherited factor inducing sterility in *D. melanogaster. Drosophila Inf. Serv.* **46,** 54.

Picard, G., and Pelisson, A. (1979). Non-Mendelian female sterility in *Drosophila melanogaster:* characterization of the non inducer chromosomes of inducer strains. *Genetics* **91,** 473–489.

Picard, G., and Wolstenholme, D. R. (1980). *SF* sterility in *Drosophila melanogaster:* a comparative study of mitochondrial DNA molecules from flies of inducer and reactive strains. *Biol. Cell.* **38,** 157–162.

Picard, G., Bucheton, A., Lavige, J. M., and Fleuriet, A. (1972). Contribution à l'étude d'un phénomène de stérilité à déterminisme non mendelien chez *Drosophila melanogaster. C. R. Hebd. Seances Acad. Sci. Ser. D* **275,** 933–936.

Picard, G., Bucheton, A., Lavige, J. M., and Pelisson, A. (1976). Répartition géographique des trois types de souche impliquées dans un phénomène de sterilité à déterminisme non mendelien chez *Drosophila melanogaster. C. R. Hebd. Seances Acad. Sci. Ser. D* **282,** 1813–1816.

Picard, G., Lavige, J. M., Bucheton, A., and Bregliano, J. C. (1977). Non-Mendelian female sterility in *Drosophila melanogaster:* physiological pattern of embryo lethality. *Biol. Cell.* **29,** 89–98.

Picard, G., Bregliano, J. C., Bucheton, A., Lavige, J. M., Pelisson, A., and Kidwell, M. G. (1978). Non-Mendelian female sterility and hybrid dysgenesis in *Drosophila melanogaster. Genet. Res.* **32,** 275–287.

Proust, J., and Prudhommeau, C. (1982). Hybrid dysgenesis in *Drosophila melanogaster.* I. Further evidence for and characterization of the mutator effect of the inducer–reactive interaction. *Mutat. Res.* **95,** 225–235.

Raymond, J. D., and Simmons, M. J. (1981). An increase in the X-linked lethal mutation rate associated with an unstable locus in *Drosophila melanogaster. Genetics* **98,** 291–302.

Rubin, G., Kidwell, M. G., and Bingham, P. A. (1982). The molecular basis of *P–M* hybrid dysgenesis: the nature of induced mutations. *Cell* **29,** 987–994.

Sainsart-Chanet, A. (1976). Gene controlled selection of mitochondria in *Paramecium. Mol. Gen. Genet.* **145,** 23–30.

Schaefer, R. E., Kidwell, M. G., and Fausto-Sterling, A. (1979). Hybrid dysgenesis in *Drosophila melanogaster:* morphological and cytological studies of ovarian dysgenesis. *Genetics* **92,** 1141–1152.

Simmons, M. J., and Lim, J. K. (1980). Site-specificity of mutations arising in dysgenic hybrids of *Drosophila melanogaster. Proc. Natl. Acad. Sci. USA.* **77,** 6041–6046.

Simmons, M. J., Johnson, N. A., Fahey, T. M., Nellett, S. M., and Raymond, J. D. (1980). High mutability in male hybrids of *Drosophila melanogaster. Genetics* **96,** 479–490.

Sinclair, D. A. R., and Green, M. M. (1979). Genetic instability in *Drosophila melanogaster:* the effect of male recombination (MR) chromosomes in females. *Mol. Gen. Genet.* **170,** 219–224.

Slatko, B. E. (1978a). Parameters of male and female recombination influenced by the *T-007* second chromosome in *Drosophila melanogaster*. *Genetics* **90,** 257–276.

Slatko, B. E. (1978b). Evidence for newly induced genetic activity responsible for male recombination induction in *Drosophila melanogaster*. *Genetics* **90,** 105–124.

Slatko, B. E., and Hiraizumi, Y. (1975). Elements causing male crossing-over in *Drosophila melanogaster*. *Genetics* **81,** 313–324.

Sochacka, J. H. M., and Woodruff, R. C. (1976). Induction of male recombination in *Drosophila melanogaster* by injection of extracts of flies showing male recombination. *Nature (London)* **262,** 287–289.

Sonneborn, T. M., and Lesuer, A. (1948). Antigenic characters in *Paramecium aurelia* (variety 4): determination inheritance and induced mutations. *Am. Nat.* **82,** 69–70.

Stamatis, N., Yannopoulos, G., and Pelecanos, M. (1981). Comparative studies of two male recombination factors (MRF) isolated from a Southern Greek *Drosophila melanogaster* population. *Genet. Res.* **38,** 125–136.

Sved, J. A. (1976). Hybrid dysgenesis in *Drosophila melanogaster:* a possible explanation in terms of spatial organization of chromosomes *Aus. J. Biol. Sci.* **29,** 375–388.

Sved, J. A. (1978). Male recombination in dysgenic hybrids of *Drosophila melanogaster:* chromosome breakage or mitotic crossing-over? *Aus. J. Biol. Sci.* **31,** 303–309.

Sved, J. A. (1979). The "hybrid dysgenesis" syndrome in *Drosophila melanogaster*. *BioScience* **29,** 659–664.

Sved, J. A., Murray, D. C., Schaefer, R. E., and Kidwell, M. G. (1978). Male recombination is not induced in *Drosophila melanogaster* by extracts of strains with male recombination potential. *Nature (London)* **257,** 457–458.

Thompson, J. N., Jr., and Woodruff, R. C. (1978). Chromosome breakage: a possible mechanism for diverse genetic events in outbred populations. *Heredity* **40,** 153–157.

Thompson, J. N., Jr., Woodruff, R. C., and Schaefer, G. B. (1978). An assay of somatic recombination in male recombination lines of *Drosophila melanogaster*. *Genetica (The Hague)* **49,** 77–80.

Wickner, R. B. (1974). Killer character of *Saccharomyces cerevisiae:* curing by growth at elevated temperature. *J. Bacteriol.* **117,** 1356–1357.

Wickner, R. B. (1978). Twenty-six chromosomal genes needed to maintain the killer double-stranded RNA plasmid of *Saccharomyces cerevisiae*. *Genetics* **88,** 419–425.

Woodruff, R. C., and Bortolozzi, J. (1976). Spontaneous recombination in males of *Drosophila simulans*. *Heredity* **37,** 295–298.

Woodruff, R. C., and Thompson, J. N., Jr. (1977). An analysis of spontaneous recombination in *Drosophila melanogaster* males: isolation and characterization of male recombination lines. *Heredity* **38,** 291–307.

Woodruff, R. C., Slatko, B. E., and Thompson, J. N., Jr. (1978). Lack of an interchromosomal effect associated with spontaneous recombination in males of *Drosophila melanogaster*. *Ohio J. Sci.* **78,** 310–317.

Woodruff, R. C., Slatko, B. E., and Thompson, J. N., Jr. (1982). Factors affecting mutation rate in natural populations. *In* "The Genetics and Biology of *Drosophila*" (M. Ashburner, H. L. Carson, and J. N. Thompson, Jr., eds.), Vol. 3C. Academic Press, London and New York.

Yannopoulos, G. (1978a). Progressive resistance against the male recombination factor 31.1 MRF acquired by *Drosophila melanogaster*. *Experientia* **34,** 1000–1001.

Yannopoulos, G. (1978b). Studies on the sterility induced by the male recombination factor 31.1 MRF in *Drosophila melanogaster*. *Genet. Res.* **32,** 239–247.

Yannopoulos, G. (1979). Ability of the male recombination factor 31.1 MRF to be transposed to another chromosome in *Drosophila melanogaster*. *Mol. Gen. Genet.* **176,** 247–253.

Yannopoulos, G., and Pelecanos, M. (1977). Studies on male recombination in a southern Greek *Drosophila melanogaster* population. (a) Effect of temperature. (b) Suppression of male recombination in reciprocal crosses. *Genet. Res.* **29,** 231–238.

Yannopoulos, G., and Zacharopoulov, A. (1980). Studies on the chromosomal rearrangements induced by the male recombination factor 31.1 MRF in *Drosophila melanogaster*. *Mutat. Res.* **73,** 81–92.

Young, M. W., and Schwartz, H. E. (1981). Nomadic gene families in *Drosophila*. *Cold Spring Harbor Symp. Quant. Biol.* **45,** 629–640.

Zalokar, M. (1976). Autoradiographic study of protein and RNA formation during early development of *Drosophila* eggs. *Dev. Biol.* **49,** 425–437.

CHAPTER 10

Retroviruses

HAROLD E. VARMUS

I. Introduction

Retroviruses compose a large class of animal viruses united by a single principle: Their single-stranded RNA genomes are replicated through double-stranded DNA intermediates. Despite this simple unifying notion, retroviruses are rich in diversity and experimental opportunity. The biological manifestations of infection with these agents include many kinds of tumors and other diseases; the viruses originate in many animal species and can be transmitted both by conventional horizontal means and by inheritance through the germ line; and the unusual form of parasitism that retroviruses exhibit in their host cells makes them useful reagents for investigating gene regulation and expression in animal cells. For these and other reasons, retroviruses have acquired a wide following in the scientific community and studies of them occupy a large berth in journals and library shelves.

It is, however, not apparent from these few statements why retroviruses are

Mobile Genetic Elements

ISBN 0-12-638680-3

entitled to inclusion in a volume devoted to transposable elements. There are now several reasons to support this editorial decision:

1. The structure of retroviral DNA bears an uncanny resemblance to that of transposable elements of bacteria, yeast, and insects. Features of the integrated (proviral) form of viral DNA suggest that similar mechanisms may be employed for both transposition of those elements and the integration of retroviral DNA.

2. Although it is conventional for virologists to view proviruses as intermediates in the replication of retroviruses, it may also be instructive to view the viruses themselves as intermediates in the transposition of viral genes from one proviral integration site to another. Thus, proviruses can be considered as transposable genetic elements in the broad sense, even though it must be stressed that proviruses have not been shown to undergo transposition in the manner understood for the structurally homologous elements in bacteria. It is possible, however, that the mechanism of transposition employed by retroviral proviruses may be shared by movable, apparently nonviral elements of other organisms (e.g., the *copia* elements of *Drosophila*).

3. Proviruses exhibit many of the functional properties expected of other transposable elements: insertion mutagenesis; excision by recombination between terminally repeated sequences; activation of the expression of flanking host genes; transduction of cellular genes; and deletion formation.

My purpose in this chapter is twofold: (*1*) to introduce the outsider to the cardinal features of retroviruses that govern their biological behavior and guide experimental manipulation and (*2*) to present a detailed account of those properties of retroviruses and their proviruses relevent to a perception of them as movable, mutable, and mutagenic agents. The discussion of the principles of retrovirology will be kept to the minimum necessary to follow the ensuing description of retroviruses as transposable elements. Interested readers are advised to consult the recently published monograph on RNA tumor viruses (Weiss *et al.*, 1982) for a thoroughly detailed and fully bibliographed account of the tenets of this field.

II. The Properties of Retroviruses

A. Discoveries and Principles of Classification

The study of retroviruses has a long and rich history that antedates any sophisticated notions of what animal viruses are or how they work. Retroviruses were first encountered as filterable, infectious agents capable of inducing tumors in experimental animals (Gross, 1970). The first such agent to be rigorously studied, a virus isolated from a spontaneous connective tissue tumor (sarcoma) in a domestic chicken, is named for Peyton Rous, who discovered it in 1910 (Rous, 1911); it has served as the most useful single experimental reagent throughout the growth of this science. Among the other early isolates first noted as oncogenic agents and later proved to belong to this virus class were viruses capable of inducing leukosis in chickens (Ellerman and Bang, 1908), mammary carcinomas in laboratory mice (Bittner, 1936), and leukemia in newborn mice (Gross, 1951).

1. Common Properties

Electron microscopy of virus particles provided the first rational basis for grouping these viruses together and for distinguishing them from other kinds of tumor viruses (Bernhard, 1960; Teich, 1982). However, the development of biochemical techniques to analyze viruses was required for a coherent system of taxonomy. Although the primary criterion for admission of a virus to the family of Retroviridae is replication of an RNA genome through a DNA intermediate, there are a number of additional features that are inevitably associated with members of this group. Foremost among these are: (a) *morphological characteristics:* despite subtle variations in virion structure, all retroviruses are approximately 80–120 nm in diameter and are composed of nucleoprotein cores (ca 40–70 nm in diameter) enveloped by plasma membrane derived from the host cell during virus release; (b) *biochemical composition:* approximately 60% protein, 30% lipid, 5% carbohydrate, and 1% RNA; (c) *the nature of their genome:* a complex of two subunits of single-stranded RNA and low molecular weight cellular RNAs (see Section II,B); (d) the *gene products required for replication:* several core proteins produced by a gene called *gag;* an RNA-directed DNA polymerase (reverse transcriptase) encoded in the *pol* gene; and glycoproteins synthesized under the direction of a viral *env* gene (Baltimore, 1975); and (e) certain *physiochemical properties:* buoyant density of approximately 1.16–1.18 g/ml in CsCl; sensitivity to heat, lipid solvents, and detergents; and relative resistance to UV and X irradiation.

Assiduous work during the past couple of decades has uncovered hundreds of viruses that belong to the retroviral family (Teich, 1982), so that the current challenge to the taxonomist is one of distinguishing among members of this group in a rational and comprehensible fashion. In fact, there have proved to be many useful ways to categorize retroviruses, some of which are germane to the central concerns of this chapter.

2. The Natural Hosts

The simplest organizing principle depends upon the *host of origin* for each retroviral isolate. Retroviruses have been discovered in a wide range of vertebrates, including snakes, fish, birds, and many mammalian orders, and there is tentative evidence for retroviruses in tapeworms and insects (Teich, 1982). The most useful experimental agents have come from chickens, mice, cats, cows,

Table I Properties of Some Commonly Studied Retroviruses

Viruses (classes and strains)[a]	Abbreviations	Species of origin	Exogenous[b]	Endogenous[c]	Transforming[d]	Tumors[e]	Other diseases[f]
Avian acute leukemia viruses	AcLVs	Chicken					
Erythroblastosis virus	AEV	Chicken	+		+	+ (S)	
Myeloblastosis virus	AMV	Chicken					
Myelocytomatosis virus	MC29	Chicken					
Avian leukosis viruses	ALVs	Chicken					
Rous-associated virus-0	RAV-0	Chicken		+	–	–	
Rous-associated viruses-1, -2, etc.	RAV-1, -2, etc.	Chicken	+		–	+ (L)	A, O
Myeloblastosis-associated viruses	MAV-1, MAV-2	Chicken	+		–	+ (L)	O
Avian sarcoma viruses	ASVs	Chicken					
Esh	ESV	Chicken					
Fujinami	FSV	Chicken					
Poultry Research Center	PRC II, PRC IV	Chicken					
Rous	RSV	Chicken	+		+	+ (S)	
Bratislava	B77-RSV	Chicken					
Bryan high titer	BH-RSV	Chicken					
Prague	PR-RSV	Chicken					
Schmidt Ruppin	SR-RSV	Chicken					
Baboon endogenous virus	BEV	Baboon		+	–	–	
Bovine leukemia virus	BLV	Cow	+		–	+ (L)	
Equine infectious anemia virus	EIAV	Horse	+		–	–	A, Ar
Feline leukemia viruses	FeLV	Cat	+		+	+ (L)	A
Feline sarcoma viruses	FeSV	Cat					
Gardner–Arnstein	GA-FeSV	Cat	+		+	+ (S)	
McDonough	SV-FeSV	Cat					
Snyder–Theilen	ST-FeSV	Cat					
Gibbon ape leukemia virus	GALV	Gibbon	+		–	+ (L)	

Human T-cell leukemia viruses[g]	HTLV	Human	+		?	+ (L)(?)	?
Murine leukemia viruses	MLV	Mouse					
Abelson	Ab-MLV	Mouse	+		+	+ (S)	
Friend	Fr-MLV	Mouse	+		−	+ (S)	A, E
Moloney	Mo-MLV	Mouse	+		−	+ (L)	
Rauscher	Ra-MLV	Mouse	+		−	+ (L)	
Mink focus forming	MCF-MLV	Mouse	+		−	+ (L)	
Spleen focus-forming virus	SFFV	Mouse	+		−	+ (S)	A, E
Xenotropic	X-MLV	Mouse		+	−	−	
Murine mammary tumor viruses	MMTV	Mouse	+/−	+/−	−	+ (L)	
Murine sarcoma viruses	MSV	Mouse, rat	+		+	+ (S)	
Harvey	Ha-MSV	Rat					
Kirsten	Ki-MSV	Rat					
Moloney	Mo-MSV	Mouse					
Rat leukemia virus	RaLV	Rat		+	−	−	
Rat sarcoma virus	RaSV	Rat	+		+	+ (S)	
RD 114 virus	RD114	Cat		+	−	−	
Reticuloendotheliosis viruses	REVs	Birds	+		+/−	+/−	
Chicken syncytial virus	CSV	Chicken	+		−	+ (L)	Various
Reticuloendotheliosis virus-A	REV-A	Turkey	+		−	+ (L)	
Reticuloendotheliosis virus, strain T	REV-T	Turkey	+		+	+ (S)	
Spleen necrosis virus	SNV	Turkey	+		−	−	
Simian sarcoma virus	SSV	Wooley monkey	+		+	+ (S)	
Viper retrovirus	VRV	Russell's viper	?		?	?	
Visna virus	Visna	Sheep	+		−	−	N, P

[a] Virus groups are listed alphabetically, with significant strains and substrains as examples. (Modified from Teich, 1982, Table 2.3)
[b] + means that some component of the viral genome is not present in the germ line of the species of origin in proviral form.
[c] + means that complete, infectious genomes are present in the germ line of some animals.
[d] + means that the virus can regularly alter behavior of cultured cells to a neoplastic phenotype; all transforming viruses listed here carry an oncogene derived from host cells (see Table III).
[e] + means that the virus can induce some kind of malignant tumor after short (S) or long (L) latency.
[f] A, anemia; Ar, arthritis; E, erythroblastosis; N, neurological disease; P, pneumonia; O, osteopetrosis.
[g] The oncogenicity of HTLV has not been documented, but its presence is strongly associated with T-cell lymphoma (Poiesz *et al.*, 1980; Hinuma *et al.*, 1981).

and lower primates and are generally named in a manner than reflects their origins; Table I provides a list of some of the most commonly employed viruses.

3. Retroviral Diseases

The first retroviruses to be sighted were those competent to produce tumors, and the *pathological consequences* of infection of animals or cultured cells remain important classifying characteristics (Gross, 1970; Teich, 1982). As indicated in Table I, retroviruses can cause a considerable number of disorders in animals, or no disease at all. Upon infection of cultured cells—generally fibroblasts or hematopoietic cells—several retroviruses are capable of transforming the behavior of their hosts to a phenotype resembling that of a tumor cell (neoplastic transformation), but the vast majority of retroviruses have little or no effect upon the phenotype of the host cells, even though they may be oncogenic in animals. A few viruses induce cytopathic effects in cultured cells (vacuolization, syncytia formation, or even cell death), although the effects are usually confined to certain types of host cell.

4. Morphological Types

One of the earliest efforts at classification of retroviruses took advantage of consistent if subtle *morphological differences* perceived during electron microscopic examination of viruses and virus-producing cells (Bernhard, 1960; Teich, 1982). Although all extracellular retroviruses appear to acquire an envelope by budding through the cellular plasma membrane, they differ in the manner by which the core is assembled, in the position of the core within the mature particle, and in the prominence of surface spikes formed by envelope glycoproteins. Thus, the *C-type particles*—by far the most common structural form, accounting for the frequent misuse of the term as a synonym for all retroviruses—assemble their nucleocapsids just beneath the plasma membrane at the site of budding; the mature form has a dense, centrally located nucleoid and spikes that are visualized only with difficulty. In contrast, the *B-type particles*—represented primarily by the mouse mammary tumor virus (MMTV)—form a complete nucleocapsid at some distance from the plasma membrane, and the finished extracellular particles have eccentrically located cores and prominent spikes. *A-type particles* are intracellular, nonenveloped forms present either free in the cytoplasm or in intracisternal spaces; they are noninfectious but in some cases are probably precursors to infectious extracellular particles, particularly of the B type. *D-type particles*—exemplified by a number of primate retroviruses—have properties of both B- and C-type particles.

5. Modes of Transmission

Retroviruses are transmitted in nature by a variety of routes. Many retroviruses, like animal viruses generally, are transmitted only as *exogenous agents* by contact between an infected and a susceptible animal, by congenital infections (via a competent virus carried in the milk or gametes), or by experimental inoculation. In addition, retroviruses are often transmitted genetically as *endogenous agents* through proviruses present in the germ line of fertile animals. These endogenous proviruses are thought to have been established in the germ line by rare infectious events, and they frequently share interesting peculiarities (lack of oncogenicity or genetic defectiveness). These issues will be more extensively discussed in Section II,F.

6. Host Range

Retroviruses gain entry to cells by a specific interaction between their envelope glycoproteins and cellular receptors; within the cell, they are dependent upon host mechanisms for full expression of viral genes. The varied capacities of cells from different species and different lineages to synthesize suitable receptors and/or create a favorable environment for viral replication determine the host range of each virus isolate (Weiss, 1982). Exogenous viruses (or endogenous viruses that can replicate in their host of origin) are called *ecotropic,* and those viruses transmitted as endogenous proviruses in one host but able to replicate only in cells of another species are called *xenotropic.* Some murine viruses that grow in both mouse and foreign cell types have been assigned to two additional classes, *amphotropic* and *dualtropic,* that appear to use different host receptors. The major host range classes can often be usefully subdivided, for example, according to the capacity of ecotropic viruses to infect cells of a defined genotype, of xenotropic viruses to infect cells from certain foreign species, or of either type to infect cells from certain tissues. Last, cells can categorized as *permissive* or *nonpermissive* hosts for the replication of certain retroviruses, depending upon whether they permit the production of infectious virus after the successful introduction of a replication-competent genome into the cell.

Three of the mechanisms that govern the viral host range have been scrutinized genetically, biochemically, or both; because these mechanisms will figure in our later discussions, they deserve a fuller description here.

1. Entry of viruses of the avian sarcoma-leukosis class into chicken (and other avian) cells is determined by interactions between the products of a polymorphic viral locus (*env*) and the products of several cellular loci called *tva, tvb, tvc,* and *tve* (Vogt and Ishizaki, 1966; Weiss, 1982). The alleles found at the *env* locus (*env*-A, *env*-B, *env*-C, etc.) code for viral envelope glycoproteins that apparently interact specifically with as yet poorly characterized host membrane proteins encoded by dominant alleles at the *tv* loci. The presence of an appropriate membrane receptor (e.g., the product of a dominant *tva* allele for the product of *env*-A) appears to be a virtually absolute requirement for penetration of the host cell by the virus, because virus titers are reduced by as much as six logs when assayed in resistant host cells (e.g., *env*-A virus in cells homozygous for the recessive *tva* allele). Competition among coinfecting viruses of the same subgroup for suitable receptors accounts for virus interference in this system.

2. Resistance of certain mouse cells to replication by certain murine leukemia viruses is determined in part by alleles at a locus termed *Fv-1* (Pincus *et al.*, 1971; Jolicoeur, 1979). N-tropic viruses (defined by their capacity to replicate in prototypic NIH-3T3 cells) and B-tropic viruses (defined by their capacity to replicate in BALB/c cells) demand homozygosity (*Fv-1*^{nn} or *Fv-1*^{bb}, respectively) at the *Fv-1* locus for efficient growth. Thus, in this system, resistance is the dominant character (e.g., *Fv-1*^{nb} hosts restrict the growth of both N- and B-tropic viruses), but resistance is not absolute and can be overcome at high multiplicities at infection. Although the precise mechanism of restriction is not known, it appears to operate at some step in the virus life cycle subsequent to virus entry and has the effect of preventing integration of viral DNA (Jolicoeur and Baltimore, 1976a). Thus, the products of alleles conferring resistance must interfere with either the correct synthesis of viral DNA or its integration into the host chromosome.

3. The nonpermissivity of mammalian cells for the replication of avian sarcoma-leukosis viruses is absolute and may reflect an inability of the host cell to process or assemble viral structural proteins (Eisenman and Vogt, 1978). The avian viruses generally enter their mammalian host cells with low efficiency (Altaner and Temin, 1970), and avian proviral DNA, once established, is expressed less efficiently than in avian hosts (Quintrell *et al.*, 1980). However, all of the necessary mRNAs and primary products of translation are synthesized in the mammalian hosts, although no infectious virus can be detected in the culture medium. Viral structural proteins do not undergo their normal cleavages in these cells, but it is not certain whether this phenomenon is the cause or a secondary consequence of the failure to assemble virus particles.

7. Genomic Content of Replication Genes

The genetic composition of retroviruses also provides a useful tool for classification (Coffin, 1982a). As described in more detail below, replication of retroviruses requires the participation of three viral genes—*gag, pol,* and *env*—and viruses carrying all three of these in functional form are called *replication-competent*. The functional similarities and arrangements of these three genes strongly suggest that all retroviruses from a wide range of species are descended from a single progenitor virus, rather than being independently generated within individual animal species or genera. Nevertheless, the divergence among contemporary virus isolates (as measured by molecular hybridization or immunological tests or by direct sequencing of proteins or nucleic acids) is as great as, and often greater than, the divergence among typical cellular genes in their respective natural hosts. On the one hand, such tests can readily distinguish agents found within a single species: Chickens, mice, cats, and some primates are hosts for two or more groups of viruses that are virtually unrelated by such criteria (Teich, 1982). On the other hand, genomic relatedness has frequently established genealogical connections between viruses isolated from widely divergent species. In such cases, it has been proposed that the viruses may have been transmitted between species in the relatively recent past (Todaro, 1980).

Viruses with mutations affecting one or more of the three common genes are *replication-defective;* this term is most frequently applied to those viruses with large deletions or substitutions in coding domains for structural proteins (Coffin, 1982a; Linial and Blair, 1982). Because replication-competent viruses can generally provide the gene products needed by closely related, replication-defective viruses, the former are sometimes referred to as *helper viruses*. The phenotypically mixed viruses that result from these mixed infections are described in a later section (Section II,D,2).

8. *onc* GENES

Several of the most extensively studied retroviruses carry sequences closely related to those of conserved cellular genes (e.g., Stehelin *et al.*, 1976; Spector *et al.*, 1978; Roussel *et al.*, 1979; Goff *et al.*, 1980; DeFeo *et al.*, 1980). Because the presence of these sequences correlates with the capacity of the virus to transform cultured cells and induce tumors efficiently in animals, the sequences have been generically called viral *onc* (or v-*onc*) regions, and their cellular homologues are called c-*onc* regions. The presence of such sequences usefully divides retroviruses into those that are v-*onc*$^+$ and those that are v-*onc*$^-$, but it is important to keep in mind that many of the v-*onc*$^-$ viruses are nevertheless able to induce a wide variety of malignant tumors, albeit with a considerable latency period and with reduced efficiency in most cases (Teich *et al.*, 1982). Viruses lacking v-*onc* have not yet been found to transform cultured cells to a neoplastic phenotype. Although tumorigenic v-*onc*$^-$ viruses are commonly replication-competent, almost all of the v-*onc*$^+$ viruses are replication-defective. This is so because the cellular genetic information (*onc*) was acquired at the expense of coding sequences for one or more of the viral replication genes (see Section II,B,3 and Fig. 2); only in certain strains of Rous sarcoma virus (RSV) is a v-*onc* gene accompanied by a full complement of replication genes.

Using biochemical and immunological criteria, more than a dozen distinct cellular genes have been identified as sources of v-*onc* genes present in the genomes of numerous transforming retroviruses (Bishop and Varmus, 1982). The nature of the *onc* element carried by a retrovirus now provides one of the most useful means of classification. To distinguish among the many *onc* sequences, they have been assigned trivial names based upon the names of viruses in which they were originally identified in each animal species (Coffin *et al.*, 1981). A current list of these sequences and their protein products is provided in Table II. It is apparent that several genes are represented in multiple virus isolates from the same species; in a few cases (v-*bas* and v-*Ha-ras;* v-*fps* and v-*fes*), related genes in different species have been the source of viral oncogenes (Shibuya *et al.*, 1980; Andersen *et al.*, 1981).

Table II Oncogenes and Their Products

onc[a]	Viral isolates[b]	Virus strains[c]	v-*onc* product[d]	Tyrosine kinase[e]	Species of origin[f]	c-*onc* expression[g]	c-*onc* product[g]
src	>2	Rous sarcoma virus	$pp60^{v\text{-}src}$	+	Chicken	+	$pp60^{c\text{-}src}$
fps[h]	>3	Fujinami sarcoma virus	$P140^{gag\text{-}fps}$	+	Chicken		
yes	2	Yamaguichi-73 sarcoma virus	$P90^{gag\text{-}yes}$	+	Chicken	+	?
ski	3	SKI 770 avian sarcoma virus	?	?	Chicken	?	?
ros	1	URI virus	$P68^{gag\text{-}ros}$	+	Chicken	+	?
myc	4	Myelocyctomatosis virus MC29	$P110^{gag\text{-}myc}$	−	Chicken	+	?
myb	2	Avian myeloblastosis virus	$p45^{v\text{-}myb}$	−	Chicken	+	?
erb-A	1	Avian erythroblastosis virus	$P75^{gag\text{-}erbA}$	−	Chicken	+	?
erb-B	2	Avian erythroblastosis virus	$gP65^{v\text{-}erbB}$	−	Chicken	+	?
rel	1	Reticuloendothesiosis virus, strain T	?	?	Turkey	+	?
abl	1	Abelson murine leukemia virus	$P120^{gag\text{-}abl}$	+	Mouse	+	$p150^{c\text{-}abl}$
mos[h]	2	Moloney murine sarcoma virus	$P37^{v\text{-}mos}$	−	Mouse	−	—
fos	2	FBJ osteosarcoma virus	$p55^{v\text{-}fos}$	−	Mouse	+	?
bas[h]	1	BALB murine sarcoma virus	$P21^{v\text{-}bas}$	?	Mouse	+	$p21^{c\text{-}bas}$
ras	>3	Harvey murine sarcoma virus	$p21^{Ha\text{-}ras}$	−	Rat	+	$p21^{c\text{-}ras}$
		Kirsten murine sarcoma virus	$p21^{Ki\text{-}ras}$				
fes[h]	>2	Snyder-Theilen feline sarcoma virus	$P85^{gag\text{-}fes}$	+	Cat	+	$p92^{c\text{-}fes}$
		Gardner-Arnstein feline sarcoma virus	$P95^{gag\text{-}fes}$	+			
fms	1	McDonough feline sarcoma virus	$gP180^{gag\text{-}fms}$	?	Cat	+	?
sis	1	Simian sarcoma virus	?	?	Wooley monkey	+	?

[a] The three-letter names for various *onc* sequences are derived according to rules described in Coffin *et al.* (1981).
[b] The minimum number of probably independent virus isolates carrying the same or a closely related *onc* sequence.
[c] One or two examples of actively studied viral strains carrying each *onc* element.
[d] The probable protein products of *onc* genes from each virus strain were identified as reviewed in Bishop and Varmus (1982) and Bishop (1983).
[e] + signifies evidence for a protein kinase activity specific for tyrosine residues associated with the v–*onc* protein. − implies evidence against tyrosine kinase activity; ? means insufficient evidence.
[f] The species in which the listed virus strains were isolated.
[g] Evidence for expression and products of c–*onc* genes are summarized in Bishop and Varmus (1982) and Bishop (1983).
[h] *fps* is related to *fes* (Shiyuba *et al.*, 1980) and *bas* is related to *ras* (Andersen *et al.*, 1981).

B. Characteristics of Retroviral Genomes

1. STRUCTURAL OF THE VIRAL GENOMIC COMPLEX

All of the viral information carried in a retrovirus particle is contained within a single-stranded piece of RNA, ranging in length among virus strains from about 3.5 to 9.5 kb and exhibiting the structural features of eukaryotic mRNA. In mature virions, however, the viral genome is normally found as a 60–70 S aggregate composed of two identical 30–35 S viral RNA subunits and multiple low molecular weight RNAs of host origin (Coffin, 1982a). The presence of two viral subunits in each genomic complex implies that retroviruses are diploid, an oddity among animal viruses. Furthermore, virions produced during mixed infection by closely related viruses are likely to be heterozygotes, and these, in turn, may be intermediates in viral recombination (see Section II,D,2). Based upon thermal denaturation and electron microscopic studies, the two subunits are believed to be joined by hydrogen bonds, involving sequences located close to the 5′-end of each subunit (Kung *et al.*, 1975; Bender and Davidson, 1976; Bender *et al.*, 1978). However, the position of the linkage has not been precisely defined, the complex has not been reassembled *in vitro* from denatured subunits, and a role for the dimer linkage in the virus life cycle has not been established. Variable amounts of host 4 and 5 S RNAs are associated with the 60–70 S complex by virtue of weak hydrogen bonding; additional 4, 5, 7, and ribosomal 18 and 28 S RNA are also found free within virus particles (Coffin, 1982a).

Although the selection of host RNAs for packaging is not random, the mechanism and purpose for the inclusion of most of these species are not known. The single exception is a host tRNA tightly associated with each 30–35 S RNA subunit (Taylor, 1977). Generally, 18 bases at the 3′-end of this tRNA are base-paired with a viral subunit about 100–180 bases from its 5′-terminus (see Fig. 1

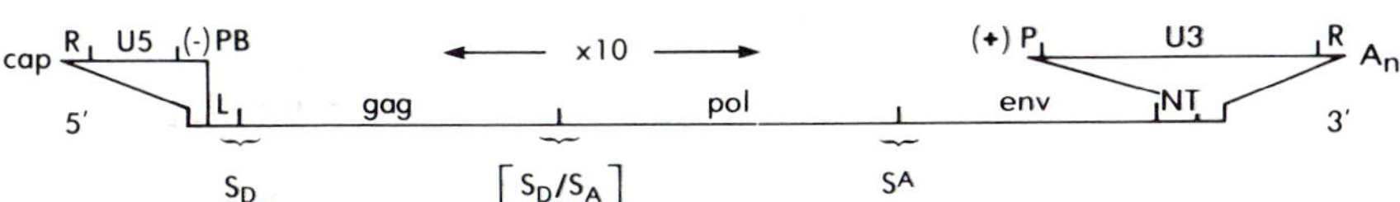

Fig. 1 Anatomy of an RNA subunit of a replication-competent retrovirus. The important structural and genetic domains of a typical single-stranded RNA subunit (ca. 8–9 kb) of a replication-competent retrovirus are indicated by the following notations: cap, 5′-"cap" nucleotide; R, terminal redundancy; U5, sequence unique to the 5′-terminus, present twice in viral DNA; (−)PB, binding site for (−)DNA strand primer, a host tRNA; L, leader region, S_D, donor splice site; S_A, acceptor splice site; *gag*, coding region for nucleocapsid proteins; *pol*, coding region for DNA polymerase; *env*, coding region for envelope glycoproteins; NT, nontranslated region between *env* and U3; (+)P, polypurine tract that is the probable primer for the (+)DNA strand; U3, sequence unique to the 3′-terminus, present twice in viral DNA; A_n, tract of poly(A) at the 3′-terminus. For clarity, terminal regions are shown at 10× expanded scale. Lengths of all regions are approximate and vary among retroviruses (e.g., see Table III for variations in R, U5, and U3 sequences). The bracketed S_D/S_A refers to the proposed splicing sites used to generate the infrequent mRNA for Pr180$^{gag\text{-}pol}$ (see text).

Table III Important Sequences in and around LTRs[a]

Virus	Putative (+) strand primer sequence	U3 IR sequence[b]	U3 Length[c]	U3 CCAAT box
RSV	. . .AGGGAGGGGGA	AATGTAGTCTTATGC. . .	230	. . .CCACT. . . (24)
ev-1	. . .AGGGAGGGGGA	AATCTAGTC. . .	172	. . .CATAT. . . (23)
SNV	. . .AGTGGGG	AATGT. . .	369	. . .ACATT. . . (34)
MoMLV-MSV	. . .AGAAAAAGGGGGG	AATGAAAGACCCC. . .	371–442	. . .CCAAT. . . (45)
MMTV	. . .AAAAAGAAAAAAGGGGGA	AATGCCGC. . .	1192	. . .CCAAAT. . . (29)
	↑ Initiation site for (+) DNA	↑ Integration site		

[a] Sequences are abstracted from published work of Swanstrom *et al.* (1981) (RSV), Hishinuma *et al.* (1981) (*ev*-1), Shimotohno *et al.* (1980) (SNV), Dhar *et al.* (1980) and Van Beveren *et al.* (1980) (MoMLV-MSV) and Donehower *et al.* (1981) and Majors and Varmus (1981) (MMTV). For a description of functional and structural properties of LTRs, see Sections II,B and C.

and Table III). This tRNA species serves as primer for the first (minus) strand of viral DNA during reverse transcription. It also appears to bind to reverse transcriptase in some cases, and this affinity may direct the inclusion of the correct tRNA in virus particles (Panet *et al.*, 1975).

2. Structure and Coding Potential of the Typical Subunit of Viral RNA

The genetic organization and the biochemical features of a single subunit of viral RNA are illustrated in Fig. 1. Many of these landmarks were defined by correlating genetic properties of virus stocks, particularly deletion mutations, with structural studies [initially, oligonucleotide mapping of RNA; later, heteroduplex formation, restriction endonuclease mapping, and sequencing of DNA (Coffin, 1982a)]. Recently, at least three viral genomes [those of PRC-RSV (Schwartz *et al.*, 1982), Mo-MLV (Shinnick *et al.*, 1981), and Mo-MSV (Van Beveren *et al.*, 1981a; Reddy *et al.*, 1981)] have been sequenced in their entirety, substantial portions of other genomes (RAV-O, MMTV) have been sequenced, and considerable amounts of corresponding amino acid sequence have also been determined (Hughes, 1982). The generalized picture of a replication-competent *onc*$^-$ virus that emerges from these studies can be summarized as follows, starting from the 5′-terminus of the RNA:

a. The 5′-Cap. The 5′-end of the viral subunit, like the 5′-ends of most mRNAs, is adorned with an inverted 7-methylguanyl nucleotide (the "cap" nucleotide) not encoded in the viral genome.

b. The R Sequence. The penultimate nucleotide (also methylated) is the first base in a sequence of about 10–80 nucleotides (nt) repeated at both ends of the RNA. [The sequence adjoins but does not include those noncoded bases, the

U3			R	U5		
TATA box		Poly(A) signal	length[c]	Length[c]	IR sequence[b]	Binding site for tRNA primer
. . .TATTTAG. . .	(16)	. . .AATAAA (1)	21	80	GCAGAAGGCTTCATT	TGGTGACCCCGACGTGAT. . .
. . .TATATAA. . .	(16)	. . .AATAAA (1)	21	80	GGCTTCATT	TGGTGACCCCGACGTGAT. . .
. . .TATAAG. . .	(21)	(53) AATAAA. . . (21). . .	80	100	ACATT	TGGGGGCTCGTCCGGGAT. . .
. . .AATAAAAG. . .	(21)	(47) AATAAA. . . (16). . .	70	75	GGGGTCTTTCATT	TGGGGGCTCGTCCGGGAT. . .
. . .TATAAAAG. . .	(15)	. . . AGTAAA(3)	13	122	GCGGCAGC	TGGCGCCCGAACAGGGAC. . .
		Cap site (initiation of viral RNA)	Polyadenylation site		Integration site	Initiation site for (−) DNA

[b] Inverted repeat sequence at the ends of LTRs.

[c] Number (in base pairs) indicates the length of the entire U3, R, or U5 sequence. Numbers in parenthesis indicate base pairs between illustrated sequences or between sequences and boundaries.

5′-cap and 3′-poly(A), added posttranscriptionally.] This direct repeat, called *R*, figures prominently in the mechanism of reverse transcription. One copy of R is present within the long terminal repeat (LTR) at each end of proviral DNA (Table III).

c. The U5 Sequence. The R sequence is followed by a sequence of 80–100 nt called *U5* that occurs only once in viral RNA but twice in viral DNA as a component of the LTR. This sequence tends to be quite highly conserved among virus strains but in no case encodes viral proteins. Ribosomes have been found to protect a portion of RSV RNA in the U5 domain (Darlix *et al.*, 1979), but there is little reason to believe that this binding is functionally significant, because multiple unused AUGs lie between the binding site and the AUG for *gag* (Swanstrom *et al.*, 1982a).

d. The Primer Binding Site (PBS). The 3′-boundary of U5 is determined by the initiation site for synthesis of the first strand of viral DNA, and the initiating site, in turn, is fixed by the site at which the tRNA primer is bound to the RNA. The PBS is 18 nt in length, extending to the position of the first modified base from the 3′-end of the priming tRNA in the several cases examined.

e. The Leader Region (L). Following the PBS, there is a sequence of variable length, probably a few hundred nucleotides in most cases, preceding the start of the first structural gene, *gag*. This noncoding 5′-region is sometimes referred to by the letter L, although the *gag* leader sequence includes R, U5, and PBS, in addition to L. In some retroviral genomes, the splice donor site (S_D) for generation of subgenomic mRNAs lies within the L domain (J. Majors, unpublished); in other genomes (e.g., RSV), the donor site lies within *gag* (Hackett *et al.*, 1982). The L sequence may also constitute part of the signal that facilitates packaging of viral RNA into particles (Shank and Linial, 1980).

f. gag. The *gag* region (ca. 2.0 kb) encodes a polyprotein of 65,000–76,000 MW, later cleaved to form the four or five components that make up most of the core of the virus particle (Dickson *et al.*, 1982). The full functional significance of the *gag* proteins has yet to be understood with regard either to the architecture of the virions or to the possible regulatory roles of these proteins in packaging, uncoating, reverse transcription, and processing of viral RNA and proteins. The *gag* polyprotein is probably synthesized from an mRNA structurally indistinguishable from the virion subunit, but in the infected cell the RNA destined for progeny virions and *gag* mRNA have distinctly different half-lives and may be in physically separated pools (Levin and Rosenak, 1976). Furthermore, some retroviruses (e.g., Mo-MLV) produce a second *gag* polyprotein, probably from an mRNA generated by an uncommon splicing event. This *gag* product can be found as a glycosylated surface membrane protein, although its function in the virus life cycle is not known (Edwards and Fan, 1979).

g. pol. The *pol* region (ca. 2.9 kb) encodes at least one-half of a polyprotein also containing most or all of the amino acid sequence encoded in *gag*. The *gag-pol* polyprotein (ca. 160,000–180,000 MW) is the precursor to the mature form of reverse transcriptase, the properties of which are described in more detail below (Section II,C,1). *pol* also encodes an endonuclease, considered below, that could play a role in DNA synthesis or integration. Because the reading frames for *gag* and *pol* are different in the RSV genome, it is likely that a splicing event produces the mRNA required for synthesis of the *gag-pol* protein (Schwartz *et al.*, 1982). At least 10-fold more *gag* than *gag-pol* protein is synthesized in infected cells; hence, the predicted splice is not commonly made, and the amount of *pol* mRNA has been too small to permit direct physical characterization. As a result, the precise relationship of *gag* to *pol* at their junction is uncertain and probably varies among retroviruses. For example, in the RSV genome, the termination codon for *gag* is followed by in-frame terminators, and the start of the coding sequence for mature DNA polymerase is about 20 nt downstream in the −1 reading frame (Schwartz *et al.*, 1982). In the Mo-MLV genome, the open reading frame for *pol* directly follows the single *amber* termination codon for *gag* (Shinnock *et al.*, 1981). This structure permits enhanced synthesis of the *gag-pol* polyprotein in reticulocyte extracts to which *amber* suppressor tRNA has been added (Philipson *et al.*, 1978). Although a suppression mechanism might also operate *in vivo,* it seems more likely that the termination codon is normally bypassed by splicing, but the acceptor site (the beginning of the *pol* domain) has not been defined.

h. env. The third coding domain, *env,* is approximately 1.8 kb in length and also encodes a polyprotein (ca. 60,000–68,000 MW). A signal peptide is removed from the primary translation product, and the *env* protein is also glycosylated and cleaved to form the major components of the viral envelope (Dickson *et al.*, 1982). These proteins determine the host range of the virus and are the targets for neutralizing antibodies. In the two strains of nondefective retroviruses sequenced to date, the 3′-end of *pol* overlaps the 5′-end of *env*, and their proteins are encoded in different reading frames (Shinnick *et al.*, 1981; Schwartz *et al.*, 1982). Because *env* is expressed via a subgenomic, spliced mRNA, a splice acceptor site (S_A) precedes the translated domain. In the case of RSV, it appears that the splicing event joins six codons from the 5′-end of *gag* to a site five codons within the open reading frame in *env;* thus, the primary translation product is encoded in both *gag* and *env* (Schwartz *et al.*, 1982; Hackett *et al.*, 1982). The position of the acceptor splice site has not been defined for MLV, but a conventional situation obtains for MMTV: A leader sequence lacking any part of *gag* is joined in *env* mRNA just upstream from the start of the *env* open reading frame (J. Majors, unpublished).

i. 3′-NT. The noncoding region on the 3′-side of *env* varies in length among retroviruses and can be subdivided into the portion that appears in the LTR (U3 and R) and the portion that does not. The latter portion, labeled 3′-NT in Fig. 1, lacks substantial open reading frames and has only one known function: The 10–20 nt directly preceding the U3 boundary constitutes a purine-rich tract, highly conserved among retroviruses, that appears to serve as primer for synthesis of plus strand DNA (see Section II,C,3 and Tables III and V).

j. U3. U3 is the unique sequence near the 3′-end of viral RNA that appears twice in viral DNA as a component of the LTR. The first 5–23 nt of this sequence are inversely repeated at the 3′-end of the U5 sequence to form inverted repeats within each LTR. The U3 sequence ranges in size from about 170 to 1250 nt among strains of retroviruses and, as a result of deletions and duplications, also varies appreciably in length among isolates of a single virus strain (Van Beveren *et al.*, 1982; Shimotohno and Temin, 1982; Ju *et al.*, 1980). The 3′-NT–U3 boundary is determined by the site of initiation of plus strand DNA; U3 always begins with the sequence AAUG, which also forms part of the inverted repeat within LTRs (Table III). The major functions of the U3 sequence concern the regulation of synthesis and processing of viral RNA; these functions are reviewed in Section II,C,5. U3, like the 3′-NT region, is usually devoid of open reading frames, but in the case of MMTV, with a U3 sequence of about 1250 nt, there is a translatable sequence sufficient to encode a protein of 36,000 MW beginning with the AUG at positions 2–4 (Dickson and Peters, 1981; Donehower *et al.*, 1981; Kennedy *et al.*, 1982; J. Majors, unpublished; E. Buetti and H. Diggelmann, personal communication). Although the anticipated protein has been synthesized *in vitro*, the protein has yet to be observed in infected cells.

k. R. The final heteropolymeric domain in viral RNA has the same sequence, R, found at the 5′-terminus. Because the length of R is determined by the site of polyadenylation, there may be heterogeneity in the length of R at the 3′-terminus [e.g., as in RSV, with lengths of 16, 19, and 21 nt (Schwartz *et al.*,

1977)]. The polyadenylation signal itself is generally in U3, except in instances in which R is significantly longer than 20 nt (e.g., MLV and SNV; see Table III and Section II,C,5,*c*).

l. Poly(A). All retroviral genomes and mRNAs analyzed to date conclude with a poly(A) tract (generally 100–200 nt in length). The poly(A) is added posttranscriptionally, presumably by cellular polynucleotide synthetases; in other words, the poly(A) is not copied during the synthesis of viral DNA and hence does not appear in the proviral DNA template for synthesis of viral RNA.

3. Structures of Variant Genomes

Retroviruses are genetically unstable, and abnormal genomes, particularly deletion mutants, arise at high frequency (Coffin, 1979). Although there is little in the viral genome that is not necessary for replication, the major coding domains, *gag, pol,* and *env,* can be complemented in trans and hence are not absolutely required for multiple rounds of infection, provided that a compatible helper virus is available to supply missing functions. On the other hand, lesions affecting the DNA priming sites and essential regions of the LTR would not be expected to be complemented by companion viruses; genomes bearing such lesions are nonconditionally defective and can be identified only if they exist in proviral form.

Surely the most interesting of the variant genomes—and those to which most attention has been given—are those that carry *onc* sequences derived from cellular genes (see Section II,A,8). Several versions of *onc*$^+$ genomes have been encountered (Fig. 2), differing in the competence of the genome to direct replication and in the manner in which the *onc* sequences are expressed (Coffin, 1982a).

1. In the simplest case, represented thus far only by strains of RSV, the *onc* region (*src*) has been introduced into a helper virus genome between *env* and U3; hence, the genome is replication-competent (Fig. 2B). Expression of *src* requires production of a spliced subgenomic mRNA, using the donor site also employed for *env* mRNA and an acceptor site derived from the cellular sequence acquired during the transduction of *src* (Hackett *et al.*, 1982; R. Swanstrom and R. Parker, unpublished).

2. Of the other *onc*$^+$ genomes, all of which are replication-defective, the most common arrangement has the 5′-boundary of the *onc* sequence fused in frame to the coding domain for *gag* (Fig. 2D). The acquisition of *onc* sequences has occurred at the expense of viral structural genes, not uncommonly with the loss of at least part of all three major coding domains. Expression of the *onc* sequence in such cases occurs by synthesis of a hybrid protein (*gag-onc*), apparently from an mRNA structurally equivalent to genomic RNA.

3. In several cases, the *onc* sequence is positioned closer to the 3′-terminus, sparing *gag* and *pol* but sacrificing *env* (Fig. 2C). Synthesis of the *onc* protein most likely occurs via a spliced subgenomic mRNA.

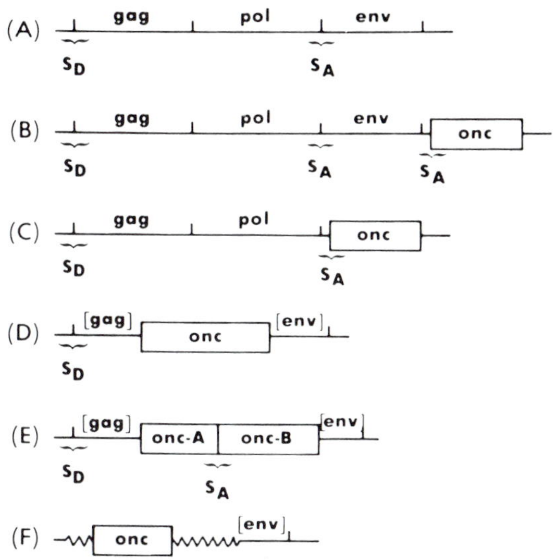

Fig. 2 Naturally occurring types of genomic variations produced by the inclusion of *onc* genes within viral RNA. Each line illustrates an RNA subunit representative of the indicated virus group. Symbols are as described for Fig. 1, with bracketed names indicating incomplete coding domains. For simplicity, terminal sequences are not identified. (A) Replication-competent genome without *onc*, as in Fig. 1. (B) Replication-competent genome with *onc;* only certain strains of Rous sarcoma examplify this type. (C) Replication-defective genome with the *onc* gene expressed via a subgenomic mRNA; examples include avian myeloblastosis virus, Moloney murine sarcoma virus, and the Bryan strain of Rous sarcoma virus. (D) Replication-defective genome with the *onc* gene expressed by synthesis of the *gag-onc* fusion protein. Examples include Fujinami, Y73, UR-1, and PRC II avian sarcoma viruses, the avian leukemia virus MC-29, Abelson murine leukemia virus, and the feline sarcoma viruses. (E) A replication-defective genome containing an *onc* element composed of two coding domains expressed by the mechanisms illustrated for both (C) and (D); the only known example is the avian erythroblastosis virus. (F) A replication-defective genome composed of two retrovirus-like components (straight and serrated lines) and an *onc* gene; examples include the Kirsten and Harvey strains of murine sarcoma virus.

4. In at least one instance (avian erythroblastosis virus), the *onc* region (*erb*) appears to have two functional domains (*erb*-A and *erb*-B), independently expressed by two of the mechanisms described above (Fig. 2E). Thus, *erb*-A is translated as part of a *gag-erb*-A fusion protein and *erb*-B is translated independently from a spliced, subgenomic mRNA.

5. The genomes of Harvey and Kirsten strains of murine sarcoma virus (MSV) present another special case. Here the oncogene (*Ha-ras* or *Ki-ras*) is joined to sequences from *two* sets of *onc*$^-$ genomes: that of the Harvey or Kirsten strain of MLV and that of the endogenous, virus-like (VL) 30 S RNA of rats (Ellis *et al.*, 1980; Young *et al.*, 1980; Fig. 2F). The transforming genomes, which arose during propagation of Ha- or Ki-MLV in rats expressing VL 30 S genomes, thus represent the products of particularly complex recombinational events. The *ras* sequences are apparently located on the 5′-side of any residual viral coding

domains and are expressed as unfused proteins, perhaps by genome-sized mRNA.

C. The Retrovirus Life Cycle

For several years prior to the discovery of reverse transcriptase and unequivocal demonstration of the provirus, it was appreciated that retroviruses did not behave like other animal viruses with RNA genomes: Virus replication was sensitive to inhibitors of both DNA synthesis (e.g., cytosine arabinoside) and DNA-templated RNA synthesis (e.g., actinomycin D) (Bader, 1965; Temin, 1963, 1967). Although these findings were interpreted by some to imply a dependence upon synthesis of cellular DNA and RNA, Temin's proposal—that the RNA genome was converted to a DNA provirus that then served as a template for synthesis of viral RNA (Temin, 1964)—was ultimately found to be correct. The decisive support for the *provirus hypothesis,* as it was called, came from several directions: biological experiments in which a form of the viral genome appeared to be labeled with BUdR (Boettiger and Temin, 1970; Balduzzi and Morgan, 1970); identification of an RNA-directed DNA polymerase activity in virions (Baltimore, 1970; Temin and Mizutani, 1970); isolation of temperature-sensitive (*ts*) and nonconditional mutants in the viral *pol* gene (Hanafusa and Hanafusa, 1971; Linial and Mason, 1973); demonstration that DNA from infected cells was infectious when introduced into new cells (Hill and Hillova, 1972); and physical characterization, isolation, and finally molecular cloning of proviral DNA (reviewed by Varmus and Swanstrom, 1982).

The profound biological implications of DNA synthesis from an RNA template have focused most experimental attention upon that phase of the life cycle, but earlier and later events during replication also offer considerable insight into the behavior of eukaryotic cells. In this section, the life cycle is viewed from the vantage point of the virologist, considering in classic order the early events (adsorption, penetration, and uncoating), the middle phase (replication and expression of the genome), and the late events (viral assembly, release, and maturation). In Section III, the sequence will be permuted to view the events from the perspective of the student of DNA transposition, starting with a provirus and ending with a new provirus in a new cell. Before considering the life cycle, however, a brief digression is required to describe some of the biochemical properties of the products of *pol,* because these proteins are so intimately involved in replicative events.

1. The Products of *pol*

Several enzymatic activities have been associated with the products of retroviral *pol* genes: (*1*) a DNA polymerase activity capable of using RNA or DNA templates; (*2*) an RNase H activity; and (*3*) a DNA endonuclease activity (Ver-

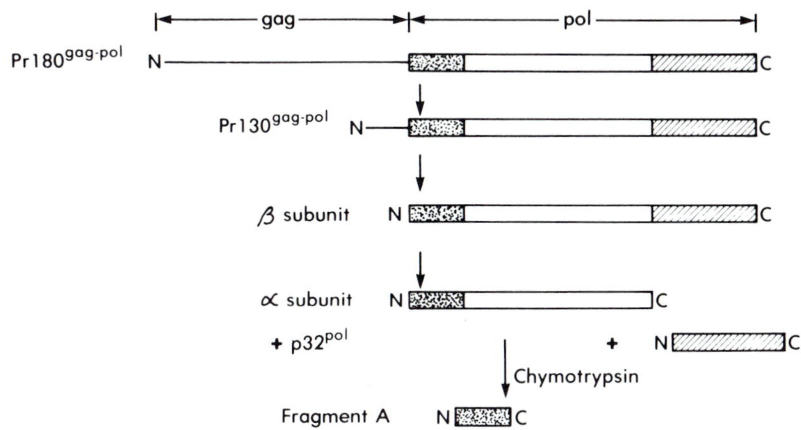

Fig. 3 Derivation of components of the avian sarcoma-leukosis virus *pol* protein from the primary translation product. The coding domains for *gag* and *pol* collaborate to produce the polyprotein precursor Pr180$^{gag\text{-}pol}$. Sequential cleavages, probably mediated by the protease activity in P15^{gag}, generate the intermediate Pr130$^{gag\text{-}pol}$ and the β subunit of the mature polymerase (Eisenman *et al.*, 1980). Further cleavage of the β subunit produces the α subunit and the viral endonuclease p32^{pol}. (The substrate for this cleavage may be a dimer of β-subunits, yielding a complex of α and β subunits and one molecule of p32^{pol}.) Treatment of the α subunit with chymotrypsin *in vitro* produces fragment A, a MW 24,000 peptide with ribonuclease H activity (Lai and Verma, 1978).

ma, 1977; Gerard and Grandgenett, 1980). All functions but the last have been assigned roles in the virus life cycle, as described in later sections.

The structure and biogenesis of the proteins responsible for these activities have been most thoroughly studied with the avian leukosis and sarcoma viruses (ALSVs). The ALSV *pol* proteins are derived from a polyprotein precursor, Pr180$^{gag\text{-}pol}$, by a series of cleavages, at least some of which are mediated by a product of *gag* (p15^{gag}) (Moelling *et al.*, 1980; Fig. 3). The mature enzyme, as isolated from virions, is composed of a 1 : 1 complex of two subunits called β (95,000 MW) and α (65,000 MW). The two subunits are coextensive at their amino termini (Copeland *et al.*, 1980); all of the α peptides are also found in β (Gibson and Verma, 1974). The α subunit can catalyze both DNA synthesis and hydrolysis of RNA in an RNA–DNA hybrid; the β subunit demonstrates these activities as well and appears to enhance the stability of the interaction of the enzyme complex with its polymeric templates and substrates (Grandgenett and Green, 1974). The complex also demonstrates a DNA endonuclease activity assigned to the carboxyl-terminal domain, because the activity is present in a 32,000 MW carboxy-teminal cleavage products of β, called the p32 endonuclease (Grandgenett *et al.*, 1978, 1980). The active site for DNA synthesis has not been localized within the α subunit, but a 24,000 MW product of chymotrypsin digestion displays RNase H activity in the absence of polymerizing activity (Lai and Verma, 1978).

Less is known about the protein products of the *pol* genes of other retroviruses. The murine leukemia virus (MLV) polymerase is also derived from a $Pr180^{gag\text{-}pol}$, but the mature form in virions is a monomer of 70,000 MW with an affiliated RNase H activity (Verma, 1975). A DNA endonuclease of 40,000 MW is present in MLV particles, but the manner in which it is generated is not known (Kopchick *et al.*, 1981).

Both *ts* and nonconditional mutants in *pol* have been studied (Linial and Blair, 1982). The nonconditional mutants are defective for replication unless complemented by phenotypic mixing with a pol^+ helper virus (Hanafusa and Hanafusa, 1971). The *ts* mutants produce enzymatic complexes that display thermolabile function *in vitro* and show impaired DNA synthesis and reduced infectivity when used to infect cells at nonpermissive temperatures (Verma *et al.*, 1974, 1976). Mutations in *pol* examined to date always affect the DNA polymerase function and generally all functions (Golomb *et al.*, 1981); mutations specifically affecting endonuclease or ribonuclease H activities would be of considerable interest.

The biochemical functions of the *pol* products have been examined in three ways: using enzymes purified from virions in *reconstructed reactions;* using virions treated with nonionic detergents to stimulate *endogenous reactions;* and using materials from *infected cells* to deduce enzymatic behavior *in vivo*. Most attention has been given to the DNA polymerase activity in each context.

The ease of solubilization of retroviral polymerases, compared to other virion polymerases, has greatly aided their purification and analysis (Verma, 1977). However, purified reverse transcriptases have not proved capable of copying their natural templates into DNA resembling that found in infected cells. The purified enzymes are active on both RNA and DNA templates and, like other DNA polymerases, exhibit an absolute requirement for a primer. Although host tRNAs are natural primers for reverse transcriptases, the enzymes display a strong preference for deoxyribonucleotide primers when copying their favored templates, homopolymeric ribonucleotides, in reconstructed reactions. The purified enzymes will also copy DNA with great efficiency, using DNA primers.

The major product of synthesis by purified enzymes on single-stranded RNA templates is complementary, single-stranded DNA (cDNA). With retroviral template-primer, most of the cDNA is a run-off product (strong stop cDNA) of 100–180 nt (Haseltine *et al.*, 1976; Haseltine and Kleid, 1978), reflecting the position of the primer close to the 5′-end of the viral subunit (Fig. 1). Purified reverse transcriptase is commonly used in molecular cloning experiments to generate copies of eukaryotic mRNAs, using oligo(dT) to prime synthesis on poly(A) tracts at their 3′-ends. The cloning procedures are often facilitated by the ability of the viral polymerase to synthesize a second DNA strand when the cDNA forms a self-priming "hairpin" by base-pairing between its 3′-end and an internal region with sufficient homology (Efstratiatis *et al.*, 1976).

By careful manipulation of reaction conditions, several investigators have succeeded in synthesizing full-length, infectious linear and open circular DNA in endogenous reactions. The use of mellitin or precise concentrations of detergent to permeabilize virions and attention to divalent cation concentrations have been particularly important (Junghans *et al.*, 1975; Rothenberg *et al.*, 1977; Boone and Skalka, 1980, 1981a). Synthesis under these conditions proceeds as fast as or faster than that observed in infected cells, and the full-length products appear to be structurally and biologically indistinguishable from those in infected cells. It is not known whether the endogenous reaction contains specific protein factors that are required for correct synthesis and are lost during purification of the enzyme, or whether some topological feature of the viral core permits a faithful rendering of the synthetic process.

The *RNase H* activity associated with reverse transcriptase is generally measured using an artificial substrate composed of a radiolabeled ribonucleotide homopolymer annealed to a complementary deoxynucleotide polymer (Moelling *et al.*, 1971; Leis *et al.*, 1973). The viral enzyme, unlike the RNase H activity isolated from *Escherichia coli,* will not digest RNA that lacks free ends (e.g., RNA in a Col El plasmid); rather, retroviral RNase H appears to act as an exonuclease, working in either the 3′ to 5′ or the 5′ to 3′ direction to generate oligoribonucleotides 6–12 nt in length. However, its suggested role in DNA synthesis requires that it ignore the nontranscribed poly(A) tract at the 3′-end of the virion subunit to digest heteropolymeric RNA found in hybrid form after DNA synthesis. Other proposed functions for RNase H include removal of part or all of the 5′-R sequence after synthesis of strong stop cDNA and generation of oligomers of genomic RNA to act as primers for synthesis of the second strand of DNA. Evidence for both of these functions has been obtained through experiments with reconstructed and endogenous reactions (Collett *et al.*, 1978; Friedrich and Moelling, 1979; Olsen and Watson, 1980).

The *DNA endonuclease* activity associated with reverse transcriptase has been most extensively studied with materials isolated from avian myeloblastosis (AMV) virions. The activity associated with the α–β complex is usually assayed for its ability to nick superhelical, double-stranded, circular (form I) DNA in the presence of Mn^{2+}; under these conditions, but less efficiently, it will also linearize form I DNA or nick linear duplex DNA or single-stranded RNA or DNA (Golomb and Grandgenett, 1979; Grandgenett *et al.*, 1980). The endonuclease activity of the α–β complex presumably resides in the p32 region of the β subunit (see above), although p32 also exhibits some activity in the presence of Mg^{2+}. The p32 product binds with high affinity to nucleic acid (Grandgenett *et al.*, 1978) and is alleged to bind with some specificity to sequences within cloned RSV LTRs (Misra *et al.*, 1982). The endonuclease could function during either DNA synthesis or, more likely, integration (e.g., by cleaving either viral or host DNA; see below), but no direct biochemical or genetic support for these proposals is available. The functions of other enzymatic activities found in virions—a topoisomerase activity (Weis and Faras, 1981; Hizi *et al.*, 1982) and an unwinding activity (Collett *et al.*, 1978)—are similarly uncertain.

2. EARLY EVENTS

Surprisingly little is known about the mechanisms that initiate the retrovirus life cycle. As discussed earlier (Section II,A,6), retroviruses enter cells with the assistance of plasma membrane receptors that interact specifically with glycoproteins in the viral envelope. Although it is often stated that *adsorption* of virus to the cell is nonspecific and that *penetration* is specifically mediated by the cellular receptors, these conclusions depend upon frail methods and should be judged accordingly. Moreover, there is recent evidence to indicate that adsorption may also occur more efficiently in the presence of appropriate receptors (Notter *et al.*, 1982). The issues are clouded by several theoretical and experimental difficulties: Most of the physical particles in any retroviral stock are noninfectious and might behave inappropriately; the biochemical nature of the receptors has not been defined; and it is not known whether the virus normally enters the cell by fusion with the plasma membrane or by engulfment and transport within cytoplasmic vesicles (Simons *et al.*, 1982). The first well-studied event in the life cycle—synthesis of the linear form of viral DNA—occurs in the cytoplasm (Varmus *et al.*, 1974), but the degree of uncoating of the virus particle and the precise location of the synthetically active complex within the cytoplasm are not known.

3. THE MIDDLE PHASE (I): SYNTHESIS AND INTEGRATION OF VIRAL DNA

Within an hour after addition of retroviruses to suitable host cells in culture, it is possible to detect newly synthesized viral DNA using molecular hybridization methods (Varmus *et al.*, 1978). The requirements for DNA synthesis include an appropriate template and primer (a viral subunit with its associated host tRNA); a functional reverse transcriptase *within* the infecting particle (*pol* mutants cannot be complemented at this step by coinfection); and a suitable cellular environment [normal synthesis does not occur in cells arrested by starvation (Fritsch and Temin, 1977; Varmus *et al.*, 1977)].

During the first few hours of infection, the number of viral DNA molecules is generally less than 10, corresponding roughly to the multiplicity of infection. Within 1–2 days the number of molecules per cell may rise to over 100 (Khoury and Hanafusa, 1976), but this is generally attributed to a secondary round of infection rather than to semiconservative replication, because there is no evidence that unintegrated viral DNA constitutes a replicon. The major product of viral DNA synthesis in the cytoplasm is a linear duplex, slightly longer than a subunit of viral RNA by virtue of the duplication of U3 and U5 that generates a copy of the LTR unit at each end (Shank *et al.*, 1978a; Hsu *et al.*, 1978; Fig. 4). Synthesis of viral DNA is rather slow; it takes 4 hr or more to complete a 10-kb molecule (Varmus *et al.*, 1978). The first (minus) strand appears to be made continuously, using host tRNA as a single primer to initiate synthesis at the 3′-

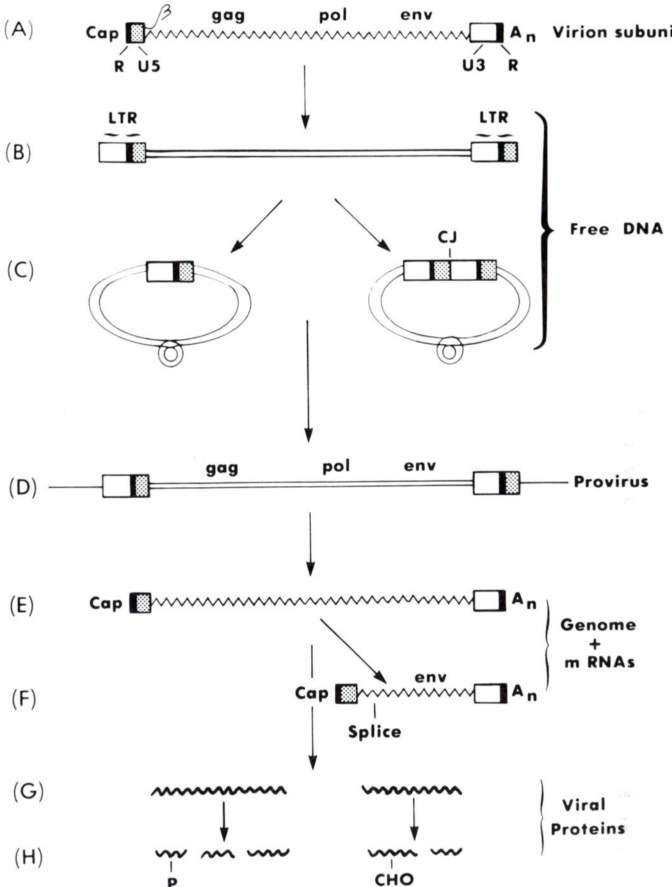

Fig. 4 The life cycle of retroviruses. (A) One of the two identical subunits of a viral RNA genome with its major structural and genetic features, as in Fig. 1. (B) The primary product of reverse transcription, linear duplex DNA, with its LTRs composed of U3, R, and U5. (C) The major forms of closed circular DNA, with one or two copies of the LTR. (D) Proviral DNA. (E, F) Genomic and mRNAs, derived from the primary transcript by capping, polyadenylation, and (in the case of subgenomic mRNA) splicing; the site at which the 5′- and 3′-domains of subunit RNA are joined to form *env* mRNA is indicated in (F). (G, H) The polyproteins synthesized from viral mRNAs and their mature products after cleavage and, in some cases, glycosylation (CHO) or phosphorylation (P).

boundary of U5. The second (plus) strand seems to be made continuously by some retroviruses and discontinuously by others (Kung *et al.*, 1981; Gilboa *et al.*, 1979); but in all cases, the first plus strand fragment to be synthesized is the most critical (Varmus *et al.*, 1978; Mitra *et al.*, 1979), and it is primed at a specific site, the 5′-boundary of U3.

The manner in which primary events and transfers of nascent strands between templates conspire to produce a linear duplex with LTRs is diagrammed in Fig. 5

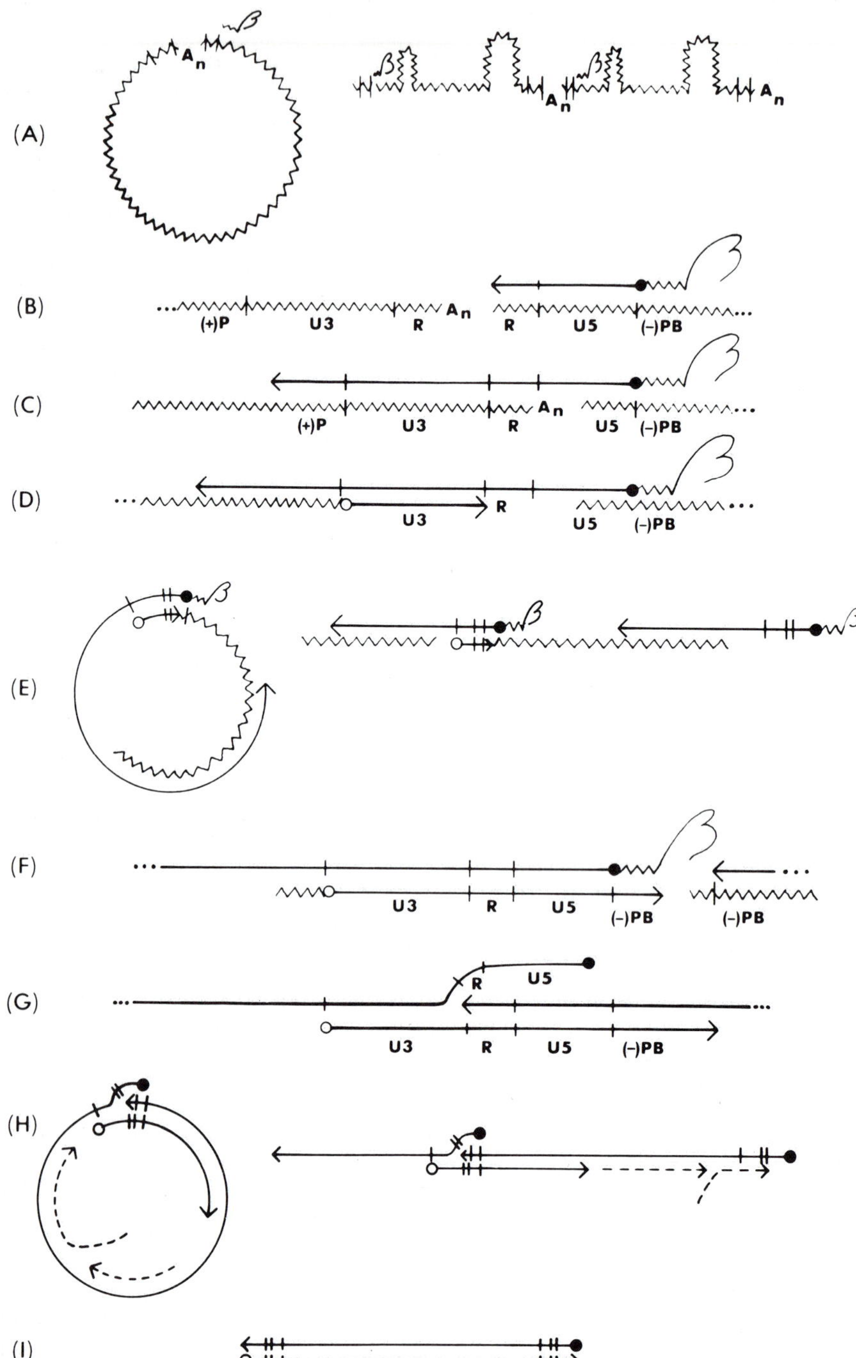

Fig. 5 The synthesis of retroviral DNA. The major steps in synthesis of viral DNA are outlined at two scales. The general topography is shown at a reduced scale [lines (A), (E), (H), and (I)], with the figures on the left illustrating the use of a single viral subunit to generate viral DNA and those on the

(Gilboa *et al.*, 1979; Varmus and Swanstrom, 1982). Synthesis of the minus strand must come to a temporary halt when the polymerase reaches the 5′-end of its RNA template 100–180 nt from the PBS. Removal of some of the RNA in the newly formed RNA:DNA hybrid, using the RNase H activity associated with the polymerase, produces a free nascent DNA strand, with a complementary copy of R, capable of annealing to the R sequence at the 3′-end of an RNA subunit. In this position, the nascent minus strand can prime synthesis along the full length of the RNA subunit. Before the minus strand is completed, however, plus strand synthesis is initiated at the U3 boundary. Although the primer for the second strand has not been identified, it is likely to consist of an oligoribonucleotide product of RNase H digestion. In all retroviruses examined, the plus strand priming site is purine-rich and the plus strand (i.e., U3) begins with the sequence AATG (Mitra *et al.*, 1982; Table III). The mechanisms of these intriguingly specific events are not yet understood.

Extension of the plus strand, like the initial phase of minus strand synthesis, is frustrated by an exhaustion of template. After copying the 300–1400 nt at the 5′-end of the minus strand, the polymerase appears to extend the plus strand about another 18 nt by copying the tRNA primer up to the position of its first modified base ($m^1{}_A$) (Gilboa *et al.*, 1979; Taylor and Hsu, 1980; Swanstrom *et al.*, 1982b). Once the genome-sized minus strand has read through the PBS, the stage is set for another transfer of strands between templates. In this instance, the base

---

right illustrating the use of two viral subunits; other variations employing two subunits are also possible and may contribute to viral recombination (see text and Fig. 9). The priming events for minus and plus strands of viral DNA (B and D) and the two transfers of nascent strands between templates (B–C and F–G), the pivotal moments in reverse transcription, are pictured on an expanded scale. RNA is shown as wavy lines; DNA as straight lines, with arrows denoting the direction of synthesis; the 5′-end of plus strand DNA is indicated by a closed circle and the 5′-end of plus strand DNA by an open circle. (−)PB is the tRNA primer binding site, and (+)P is the putative plus strand primer sequence. The short vertical lines designate the boundaries of U3, R, and U5 in all panels. (A) Template:primer positioned to permit the first transfer to occur between the ends of the same subunit (left) or between the ends of different subunits (right). (B) A nascent minus strand copy of R–U5, primed by host tRNA, is extended to the extreme 5′-end of its template. (C) After removal of the 5′-end of the primary template, the nascent minus strand has base-paired with the R sequence at the 3′-terminus of the same or a companion RNA subunit and is extended along its secondary template of viral RNA. (D) Synthesis of plus strand DNA commences from a priming site at the boundary of U3. (E) The products synthesized to the point shown in line (D) are redrawn at reduced scale to show two possible configurations of templates and products; note that much of the RNA subunit has been removed, presumably by RNase H. (F) The plus strand is extended through a portion of the tRNA sequence originally bound to viral RNA; the minus strand concurrently elongates toward the 5′-end of viral RNA, into or beyond the (−)PB site. (G) The nascent minus strand in (F) has base-paired with the plus strand, using complementary sequences from the (−)PB region. The plus strand is then extended on its second template, minus strand DNA, and the minus strand is extended by displacement synthesis along its third template, plus strand DNA. (H) The events in (G) are shown at a reduced scale, again indicating the use of either one RNA subunit (left) or two (right). Dashed lines indicate the possibility of multiply initiated internal plus strands; these may displace adjacent fragments during synthesis and play a role in viral recombination (cf. Fig. 9). (I) The completed product, duplex linear DNA with LTRs.

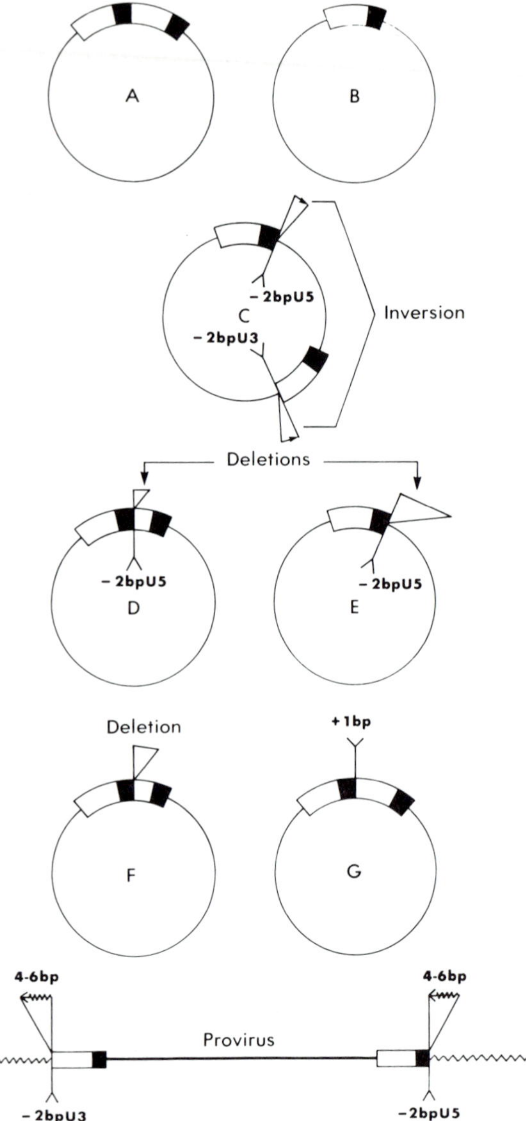

Fig. 6 Varieties of closed circular retroviral DNA identified by analysis of molecularly cloned molecules. The figure illustrates the structure of types of monomeric circular DNA, as deduced from restriction mapping and nucleotide sequencing of molecules cloned from cells infected with retroviruses. For comparison, an integrated provirus is shown at the bottom. Open boxes indicate U3 regions; closed boxes, U5 regions. Expanded regions with arrows represent short duplicated sequences, with the arrows indicating the direction of the repeated sequence; inversions, insertions, and deletions are so labeled. (A) Circular DNA with two complete LTRs. (B) Circular DNA with one copy of the LTR. (C) Circular DNA resulting from a probable autointegration event (see the text for discussion of the precursor and mechanism). (D, E) Circular DNA lacking 2 bp from one LTR and

pairing involves the PBS, with the 3′-end of the long minus strand, freed of RNA template, now capable of annealing to the 3′-end of the short plus strand. When this transfer has been made, the minus strand can be extended another several hundred nucleotides by copying the rest of the plus strand, and the plus strand can be extended several thousand nucleotides by copying the rest of the minus strand. [In some cases, however, multiple internal priming sites exist for plus strands, and little if any full-length plus strand is made (Kung *et al.*, 1981).] The final product is a blunt-ended linear duplex with two copies of the U3-R-U5 unit that constitutes the LTR.

Although this scheme for synthesis of retroviral DNA has considerable experimental support, there are a few interesting uncertainties:

1. What is the role of the dimeric genome in DNA synthesis? Does each transfer involve two subunits or one? Can two complete linear molecules be synthesized from each dimer?
2. How is the poly(A) tract eliminated to allow RNase H access to the heteropolymeric RNA:DNA hybrid?
3. What is required for displacement synthesis by viral DNA polymerase? The enzyme must read the PBS from genomic RNA despite the presence of tRNA and must copy the 5′-end of plus strand DNA even though it forms a duplex with the 5′-end of minus strand DNA.
4. How is the precise priming of the first plus strand achieved? What is the primer and how is it generated?
5. Are the predicted ends of linear DNA correct? Only in one case has an apparently intact (albeit noninfectious) linear molecule been cloned from infected cells using a prokaryotic host–vector system (Scott *et al.*, 1981). In this instance (linear DNA of gibbon ape leukemia virus, or GALV), the sequences predicted to be at the ends of the LTR were present at the junctions of viral and vector DNA produced by blunt end ligation *in vitro*.

Although the biological fate of a considerable portion of linear DNA is uncertain, pulse-chase experiments with BUdR and thymidine have shown that some of the linear DNA is transported to the nucleus (by unknown mechanisms) and that some of the transported DNA is converted to closed circular forms in the nucleus within 6–8 hr after infection (Shank and Varmus, 1978). It is not known whether linear DNA can be integrated directly (see below). The amount of circular DNA generally increases for 12–48 hr, with the absolute amounts varying from 0.1 to 5.0 copies per infected cell in different systems (Gianni *et al.*, 1975; Guntaka *et al.*, 1976).

---

some or all of the other LTR; these may also arise from autointegration events (e.g., involving only one end of linear DNA). (F) Circular DNA with one complete and one partially deleted LTR; errors in DNA synthesis (e.g., choice of the priming site for plus strand DNA) or circularization may produce these forms. (G) Circular DNA resembling that in (A) save for an additional unexplained base pair at the point at which the ends of linear DNA have been joined to form circular DNA.

Two major forms of monomeric, closed circular DNA have been regularly observed in acutely infected cells by gel electrophoresis and molecular hybridization (Shank *et al.*, 1978a,b; Yoshimora and Weinberg, 1979): The more abundant form contains one copy of the LTR unit and the other contains two copies of the LTR, usually joined in tandem (Figs. 4 and 6). The circles probably arise by multiple mechanisms because analysis of molecules cloned in bacteria has revealed several subtle but significant variations (Fig. 6). Circles with one complete LTR (Fig. 6B) could arise by homologous recombination between the LTRs at the ends of linear DNA. When two complete and identical LTRs are present in tandem, the blunt ends of linear DNA may have been directly joined (Fig. 6A; Swanstrom *et al.*, 1981; Shoemaker *et al.*, 1980; Ju and Skalka, 1980; Donehower *et al.*, 1981). Occasionally, a single extra nucleotide of uncertain origin is observed at the circle junction (Fig. 6G; Van Beveren *et al.*, 1982). In several cases, nucleotides are missing from one or both LTRs at the site of joining (Fig. 6D–F; Ju and Skalka, 1980; Shoemaker *et al.*, 1981; Highfield *et al.*, 1980). In some of these instances, the aberrations appear to reflect an autointegration event, because 2 bp are missing from one LTR (Fig. 6D,E; Swanstrom *et al.*, 1981; Ju and Skalka, 1980; Shoemaker *et al.*, 1981). (Loss of 2 bp from each LTR is characteristic of the integrated form of retroviral DNA; see below.) There is further evidence that unintegrated DNA can undergo changes akin to those of integration into itself. A number of circular molecules bear two LTRs in inverted orientation, separated by an inverted viral sequence of variable length (Fig. 6C); the stigmata of retroviral integration (loss of 2 bp from the LTRs and a short duplication of the target site) are present in these molecules (Shoemaker *et al.*, 1980, 1981; Van Beveren *et al.*, 1982). In addition to the monomeric forms, small amounts of dimeric and trimeric closed circular molecules have been detected in cells acutely infected by RSV, with the monomers arranged head to tail within the oligomeric structures (Goubin and Hill, 1979; Kung *et al.*, 1980). It is not known how the dimers and trimers arise, although it is likely that they are the products of joining rather than replication of monomers. One unusual cloned fragment of circular DNA may have been generated by an autointegration event involving a dimeric precursor (Shoemaker *et al.*, 1981). In none of the postulated autointegration events is it clear whether linear or circular forms are integrating into themselves.

4. The Middle Phase (II): Integration of Viral DNA

Every successfully infected cell ultimately acquires at least one and as many as 10 or more integrated (proviral) forms of viral DNA. However, the kinetics, substrates, and mechanisms of integration are poorly understood.

a. Kinetics. Small amounts of integrated viral DNA have been reported as early as 8 hr after infection (Varmus *et al.*, 1973), although it is likely that the amount increases for at least 24–48 hr. A single provirus encoding one subunit of a replication-competent genome is sufficient to direct the production of high titers of infectious virus in a suitable host. The accompanying synthesis of envelope glycoproteins ultimately saturates plasma membrane receptors and effectively prevents superinfection by viruses competing for the same receptors (see Section II,A,6). This barrier (viral interference) tends to limit the number of proviruses of related viruses in any one cell. There are, nevertheless, many ways in which multiple homologous proviruses can be established in a single cell. Initial infection at a high multiplicity of virus to cells or repeated exposure to virus before the establishment of viral interference can facilitate the introduction of multiple copies of viral DNA. In many cases, the initially integrated provirus is not competent to produce viral glycoprotein—either because it is genetically defective (as in the case of many transforming viruses) or because its site of integration is not conducive to viral gene expression—and the cell thus remains susceptible to superinfection.

There is considerable evidence for repeated cycles of "intracellular reinfection" in certain chronically infected cells (Guntaka *et al.*, 1976; Varmus and Shank, 1976; Ringold *et al.*, 1978); in these situations, unintegrated DNA appears to be synthesized from RNA templates, perhaps in nascent virus particles, without a requirement for cell-to-cell spread. Integration of such DNA could augment the number of proviruses despite surface interference.

b. Substrates and Catalysts. There is evidence favoring the integration of viral DNA into domains of cellular DNA that have replicated during infection, although it is uncertain whether this means that the S phase is necessary for synthesis of competent viral DNA or that the topology of replicating DNA provides an advantageous target for the integrative process (Varmus *et al.*, 1977; Humphries *et al.*, 1981). A more fundamental uncertainty surrounds the question of the viral substrate: It is still not known which of the free forms of viral DNA is the proximal precursor to the provirus. Furthermore, no specific viral or cellular functions have been implicated in the integrative process. The viral endonuclease activity, encoded in *pol* and capable of nicking superhelical closed circular DNA, could be imagined to have a role in integration, but the only mutations affecting the endonuclease also affect the DNA polymerase activity (Golomb *et al.*, 1981). Likewise, the products of alleles at the murine *Fv-1* locus may directly influence integration, but the phenotype could also be explained by subtle influences upon synthesis of viral DNA (see Section II,B,6).

c. Structure and Location. Despite the mechanistic uncertainties, proviral structure and location have been carefully defined (Varmus and Swanstrom, 1982). Although viral DNA can be inserted at any of a large number of cellular sites, perhaps at any site (see below), the insertion always seems to occur at two specific sites in the viral DNA 2 bp from the ends of each LTR. Regardless of the origins or genetic competence of a retroviral provirus, its structure conforms to a few simple rules resembling those that govern the structure of many transposable elements (Fig. 4D; see Section III,B and Fig. 11). (*1*) A central coding domain that may encompass as many as four genetically distinguishable regions is flanked on both sides by an identical sequence of several hundred base pairs (LTRs). The structure of the provirus thus approximates that of linear DNA, with the genes flanked by LTRs or, viewed slightly differently, with a complete copy of viral RNA flanked upstream by the U3 sequence and downstream by the U5 sequence (Hughes *et al.,* 1978; Sabran *et al.,* 1979). (*2*) Proviruses terminate with short (3–22 bp) inverted repeat sequences, as each LTR also concludes with inverted repeats (Fig. 8, Tables III and V). However, comparisons of the various intracellular forms of viral DNA have revealed that 2 bp are lost from each LTR at some time prior to or during the integrative process; the additional 2 bp usually extend the length of the inverted repeat in unintegrated DNA (Table III). (*3*) Each provirus is flanked on both sides by the same short sequence (4, 5, or 6 bp) from host DNA; this sequence, which differs from provirus to provirus (Table IV), is apparently duplicated during the integrative process because it is present only once in the unoccupied integration site (Majors and Varmus, 1981; Hughes *et al.,* 1981a; Hishinuma *et al.,* 1981; Van Beveren *et al.,* 1982). The size of the duplicated host sequence appears to be determined by the virus in question (Table IV), so it is possible that the duplication is generated by a staggered cleavage executed at the host integration site by a viral endonuclease, for example, the endonuclease activity encoded in *pol* (see Section II,C,1). (*4*) Aside from the short duplication generated during integration and the interruption imposed by the provirus itself, no other aberrations of the host chromosome are observed in the regions surrounding the majority of proviruses. (*5*) Analysis of cells with multiple homologous proviruses (e.g., Hughes *et al.,* 1978) has failed to reveal evidence for close linkage of those proviruses or for integrated units larger than the prototypic subunit-length provirus. (The sole observed exception to this statement, considered in Section III,D,3, resulted from a duplication event subsequent to the initial insertion.) The absence of tandem arrays of retroviral DNA indicates that sequential integration events do not occur in the large homologous regions provided by preexisting proviruses and that the dimers and trimers of closed circular DNA, present at low concentrations in acutely infected cells, are rarely, if ever, used as substrates for integrative recombination. In contrast, the integrated DNAs of polyoma and simian virus-40 (SV40) viruses are often found in tandem arrays, most likely as a consequence of using multimeric intermediates, formed during replication, as immediate precursors to integrated DNA (Weinberg, 1980).

There are several implications of these fundamental principles of retroviral structure. First, two highly specific sites, each 2 bp from opposite ends of the LTRs, are joined directly to host DNA and are presumably involved in the enzymology of integration. It is not known whether these sites are selected by a mechanism that measures from the ends of unintegrated linear DNA or by a mechanism that recognizes defined sequences in a circular form bearing either one or two LTRs. Figure 7 illustrates diagrammatically how the three major forms of unintegrated DNA could be joined to host DNA at the site of a staggered cleavage. No information is available about the enzymes responsible for the postulated cleavages and ligations. Second, the organization of proviral DNA is admirably suited to facilitate the synthesis of viral RNA. The provirus includes a complete copy of a subunit of the viral genome downstream (in a transcriptional sense) from a redundant copy of the U3 sequence and upstream from an extra copy of the U5 sequence. This format allows the virus to provide its own signals for initiating and terminating transcripts in repeated regions that do not have to be converted to RNA. Third, the overall design of proviral DNA, the precision with which viral sequences are joined to host DNA, and the minimal distortion of the host chromosome at the insertion site argue that retroviruses evolved with the intention of integrating their genomes into host DNA. These features set retroviruses apart from the other animal viruses—most of the known DNA viruses—that often introduce their genome into cellular genomes, but do so in a manner that is inefficient, nonspecific with respect to the viral sequences linked to host DNA, and usually disruptive to both host and viral chromosomes in the vicinity of the integration site (Weinberg, 1980; Tooze, 1980). It is likely that the mechanisms involved in the integration of animal virus genomes other than retroviral proviruses were not specifically designed for this purpose, and that the processes are akin to the assimilation of DNA introduced into cells by microinjection (Folger *et al.*, 1982) or as coprecipitates with calcium phosphate (Perucho *et al.*, 1980).

Although proviruses exhibit great regularity with respect to the position within the viral genome at which linkage to cellular DNA occurs, the evidence to date is consistent with the proposal that integration sites within the host genome are chosen at random. This hypothesis is obviously difficult to substantiate without detailed study of a daunting number of proviruses of various types. However, two types of analysis fail to offer support for the contrary view that proviruses prefer certain sites over others.

1. The *restriction maps* of the cellular DNA that surrounds many types of proviruses are sufficiently dissimilar to show that different regions of the host genome have been employed as integration sites. In the most extensive surveys within a homogeneous experimental system, however, only about 20 insertions were investigated, and of those only about 10 were mapped with confidence; a simple arithmetic analysis suggests that there is at least a 75% probability of there being more than 40 integration sites for the virus in question (RSV) in that cell type (rat) (Hughes *et al.*, 1978). In subsequent sections (II,E,2 and III,C,1,b), we will consider a situation that seems at first glance to elicit contrary results: In certain virus-induced tumors, proviruses are frequently found at the same cellular locus (the cellular homologue of a viral transforming gene) in each tumor (Neel *et al.*, 1981; Payne *et al.*, 1981; Hayward *et al.*, 1981). However, it is likely that the apparent concordance of integration sites reflects strong selec-

Table IV Integration Sites

Virus[a]	Cellular or molecular clone[b]	Host for integration event[c]	Sequences at integration sites[d]	Type of molecule resulting from integration event[e]	Reference[f]
			AA AG GC (ev-1)		1
RSV/ev1[g]			U5. . . GAAGCAGAAGGCTTCATT AATGTAGTCTTATGCAAT . . . U3	—	2
	NRK-2	Rat	GCCCTTCAGGGTCTA GCCTTA GCCTTTCTCTTAGCC	Provirus	3
	NRK-4	Rat	CTGTGG AAAAGTGTGAGAAAA	Provirus	3
	ev-1	Chicken	CACAGTAAAAAGAGA ACCGTT GACAGATATTAAAGT	Endogenous provirus	1
	SRA-1	Quail	ATGAGTTAGCAACAT GCCTTA CCAGGAGAGAAAAAG	Circle type D	2
SNV			U5. . .GTACTTCGGTACAACATT AATGTGGGAGGGAGCTCT. . .U3	—	4
	14–44	Chicken	AAATAAAAGATACAA AAAAT CCGATTTACCCAGGCA	Provirus	4
	44	Chicken	GTCTCTGGGGGAACA TTAAT AATGCCAGCAGGATTT	Provirus	5
	76	Chicken	TTGGAGAACTTGCTG TTTCT AAGGGAATAACCAGGT	Provirus	5
	32	Chicken	AATCAGTTGGTGCTG CTAAC GCCTTGCTTTTACTCA	Provirus	5
	60	Chicken	TTACAAACATGCCTG GATAC GAGCCATTTATATCCC	Provirus	5
	36	Chicken	TCTCCAACCCGAGAC CCTTC CCCCAGTGGGGCAGCC	Provirus	5
MLV/MSV			U5. . .CAGCGGGGGTCTTTCATT AATGAAAGACCCCACCTG. U3	—	6–9
	1G	Mouse	TCCTTCTGTTTCTCTC CTGA TACGTTCCTCTCTTTG	Circle type C	9,10
	G2	Mouse	CCAGTAAGCTCTACCG GTTG TCTCACACCCCCAATA	Circle type C	10
	A2	Mouse	TAGGTGATGAGGTCTC GGTT AAAGGTGCCGTCTCGC	Circle type C	10
	3S[h]	Mouse	AACATTTAGCTTCTTA CCTT CTGCCATCTTTAGGGC	Circle type E	10
	15G	Mouse	CAACATCCAGGGCTTA CATT TGGGATCCGACTCAAT	Circle type E	10
	H5	Mouse	ACTACATGCTGAACCG GGAT CGAACACTCAAAAATA	Circle type E	10
	J2	Mouse	? AAGA CCACCCCCTTCCGACA	Circle type E	10
	C7	Mouse	? CCAC CTAGTCCACTCTCGCC	Circle type E	10
	J7	Mouse	? AGCC CTTTGTACACCCTAAG	Circle type F	10

	ZIP	Rat	TTATTTTTGACCTCTA	CATG	CAGCACGTATCAAAAT	Provirus	10
	ZAP	Rat	CCCGCCCCTGGGTACA	CTGG	GAGAATCCACTGAACA	Provirus	10
	m1	Mink	TCCCCTCATAGATATA	AACG	CTAGTGCTGACCTAGT	Provirus	6
	HT1	Mink	GTC	CCCC	TGA	Provirus	11
	AKR623	Mouse	CAATTTCT	ACAT	GAATATGC	Provirus	12
	AKR614	Mouse	AGGAAATT	GTGAC	AATCTCCC	Provirus	13
	MOV-1	Mouse	CTTGTCCTGGAGAAAA	TTAG	?	Provirus	7
	MSV-12	Mouse	GAGA	CGAG	AACC	Circle type C	12
	SFFV	Mouse	TTCTTC	ATCC	TACTCAGTTACTCTTT	Provirus	14
MMTV			U5. . .GGCCGACTGCGGCAGC		AATGCCGCGCCTGCAG. . .U3		15, 16
	8	Rat	GGGGCCTAGTTCACT	GTAAGG	GATGCCCCAGACCA	Provirus	16, 17
	15	Rat	TATACATATGCTTAT	CAAAAG	CACTAGTG	Provirus	16, 17
	34	Rat	CCACTTTCAGACCTA	GAGGTT	GTTTCATGATTCCA	Provirus	16, 17
	42a-1	Mouse		CAGTTT	TTTGCTCCTTTTTCT	Provirus	15
	3a1-3	Mouse		GCAGAA	GATAATGCTCTCACA	Provirus	15
	2a1-4	Mouse		GAGAGC	CACAGTGAACCTCAT	Provirus	15
	Unit II/GR40	Mouse	GTTCTATTAGCTTC	TTGTAC	CTTAATGTCCA	Endogenous provirus	18, 19
	Unit V	Mouse	TTTTCTATCTTGTTG	ATTTTC	TCTAAGAACCAGATC	Endogenous provirus	20

[a] The sequences of the 3′-end of the LTR (U5) and the 5′-end of the LTR (U3) are shown for each virus as they would appear at the junction of the LTRs in circular DNA of type A (Fig. 6). This format is used for convenience and is not meant to imply that this species of integrated DNA is the substrate for integration (see text and Fig. 7).

[b] The infected cell line or the molecular clone used as a source of DNA for sequencing integration sites is provided for reference.

[c] The host cell in which the integration event occurred is listed for each molecule analyzed.

[d] Beneath the relevant sequences from the viral LTRs are listed the sequences predicted from analysis of host–viral joints to constitute the native insertion site. In each case, the sequence believed to be duplicated by the insertion is boxed, although only in a few cases has the duplicated sequence been directly shown to be present in a single copy in the native site.

[e] The types of circular molecules are illustrated in Fig. 6.

[f] References: 1, Hishinuma *et al.* (1981); 2, Swanstrom *et al.* (1981); 3, Hughes *et al.* (1981a); 4, Shimotohno *et al.* (1980); 5, Shimotohno and Temin (1980); 6, Dhar *et al.* (1980); 7, Van Beveren *et al.* (1980); 8, Sutcliffe *et al.* (1980); 9, Shoemaker *et al.* (1980); 10, Shoemaker *et al.* (1981); 11, McClements *et al.* (1981); 12, Van Beveren *et al.* (1981); 13, Lowy *et al.* (1980); 14, Clark and Mak (personal communication); 15, Donehower *et al.* (1981); 16, Majors and Varinus (1981); 17, J. E. Majors (unpublished); 18, Kennedy *et al.* (1982); 19, L. A. Donehower (unpublished); 20, G. L. Hager (unpublished).

[g] The differences between RSV and ev^- are indicated above the U3 sequence of RSV.

[h] This molecule may be derived from a dimeric percursor (see text).

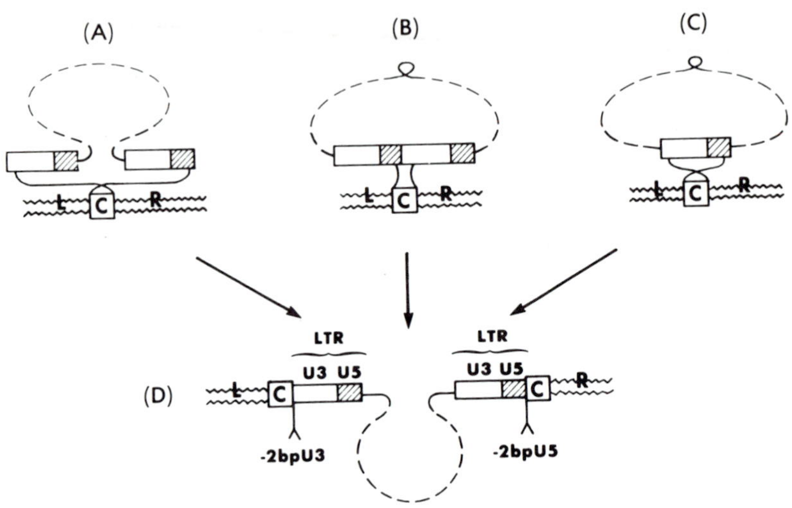

Fig. 7 Models for integration of retroviral DNA. Each diagram indicates the cleavages and ligations required to produce a provirus (D, at bottom) from each of the major forms of unintegrated DNA. The final product is flanked by cellular DNA, including a short duplicated cellular sequence (box C) and terminated by LTRs lacking 2 bp at each end. LTR sequences are represented as open (U3) and closed (U5) boxes, and other regions of viral DNA as dashed lines; cellular DNA is shown as wavy double lines (L, left; R, right flanking DNA). (A) Linear DNA as proximal precursor: Sites in viral DNA 2 bp from the ends of linear DNA are joined to cellular DNA at staggered nicks at either end of sequence C. (B) Circular DNA with two LTRs as proximal precursor: Sites in viral DNA 2 bp one each side of the junction point of the two LTRs are joined to cellular DNA at staggered nicks. (C) Circular DNA with one LTR as proximal precursor: Sites produced in viral DNA by staggered nicks 2 bp from the ends of the single LTR are joined to host DNA at staggered nicks at each end of sequence C; in this case, gap filling by host DNA polymerases would serve to duplicate the LTR (minus 2 bp at each end) as well as C.

tion for the rare infected cell bearing a provirus at a site in which the tumorigenic process can be initiated. In another experimental context, in which it is possible to select for proviral insertion mutants in a defined genetic target in cultured cells, the frequency with which such mutants arise approximates the rate expected on the basis of random insertions (Varmus *et al.*, 1981a; Section III,C,1,a).

2. Determinations of the *nucleotide sequence* of host DNA at host–proviral junctions have failed to reveal significant homology either between viral and host sequences or among host sequences; similar conclusions have been drawn from the study of cloned circular products of presumed autointegration events (Table IV). Again, the number of integration sites studied in this arduous manner is small within each experimental system. Only a much more extensive survey is likely to establish whether preferred sites exist.

Two kinds of experiments are often cited in support of nonrandom integration. First, there is genetic and biochemical evidence that argues that a primate retrovirus endogenous to baboons (BEV) preferentially inserts its provirus into chromosome 6 upon infection of human cells (Lemons *et al.*, 1978). By restriction mapping, however, the BEV proviruses appear to enter many sites in host genomes (J. C. Cohen, personal communication). It is possible that BEV poses a special case in which integration is confined to multiple sites only or mainly on a single chromosome. In other systems, many if not all chromosomes appear to be suitable targets for integration of both exogenous and endogenous proviruses (Hughes *et al.*, 1981b; Coffin, 1982b). Second, a single size class of *Eco*RI restriction fragments of DNA from cells chronically infected with avian reticuloendotheliosis viruses (REVs) appeared to be infectious, suggesting that infectious proviruses were present only in a distinct subclass of host sequences distinguished by the distance between *Eco*RI sites (Battula and Temin, 1977, 1978). More recently, a collection of cloned infectious REV proviruses failed to reveal the expected homologies in flanking cellular sequences (O'Rear *et al.*, 1980). It is possible that the initial infectivity studies were misleading because of the small number of infectious events scored. Physical analyses of the total REV proviral population (as well as several cloned proviruses) and sequence determinations at host–viral junctions have also failed to demonstrate a proclivity for any integration sites (Keshet and Temin, 1978; Shimotohno *et al.*, 1980).

5. THE MIDDLE PHASE (III): MAINTENANCE AND EXPRESSION OF THE PROVIRUS

a. Perpetuating the Provirus. Once established within host chromosomes, proviral DNA is maintained and distributed equally to daughter cells as a result of its replication as part of host replicons. These features account for the persistence of retroviral infection and for the ability of most or all progeny cells to express viral genes and produce infectious virus. Thus, retroviral infection implies stable genetic changes—both the acquisition of viral coding and regulatory domains and the interruption of a host chromosome at the insertion site. These alterations can occur in somatic or germinal cells, establishing in the latter case proviruses considered endogenous to those offspring that inherit the appropriate chromosome. The association of provirus and host DNA ends only with the death of the infected animals or cells, with chromosomal loss or deletion mutations (see Section III,C,3), or with failure to transmit a chromosome containing an endogenous provirus to subsequent generations.

b. Determinants of Transcriptional Activity. Fulfillment of the virus life cycle depends upon transcription of the provirus to form progeny genomes and mRNAs. Under ideal circumstances, this is the stage of the life cycle at which amplification of the genome occurs: A single provirus can be the source of thousands of copies of viral RNA per cell, resulting in the production of as many as 10^3–10^4 virus particles per cell per day. However, it is now apparent that many factors, including the chromosomal location of the provirus, the inherent strength of its promoter, the physiological state of the cell, and specific regulators, can markedly affect the transcriptional activity of endogenous or exogenous proviruses (Varmus and Swanstrom, 1982).

It is generally assumed that synthesis of viral RNA follows (and is dependent upon) integration of viral DNA, but this point has never been rigorously established. Kinetic studies of the appearance of new RNA (e.g., Schincariol and Joklik, 1973) are consistent with synthesis from integrated templates but do not exclude transcription from unintegrated ones as well. Perhaps more telling, cells that restrict integration of viral DNA by virtue of *Fv-l* alleles (Section II,A,6) also restrict synthesis of viral RNA (Jolicoeur and Baltimore, 1976b); but it can be argued that the free viral DNA is defective for both integration and transcription. Although the transcriptional status of free DNA is in doubt, there is no doubt that integrated DNA is the usual template for RNA synthesis: Most infected cells expressing viral genes do not carry detectable amounts of unintegrated DNA, but [with the possible exception of cells infected with visna virus (A. Haase, personal communication)], integrated genomes are inevitably encountered.

Transcription appears to be catalyzed by host RNA polymerase II: It is highly sensitive to the inhibiting effects of α-amanitin (Jaquet *et al.*, 1974; Rymo *et al.*, 1974); the initiation site appears to be preceded by a sequence resembling TATAAA, found in many other eukaryotic transcriptional units dependent upon RNA polymerase II (Table III); and correct initiation occurs on cloned viral DNA templates in soluble systems that transcribe other templates with RNA polymerase II (Yamamoto *et al.*, 1980a). In general, transcriptionally active proviruses appear to be located in DNAse I-sensitive regions of chromatin, and they are usually hypomethylated; conversely, unexpressed proviruses are frequently located in DNase I-resistant domains and are hypermethylated (Panet and Cedar, 1977; Cohen, 1980; Breindl *et al.*, 1980; Harbers *et al.*, 1981a). Although these findings conform to current dogmas, it is difficult to say whether such correlations imply determinants or consequences of transcriptional activity. In at least one case, however, the use of 5-azacytidine to inhibit methylation at CpG dinucleotides has been followed by enhanced transcription and increased DNAse I sensitivity of a provirus (Groudine *et al.*, 1981).

There are many observations suggesting that complex and interesting controls govern transcription of proviruses. In some instances, proviruses of different retroviruses appear to possess promoters of different strengths (Tsichlis and Coffin, 1980). But even apparently identical or very similar proviruses may be expressed with differing efficiencies in the same cell type, suggesting that the nature of the integration site may be an important determinant (Dina and Penhoet, 1978; Fan *et al.*, 1978; Jahner and Jaenisch, 1980). Moreover, marked changes, increases or decreases, in the transcriptional activity of a provirus may occur with the passage of time and without apparent cause (Deng *et al.*, 1974; Porzig *et al.*, 1979; Turek and Oppermann, 1980). Various chemical inducers (e.g., halogenated pyrididines) and hormones (e.g., glucocorticoids) may en-

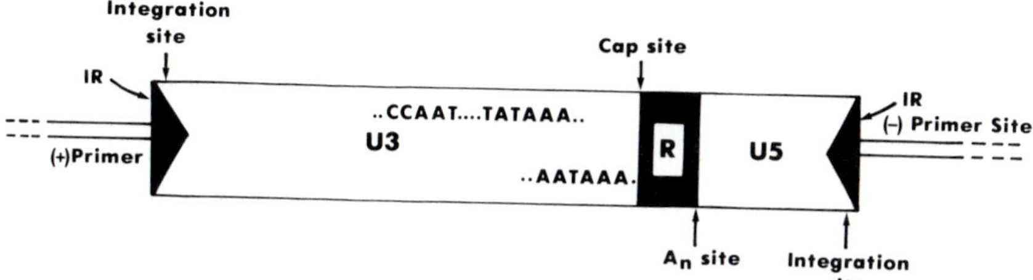

Fig. 8 Anatomy of an LTR. The important features of the U3, R, U5, and flanking viral sequence of LTRs are shown in graphic form here and are presented in greater detail in Table III. Features that pertain to a 5′-LTR are shown at the top of the figure, and those that are relevant to a 3′-LTR are shown at the bottom. The viral sequence at the 5′-boundary of U3 is the sequence of the putative primer for plus DNA, and the sequence of the 3′-boundary of U5 is the binding site for the tRNA primer for minus DNA. The LTRs terminate with short inverted repeats (IRs). The integration sites are 2 bp from each boundary. The sequences resembling CCAAT and TATAAA probably influence the initiation of RNA synthesis (at the cap site), and the nonoverlapping sequence, usually AATAAA, probably determines the polyadenylation (A_n) site.

hance transcriptional activity (Besmer *et al.*, 1975; Ringold *et al.*, 1975). In the best-studied case, glucocorticoids produce a rapid, direct effect upon transcription of MMTV proviruses (Ringold *et al.*, 1977; Young *et al.*, 1977; see below).

c. Structural and Functional Properties of LTRs Affecting Gene Expression. There is now considerable evidence from both structural and functional analyses that most of the regulatory apparatus influencing retroviral replication resides within the LTRs. The importance of sequences at and around the boundaries of LTRs during the synthesis and integration of viral DNA was discussed earlier and is summarized in Figs. 4, 5, and 8 and in Table III; here, we shall be concerned principally with the impact of LTRs upon expression of viral genes in proviral DNA. At least four attributes of LTRs seem to be relevant to this discussion: (*1*) the capacity to determine the site of initiation of RNA synthesis by RNA polymerase II; (*2*) responsiveness to factors that regulate transcriptional activity; (*3*) the ability to enhance the use of transcriptional promoters, including heterologous promoters, in a nonpolar fashion; and (*4*) the presence of instructions for RNA processing, particularly polyadenylation.

1. Nucleotide sequencing of many types of retroviral LTRs has confirmed the considerable heterogeneity of size and sequence of the three domains (U3, R, and U5) previously deduced from annealing tests and restriction endonuclease maps (see also Section II,B,2). Nonetheless, virtually all LTRs contain two short sequences believed to determine the initiation site for transcription by eukaryotic RNA polymerase II (Pol II) and to influence the efficiency of initiation (Breathnach and Chambon, 1981). A sequence containing the TATAA refrain, or a very similar sequence, is generally present about 25–30 nt upstream from the cap site (equivalent to the initiation site), and the TATAA "box" is usually preceded by a sequence resembling CCAAT an additional 30–40 nt upstream

(Fig. 8, Table III). Studies of transcription in cell-free systems (Yamamoto *et al.*, 1980a; Fuhrman *et al.*, 1981; Ostrowski *et al.*, 1981) and in cells into which cloned DNA has been introduced by transfection or microinjection (Blair *et al.*, 1980, 1981; Lee *et al.*, 1981; P. Luciw, unpublished) confirm that a solitary LTR is sufficient to cause initiation of Pol II transcripts at or very near the cap site. However, construction of mutants *in vitro* will be required to assess the relative importance of the putative signals for initiation. Surprisingly, the RSV LTR also appears to act as a transcriptional promoter in *E. coli,* with initiation occurring near or at the "correct" site (Guntaka and Mitsialis, 1980).

2. Although most cases of transcriptional variation noted above (Section II,C,5,*b*) are not understood, it is now apparent that regulation of synthesis of MMTV RNA by glucocorticoids is mediated by interactions involving viral DNA, rather than flanking cellular DNA (Ucker *et al.*, 1981), and an activated hormone–receptor complex. The best evidence for this view comes from experiments in which portions of cloned MMTV DNA have been delivered to cells by transfection procedures, sometimes linked to heterologous genes (Lee *et al.*, 1981; Huang *et al.*, 1981; Buetti and Diggelman, 1981; Hynes *et al.*, 1981). The most refined analysis to date indicates that only 190 nt on the 5′-side of the cap site, plus R and part of U5, suffice to produce a dramatic increase in transcripts from the downstream domain (J. Majors, unpublished). Studies with purified hormone–receptor complex suggest that the 3′-portion of the MMTV LTR provides a binding site for the complex (Govindan *et al.*, 1982; H. Govindan and F. Payvar, personal communications), although the interactions among the MMTV DNA, the hormone–receptor complex, and Pol II have yet to be precisely defined. The protein encoded in the open reading frame of the MMTV U3 region (see Section II,B,2) cannot be required for the hormone response: The deletions that remove all but 190 bp of U3 without affecting induction also eliminate the entire open reading frame.

3. Components within some retroviral LTRs have a still poorly understood but immensely interesting ability to influence the activity of nearby heterologous promoters. It is also likely, albeit unproven, that the same property influences the strength of the LTR's own promoter. Several phenomena have been recorded as examples of this property: A portion of U3 from an MSV LTR can restore expression of early genes to a deletion mutant of SV40 (Levinson *et al.*, 1982; M. Botchan, personal communication) and can improve the transcriptional efficiency of the MMTV LTR (G. Hager, personal communication); the efficiency of transformation of cells to a thymidine kinase$^+$ (tk$^+$) phenotype is enhanced 20-fold when microinjected plasmids containing the complete herpes *tk* gene in either transcriptional orientation also include an RSV LTR (P. Luciw and M. Capecchi, unpublished); and the efficiency of expression of cellular genes can be augmented manyfold by the presence of a nearby provirus, even when the LTR is not positioned appropriately to act as a promoter for the cellular gene (Payne *et al.*, 1982; see Sections II,E,2 and III,C,1,*b*). It is not yet clear whether these effects are due to the same mechanism or are related to similar enhancing proper-

ties assigned to the 72-bp repeat near the origin of SV40 DNA (Banerjee *et al.*, 1981; Benoist and Chambon, 1981; Lee *et al.*, 1981; M. Fromm and P. Berg, personal communication; Gruss *et al.*, 1981) and some other eukaryotic viruses and genes (M. Botchan, T. Maniatis, R. Kamen, and P. Luciw, personal communications).

4. The sequence implicated as a signal for polyadenylation in eukaryotic cells, AATAAA (Proudfoot and Brownlee, 1974; Fitzgerald and Shenk, 1981), is usually found about 20 nt upstream from one or more CA dinucleotides at which poly(A) is added to viral RNA posttranscriptionally (Table III). [However, in one case (MMTV) a slightly different sequence (AGTAAA) determines the polyadenylation site (Table III).] The polyadenylation signal is usually in U3 (as shown in Fig. 8), but in viruses with large R sequences (Table III), the signal is within R and must be ignored when the 5′-LTR is being transcribed. A model involving the secondary structure of nascent RNA has been invoked to explain this phenomenon (Benz *et al.*, 1980).

d. Products of RNA Synthesis and Their Functions. The primary product of transcription is almost certainly initiated at the site within the 5′-LTR that corresponds to the 5′-boundary of the R sequence and hence at the nucleotide to which the cap nucleotide is added posttranscriptionally (Table III). Primary transcripts are approximately the length of a subunit of viral RNA, but the sites and determinants of termination of transcription are not yet defined. There is both direct evidence (Yamamoto *et al.*, 1980b; Ucker, 1981) and precedent (Nevins, 1982) for believing that transcription can extend some distance beyond the sequence to which the poly(A) tract is added (the 3′-end of the R sequence derived from the 3′-LTR).

Transcripts of subunit length can be synthesized and polyadenylated in the nucleus within 10 min (Robertson and Varmus, 1981; Ucker, 1981). Some transcripts are then further processed, most significantly by splicing, probably before transport to the cytoplasm (Weiss *et al.*, 1977; Cordell *et al.*, 1978; Rothenberg *et al.*, 1978; Stacey and Hanafusa, 1978). Because the half-lives for genomic RNA and the structurally similar *gag* mRNA are demonstrably different (Levin and Rosenak, 1976), it is likely that "pregenomes" and mRNAs are assigned to different intracellular pools, but the manner in which this occurs is not known.

Approximately one-half of the intracellular viral RNA serves as messenger during productive infection. (The structures of the various mRNAs were described previously in Section II,B,2) Generally, about one-half of the mRNA is *gag* mRNA, less than one-half is *env* mRNA, and less than one-tenth is *pol* mRNA, but these proportions may depend upon the nature of the host cell. For example, in avian cells productively infected by RSV, about one-fourth of the large amount of viral RNA is *src* mRNA, whereas in infected, nonpermissive mammalian cells, well over one-half of the 10- to 100-fold lower amounts of viral RNA is composed of *src* mRNA (Quintrell *et al.*, 1980).

Although it is inappropriate here to relate the intricate details of retroviral protein synthesis, the principles governing this phase of the life cycle should be briefly summarized. In general, the second half of the retrovirus life cycle capitalizes upon the potential of the host cell to synthesize and process both viral RNA and viral protein. To maximize the protein products of a single transcriptional unit, retroviruses employ two strategies familiar to students of eukaryotic viruses: (*1*) multiple mRNAs with common leader sequences and 3′-termini are produced by splicing of a single primary transcript, thereby facilitating translation of internal coding domains (cf. Fig. 4); (*2*) the primary products of translation of *gag, pol,* and *env* mRNAs are polyproteins, subsequently cleaved and processed to generate mature virion proteins. A full discussion of these events can be found in treatises by Varmus and Swanstrom (1982) and Dickson *et al.* (1982).

6. Late Events

Retroviruses, like many other enveloped viruses, assemble at the plasma membrane and exit from the cell by a budding process that does not overtly damage the integrity of the membrane. Although these events have been frequently observed in the electron microscope, there is little biochemical insight into the mechanisms of assembly and release.

Studies of deletion mutants suggest that a recognition site for inclusion of viral RNA into particles exists within a domain of several hundred bases that includes L and sometimes the 5′-end of *gag* (Shank and Linial, 1980; K. Toyoshima, R. Mulligen, and H. Temin, personal communications). Because, with a few exceptions, subgenomic mRNAs are not included in virus particles at a frequency commensurate with their intracellular abundance, the packaging signal presumably includes some sequences that are usually eliminated by splicing. Alternatively, most mRNA pools may not be physically accessible to the packaging events.

Formation of the nucleoprotein core of the virus is thought to involve the gathering of "immature" components: viral RNA subunits not yet joined to form dimers; uncleaved *gag* and *gag-pol* polyproteins; and various host RNAs, including the tRNA primer not yet bound to viral subunits. How these components coalesce is still a mystery, but regions of the *gag* polyprotein have a strong affinity for nucleic acids and can be imagined to initiate an interaction between viral RNA and *gag* polyproteins. Similarly, the *pol* region of the *gag-pol* polyprotein may expedite the inclusion of the subset of host RNAs, because binding of tRNA to reverse transcriptase can be measured in some instances (Panet *et al.*, 1975). Whether cores form throughout the cytoplasm (as in the case of B-type particles) or just beneath the plasma membrane (C-type particles), they commonly affiliate with a region of the membrane enriched with viral glycoproteins, suggesting specific attractions between proteins encoded in *gag* and *env* (Dickson *et al.*, 1982).

After budding, the released particles undergo a maturation process that can be monitored in several ways: by electron microscopic changes, such as condensation of the core structure (Yoshinaka and Luftig, 1977a); by the appearance of 60–70 S dimers, with tRNA positioned to serve as primer for DNA synthesis (Canaani and Duesberg, 1972); and by the cleavage of *gag* and *gag-pol* proteins to generate the low molecular weight products of *gag* and the enzymatically active products of *pol* (Witte and Baltimore, 1978). Although the precise sites of *gag* cleavage have been identified in a few cases, the mechanisms remain controversial. In the case of RSV and closely related viruses, the carboxy-terminal component of *gag*, p15, is a protease capable of catalyzing the appropriate scissions in *gag* and *gag-pol* proteins once it has itself been released, perhaps by a cellular protease (von der Helm, 1977). However, virus-coded proteases have not yet been found in other cases (e.g., MLV) in which processing of *gag* products may be performed by proteases of host origin (Yoshinaka and Luftig, 1977b).

D. Genetic Changes and Interactions

Retroviruses are noted for their genetic and phenotypic plasticity, to which many of their fundamental characteristics contribute: their diploid genomes; their unusual life cycle, including integration of their genomes into host chromosomes; the mechanism and fidelity of reverse transcription; and their ability to exchange components with unrelated retroviruses and other enveloped viruses (for reviews see Vogt, 1977; Linial and Blair, 1982). The roles of such factors in the production of retroviral diversity can be most readily summarized by considering first those changes that arise during solitary propagation of a cloned virus and then those that arise during mixed infection of cells by certain combinations of viruses.

1. Changes Arising during the Solitary Life Cycle

There are multiple opportunities for the generation of viral mutants during infection by single virus particles, but it is generally assumed that mutations arise most frequently during reverse transcription. There may be several reasons for mutability at this step. First, reverse transcriptase is able to transfer nascent DNA strands from one template to another, using homologous regions of the two templates (see Section II,C,3). Inappropriate transfers between short regions of homology on the two subunits of viral RNA may be responsible for the high frequency with which deletions (and, less commonly, duplications) are found in viral genomes (Coffin, 1979). Some of the resulting lesions can be quite bizarre. For example, a recently described mutant of spleen necrosis virus has a short insertion of part of U5 and part of the PBS sequence at the site of a small deletion in the middle of *gag;* this is likely to have resulted from an erroneous "second jump" between templates during DNA synthesis, perhaps encouraged by a failure to copy the PBS completely (O'Rear and Temin, 1982). Second, reverse transcriptase has no known DNA exonuclease activity (and thus no "proofreading" capacity), and it has been shown to exhibit less fidelity during DNA synthesis *in vitro* than some other DNA polymerses (Weymouth and Loeb, 1978; Gopinathan *et al.*, 1979). These properties could clearly contribute to the generation of point mutations during viral replication. However, neither a rigorous assessment of the frequency with which point mutations arise during natural infection nor a careful comparison of eukaryotic cell DNA polymerases and reverse transcriptases for faithful copying of appropriate templates has yet been made.

There are three steps other than reverse transcription at which mutations can be envisioned to occur: (*1*) integration of viral DNA into host chromosomes, (*2*) replication and maintenance of the provirus, and (*3*) synthesis of viral RNA.

1. Because the regions of viral DNA involved in the integrative steps are the copies of U3 and U5 that are not copied into genomic RNA (cf. Fig. 4), lesions occurring in these areas during integration will not appear in the genome in subsequent rounds of virus replication. When deletions involving LTRs are identified in proviral form (Section III,C,4), it is not usually possible to say whether the damage occurred before or after integration. During natural infection, however, there is little evidence for grossly aberrent integration (permuted DNA or DNA lacking an LTR); in contrast, retroviral DNA delivered by microinjection or as calcium phosphate precipitates is commonly permuted (if delivered as a circle) or truncated (if delivered as a linear duplex) by integration (Chang *et al.*, 1980; Copeland *et al.*, 1981; Folger *et al.*, 1982; P. Luciw, unpublished).

2. Viral mutations that occur within integrated proviruses might be expected to arise at the frequency with which any cellular gene mutates, but there are few studies to date that address this question. Under special circumstances it is possible to identify mutations directly afflicting proviral DNA—for example, when cells are transformed by a single *onc*$^+$ provirus that does not enter the virus life cycle either because it is replication-defective or because the host cell is nonpermissive. In such cases, the frequency of *onc* mutations that revert the cellular phenotype to normal appears to be low, as judged by the prevalence of reverted cells (10^{-5}–10^{-7}), but the differential growth capacities of the wild-type and mutant cells and the crude selection procedures used to isolate mutants make the numbers imprecise (Varmus *et al.*, 1981b). Moreover, the frequency of somatic mutation of cellular genes in cultured cells is not well established and probably varies among genes; for instance, it is probable that immunoglobulin genes experience a particularly high rate of somatic mutation (Baltimore, 1981a).

The issue of proviral mutability is of interest in view of the possibility that structural homologies with transposable elements imply the existence of the kinds of genetic lability associated with those elements. Some of the observed proviral lesions that may be manifestations of such lability (large deletions or excision by recombination between LTRs) will be considered in a later section (III,C). In one experimental system (a nonpermissive rat cell transformed by a single RSV provirus), several proviral mutants have been isolated (Varmus *et al.*, 1981b; Oppermann *et al.*, 1981) and a few analyzed by nucleotide sequencing. One mutation was found to be a +1 frameshift (GAAG to GAAAG) within the *src* coding region, and a rare secondary mutation, shown to be a duplication of about 242 bp that included the primary mutation, suppressed the mutant phenotype by restoring the correct reading frame (G. Mardon, unpublished). This mutant, however, appears to be unusual and was selected for analysis because of the unexpected proteins produced by the mutant and back-mutant genes (Oppermann *et al.*, 1981). The majority of *src* mutants encountered in the survey synthesized nonfunctional proteins of normal size and probably resulted from simple base substitutions. About one-third of the mutants produced structurally abnormal proteins without appreciable deletions or duplications, implying small frameshift or nonsense mutations. However, no cellular gene mutants, other than those arising at immunoglobulin and globulin gene loci, have been collected and analyzed in sufficient numbers to judge whether the kinds of mutations arising in proviral DNA are typical eukaryotic mutations.

3. Synthesis of retroviral RNA should be no more error-prone than synthesis of any cellular mRNA by RNA polymerase II. Of course, errors by eukaryotic RNA polymerases are difficult to measure because they usually have no genetic consequence; however, mistakes in the synthesis of retroviral RNA genomes can be perpetuated by reverse transcription in the next infectious cycle. There is, at present, no obvious way to distinguish between errors generated during reverse transcription and those generated during RNA synthesis.

An additional example of genetic change occurs extremely rarely when retroviruses are passaged through animal hosts: the acquisition of host genes (c-*onc*) that function as viral transforming genes (v-*onc*). The possible mechanisms by which this rare but important transduction of host information occurs will be considered in a later section (III,C,5).

2. Changes Arising during Mixed Infection

a. Heterozygosity. Because retroviruses have diploid genomes, it is reasonable to suppose that cells coinfected with closely related but distinguishable viruses (e.g., two mutants of a single parent) will produce heterozygotic particles at high frequency. This view is so widely accepted that it forms the basis for the prevalent explanation of retroviral recombination (see below). Nevertheless, the direct evidence for heterozygotes is still rather meager and, perhaps more importantly, the requirements for formation of heterologous dimers have not been defined. The best evidence for heterozygotes is derived from genetic analyses of the progeny of mixed infection: Genetic markers peculiar to each of the partners in the initial infection can be transferred together in a second round of infection but later segregate independently (Weiss *et al.*, 1973; Vogt, 1977). The major drawback in these studies is that of distinguishing virus aggregates from true heterozygotes. The seemingly obvious solution is to identify heterozygotes by microscopic inspection of dimers formed between genomes with differing physical characteristics. Efforts to do this have thus far been unsuccessful (Maisel *et al.*, 1978), but they have largely been made with coinfecting viruses that are not known to recombine and hence may not be capable of producing heterodimers.

b. Recombination. A striking feature of retroviruses is the high frequency with which genetic recombinants form after dual infection of closely related viruses. The frequency is often so high [recombinants may account for as much as 10–50% of the progeny virus after a few rounds of mixed infection (Vogt, 1971, 1977; Kawai and Hanafusa, 1972)] that this feature was originally construed as evidence in favor of a segmented haploid genome able to reassort subunits during mixed infection. The frequency of recombination has also confounded tests for complementation in some cases; for example, it was initially thought that RSV might contain four transforming genes, based upon the division of mutants into "complementation groups" that proved to reflect recombination frequencies (Wyke *et al.*, 1975). Perhaps because the frequency of recombination is so high, it has not been possible to derive a genetic map for any retrovirus in this way; in fact, it is still not certain whether the recombinational map is linear or circular (e.g., whether *gag* and *src* are as closely linked as *pol* and *env* in the genetic map of RSV).

Considerable speculation and indirect data surround the issue of the mechanism of recombination. Although mechanisms that involve crossings-over between unintegrated or integrated forms of viral DNA have not been rigorously excluded, the weight of argument favors schemes that emphasize a requirement for heterozygote intermediates (Fig. 9). The essence of such models is that any two subunits that can form heterodimers will be subject to frequent recombination, often involving multiple crossovers, during a subsequent round of DNA synthesis. In one version (Coffin, 1979), recombination is dependent upon the ability of reverse transcriptase to guide nascent DNA strands from template to template (copy choice). In this case, the frequency will be enhanced by nicks in the genome, demanding transfer of nascent strands between templates, and by a high degree of homology between the components of the heterodimer. In a second version (Hunter, 1978), also outlined in Fig. 9, genetic exchange would occur by exchange of segmented plus strands of DNA. This model is supported by the observation *in vitro* of multiple plus strands synthesized by displacement and capable of invasion among intermediates of avian virus DNA (Boone and Skalka, 1981b; Junghans *et al.*, 1982); the model also makes some genetic

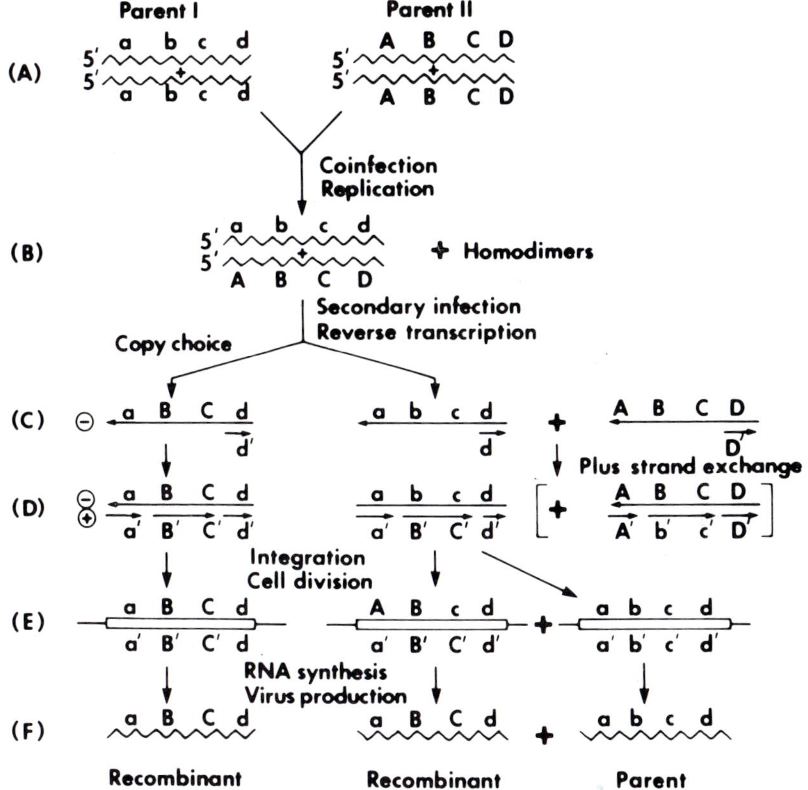

Fig. 9 Possible mechanisms for high-frequency interviral recombination. Two leading contenders for the mechanism of retroviral recombination, both requiring heterozygotic intermediates, are displayed using closely related parental viruses (I and II) with genomic domains represented as a, b, c, d or A, B, C, D. RNA forms of the genome are presented as wavy lines, DNA forms as straight lines; integrated proviruses are shown as boxes flanked by a single line representing cellular DNA. (A) The homodimeric genomes of viruses I and II. (B) Cells coinfected with the two parents will product progeny with the illustrated heterodimers, as well as homodimers; the latter are not considered further. (C–F) Infection of new cells by heterozygotes could lead to recombination by at least two pathways. (*1*) The pathway on the left shows the consequences of a "copy choice" mechanism, in which reverse transcriptase uses both subunits of RNA to synthesize a recombinant minus strand of DNA [line (C)]; the plus strands [line (D)], the provirus [line (E)], and the progeny RNA [line (F)] will each reflect the initial recombination events, which in the case presented include multiple crossings-over. Note that neither the reciprocal recombinants nor the parental genomes would result from a single infection producing recombinants. (*2*) The pathway on the right shows the steps in a "plus strand exchange" mechanism. Minus strands are not recombinant [line (C)], but exchanges during or after synthesis of plus strands would produce DNA heteroduplexes [line (D)]; reciprocal heteroduplexes (in brackets) may be formed but are not considered further here. Integration of a heteroduplex followed by DNA replication and cell division would produce cells with either a recombinant provirus or a parental provirus [line (E)]; these, in turn, would produce recombinant or parental viruses. Note that clones derived from cells infected with heterozygotes could produce both recombinant and parental viruses, as observed by Wyke and Beamand (1979).

predictions (see legend of Fig. 9) for which there is modest support (Wyke and Beamand, 1979).

An unresolved but important issue concerns the degree of relatedness required for recombination. It is possible, for example, that considerable homology near the 5′-ends of two RNAs is necessary for heterodimer formation, but that apparently unrelated domains can recombine, using very short stretches of homologous sequences, within heterozygotic particles. This point will be considered further in the discussion of transduction of cellular genes by retroviruses (Section III,C,5).

c. Phenotypic Mixing and Complementation. When engaged in mixed infection, retroviruses are capable of nongenetic as well as genetic interactions with coinfecting viruses. In the most common form of nongenetic interaction, one virus provides structural proteins for another; envelope glycoproteins are most easily recognized as having been phenotypically mixed because they are the targets for neutralizing antisera and the determinants of host range (Hanafusa *et al.,* 1964; Ishizaki and Vogt, 1966; Vogt and Ishizaki, 1966; Weiss, 1982). Phenotypic mixing of envelope components is not subject to the constraints of homology placed upon genetic interactions: Retroviruses from widely different species and with no apparent homology to each other can still engage in phenotypic mixing. Moreover, retroviruses can exchange envelope glycoproteins with some members of the rhabdovirus family, such as vesicular stomatitis virus (VSV) (Zavada, 1972; Weiss, 1982). In general, such exchanges are reciprocal, although it is easier to detect pseudotypes containing VSV genomes and retrovirus coat proteins because of the simpler biological assay for VSV. Phenotypic mixing involving products of genes other than *env* appears to demand close relations between the viruses. For example, only members of the avian leukosis-sarcoma virus family can supply functionally compatible reverse transcriptase to a *pol*$^-$ member of the same family (Vogt, 1977).

E. Retroviral Pathogenesis

Retroviruses induce a wide variety of diseases (Section II,A,3, Table I), but efforts to determine pathogenetic mechanisms have been focused largely upon neoplastic diseases, and only neoplasms will be considered here. For a broader perspective on retroviral diseases, the reader is referred to Teich *et al.* (1982) and Gross (1970).

1. PATHOGENESIS BY v-*onc*$^+$ VIRUSES

It is convenient to discuss separately the oncogenic viruses that carry an *onc* sequence (v-*onc*$^+$) and those that do not (v-*onc*$^-$) (see Section II,A,8). Although in most cases genetic proof is not yet available, it is commonly assumed that tumor induction by v-*onc*$^+$ viruses is due to proteins partially or completely encoded by v-*onc* (reviewed by Bishop and Varmus, 1982). These proteins are generally not present in virus particles and hence have been identified by immunoprecipitation of labeled proteins from extracts of infected cells or by translation of viral RNA *in vitro*. Appropriate antisera have been easily procured in the several cases in which v-*onc* sequences are expressed as hybrid proteins also containing *gag* determinants (Fig. 2D, E), because antisera raised against viral structural proteins can also precipitate the hybrid proteins. However, it has generally been difficult to obtain antisera reactive with *onc* proteins synthesized independently of viral structural genes (Fig. 2B, C, E, F) or antisera reactive with v-*onc* antigens in hybrid proteins. Sera from animals bearing virus-induced tumors, particularly tumors observed to regress, have been found active against the products of *src* (Brugge and Erikson, 1977) and *ras* (Shih *et al.*, 1979), and useful antisera against *mos erb*-B, and *myb* proteins have been generated by using synthetic peptides (predicted from nucleic acid sequences) as immunogens (Papkoff *et al.*, 1981). However, the products of some oncogenes (or peptides synthesized in bacteria carrying cloned v-*onc* DNA) have been examined only by *in vitro* translation, a method insufficient to establish the nature of the true gene product as it exists in the cell.

Insight into pathogenesis under the direction of v-*onc* sequences has come from efforts to define the structures, biochemical functions, and intracellular locations of the gene products and to correlate them with the important characteristics of the cancer cell. The goal of such labors has been to perceive that property (or properties) of an oncogenic protein responsible for initiating and maintaining the complex altered phenotype. Progress in this area has been greatly expedited by the capacity of all v-*onc*$^+$ viruses to transform at least some type of cultured cell (see Section II,A,3). Although a detailed discussion of this topic is clearly beyond the confines of this chapter, a few central points should be summarized.

1. The products of several, but far from all, oncogenes are able to transfer γ-phosphate from ATP (or GTP) donors to tyrosine residues in suitable target proteins (Collett and Erikson, 1978; Levinson *et al.*, 1978; Witte *et al.*, 1980; Hunter and Sefton, 1980). This is an unusual form of phosphorylation in eukaryotic cells—over 99.9% of phosphoamino acid is in the form of phosphoserine or phosphothreonine—yet the products of *src, yes, fps, ros, abl, fes,* and *fms* are able to catalyze this reaction *in vitro* (Bishop and Varmus, 1982; Table II). Several observations sustain the argument that this activity is operative *in vivo* and contributory to the transformed state: transformation-defective mutants have impaired kinase activity; the amount of phosphotyrosine in total cellular protein is increased after viral transformation in all cases save *fms;* and at least one growth-stimulating hormone (epidermal growth factor) also provokes tyrosine phosphorylation. Several phosphotyrosine-containing proteins—some previously known (e.g., vinculin), some identified by physical association with

onc proteins, and some found only by diligent analysis of two-dimensional gels—have been identified, but the relationship between these phosphorylations and transformation is still obscure.

2. The products of at least some v-*onc* genes (*src, ras,* and *abl* have been studied to date) are found mainly in the cytoplasm, particularly in the plasma membrane (apparently as integral membrane proteins), rather than in the nucleus (Willingham *et al.*, 1979, 1980; Courtneidge *et al.*, 1980; Witte *et al.*, 1979). Although at first it may seem surprising not to find transforming proteins in the nucleus, where genes are replicated and expressed and where the "tumor antigens" of several DNA viruses had previously been located, it is now recognized that cytoplasmic events, particularly events at the plasma membrane, can determine the growth properties and morphology of a cell; moreover, there is increasing evidence favoring a cytoplasmic locus for transformation by DNA viruses as well (Ito *et al.*, 1977; Treisman *et al.*, 1981).

3. The target cells for transformation by each virus appear to be restricted to members of certain cell lineages, both *in vivo* and *in vitro* (e.g., Graf and Beug, 1978; Beug *et al.*, 1979), and at least some of the consequences of transformation can be likened to an arrest of differentiation. Because v-*onc* genes are all derived from highly conserved cellular genes, most of which are expressed in at least some tissues at a generally low level, it is tempting to speculate that oncogenesis by v-*onc*$^+$ viruses is, in some sense, a manifestation of differentiation gone awry. Excessive production of v-*onc* protein, structural differences between v-*onc* and c-*onc* products (resulting either from mutation or from fusion of v-*onc* peptides with viral structural peptides), or differences in cellular location could lead to enzymatic action (e.g., phosphorylation) upon the "wrong" substrates or to crucial imbalances that would derail normal regulatory controls.

2. PATHOGENESIS BY V-*onc*$^-$ VIRUSES

It has been much more difficult to fathom the mechanisms by which viruses lacking *onc* genes induce tumors: These viruses do not transform cultured cells and hence must be studied in the more complicated environment of a whole animal; their genomes do not generally contain coding domains other than those for the three viral structural genes; and tumor induction is slow and inefficient, often requiring 6–12 months to produce tumors in a majority of animals. Most of the v-*onc*$^-$ viruses are replication-competent, with the advantage that they can be biologically cloned; on the other hand, it may be difficult to differentiate between features of the viral genome required for efficient replication and those required specifically for tumorigenesis, because virus replication and spread are so tightly linked to tumor production. Moreover, the many cycles of replication are likely to produce a large number of genetic variants or recombinants, particularly when the initial virus stock is not clonal or when the host harbors endogenous proviruses homologous to the agent under study.

Two hypotheses have dominated recent discussion of pathogenesis by viruses lacking *onc* genes: (*1*) that oncogenesis is dependent upon the generation of abnormal (e.g., recombinant) viral structural genes with oncogenic potential and (*2*) that oncogenesis is dependent upon the capacity of proviral insertions to mutate or alter the expression of cellular genes. Before considering briefly the evidence for these hypotheses, it is prudent to note that the hypotheses are not mutually exclusive, that kinetics of tumor induction suggest that multiple events may be required, and that a full description of the mechanism must explain the choice of target cell as well as neoplastic transformation per se.

Several features of leukemogenesis by various MLVs argue for a role for recombinant *env* genes. Prominent among these is the repeated isolation of unusual recombinant viruses from the preleukemic thymus of mice infected with certain ecotropic strains of MLV (Hartley *et al.*, 1977). The recombinant viruses have an amphotropic host range (growing in both mouse and foreign cell types), induce foci of altered growth patterns in mink cells (hence the name MCF, or mink cell focus-forming, virus), and sometimes accelerate the appearance of leukemia when inoculated into uninfected animals. Analysis by peptide mapping, oligonucleotide fingerprinting, restriction endonuclease mapping, and heteroduplex formation indicates that the *env* gene of the MCF viruses resembles an amalgam of ecotropic and xenotropic portions that could have united by recombination between an infecting ecotropic virus and an endogenous xenotropic virus (Elder *et al.*, 1977; reviewed by Teich *et al.*, 1982; Coffin, 1982a). Recent work, however, favors the presence of proviruses with MCF-like *env* genes in the germ line (Chattopadhyay *et al.*, 1982); it is probable that genomes encoded by such proviruses recombine with ecotropic genomes to produce MCF viruses. The hypothesis that MCF rather than ecotropic viruses are the proximal causative agents of thymic leukemia has been strengthened by the finding that the tumors are clonal with respect to MCF but not ecotropic proviruses (Yoshimora and Breda, 1981; van der Putten *et al.*, 1981; Chattopadhyay *et al.*, 1982).

Further evidence implicating the *env* gene in leukemogenesis comes from molecular cloning of the genomes of the replication-competent Friend MLV and the replication-defective spleen focus-forming virus (SFFV). Both of these viruses, unlike most v-*onc*$^-$ viruses, can induce tumors quite rapidly under appropriate conditions; the former contains an MCF-like *env* gene and the latter contains a recombinant *env* gene that encodes an atypically small major glycoprotein (Racevskis and Koch, 1977; Dresler *et al.*, 1979; Troxler *et al.*, 1977; Ruscetti *et al.*, 1979). It appears from properties of fragments of cloned DNA that the oncogenic potential of the viruses resides in the 3′-half of the RNA genome, presumably in the *env* gene (Oliff *et al.*, 1980; Linemeyer *et al.*, 1981).

Although such findings implicate *env* in some aspect of leukemogenesis, they are not mechanistically illuminating. It is uncertain whether the recombinant *env* genes are required for growth to high titer, for entry into a small cohort of target cells, or for effecting the transformation itself. It has been argued that *env* gene products could instigate tumorous growth by direct effects upon cell membranes or by stimulating the unrestrained growth of lymphocytes with specific receptors for certain *env* glycoproteins (McGrath and Weissman, 1979). The significance of such mechanistic possibilities has not been established.

In the last few years, considerable attention has been given to the possibility that tumor induction by some v-*onc*$^-$ viruses is dependent upon insertion of proviral DNA at certain critical sites in the host genome; the effect of such insertions could be to interrupt coding domains or to alter gene expression. The recognition that most tumors induced by v-*onc*$^-$ viruses are clonal (Steffen and Weinberg, 1978; Cohen *et al.*, 1979) provided experimental access to this hypothesis, because it is feasible to examine the structure and position of proviruses and the nature of RNA transcripts from the interrupted domain in cell populations dominated by a single infected cell.

The most persuasive evidence for an insertional mechanism of oncogenesis comes from studies of B (or bursal) lymphomas induced in chickens by infection with ALVs. The central features of the story are as follows: The tumors are clonal, arising in the bursa of Fabricius several months after the infection of newborn animals. All tumors contain at least part of an ALV provirus, but the proviruses are frequently truncated by deletions (Payne *et al.*, 1981; Neel *et al.*, 1981; Fung *et al.*, 1981; see Section III,C,4), and viral genes are not expressed; this implies that virus-encoded proteins are not required for maintenance of the tumor phenotype. The great majority of tumors have proviral insertions in the c-*myc* locus (Hayward *et al.*, 1981), the host gene from which the v-*myc* gene of MC29, an acute defective leukemia virus, was derived (Table II). Although the proviral DNA may be positioned within the c-*myc* region in various ways (Hayward *et al.*, 1981; Payne *et al.*, 1982; see Section III,C,1,*b*), the accompanying effect is always a marked increase in the concentration of c-*myc* mRNA. The structure and function of the c-*myc* protein have yet to be determined, and there is no genetic support for the presumption that enhanced expression of c-*myc* is directly responsible for tumor formation. In fact, there is reason to suppose that additional steps are required for full expression of the tumor phenotype: DNA from the lymphomas transforms an established line of mouse fibroblasts to a malignant phenotype, but the transforming activity appears not to reside in the interrupted c-*myc* locus (Cooper and Neiman, 1980, 1981).

Experiments with other v-*onc*$^-$ viruses offer further support for the hypothesis that the site of integration is an important determinant of oncogenesis. Bursal lymphomas induced by two other chicken viruses—a reliculoendotheliosis virus (chicken syncytial virus, or CSV) unrelated to ALV and a myeloblastosis-associated virus (MAV) differing from ALV in the U3 domain—also contain proviral insertions in c-*myc* (Noori-Daloii *et al.*, 1981; D. Westaway and C. Moscovici, unpublished). The interrupted c-*myc* locus in some CSV-induced tumors appears to be amplified severalfold, although the effect of gene amplification upon gene expression has not been directly measured. The observation that three different viruses induce the same type of tumor through insertions affecting the same locus suggests that overexpression of c-*myc* may be specifically required for B cell lymphomas, at least in birds. This view is strengthened by the finding that renal tumors induced by MAV do not bear insertions in c-*myc* (D. Westaway and C. Moscovici, unpublished). In addition, there is recent evidence for integration sites that predispose to the development of mammary carcinomas after infection by MMTV. Over half of a large group of virus-induced mammary tumors have been found to contain an MMTV provirus within the same 20-kb domain of the mouse genome (Nusse and Varmus, 1982), a coincidence too improbable to explain by chance. Furthermore, a transcriptional unit positioned amidst the integration sites is active in the tumors, but not in normal mammary glands.

F. Endogenous Proviruses

Among the most extraordinary properties of retroviruses is the frequent and widespread presence of their proviruses in the germ lines of apparently normal animals. Many clues led to the hypothesis that retroviral genomes were often endogenous to vertebrate genomes and genetically transmitted: the appearance of viruses and sometimes virus-induced diseases in ostensibly uninfected animals (Gross, 1970); the detection of viral antigens in normal embryos (Dougherty *et al.*, 1967; Huebner and Todaro, 1969) or of physical particles in normal placentas (Kalter *et al.*, 1973); Mendelian segregation of the capacity to synthesize virus-related proteins in normal embryo cells (Payne and Chubb, 1968); the release of viruses from cultured embryo fibroblasts spontaneously (Livingstone and Todaro, 1973) or after induction with chemical agents, such as halogenated pyridimines (Lowy *et al.*, 1971); and the measurement of virus-specific DNA in chromosomal DNA from normal animals and embryos by molecular hybridization (reviewed by Coffin, 1982b).

Initially, these observations were confined to those endogenous viruses closely related to the most intensively studied laboratory agents—ALVs, MLVs, and MMTVs. Although these remain the best-studied groups of endogenous agents, endogenous genomes have now been recognized in a vast number of vertebrates (Teich, 1982; Table I) as a result of many technical advances: appropriate indicator cells for growing viruses with limited host range; sensitive assays for viral antigens; and restriction endonuclease mapping, molecular cloning, and low stringency hybridization procedures for detecting distantly homologous sequences. The features of endogenous viruses that enforce the need for these methods—their xenotropism, their defectiveness for replication, their partial homology with viruses transmitted exogenously in the same or other species, and their variable levels of gene expression—are common ones and will be discussed at greater length below. It is worth emphasizing at this point, however, that although endogenous viruses are frequently related to common horizontally transmitted agents, they generally have features peculiar to them as a class. Thus, the endogenous viruses of chickens can be differentiated from the exogenous viruses by their *env*-determined subgroup (E), their U3 sequence, and their lack of pathogenicity (Coffin, 1982b). Similarly, endogenous feline leukemia virus (FeLV)-like proviruses can be distinguished from horizontally transmitted FeLV by their U3 domains (Casey *et al.,* 1981). Many endogenous viruses of mice, cats, and primates would have been difficult to discern without the recognition that they may be replication-incompetent or able to replicate only in foreign hosts. Endogenous proviruses are sometimes recognized as such only by their structure and by some distant homologies with competent retroviruses. The most provocative example of this situation is described by Martin *et al.* (1981), who used cloned endogenous MLV DNA to isolate endogenous proviral DNA from the African green monkey genome; the cloned monkey DNA, in turn, was used to identify putative proviral DNA in the human germ line. Streeck (1982) has identified an LTR-like sequence in a repeated satellite component of bovine DNA. The progenitor of this sequence is not known, but it is likely that a provirus was inserted in the satellite DNA and partly excised by homologous recombination between its LTRs (see Fig. 13B) before amplification of the satellite.

Retroviruses remain unique among animal viruses in their capacity to be transmitted through the germ line. This property of retroviruses has fostered the notion that proviruses are normal cellular components (or recent derivatives thereof) and has stimulated speculation about their roles in development, evolution, and disease. Such views should be treated with caution, however, because there is now considerable evidence (reviewed briefly below and more extensively in Coffin, 1982b) suggesting that endogenous proviruses are most commonly the consequence of rare and usually recent germ line infections. According to the latter perspective, endogenous proviruses are distinctly viral rather than cellular in character, probably useless and adventitious components of germ lines in most cases, and well tolerated only if relatively innocuous. The most dramatic supporting evidence for these conjectures is the complete absence of certain types of chicken or murine retroviruses from the germ lines of a minority of apparently healthy animals (Cohen and Varmus, 1979; Astrin *et al.,* 1979; Frisby *et al.,* 1979; Steffen *et al.,* 1980).

1. Distribution and Structure

The contention that endogenous proviruses are the residua of relatively recent germ line infections is most strongly supported by two kinds of observations: (*1*) endogenous proviruses have all of the structural properties of proviruses acquired by experimental infection and (*2*) the distribution of endogenous proviruses usually shows poor concordance with evolutionary relationships among the hosts.

In general, endogenous proviruses conform to the organizational rules described earlier for exogenous proviruses. Most carry the three replication genes, *gag, pol,* and *env,* in their usual order and without interruptions in coding sequences, and the structural genes are flanked by LTRs; in the few cases in which the nucleotide sequences are known (Hishinuma *et al.,* 1981; Kennedy *et al.,* 1982; L. Donehower and G. Hager, personal communication), the LTRs are identical, concluded with inverted repeats, and flanked by short duplicated sequences of host DNA; and proviruses of the same type may be found at many different sites in the host genome. A significant proportion of endogenous proviruses bear mutations, commonly deletions that impair their ability to replicate. For example, more than one-half of the mapped endogenous proviruses of white leghorn chickens have lost various components (the 5′-LTR, part of *gag-pol,* or virtually all of the coding domain) or have less obvious lesions that incapacitate replicative genes (Hayward *et al.*, 1980; Hughes *et al.*, 1981c). Similar lesions have been encountered among exogenous proviruses (Section III,C,4); it is probable that the greater frequency of mutated endogenous elements reflects selection pressures against functional proviruses (see below).

The distribution of endogenous proviruses of any single type can vary dramatically both within species and among closely related species. For example, strains of white leghorn chickens or laboratory mice each carry a different set of endogenous proviruses related to ALVs or MLVs and MMTVs, respectively (Astrin, 1978; Cohen and Varmus, 1979; Hughes *et al.*, 1979; Dolberg *et al.*, 1981; Steffen *et al.*, 1980). Outbred animals display even greater heterogeneity with respect to the numbers and chromosomal sites of endogenous proviruses (Cohen and Varmus, 1979; Hughes *et al.*, 1981d). Comparison of animal species frequently reveals a pattern of proviral distribution in conflict with phylogenetic relationships. Thus, only a few species of cats contain the endogenous virus, RD-114, that is closely related to (and presumably derived from) an endogenous virus widespread in baboons (BEV) (Benveniste and Todaro, 1974). Similarly, ALVs are present in only one of three species of wild jungle fowl closely related to domestic chickens, but they are frequent in the further-removed ring-necked pheasants (Frisby *et al.*, 1979; Hughes *et al.*, 1981d).

Several factors seem likely to affect the absolute numbers of endogenous proviruses and their distribution among related animals:

1. The time of germ line infection during the history of a species appears to have a profound effect upon proviral distribution. Thus, in the case of endogenous ALV, ecotropic MLV, and MMTV, infection long after speciation seems the most probable explanation for the extreme heterogeneity in proviral carriage found among members of the same species. In contrast, endogenous viruses (such as BEV) that appear to be more uniformly distributed among members of one (and perhaps several related) species (Benveniste *et al.*, 1974) may have entered the germ line before speciation.
2. The frequency of infection can obviously influence the number of germ line copies. The frequency, in turn, will be influenced by the propensity of each virus to infect germinal cells and by the degree of exposure of those cells to the virus. There is recent evidence to suggest that in laboratory animals high-titer viruses may enter the germ line within periods compatible with experimental observation (Rowe and Kozak, 1980; Quint *et al.*, 1982; W. Herr, personal communication). Presumably, this high frequency does not accurately reflect the situation in the wild.

3. Endogenous proviruses have been shown to segregate like stable Mendelian markers in genetic crosses and during inbreeding procedures, with preservation of their physical organization as perceived from restriction mapping (Cohen and Varmus, 1979; Astrin *et al.*, 1980; Quint *et al.*, 1982). This implies both that they can be lost in the segregation process if not distributed to viable offspring and that they are not prone to unexpected genetic aberrations (e.g., transposition, amplification, rearrangement, deletion) at a frequency greater than one per 100 or more generations.

4. The numbers of proviruses may depend upon the genetic competence of viral DNA. Although apparently stable during periods of experimental observation (up to 60 years), endogenous proviruses are often first encountered as imperfect elements (see above and Section III,C,4). Furthermore, other types of mutations frequently afflict proviruses and curtail their ability to replicate (see below). Such lesions can have various effects upon the germ line content of proviruses: Complete deletions obviously reduce the load of proviral DNA to be replicated in the chromosome; mutations that render proviruses replication-deficient impair their ability to be amplified; and mutations that affect the expression of *env* products will abort viral interference and thus increase the chances of superinfection.

5. Selective pressures that act upon the evolution of cellular genomes presumably can also act upon endogenous proviruses. This is particularly likely to be the case when a provirus has been inserted within, and has thereby inactivated a necessary or influential gene (see Section III,C,1), or when a provirus has oncogenic or other pathogenetic potential. The latter may explain the absence of v-*onc*$^+$ viruses from the germ line of any tested animals and the rarity of oncogenic viruses in the germ lines. (The only examples of oncogenic viruses—some strains of MMTV and MLV—have been encountered in laboratory mouse colonies in which there may have been selection for, rather than against, the oncogenic phenotype.) Presumably, virus expression and virus spread may also be detrimental to the host, accounting for the high frequency of xenotropic and mutated proviruses in germ lines.

Jaenisch and his colleagues have directly demonstrated the feasibility of infecting the germ lines of mice by establishing the genome of an exogenous virus, the Moloney strain of MLV, as an endogenous provirus after infection of preimplantation embryos (Jaenisch, 1976; Jaenisch *et al.*, 1980, 1981); more recently, they have introduced endogenous proviruses by microinjection of fertilized eggs with cloned proviral DNA (Harbers *et al.*, 1981b). Some of these proviruses, such as Mo-MLV introduced by neonatal infection, place the animals at high risk of thymic leukemia (see Section II,E,2). Presumably, such proviruses would be selected against in the wild.

6. It may be that some classes of endogenous proviruses utilize mechanisms other than simple germ line infection to achieve amplified representation. For example, the genomes of mice contain hundreds of proviral copies of the genome of intracytoplasmic A-type particles (IAPs; Leuders and Kuff, 1977; Ono *et al.*, 1980) and at least tens of copies of the 30 S virus-like (VL30) RNA Keshet *et al.*,

1980), yet there is no evidence that either IAPs or VL30-containing particles are replication-competent. Germ line infection by pseudotypes carrying these defective genomes, multiple rounds of intracellular infection, or other mechanisms of gene amplification and transposition must account for the large numbers of proviral copies. It is also conceivable that the large numbers reflect some benefits conferred on the host by these endogenous elements.

2. FUNCTIONS

Regardless of whether endogenous proviruses provide any service to their hosts, it is certain that they exhibit varied levels of expression, ranging from complete silence in most or all cell types, through expression of only selected viral genes, to synthesis of complete virus particles that may be noninfectious or infections, ecotropic or xenotropic, and oncogenic or apparently harmless.

Without doubt, the most important determinant of expression of endogenous viral genes in any host is the inherited collection of endogenous proviruses; no host genes have been reliably implicated as trans-active regulators of viral gene expression, although some do affect virus spread (see Section II,A,6). Each proviral unit seems to have its characteristic pattern of expression, determined both by its genetic competence and by the context of flanking host DNA. Among the endogenous proviruses (*ev*) of chickens, for example, there is a provirus (*ev*-3) that lacks about 1 kb around the *gag-pol* junction and can make only a nonfunctional *gag-pol* fusion protein; a provirus (*ev*-6) that lacks the 5′-LTR, *gag,* and most or all of *pol,* and expresses *env* under the apparent control of a flanking host promoter; a provirus (*ev*-5) that is similar to *ev*-6 but not expressed; a provirus (*ev*-1) that appears structurally normal but is inefficiently expressed and contains a defective *pol* gene; and a provirus (*ev*-2) that encodes a fully infectious virus (RAV-0) but is expressed at an extremely low level (Hayward *et al.*, 1980; Hughes *et al.*, 1981c; Baker *et al.*, 1981).

Although the functional behavior of each endogenous provirus seems to be characteristic, the regulation can be perturbed. Expression of *ev*-1 is induced by treatment of embryo cells with the demethylating agent 5-azacytidine (Groudine *et al.*, 1981), and various endogenous MLV proviruses can be induced with halogenated pyrimidines, inhibitors of protein synthesis, and other agents (Aaronson and Stephenson, 1976). Furthermore, the inherent efficiency of expression of each provirus may be strongly influenced by its chromosomal site. This is particularly well documented for *ev*-2, because RAV-0, the product of *ev*-2, replicates quite efficiently when RAV-0 proviruses are randomly inserted in the genomes of cells to which the virus has spread (Jenkins and Cooper, 1980; Humphries *et al.*, 1979).

The consequences of endogenous viral gene expression are largely unknown. In a few cases (the *mtv*-1, *mtv*-2, and *akv* proviruses of mice), tumor development results from expression and virus spread (Coffin, 1982b). It is possible that expression of certain endogenous viruses can protect the host from deleterious infection by exogenous viruses (Crittenden, 1981) or that partial expression of endogenous proviruses can prevent spread of replication-competent endogenous viruses (Robinson *et al.*, 1981). Despite some evidence that expression may be coincident with developmental stages (Dell Villano *et al.*, 1975), the notion that endogenous proviruses are significant determinants of development should be viewed with caution.

3. Origins of Retroviruses

As a family of viruses with extraordinarily similar features, retroviruses seem likely to have descended from a single ancestor, no matter how widely the progeny are now dispersed. The origins of that ancestor pose the same problem that perplexes students of any other large class of viruses, from small RNA viruses to large DNA viruses: Did the forefather arise from some cellular elements, from some other virus, or from both? In the case of retroviruses, the inclination to seek their origins in host genomes is particularly strong, because the viruses are commonly encountered in the germ line, because proviruses have strong structural similarities to transposable elements dispersed throughout living systems (see Section III,B), and because reverse transcription may be a mechanism generally operative in the relocation of cellular genes in higher eukaryotes (see Section III,A). Temin (1980) has argued most strongly for the view that retroviruses evolved from cellular elements. In one possible sequence of events, a set of primordial IS-like transposible elements might have surrounded a host gene for a DNA polymerase. Amplification of the gene (*pol*) and its flanking elements (LTRs) would be favored by mutations that rendered the polymerase capable of synthesizing DNA from RNA transcripts of the region. Further rapid evolution toward a true retrovirus would occur with subsequent rounds of reverse transcription, reinsertion, and finally, incorporation of the forebears of viral structural and transforming genes.

Although this hypothesis has its attractions, there is little direct evidence to favor it. Attempts to identify the postulated forerunners of either LTRs or *pol* have not been conclusive, although it is possible to rationalize such failures as evidence that contemporary retroviruses and the cellular elements that gave rise to them have now considerably diverged. For the moment, it seems most sensible to consider exogenous retroviruses to be of uncertain provenance and endogenous proviruses to be the consequences of germ line infections; only viral transforming genes (v-*onc* genes) seem certain to be derived from cellular genes (see Sections II,A,8 and III,C,5).

III. Proviruses as Movable Genetic Elements

A. The Transposition Cycle

In the tryptich of the virus life cycle, the provirus forms the centerpiece, the goal toward which the initiating events—virus entry, DNA synthesis, and integration—are directed, and the source from which the culminating events—ex-

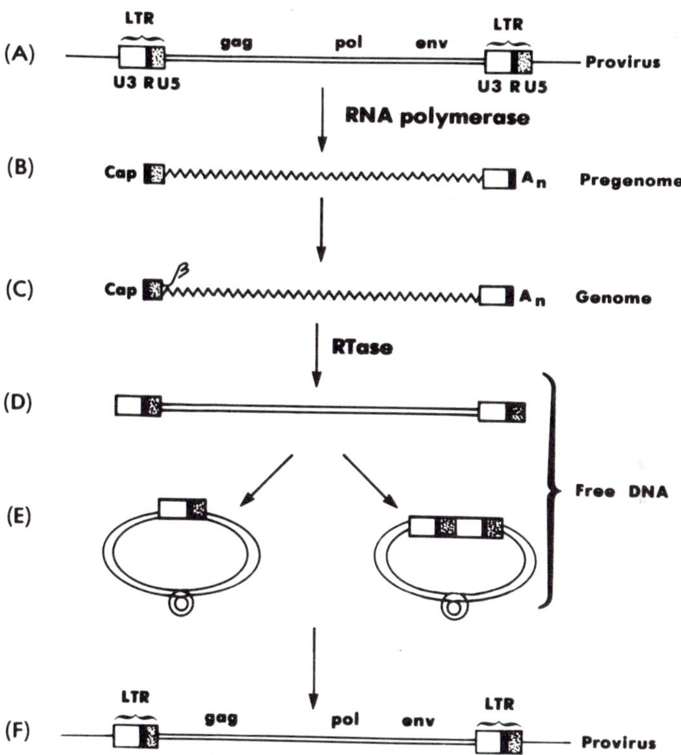

Fig. 10 Transposition of proviruses. The replication cycle (Fig. 4) has been redrawn to demonstrate how retrovirus particles can act as intermediates in the transposition of proviruses. The transposition cycle begins with a provirus [line (A), symbols as in Fig. 4] that is transcribed into RNA [line (B)], some of which is packaged into virions [line (C)] and used as template for synthesis of unintegrated forms of viral DNA [lines (D) and (E)] during a subsequent round of infection. Integration of an unspecified form produces a new provirus [line (F)] at a new site in the infected cell. Occasionally, this cycle can occur within the cell carrying the original provirus (see text).

pression of viral genes, virus production, and transformation—must flow. By imposing a shift of frame, the series of events that constitutes the life cycle can be viewed as a mechanism of DNA transposition (Fig. 10). A single provirus in a single cell can engineer the production of thousands of infectious virus particles, and these particles can introduce the viral genome into new cells in which new viral DNA can then be synthesized and integrated. Seen in this light, the provirus is both the beginning of the transposition process and its termination, with the virus its intermediate.

Viewed as transposable elements, proviruses are extraordinarily proficient and well adapted to their task, much like the prophages of temperate bacteriophage. The status of genetic elements perceived as both transposable elements and viruses is necessarily a privileged one: These elements can persist either as a component of the host chromosome or as the genome of an extracellular particle capable of resisting environmental factors and infecting new host cells. For viral transposable elements, the act of transposition is generally frequent, almost always intercellular, and sometimes interspecies. Moreover, the production of many virus particles from a single provirus (or prophage) can engender an abrupt amplification of the number of copies of an element within a population of cells simply by recruiting a large number of cells as infected hosts. The adaptation of such elements to survive as viruses as well as inserted components of chromosomes usually requires that they acquire the coding potential for viral structural proteins (e.g., *gag* and *env* in the case of retroviruses), as well as coding domains (e.g., *pol*) and noncoding regions (e.g., LTRs) intimately involved in the transposition process.

Recent structural studies of vertebrate genomes suggest that retroviruses may employ a mode of transposition much more widely used than previously supposed for amplification, alteration, and relocation of several chromosomal components. A growing list of genetic units shows some evidence of having been established by reverse transcription of RNAs transcribed from preexisting units. Some pseudogenes of immunoglobulin, tubulin, globin, and metallothenin genes lack introns, are flanked by direct repeats, end with an oligo(dA) tract, and are unlinked to their functional counterparts (Leder *et al.*, 1981; Hollis *et al.*, 1982; P. Sharp and M. Karin, personal communications). ''Alu''-type sequences, dispersed widely and abundantly in the genome, and pseudogenes for small nuclear RNAs are both often flanked by direct repeats and concluded at one end with oligo(dA) (Van Arsdell *et al.*, 1981; Jagadeeswaran *et al.*, 1981; Schon *et al.*, 1981). The hepatitis B viruses may represent another member of the class of eukaryotic genetic elements that uses some component(s) of the retrovirus life cycle: Recent studies by Summers and Mason (1982) indicate that these viruses may replicate by reverse transcription of an amplified RNA intermediate, with packaging of incomplete duplex DNA products in progeny virions. In this case, although replication appears to be efficient and is probably mediated by a virus-coded, RNA-directed DNA polymerase, there is no evidence for a precise mechanism for integration. Viewed in comparison either with those cellular genes (and pseudogenes) that may occasionally be established by reverse transcription or with other viral genomes that may replicate their DNA through RNA intermediates, retroviral genomes seem to have achieved a high level of genetic competence and structural design.

B. Structural Homologies with Transposable Elements

One of the conclusions readily drawn from perusal of this chapter is that the several transposable elements within a single class of organisms are likely to exhibit more structural diversity than certain elements widely scattered throughout the evolutionary tree. Simple inspection of structural features assigns retroviruses to that class of mobile elements that includes Tn*9* (and others) in bacteria;

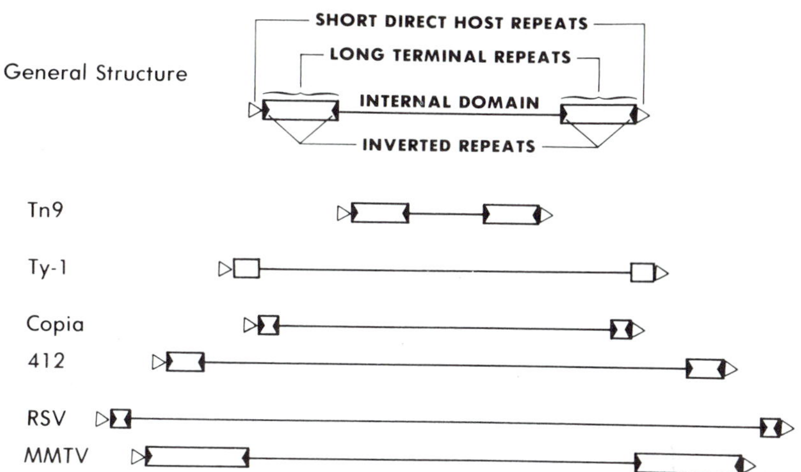

Fig. 11 Structurally related nomadic elements of prokaryotic and eukaryotic organisms. Several examples of transposable elements with particularly strong resemblances to proviruses—Tn*9* of *Escherichia coli,* Ty-1 of *Saccharomyces,* and elements *copia and* 412 of *Drosophila melanogaster*—are compared with retroviral proviruses of RSV and MMTV. Large boxes represent direct LTRs within the nomadic elements; open triangles represent short repeats of host DNA at insertion sites; solid triangles denote inverted repeats. The sequences denoted by triangles are not drawn to scale. Further details are presented in the text and Table V.

Ty-1 in yeast; and copia, 412, and several others in *Drosophila* (Fig. 11). All of these share an internal coding domain of one to several kilobase pairs; large terminal direct repeat units of several hundred base pairs, usually with short inverted terminal repeats; and short flanking direct repeats of host DNA duplicated during insertion (Table VA). Some of these features distinguish this class from other groups of mobile elements, such as the classes that include Tn*3* of *E. coli* and P factors of *Drosophila,* the variable surface glycoprotein genes of trypanosomes, mating type casettes of yeast, immunoglobulin genes, or the inverting elements in *Salmonella* or Mu-1 phage.

Many authors have previously commented upon the structural similarities among members of the Tn*9* class of mobile elements. In Fig. 11 and Table VA, the major features of several members of this class are compared, and in Table VB some of the homologies described in nucleotide sequence are listed for eukaryotic members. Some of these homologies, particularly the TG . . . CA format at the ends of the entire units and the nucleotide sequences at the ends of the internal domains, where retroviral DNA synthesis initiates, are extraordinarily provocative. At the least, they testify to the correctness of assigning proviruses to this class of mobile elements, regardless of the currently perceived differences in modes of transposition. More important but less clear is the functional significance of the structural similarities. Are the similarities vestiges of a common predecessor? If so, have the vestiges persisted *despite* the development

Table V Similarities among Structurally Related Transposable Elements

A. Organism	Element	Internal domain[a]	Long terminal repeat[b]	Inverted repeat[c]	Host duplication[d]
Escherichia coli	Tn9	1.1	768	18/23	9
Saccharomyces cerevisiae	Ty-1	5.1	334,338	2/2	5
Drosophila melanogaster	*copia*	4.5	276	13/17	5
	297	5.7	412	2/2 or 3/3	4 or 6
	412	6.5	481,571	5/6	4
	Mdg 1	6.3	442,444	6/6	4
	B104	7.9	429	3/3	5
Aves	ASVs, RAVs	3.0–8.6	270–350	10/13 (12/15)	6
	REVs	5.0–8.0	569	3/3 (5/5)	5
Mus	MLV/MSVs	4.5–7.8	515–588	11/11 (13/13)	4
	MMTV	6.2	1330	6/6 (6/8)	6

B. Element	5′-end of LTR[e]	5′-end of internal domain[f]	3′-end of internal domain[g]	3′-end of LTR[h]	Reference[i]
Ty-1	TGTTGGAATA	TGGTAGCGCCTGTGCTTC	GGGTGGTA	ATTTCTCA	1
copia	TGTTGGAATA	GGTTATGGGCCCAGTGCA	GAGGGGCG	TTACAACA	2
412	TGTACT	TGGCGACCGTGACAGTCGTC	AAAAGGAGGGAGA	ATTACA	3
Mdg1	TGTAGT	TGGCGACCGTGACAAAGGAT	AAAAGGAGGGAGA	ACTACA	4
297	GTGACGTATTT	TGG	AAGGGAAGGGGA	AATTTTAC	5
B104	TGTTCA	TTTGGTCAATCGA		ATTTTACA	6
RSV	(AA)TGTAGTCTTATGC	TGGTGACCCCGACGTGAT	AGGGAGGGGGA	GCAGAAGGCTTCA(TT)	7
SNV	(AA)TGT	TGGGGGCTCGTCCGGGAT	AGTGGGG	GTACAACA(TT)	8
MLV/MSV	(AA)TGAAAGACCCC	TGGGGGCTCGTCCGGGAT	AGAAAAAGGGGGG	GGGGTCTTTCA(TT)	9
MMTV	(AA)TGCCGC	TGGCGCCCGAACAGGGAC	AAAAAGAAAAAAGGGGA	GCGGCA(GC)	19

[a] The lengths of the domain between long terminal repeats (LTRs) are given in kilobase pairs.
[b] The lengths of sequences in LTRs from examples of each listed element are given in base pairs.
[c] The lengths of inverted repeated sequences at the ends of LTRs are presented as the fraction of perfectly matched base pairs. For the proviruses, the lengths of the inverted repeats in the unintegrated DNA forms (2 bp longer) are shown in parentheses.
[d] The lengths of the host sequence at the integration site, duplicated during insertion to flank the element, are given in base pairs.
[e] Sequences at the 5′-ends of the LTR showing homology among elements. For the proviruses, the two bases present only in unintegrated DNA are shown in parentheses, and the entire inverted repeat sequence is provided.
[f] Sequences at the 5′-end of the internal domain that could provide a binding site for tRNA. The proviruses are oriented as elsewhere in this chapter, so that the direction of transcription is left to right; for the other elements, the orientation is arbitrarily selected to maximize the similarities among the elements. For some of the *Drosophila* elements, most if not all transcription would proceed from right to left.
[g] Purine-rich sequences at the other end of the internal domain, homologous to the putative primer for the second strand of retroviral DNA.
[h] Homologous sequences at the 3′-ends of LTRs. Again, the two bases present only in unintegrated forms of retroviral DNA are shown in parentheses.
[i] References: 1, Farabaugh and Fink (1980); 2, Levis *et al.* (1980); 3, Will *et al.* (1982); 4, Kulgushkin *et al.* (1981); 5, Ikenaga and Saigo (1982); 6, G. Scherer *et al.* (1982); 7, Swanstrom *et al.* (1981); 8, Shimotohno *et al.* (1980); 9, Dhar *et al.* (1980); 10, Majors and Varmus (1981).

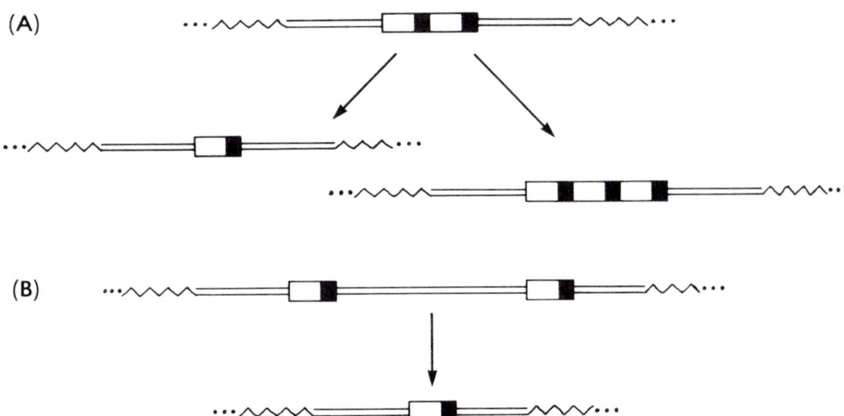

Fig. 12 Unusual behavior of retroviral DNA during propagation in prokaryotic hosts. (A) Changes in numbers of adjacent LTRs. Cloning of closed circular DNA containing two LTRs in phage or plasmid vectors in *Escherichia coli* leads to segregation of molecules containing one or three LTRs (Hager *et al.*, 1979). Viral DNA is shown as double lines, with open (U3) and closed (U5) boxes forming LTRs; vector DNA is represented as wavy lines. (B) Excision of proviral DNA by homologous recombination. Cloning of proviral DNA in *E. coli* host–vector systems leads to occasional loss of most of the proviral DNA by apparently homologous recombination between LTRs, leaving a single LTR flanked by the sequences originally flanking the provirus (double line) in its animal host (McClements *et al.*, 1980).

of different modes of behavior or *because* of the retention of some common functional attributes? Or are the similar sequences and structures the products of convergent evolution, with features that prove useful for some common (and perhaps incompletely perceived) aspects of their overtly different transpositional mechanisms?

One potentially provocative way to assess the structural homologies is to propagate one member of the class in the host cells of another member. Naturally, this has been done most frequently during the cloning of various elements in *E. coli* host–vector systems. To date, there have been no reports of transposition of retroviral DNA during its growth in coliform hosts, but two types of significant genetic change have been regularly observed (Fig. 12). First, attempts to amplify intact proviral DNA in bacteriophage vectors regularly generate deletion mutants that retain only one LTR and flanking host DNA (McClements *et al.*, 1980; O'Rear *et al.*, 1980; Fig. 12B). Because the vectors are *int*$^-$, *xis*$^-$, and *red*$^-$, this reaction appears to be independent of the λ recombination system; thus, it resembles both the "*cam* cutout" reaction, in which Tn*9* inserted in λ phage is deleted by recombination events, leaving only one IS*1* sequence (MacHattie and Jackowski, 1977), and excision of λ prophage inserted between IS*1* elements (Shapiro and MacHattie, 1979). None of these reactions appears to require the host *rec* A function (Shapiro and MacHattie, 1979; G. Vande Woude, personal communication), although similar deletions have not been reported when proviruses are grown in plasmid vectors in *rec* A^- hosts. The deletions could be due to recombination between the terminally repeated sequences (LTRs or IS*1*s) on a single molecule or to unequal crossing-over between two phage genomes. The expected reciprocal product of unequal crossing-over (phage with a duplicated provirus) is rarely seen, perhaps because of packaging constraints, although there is one reported instance of duplicated spleen necrosis virus (SNV) proviruses that might have arisen from unequal crossing-over (O'Rear *et al.*, 1980). These proviruses, however, could also have been duplicated in their original host cell, as shown for an RSV provirus (see Section III,C,3 and Fig. 13C). Unequal crossing-over is the likely mechanism for another frequently observed change during the propagation of cloned fragments of circular retroviral DNA in bacteriophage vectors: the variation in number of tandem LTRs (Fig. 12A). In the reported cases, recombinant molecules originally containing two tandem LTRs seem to generate molecules with either three tandem copies or a solitary LTR (Hager *et al.*, 1979; S. Hughes, unpublished).

Other chapters in this volume consider the possibility that members of the Tn*9* structural class other than retroviral proviruses may transpose via RNA intermediates. It is sufficient to reiterate here that the evidence remains weak. The precise identity between pairs of long terminal repeats in each sequenced element of Ty1 (Farabaugh and Fink, 1980; Gafner and Phillipsen, 1980), *copia* (Levis *et al.*, 1980), 412 (Will *et al.*, 1982), and *Mdg*-1 (Kulgushkin *et al.*, 1981), despite differences between repeats in separate elements of the same type, can be interpreted to mean that the two LTRs are generated from a single LTR during each transposition event. This could occur, as it does for proviruses, by DNA synthesis from an RNA intermediate with only one copy of the sequences to be repeated, or it could occur, as it does during replicon fusion, by DNA replication during insertion of a circular element containing a single LTR (Shapiro, 1979). It is also possible that identity between LTRs could be maintained by a process unrelated to transposition, such as gene conversion (Baltimore, 1981b). [Gene conversion is a tempting explanation for identical LTRs in endogenous proviruses (Hishinuma *et al.*, 1981; Kennedy *et al.*, 1982; G. Hager, personal communication), but it is uncertain whether those proviruses are sufficiently ancient to have accumulated spontaneous mutations.] The significance of identical LTRs may be further complicated by the single sequenced example of *Mdg*-3 of *Drosophila,* an element very similar to members of the Tn*9* class but with inverted terminal repeats (18 bp) that are adjacent to, not part of, identical, direct subterminal repeats (268 bp) (Bayev *et al.*, 1980). Thus, *Mdg*-3 may be required to transpose its termini independently, rather than by generating two from one.

There are, however, other kinds of evidence to suggest that reverse transcription might be involved in the transposition of some members of this structural class. In the cases of 412 and *Mdg*-1, the sequences flanking the inner boundaries of the terminal direct repeats include a possible site for tRNA binding (and hence priming of DNA synthesis) on one side and a polypurine tract (and hence a possible second strand priming site) on the other (Table VB). However, it is surprising to find that the only known transcripts of 412 and the abundant transcripts of *Mdg*-1 are *complementary* to the strand that would be predicted to act as template for reverse transcription regulated by the homologous signals (Finnegan *et al.*, 1978; Ilyin *et al.*, 1980). For *copia*, the case may be somewhat stronger, despite the less suggestive sequences at the boundaries of the internal domain (Table VB), because the abundant RNA species has the predicted polarity (Flavell *et al.*, 1981) and, more important, because unintegrated forms of *copia* DNA have been found, cloned, analyzed structurally, and shown to resemble retroviral circles bearing one and two copies of the LTR (Flavell and Ish-Horowicz, 1981). The only obvious structural difference is that retroviral circles with two LTRs include the 4 bp of viral sequence missing from the ends of proviral DNA, whereas the corresponding *copia* circles have only a single extra A : T base pair of uncertain origin (A. Flavell, personal communication). Although there is some marginal evidence for particles with reverse transcriptase activity in cultured *Drosophila* cells (Heine *et al.*, 1980), there is no direct information about the manner in which the circular *copia* DNA is generated. Last, there have been suggestions that Ty-1 may transpose via RNA intermediates, because the major RNA transcripts are terminally redundant (J. Elder and R. Davis, personal communication) and inner boundaries of the δ units are again similar to retroviral priming sites (Table VB; Eibel *et al.*, 1981). Furthermore, at least one size class of unintegrated Ty-1-specific circular DNA has been detected in certain yeast strains (Newlon *et al.*, 1982).

The members of a structurally related group of transposable elements would be expected to share some important component of their transposition mechanism. As members of the Tn9 class of transposable elements, retroviral proviruses probably require for transposition the structural properties that admit them to this class, even though the manner in which they transpose seems to set them off from the other members. The significance of those structural properties will be apparent only when the mechanisms of proviral integration and transposition are more clearly understood.

C. Mutational Capacities

Like all other transposable elements, regardless of structural class, retroviral proviruses display mutational properties. Some of these are predictable as inherent features of DNA that can insert randomly within host genomes, whereas others are more difficult to explain.

1. Insertion Mutagenesis

a. Inactivating, Recessive Lesions. Any unit of DNA that can be transposed to a new site at random is expected to make interruptions in cellular genes at a frequency dependent upon the size of the genome and the efficiency of transposition. Because most retroviruses generally introduce no more than a few copies of proviral DNA into any single cell and because vertebrate genomes are diploid and large (ca. 4×10^6 kbp), insertional mutations are expected to be rare. The precise frequency has been difficult to determine, because appropriate selection methods are not available for mutations of many genes and because reagents are lacking for validation of the nature of the mutations in most genes for which selection methods do exist. Naturally, the matter may be complicated by the variable sizes of genetic units, by the uncertain consequences of insertion in noncoding regions (e.g., introns), by the presence of multiple copies of some genes, and by the possibility that insertion is not strictly random within the host genome.

Inactivating, insertional mutations by proviral DNA have been most fully examined in an experimental system using one provirus as the target for mutation and another as the mutagen (Varmus *et al.*, 1981a; Fig. 13A). The host cell for mutagenesis was derived from a continuous line of rat fibroblasts by morphological transformation with RSV. The transformed cell (called B31) carried only a single RSV provirus (and hence was haploid for the mutational target); being nonpermissive for replication by RSV, the line was unlikely to accumulate additional RSV proviruses. Moreover, the transformed phenotype was particularly stable in the B31 line: Unlike most RSV-transformed mammalian cells, which revert frequently to a normal phenotype by modulating transcriptional activity (Section II,C,5,*c*), the prevalence of revertants of B31 cells was about 10^{-5}, and the responsible lesions were loss of the entire provirus, small mutations (frameshifts and base substitutions) in *src*, or deletions affecting the 5′-LTR (Varmus *et al.*, 1981b; Oppermann *et al.*, 1981; see below). After infection with Mo-MLV, a nontransforming retrovirus for which rat cells are a permissive host, all the B31 cells acquired one to five MLV proviruses; two insertion mutations were then identified by isolating and screening over 60 morphological revertants for disruptions in the RSV provirus. The insertion mutation rate appeared to be between 10^{-6} and 10^{-7}, in accord with the principle of random insertion of the superinfecting provirus.

Both mutations were found to have interrupted the RSV provirus outside the *src* coding domain; in both cases, the MLV DNA was between the splice donor and acceptor sites for *src* mRNA, in the same transcriptional orientation as the RSV DNA. Examination of viral RNA in the two lines showed that a normal amount of RNA annealed with probe for the U5 domain of RSV DNA, suggesting that the promotional activity of the RSV 5′-LTR had not been affected, but transcripts appeared not to extend beyond the MLV 3′-LTR. (Notably, the MLV 5′-LTR did not serve to polyadenylate RNAs initiated in the RSV 5′-LTR, nor did the MLV 3′-LTR promote transcription of *src*.) Thus, these mutations are polar intron insertions affecting synthesis or processing of RSV RNA. As discussed below (Section III,C,2), one of the two mutants is unstable, converting to a wild-type phenotype after recombination between the two MLV LTRs.

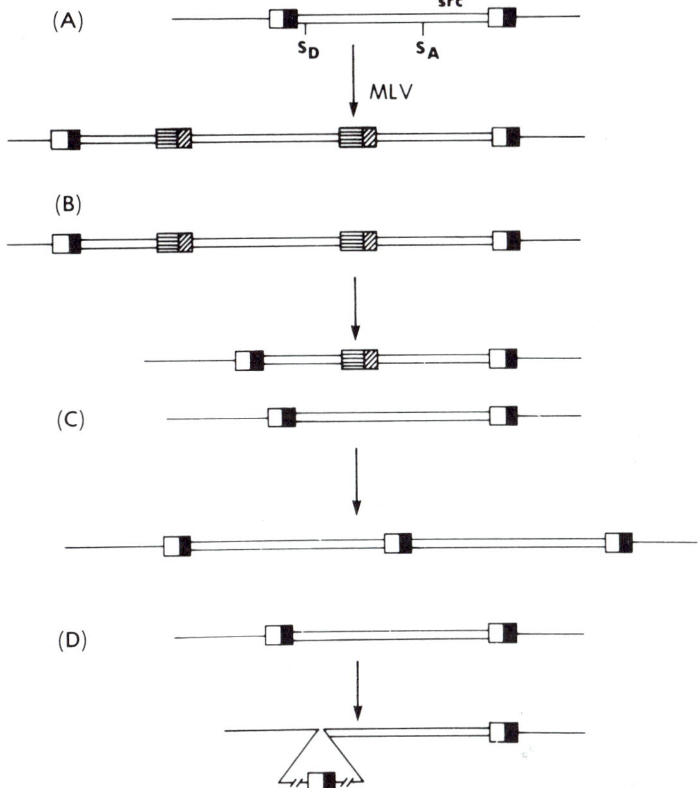

Fig. 13 Unusual behavior of proviruses in infected animal cells. (A) Insertion mutagenesis. In the illustrated case, a provirus of RSV is the genetic target for mutation, and a provirus introduced by infection with MLV is the mutagen (Varmus *et al.*, 1981a). The insertion occurred between the donor and acceptor splice points for *src* mRNA and altered the phenotype of the RSV-transformed rat cell by prohibiting the expression of *src*. The RSV provirus is drawn as a double line, with LTRs composed of open (U3) and closed (U5) boxes; the MLV provirus is drawn as a double line, with LTRs composed of horizontally (U3) and diagonally (U5) striped boxes. (B) Proviral excision by homologous recombination between LTRs. An MLV provirus, inserted within an RSV provirus as described for panel (A), is largely eliminated at low frequency by recombination between its LTRs, leaving behind a single MLV LTR within an RSV provirus now able to express *src* at normal levels (Varmus *et al.*, 1981a). (C) Proviral duplication. An apparently normal RSV provirus in a transformed quail cell was found duplicated as diagrammed in a subclone of the original transformant (Hsu *et al.*, 1981). The mechanism of duplication is not known (see text). (D) Deletions encompassing the 5′-LTR. In several instances, the RSV provirus shown in panel (A) has undergone deletion removing the 5′-LTR and variable amounts of adjacent cellular and viral DNA (Varmus *et al.*, 1981a). Similar truncated proviruses are frequently observed endogenous to germ lines (Hughes *et al.*, 1981a; Hayward *et al.*, 1980) and in retrovirus-induced tumors (Payne *et al.*, 1981), but the structure of the primary proviral insert in these cases is not known.

During a survey of many mouse strains for endogenous, ecotropic MLV proviruses, Jenkins *et al.* (1981) noted an extra provirus in DBA/2J mice bearing the *dilute* coat color allele, a recessive lesion mapped to chromosome 9. This provirus was missing from mice descended from spontaneous revertants of *dilute* and homozygous for the revertant allele. The presumed mutagenizing provirus has been cloned with its flanking DNA, and a residual LTR has been found in the homologous region of the revertant genome (N. Jenkins and N. Copeland, personal communication). Two mutations arising in cell culture in a mouse light-chain immunoglobulin gene have been identified as insertions of IAP DNA (Section II,F,1; M. Shulman, personal communication).

An attractive approach to studying retroviral insertion mutants is to use genetic targets for which the host cell is hemizygous (e.g., X-linked genes). Preliminary attempts to produce hypoxanthine-guanosine phosphoribosyl transferase mutants by infection with MLV seemed to raise the frequency of mutations over the background rate (M. Capecchi, personal communication), but the nature of the lesions has not yet been documented. Theoretically, proviral insertion mutations could be used to isolate host genes that are difficult to clone with standard techniques. However, the probability of making the necessary double mutations is too low to permit this approach for autosomal genes, unless one member of the pair can be inactivated by another means without raising the frequency of spontaneous mutations to an unacceptable level.

b. Dominant, Activating Insertion Mutants. Retroviral proviruses, like several other transposable elements, have the capacity to augment the expression of genes adjacent to an integration site. These phenomena have usually been ascribed to one or more of the functions attributed to LTRs: the ability to promote transcription, to polyadenylate transcripts, or to enhance transcription from linked promoters in a manner independent of orientation (Section II,C,4,*d*). All of the well-studied mutants of this type are derived from bursal lymphomas induced by ALV (see Section II,E,2); the selected phenotype (tumorous growth) has not been rigorously demonstrated to depend upon the observed molecular lesions (enhanced transcription of a c-*myc* locus bearing a proviral insertion).

Figure 14 illustrates the arrangements of proviral and c-*myc* DNA that have been encountered in tumors showing augmented levels of c-*myc* RNA. In the most common case (Fig. 14I), proviral DNA is positioned on the 5′-side of the two known exons of c-*myc,* in the same transcriptional orientation, generally 1–5 kb from the start of the first exon, so that the 3′ LTR can serve as a promoter for c-*myc* RNA (Hayward *et al.*, 1981). The resulting transcripts anneal with probes for U5 but not U3, confirming the prediction that the transcripts are initiated in the LTR. In several cases, the 5′ LTR and sometimes additional ALV DNA has been deleted; this may potentiate the use of the 3′-LTR as a promoter (Payne *et al.*, 1981; Neel *et al.*, 1981). In one instance, a proviral insert has been discovered on the 3′-side of the major c-*myc* exons without evidence for an insertion upstream from c-*myc* (Fig. 14II; Payne *et al.*, 1982). The provirus is in the same

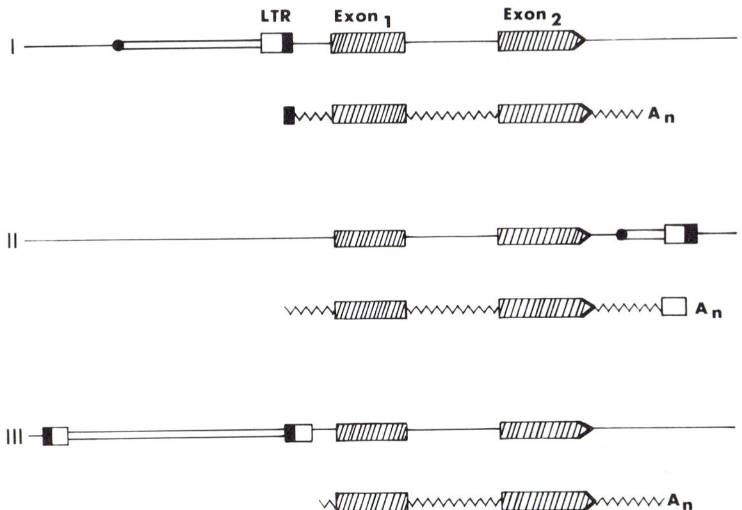

Fig. 14 Arrangements of proviral DNA and a cellular oncogene associated with enhanced expression of the oncogene. Panels I, II, and III depict the three configurations in which ALV DNA has been found in DNA from chicken bursal lymphomas in the region of c-*myc*, the cellular homologue of a viral transforming gene (v-*myc*) (Payne *et al.*, 1982). The two known exons of c-*myc* are denoted by striped boxes, with the shape of the second exon indicating the direction of synthesis of normal c-*myc* mRNA. Proviral DNA is shown as a double line, with open and closed boxes representing U3 and U5 regions, respectively; cellular DNA is shown as a single line. Junctions formed by deletions that probably occurred subsequent to proviral integration are indicated by solid circles. The direction of synthesis of each RNA species is indicated by the position of the polyadenylate tract (A_n) at the 3′-terminus. The predicted primary transcripts containing c-*myc* sequences are illustrated for convenience, but the observed abundant c-*myc* RNA in bursal tumors is a processed species in which the region between $exon_1$ and $exon_2$ has probably been removed by splicing. The positions of integration, initiation, and polyadenylation sites are approximate; the drawing is not to scale.

transcriptional orientation as c-*myc*, and a large deletion, affecting both cellular and ALV DNA, has spared only about 0.7 kb of ALV sequence, including the 3′-LTR. The LTR appears to provide the polyadenylation site for the abundant c-*myc* transcripts; the site at which transcription is initiated has not been determined. In several additional tumors, ALV proviruses have been found on the 5′-side of c-*myc*, usually about 1 kb from the 5′-boundary of the first exon, in the opposite transcriptional orientation (Fig. 14 III; Payne *et al.*, 1982). Again, the level of c-*myc* RNA is augmented, but the stable transcripts do not contain ALV sequences. The initiation site and the mechanism of enhanced expression are not known.

These phenomena are analogous to a number of phenomea observed with transposable elements in other organisms. For example, insertions of IS2 or IS*3* can provide a strong promoter for *E. coli* operons (reviewed by Kleckner, 1981), and the solitary δ sequences remaining after excision of Ty-1 from polar insertion mutants in *his*-4 restore expression of *his*-4, presumably by serving as a promoter site (Roeder and Fink, 1980; G. Fink, personal communication). Another comparison with Ty-1 is of perhaps greater interest: A large number of Ty-1 insertion mutants that diaplay enhanced expression of a flanking gene are due to insertions of Ty-1 on the 5′-side of the gene in the opposite transcriptional orientation (Errede *et al.,* 1980; Williamson *et al.,* 1981; S. Scherer *et al.,* 1982).

2. Proviral Excision

Although there are several reported instances in which an entire provirus has been lost from infected, cultured cells (Frankel *et al.,* 1976; Trainor and Reitz, 1979; Yang *et al.,* 1979; Varmus *et al.,* 1981b), in none of these cases has the mechanism of elimination been ascertained. Chromosomal loss, large deletions, and even precise excision remain possible explanations. (Precise excision would be difficult to prove unless the insertion had occurred originally on a marked chromosome, allowing a distinction between precise excision and loss of one member of a chromosomal pair.)

Solitary LTRs, suggestive of elimination of most of a provirus by recombination between 5′- and 3′-LTRs, have been observed in a variety of contexts: RAV-0-related (Hughes *et al.,* 1981c) and FeLV-related (Casey *et al.,* 1981) LTRs in the germ lines of certain chickens and cats; an FeSV LTR in a morphological revertant of an FeSV-transformed cell line (Donner *et al.,* 1980); an MLV LTR in revertants of the *dilute* mutation (Section III,C,1,*a*); an ALV LTR in a bursal lymphoma (Payne *et al.,* 1981; D. Westaway, unpublished); and an MLV LTR in retransformed cells derived from an MLV insertion mutants of B31 cells described above (Section III,C,1,*a*; Varmus *et al.,* 1981a). Only in the last situation has the excision phenomenon been reproducibly observed (Fig. 13B). Because the excision permits reexpression of the transformed phenotype, foci are readily scored; the frequency of recombination is estimated to be 10^{-7}–10^{-8} per cell per generation. Several of the solitary MLV LTRs that remain have been mapped: In all but one case, the LTRs appear to be identical to those in the mutating provirus (S. Ortiz, unpublished). The one exceptional LTR has suffered a deletion of about 150 bp. No information is available about viral or cellular factors that may mediate the excision events. Nucleotide sequencing of a solitary ALV LTR from a lymphoma reveals the expected features of the excision product; the LTR lacks 2 bp at each end and is flanked by a 6-bp duplication of host DNA (D. Westaway, unpublished).

Recombination between LTRs is one of the few examples of homologous recombination in somatic cells of higher eukaryotes. Homologous recombination has also been observed within tandem repeats of free (Wake and Wilson, 1980) and integrated (Basilico *et al.,* 1979; Botchan *et al.,* 1979; Lania *et al.,* 1982) papovavirus genomes; the frequency is dependent upon the presence of a product of a viral early gene (large T antigen), although in one case recombination appears to occur occasionally in the absence of any normal T antigen (Lania *et al.*, 1982). In the reaction involving retroviral LTRs, it is not clear whether recombination can occur randomly within the LTRs, or whether a defined portion of sequences constitutes a favored site [as is the case, for example, in the resolution of Tn*3* cointegrates (Reed, 1981; Reed and Grindley, 1981)]. It is also unknown whether proviral products (e.g., analogues of resolvase) or host genes (e.g., analogues of *Spm* of yeast; see Chapter 7) have any role in the process. Other structurally similar transposable elements undergo similar reactions [e.g., the *cam* cutout reaction of Tn*9* (MacHattie and Jackowski, 1977); excision of Ty-1 by recombination between δs (Roeder and Fink, 1980)], but there appears to be considerable variation in frequency; for example, the related elements of *Drosophila* rarely or never appear to generate solitary repeat units in this fashion (Levis *et al.*, 1980).

Regardless of the mechanism, the excision reaction can have important consequences other than reversion of a mutation. It may, for example, be a means for restraining the accumulation of transposable elements, such as proviruses, in the germ line. On the other hand, it is conceivable that the second excision product, a free circle with a single LTR, may be able to reinsert in a new chromosomal site; as yet, there have been no unambiguous tests of this possibility. Last, the repeat unit remaining behind in the chromosome may be more likely than the repeats at the ends of the complete element to influence the expression of adjacent cellular sequences. This has been documented in the case of Ty-1 (Chapter 7) and seems likely in the case of some ALV proviruses in bursal lymphomas (see above).

3. Proviral Duplication

The failure to find tandem proviruses in infected cells is one of the reasons for concluding both that proviruses do not preferentially use homologous sites for integrative recombination and that the integrating species of viral DNA is monomeric. In the one instance in which a tandem arrangement has been found, a single normal RSV provirus was observed to undergo a duplication during prolonged passage of its host cells (quail) (Hsu *et al.*, 1981; Fig. 13C). The resulting tandem structure has terminal LTRs, plus a single LTR unit in the center separating two complete internal domains, all in the same apparent orientation. The mechanism by which the duplication occurred is not known, although unequal crossing-over between the 5′- and 3′-LTRs of the original provirus during mitosis could explain the observation. A similar unexplained phenomenon has been noted for Tn*9* inserted in λ phage (MacHattie and Jackowski, 1977). The duplicated provirus appears to produce a moderate amount of RNA the size of the entire 19-kb unit, as well as abundant normal RSV mRNAs and perhaps minor spliced products of the 19-kb transcript. Thus, as in the case of the B31 cell insertion mutants discussed above, an LTR in the middle of a proviral unit flanked by LTRs does not direct polyadenylation of all of its transcripts.

4. Deletion Formation

Among prokaryotic transposable elements, deletions are common and frequently occur at specific sites within or adjacent to the elements (Kleckner, 1977, 1981). Deletion mutants are frequently encountered among proviruses, but the vast majority are presumed to represent the products of deletions generated during reverse transcription (or at some stage in the life cycle other than propagation of integrated DNA), because the deletions do not cross LTR boundaries (see Coffin, 1979). However, a number of deletion mutants have been identified with lesions that affect LTR boundaries and flanking cellular DNA; those that have been most fully characterized are the several endogenous, RAV-0-related proviruses of chickens (Hayward *et al.*, 1980; Hughes *et al.*, 1981c); some ALV proviruses in the c-*myc* locus in bursal lymphomas (Payne *et al.*, 1981, 1982; Neel *et al.*, 1981; Fung *et al.*, 1981); and truncated proviruses in a few morphological revertants of the B31 cell line (Fig. 13D). All of the observed deletions seem to affect the 5′-LTR rather than the 3′ LTR, although in each context there may be selection pressures favoring mutants without a 5′-LTR. Usually, the loss of the 5′-LTR is accompanied by an absence of transcripts of the provirus, as expected for promoter deletions. However, at least one endogenous chicken provirus (*ev*-6) is expressed under the apparent control of an adjacent cellular promoter (Hayward *et al.*, 1980; Baker *et al.*, 1981); it is also possible that deletion of the 5′-LTR favors the use of the 3′-LTR as a promoter of flanking cellular genes (e.g., in bursal lymphomas).

It has been impossible thus far to judge whether proviral deletion mutations are more common than similar lesions affecting vertebrate chromosomal DNA generally. For endogenous proviruses and proviruses in tumors, it is not even possible to estimate the absolute frequency of deletion formation. For the B31 cell mutants, the frequency appears to be about 10-fold less than that of point mutations in *src* (i.e., the prevalence of deletion mutants is about 10^{-6}; Varmus *et al.*, 1981a), but there is no available estimate for the frequency of large deletions elsewhere in the genome (other than the specialized immunoglobulin locus). If the structure of the RSV provirus (or some property of the superinfecting MLV also present during deletion formation) influences the deletions in B31 cells, this is not apparent from physical maps of the affected proviruses: The deletions vary from 1.5 to over 20 kb in length and join different sites in viral DNA to different sites in host DNA in most if not all cases. It is likely that deletions in circular molecules from acutely infected cells (Section II,C,3 and Fig. 6) are the consequences of mechanisms that specifically affect unintegrated rather than proviral forms.

5. Transduction

Retroviruses are admirably designed to serve as vectors of genetic information from one cell to another. This property is most dramatically displayed by the v-*onc*$^+$ viruses (Fig. 2; Section II,B,3), which transport transforming genes of cellular origin into newly infected cells and express them under the control of viral regulatory signals. Recent efforts to generate viral vectors carrying other heterologous genes by DNA recombination *in vitro* (Shimotohno and Temin, 1981; Wei *et al.*, 1981) also substantiate the utility of retroviruses as conveyers of genes.

How have the naturally occuring (v-*onc*$^+$) recombinant viruses transduced cellular information? Because the generation of such viruses is a rare event not readily reproducible under controlled conditions, only indirect evidence bears upon the question of mechanism. However, it is apparent from the structure of the transforming virus genomes that at least two recombination events must have occurred in each case: one to join viral sequences to the 5′-side of a cellular insert and one to join viral sequences to the 3′-side. These requirements are implicit in the retrovirus life cycle, because viral sequences essential for synthesis of DNA with LTRs must be on both sides of any transduced sequence that is to be repeatedly replicated. In addition, there must be a step in which any introns between coding exons of c-*onc* genes are removed during generation of v-*onc* genes, because v-*onc* coding domains are invariably uninterrupted. Despite the apparent odds against the collusion of such rare events, v-*onc*$^+$ viruses have arisen on numerous occasions (see Table II); in the mechanism proposed below, two factors can be viewed as potentiators of the recombination events—the use of a viral packaging signal to augment the population of heterozygote particles and the tumorous expansion of the cellular clone in which the subsequent events will occur.

The transduction process is now thought to begin with the integration of the provirus of a replication-competent v-*onc*$^-$ virus on the 5′-side of the cellular oncogene destined to supply v-*onc* sequences (Fig. 15). The precedent of ALV insertions near c-*myc* suggests that such insertions can lead to the production of a tumor in which large numbers of cells (10^8 or more) contain the provirus linked to the oncogene. Deletions that remove the 3′-LTR and variable amounts of adjacent viral and cellular DNA would create an opportunity for synthesizing RNA that is initiated at the cap site in the 5′-LTR and extended through the putative packaging signal and the residual viral coding domains into the remaining exons and introns of the cellular gene. The only information presently accessible about the boundaries of the postulated deletions resides in the nucleotide sequences of cloned c-*onc* genes and of viral genomes (both the presumed transducing agent and the resulting recombinant). Some of the relevant data are now available for at least three oncogenes. (*1*) The deletion on the 5′-side of *abl* during the generation of Abelson MLV apparently joined the coding sequence in *gag* to the middle of an exon in c-*abl*, with preservation of the reading frame; the deletion resulted from recombination within a 4-bp homologous sequence (D. Baltimore, personal communication). Presumably, other viruses that carry *gag*-*onc* fusions (Fig. 2D) have required similar deletions, although nothing is known about homology at the site of deletion formation. (*2*) Production of Moloney

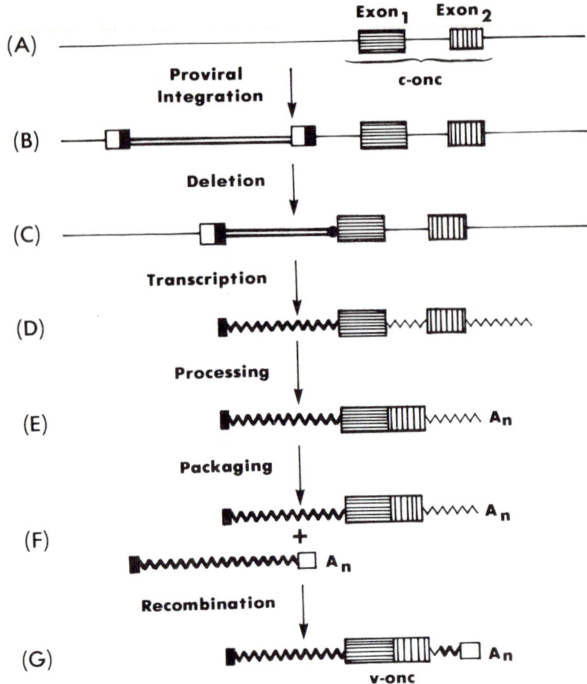

Fig. 15 A possible mechanism for transduction of a cellular oncogene by a nontransforming retrovirus (R. Swanstrom, unpublished). The horizontally and vertically striped boxes in (A) represent exons in a cellular gene (c-*onc*) destined to be a precursor to a viral transforming gene [v-*onc;* (G)]. (B) Proviral DNA of a retrovirus lacking *onc* integrates on the 5′-side of c-*onc* genes in the same transcriptional orientation. (Viral DNA is indicated by double lines, U3 as open boxes, U5 as closed boxes, and host DNA as a single line.) (C) A deletion removes a 3′-portion of proviral DNA and adjacent cellular DNA (the point of rejoining is indicated by a closed circle). The figure is purposefully ambiguous about whether the deletion ends in a coding or noncoding sequence of c-*onc;* see text. (D) Transcription from the remaining retroviral LTR produces a primary transcript linking sequences from the 5′-end of the viral genome [at least R, U5, (−)PB, and L (cf. Fig. 1); heavy wavy line] to c-*onc* sequences. (E) The primary transcript is polyadenylated and spliced to remove c-*onc* introns. (F) The processed RNA is packaged in retroviral particles, forming heterozygotic dimers with a subunit of wild-type viral RNA. (G) Recombination during reverse transcription (cf. Fig. 9) occurs upon infection of neighboring cells, yielding a recombinant genome; because the components of the heterodimer are only partially related, the frequency of recombination is expected to be relatively low (see text).

MSV appears to have involved a deletion that fused sequences near the 5′-end of *env* to sequences near the 5′-end of the only known exon of c-*mos;* 2 bp of homology are present at the deletion boundaries (Van Beveren *et al.*, 1981b). (*3*) Deletions likely to have occurred during the derivation of two strains of RSV have fused different noncoding sequences from the 3′-side of *env* to an intron upstream from a 5′-noncoding exon of c-*src* (R. Parker and R. Swanstrom, unpublished). (This construction has the advantage of retaining a normal S_A on the 5′-side of v-*src* while retaining all the coding domains for viral replication genes and for c-*src*.) A similar situation has been described for avian myeloblastosis virus (K. H. Klempnauer and J. M. Bishop, personal communication); again intron sequence from the oncogene (*myb*) precedes coding *myb* sequences in the viral genome.

The primary transcript made from the template of fused provirus and c-*onc* is then presumably processed to remove any complete introns by splicing. If the cell in which these events have occurred contains a wild-type helper virus, heterozygous virus is likely to be produced by the efficient packaging of the hybrid RNA, which contains the packaging signal as well as *onc* sequences. The next necessary step—the joining of viral sequences downstream from *onc*—can presumably employ the mechanisms proposed for high-frequency viral recombination (Section II,D,2; Fig. 9), but the efficiency is likely to be severely impaired by the lack of homology at the recombination sites. There is, however, direct experimental evidence for the occasional occurrence of the necessary joinings. Goldfarb and Weinberg (1981) and Goff *et al.* (1982) have shown that retroviruses capable of efficient replication in the company of a helper virus can be recovered from cells transformed by DNA fragments containing viral sequences (an LTR and variable coding regions) only on the 5′-side of v-*onc*. Several different sites were used for recombination between the 3′-end of an accompanying helper virus genome and sequences on the 3′-side of v-*onc*. It is not yet known whether any sequence homology was required for these rare recombinations.

Although the model described has several attractive features, it is not yet possible to exclude the formerly popular view that transduction of *onc* genes results from recombination between retroviral subunits and adventitiously packaged c-*onc* mRNAs. It is apparent that the probability of such a phenomenon must be very low: The number of c-*onc* mRNAs is generally low per cell, the efficiency of packaging host mRNA is extremely poor, and both recombinations are likely to be nonhomologous. More telling, perhaps, is the evidence that intron sequences from c-*src* and c-*myb* are present in viral genomes; intron sequences would not be anticipated in packaged c-*onc* mRNAs.

IV. Prospects

It is apparent from this lengthy treatise that retroviruses are instructive about matters far beyond the confines of their own life cycles: eukaryotic gene expression and regulation, cancer, mutagenesis, transposition, protein processing, membrane assembly, and others. Certain questions seem likely to receive the greatest attention from retrovirologists in the near future: What are the biochemical components of the integration mechanism? What features of viral and flanking host DNA contribute to the regulation of viral gene expression? What are the mutational effects of infection by retroviruses? What are the biochemical properties of products of transforming genes? What is the role of c-*onc* genes in various kinds of animal and human malignancy? What are the prospects for the use of retroviruses as experimental (and even therapeutic) vectors of genetic information into somatic or germinal cells? The answers to such questions promise to be useful to students of many aspects of contemporary biology.

Acknowledgments

I wish to thank numerous colleagues for their open exchange of unpublished results and Bertha Cook for her patient stenography. Work in this laboratory is supported by grants from the American Cancer Society and the National Institutes of Health. Some of the figures shown here are modified versions of those previously prepared for a short review in *Science (Washington, D.C.)* [**216,** 812–820 (1982)].

References

Aaronson, S. A., and Stephenson, J. R. (1976). Endogenous type-C RNA viruses of mammalian cells. *Biochim. Biophys. Acta* **458,** 323–354.

Altaner, C., and Temin, H. M. (1970). Carcinogenesis by RNA sarcoma viruses. XII. A quantitative study of infection of rat cells *in vitro* by avian sarcoma viruses. *Virology* **40,** 118–134.

Andersen, P. R., Devare, S. G., Tronick, S. R., Ellis, R. W., Aaronson, S. A., and Scolnick, E. M. (1981). Generation of BALB-MuSV and Ha-MuSV by type C virus transduction of homologous transforming genes from different species. *Cell* **26,** 129–134.

Astrin, S. M. (1978). Endogenous viral genes of white leghorn chickens: common site of residence and sites associated with specific phenotypes of viral gene expression. *Proc. Natl. Acad. Sci. USA* **75,** 5941–5945.

Astrin, S. M., Buss, E. G., and Hayward, W. S. (1979). Endogenous viral genes are non-essential in the chicken. *Nature (London)* **282,** 339–341.

Astrin, S. M., Robinson, H. L., Crittenden, L. B., Buss, E. G., Wyban, J., and Hayward, W. S. (1980). Ten genetic loci in the chicken that contain structural genes for endogenous avian leukosis viruses. *Cold Spring Harbor Symp. Quant. Biol.* **44,** 1105–1109.

Bader, J. P. (1965). The requirement for DNA synthesis in growth of Rous sarcoma and Rous-associated viruses. *Virology* **26,** 253–261.

Baker, B., Robinson, H. L., Varmus, H. E., and Bishop, J. M. (1981). Analysis of endogenous avian retrovirus DNA and RNA: viral and cellular determinants of retrovirus gene expression. *Virology* **144,** 8–22.

Balduzzi, P., and Morgan, H. R. (1970). Mechanisms of oncogenic transformation by Rous sarcoma virus. I. Intracellular inactivation of cell-transforming ability of Rous sarcoma virus by 5-bromodeoxyuridine and light. *J. Virol.* **5,** 470–478.

Baltimore, D. (1970). RNA-dependent DNA polymerase in virions of RNA tumour viruses. *Nature (London)* **226,** 1209–1211.

Baltimore, D. (1975). Tumor viruses: 1974. *Cold Spring Harbor Symp. Quant. Biol.* **39,** 1187–1200.

Baltimore, D. (1981a). Gene conversion: some implications for immunoglobulin genes. *Cell* **24,** 592–594.

Baltimore, D. (1981b). Somatic mutation gains its place among the generators of diversity. *Cell* **26,** 295–296.

Banerji, J., Rusconi, S., and Schaffner, W. (1981). Expression of a β-globin gene is enhanced by remote SV40 DNA sequences. *Cell* **27,** 299–308.

Basilico, C., Gattoni, S., Zouaias, D., and Della Valle, G. (1979). Loss of integrated viral DNA sequences in polyoma transformed cells is associated with an active viral A function. *Cell* **17,** 645–659.

Battula, N., and Temin, H. M. (1977). Infectious DNA of spleen necrosis virus is integrated at a single site in the DNA of chronically infected chicken fibroblasts. *Proc. Natl. Acad. Sci. USA* **74,** 281–285.

Battula, N., and Temin, H. M. (1978). Sites of integration of infectious DNA of avian reticuloendotheliosis virus in different avian cellular DNAs. *Cell* **13,** 387–398.

Bayev, A. A., Jr., Krayev, A. S., Lyubomirskaya, N. V., Ilyin, Y. V., Skryabin, K. G., and Georgiev, G. P. (1980). The transposable element Mdg3 in *Drosophila melanogaster* is flanked with perfect direct and mismatched inverted repeats. *Nucleic Acids Res.* **8,** 3263–3286.

Bender, W., and Davidson, N. (1976). Mapping of poly(A) sequences in the electron microscope reveals unusual structure of type C oncornavirus RNA molecules. *Cell* **7,** 595–607.

Bender, W., Chien, Y.-H., Chattopaddhyay, S., Vogt, P. K., Gardner, M. B., and Davidson, N. (1978). High-molecular-weight RNAs of AKR, NZB, and wild mouse viruses and avian reticuloendotheliosis virus all have similar dimer structures. *J. Virol.* **25,** 888–896.

Benoist, C., and Chambon, P. (1981). *In vivo* sequence requirements of the SV40 early promoter region. *Nature (London)* **290,** 304–310.

Benveniste, R. E., and Todaro, G. J. (1974). Evolution of C-type viral genes: inheritance of exogenously acquired viral genes. *Nature (London)* **252,** 456–459.

Benveniste, R. E., Heinemann, R., Wilson, G. L., Callahan, R., and Todaro, G. J. (1974). Detection of baboon type C viral sequences in various primate tissues by molecular hybridization. *J. Virol.* **14,** 56–67.

Benz, E. W., Jr., Wydro, R. W., Nadal-Ginard, B., and Dina, D. (1980). Moloney murine sarcoma proviral DNA is a transcriptional unit. *Nature (London)* **288,** 665–669.

Bernhard, W. (1960). The detection and study of tumor viruses with the electron microscope. *Cancer Res.* **20,** 712–727.

Besmer, P., Smotkin, D., Haseltine, W., Fan, H., Wilson, A. T., Paskind, M., Weinberg, R., and Baltimore, D. (1975). Mechanism of induction of RNA tumor viruses by halogenated pyrimidines. *Cold Spring Harbor Symp. Quant. Biol.* **39,** 1103–1107.

Beug, H., von Kirchbach, A., Doderlein, G., Conscience, J.-F., and Graf, T. (1979). Chicken hematopoietic cells transformed by seven strains of defective avian leukemia viruses display three distinct phenotypes of differentiation. *Cell* **18,** 375–390.

Bishop, J. M. (1983). Cellular oncogenes and retroviruses. *Annu. Rev. Biochem.* (in press).

Bishop, J. M., and Varmus, H. E. (1982). Functions and origins of retroviral transforming genes. *In* "Molecular Biology of Tumor Viruses. RNA Tumor Viruses" (R. A. Weiss, N. Teich, H. E. Varmus, and J. M. Coffin, eds.), pp. 999–1108. Cold Spring Harbor Lab., Cold Spring Harbor, New York.

Bittner, J. J. (1936). Some possible effects of nursing on the mammary gland tumor incidence in mice. *Science (Washington, D.C.)* **84,** 162.

Blair, D. G., McClements, W. L., Oskarsson, M. K., Fischinger, P. J., and Vande Woude, G. F. (1980). Biological activity of cloned Moloney sarcoma virus DNA: terminally redundant sequences may enhance transformation efficiency. *Proc. Natl. Acad. Sci. USA* **77,** 3504–3508.

Blair, D. G., Oskarsson, M., Wood, T. G., McClements, W. L., Fischinger, P. J., and Vande Woude, G. G. (1981). Activation of the transforming potential of a normal cell sequence: a molecular model for oncogenesis. *Science (Washington, D.C.)* **212,** 941–943.

Boettiger, D., and Temin, H. M. (1970). Light inactivation of focus formation by chicken embryo fibroblasts infected with avian sarcoma virus in the presence of 5-bromodeoxyuridine. *Nature (London)* **228,** 622–624.

Boone, L. R., and Skalka, A. (1980). Two species of full-length cDNA are synthesized in high yield by mellitin-treated avian retrovirus particles. *Proc. Natl. Acad. Sci. USA* **77,** 847–851.

Boone, L. R. and Skalka, A. (1981a). Viral DNA synthesized *in vitro* by avian retrovirus particles permeabilized with melittin. I. Kinetics of synthesis and size of minus- and plus-strand transcripts. *J. Virol.* **37,** 109–116.

Boone, L. R. and Skalka, A. (1981b). Viral DNA synthesized *in vitro* by avian retrovirus particles permeabilized with melittin. II. Evidence for a strand displacement mechanism in plus-strand synthesis. *J. Virol.* **37,** 117–126.

Botchan, M., Topp, W., and Sambrook, J. (1979). Studies on excision of SV40 from cellular chromosomes. *Cold Spring Harbor Symp. Quant. Biol.* **43,** 709–719.

Breathnach, R., and Chambon, P. (1981). Organization and expression of eucaryotic split genes coding for proteins. *Annu. Rev. Biochem.* **50,** 349–384.

Breindl, M., Bacheler, L., Fan, H., and Jaenisch, R. (1980). Chromatin conformation of integrated Moloney leukemia virus DNA sequences in tissues of BALB/Mo mice and in virus-infected cell lines. *J. Virol.* **34,** 373–382.

Brugge, J. S., and Erikson, R. L. (1977). Identification of a transformation-specific antigen induced by an avian sarcoma virus. *Nature (London)* **269,** 346–348.

Buetti, E., and Diggelmann, H. (1981). Cloned mouse mammary tumor virus DNA is biologically active in transfected mouse cells and its expression is stimulated by glucocortoid hormones. *Cell* **23,** 335–346.

Canaani, E., and Duesberg, P. (1972). Role of subunits of 60 to 70 S avian tumor virus ribonucleic acid in its template activity for the viral deoxyribonucleic acid polymerase. *J. Virol.* **10,** 23–31.

Casey, U. W., Roach, A., Mullins, J. I., Burck, K. B., Nicolson, M. O., Gardner, M. B., and Davidson, N. (1981). The U3 portion of feline leukemia virus DNA identifies horizontally acquired proviruses in leukemia cells. *Proc. Natl. Acad. Sci. USA* **78,** 7778–7782.

Chang, E. H., Mryoak, J. M., Wei, C.-M., Shih, T. Y., Shober, R., Cheung, H. L., Ellis, R. W., Hager, G. L., Scolnick, E. M., and Lowy, D. R. (1980). Functional organization of the Harvey murine sarcoma virus genome. *J. Virol.* **35,** 76–92.

Chattopadhyay, S. K., Cloyd, M. W., Linemeyer, D. L., Lander, M. R., Rands, E., and Lowy, D. R. (1982). Cellular origin and role of mink cell focus-forming viruses in murine thymic lymphomas. *Nature (London)* **295,** 25–30.

Coffin, J. M. (1979). Structure, replication, and recombination of retrovirus genomes: some unifying hypotheses. *J. Gen. Virol.* **42,** 1–26.

Coffin, J. M. (1982a). Structure of the retroviral genomes. *In* "Molecular Biology of Tumor Viruses. RNA Tumor Viruses" (R. A. Weiss, N. Teich, H. E. Varmus, and J. M. Coffin, eds.), pp. 261–368. Cold Spring Harbor Lab., Cold Spring Harbor, New York.

Coffin, J. M. (1982b). Endogenous proviruses. *In* "Molecular Biology of Tumor Viruses" (R. A. Weiss, N. Teich, H. E. Varmus, and J. M. Coffin, eds.), pp. 1109–1204. Cold Spring Harbor Lab., Cold Spring Harbor, New York.

Coffin, J. M., Varmus, H. E., Bishop, J. M., Essex, M., Hardy, W. D., Martin, G. S., Rosenberg, N. E., Scolnick, E. M., Weinberg, R. A., and Vogt, P. D. (1981). A proposal for naming host cell-derived inserts in retrovirus genomes. *J. Virol.* **49,** 953–957.

Cohen, J. C. (1980). Methylation of milk-borne and genetically transmitted mouse mammary tumor virus proviral DNA. *Cell* **19,** 653–662.

Cohen, J. C., and Varmus, H. E. (1979). Endogenous mammary tumour virus DNA varies among wild mice and segregates during inbreeding. *Nature (London)* **278,** 418–423.

Cohen, J. C., Shank, P. R., Morris, V. L., Cardiff, R., and Varmus, H. E. (1979). Integration of the DNA of mouse mammary tumor virus in virus-infected normal and neoplastic tissue of the mouse. *Cell* **16,** 333–346.

Collett, M. S., and Erikson, R. L. (1978). Protein kinase activity associated with the avian sarcoma virus *src* gene product. *Proc. Natl. Acad. Sci. USA* **75,** 2021–2024.

Collett, M. S., Dierks, P., Parsons, J. T., and Faras, A. J. (1978). RNase H hydrolysis of the 5′ terminus of the avian sarcoma virus genome during reverse transcription. *Nature (London)* **272,** 181–183.

Collett, M. S., Leis, J. P., Smith, M. S., and Faras, A. J. (1978). Unwinding-like activity associated with avian retrovirus RNA-directed DNA polymerase. *J. Virol.* **26,** 498–509.

Cooper, G. M., and Neiman, P. E. (1980). Transforming genes of neoplasms induced by avian lymphoid leukosis viruses. *Nature (London)* **287,** 659–660.

Cooper, G. M., and Neiman, P. E. (1981). Two distinct candidate transforming genes of lymphoid leukosis virus-induced neoplasms. *Nature (London)* **292,** 857–858.

Copeland, T. D., Grandgenett, D. P., and Oroszlan, S. (1980). Amino acid sequence analysis of reverse transcriptase subunits from avian myeloblastosis virus. *J. Virol.* **36,** 115–119.

Copeland, N. G., Jenkins, N. A., and Cooper, G. M. (1981). Integration of Rous sarcoma virus DNA during transfection. *Cell* **23,** 51–60.

Cordell, B., Weiss, S. R., Varmus, H. E., and Bishop, J. M. (1978). At least 104 nucleotides are transposed from the 5′ terminus of the avian sarcoma virus genome to the 5′ termini of smaller viral mRNAs. *Cell* **15,** 79–91.

Courtneidge, S. A., Levinson, A. D., and Bishop, J. M. (1980). The protein encoded by the transforming gene of avian sarcoma virus ($pp60^{src}$) and homologous protein in normal cells ($pp60^{proto\text{-}src}$) are associated with the plasma membrane. *Proc. Natl. Acad. Sci. USA* **77,** 3783–3787.

Crittenden, L. B. (1981). Exogenous and endogenous leukosis virus genes. *Avian Pathol.* **10,** 101–112.

Darlix, J.-L., Spahr, P.-F., Bromley, P. A., and Jaton, J.-C. (1979). *In vitro,* the major ribosome binding site on Rous sarcoma virus RNA does not contain the nucleotide sequence coding for the N-terminal amino acids of the *gag* gene product. *J. Virol.* **29,** 597–611.

DeFeo, D., Gonda, M. A., Young, H. A., Chang, E. H., Lowy, D. R., Scolnick, E. M., and Ellis, R. W. (1981). Analysis of two divergent rat genomic clones homologous to the transforming gene of Harvey murine sarcoma virus. *Proc. Natl. Acad. Sci. USA* **78,** 3328–3332.

Del Villano, B. C., Nave, B., Croker, B. P., Lerner, R. A., and Dixon, F. J. (1975). The oncornavirus glycoprotein gp69/71: a constituent of the surface of normal and malignant thymocytes. *J. Exp. Med.* **141,** 172–187.

Deng, C. T., Boettiger, D., Macpherson, I., and Varmus, H. E. (1974). The persistance and expression of virus-specific DNA in revertants of Rous sarcoma virus-transformed BHK-21 cells. *Virology* **62,** 512–521.

Dhar, R., McClements, W. L., Enquist, L. W., and Vande Woude, G. F. (1980). Nucleotide sequences of integrated Moloney sarcoma provirus long terminal repeats and their host and viral junctions. *Proc. Natl. Acad. Sci. USA* **77,** 3937–3941.

Dickson, C., and Peters, G. (1981). Protein-coding potential of mouse mammary tumor virus genome RNA as examined by *in vitro* translation. *J. Virol.* **37,** 36–47.

Dickson, C., Eisenman, R., Fan, H., Hunter, E., and Teich, N. (1982). Protein biosynthesis and assembly. *In* "Molecular Biology of Tumor Viruses. RNA Tumor Viruses" (R. A. Weiss, N. Teich, H. E. Varmus, and J. M. Coffin, eds.), pp. 513–648. Cold Spring Harbor Lab., Cold Spring Harbor, New York.

Dino, D., and Penhoet, E. E. (1978). Viral gene expression in murine sarcoma virus (murine leukemia virus)-infected cells. *J. Virol.* **27,** 768–775.

Dolberg, D. S., Bacheler, L. T., and Fan, H. (1981). Endogenous type C retroviral sequences of mice are organized in a small number of virus-like classes and have been acquired recently. *J. Virol.* **40,** 96–106.

Donehower, L. A., Huang, A. L., and Hager, G. L. (1981). Regulatory and coding potential of the mouse mammary tumor virus long terminal redundancy. *J. Virol.* **37,** 226–238.

Donner, L., Turek, L. P., Ruscetti,.S. K., Fedele, L. A., and Sherr, C. J. (1980). Transformation-defective mutants of feline sarcoma virus which express a product of the viral *src* gene. *J. Virol.* **35,** 129–140.

Dougherty, R. M., Di Stefano, H. S., and Roth, F. K. (1967). Virus particles and viral antigens in chicken tissues free of infectious avian leukosis virus. *Proc. Natl. Acad. Sci. USA* **58,** 808–817.

Dresler, S., Ruta, M., Murray, M. J., and Kabat, D. (1979). Glycoprotein encoded by the Friend spleen focus-forming virus. *J. Virol.* **30,** 564–575.

Dunsmuir, P., Brorein, W. J., Jr., Simon, M. A., and Rubin, G. M. (1980). Insertion of the *Drosophila* transposable element *copia* generates a 5 base pair duplication. *Cell* **21,** 575–580.

Edwards, S. A., and Fan, H. (1979). *gag*-related polyproteins of Moloney murine leukemia virus: evidence for independent synthesis of glycosylated and unglycosylated forms. *J. Virol.* **30,** 551–563.

Efstratiatis, A., Kafatos, F. C., Maxam, A. M., and Maniatis, T. (1976). Enzymatic *in vitro* synthesis of globin genes. *Cell* **7,** 279–288.

Eibel, H., Gafner, J., Stotz, A., and Philippsen, P. (1981). Characterization of the yeast mobile element Ty1. *Cold Spring Harbor Symp. Quant. Biol.* **45,** 609–617.

Eiden, J. J., Quade, K., and Nichols, J. L. (1976). Interaction of tryptophan transfer RNA with Rous sarcoma virus 35 S RNA. *Nature (London)* **259,** 245–247.

Eisenman, R. N., and Vogt, V. M. (1978). The biosynthesis of oncovirus proteins. *Biochim. Biophys. Acta* **473,** 187–239.

Eisenman, R. N., Mason, W. S., and Linial, M. (1980). Synthesis and processing of polymerase proteins of wild-type and mutant avian retroviruses. *J. Virol.* **36,** 62–78.

Elder, J. H., Gautsch, J. W., Jensen, F. C., Lerner, R. A., Hartley, J. W., and Rowe, W. P. (1977). Biochemical evidence that MCF murine leukemia viruses are envelope (*env*) gene recombinants. *Proc. Natl. Acad. Sci. USA* **74,** 4676–4680.

Ellermann, V., and Bang, O. (1908). Experimentelle Leukamie bei Huhnern. *Zentralbl. Bakteriol. Parasitenkde. Infektionskr.* **46,** 595.

Ellis, R. W., DeFeo, D., Maryak, J. M., Young, H. A., Shih, T. Y., Chang, E. H., Lowy, D. R., and Scolnick, E. M. (1980). Dual evolutionary origin for the rat genetic sequences of Harvey murine sarcoma virus. *J. Virol.* **36,** 408–420.

Errede, B., Cardillo, T. S., Sherman, F., Duboi, E., Deschamps, J., and Wiame, J.-M. (1980). Mating signals control expression of mutations resulting from insertion of a transposable repetitive element adjacent to diverse yeast genes. *Cell* **25,** 427–436.

Fan, H., Jaenisch, R., and MacIsaac, P. (1978). Low-multiplicity infection of Moloney murine leukemia virus in mouse cells: effect on number of viral DNA copies and virus production in producer cells. *J. Virol.* **28,** 802–809.

Farabaugh, P. J., and Fink, G. R. (1980). Insertion of the eukaryotic transposable element Ty1 creates a 5-base pair duplication. *Nature (London)* **286,** 352–356.

Finnegan, D. J., Rubin, G. M., Young, M. W., and Hogness, D. S. (1978). Repeated gene families in *Drosophila melanogaster*. *Cold Spring Harbor Symp. Quant. Biol.* **42,** 1053–1063.

Fitzgerald, M., and Shenk, T. (1981). The sequence 5′-AAUAAA-3′ forms part of the recognition site for polyadenylation of late SV40 mRNAs. *Cell* **24,** 251–260.

Flavell, A. J., and Ish-Horowicz, D. (1981). Extra chromosomal circular copies of the eucaryotic transposable element *copia* in *Drosophila* cells. *Nature (London)* **292,** 591–594.

Flavell, A. J., Levis, R., Simon, M. A., and Rubin, G. M. (1981). The 5′ termini of RNAs encoded by the transposable element *copia*. *Nucl. Acids Res.* **9,** 6279–6291.

Folger, K. R., Wong, E. A., Wahl, G., and Capecchi, M. R. (1982). Patterns of integration of DNA microinjected into cultured mammalian cells: evidence for homologous recombination between injected plasmid DNA molecules. *Mol. Cell. Biol.* **2,** 1372–1387.

Frankel, A. E., Haapala, D. K., Neubauer, R. L., and Fischinger, P. J. (1976). Elimination of the sarcoma genome from murine sarcoma virus transformed cat cells. *Science (Washington, D.C.)* **191,** 1264–1266.

Friedrich, R., and Moelling, K. (1979). Effect of viral RNase H on the avian sarcoma viral genome during early transcription *in vitro*. *J. Virol.* **31,** 630–638.

Frisby, D. P., Weiss, R. A., Roussel, M., and Stehelin, D. (1979). The distribution of endogenous retrovirus sequences in the DNA of galliform birds does not coincide with avian phylogenetic relationships. *Cell* **17,** 623–634.

Fritsch, E., and Temin, H. M. (1977). Inhibition of viral DNA synthesis in stationary chicken embryo fibroblasts infected with avian retroviruses. *J. Virol.* **24,** 461–469.

Fuhrman, S. A., Van Beveren, C., and Verma, I. M. (1981). Identification of an RNA polymerase II initiation site in the long terminal repeat of Moloney murine leukemia viral DNA. *Proc. Natl. Acad. Sci. USA* **78,** 5411–5416.

Fung, Y. K. T., Fadly, A. M., Crittenden, L. B., and Kung, H.-J. (1981). On the mechanism of retrovirus-induced avian lymphoid leukosis: deletion and integration of the proviruses. *Proc. Natl. Acad. Sci. USA* **78,** 3418–3422.

Gafner, J., and Phillipsen, P. (1980). The yeast transposon Ty1 generates duplications of target DNA on insertion. *Nature (London)* **286,** 414–418.

Gerard, G. F., and Grandgenett, D. P. (1980). Retrovirus reverse transcriptase. *In* "Molecular Biology of RNA Tumor Viruses" (J. Stephenson, ed.), pp. 345–394. Academic Press, New York.

Gianni, A. M., Smotkin, D., and Weinberg, R. A. (1975). Murine leukemia virus: detection of unintegrated double-stranded DNA forms of the provirus. *Proc. Natl. Acad. Sci. USA* **72,** 447–451.

Gibson, W., and Verma, I. M. (1974). Studies on the reverse transcriptase RNA tumor viruses. Structural relatedness of two subunits of avian RNA tumor viruse. *Proc. Natl. Acad. Sci. USA* **71,** 4991–4994.

Gilboa, E., Mitra, S. W., Goff, S., and Baltimore, D. (1979). A detailed model of reverse transcription and tests of crucial aspects. *Cell* **18,** 93–100.

Goff, S. P., Gilboa, E., Witte, O. N., and Baltimore, D. (1980). Structure of the Abelson murine leukemia virus genome and the homologous cellular gene: studies with cloned viral DNA. *Cell* **22,** 777–785.

Goff, S. P., Tabin, C. J., Wang, J. Y.-J., Weinberg, R., and Baltimore, D. (1982). Transfection of fibroblasts by cloned Abelson murine leukemia virus DNA and recovery of transmissible virus by recombination with helper virus. *J. Virol.* **41,** 271–285.

Goldfarb, M. P., and Weinberg, R. A. (1981). Generation of novel, biologically active Harvey sarcoma virus via apparent illegitimate recombination. *J. Virol.* **38,** 136–150.

Golomb, M., and Grandgenett, D. P. (1979). Endonuclease activity of purified RNA directed DNA polymerase from AMV. *J. Biol. Chem.* **254,** 1606–1613.

Golomb, M., Grandgenett, D. P., and Mason, W. (1981). Virus-coded DNA endonuclease from avian retrovirus. *J. Virol.* **38,** 548–555.

Gopinathan, K. P., Weymouth, L. A., Kunkel, T. A., and Loeb, L. A. (1979). Mutagenesis *in vitro* by DNA polymerase from an RNA tumor virus. *Nature (London)* **278,** 857–859.

Goubin, G., and Hill, M. (1979). Monomer and multimer covalently closed circular forms of Rous sarcoma virus DNA. *J. Virol.* **29,** 799–804.

Govindan, M. V., Spiess, E., and Majors, J. E. (1982). Purified glucocorticoid receptor-hormone complex from rat liver cytosol binds specifically to cloned mouse mammary tumor virus long terminal repeats *in vitro*. *Proc. Natl. Acad. Sci. USA* **79,** 5157–5161.

Graf, T. and Berg, H. (1978). Avian leukemia viruses. Interaction with their target cells *in vivo* and *in vitro*. *Biochim. Biophys. Acta* **516,** 269–299.

Grandgenett, D. P., and Green, M. (1974). Different mode of action of ribonuclease H in purified α and β riboinucleic acid-directed deoxyribonucleic acid polymerase from avian myeloblastosis virus. *J. Biol. Chem.* **294,** 5148–5152.

Grandgenett, D. P., Vora, A. C., and Schiff, R. D. (1978). A 32,000-dalton nuclecic acid-binding protein from avian retrovirus cores possesses DNA endonuclease activity. *Virology* **89,** 119–132.

Grandgenett, D. P., Golomb, M., and Vora, A. C. (1980). Activation of an Mg^{+2}-dependent DNA endonuclease of avian myeloblastosis virus $\alpha\beta$ DNA polymerase by *in vitro* proteolytic cleavage. *J. Virol.* **33,** 264–271.

Gross, L. (1951). "Spontaneous" leukemia developing in C3H mice following inoculation, in infancy, with Ak-leukemic extracts or Ak-embryos. *Proc. Soc. Exp. Biol. Med.* **76,** 27–32.

Gross, L. (1970). "Oncogenic Viruses." Pergamon, Elmsford, New York.

Groudine, M., Eisenman, R., and Weintraub, H. (1981). Chromatin structure of endogenous retroviral genes and activation by an inhibitor of DNA methylation. *Nature (London)* **292,** 311–317.

Gruss, P., Dhar, R., and Khoury, G. (1981). Simian virus 40 tandem repeated sequences as an element of the early promoter. *Proc. Natl. Acad. Sci. USA* **78,** 943–947.

Guntaka, R. V., and Mitsialis, S. A. (1980). Cloning of avian tumor virus DNA fragments in plasmid pBR322: evidence for efficient transcription in *E. coli* from a virus-coded promoter. *Gene* **12,** 113–121.

Guntaka, R. V., Richards, O. C., Shank, P. R., Kung, H.-J., Davidson, N., Fritsch, E., Bishop, J. M., and Varmus, H. E. (1976). Covalently closed circular DNA of avian sarcoma virus: purification from nuclei of infected quail tumor cells and measurement by electron microscopy and gel electrophoresis. *J. Mol. Biol.* **106,** 337–357.

Hackett, P. B., Swanstrom, R., Varmus, H. E., and Bishop, J. M. (1982). The leader sequence of the subgenomic messenger RNAs of Rous sarcoma virus is approximately 390 nucleotides. *J. Virol.* **41,** 527–534.

Hager, G. L., Chang, E. H., Chan, H. W., Garon, C. F., Israel, M. A., Martin, M. A., Scolnick, E. M., and Lowy, D. R. (1979). Molecular cloning of the Harvey sarcoma virus closed circular DNA intermediates: initial structural and biological characterization. *J. Virol.* **31,** 795–809.

Hanafusa, H., and Hanafusa, T. (1971). Noninfectious RSV deficient in DNA polymerase. *Virology* **43,** 313–316.

Hanafusa, H., Hanafusa, T., and Rubin, H. (1964). Analysis of the defectiveness of Rous sarcoma virus. II. Specifications of RSV antigenicity by helper virus. *Proc. Natl. Acad. Sci. USA* **51,** 41–48.

Harbers, K., Schnieke, A., Stuhlmann, H., Jahner, D., and Jaenisch, R. (1981a). DNA methylation and gene expression: endogenous retroviral genome becomes infectious after molecular cloning. *Proc. Natl. Acad. Sci. USA* **78,** 7609–7613.

Harbers, K., Jahner, D., and Jaenisch, R. (1981b). Microinjection of cloned retroviral genomes into mouse zygotes: integration and expression in the animal. *Nature (London)* **293,** 540–542.

Hartley, J. W., Wolford, N. K., Old, L. J., and Rowe, W. P. (1977). A new class of murine leukemia virus associated with development of spontaneous lymphomas. *Proc. Natl. Acad. Sci. USA* **74,** 789–792.

Haseltine, W. A., and Kleid, D. G. (1978). A method for classification of 5′ termini of retroviruses. *Nature (London)* **273,** 358–364.

Haseltine, W. A., Kleid, D. G., Panet, A., Rothenberg, E., and Baltimore, D. (1976). Ordered transcription of RNA tumor virus genomes. *J. Mol. Biol.* **106,** 109–131.

Hayward, W. S., Braverman, S. B., and Astrin, S. M. (1980). Transcriptional products and DNA structure of endogenous avian proviruses. *Cold Spring Harbor Symp. Quant. Biol.* **44,** 1111–1121.

Hayward, W. S., Neel, B. G., and Astrin, S. M. (1981). Activation of a cellular *onc* gene by promoter insertion in ALV-induced lymphoid leukosis. *Nature (London)* **290,** 475–480.

Heine, C. W., Kelly, D. C., and Avery, R. J. (1980). The detection of intracellular retrovirus-like entities in *Drosophila melanogaster* cell cultures. *J. Gen. Virol.* **49,** 385–395.

Highfield, P. E., Rafield, L. F., Gilmer, T. M., and Parsons, J. T. (1980). Molecular cloning of avian sarcoma virus closed circular DNA: structural and biological characterization of three recombinant clones. *J. Virol.* **36,** 271–279.

Hill, M., and Hillova, J. (1972). Virus recovery in chicken cells tested with Rous sarcoma cell DNA. *Nature (London) New Biol.* **237,** 35–39.

Hinuma, Y., Nagata, K., Hanaoka, M., Nakai, M., Matsumoto, T., Kinoshita, K. I., Shiradawa, S., and Miyoshi, I. (1981). Adult T-cell leukemia: antigen in an ATL cell line and detection of antibodies to the antigen in human sera. *Proc. Natl. Acad. Sci. USA* **78,** 6476–6480.

Hishinuma, F., DeBona, P. J., Astrin, S., and Skalka, A. M. (1981). Nucleotide sequence of the acceptor site and termini of integrated avian endogenous provirus *ev*-1: integration creates a 6 bp repeat of host DNA. *Cell* **23,** 155–164.

Hizi, A., Gazit, A., Geithmann, D., and Yaniv, A. (1982). DNA-processing activities associated with the purified α, β_2, and $\alpha\beta$ molecular forms of avian sarcoma virus RNA-dependent DNA polymerase. *J. Virol.* **41,** 974–981.

Hollis, G. F., Hinter, P. A., McBride, O. W., Swan, D., and Leder, P. (1982). Processed genes: a dispersed human immunoglobulin gene bearing evidence of RNA-type processing. *Nature (London)* **296,** 321–325.

Hsu, T. W., Sabran, J. L., Mark, G. E., Guntaka, R. V., and Taylor, J. M. (1978). Analysis of unintegrated avian RNA tumor virus double-stranded DNA intermediates. *J. Virol.* **28,** 810–818.

Hsu, T. W., Taylor, J. M., Aldrich, C., Townsend, J. B., Seal, G., and Mason, W. S. (1981). Tandem duplication of the proviral DNA in avian sarcoma virus-transformed quail clone. *J. Virol.* **38,** 219–223.

Huang, A. L., Ostrowski, M. C., Berard, D., and Hager, G. L. (1981). Glucocorticoid regulation of the Ha-MuSV p21 gene conferred by sequences from mouse mammary tumor virus. *Cell* **27,** 245–256.

Huebner, R. J., and Todaro, G. J. (1969). Oncogenes of RNA tumor viruses as determinants of cancer. *Proc. Natl. Acad. Sci. USA* **64,** 1087–1094.

Hughes, S. H. (1982). *In* "Molecular Biology of Tumor Viruses. RNA Tumor Viruses" (R. A. Weiss, N. Teich, H. E. Varmus, and J. M. Coffin, eds.), pp. 1285–1386. Cold Spring Harbor Lab., Cold Spring Harbor, New York.

Hughes, S. H., Shank, P. R., Spector, D. H., Kung, H.-J., Bishop, J. M., Varmus, H. E., Vogt, P. K., and Breitman, M. L. (1978). Proviruses of avian sarcoma virus are terminally redundant, co-extensive with unintegrated liner DNA and integrated at many sites. *Cell* **15,** 1397–1410.

Hughes, S. H., Payvar, F., Spector, D., Schimke, R. T., Robinson, H. L., Payne, G. S., Bishop, J. M., and Varmus, H. E. (1979). Heterogeneity of genetic loci in chickens: analysis of endogenous viral and nonviral genes by cleavage of DNA with restriction endonucleases. *Cell* **18,** 347–360.

Hughes, S. H., Mutschler, A., Bishop, J. M., and Varmus, H. E. (1981a). A Rous sarcoma virus provirus is flanked by short direct repeats of a cellular DNA sequence present in only one copy prior to integration. *Proc. Natl. Acad. Sci. USA* **78,** 4299–4303.

Hughes, S. H., Vogt, P. K., Stubblefield, E., Bishop, J. M., and Varmus, H. E. (1981b). Integration of avian sarcoma virus DNA in chicken cells. *Virology* **108,** 208–221.

Hughes, S. H., Toyoshima, K., Bishop, J. M., and Varmus, H. E. (1981c). Organization of the endogenous proviruses of chickens: implications for origin and expression. *Virology* **108,** 189–207.

Hughes, S. H., Vogt, P. K., Bishop, J. M., and Varmus, H. E. (1981d). Endogenous proviruses of random-bred chickens and ring-necked pheasants: analysis with restriction endonucleases. *Virology* **108,** 222–229.

Humphries, E. H., Glover, C., Weiss, R. A., and Arrand, J. R. (1979). Differences between the endogenous and exogenous DNA sequences of Rous-associated virus-0. *Cell* **18,** 803–815.

Humphries, E. H., Glover, C., and Reichmann, M. E. (1981). Rous sarcoma virus infection of synchronized cells establishes provirus integration during S phase DNA synthesis prior to cell division. *Proc. Natl. Acad. Sci. USA* **78,** 2601–2605.

Hunter, E. (1978). The mechanism for genetic recombination in the avian retroviruses. *Curr. Top. Microbiol. Immunol.* **79,** 295–309.

Hunter, T., and Sefton, B. M. (1980). Transforming gene product of Rous sarcoma virus phosphorylates tyrosine. *Proc. Natl. Acad. Sci. USA* **77,** 1311–1115.

Hynes, N. E., Kennedy, N., Rahmsdorf, V., and Groner, B. (1981). Hormone responsive expression of an endogenous proviral gene of mouse mammary tumor virus after molecular cloning and gene transfer into cultured cells. *Proc. Natl. Acad. Sci. USA* **78,** 2038–2042.

Ikenga, H., and Saigo, K. (1982). Insertion of a movable genetic element, *297,* into the T-A-T-A box for the H3 histone gene in *Drosophila melanogaster. Proc. Natl. Acad. Sci. USA* **78,** 4143–4147.

Ilyin, Y. V., Chmeliauskaite, V. G., and Georgiev, G. P. (1980). Double stranded sequences in RNA of *Drosophila melanogaster:* relation to mobile dispersed genes. *Nucleic Acids Res.* **8,** 3439–3457.

Ishizaki, R., and Vogt, P. K. (1966). Immunological relationships among envelope antigens of avian tumor viruses. *Virology* **30,** 375–387.

Ito, Y., Brocklehurt, J. R., and Dulbecco, R. (1977). Virus-specific proteins in the plasma membrane of cells lytically infected or transformed by polyoma virus. *Proc. Natl. Acad. Sci. USA* **74,** 4666–4670.

Jaenisch, R. (1976). Germ line integration and Mendelian transmission of the exogenous Moloney leukemia virus. *Proc. Natl. Acad. Sci. USA* **73,** 1260–1264.

Jaenisch, R., Jahner, D., and Grotkopp, D. (1980). Derivation of three mouse strains carrying Moloney leukemia virus in their germ line at different genetic loci. *ICN-UCLA Symp. Mol. Cell. Biol.* **18,** 265–279.

Jaenisch, R., Janer, D., Nobis, P., Simon, I., Lohler, J., Harbers, K., and Grotkopp, G. (1981). Chromosomal position and activation of retroviral genomes inserted into the germ line of mice. *Cell* **24,** 519–529.

Jagadeeswaran, P., Forget, B. G., and Weissman, S. M. (1981). Short interspersed repetitive DNA elements in eucaryotes: transposable DNA elements generated by reverse transcription of RNA *pol* III transcripts. *Cell* **26,** 141–142.

Jahner, D., and Jaenisch, R. (1980). Integration of Moloney leukemia virus into the germ line of mice: correlation between site of integration and virus activation. *Nature (London)* **287,** 456–457.

Jaquet, M., Groner, Y., Monroy, G., and Hurwitz, J. (1974). The *in vitro* synthesis of avian myeloblastosis virus RNA sequences. *Proc. Natl. Acad. Sci. USA* **71,** 3045–3049.

Jenkins, N. A., and Cooper, G. M. (1980). Integration, expression, and infectivity of exogenously acquired proviruses of Rous-associated virus-0. *J. Virol.* **36,** 684–691.

Jenkins, N. A., Copeland, N. G., Taylor, B. A., and Lee, B. K. (1981). Dilute (d) coat colour mutation of DBA/2J mice is associated with the site of integratio of an ecotropic MuLV genome. *Nature (London)* **293,** 370–374.

Jolicoeur, P. (1979). The *Fv*-1 gene of the mouse and its control of murine leukemia virus replication. *Curr. Top. Microbiol. Immunol.* **86,** 67–122.

Jolicoeur, P., and Baltimore, D. (1976a). Effect of *Fv*-1 gene product on proviral DNA formation and integration in cells infected with murine leukemia viruses. *Proc. Natl. Acad. Sci. USA* **73,** 2236–2240.

Jolicoeur, P., and Baltimore, D. (1976b). Effect of *Fv*-1 gene product on synthesis of N-tropic and B-tropic murine leukemia viral RNA. *Cell* **7,** 33–39.

Ju, G., and Skalka, A. M. (1980). Nucleotide sequence analysis of the long terminal repeat (LTR) of avian retroviruses: structural similarities with transposable elements. *Cell* **22,** 379–386.

Ju, G., Boone, L., and Skalka, A. M. (1980). Isolation and characterization of recombinant DNA clones of avian retroviruses: size heterogeneity and instability of the direct repeat. *J. Virol.* **33,** 1026–1033.

Junghans, R. P., Duesberg, P. H., and Knight, C. A. (1975). *In vitro* synthesis of full-length DNA transcripts of Rous sarcoma virus RNA by viral DNA polymerase. *Proc. Natl. Acad. Sci. USA* **72,** 4895–4899.

Junghans, R. P., Boone, L. R., and Skalka, A. M. (1982). Retroviral DNA structures: displacement–assimilation model of recombination. *Cell* **30,** 53–62.

Kalter, S. S., Helmke, R. J., Panigel, M., Heberling, R. L., Felsburg, P. J., and Axelrod, L. R. (1973). Observations of apparent C-type particles in baboon (*Papio cynocephalus*) placenta. *Science (Washington, D.C.)* **179,** 1332–1333.

Kawai, S., and Hanafusa, H. (1972). Genetic recombination with avian tumor virus. *Virology* **49,** 37–44.

Kennedy, N., Knedlitschek, Groner, B., Hynes, N. E., Herrlich, P., Michalides, R., and Van Ooyen, A. J. J. (1982). Long terminal repeats of endogenous mouse mammary tumor virus contain a long open reading frame which extends into adjacent sequences. *Nature (London)* **295,** 622–624.

Keshet, E., and Temin, H. M. (1978). Sites of integration of reticuloendotheliosis virus in chicken DNA. *Proc. Natl. Acad. Sci. USA* **75,** 3372–3376.

Keshet, E., Shaul, Y., Kaminchik, J., and Aviv, H. (1980). Heterogeneity of "virus-like" genes encoding retrovirus-associated 30S RNA and their organization within the mouse genome. *Cell* **20,** 431–439.

Khoury, A. T., and Hanafusa, H. (1976). Synthesis and integration of viral DNA in chicken cells at different times after infection with various multiplicities of avian oncornavirus. *J. Virol.* **18,** 383–400.

Kleckner, N. (1977). Translocatable elements in procaryotes. *Cell* **11,** 11–23.

Kleckner, N. (1981). Transposable elements in prokaryotes. *Annu. Rev. Genet.* **15,** 341–404.

Kopchick, J. J., Harless, J., Geissner, B. S., Killam, R., Hewitt, R. R., and Arlinghaus, R. B. (1981). Endodeoxyribonuclease activity associated with Rauscher murine leukemia virus. *J. Virol.* **37,** 274–283.

Kulgushkin, V. V., Ilyin, Y. V., and Georgiev, G. P. (1981). Mobile dispersed genetic element mdgl of *Drosophila melanogaster:* nucleotide sequence of long terminal repeats. *Nucleic Acids Res.* **9,** 3451–3464.

Kung, H.-J., Bailey, J. M., Davidson, N., Vogt, P. K., Nicolson, M. O., and McAllister, R. M.

(1975). Electron microscope studies of tumor virus RNA. *Cold Spring Harbor Symp. Quant. Biol.* **39,** 827–834.

Kung, H.-J., Shank, P. R., Bishop, J. M., and Varmus, H. E. (1980). Identification and characterization of dimeric and trimeric circular forms of avian sarcoma virus-specific DNA. *Virology* **103,** 425–433.

Kung, H.-J., Fung, Y. K., Majors, J. E., Bishop, J. M., and Varmus, H. E. (1981). Synthesis of plus strands of retroviral DNA in cells infected with avian sarcoma virus and mouse mammary tumor virus. *J. Virol.* **37,** 127–138.

Lai, M.-H. T., and Verma, I. M. (1978). Reverse transcriptase of RNA tumor viruses. V. *In vitro* proteolysis of reverse transcriptase from avian myeloblastosis virus and isolation of a polypeptide manifesting only RNase H activity. *J. Virol.* **25,** 652–663.

Lania, L., Boast, S., and Fried, M. (1982). Excision of polyoma virus genomes from chromosomal DNA by homologous recombination. *Nature (London)* **295,** 349–351.

Leder, A., Swan, D., Ruddle, F., D'Eustachio, P., and Leder, P. (1981). Dispersion of α-like globin genes of the mouse to three different chromosomes. *Nature (London)* **293,** 196–199.

Lee, F., Mulligan, R., Berg, P., and Ringold, G. (1981). Glucocorticoids regulate expression of dihydrofolate reductase cDNA in mouse mammary tumour virus chimaeric plasmids. *Nature (London)* **294,** 228–231.

Leis, J. P., Berkower, I., and Hurwitz, J. (1973). Mechanism of action of ribonuclease H isolated from avian myeloblastosis virus and *Escherichia coli*. *Proc. Natl. Acad. Sci. USA* **70,** 466–470.

Lemons, R. S., Nash, W. G., O'Brien, S. J., Benveniste, R. E., and Sherr, C. J. (1978). A gene (*Bevl*) on human chromosome 6 is an integration site for baboon type C DNA provirus in human cells. *Cell* **14,** 995–1005.

Levin, J. G., and Rosenak, M. J. (1976). Synthesis of murine leukemia virus proteins associated with virions assembled in actinomycin D-treated cells: evidence for the persistence of viral messenger RNA. *Proc. Natl. Acad. Sci. USA* **73,** 1154–1158.

Levinson, A. D., Oppermann, H., Levintow, L., Varmus, H. E., and Bishop, J. M. (1978). Evidence that the transforming gene of avian sarcoma virus encodes a protein kinase associated with a phosphoprotein. *Cell* **15,** 561–572.

Levinson, B., Khoury, G., Vande Woude, G., and Gruss, P. (1982). Activation of SV40 genome by 72-base pair tandem repeats of Moloney sarcoma virus. *Nature (London)* **295,** 568–572.

Levis, R., Dunsmuir, P., and Rubin, G. M. (1980). Terminal repeats of the *Drosophila* transposable element *copia:* nucleotide sequence and genomic organization. *Cell* **21,** 581–588.

Linemeyer, D. L., Ruscetti, S. K., Scolnick, E. M., Evans, L. H., and Duesberg, P. H. (1981). Biological activity of the spleen focus-forming virus is encoded by a molecularly cloned subgenomic fragment of spleen focus-forming virus DNA. *Proc. Natl. Acad. Sci. USA* **78,** 1401–1405.

Linial, M., and Blair, D. (1982). Genetics of retroviruses. *In* "Molecular Biology of Tumor Viruses. RNA Tumor Viruses" (R. A. Weiss, N. Teich, H. E. Varmus, and J. M. Coffin, eds.), pp. 649–784. Cold Spring Harbor Lab., Cold Spring Harbor, New York.

Linial, M., and Mason, W. S. (1973). Characterization of two conditional early mutants of Rous sarcoma virus. *Virology* **53,** 258–273.

Livingston, D. M., and Todaro, G. J. (1973). Endogenous type C virus from a cat cell clone with properties distinct from previously described feline type C virus. *Virology* **53,** 142–151.

Lowy, D. R., Rowe, W. P., Teich, N., and Hartley, J. W. (1971). Murine leukemia virus: high-frequency activation *in vitro* by 5'iododeoxyuridine and 5'bromodeoxyuridine. *Science (Washington, D.C.)* **174,** 155–156.

Lowy, D. R., Rands, E., Chattopadhyay, S. K., Garon, C. F., and Hager, G. L. (1980). Molecular cloning of infectious integrated murine leukemia virus DNA from infected mouse cells. *Proc. Natl. Acad. Sci. USA* **77,** 614–618.

Lueders, K. K., and Kuff, E. L. (1977). Sequences associated with intracisternal A particles are reiterated in the mouse genome. *Cell* **12,** 963–972.

MacHattie, L. A., and Jackowski, J. B. (1977). Physical structure and deletion effects of the chloramphenicol resistance element Tn9 in phage lambda. *In* "DNA Insertion Elements, Plasmids and Episomes" (A. I. Bukhari,) pp. 219–228. Cold Spring Harbor Lab., Cold Spring Harbor, New York.

Maisel, J., Bender, W., Hu, S., Duesberg, P. H., and Davidson, N. (1978). Structure of 50 to 70 S RNA from Moloney sarcoma virus. *J. Virol.* **25,** 384–394.

Majors, J. E., and Varmus, H. E. (1981). Nucleotide sequences at host–proviral junctions for mouse mammary tumour virus. *Nature (London)* **289,** 253–258.

Martin, M. A., Bryan, T., Rasheed, S., and Khan, A. S. (1981). Identification and cloning of endogenous retroviral sequences present in human DNA. *Proc. Natl. Acad. Sci. USA* **78,** 4892–4896.

McClements, W. L., Enquist, L., Oskarsson, M., Sullivan, M., and Vande Woude, G. F. (1980). A frequent site specific deletion of coliphage lambda murine sarcoma virus recombinants and its use in the identification of a retrovirus integration site. *J. Virol.* **35,** 488–497.

McClements, W. L., Dhar, R., Blair, D. G., Enquist, L., Oskarsson, M., and Vande Woude, G. F. (1981). The long terminal repeat of Moloney sarcoma provirus. *Cold Spring Harbor Symp. Quant. Biol.* **45,** 699–706.

McGrath, M. S., and Weissman, I. L. (1979). AKR leukemogenesis: identification and biological significance of thymic lymphoma receptors for AKR retroviruses. *Cell* **17,** 65–75.

Misra, T. P., Grandsenett, D. P., and Parsons, J. T. (1982). Avian retrovirus pp32 DNA-binding protein. I. Recognition of specific sequences on retrovirus DNA terminal repeats. *J. Virol.* **44,** 330–343.

Mitra, S. W., Goff, S., Gilboa, E., and Baltimore, D. (1979). Synthesis of a 600-nucleotide-long plus-strand DNA by virions of Moloney murine leukemia virus. *Proc. Natl. Acad. Sci. USA* **76,** 4355–4359.

Mitra, S. W., Chow, M., Champoux, J., and Baltimore, D. (1982). Synthesis of murine leukemia virus plus strong stop DNA initiates at a unique site. *J. Biol. Chem.* (in press).

Moelling, K., Bolognsi, D. P., Bauer, H., Busen, W., Plassmann, H. W., and Hausen, P. (1971). Association of viral reverse transcriptase with an enzyme degrading the RNA moiety of RNA–DNA hybrids. *Nature (London) New Biol.* **234,** 240–243.

Moelling, K., Scott, A., Dittmar, K. E. J., and Owada, M. (1980). Effect of p15-associated protease from an avian RNA tumor virus on avian virus-specific polyprotein precursors. *J. Virol.* **33,** 680–688.

Neel, B. G., Hayward, W. S., Robinson, H. L., Fang, J., and Astrin, S. M. (1981). Avian leukosis virus-induced tumors have common proviral integration sites and synthesize discrete viral RNAs: oncogenesis by promoter insertion. *Cell* **23,** 323–334.

Nevins, J. R. (1982). Adenovirus gene expression: control at multiple steps of mRNA biogenesis. *Cell* **28,** 1–2.

Newlon, C. S., Devenish, R. J., and Lipchitz, L. R. (1982). Mapping autonomously replicating segments on a circular derivative of chromosome III. *In* Proceedings of the Berkeley Workshop on Recent Advances in Yeast Molecular Biology: Recombinant DNA" (in press).

Noori-Daloii, M. R., Swift, R. A., Kung, H.-J., Crittenden, L. B., and Witter, R. L. (1981). Specific integration of REV proviruses in avian bursal lymphomas. *Nature (London)* **294,** 574–575.

Notter, M. F. D., Leary, J. F., and Balduzzi, P. C. (1982). Adsorption of Rous sarcoma virus to genetically susceptible and resistant chicken cells studied by laser flow cytometey. *J. Virol.* **41,** 958–964.

Nusse, R., and Varmus, H. E. (1982). Many tumors induced by the mouse mammary tumor virus contain a provirus integrated in the same region of the host genome. *Cell* **31,** 99–109.

Oliff, A., Linemeyer, D., Ruscetti, S., Lowe, R., Lowy, D. R., and Scolnick, E. (1980). Subgenomic fragment of molecularly cloned Friend murine leukemia virus DNA contains the gene(s) responsible for Friend murine leukemia virus-induced disease. *J. Virol.* **35,** 924–936.

Olsen, J. C., and Watson, K. F. (1980). Avian retrovirus RNA-directed DNA synthesis by purified reverse transcriptase. Covalent linkage of RNA to plus strand DNA. *Biochem. Biophys. Res. Commun.* **97,** 1376–1383.

Ono, M., Cole, M. D., White, A. T., and Huang, R. C. C. (1980). Sequence organization of cloned intracisternal A particle genes. *Cell* **21,** 465–474.

Oppermann, H., Levinson, A. D., and Varmus, H. E. (1981). The structure and protein kinase activity of proteins encoded by nonconditional mutants and back mutants in the *src* gene of avian sarcoma virus. *Virology* **108,** 47–70.

O'Rear, J. J., and Temin, H. M. (1982). Spontaneous changes in nucleotide sequences in proviruses of spleen necrosis virus, an avian retrovirus. *Proc. Natl. Acad. Sci. USA* **79,** 1230–1234.

O'Rear, J. J., Mizutani, S., Hoffman, G., Fiandt, M., and Temin, H. M. (1980). Infectious and noninfectious recombinant clones of the provirus of SNV differ in cellular DNA and are apparently the same in viral DNA. *Cell* **20,** 423–430.

Ostrowski, M. C., Berard, D., and Hager, G. L. (1981). Specific transcriptional initiation *in vitro* on murine type-C retrovirus promoters. *Proc. Natl. Acad. Sci. USA* **78,** 4485–4489.

Panet, A., and Cedar, H. (1977). Selective degradation of integrated murine leukemia proviral DNA by deoxyribonucleases. *Cell* **11,** 933–940.

Panet, A., Haseltine, W. A., Baltimore, D., Peters, G., Harada, F., and Darlberg, J. E. (1975). Specific binding of tryptophan transfer RNA to avian myeloblastosis virus RNA-dependent DNA polymerase (reverse transcriptase). *Proc. Natl. Acad. Sci. USA* **72,** 2535–2539.

Papkoff, J., Lai, M. H.-T., Hunter, T., and Verma, I. M. (1981). Analysis of transforming proteins from Moloney murine sarcoma virus. *Cell* **27,** 109–120.

Payne, L. N., and Chubb, R. C. (1968). Studies on the nature and genetic control of an antigen in normal chick embryos which reacts in the COFAL test. *J. Gen. Virol.* **3,** 379–391.

Payne, G. S., Courtneidge, S. A., Crittenden, L. B., Fadly, A. M., Bishop, J. M., and Varmus, H. E. (1981). Analysis of avian leukosis virus DNA and RNA in bursal tumors: viral gene expression is not required for maintenance of the tumor state. *Cell* **23,** 311–322.

Payne, G. S., Bishop, J. M., and Varmus, H. E. (1982). Multiple arrangements of viral DNA and an activated host oncogene in bursal lymphomas. *Nature (London)* **295,** 209–213.

Perucho, M., Hanahan, D., and Wigler, M. (1980). Genetic and physical linkage of exogenous sequences in transformed cells. *Cell* **22,** 309–317.

Philipson, L., Andersson, P., Olshevsky, U., Weinberg, R., Baltimore, D., and Gesteland, R. (1978). Translation of MuLV and MSV RNA6s in nuclease-treated reticulocyte extracts: enhancement of the *gag-pol* polypeptide with yeast suppressor tRNA. *Cell* **13,** 189–199.

Pincus, T., Hartley, J. W., and Rowe, W. P. (1971). A major genetic locus affecting resistance to infection with murine leukemia viruses. I. Tissue culture studies of naturally occurring viruses. *J. Exp. Med.* **133,** 1219–1233.

Poiesz, B. J., Ruscetti, F. W., Gazdar, A. F., Bunn, P. A., Minna, J. D., and Gallo, R. C. (1980). Detection and isolation of type C retrovirus particles from fresh and cultured lymphocytes of a patient with cutaneous T-cell lymphoma. *Proc. Natl. Acad. Sci. USA* **77,** 7415–7419.

Porzig, K. J., Robbins, K. C., and Aaronson, S. A. (1979). Cellular regulation of mammalian sarcoma virus expression: a gene regulation model for oncogenesis. *Cell* **16,** 875–884.

Proudfoot, N. J., and Brownlee, G. G. (1974). Sequence analysis at the 3′ end of globin mRNA shows homology with immunoglobulin light chain messenger mRNA. *Nature (London)* **252,** 359–362.

Quint, W., van der Putten, H., Janseen, F., and Berns, A. (1982). Mobility of endogenous ecotropic murine leukemia viral genomes within mouse chromosomal DNA and integration of a mink cell focus-forming virus-type recombinant provirus in the germ line. *J. Virol.* **41,** 901–908.

Quintrell, N., Hughes, S. H., Varmus, H. E., and Bishop, J. M. (1980). Structure of viral DNA and RNA in mammalian cells infected with avian sarcoma virus. *J. Mol. Biol.* **143,** 363–393.

Racevskis, J., and Koch, G. (1977). Viral protein synthesis in Friend erytroleukemia cell lines. *J. Virol.* **21,** 328–337.

Reddy, E. P., Smith, M. J., and Aaronson, S. A. (1981). Complete nucleotide sequence and organization of the Moloney murine sarcoma virus genome. *Science (Washington, D.C.)* **214,** 445–449.

Reed, R. (1981). Resolution of cointegrates between transposons and Tn3 defines the recombination site. *Proc. Natl. Acad. Sci. USA* **78,** 3428–3432.

Reed, R. R., and Grindley, N. D. F. (1981). Transposon-mediated site-specific recombination *in vitro:* DNA cleavage and protein–DNA linkage at the recombination site. *Cell* **25,** 721–728.

Ringold, G. M., Yamamoto, K. R., Tomkins, G. M., Bishop, J. M., and Varmus, H. E. (1975). Dexamethasone-mediated induction of mouse mammary tumor virus RNA: a system for studying glucocorticoid action. *Cell* **6,** 299–305.

Ringold, G. M., Yamamoto, K. R., Bishop, J. M., and Varmus, H. E. (1977). Glucocorticoid-stimulated accumulation of mouse mammary tumor virus RNA: increased rate of synthesis of viral RNA. *Proc. Natl. Acad. Sci. USA* **74,** 2879–2883.

Ringold, G. M., Shank, P. R., and Yamamoto, K. R. (1978). Production of unintegrated mouse mammary tumor virus DNA in infected rat hepatoma cells is a secondary action of dexamethasone. *J. Virol.* **26,** 93–101.

Robertson, D. L., and Varmus, H. E. (1979). Structural analysis of the intracellular RNAs of murine mammary tumor virus. *J. Virol.* **40,** 673–682.

Robertson, D. L., and Varmus, H. E. (1981). Dexamethasone induction of the intracellular RNAs and proteins of mouse mammary tumor virus. *J. Virol.* **40,** 673–682.

Robinson, H. L., Astrin, S. M., Senior, A. M., and Salazar, F. H. (1981). Host susceptibility to endogenous viruses: defective glycoprotein-expressing proviruses interfere with infections. *J. Virol.* **40,** 745–751.

Roeder, G. S., and Fink, G. R. (1980). DNA rearrangements associated with a transposable element in yeast. *Cell* **21,** 239–250.

Rothenberg, E., Smotkin, D., Baltimore, D., and Weinberg, R. A. (1977). *In vitro* synthesis of infectious DNA of murine leukemia virus. *Nature (London)* **269,** 122–126.

Rothenberg, E., Donoghue, D. J., and Baltimore, D. (1978). Analysis of a 5′ leader sequence on murine leukemia virus 21S RNA: heteroduplex mapping with long reverse transcriptase products. *Cell* **13,** 435–451.

Rous, P. (1911). A sarcoma of the fowl transmissible by an agent separable from the tumor cells. *J. Exp. Med.* **13,** 397–413.

Roussel, M., Saule, S., Lagrou, C., Rommens, C., Beug, H., Graf, T., and Stehelin, D. (1979). Three new types of viral oncogene of cellular origin specific for haematopoietic cell transfromation. *Nature (London)* **281,** 452–455.

Rowe, W. P., and Kozak, C. A. (1980). Germ-line reinsertions of AKR murine leukemia virus genomes in Akv-1 congenic mice. *Proc. Natl. Acad. Sci. USA* **77,** 4871–4874.

Ruscetti, S. K., Linemeyer, D., Field, J., Troxler, D., and Scolnick, E. M. (1979). Characterization of a protein found in cells infected with the spleen focus-forming virus that shares immunological cross-reactivity with the gp70 found in mink cell focus-inducing virus particles. *J- Virol.* **30,** 787–798

Rymo, L., Parsons, J. T., Coffin, J. M., and Weissmann, C. (1974). *In vitro* synthesis of Rous sarcoma virus-specific RNA is catalyzed by a DNA-dependent RNA polymerase. *Proc. Natl. Acad. Sci. USA* **71,** 2782–2786.

Sabran, J. L., Hsu, T. W., Yeater, C., Kaji, A., Mason, W. S., and Taylor, J. M. (1979). Analysis of integrated avian RNA tumor virus DNA in transformed chicken, duck, and quail fibroblasts. *J. Virol.* **29,** 170–178.

Scherer, S., Mann, C., and Davis, R. W. (1982). Reversion of a promoter deletion in yeast. *Nature* **298,** 815–818.

Schincariol, A. L., and Joklik, W. K. (1973). Early synthesis of virus-specific RNA and DNA in cells rapidly transformed with Rous sarcoma virus. *Virology* **56,** 532–548.

Scherer, G., Tschndi, C., Perera, J., DeLuiss, H., and Pirrotta, V. (1982). *B104,* a new dispersed repeat gene family in *Drosophila melanogaster* and its analogies with retroviruses. *J. Mol. Biol.* **157,** 435–451.

Schon, E. A., Cleary, M. L., Haynes, J. R., and Lingrel, J. B. (1981). Structure and evolution of goat α, β^c- and β^a-globin genes: three developmentally regulated genes contain inserted elements. *Cell* **27,** 359–370.

Schwartz, D. E., Zamecnik, P. C., and Weith, H. L. (1977). Rous sarcoma virus genome is terminally redundant: the 3′ sequence. *Proc. Natl. Acad. Sci. USA* **74,** 994–998.

Schwartz, D., Tizard, R., and Gilbert, W. (1982). Nucleotide sequence of Rous sarcoma virus. *Cell* (in press).

Scott, M. L., McKereghan, K., Kaplan, H. S., and Fry, K. E. (1981). Molecular cloning and partial characterization of unintegrated linear DNA from gibbon ape leukemia virus. *Proc. Natl. Acad. Sci. USA* **78,** 4213–4217.

Shank, P. R., and Linial, M. (1980). Avian oncornavirus mutant (Se21Q1b) deficient in genomic RNA: characterization of a deletion in the provirus. *J. Virol.* **36,** 450–456.

Shank, P. R., and Varmus, H. E. (1978). Virus-specific DNA in the cytoplasm of avian sarcoma virus-infected cells is a precursor to covalently closed circular viral DNA in the nucleus. *J. Virol.* **25,** 104–114.

Shank, P. R., Hughes, S. H., Kung, H.-J., Majors, J. E., Quintrell, N., Guntaka, R. V., Bishop, J. M., and Varmus, H. E. (1978a). Mapping unintegrated avian sarcoma virus DNA: termini of linear DNA bear 300 nucleotides present once or twice in two species of circular DNA. *Cell* **15,** 1383–1395.

Shank, P. R., Cohen, J. C., Varmus, H. E., Yamamoto, K. R., and Ringold, G. M. (1978b). Mapping of linear and circular forms of mouse mammary tumor virus DNA with restriction endonucleases: evidence for a large specific deletion occurring at high frequency during circularization. *Proc. Natl. Acad. Sci. USA* **75,** 2112–2116.

Shapiro, J. A. (1979). Molecular model for the transposition and replication of bacteriophage Mu and other transposable elements. *Proc. Natl. Acad. Sci. USA* **76,** 1933–1937.

Shapiro, J. A., and MacHattie, L. A. (1979). Integration and excision of prophage λ mediated by the IS1 element. *Cold Spring Harbor Symp. Quant. Biol.* **43,** 1135–1142.

Shibuya, M., Hanafusa, T., Hanafusa, H., and Stephenson, J. R. (1980). Homology exists among the transforming sequences of avian and feline sarcoma viruses. *Proc. Natl. Acad. Sci. USA* **77,** 6536–6540.

Shih, T. Y., Weeks, M. O., Young, H. A., and Scolnick, E. M. (1979). Identification of a sarcoma virus-coded phosphoprotein in nonproducer cells transformed by Kirsten or Harvey murine sarcoma virus. *Virology* **96,** 64–79.

Shimotohno, D., and Temin, H. M. (1980). No apparent nucleotide sequence specificity in cellular DNA juxtaposed to retrovirus proviruses. *Proc. Natl. Acad. Sci. USA* **77,** 7357–7361.

Shimotohno, K., and Temin, H. M. (1981). Formation of infectious progeny virus after insertion of herpes simplex thymidine kinase gene into DNA of an avian retrovirus. *Cell* **26,** 67–78.

Shimotohno, K., and Temin, H. M. (1982). Spontaneous variation and synthesis in the U3 region of the long terminal repeat of an avian retrovirus. *J. Virol* **41,** 163–171.

Shimotohno, K., Mizutani, S., and Temin, H. M. (1980). Sequence of retrovirus provirus resembles that of bacterial transposable elements. *Nature (London)* **285,** 550–554.

Shimotohno, K., (1981).

Shinnock, T. M., Lerner, R. A., and Sutcliffe, J. G. (1981). Nucleotide sequence of Moloney murine leukemia virus. *Nature (London)* **293,** 543–548.

Shoemaker, C., Goff, S., Gilboa, E., Paskind, M., Mitra, S. W., and Baltimore, D. (1980). Structure of a cloned circular Moloney murine leukemia virus molecule containing an inverted segment: implications for retrovirus integration. *Proc. Natl. Acad. Sci. USA* **77,** 3932–3936.

Shoemaker, C., Hoffman, J., Goff, S. P., and Baltimore, D. (1981). Intramolecular integration within Moloney murine leukemia virus DNA. *J. Virol.* **40,** 164–172.

Simons, K., Garoff, H., and Helenius, A. (1982). How an animal virus gets into and out of its host cell? *Sci. Am.* **246,** 58–97.

Spector, D. H., Varmus, H. E., and Bishop, J. M. (1978). Nucleotide sequences related to the transforming gene of avian sarcoma virus are present in the DNA of uninfected vertebrates. *Proc. Natl. Acad. Sci. USA* **75,** 4102–4106.

Stacey, D. W., and Hanafusa, H. (1978). Nuclear conversion of microinjected avian leukosis virion RNA into a envelope-glycoprotein messenger. *Nature (London)* **273,** 779–782.

Steffen, D., and Weinberg, R. A. (1978). The integrated genome of murine leukemia virus. *Cell* **15,** 1003–1010.

Steffen, D. L., Bird, S., and Weinberg, R. A. (1980). Evidence for the Asiatic origin of endogenous AKR-type murine leukemia proviruses. *J. Virol.* **35,** 824–835.

Stehelin, D., Varmus, H. E., Bishop, J. M., and Vogt, P. K. (1976). DNA related to the transforming gene(s) of avian sarcoma viruses is present in normal avian DNA. *Nature (London)* **260,** 170–173.

Streeck, R. E. (1982). A multicopy insertion sequence in the bovine genome with structural homology to the long terminal repeats of retroviruses. *Nature* **298,** 767–769.

Summers, J., and Mason, W. S. (1982). Replication of the genome of a hepatitis B-like virus by reverse transcription of an RNA intermediate. *Cell* **29,** 403–415.

Sutcliffe, J. G., Shinnick, T. M., Verma, I. M., and Lerner, R. A. (1980). Nucleotide sequence of Moloney leukemia virus: 3′ end reveals details of replication, analogy to bacterial transposons, and an unexpected gene. *Proc. Natl. Acad. Sci. USA* **77,** 3302–3306.

Swanstrom, R., DeLorbe, W. J., Bishop, J. M., and Varmus, H. E. (1981). Nucleotide sequence of cloned unintegrated avian sarcoma virus DNA: viral DNA contains direct and inverted repeats similar to those in transposable elements. *Proc. Natl. Acad. Sci. USA* **78,** 124–128.

Swanstrom, R., Varmus, H. E., and Bishop, J. M. (1982a). Nucleotide sequence of the 5′ noncoding region and part of the *gag* gene of Rous sarcoma virus. *J. Virol.* **41,** 535–541.

Swanstrom, R., Bishop, J. M., and Varmus, H. E. (1982b). Structure of a replication intermediate in the synthesis of Rous sarcoma virus DNA *in vivo*. *J. Virol.* **42,** 337–341.

Taylor, J. M. (1977). An analysis of the role of tRNA species as primers for the transcription into DNA of RNA tumor virus genomes. *Biochim. Biophys. Acta* **473,** 57–71.

Taylor, J. M., and Hsu, T. W. (1980). Reverse transcription of avian sarcoma virus RNA into DNA might involve copying of the tRNA primer. *J. Virol.* **33,** 531–534.

Teich, N. (1982). Taxonomy of retroviruses. *In* "Molecular Biology of Tumor Viruses. RNA Tumor Viruses" (R. A. Weiss, N. Teich, H. E. Varmus, and J. M. Coffin, eds.), pp. 25–207. Cold Spring Harbor Lab., Cold Spring Harbor, New York.

Teich, N., Bernstein, A., Mak, T., Wyke, J., and Hardy, W. (1982). Pathogenesis of retrovirus-induced disease. *In* "Molecular Biology of Tumor Viruses. RNA Tumor Viruses" (R. A. Weiss, N. Teich, H. E. Varmus, and J. M. Coffin, eds.), pp. 785–998. Cold Spring Harbor Lab., Cold Spring Harbor, New York.

Temin, H. (1963). The effects of actinomycin D on growth of Rous sarcoma virus *in vitro*. *Virology* **20,** 577–582.

Temin, H. M. (1964). Nature of the provirus of Rous sarcoma. *Natl. Cancer Inst. Monogr.* **17,** 557–570.

Temin, H. M. (1967). Studies on carcinogenesis by avian sarcoma virus. V. Requirement for new DNA synthesis and for cell division. *J. Cell. Physiol.* **69,** 53–64.

Temin, H. M. (1980). Origin of retroviruses from cellular moveable genetic elements. *Cell* **21,** 599–600.

Temin, H. M., and Mizutani, S. (1970). RNA-directed DNA polymerase in virions of Rous sarcoma virus. *Nature (London)* 1211–1213.

Todaro, G. (1980). Interspecies transmission of mammalian retroviruses. *In* "Molecular Biology of RNA Tumor Viruses" (J. Stephenson, ed.), pp. 46–77. Academic Press, New York.

Tooze, J. "Molecular Biology of Tumor Viruses. DNA Tumor Viruses." Cold Spring Harbor Lab., Cold Spring Harbor, New York.

Trainor, C. D., and Reitz, M. S., Jr. (1979). Loss of proviral DNA sequences in a revertant of Kirsten sarcoma virus-transformed murine fibroblasts. *J. Gen. Virol.* **44,** 245–249.

Treisman, R., Novak, U., Favaloro, J., and Kamen, R. (1981). Transformation of rat cells by an altered polyoma virus genome expressing only the middle-T protein. *Nature (London)* **292,** 595–600.

Troxler, D. H., Lowy, D., Howk, R., Young, H., and Scolnick, E. M. (1977). Friend strain of spleen focus-forming virus is a recombinant between ecotropic murine type C virus and the *env* gene region of xenotropic type C virus. *Proc. Natl. Acad. Sci. USA* **84,** 4671–4675.

Tsichlis, P. N., and Coffin, J. M. (1980). Recombinants between endogenous and exogenous avian tumor viruses: role of the C region and other portions of the genome in the control of replication and transformation. *J. Virol.* **33,** 238–249.

Turek, L. P., and Oppermann, H. (1980). Spontaneous conversion of nontransformed avian sarcoma virus-infected rat cells to the transformed phenotype. *J. Virol.* **35,** 466–478.

Ucker, D. (1981). Ph.D. Dissertation, Univ. California, San Francisco.

Ucker, D. S., Ross, S. R., and Yamamoto, K. R. (1981). Mammary tumor virus DNA contains sequences required for its hormone-regulated transcription. *Cell* **27,** 257–266.

Van Arsdell, S. W., Denison, R. A., Bernstein, L. B., Weiner, A. M., Manser, T., and Gestland, R. F. (1981). Direct repeats flank three small nuclear RNA pseudogenes in the human genome. *Cell* **26,** 11–18.

Van Beveren, C., Goddard, J. C. G., Berns, A., and Verma, I. M. (1980). Structure of Moloney murine leukemia viral DNA: nucleotide sequence of the 5′ long terminal repeat and adjacent cellular sequence. *Proc. Natl. Acad. Sci. USA* **77,** 3307–3311.

Van Beveren, C., van Straaten, F., Galleshaw, J. A., and Verma, I. M. (1981a). Nucleotide sequence of the genome of a murine sarcoma virus. *Cell* **27,** 97–108.

Van Beveren, C., Galleshaw, J. A., Jonas, V., Berns, A. J. M., Doolittle, R. F., Donoghue, D. J., and Verma, I. M. (1981b). Nucleotide sequence and formation of the transforming gene of a mouse sarcoma virus. *Nature (London)* **289,** 258–262.

Van Beveren, C., Rands, E., Chattopadhyay, S. K., Lowy, D. R., and Verma, L. M. (1982). Long terminal repeat of murine retroviral DNAs: sequence analysis, host–proviral junctions and preintegration site. *J. Virol.* **41,** 542–556.

van der Putten, H., Quint, W., van Raaij, J., Maandag, E. R., Verma, I. M., and Berns, A. (1981). M–MLV-induced leukemogenesis: integration and structure of recombinant proviruses in tumors. *Cell* **24,** 729–740.

Varmus, H. E., and Shank, P. R. (1976). Unintegrated viral DNA is synthesized in the cytoplasm of avian sarcoma virus-transformed duck cells by viral DNA polymerase. *J. Virol.* **18,** 567–573.

Varmus, H. E., and Swanstrom, R. (1982). Replication of retroviruses. *In* "Molecular Biology of Tumor Viruses. RNA Tumor Viruses" (R. A. Weiss, N. Teich, H. E. Varmus, and J. M. Coffin, eds.), pp. 369–512. Cold Spring Harbor Lab., Cold Spring Harbor, New York.

Varmus, H. E., Bishop, J. M., and Vogt, P. K. (1973). Appearance of virus-specific DNA in mammalian cells following transformation by Rous sarcoma virus. *J. Mol. Biol.* **74,** 613–626.

Varmus, H. E., Guntaka, R. V., Fan, W. J. W., Heasley, S., and Bishop, J. M. (1974). Synthesis of viral DNA in the cytoplasm of duck embryo fibroblasts and in enucleated cells after infection by avian sarcoma virus. *Proc. Natl. Acad. Sci. USA* **71,** 3874–3878.

Varmus, H. E., Padgett, T., Healsey, S., Simon, G., and Bishop, J. M. (1977). Cellular functions are required for the synthesis and integration of avian sarcoma virus-specific DNA. *Cell* **11,** 307–319.

Varmus, H. E., Healsey, S., Kung, H.-J., Oppermann, H., Smith, V. C., Bishop, J. M., and Shank, P. R. (1978). Kinetics of synthesis, structure and purification of avian sarcoma virus-specific DNA made in the cytoplasm of acutely infected cells. *J. Mol. Biol.* **120,** 55–82.

Varmus, H. E., Quintrell, N., and Oritz, S. (1981a). Retroviruses as mutagens: insertion and excision of a nontransforming provirus after expression of a resident transforming provirus. *Cell* **25,** 23–36.

Varmus, H. E., Quintrell, N., and Wyke, J. (1981b). Revertants of an ASV-transformed rat cell line have lost the complete provirus or sustained mutations in *src*. *Virology* **108,** 28–46.

Verma, I. M. (1975). Studies on reverse transcriptase of RNA tumor viruses. III. Properties of purified Moloney murine leukemia virus DNA polymerase and associated RNase H. *J. Virol.* **15,** 843–854.

Verma, I. M. (1977). The reverse transcriptase. *Biochim. Biophys. Acta* **473,** 1–38.

Verma, I. M., Mason, W. S., Drost, S. D., and Baltimore, D. (1974). DNA polymerase activity from two temperature-sensitive mutants of Rous sarcoma virus is thermolabile. *Nature (London)* **251,** 27–31.

Verma, I. M., Varmus, H. E., and Hunter, E. (1976). Characterization of "early" temperature-sensitive mutants of avian sarcoma viruses: biological properties, thermolability of reverse transcriptase *in vitro* and synthesis of viral DNA in infected cells. *Virology* **74,** 16–29.

Vigne, R., Breitman, M. L., Moscovici, C., and Vogt, P. K. (1971). Genetically stable reassortment of markers during mixed infection with avian tumor viruses. *Virology* **46,** 947–952.

Vogt, P. K. (1971). Genetically stable reassortment of markers during mixed infection with avian tumor viruses. *Virology* **46,** 947–952.

Vogt, P. K. (1977). Genetics of RNA tumor viruses. *In* "Comprehensive Virology" (H. Fraenkel-Conrat and R. R. Wagner, eds.), Vol. 9, pp. 341–455. Plenum, New York.

Vogt, P. K., and Ishizaki, R. (1966). Patterns of viral interference in the avian leukosis and sarcoma complex. *Virology* **30,** 368–374.

von der Helm, K. (1977). Cleavage of Rous sarcoma virus polyprotein precursor into internal structural proteins *in vitro* by viral protein p15. *Proc. Natl. Acad. Sci. USA* **74,** 911–915.

Wake, C. T., and Wilson, J. H. (1980). Defined oligomeric SV40 DNA: a sensitive probe of general recombination in somatic cells. *Cell* **21,** 141–148.

Wei, C.-M., Gibson, M., Spear, P. G., and Scolnick, E. M. (1981). Construction and isolation of a transmissible retrovirus containing the *src* gene of Harvey murine sarcoma virus and the thymidine kinase gene of herpes simplex virus type 1. *J. Virol.* **39,** 935–944.

Weinberg, R. A. (1980). Integrated genomes of animal viruses. *Annu. Rev. Biochem.* **49,** 197–226.

Weis, J. H., and Faras, A. J. (1981). DNA topoisomerase activity associated with Rous sarcoma virus. *Virology* **114,** 563–566.

Weiss, R. A. (1982). Experimental biology and assay of retroviruses. *In* "Molecular Biology of Tumor Viruses. RNA Tumor Viruses" (R. A. Weiss, N. Teich, H. E. Varmus, and J. M. Coffin, eds.), pp. 209–260. Cold Spring Harbor Lab., Cold Spring Harbor, New York.

Weiss, R. A., Mason, W. S., and Vogt, P. K. (1973). Genetic recombinants and heterozygotes derived from endogenous and exogenous avian RNA tumor viruses. *Virology* **52,** 535–552.

Weiss, S. R., Varmus, H. E., and Bishop, J. M. (1977). The size and genetic composition of virus-specific RNAs in the cytoplasm of cells producing avian sarcoma-leukemia viruses. *Cell* **12,** 983–992.

Weiss, R. A., Teich, N., Varmus, H. E., and Coffin, J. M. (eds.) (1982). "Molecular Biology of Tumor Viruses. RNA Tumor Viruses." Cold Spring Harbor Lab., Cold Spring Harbor, New York.

Weymouth, L. A., and Loeb, L. A. (1978). Mutagenesis during *in vitro* DNA synthesis. *Proc. Natl. Acad. Sci. USA* **75,** 1924–1928.

Will, B. M., Bayev, A. A., and Finnegan, D. J. (1982). Nucleotide sequence of terminal repeats of 412 transposable elements of *Drosophila melanogaster*. A similarity to proviral long terminal repeats and its implications for the mechanism of transposition. *J. Mol. Biol.* **153,** 897–916.

Williamson, V. M., Young, E. T., and Ciriacy, M. (1981). Transposable elements associated with constitutive expression of yeast alcohol dehydrogenase II. *Cell* **23,** 605–614.

Willingham, M. C., Jay, G., and Pastan, I. (1979). Localization of the ASV *src* gene product to the plasma membrane of transformed cells by electron microscopic immunocytochemistry. *Cell* **18,** 125–134.

Willingham, M. C., Pastan, I., Shih, T. Y., and Scolnick, E. M. (1980). Localization of the *src* gene product of the Harvey strain of MSV to the plasma membrane of transformed cells by electron microscopic immunocytochemistry. *Cell* **19,** 1005–1014.

Witte, O. N., and Baltimore, D. (1978). Relationship of retrovirus polyprotein cleavages to virion maturation studied with temperature-sensitive murine leukemia virus mutants. *J. Virol.* **26,** 750–761.

Witte, O. N., Rosenberg, N., and Baltimore, D. (1979). Preparation of syngeneic tumor regressor serum reactive with the unique determinants of the Abelson murine leukemia virus-encoded P120 protein at the cell surface. *J. Virol.* **31,** 776–784.

Witte, O. N., Dasgupta, A., and Baltimore, D. (1980). Abelson murine leukemia virus protein is phosphorylated *in vitro* to form phosphotyrosine. *Nature (London)* **283,** 826–831.

Wyke, J. A., and Beamand, J. A. (1979). Genetic recombination in Rous sarcoma virus: the genesis of recombinants and lack of evidence for linkage between *pol, env* and *src* genes in three factor crosses. *J. Gen. Virol.* **43,** 349–364.

Wyke, J. A., Bell, J. G., and Beamand, J. A. (1975). Genetic recombination among temperature-sensitive mutants of Rous sarcoma virus. *Cold Spring Harbor Symp. Quant. Biol.* **39,** 897–905.

Yamamoto, T., de Crombrugghe, B., and Pastan, I. (1980a). Identification of a functional promoter in the long terminal repeat of Rous sarcoma virus. *Cell* **22,** 787–797.

Yamamoto, T., Jay, G., and Pastan, I. (1980b). Unusual features in the nucleotide sequence of a cDNA clone derived from the common region of avian sarcoma virus messenger RNA. *Proc. Natl. Acad. Sci. USA* **77,** 176–180.

Yang, Y. H. J., Rhim, J. S., Rasheed, S., Klement, V., and Roy-Burman, P. (1979). Reversion of Kirsten sarcoma virus transformed human cells: elimination of the sarcoma virus nucleotide sequences. *J. Gen. Virol.* **43,** 447–451.

Yoshimora, F. K. and Breda, M. (1981). Lack of an AKR ecotropic provirus amplification in AKR leukemic thymuses. *J. Virol.* **39,** 808–815.

Yoshimora, F. K., and Weinberg, R. A. (1979). Restriction endonuclease cleavage of linear and closed circular murine leukemia viral DNAs: discovery of a smaller circular form. *Cell* **16,** 323–332.

Yoshinaka, Y., and Luftig, R. B. (1977a). Murine leukemia virus morphogenesis: cleavage of P70 *in vitro* can be accompanied by a shift from a concentrically coiled internal stand (''immature'') to a collapsed (''mature'') form of the virus core. *Proc. Natl. Acad. Sci. USA* **74,** 3446–3450.

Yoshinaka, Y., and Luftig, R. B. (1977b). Properties of a P70 proteolytic factor of murine leukemia viruses. *Cell* **12,** 709–719.

Young, H. A., Shih, T. Y., Scolnick, E. M., and Parks, W. P. (1977). Steroid induction of mouse mammary tumor virus: effect upon synthesis and degradation of viral RNA. *J. Virol.* **21,** 139–146.

Young, H. A., Gonda, M. A., DeFeo, D., Ellis, R. W., Nagashima, K., and Scolnick, E. M. (1980). Heteroduplex analysis of cloned rat endogenous replication-defective (30S) retrovirus and Harvey murine sarcoma virus. *Virology* **107,** 89–99.

Zavada, J. (1972). VSV pseudotype particles with the coat of avian myeloblastosis virus. *Nature (London) New Biol.* **240,** 122–124.

CHAPTER 11

Agrobacterium Tumor Induction

PATRICIA ZAMBRYSKI
HOWARD M. GOODMAN
MARC VAN MONTAGU
JEFF SCHELL

Mobile Genetic Elements

ISBN 0-12-638680-3

I. Crown Gall: Agrobacterium*–Plant-Cell Interaction*

Crown gall is a disease that affects most dicotyledonous plants (for recent reviews see Braun, 1982; Nester and Kosuge, 1981; Van Montagu and Schell, 1981; Kahl and Schell, 1982; Schell *et al.*, 1981, 1982). The inducing agent is a soil bacterium, *Agrobacterium tumefaciens*. Although the host range is broad, it does not include monocotyledonous plants. However, with increased knowledge of the process of crown gall tumor formation, monocots can be expected to become susceptible to genetic exchanges similar to those induced by *Agrobacterium*. The disease represents the first documented example of a parasitic interaction between prokaryotes and eukaryotes whereby a pathogen (*Agrobacterium*) genetically determines the infected host to express functions that benefit the survival of the parasite. Although the bacteria are required for tumor induction and are mostly found in association with crown gall-harboring plants in nature, they are not necessary for tumor maintenance and growth (Braun, 1953). In fact, sterile crown gall tissues can readily be cultured indefinitely on simple media lacking any added growth hormones. Crown gall plant cells therefore are persistently altered cells with a capacity for autonomous growth in the absence of any outside stimulation by the inciting bacteria. The normal mechanisms controlling growth and differentiation of plant cells are altered in these transformed cells and, as such, crown gall cells may be compared with animal tumor cells.

Braun and Mandle (1948) were the first to conclude from their observations that a factor (tumor-inducing principle or TIP factor) is transmitted from the inciting bacteria to the host cells, resulting, in a relatively short time (about 36 hr), in a stable transformation of the plant cells into crown gall tumor cells. The cells acquire a capacity for autonomous growth, possibly because they persistently synthesize or otherwise become independent of the growth-regulating substances and metabolites that are externally required for cellular growth and division of normal, untransformed cells.

Because of its possible relation to animal cancers and its value as an ideal experimental model system to study the control mechanisms underlying cellular growth and differentiation, crown gall has been extensively and continuously studied ever since the beginning of this century (Smith, 1916; Jensen, 1918; Braun and White, 1943; Braun, 1982). Until about 1970, most of the crown gall research was centered on the study of the properties of transformed plant cells. It is largely because some of the attention shifted to the study of the biological and genetic properties of the inciting bacteria that the nature of the tumor-inducing principle was discovered and the biological significance of the crown gall phenomenon could be understood.

The observations of Morel and his collaborators led to a more fundamental implication of the inciting bacteria: These authors (Petit *et al.*, 1970) demonstrated that there are at least two different types of crown gall-inducing bacteria that induce tumors with different properties. One type of *Agrobacterium* strain induces crown gall tumors in which *N*-α-(D-1-carboxyethyl)-L-arginine (octopine) was synthesized; the other type induces tumors containing *N*-α-(1,3-dicarboxypropyl)-L-arginine (nopaline). The important fact here is that the type of arginine derivative synthesized in the tumor was found to be specified by the particular *Agrobacterium* strain used to incite the tumor and to be independent of the host plant on which the tumors were induced.

It was furthermore demonstrated that *Agrobacterium* strains that induce the synthesis of octopine in crown gall can selectively use this product, but not nopaline, as the sole carbon and/or nitrogen energy source, whereas *Agrobacterium* strains that induce the synthesis of nopaline in crown gall can selectively use it, but not octopine, as the carbon and/or nitrogen energy source. Thus, a genetic linkage between oncogenicity and opine metabolism was established, and the hypothesis was proposed that genetic information could be somehow transferred from bacterium to plant. The name *opine* was subsequently proposed (Tempé and Schell, 1977; Tempé *et al.*, 1978) to describe products, specifically synthesized by crown gall plant cells, that can be used by agrobacteria as specific growth substances.

The identification of the nature and function of the elusive tumor-inducing principle resulted primarily from experiments designed to identify the genetic determinants in *Agrobacterium* responsible for its tumor-inducing capacity. Thus, it was first reported in 1973 (see Schell, 1975), that the tumor-inducing principle of *Agrobacterium* was carried by extrachromosomal DNA elements of the plasmid type and not of the lysogenic type, as had been previously reported.

The experimental demonstration of the involvement of tumor-inducing (Ti) plasmids (Zaenen *et al.*, 1974; Van Larebeke *et al.*, 1974, 1975; Watson *et al.*, 1975; Engler *et al.*, 1975; Bomhoff *et al.*, 1976) relied on the analysis of some of the many observations previously published about the nature of the tumor-inducing principle. Thus, the observations by Hamilton and Fall (1971) that certain strains of crown gall bacteria, such as strain C58, irreversibly lost their tumor-inducing capacity when grown at 37°C were confirmed and shown to be due to the irreversible loss (by the nononcogenic bacteria) of their large extrachromosomal Ti plasmids. The remarkable observations of Kerr (1969, 1971), who discovered that oncogenicity could be transferred from one strain of *Agrobacterium* to another by inoculating both strains, together or in succession, into the same susceptible plant, were also confirmed and shown to be due to the transmission of Ti plasmids between *Agrobacterium* strains. The discovery of Ti plasmids and of their central role in the crown gall phenomenon has led to the formulation of a general concept describing and explaining this neoplastic transformation of plant cells. Thus, the demonstration that the genes controlling opine catabolism in *Agrobacterium* and the genes determining opine synthesis in transformed plant cells are both localized on Ti plasmids (Bomhoff *et al.*, 1976; Kerr and Roberts, 1976; Genetello *et al.*, 1977; Kerr *et al.*, 1977; Montoya *et al.*, 1977; Hooykaas *et al.*, 1977; Van Larebeke *et al.*, 1977; Schell *et al.*, 1977; Holsters *et al.*, 1978; Klapwijk *et al.*, 1978) explained the linkage between opine metabolism and oncogenicity and provide genetic evidence for the involvement of the Ti plasmid in a mechanism of DNA transfer from bacterium to plant. The obvious way to explain the role played by Ti plasmids in oncogenicity and in opine synthesis was to assume that part or all of the Ti plasmid was somehow transferred to—and maintained and expressed in—the transformed plant cells. The experiments demonstrating that this assumption is correct were based on a detailed genetic analysis of the Ti plasmid and on the demonstration, by means of DNA–DNA hybridizations, of the physical presence of a Ti plasmid DNA fragment (the so-called T-DNA) in the nucleus (Willmitzer *et al.*, 1980; Chilton *et al.*, 1980) of transformed plant cells. These observations will be described in more detail in the following sections and are the reasons why *Agrobacterium* is now considered a "natural genetic engineer" because it transfers DNA to the plant cell; this DNA becomes stably integrated into the nuclear DNA of the plant cell genome, and the plant cell is transformed to a crown gall tumor cell. The transferred DNA (T-DNA), which is part of the Ti plasmid found, contains genetic information for the synthesis of opines and, in addition, encodes information for plant regulatory functions that are directly or indirectly responsible for the neoplastic growth pattern. Due to the gene transfer mechanism determined by Ti plasmids, agrobacteria have the capacity genetically to force plant cells to divert some of their metabolism to produce compounds (opines) that these bacteria are selectively equipped to catabolize. The transferred plasmid genes coding for "opine synthase" enzymes are linked, and therefore cotransferred with other genes that force the transformed cells into a tumorous mode of growth, thus establishing relatively large tissues (crown galls) of opine-producing cells. The term *genetic colonization* was proposed to describe this type of parasitic interaction (Schell *et al.*, 1979a).

The crown gall phenomenon indeed illustrates and documents a kind of parasitism and/or symbiosis in which one of the partners not only adapts itself genetically to the properties of the other partner (e.g., the host) but also specifically changes some of the host's properties by inserting a set of genes into its genome. In view of the sophistication of the Ti plasmid as a natural gene vector, it would seem reasonable to assume that similar gene transfer phenomena may account for other types of parasitic and symbiotic interactions. The induction of nodules on leguminous plants by *Rhizobium* strains is a prime example of another instance in which a gene transfer mechanism may be involved.

Recent evidence indicates that large plasmids in *Rhizobium* play an essential role in nodulation and, at least in some instances, carry the nitrogenase genes (for recent review see Nuti *et al.*, 1982). In view of the fact that nodules contain active bacterioids (large fused aggregates of bacteria), it is conceivable that bacterial (or plasmid) genes involved in the control of nodular growth remain in the bacterioid and exert their influence via products released from the bacterioids. On the other hand, it is equally conceivable that some bacterial genes involved in nodule formation and maintenance are first transferred to plant cells. Experiments to test these possibilities are in progress.

Theoretically, the galls formed on plants under the influence of insects (Rohfritsch and Shorthouse, 1981) could also conceivably involve a gene transfer mechanism. However, the fact that gall cell proliferation ceases once the gall-forming insect is withdrawn or killed, and that *in vitro* cultures of galls require the same growth substances and medium as do normal tissues (Rohfritsch, 1971), argues against such a hypothesis.

Possibly the strongest circumstantial evidence in favor of natural gene transfer in animals was recently presented by Martin and Fridovich (1981). In this case, it is supposed that the symbiotic bacteria carry and express a gene derived from the fish host. The argument is that *Photobacter leiognathi* was found to contain a copper- and zinc-containing superoxide dismutase, the amino acid composition of which was more closely related to the corresponding fish enzymes than it was to any other described bacterial CuZn superoxide dismutase. Finally, one can find in the literature a number of examples in which bacteria were found to synthesize peptides resembling animal hormones (e.g., see Cohen and Strampp, 1976); some of these may well result from a natural gene transfer phenomenon.

II. Genetic and Functional Characteristics of Different Types of Ti Plasmids

A. Taxonomic Classification of *Agrobacterium*

The biological and economically important family of the Rhizobiaceae groups comprises only two genera, *Rhizobium* and *Agrobacterium*. Officially, the latter is still considered to contain four species: *A. tumefaciens, A. radiobacter, A. rhizogenes,* and *A. rubi* (Kersters and De Ley, 1982). This classification is mostly of interest to plant pathologists. The properties characterizing and distinguishing each of the species are determined by the presence or absence of characteristic Ti plasmids. However, transfer of these plasmids from *A. tumefaciens* or *A. rhizogenes* into *A. radiobacter* will give the latter all properties of the donor strain (Van Larebeke *et al.*, 1975; Thomashow *et al.*, 1980c); thus, this classification is controversial. A new classification into biotypes 1, 2, and 3 has been proposed (Kerr and Panagopoulos, 1977; Kersters and De Ley, 1982) and is based on more traditional taxonomic criteria such as fermentation capacity. In this classification, each of the classes would contain pathogenic and nonpathogenic strains. With this classification there is a corresponding regional specificity: biotype 1, predominant for Western Europe and the United States, biotype 2 for Australia, and biotype 3 for Southeastern Europe.

The typing of strains that should be considered for the genus *Rhizobium* are also a matter of debate. The actual classification is largely based on the plant species with which the *Rhizobium* strains interact as symbionts (Elkan, 1981). Many of these properties are plasmid-encoded, and because plasmids are widely exchanged, one wonders if it is correct to make a taxonomic distinction between some *Rhizobium* strains and *Agrobacterium* strains. Indeed, transfer of a Ti plasmid from *Agrobacterium* into some *Rhizobium* strains allows them to induce crown gall (Hooykaas *et al.*, 1977; Schilperoort *et al.*, 1979).

A wide variety of natural isolates of *Agrobacterium* have been observed. However, only a few strains are actively used by investigators. These strains have been chosen by their Ti plasmids. A summary of these Ti plasmids and their properties follows.

B. The Study of the Ti Plasmids

Ti plasmids of different natural isolates of *Agrobacterium* have been characterized according to length (Zaenen *et al.*, 1974), by pattern of digestion by restriction endonucleases (Schell *et al.*, 1979b; Gordon *et al.*, 1979), or by direct DNA–DNA hybridization between different strains (Currier and Nester, 1976). These physical studies can be correlated with a biological classification, that is, according to the types of opine(s) specified and the tumor morphology on some characteristic test plants (Bomhoff *et al.*, 1976; Ellis and Murphy, 1981; Guyon *et al.*, 1980).

So far, four major classes of Ti plasmids can be considered. The two most studied classes are as follows. (*1*) Nopaline plasmids specify the synthesis of nopaline in crown gall tumor cells and encode catabolic enzymes for nopaline utilization in bacteria. Nopaline plasmids also encode the synthesis of enzymes for ornithine metabolism and for another more recently discovered opine, called *agrocinopine* (Ellis and Murphy, 1981). Nopaline tumors can differentiate into leaf-like structures that develop into abnormally structured shoots. Hence, these tumors have been called *teratomas* (for a review see Braun, 1969). (*2*) Octopine plasmids specify the synthesis of octopine in crown gall cells and its catabolism in bacteria. Octopine plasmids also encode the synthesis and catabolism of agropine (Firmin and Fenwick, 1978). Wild-type octopine tumors are usually undifferentiated and do not form teratoma shoots but can, on some test plants, develop adventitious roots. Figure 1 illustrates a nopaline teratoma of tobacco containing many plant-like structures. The original teratoma was cloned in tissue culture through plant protoplasts, and the cloned tissue was grafted onto a different host species of tobacco. From such grafts, it is possible to obtain normal regenerating shoots (Braun and Wood, 1976) that still contain Ti plasmid-specific sequences (Lemmers *et al.*, 1980).

The third class of Ti plasmids encodes only agropine metabolism (Guyon *et*

Fig. 1 A nopaline teratoma induced on *Nicotiana tabacum* cultivar. W38 by *A. tumefaciens* T37. The original teratoma was cloned via plant protoplasts and subsequently grafted onto the base of *N. tabacum* var. *xanthii,* as shown here.

al., 1980) and a second type of opine called *agrocinopine C* (Ellis and Murphy, 1981), and was formally classified as a null-type (without octopine or nopaline) plasmid. Agropine tumors often develop less well, do not show any form of differentiation, and die early.

The fourth class of Ti plasmids (White and Nester, 1980) give *Agrobacterium* the capacity to induce the hairy root or "woolly knot" disease on some plant species and regular crown galls on others (De Cleene and De Ley, 1981). These plasmids encode agropine (Tepfer and Tempé, 1981) and have also been called *Ri plasmids* (Chilton *et al.*, 1982). Some segments of the *Rhizogenes* Ti plasmids seem to be derived from nopaline plasmids and others from octopine plasmids (Willmitzer *et al.*, 1982b).

The major functions of the Ti plasmids (reviewed in Holsters *et al.*, 1982a) include crown gall tumor induction, specificity of opine synthesis in transformed plant cells, and catabolic functions for the utilization of specific opines, as described above. In addition, there are other functions that regulate replication (Koekman *et al.*, 1980; Engler *et al.*, 1981), genetic transfer between agrobacteria (Kerr *et al.*, 1977; Genetello *et al.*, 1977; Petit *et al.*, 1978), sensitivity to particular bacteriophages (Holsters *et al.*, 1980), and the synthesis of bacterial compounds such as agrocin 84, which are toxic to other soil bacteria (Kerr and Htay, 1974; Engler *et al.*, 1975).

C. The Genetic Map of Ti Plasmids

The conclusion that a property of an *A. tumefaciens* strain is Ti plasmid-encoded could be made if the loss of this property is concurrent with the curing of the Ti plasmid or if the transfer of the Ti plasmid into a new host introduces the property under study (Van Montagu and Schell, 1979). These transfer studies are relevant because several catabolic functions that were shown to be Ti plasmid-encoded are for some *Agrobacterium* strains additionally chromosomally encoded.

The mapping of these functions on the plasmids became possible through the development of the transposon insertion mutagenesis techniques (Hernalsteens *et al.*, 1978; Holsters *et al.*, 1978; Garfinkel and Nester, 1980; Ooms *et al.*, 1980; De Greve *et al.*, 1981) or deletion mutagenesis (Koekman *et al.*, 1979; Holsters *et al.*, 1980). All the mutations were localized on the physical maps for nopaline (Depicker *et al.*, 1980) or octopine (De Vos *et al.*, 1981) Ti plasmids, thus establishing functional genetic maps for these plasmids. A summary of these results is diagrammed in Fig. 2.

A number of general conclusions can be derived from this genetic analysis of Ti plasmids. Mutants of all the expected Ti-determined phenotypes have been found and assigned to specific regions of the Ti plasmid maps. Several regions of Ti plasmids must be directly or indirectly involved in determining the different steps leading to neoplastic transformation. Indeed, mutants affecting so-called Onc (oncogenicity) functions are distributed over about one-half of the Ti plasmid map; the other half does not appear to carry any genes involved in transformation. No Onc$^-$ mutants were observed in this region of various Ti plasmids (right half of the maps in Fig. 2), and large deletions, eliminating most of this region, were obtained that do not affect crown gall formation by these plasmids (Koekman *et al.*, 1979).

The most important conclusion of this work was that the non-T-DNA part of the Ti plasmid also contained extensive regions essential for tumor induction. These so-called Onc regions seem to be conserved among nopaline and octopine plasmids, because Onc$^-$ insertion mutants were mapped in each of the four areas of homology between octopine and nopaline plasmids. Electron microscopic heteroduplex analyses reveal blocks of sequence equal to a total homology of approximately 30% (Engler *et al.*, 1981); these blocks of homology are contained in four regions, A, B, C, and D, diagrammed in black in Fig. 2. Some of these regions possibly encode functions essential for the transfer of the T-DNA into the plant nucleus. Other functions may interfere with the balance of growth factors (plant hormones) of the infected tissue and may therefore be essential for the stimulation of cell proliferation. It has been shown that *A. tumefaciens* strains excrete trans-zeatin and that this production is Ti plasmid-determined (Kaiss-Chapman and Morris, 1977; Claeys *et al.*, 1978; McCloskey *et al.*, 1980). Ti plasmid deletion and insertion mutants that no longer mediate trans-zeatin synthesis confer an Onc$^-$ phenotype (R. Morris and E. Messens, personal communication).

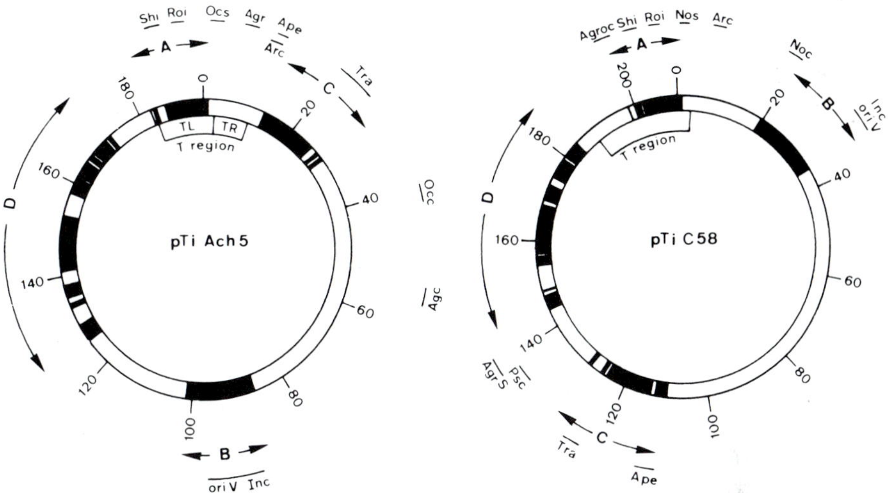

Fig. 2 Functional maps of standard octopine (Ach5) and nopaline (C58) Ti plasmids. The four areas of homology, A, B, C, and D, are indicated by the black regions. The locations of different functional regions common to both plasmids are also indicated: Shi, shoot induction; Roi, root induction; Tra, transfer function; Inc, incompatibility; Ape, phage exclusion; OriV, origin of replication; Arc, arginine catabolism. The regions that are specific for each type of plasmid are also indicated: Nos, nopaline synthesis; Noc, nopaline catabolism; Agroc, agrocinopine synthesis; Psc, phosphorylated sugar catabolism; Agr^S, agrocin sensitivity; Ocs, octopine synthesis; Occ, octopine catabolism; Agr, agropine synthesis; Agc, agropine catabolism. The numbers indicate the relative size of the plasmids in kilobase pairs. The two T-DNA regions are indicated at the top of the diagrams.

Interestingly some Onc mutants (both inside and outside the T region) suggest a host-specific interaction. For example, some mutants affect functions that are dispensable for tumor formation on *Kalanchoë* but not on tobacco or potatoes. There is no information about the products encoded by Onc mutants outside of the T region except that several of these DNA segments, when cloned in *Escherichia coli* plasmids, synthesize proteins in minicells (P. Dhaese, unpublished data).

Ti plasmids, like most large bacterial plasmids, were found to be conjugative or autotransferable (Kerr *et al.*, 1977; Genetello *et al.*, 1977; Holsters *et al.*, 1978), that is, they promote their own transfer from one bacterial host to another. It was therefore to be expected that Ti plasmids would carry an elaborate set of genes (so-called *tra* genes) to control this transfer. Tra^- plasmid mutants were obtained by deletion and insertion mutagenesis and mapped in at least two different regions of nopaline Ti plasmids (Holsters *et al.*, 1980). It had been suggested that some of the *tra* functions essential for bacterial conjugation would also be involved in the transfer of plasmid DNA to plant cells (Tempé *et al.*,

1977). However, most of the Tra$^-$ mutations obtained thus far are still oncogenic, indicating that the transfer of plasmid DNA to plants occurs normally, despite the fact that bacterial conjugation is impaired.

The catabolic functions of Ti plasmids provide *Agrobacterium* with the capacity to use opines as the sole carbon and/or nitrogen energy source. These functions have been mapped (see Fig. 2), and the plasmid DNA fragments thus identified have been isolated and propagated by molecular cloning techniques. Their expression as pBR322 mini-Sa clones (for the vector description see Leemans *et al.*, 1981) in Ti plasmid-free *Agrobacterium* was studied (E. Van Haute, unpublished results) and points to the existence of a complex operon structure possibly containing both positive and negative controlling elements. Such an operon probably consists of genes for a permease (Klapwijk *et al.*, 1977), the octopine or nopaline dehydrogenase, and enzymes involved in the further degradation of arginine and ornithine (Ellis *et al.*, 1979).

Mutants defective in the catabolism of agropine map in a Ti plasmid segment distinct from the segment involved in octopine catabolism. Likewise, the genes involved in the uptake and degradation of agrocinopine (Ellis and Murphy, 1981) by nopaline-type Ti plasmids map in a segment opposite the segment involved in nopaline catabolism. The degradation of the different opines is therefore controlled by a different set of genes, and each Ti plasmid may harbor catabolic functions for more than one type of opine. Plasmids that code for the degradation of a given opine, but do not harbor genes specifying the synthesis of this opine in their T region, have been observed and can usually be regarded as naturally occurring mutant plasmids. Interestingly, the genes in the T region involved in opine synthesis and the various genes outside the T region involved in opine and amino acid catabolism have different (nonhomologous) DNA sequences in different Ti plasmid types. On the other hand, most of the *tra* functions (involved in the conjugative transfer of Ti plasmids) are localized in DNA sequences conserved among different types of Ti plasmids. The Ti plasmids studied thus far turn out to be repressed for autotransferability, that is, the conjugative *Tra* functions are normally repressed but are induced by the presence of specific opines, thus further stressing the central role of opine synthesis and catabolism in the evolution of Ti plasmids (Petit *et al.*, 1978; Klapwijk *et al.*, 1978; Kerr and Ellis, 1982).

Another argument in favor of the generality of the opine concept stems from the study of the hairy root tumors induced by *A. rhizogenes* strains. The disease is characterized by abundant proliferation of roots at the wound sites. Large Ti-like plasmids were shown to be responsible for the rhizogenicity (Moore *et al.*, 1979; White and Nester, 1980) because the transfer of *rhizogenes* plasmids to a Ti plasmidless *A. tumefaciens* strain resulted in receptor strains that induce hairy root disease. These *rhizogenes* plasmids are compatible with Ti plasmids (Costantino *et al.*, 1980) and show only a limited base sequence homology with Ti plasmids (White and Nester, 1980). The fact that the T-DNA of a *rhizogenes* was shown to contain only the TR segment (see Section III,A, B) of an octopine T-

DNA (Willmitzer *et al.*, 1982b) is of particular interest. Despite these differences with Ti plasmids, *A. rhizogenes* strains were shown to be capable of utilizing the opine agropine, and J. Tempé (unpublished data) has shown that hairy roots in axenic culture do produce agropine. Again, opines are really a basic part of the plant cell transformations induced by bacterial plasmid genes.

III. T-DNA Transfer from the Ti Plasmid to the Nucleus of Plant Cells

A. The T Region of Ti Plasmids

The T region is defined as that segment of the Ti plasmid that is homologous to sequences present in crown gall cells. The sequences that are transferred from the Ti plasmid to the plant and determine tumorous growth have been called *T-DNA*. The T region of octopine and nopaline Ti plasmids has been studied in detail, both physically and functionally, and Fig. 3 summarizes the results. The T region is only a portion of the entire plasmid, roughly 23 kb in size (Lemmers *et al.*, 1980; Thomashow *et al.*, 1980a; De Beuckeleer *et al.*, 1981; Engler *et al.*, 1981). Southern blotting and cross-hybridization of restriction endonuclease digests of the two types of plasmids, as well as electron microscopic heteroduplex analyses, have revealed that 8–9 kb of the T-DNA regions are conserved (Chilton *et al.*, 1978; Depicker *et al.*, 1978; Engler *et al.*, 1981); these sequences are represented as the shaded areas in the figure.

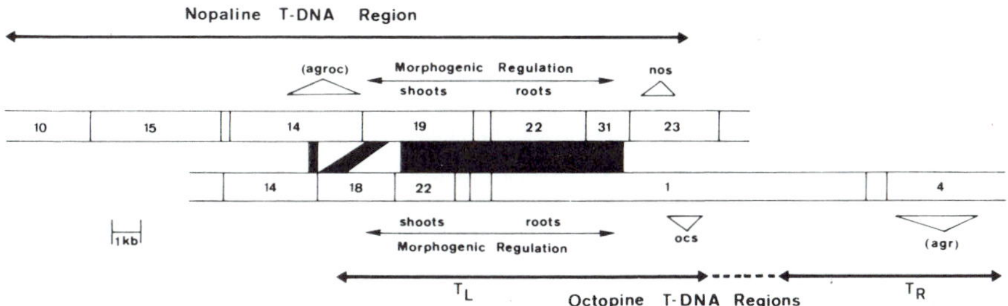

Fig. 3 The T regions of the nopaline and octopine Ti plasmids. The T regions of the two types of plasmids are aligned; shaded areas indicate the common DNA. The numbers refer to fragments produced following digestion with restriction endonuclease *Hin*dIII (Depicker *et al.*, 1980; De Vos *et al.*, 1981). The octopine T-DNA region is shown as two interrupted regions because sometimes only the left part is found in transformed plant cell DNA (see text). The abbreviations refer to regions controlling the synthesis of opines: nos, nopaline synthesis; ocs, octopine synthesis; agr, agropine synthesis; agroc, agrocinopine synthesis. The locations of agr and agroc are preliminary assignments (J. Ellis, H. Joos, and J. Leemans, unpublished results).

The T region has been subjected to intense mutagenesis in the hope of defining the functions encoded by this DNA. In addition, such analyses establish regions of similarity between octopine and nopaline T-DNA (Holsters *et al.*, 1980; De Greve *et al.*, 1981; Ooms *et al.*, 1981; Leemans *et al.*, 1981; Garfinkel *et al.*, 1981). The mutants so far have revealed that there are regions that either (*1*) greatly affect the ability of mutant Ti plasmids to form tumors, (*2*) are not required for oncogenicity, (*3*) alter the host range, (*4*) no longer synthesize opines, or (*5*) alter the morphology of tumors to produce excessive shoots or (*6*) roots. Those mutants that allow shoot formation are especially interesting and will be exploited in the T-DNA-mediated transfer of foreign DNA to plants (see Section IV,B). Many of these phenotypes are closely related such that those mutants that regulate the morphology of transformed cells also affect either the efficiency of tumor formation or the response of different plant species.

If one looks at the functions encoded by the T-DNA, it is interesting to see that almost one-half of the T-DNA is involved in the regulation of plant growth and development; it is almost tailor-made to be situated in its host. Perhaps *Agrobacterium* has captured plant genes that it then utilizes to its advantage upon T-DNA transfer. So far, all attempts to reveal homology between the T-DNA region and plant DNA have failed. The study of the growth-regulating functions encoded by the T-DNA provides a novel probe for similar endogenous functions encoded by the plant cell itself. One can also hope to exploit the vector properties of the T-DNA to insert other plant growth regulators of choice replacing the T-DNA-encoded functions.

Studies of the T-DNA functions reveal important information on the general physiology of tumors and may help elucidate the tumorous phenotypes, but they reveal nothing about the mechanism of T-DNA transfer. In addition, a large portion of the T-DNA can either be deleted or mutated and T-DNA transfer still occurs (see Section III,E). The knowledge of T-DNA transfer consists entirely of physical studies that either determine the limits of the T-DNA borders in plant DNA by restriction endonuclease mapping and hybridization to Ti plasmid specific probes or define these borders more precisely by nucleotide sequence analysis.

B. Comparison of the T-DNA in Independent Crown Gall Lines

The transfer of DNA from *Agrobacterium* to plant cell DNA includes at least two stages: A primary interaction between bacterial and plant cell walls presumably leads to the transfer of the whole or part of the Ti plasmid to the plant cell, and a secondary reaction must involve the insertion of the T-DNA of the Ti plasmid into the plant cell genome. One possibility for elucidating this event is to compare the T region of the Ti plasmid in bacteria with the final T-DNA struc-

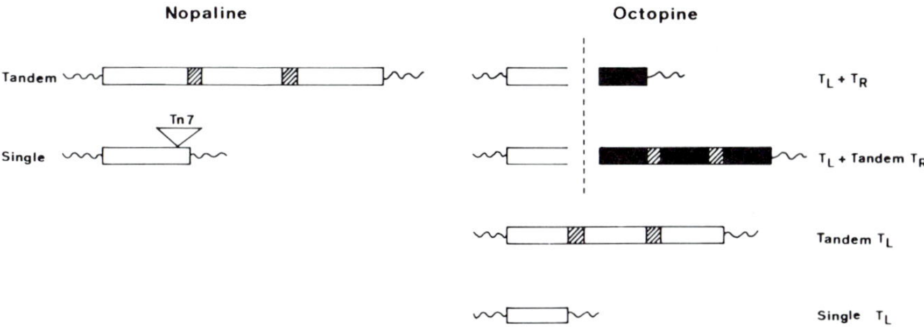

Fig. 4 The T-DNAs of nopaline- and octopine-transformed plant cell lines. All tandem T-DNAs are in direct orientation. The wavy lines indicate plant DNA sequences. The crosshatched areas indicate regions that may undergo sequence rearrangements. TL (in white) and TR (in black) refer to the left and right T-DNAs found in octopine tumor lines. It is unknown whether TL and TR are linked in the plant chromosome, as indicated by the dotted line. Multiple copies of TR have been observed only in one tumor cell line; we suggest that these are also arranged in a tandem array, although there has been no detailed analysis of this T-DNA structure.

ture after it has arrived in the plant. Figure 4 summarizes the T-DNAs of octopine and nopaline tumor lines.

Although only four nopaline tumor lines have been analyzed in detail (Lemmers *et al.*, 1980; Zambryski *et al.*, 1980, 1982; Yadav *et al.*, 1980), the data suggest that the mechanism of T-DNA transfer is rather precise because the same continuous segment of the Ti plasmid is always present. Some nopaline tumor lines appear to contain a single T-DNA copy. In two tumor lines the T-DNA is in multiple copies organized in a tandem array; this was demonstrated most clearly by molecular cloning of T-DNA fragments from transformed plant cell DNA that contains sequences derived from the right and left ends of the T-DNA region of the Ti plasmid. To date we have more detailed information on the boundaries of nopaline T-DNAs because the nucleotide sequence of these borders has been determined by comparison with the nucleotide sequence of the same region of the nopaline Ti plasmid (see Section III,C). Sequence analysis of left–right junctional clones derived from the tandemly organized T-DNAs has revealed that there can be reorganization of the sequences contained at the ends of the T-DNA; this reorganization is indicated by the crosshatched areas in Fig. 2. This reorganization is limited to the ends of the T-DNA. For example, in one clone there are 136 bp at the left–right junction that are not continuous with the ends of the T-DNA; however, almost all of these 136 bp can be found to be derived from sequences within 50 bp of the ends of the T-DNA region of the Ti plasmid (Zambryski *et al.*, 1980).

Thus, the nopaline T-DNA as shown here is of two forms: a tandem array with the altered junctional sequence indicated by the crosshatched region, and a single T-DNA copy. As an example of a single T-DNA copy, we employ the tumor line that was induced using a Ti plasmid containing the transposon Tn*7* in the right end of the T region (Lemmers *et al.*, 1980; Hernalsteens *et al.*, 1980; Holsters *et al.*, 1980). In this tumor line, T-DNA is enlarged by the exact size of 14.25 kb of Tn*7*, and molecular cloning suggests that the borders of this T-DNA occur in the same region of Ti plasmid sequences as observed for wild-type nopaline Ti plasmids (Holsters *et al.*, 1982b). The generation of tandem copies of T-DNA in some nopaline tumor lines and the integration of internally enlarged T-DNAs with similar boundaries suggest that the ends of the T-DNA in nopaline tumor lines are well defined.

Similar data obtained for several octopine tumor lines suggest that the octopine T-DNA is more variable (Thomashow *et al.*, 1980a,b; De Beuckeleer *et al.*, 1981). A left T-DNA region (TL), containing the sequence also conserved in the nopaline T-DNA region, is always present; this region is usually 12 kb in size, but at least one *Petunia* tumor line contains a TL shortened at the right end by about 4 kb (De Beuckeleer *et al.*, 1981). In addition, there is often a 6–7-kb right T-DNA region (TR) that contains sequences that are adjacent but not contiguous in the octopine Ti plasmid, and in one tumor line TR is amplified (Merlo *et al.*, 1980; Thomashow *et al.*, 1980a). Recent observations indicate that TL can also be a part of a tandem array (M. Holsters, unpublished results). The dotted vertical line in Fig. 4 indicates that it is unknown whether TL and TR are linked together in the plant chromosome. Both the interruption of Ti plasmid sequences in the generation of TR and TL and the existence of a tandem TL and an amplified TR indicate that reorganization of octopine T-DNA sequences also occurs. Furthermore, even a tumor line that contains a single TL (Thomashow *et al.*, 1980b) has sequences at the right border that are reorganized; this border sequence is homologous to an internal segment of this T-DNA. The areas between tandem octopine T-DNA copies are also crosshatched in Fig. 4 to suggest rearrangements analogous to those found in nopaline T-DNA.

It is obvious from even this simplified summary that the T-DNAs of nopaline and octopine tumor lines are not identical. However, there are some similarities: Both T-DNAs contain a common and highly conserved DNA; both T-DNAs can be amplified either in whole, as in nopaline T-DNA, or in part, as in octopine T-DNA; and both appear to undergo sequence rearrangements at the borders. It is not known whether the reorganization occurs as a consequence of the initial integration of T-DNA or as a result of subsequent recombination events that lead to a stabilized structure. It is unknown whether the rearrangements are the result of plant- or Ti plasmid-specific functions, but it is not unlikely that both are involved. For example, the interruption of TR and TL in octopine T-DNA is more likely to be a property of the Ti plasmid, whereas amplification of T-DNA sequences may be due to plant functions. Site-specific amplification of DNA is a property common to prokaryotic and eukaryotic genomes, and it is not surprising that the introduction of foreign DNA into the DNA of a susceptible host results in some perturbation of the local environment.

Amplification of DNA occurs under widely different biological circumstances. For example, growth under the selective pressure of drugs or metabolic analogs results in gene duplication for prokaryotic drug resistance markers (reviewed in Anderson and Roth, 1977) or for the eukaryotic genes dihydrofolate reductase (tetrahydrofolate dehydrogenase) or aspartate carbamoyltransferase when cells are grown in analogs of these enzymes, methotrexate or PALA [*N*-(phosphoracetyl)-L-aspartase] (Alt *et al.*, 1978; Schimke *et al.*, 1979, 1980; Milbrandt *et al.*, 1981; Wahl *et al.*, 1979). Amplification can also occur as part of a normal developmental program, as in the case of ribosomal DNA (reviewed in Long and Dawid, 1980), or chorion genes in the ovarian follicle cells of *Drosophila* (Mahowald *et al.*, 1979; Spradling and Mahowald, 1980; Spradling, 1981), or the DNA puffs of the salivary glands of flies of the Sciaridae family (Breuer and Pavan, 1955; Ficg and Pavan, 1957; Rudkin and Corlette, 1957; Glover *et al.*, 1982). Perhaps the most significant analogy to the generation of tandem copies of the T-DNA is the events that occur following transformation by the DNA tumor viruses SV40 (Sambrook *et al.*, 1980; Botchan *et al.*, 1976, 1979) or adenovirus (Sambrook *et al.*, 1980); here too, it is the addition of exogenous DNA that results in duplication. Simplistically, there appear to be three different situations that can result in genetic duplications: normal developmental programs, metabolic selection pressure, or the generation of an abnormal genotype, as in the case of the T-DNA integration.

The mechanism for the generation of multiple copies of T-DNA is unknown. However, there are two likely possibilities involving either DNA replication or recombination mechanisms. Local saltatory replication has been proposed for the generation of repeated integrated copies of SV40 and adenovirus (Botchan *et al.*, 1979; Sambrook *et al.*, 1980) or for the amplification of dihydrofolate reductase genes (Schimke *et al.*, 1980). Multiple copies of T-DNA may also be regenerated by unequal crossover events between homologous repetitive sequences, as has been proposed for the evolution of multigene families (Smith, 1974).

C. Nucleotide Sequence Analysis of the Borders of the T-DNA of Nopaline Tumors

Although it is useful to compare sequences between various nucleotide stretches to describe general similarities that can lead to models for synthesis or processing of nucleic acids, we do not yet have sufficient information to make elaborate models for T-DNA transfer. However, we present the results of sequence analyses to date, along with some of the possible implications or applications that may become useful as these studies continue.

Figure 5 is a summary of the sequence of the left and right border regions of the nopaline Ti plasmid (Zambryski *et al.*, 1980, 1982). The data from the nucleotide sequences of four right and four left T-DNA borders isolated from nopaline tumor lines suggest that the mechanism of T-DNA integration and subsequent stabilization is precise on the right and less precise on the left. The

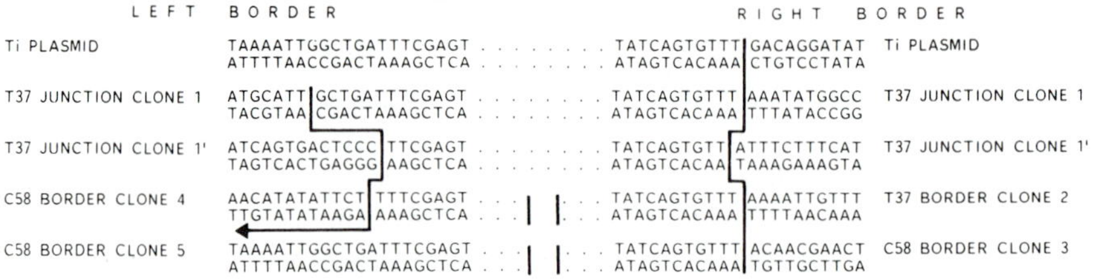

```
                        LEFT BORDER                                      RIGHT BORDER
Ti PLASMID              TAAAATTGGCTGATTTCGAGT . . . . . . . . . TATCAGTGTTT|GACAGGATAT  Ti PLASMID
                        ATTTTAACCGACTAAAGCTCA . . . . . . . . . ATAGTCACAAA|CTGTCCTATA

T37 JUNCTION CLONE 1    ATGCATT|GCTGATTTCGAGT . . . . . . . . . TATCAGTGTTT|AAATATGGCC  T37 JUNCTION CLONE 1
                        TACGTAA|CGACTAAAGCTCA . . . . . . . . . ATAGTCACAAA|TTTATACCGG

T37 JUNCTION CLONE 1'   ATCAGTGACTCCC|TTCGAGT . . . . . . . . . TATCAGTGTT|ATTTCTTTCAT  T37 JUNCTION CLONE 1'
                        TAGTCACTGAGGG|AAGCTCA . . . . . . . . . ATAGTCACAA|TAAAGAAAGTA

C58 BORDER CLONE 4      AACATATATTCT|TTTCGAGT . . . | | . . . TATCAGTGTTT|AAAATTGTTT  T37 BORDER CLONE 2
                        TTGTATATAAGA|AAAGCTCA . . . | | . . . ATAGTCACAAA|TTTTAACAAA
                        <-----------

C58 BORDER CLONE 5      TAAAATTGGCTGATTTCGAGT . . . | | . . . TATCAGTGTTT|ACAACGAACT  C58 BORDER CLONE 3
                        ATTTTAACCGACTAAAGCTCA . . . | | . . . ATAGTCACAAA|TGTTGCTTGA
```

Fig. 5 Nucleotide sequence analysis of the T-DNA borders in nopaline tumor cell DNA. The sequences at the left and right T-DNA borders from cloned DNA derived from crown gall tumor cells are compared with the sequences of the corresponding region of the Ti plasmid. The top line represents sequences from the Ti plasmid, and the corresponding sequences that occur at the left and right T-DNA borders are also aligned with the Ti-specific sequences. The sequence shown is a portion of that described by Zambryski *et al.* (1982). The vertical lines represent the point at which the clones diverge from the Ti plasmid. Clones 1, 1′, and 2 are derived from a tumor line of tobacco induced by nopaline strain T37, and clones 3, 4, and 5 are derived from another tumor line of tobacco induced by nopaline strain C58. The central portion of the T-DNA is indicated by a series of dots; however, the dots are meant to indicate the sequences in the left and right sides of the same T-DNA copy for cg clones 1 and 1′ only; clones 2, 3, 4, and 5 are not from the same T-DNA array.

arrows indicate the nucleotide sequence at which the T-DNA has been found to diverge from the Ti plasmid sequences. The T-DNA–plant DNA junction on the right has been shown to occur at exactly the same base pair with respect to Ti plasmid sequences in the T-DNA isolated from two independent tumor lines. Only a variation of a single base was found in the analysis of the right border of a junctional region containing the left–right T-DNA boundary from a T-DNA that was part of a tandem array. All the left T-DNA boundaries we have determined so far are different, as indicated. The results from the sequence analysis do not reveal how T-DNA transfer has occurred; nevertheless, the data suggest that well-defined regions of the Ti plasmid are involved in this process.

The nucleotide sequence of the T-DNA region around the borders of the nopaline Ti plasmid has also been determined (Zambryski *et al.*, 1982). It is easy to find the region of interest on the right side of the T-DNA region because the T-DNA border on this side varies over only a single base pair (as indicated in Fig. 5). On the left, we have found three borders spanning a region of 6 bp. However, there are two other left T-DNA borders that can occur farther to the left (C58 border clone 5, Fig. 5) (Yadav *et al.*, 1982). One of these has been sequenced (Yadav *et al.*, 1982) to reveal a T-DNA border 96 bp to the left of those shown here.

One can compare the sequences within 100 bp of either the right T-DNA border region or the left T-DNA border region of the nopaline Ti plasmid. There is a short direct repeat of 25 bp with four mismatched bases that is found on either side of the T-DNA region of the nopaline Ti plasmid. The 25 bp have the sequence

```
  A                    G C   G
T G G C A G G A T A T A T T G T G G T G T A A C
A C C G T C C T A T A T A A C A C C A C A T T G
  T                    C G   C
```

This sequence is found to start just 1 bp inside the T-DNA border on the right side of the nopaline Ti plasmid and 108 bp to the left of the left T-DNA border (shown in Fig. 5) on the left side of the nopaline Ti plasmid. The left T-DNA border of Yadav *et al.* (1982) in fact ends within this 22-bp sequence (to exclude the first 15 bp). A nearly identical sequence has been found close to the left border (Simpson *et al.*, 1982; M. Holsters, R. Villaroel, H. De Greve, J. Gielen, M. Van Montagu, and J. Schell, unpublished data) and the right border (M. Holsters, R. Villaroel, H. De Greve, J. Gielen, M. Van Montagu, and J. Schell, unpublished data) region of the TL-DNA of octopine Ti plasmids. The significance of the 25-bp repeat in T-DNA transfer is unknown; however, if it is involved, it is not typical of sequences used by transposable elements, because it is not always found in the transferred DNA segment. Normally repeated sequences form the exact ends of the transposable element. Future studies of mutants in which these Ti sequences are specifically altered may establish whether they are fundamental to T-DNA transfer.

There is also a sequence just adjacent to the 25-bp direct repeat at the left border of the T-DNA region of the nopaline Ti plasmid that is an exact Chi sequence ($^{\text{GCTGGTGG}}_{\text{CGACCACC}}$) (Stahl *et al.*, 1980; Smith *et al.*, 1981). Chi mainly stimulates recombination in prokaryotes and has the rather arcane property of acting to influence the recombinogenic behavior of DNA that is found at a distance from its own location. Chi has even been proposed to be involved in the synthesis of hybrid immunoglobulins (Kenter and Birshtein, 1981). Whether or not this Chi sequence plays a role in T-DNA transfer has not been determined, but its presence can alert us to the possibility of such elements that can have an influence on T-DNA integration even though they may be at a distant site.

The one general conclusion from the nucleotide sequence analysis of the T-DNA junctions is that these sequences are enriched for A-T base pairs at the immediate junctional site (Fig. 5). This high A-T content of the integration site is a more or less universal observation for most movable DNA; this includes bacteriophage λ integration into bacterial DNA (Landy and Ross, 1977; Pinkham *et al.*, 1980), the sites favored for integration of prokaryotic (Kühn *et al.*, 1979; Van Emmelo *et al.*, 1980; Heffron *et al.*, 1981; Miller *et al.*, 1980; Reed, 1981) and eukaryotic transposons (Farabaugh and Fink, 1980; Roeder *et al.*, 1980; Dunsmuir *et al.*, 1980), as well as the sites for retrovirus integration (Majors and Varmus, 1981; Hishinuma *et al.*, 1981).

The existence of sequence rearrangements at the ends of the T-DNA in nopaline tumor DNA also suggests that the observed T-DNAs were once part of a longer T-DNA copy; the rearranged DNA in the junctional sequence between tandem T-DNA copies is composed of some sequences that are found in the Ti plasmid just outside the observed T-DNA borders (Zambryski *et al.*, 1982). It is likely that some of the observed boundaries are the result of recombination and sequence rearrangements that lead to a stabilized structure. It is also likely that the observed T-DNA border structures may depend on the region of the plant genome that was a site for the T-DNA integration. To date, the T-DNA has demonstrated no particular preference for classes of plant DNA; it has been shown to be integrated adjacent to either unique or several classes of repeated sequences (Zambryski *et al.*, 1980, 1982; Thomashow *et al.*, 1980b; Yadav *et al.*, 1980). All of these studies were carried out using tumor lines of tobacco that contains more than 70% of its DNA in repeated sequences (Zimmerman and Goldberg, 1977). Not only the local DNA environment but also the general physiology of the cell may influence either the regions of the genome that are available for integration or the final stabilized T-DNA structure. For example, it is becoming established that the physiology of plant growth substances is important in the formation of crown gall tumors; there are mutants in the regions of the T-DNA controlling the ability to form roots or shoots in infected plants that are only weakly tumorigenic. However, the addition of exogenous auxins or cytokinins to such mutants at the moment of tumor induction fully restores the transforming ability of these mutant Ti plasmids (Ooms *et al.*, 1981; Van Montagu and Schell, 1981).

D. T-DNA Compared to Other Mobile Elements

A comparison of the sequences found in the region of the T-DNA boundaries reveals a lack of the properties typically observed for other transposable elements. So far, there is no evidence for sequences in inverted orientation at the ends of the T region in the Ti plasmid or the T-DNA *in planta*. None of the clones that have been isolated to date contain a complete T-DNA copy; there is therefore no evidence for the generation of short repeated sequences in the target DNA (Grindley and Sherratt, 1978; Shapiro, 1979). In view of the sequence rearrangements that occur at the ends of the T-DNA, such repeats may be difficult to observe; however, it would not be unexpected that a staggered cut is introduced into the host sequence during an initial step in integration. Perhaps the T-DNA most closely resembles other elements, both prokaryotic and eukaryotic, in the ability to induce sequence rearrangements in its neighborhood. These rearrangements have been shown at the nucleotide sequence level in many prokaryotic systems, for example, IS*1*, IS2, Tn9, Tn*3*, mu, and Tn*10* (Reif and Saedler, 1975; Nevers and Saedler, 1978; Besemer *et al.*, 1980; Ghosal *et al.*, 1979; Calos and Miller, 1980; Ohtsubo *et al.*, 1979; Faelen and Toussaint, 1978; Kleckner, 1979). Sequence rearrangements have also been observed to occur in eukaryotic systems in the integrated SV40 or adenovirus DNA in transformed animal cell lines. For example, there are tracts of viral sequences that sometimes become inverted with respect to the normal genomic sequences (Sambrook *et al.*, 1980; Deuring *et al.*, 1981), and there is also evidence that viral sequences, along with flanking cell DNA, can be amplified after integration (Sambrook *et al.*, 1980). Of course, the instability of particular loci in complex systems such as *Drosophila* (see Chapters 8 and 9) or corn (see Chapter 1) has now firmly been established to be caused by the presence of movable elements. So far, we have been able to observe these happenings only by observing the ends of the T-DNA; but in addition, there is preliminary evidence to suggest that the plant sequences near the T-DNA may also undergo reorganization (P. Zambryski, unpublished results). Here, the unique plant DNA contained in one T-DNA border clone was used as a hybridization probe either to untransformed callous DNA or to tumor DNA. Only a single DNA fragment hybridized in callous DNA, whereas this fragment plus an additional fragment was observed to hybridize in tumor DNA.

E. Transfer of the T-DNA Region in the Case of Ti Plasmids Mutated in This Region

So far, mutagenesis by the insertion of transposons within 100 bp of the left or right ends of the nopaline T-DNA region (D. Inzé, A. Caplan, and C. Genetello, unpublished results) does not affect the efficiency of T-DNA transfer of the altered Ti plasmids. The entire common DNA of the T region of either octopine (Garfinkel *et al.*, 1981; R. Deblaere, unpublished results) or nopaline (D. Inzé, G. Van den Broeck, G. Gheysen, and M. Sormann, unpublished results) Ti plasmids have been saturated by transposon insertions, and no mutant is altered in T-DNA transfer. Even large deletions within the T regions still produce transformed plant cells (Leemans *et al.*, 1982; Joos *et al.*, 1983). In fact, no transfer negative mutants in the T-DNA region have been observed to date. Only a Tn*1*-induced deletion of the right T-DNA border of a nopaline Ti plasmid has been observed to have a lowered frequency of tumor induction (Holsters *et al.*, 1980); here the right T-DNA border is formed by the Tn*1* itself (Lemmers *et al.*, 1980). On the other hand, deletion of the right border of the TL region of an octopine plasmid does not affect the oncogenicity of this plasmid (Leemans *et al.*, 1981); here it may be expected that TL and TR will be fused in tumor cell DNA. In the case of the octopine T-DNA, there are several possible borders such that the deletion of one may not have an effect. The definition of the regions that are required for effective integration of the T-DNA is important if the T-DNA is to be utilized as a vector for the transfer of foreign DNA to plants of interest.

F. Use of Ti Plasmids as Experimental Gene Vectors

Although we lack information on the exact mechanism of T-DNA transfer, we can still take advantage of the properties described above to begin to utilize the Ti plasmid as a vector for gene transfer to plants. As mentioned above, a Ti plasmid carrying the transposon Tn7 in the T-DNA region has been utilized to transform plant cells with T-DNA containing Tn7. Furthermore, this tumor cell line has

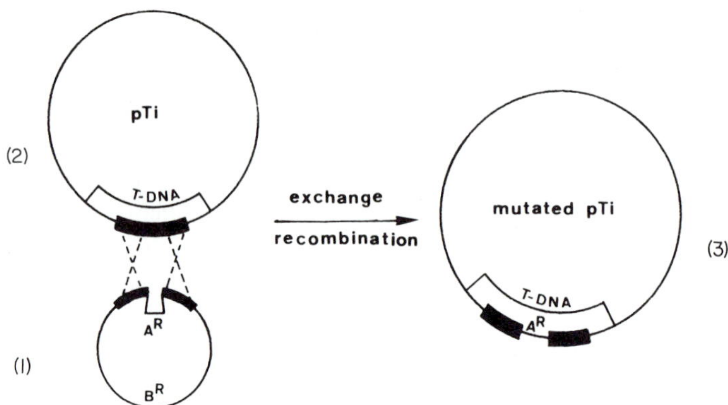

Fig. 6 Intermediate vector–Ti plasmid interaction. An intermediate vector is a pBR-derived recombinant molecule containing a segment of the T-DNA that includes the site and the sequence into which a "foreign DNA," A^R, has to be inserted in order to allow expression (1). The insertion site is flanked by a stretch of T-DNA of appropriate length. In the presence of the Ti plasmid (2), a double crossover allows an exchange of a fragment containing the insert with the unaltered homologous part of this plasmid, resulting in the formation of transfer vector (3).

been shown to synthesize transcripts specific to Tn7 (L. Willmitzer, L. Otten, and J. Schell, unpublished results). Whether or not the Tn7 can be expressed to provide functional resistance to methotrexate has not yet been firmly established, but some Tn7 containing tobacco crown gall lines were found to be capable of growing in the presence of methotrexate. No such growth was observed in normal tobacco crown gall lines. A genomic clone from this tumor line contained an intact copy of Tn7 that transposed in *E. coli* with normal frequency (2×10^{-4}). This observation clearly demonstrates that the bacterial transposon Tn7 underwent no changes during its stay in the plant genome and can be preserved both physically and functionally intact as part of the T-DNA (Holsters *et al.*, 1982b).

The expression of the T-DNA sequences is being intensely investigated. So far, information is available on the overall map position, the size, and the number of different RNAs expressed in tumor cells. Some of the most active regions of the T-DNA are those involved in opine biosynthesis; the complete nucleotide sequences of the genes encoding nopaline synthase and octopine synthase have been recently determined (Depicker *et al.*, 1982; De Greve *et al.*, 1982a). One can now begin systematically to insert genes of interest behind the nopaline or octopine promoters to assay for their expression in transformed plant cells.

So far, there is no "mini-Ti" cloning vector that would be useful for genetic and recombinant DNA manipulations and still contain all the functions required for T-DNA transfer. However, the use of "intermediate vectors" (Leemans *et al.*, 1981) provides a convenient and flexible method for the *in vitro* introduction of selected genes into the T-DNA region of the selected Ti plasmid (Fig. 6). The principle consists of utilizing a common *E. coli* cloning vehicle containing a fragment of the T-DNA region; this fragment may either be mutated or utilized to insert DNA. The intermediate vector is introduced into *Agrobacterium,* and recombination *in vivo* will transfer the DNA of interest into the appropriate site of the Ti plasmid. This technique has been successfully used to insert DNA into any part of the T-DNA region that is contained within the borders defined above.

IV. T-DNA as a New Chromosomal Locus in Transformed Plant Cells

A. Isolation of Plant Cells Harboring Partially Deleted T-DNA Segments

A detailed genetic analysis of deletion mutants of the TL region of octopine crown galls (Leemans *et al.,* 1982), when correlated with an analysis of the number, size, and map positions of TL-DNA-derived, polyadenylated transcripts (Willmitzer *et al.,* 1982a), has allowed us to assign functions to some of the seven different, well-defined transcripts coded for by TL-DNA.

The two most abundant transcripts, transcripts 7 and 3, with 670 and 1400 nucleotides, respectively, are specific for octopine tumors. Transcript 3 was found to code for the enzyme octopine synthase. The function of transcript 7 is still unknown. The other five transcripts are derived from TL-DNA sequences that are homologous with an equivalent region in the T-DNA of nopaline tumors (Willmitzer *et al.,* 1983). The sizes of transcripts 1, 2, 4, 5, and 6 are 2700, 1600, 1200, 1000, and 900 bp, respectively.

By observing the properties of crown gall tumors obtained by infection of tobacco with mutant octopine Ti plasmids carrying deletions of specified segments of the Ti region, it was possible to assign functions to each of the transcripts 1, 2, 4, and 6 and to suggest a possible function for transcript 5 (Leemans *et al.,* 1982).

The main conclusions from these studies were as follows.

1. Most, if not all, of the transcripts were expressed from individual promoters.
2. T-DNA transfer and tumorous growth are controlled by different and independently acting functions.
3. None of the TL-DNA derived transcripts are essential for T-DNA transfer.
4. Transcripts 1, 2, 4, 6, and possibly also 5 act by suppressing organ development.

5. Shoot formation (transcripts 1 and 2) and root formation (transcript 4) are suppressed by the action of different transcripts.
6. Transcripts 4 and 6 are sufficient to maintain the tumorous growth properties of transformed tobacco cells.

B. Regeneration of Fertile Plants Containing a Partially Deleted T-DNA Segment

If these conclusions regarding transcripts 1, 2, 4, 5, and 6 are correct, it follows that elimination of the coding sequences for these transcripts from the T-DNA of transformed cells should result in tobacco cells containing and expressing some of the T-DNA genes, for example, transcript 3 (thus producing octopine synthase activity), but no longer suppressed for organ development. Such mutant TL-DNAs were recently obtained (H. De Greve *et al.*, 1982b; Schell *et al.*, 1982). As predicted, they were shown to form fertile plants with normal organ development and to express transcript 3 in most, if not all, of their cells.

The occurrence of such regenerated plants was, however, a rare event because it was dependent on the formation of additional deletions in the plant by as yet unknown and thus uncontrolled mechanism(s). The first plant regenerated from a cell that had deleted most of the T-DNA, leaving only a 3.5-kb fragment in which the octopine synthase was the only gene present in intact form. This deletion occurred upon infection of the tobacco shoot with an octopine strain harboring a Tn*7* insertion in the T-DNA. Insertion mutations in this region of the T-DNA irrevocably result in a proliferation of shoots that develop from the surrounding untransformed cells. Screening these shoots for the presence of octopine synthase activity resulted in the identification of the rare regenerating transformants. Until now, nine such regenerated plants, formed from independent induced tumor lines, have been analyzed, and each time only transcript 3 was expressed.

It is clear that this information opens the way for construction of cloning vectors that will allow generation of transformed and fertile plants. However, before this can be achieved, it is essential to construct a marker gene that will allow the selection of transformed plant cells. Indeed, T-DNA-transformed plant cells are normally selected for by their hormone-independent growth characteristics. A deletion of these five important T-DNA transcripts will no longer allow this distinction. Hence, another selection scheme must be envisaged. Some bacterial antibiotic resistance genes are therefore good candidates because some have been shown to be expressed in yeast (Jimenez and Davies, 1980) or mammalian cells (Colbère-Garapin *et al.*, 1981). Now that DNA sequences of some T-DNA-encoded genes have been determined, it will be easy to link DNA-containing information for resistance with T-DNA signals for expression.

C. Mendelian Inheritance of T-DNA Linked Genes

A number of published observations (Braun and Wood, 1976; Turgeon *et al.*, 1976; Yang *et al.*, 1980) had been interpreted to indicate that T-DNA sequences in grafted teratomas were lost during meiosis in flowering grafts. This led to the important suggestion that T-DNA sequences might be excised during meiosis.

To test this possibility, sexual crosses were performed using flowering tobacco plants containing a partially deleted T-DNA and expressing transcript 3. It was found that the T-DNA was stably transmitted during meiosis, both through the pollen and through the eggs, and segregated as a single dominant Mendelian gene (Otten *et al.*, 1981).

V. General Conclusions

The crown-gall system is a well-documented, natural instance of a translocation system involving DNA transfer between prokaryotes and eukaryotes. The T-DNA is a translocating element present in a prokaryotic plasmid but consisting of DNA sequences that become functional genes in eukaryotic cells. It is interesting to speculate that other such natural instances of DNA translocations, involving widely different organisms, can be expected to occur in situations involving a close, long-term symbiotic or parasitic association between these different organisms.

References

Alt, F. W., Kellems, R. E., Bertino, J. R., and Schimke, R. T. (1978). Selective multiplication of dihydrofolate reductase genes in methotrexate-resistant variants of cultured murine cells. *J. Biol. Chem.* **253,** 1357–1370.

Anderson, R. P., and Roth, J. R. (1977). Tandem genetic duplications in phage and bacteria. *Annu. Rev. Microbiol.* **31,** 473–505.

Besemer, J., Görtz, G., and Charlier, D. (1980). Deletions and DNA rearrangements within the transposable element IS2. A model for the creation of palindromic DNA by DNA repair synthesis. *Nucleic Acids Res.* **8,** 5825–5833.

Bomhoff, G.,Klapwijk, P. M., Kester, H. C. M., Schilperoort, R. A., Hernalsteens, J. P., and Schell, J. (1976). Octopine and nopaline synthesis and breakdown genetically controlled by a plasmid of *Agrobacterium tumefaciens*. *Mol. Gen. Genet.* **145,** 177–181.

Botchan, M., Topp, W., and Sambrook, J. (1976). The arrangement of Simian Virus 40 sequences in the DNA of transformed cells. *Cell* **9,** 269–288.

Botchan, M. R., Topp, W. C., and Sambrook, J. (1979). Studies on SV40 excision from cellular chromosomes. *Cold Spring Harbor Symp. Quant. Biol.* **43,** 709–719.

Braun, A. C. (1953). Bacterial and host factors concerned in determining tumormorphology in crown gall. *Bot. Gaz. (Chicago)* **114,** 363–371.

Braun, A. C. (1969). "The Cancer Problem, a Critical Analysis and Modern Synthesis." Columbia Univ. Press, New York.

Braun, A. C. (1982). A history of the crown gall problem. *In* "Molecular Biology of Plant Tumors" (G. Kahl and J. Schell, eds.), pp. 155–208. Academic Press, New York.

Braun, A. C., and Mandle, R. J. (1948). Studies on the inactivation of the tumor-inducing principle in crowngall. *Growth* **12,** 255–269.

Braun, A. C., and White, P. R. (1943). Bacteriological sterility of tissues derived from secundary crown gall tumors. *Phytopathology* **33,** 85–100.

Braun, A. C., and Wood, H. N. (1976). Suppression of the neoplastic state with the acquisition of specialized functions in cells, tissues, and organs of crown gall teratomas of tobacco. *Proc. Natl. Acad. Sci. USA* **73,** 496–500.

Breuer, M. E., and Pavan, C. (1955). Behaviour of polytene chromosomes of Rhynchosciara angelae at different stages of larval development. *Chromosoma* **7,** 371–386.

Calos, M. P., and Miller, J. H. (1980). Molecular consequences of deletion formation mediated by the transposon Tn*9*. *Nature (London)* **285,** 38–41.

Chilton, M.-D., Drummond, M. H., Merlo, D. J., and Sciaky, D. (1978). Highly conserved DNA of Ti-plasmids overlaps T-DNA, maintained in plant tumors. *Nature (London)* **275,** 147–149.

Chilton, M.-D., Saiki, R. K., Yadav, N., Gordon, M. P., and Quetier, F. (1980). T-DNA from *Agrobacterium* Ti plasmid is in the nuclear DNA fraction of crown gall tumor cells. *Proc. Natl. Acad. Sci. USA* **77,** 4060–4064.

Chilton, M.-D., Tepfer, D. A., Petit, A., David, C., Casse-Delbart, F., and Tempé, J. (1982). *Agrobacterium rhizogenes* inserts T-DNA into the genomes of the host plant root cells. *Nature (London)* **295,** 432–434.

Claeys, M., Messens, E., Van Montagu, M., and Schell, J. (1978). GC/MS determination of cytokinins in *Agrobacterium tumefaciens* cultures. *Fresenius Z. Anal. Chem.* **290,** 125–126.

Cohen, H., and Strampp, A. (1976). Bacterial synthesis of substances similar to human chorionic gonadotrophin. *Proc. Soc. Exp. Biol. Med.* **152,** 408–410.

Colbère-Garapin, F., Horodniceanu, F., Kourilsky, P., and Garapin, A.-C. (1981). A new dominant hybrid selective marker for higher eukaryotic cells. *J. Mol. Biol.* **150,** 1–14.

Costantino, P., Hooykaas, P. J. J., den Dulk-Ras, H., and Schilperoort, R. A. (1980). Tumor formation and rhizogenicity of *Agrobacterium rhizogenes* carrying Ti-plasmids. *Gene* **11,** 79–87.

Currier, T. C., and Nester, E. W. (1976). Isolation of covalently closed circular DNA of high molecular weight from bacteria. *Anal. Biochem.* **76,** 431–441.

De Beuckeleer, M., Lemmers, M., De Vos, G., Willmitzer, L., Van Montagu, M., and Schell, J. (1981). Further insight on the transferred-DNA of octopine crown gall. *Mol. Gen. Genet.* **183,** 283–288.

De Cleene, M., and De Ley, J. (1981). The host range of infectious hairy-root. *Bot. Rev.* **47,** 147–194.

De Greve, H., Decraemer, H., Seurinck, J., Van Montagu, M., and Schell, J. (1981). The functional organization of the octopine *Agrobacterium tumefaciens* plasmid pTiB6S3. *Plasmid* **6,** 235–248.

De Greve, H., Dhaese, P., Seurinck, J., Van Montagu, M., and Schell, J. (1982a). Nucleotide sequence and transcript map of the *Agrobacterium tumefaciens* Ti plasmid-encoded octopine synthase gene. *J. Mol. Appl. Genet.* (unpublished).

De Greve, H., Leemans, J., Hernalsteens, J. P., Thia-Toong, L., De Beuckeleer, M., Willmitzer,

L., Otten, L., Van Montagu, M., and Schell, J. (1982b). Normal and fertile plants, expressing octopine synthase, regenerate from tobacco crown galls after deletion of tumour controlling functions. *Nature (London)* (unpublished).

Depicker, A., Van Montagu, M., and Schell, J. (1978). Homologous DNA sequences in different Ti-plasmids are essential for oncogenicity. *Nature (London)* **275,** 150–153.

Depicker, A., De Wilde, M., De Vos, G., De Vos, R., Van Montagu, M., and Schell, J. (1980). Molecular cloning of overlapping segments of the nopaline Ti-plasmid pTiC58 as a means to restriction endonuclease mapping. *Plasmid* **3,** 193–211.

Depicker, A., Stachel, S., Dhaese, P., Zambryski, P., and Goodman, H. M. (1982). Nopaline synthase: transcript mapping and DNA sequence. *J. Mol. Appl. Genet.* (unpublished).

De Vos, G., De Beuckeleer, M., Van Montagu, M., and Schell, J. (1981). Restriction endonuclease mapping of the octopine tumor inducing pTiAch5 of *Agrobacterium tumefaciens. Plasmid* **6,** 249–253.

Deuring, R., Winterhoff, U., Tamanoi, F., Stabel, S., and Doerfler, W. (1981). Site of linkage between adenovirus type 12 and cell DNAs in hamster tumour line CLAC3. *Nature (London)* **293,** 81–84.

Dunsmuir, P., Brorein, W. J., Jr., Simon, M. A., and Rubin, G. M. (1980). Insertion of the *Drosophila* transposable element copia generates a 5 base pair duplication. *Cell* **21,** 575–579.

Elkan, G. H. (1981). The taxonomy of *Rhizobiaceae*. *Int. Rev. Cytol. Suppl.* **13,** 1–14.

Ellis, J. G., and Murphy, P. J. (1981). Four new opines from crown gall tumours—their detection and properties. *Mol. Gen. Genet.* **181,** 36–43.

Ellis, J. G., Kerr, A., Tempé, J., and Petit, A. (1979). Arginine catabolism: a new function of both octopine and nopaline Ti-plasmids of *Agrobacterium*. *Mol. Gen. Genet.* **173,** 263–269.

Engler, G., Holsters, M., Van Montagu, M., Schell, J., Hernalsteens, J. P., and Schilperoort, R. A. (1975). Agrocin 84 sensitivity: a plasmid determined property in *Agrobacterium tumefaciens. Mol. Gen. Genet.* **138,** 345–349.

Engler, G., Depicker, A., Maenhaut, R., Villarroel-Mandiola, R., Van Montagu, M., and Schell, J. (1981). Physical mapping of DNA base sequence homologies between an octopine and a nopaline Ti-plasmid of *Agrobacterium tumefaciens*. *J. Mol. Biol.* **152,** 183–208.

Faelen, M., and Toussaint, A. (1978). Stimulation of deletions in the *Escherichia coli* chromosome by partially induced Mu*cts*62 prophages. *J. Bacteriol.* **136,** 477–483.

Farabaugh, P. J., and Fink, G. R. (1980). Insertion of the eukaryotic transposable element Ty*1* creates a 5-base pair duplication. *Nature (London)* **286,** 352–356.

Ficg, A., and Pavan, C. (1957). Autoradiography of polytene chromosomes of Rhynchosciara angelae at different stages of larval development. *Nature (London)* **180,** 983–984.

Firmin, J. L., and Fenwick, G. R. (1978). Agropine—a major new plasmid-determined metabolite in crown gall tumours. *Nature (London)* **276,** 842–844.

Garfinkel, D. J., and Nester, E. W. (1980). *Agrobacterium tumefaciens* mutants affected in crown gall tumorigenesis and octopine catabolism. *J. Bacteriol.* **144,** 732–743.

Garfinkel, D. J., Simpson, R. B., Ream, L. W., White, F. F., Gordon, M. P., and Nester, E. W. (1981). Genetic analysis of crown gall: fine structure map of the T-DNA by site-directed mutagenesis. *Cell* **27,** 143–153.

Genetello, Ch., Van Larebeke, N., Holsters, M., Depicker, A., Van Montagu, M., and Schell, J. (1977). Ti-plasmids of *Agrobacterium* as conjugative plasmids. *Nature (London)* **265,** 561–563.

Ghosal, D., Gross, J., and Saedler, H. (1979). DNA sequence of IS2-7 and generation of mini-insertions by replication of IS2 sequences. *Cold Spring Harbor Symp. Quant. Biol.* **43,** 1193–1196.

Glover, D. M., Zaha, A., Stocker, A. J., Santelli, R. V., Pueyo, M. T., de Toledo, S. M., and Lara,

F. J. S. (1982). Gene amplification in *Rhynchosciaria* salivary gland chromosomes. *Proc. Natl. Acad. Sci. USA* **79,** 2947–2951.

Gordon, M. P. (1981). Tumor formation in plants. *In* "The Biochemistry of Plants: A Comprehensive Treatise" (P. K. Stumpf and E. E. Conn, eds.), Vol. 6, "Proteins and Nucleic Acids" (A. Marcus, ed.), pp. 531–570. Academic Press, New York.

Gordon, M. P., Farrand, S. K., Sciaky, D., Montoya, A., Chilton, M.-D., Merlo, D., and Nester, E. W. (1979). The crown gall problem. *In* "The Molecular Biology of Plants" (I. Rubenstein, R. L. Phillips, C. E. Green, and B. G. Gengenbach, eds.), pp. 291–313. Academic Press, New York.

Grindley, N. D. F., and Sherratt, D. J. (1978). Sequence analysis at IS1 insertion sites: models for transposition. *Cold Spring Harbor Symp. Quant. Biol.* **43,** 1257–1261.

Guyon, P., Chilton, M.-D., Petit, A., and Tempé, J. (1980). Agropine in "null type" crown gall tumors: evidence for the generality of the opine concept. *Proc. Natl. Acad. Sci. USA* **77,** 2693–2697.

Hamilton, R. H., and Fall, M. Z. (1971). The loss of tumor initiating ability in *Agrobacterium tumefaciens* by incubation at high temperature. *Experientia* **27,** 229–230.

Heffron, F., Kostriken, R., Morita, C., and Parker, R. (1981). Tn*3* encodes a site-specific recombination system: identification of essential sequences, genes, and the actual site of recombination. *Cold Spring Harbor Symp. Quant. Biol.* **45,** 259–268.

Hernalsteens, J. P., De Greve, H., Van Montagu, M., and Schell, J. (1978). Mutagenesis by insertion of the drug resistance transposon Tn*7* applied to the Ti plasmid of *Agrobacterium tumefaciens. Plasmid* **1,** 218–225.

Hernalsteens, J. P., Van Vliet, F., De Beuckeleer, M., Depicker, A., Engler, G., Lemmers, M., Holsters, M., Van Montagu, M., and Schell, J. (1980). The *Agrobacterium tumefaciens* Ti plasmid as a host vector system for introducing foreign DNA in plant cells. *Nature (London)* **287,** 654–656.

Hishinuma, F., De Bona, P. J., Astrin, S., and Skalka, A. M. (1981). Nucleotide sequence of acceptor site and termini of integrated avian endogenous provirus *ev*I: integration creates a 6 bp repeat of host DNA. *Cell* **23,** 155–164.

Holsters, M., Silva, B., Van Vliet, F., Hernalsteens, J. P., Genetello, C., Van Montagu, M., and Schell, J. (1978). *In vivo* transfer of the Ti-plasmid of *Agrobacterium tumefaciens* to *Escherichia coli*. *Mol. Gen. Genet.* **163,** 335–338.

Holsters, M., Silva, B., Van Vliet, F., Genetello, C., De Block, M., Dhaese, P., Depicker, A., Inzé, D., Engler, G., Villarroel, R., Van Montagu, M., and Schell, J. (1980). The functional organization of the nopaline *A. tumefaciens* plasmid pTiC58. *Plasmid* **3,** 212–230.

Holsters, M., Hernalsteens, J. P., Van Montagu, M., and Schell, J. (1982a). Ti plasmids of *Agrobacterium tumefaciens,* the nature of the TIP. *In* "The Molecular Biology of Plant Tumors" (G. Kahl and J. Schell, eds.), pp. 269–298. Academic Press, New York.

Holsters, M., Villarroel, R., Van Montagu, M., and Schell, J. (1982b). The use of selectable markers for the isolation of plant-DNA/T-DNA junction fragments in a cosmid vector. *Mol. Gen. Genet.* **185,** 283–289.

Hooykaas, P. J. J., Klapwijk, P. M., Nuti, M. P., Schilperoort, R. A., and Rörsch, A. (1977). Transfer of the *Agrobacterium tumefaciens* Ti-plasmid to avirulent Agrobacteria and to *Rhizobium* ex planta. *J. Gen. Microbiol.* **98,** 477–484.

Jensen, C. O. (1918). Undersøgelser vedrørende nogle svulstlignende dannelser hos planter. *Aarsskr. K. Vet. Landbohoejsk. (Copenhagen)* **1918,** 91–143.

Jimenez, A., and Davies, J. (1980). Expression of a transposable antibiotic resistance element in Saccharomyces. *Nature (London)* **287,** 869–871.

Joos, H., Inzé, D., Caplan, A., Sormann, M., Van Montagu, M., and Schell, J. (1983). Genetic analysis of T-DNA transcripts in nopaline crown galls. *Cell* (unpublished).

Kahl, G., and Schell, J. (1982). "*Molecular Biology of Plant Tumors.*" Academic Press, New York.

Kaiss-Chapman, R. W., and Morris, R. O. (1977). Trans-zeatin in culture filtrates of *Agrobacterium tumefaciens*. *Biochem. Biophys. Res. Commun.* **76,** 453–459.

Kenter, A. L., and Birshtein, B. K. (1981). Chi, a promoter of generalized recombination in λ phage, is present in immunoglobin genes. *Nature (London)* **293,** 402–404.

Kerr, A. (1969). Transfer of virulence between isolates of *Agrobacterium*. *Nature (London)* **223,** 1175–1176.

Kerr, A. (1971). Acquisition of virulence by non-pathogenic isolation of *Agrobacterium radiobacter*. *Physiol. Plant Pathol.* **1,** 241–246.

Kerr, A., and Ellis, J. G. (1982). Conjugation and transfer of Ti plasmids in *Agrobacterium tumefaciens*. *In* "Molecular Biology of Plant Tumors" (G. Kahl and J. Schell, eds.), pp. 321–344. Academic Press, New York.

Kerr, A., and Htay, K. (1974). Biological control of crown gall through bacteriocin production. *Physiol. Plant Pathol.* **4,** 37–44.

Kerr, A., and Panagopoulos, C. G. (1977). Biotypes of *Agrobacterium radiobacter* var. *tumefaciens* and their biological control. *Phytopathol. Z.* **90,** 172–179.

Kerr, A., and Roberts, W. P. (1976). *Agrobacterium:* correlation between and transfer of pathogenicity, octopine and nopaline metabolism and bacteriocin 84 sensitivity. *Physiol. Plant Pathol.* **9,** 205–211.

Kerr, A., Manigault, P., and Tempé, J. (1977). Transfer of virulence *in vivo* and *in vitro* in *Agrobacterium*. *Nature (London)* **265,** 560–561.

Kersters, K., and De Ley, J. (1982). Family Rhizobiaceae—genus *Agrobacterium*. *In* "Bergey's Manual of Systematic Bacteriology" (N. R. Krieg, ed.), Vol. 1, 9th ed., in press. Williams & Wilkins, Baltimore, Maryland.

Klapwijk, P. M., Oudshoorn, M., and Schilperoort, R. A. (1977). Inducible permease involved in the uptake of octopine, lysopine and octopinic acid by *Agrobacterium tumefaciens* strains carrying virulence-associated plasmids. *J. Gen. Microbiol.* **102,** 1–11.

Klapwijk, P. M., Scheulderman, T., and Schilperoort, R. A. (1978). Coordinated regulation of octopine degradation and conjugative transfer of Ti-plasmids in *Agrobacterium tumefaciens:* evidence for a common regulatory gene and separate operons. *J. Bacteriol.* **136,** 775–785.

Kleckner, N. (1979). DNA sequence analysis of Tn*10* insertions: origin and role of 9 bp flanking repetition during Tn*10* translocation. *Cell* **16,** 711–720.

Koekman, B. P., Ooms, G., Klapwijk, P. M., and Schilperoort, R. A. (1979). Genetic map of an octopine Ti-plasmid. *Plasmid* **2,** 347–357.

Koekman, B. P., Hooykaas, P. J. J., and Schilperoort, R. A. (1980). Localization of the replication control region of the physical map of the octopine Ti plasmid. *Plasmid* **4,** 184–195.

Kühn, S., Fritz, H. J., and Starlinger, P. (1979). Close vicinity of IS*1* integration sites in the leader sequence of the gal operon of *E. coli*. *Mol. Gen. Genet.* **167,** 235–241.

Landy, A., and Ross, W. (1977). Viral integration and excision: structure of the lambda *att* sites. *Science (Washington, D.C.)* **197,** 1147–1160.

Leemans, J., Shaw, C., Deblaere, R., De Greve, H., Hernalsteens, J. P., Maes, M., Van Montagu, M., and Schell, J. (1981). Site-specific mutagenesis of *Agrobacterium* Ti plasmids and transfer of genes to plant cells. *J. Mol. Appl. Genet.* **1,** 149–164.

Leemans, J., Deblaere, R., Willmitzer, L., De Greve, H., Hernalsteens, J. P., Van Montagu, M., and Schell, J. (1982). Genetic identification of functions of TL-DNA transcripts in octopine crown galls. *EMBO J.* **1,** 147–152.

Lemmers, M., De Beuckeleer, M., Holsters, M., Zambryski, P., Depicker, A., Hernalsteens, J. P., Van Montagu, M., and Schell, J. (1980). Internal organization, boundaries and integration of Ti-plasmid DNA in nopaline crown gall tumours. *J. Mol. Biol.* **144,** 353–376.

Long, E. O. and Dawid, I. B. (1980). Repeated genes in eukaryotes. *Annu. Rev. Biochem.* **49,** 727–764.

Mahowald, A. P., Caulton, J. H., Edwards, M. K., and Floyd, A. D. (1979). Loss of centrioles and polyploidization in follicle cells of *Drosophila melanogaster. Exp. Cell Res.* **118,** 404–410.

Majors, J. E., and Varmus, H. E. (1981). Nucleotide sequences at host-proviral junctions for mouse mammary tumour virus. *Nature (London)* **289,** 253–257.

Martin, J. P., Jr., and Fridovich, I. (1981). Evidence for a natural gene transfer from the ponyfish to its bioluminescent bacterial symbiont *Photobacter leiognathi*. The close relationship between bacteriocuprein and the copper–zinc superoxide dismutase of teleost fishes. *J. Biol. Chem.* **256,** 6080–6089.

McCloskey, J. A., Hashizume, T., Basile, B., Ohno, Y., and Sonoki, S. (1980). Occurrence and levels of *cis*- and *trans*-zeatin ribosides in the culture medium of a virulent strain of *Agrobacterium tumefaciens. FEBS Lett.* **111,** 181–183.

Merlo, D. J., Nutter, R. C., Montoya, A. L., Garfinkel, D. J., Drummond, M. H., Chilton, M.-D., Gordon, M. P., and Nester, E. W. (1980). The boundaries and copy numbers of Ti plasmid T-DNA vary in crown gall tumors. *Mol. Gen. Genet.* **177,** 637–643.

Milbrandt, J. D., Heintz, N. H., White, W. C., Rothman, S. M., and Hamlin, J. L. (1981). Methotrexate-resistant Chinese hamster ovary cells have amplified a 135-kilobase-pair region that includes the dihydrofolate reductase gene. *Proc. Natl. Acad. Sci. USA* **78,** 6043–6047.

Miller, J. H., Calos, M. P., Galas, D., Hofer, M., Büchel, D. E., and Müller-Hill, B. (1980). Genetic analysis of transpositions in the lac region of *Escherichia coli. J. Mol. Biol.* **144,** 1–18.

Montoya, A., Chilton, M. D., Gordon, M. P., Sciaky, D., and Nester, E. W. (1977). Octopine and nopaline metabolism in *Agrobacterium tumefaciens* and crown-gall tumor cells: role of plasmid genes. *J. Bacteriol.* **129,** 101–107.

Moore, L., Warren, G., and Strobel, G. (1979). Involvement of a plasmid in the hairy root disease of plants caused by *Agrobacterium rhizogenes. Plasmid* **2,** 617–626.

Nester, E. W., and Kosuge, T. (1981). Plasmids specifying plant hyperplasias. *Annu. Rev. Microbiol.* **35,** 531–565.

Nevers, P., and Saedler, H. (1978). Mapping and characterization of an *E. coli* mutant defective in IS1-mediated deletion formation. *Mol. Gen. Genet.* **160,** 209–214.

Nuti, M., Lepidi, A., Prakash, R., Hooykaas, P., and Schilperoort, R. (1982). The plasmids of *Rhizobium* and symbiotic nitrogen fixation. *In* "Molecular Biology of Plant Tumors" (G. Kahl and J. Schell, eds.), pp. 561–589. Academic Press, New York.

Ohtsubo, H., Ohmori, H., and Ohtsubo, E. (1979). Nucleotide sequence analysis of Tn*3*(Ap): implications for insertion and deletion. *Cold Spring Harbor Symp. Quant. Biol.* **43,** 1269–1277.

Ooms, G., Klapwijk, P. M., Poulis, J. A., and Schilperoort, R. A. (1980). Characterization of Tn*904* insertions in octopine Ti plasmid mutants of *Agrobacterium tumefaciens. J. Bacteriol.* **144,** 82–91.

Ooms, G., Hooykaas, P. J., Moleman, G., and Schilperoort, R. A. (1981). Crown gall plant tumors of abnormal morphology, induced by *Agrobacterium tumefaciens* carrying mutated octopine Ti plasmids; analysis of T-DNA functions. *Gene* **14,** 33–50.

Otten, L., De Greve, H., Hernalsteens, J. P., Van Montagu, M., Schieder, O., Straub, J., and Schell, J. (1981). Mendelian transmission of genes introduced into plants by the Ti plasmids of *Agrobacterium tumefaciens. Mol. Gen. Genet.* **183,** 209–213.

Petit, A., Delhaye, S., Tempé, J., and Morel, G. (1970). Recherches sur les guanidines des tissues de crown gall. Mise en évidence d'une relation biochemique spécifique entre les souches d'*Agrobacterium* et les tumeurs qu'elles induisent. *Physiol. Veg.* **8,** 205–213.

Petit, A., Tempé, J., Kerr, A., Holsters, M., Van Montagu, M., and Schell, J. (1978). Substrate induction of conjugative activity of *Agrobacterium tumefaciens. Nature (London)* **271,** 570–572.

Pinkham, J. L., Platt, T., Enquist, L., and Weisberg, R. A. (1980). The secondary attachment site for bacteriophage λ in the *pro*A/B gene of *Escherichia coli. J. Mol. Biol.* **144,** 587–592.

Reed, R. R. (1981). Resolution of cointegrates between transposons γδ and Tn*3* defines the recombination site. *Proc. Natl. Acad. Sci. USA* **78,** 3428–3432.

Reif, H. J., and Saedler, H. (1975). IS*1* is involved in deletion formation in the gal region of *E. coli. Mol. Gen. Genet.* **137,** 17–28.

Roeder, G. S., Farabaugh, P. J., Chaleff, D. T., and Fink, G. R. (1980). The origins of gene instability in yeast. *Science (Washington, D.C.)* **209,** 1375–1380.

Rohfritsch, O. (1971). Culture *in vitro* de jeunes galles d'*Aulax glechomae* L. sur *Glechoma hederacea* L. *Marcellia* **37,** 233–239.

Rohfritsch, O., and Shorthouse, J. O. (1982). Insect galls. *In* "Molecular Biology of Plant Tumors" (G. Kahl and J. Schell, eds.), pp. 131–152. Academic Press, New York.

Rudkin, G. T., and Corlette, S. L. (1957). Disproportionate synthesis of DNA in a polytene chromosome region. *Proc. Natl. Acad. Sci. USA* **43,** 964–968.

Sambrook, J., Botchan, M., Hu, S. L., Mitchison, T., and Stringer, J. (1980). Integration of viral DNA sequences in cells transformed by adenovirus 2 or SV40. *Proc. R. Soc. Lond Ser. B* **210,** 423–435.

Schell, J. (1975). The role of plasmids in crown gall formation by *A. tumefaciens. In* "Genetic Manipulations with Plant Materials" (L. Ledoux, ed.), pp. 163–181. Plenum, New York.

Schell, J., Van Montagu, M., Depicker, A., De Waele, D., Engler, G., Genetello, C., Hernalsteens, J. P., Holsters, M., Messens, E., Silva, A., Van den Elsacker, S., Van Larebeke, N., and Zaenen, I. (1977). *Agrobacterium tumefaciens:* what segment of the plasmid is responsible for the induction of crown gall tumors? *In* "Nucleic Acids and Protein Synthesis in Plants" (L. Bogorad and J. H. Weil, eds.), pp. 329–342. Plenum, New York.

Schell, J., Van Montagu, M., De Beuckeleer, M., De Block, M., Depicker, A., De Wilde, M., Engler, G., Genetello, C., Hernalsteens, J. P., Holsters, M., Seurinck, J., Silva, B., Van Vliet, F., and Villarroel, R. (1979a). Interactions and DNA transfer between *Agrobacterium tumefaciens,* the Ti-plasmid and the plant host. *Proc. R. Soc. Lond. Ser. B* **204,** 251–266.

Schell, J., Van Montagu, M., Depicker, A., De Waele, D., Engler, G., Genetello, C., Hernalsteens, J. P., Holsters, M., Messens, E., Silva, A., Van den Elsacker, S., Van Larebeke, N., and Zaenen, I. (1979b). Crown gall: bacterial plasmids as oncogenic elements for eucaryotic cells. *In* "Molecular Biology of Plants" (I. Rubinstein, R. L. Phillips, C. E. Green, and B. G. Gengenbach, eds.), pp. 315–337. Academic Press, New York.

Schell, J., Van Montagu, M., Holsters, M., Hernalsteens, J. P., Leemans, J., De Greve, H., Willmitzer, L., Otten, L., Schröder, J., and Shaw, C. (1981a). The development of host vectors for directed gene-transfers in plants. *ICN–UCLA Symp. Mol. Cell. Biol.* **23,** 557–575.

Schell, J., Van Montagu, M., Holsters, M., Hernalsteens, J. P., Leemans, J., De Greve, H., Shaw, Ch., Zambryski, P., Goodman, H. M., Willmitzer, L., Otten, L., Schröder, J., and Schröder, G. (1982). Tumours as a result of gene transfer in plants. *In* "The Molecular Biology of Plant Development" (H. Smith and D. Grierson, eds.), *Bot. Monogr.* (Oxford). Blackwell, Oxford.

Schilperoort, R. A., Hooykaas, P. J. J., Klapwijk, P. M., Koekman, B. P., Nuti, M. P., Ooms, G., and Prakash, R. K. (1979). Characters on large plasmids in Rhizobiaceae involved in the interaction with plant cells. *In* "Plasmids of Medical, Environmental and Commercial Importance" (K. Timmis and A. Pühler, eds.), pp. 339–352, Elsevier, Amsterdam.

Schimke, R. T., Kaufman, R. J., Nunberg, J. H., and Dana, S. L. (1979). Studies on the amplification of dihydrofolate reductase genes in methotrexate-resistant cultured mouse cells. *Cold Spring Harbor Symp. Quant. Biol.* **43,** 1297–1303.

Schimke, R. T., Brown, P. C., Kaufman, R. J., McGrogan, M., and Slate, D. L. (1980). Chromosomal and extrachromosomal localization of amplified dihydrofolate reductase genes in cultured mammalian cells. *Cold Spring Harbor Symp. Quant. Biol.* **45,** 785–797.

Simpson, R., O'Hara, P. J., Montoya, A. L., Lichtenstein, C., Gordon, M. P., and Nester, E. W. (1982). DNA from the A6S/2 crown gall tumor contains scrambled Ti-plasmid sequences near its junctions with the plant DNA. *Cell* **29,** 1005–1014.

Shapiro, J. A. (1979). Molecular model for the transposition and replication of bacteriophage Mu and other transposable elements. *Proc. Natl. Acad. Sci. USA* **76,** 1933–1937.

Smith, E. F. (1916). Studies on the crown gall of plants. Its relation to human cancer. *J. Cancer Res.* **1,** 231–309.

Smith, G. P. (1974). Unequal crossover and the evolution of multigene families. *Cold Spring Harbor Symp. Quant. Biol.* **38,** 507–513.

Smith, G. R., Kunes, S. M., Schultz, D. W., Taylor, A., and Triman, K. L. (1981). Structure of Chi hotspots of generalised recombination. *Cell* **24,** 429–436.

Spradling, A. C. (1981). The organization and amplification of two chromosomal domains containing *Drosophila* Chorion genes. *Cell* **27,** 193–201.

Spradling, A. C., and Mahowald, A. P. (1980). Amplification of genes in chorion proteins during oogenesis in *Drosophila melanogaster*. *Proc. Natl. Acad. Sci. USA* **77,** 1096–1100.

Stahl, F. W., Stahl, M. M., Malone, R. E., and Crasemann, J. M. (1980). Directionality and nonreciprocality of Chi stimulated recombination in phage λ. *Genetics* **94,** 235–248.

Tempé, J., and Schell, J. (1977). Is Crown gall a natural instance of gene transfer? *In* "Translation of Natural and Synthetic Polynucleotides" (A. B. Legocki, ed.), pp. 416–420. Univ. of Agriculture, Poznan, Poland.

Tempé, J., Petit, A., Holsters, M., Van Montagu, M., and Schell, J. (1977). Thermosensitive step associated with transfer of the Ti-plasmid during conjugation: possible relation to transformation in crown gall. *Proc. Natl. Acad. Sci. USA* **74,** 2848–2849.

Tempé, J., Estrade, C., and Petit, A. (1978). The biological significance of opines. II. The conjugative activity of the Ti-plasmids of *Agrobacterium tumefaciens*. *Proc. Int. Conf. Plant Pathog. Bact., 4th* 153–160.

Tepfer, D. A., and Tempé, J. (1981). Production d'agropine par des racines formées sous l'action d'*Agrobacterium rhizogenes,* souche A4. *C. R. Hebd. Seances Acad. Sci. Ser. C* **292,** 153–156.

Thomashow, M. F., Nutter, R., Montoya, A. L., Gordon, M. P., and Nester, E. W. (1980a). Integration and organisation of Ti-plasmid sequences in crown gall tumors. *Cell* **19,** 729–739.

Thomashow, M. F., Nutter, R., Postle, K., Chilton, M.-D., Blattner, F. R., Powell, A., Gordon, M. P., and Nester, E. W. (1980b). Recombination between higher plant DNA and the Ti plasmid of *Agrobacterium tumefaciens*. *Proc. Natl. Acad. Sci. USA* **77,** 6448–6452.

Thomashow, M. F., Panagopoulos, C. G., Gordon, M. P., and Nester, E. W. (1980c). Host range of *Agrobacterium tumefaciens* is determined by the Ti plasmid. *Nature (London)* **283,** 794–796.

Turgeon, R., Wood, M. N., and Braun, A. C. (1976). Studies on the recovery of crown gall tumor cells. *Proc. Natl. Acad. Sci. USA* **73,** 3562–3564.

Van Emmelo, J., Devos, R., Ysebaert, M., and Fiers, W. (1980). Construction and characterization of a plasmid containing a nearly full-size DNA copy of satellite tobacco necrosis virus RNA. *J. Mol. Biol.* **143,** 259–271.

Van Larebeke, N., Engler, G., Holsters, M., Van den Elsacker, S., Zaenen, I., Schilperoort, R. A., and Schell, J. (1974). Large plasmid in *Agrobacterium tumefaciens* essential for crown gall-inducing ability. *Nature (London)* **252,** 169–170.

Van Larebeke, N., Genetello, C., Schell, J., Schilperoort, R. A., Hermans, A. K., Hernalsteens, J. P., and Van Montagu, M. (1975). Acquisition of tumour-inducing ability by non-oncogenic agrobacteria as a result of plasmid transfer. *Nature (London)* **255,** 742–743.

Van Larebeke, N., Genetello, C., Hernalsteens, J. P., Depicker, A., Zaenen, I., Messens, E., Van Montagu, M., and Schell, J. (1977). Transfer of Ti-plasmids between *Agrobacterium* strains by mobilization with the conjugative plasmid RP4. *Mol. Gen. Genet.* **152,** 119–124.

Van Montagu, M., and Schell, J. (1979). The plasmids of *Agrobacterium tumefaciens. In* "Plasmids of Medical, Environmental and Commercial Importance" (K. Timmis and A. Pühler, eds.), pp. 71–96. Elsevier, Amsterdam.

Van Montagu, M., and Schell, J. (1981). The Ti plasmids of *Agrobacterium. Curr. Top. Microbiol. Immunol.* **96,** 237–254.

Wahl, G. M., Padgett, R. A., and Stark, G. R. (1979). Gene amplification causes overproduction of the first three enzymes of UMP synthesis in *N*-(phosphonacetyl)-L-aspartate-resistant hamster cells. *J. Biol. Chem.* **254,** 8679–8689.

Watson, B., Currier, T. C., Gordon, M. P., Chilton, M.-D., and Nester, E. W. (1975). Plasmid required for virulence of *Agrobacterium tumefaciens. J. Bacteriol.* **123,** 255–264.

White, F. F., and Nester, E. W. (1980). Relationship of plasmids responsible for hairy root and crown gall tumorigenicity. *J. Bacteriol.* **144,** 710–720.

Willmitzer, L., De Beuckeleer, M., Lemmers, M., Van Montagu, M., and Schell, J. (1980). DNA from Ti-plasmid is present in the nucleus and absent from plastids of plant crown-gall cells. *Nature (London)* **287,** 359–361.

Willmitzer, L., Simons, G., and Schell, J. (1982a). The TL-DNA in octopine crown gall tumours codes for seven well defined polyadenylated transcripts. *EMBO J.* **1,** 139–146.

Willmitzer, L., Sanchez-Serrano, J., Buschfeld, E., and Schell, J. (1982b). DNA from *Agrobacterium rhizogenes* is transferred to and expressed in axenic hairy root plant tissues. *Mol. Gen. Genet.* **186,** 16–22.

Willmitzer, L., Dhaese, P., Schreier, P. H., Schmalenbach, W., Van Montagu, M., and Schell, J. (1983). Size, location, and polarity of T-DNA-encoded transcripts in nopaline crown gall tumors; evidence for common transcripts present in both octopine and nopaline tumors. *Cell* (unpublished).

Yadav, N. S., Postle, K., Saiki, R. K., Thomashow, M. F., and Chilton, M.-D. (1980). T-DNA of a crown gall teratoma is covalently joined to host plant DNA. *Nature (London)* **287,** 458–461.

Yadav, N. S., Vanderleyden, J., Bennett, D. R., Barnes, W. M., and Chilton, M.-D. (1982). Short direct repeats flank the T-DNA on a nopaline Ti plasmid. *Proc. Natl. Acad. Sci. USA* **79,** 6322–6326.

Yang, F., Montoya, A. L., Merlo, D. J., Drummond, M. H., Chilton, M.-D., Nester, E. W., and Gordon, M. P. (1980). Foreign DNA sequences in crown gall teratomas and their fate during the loss of the tumorous traits. *Mol. Gen. Genet.* **177,** 707–714.

Zaenen, I., Van Larebeke, N., Teuchy, H., Van Montagu, M., and Schell, J. (1974). Supercoiled circular DNA in crown gall inducing *Agrobacterium* strains. *J. Mol. Biol.* **86,** 109–127.

Zambryski, P., Holsters, M.,Kruger, K., Depicker, A., Schell, J., Van Montagu, M., and Goodman, H. M. (1980). Tumor DNA structure in plant cells transformed by *A. tumefaciens. Science (Washington, D.C.)* **209,** 1385–1391.

Zambryski, P., Depicker, A., Kruger, K., and Goodman, H. (1982). Tumor induction by *Agrobacterium tumefaciens:* analysis of the boundaries of T-DNA. *J. Mol. Appl. Genet.* **1,** 361–370.

Zimmerman, J. L., and Goldberg, R. B. (1977). DNA sequence organization in the genome of *Nicotiana tabacum. Chromosoma* **59,** 227–252.

CHAPTER 12

Phase Variation and Related Systems

MICHAEL SILVERMAN
MELVIN SIMON

I. Introduction

McClintock's work on chromosomal rearrangement in maize (McClintock, 1941, 1951) led her to the discovery of mobile genetic elements. In addition to transposition, the genetic units encoded activities that mediated a variety of rearrangements such as insertions and deletions. Furthermore, these mobile units could act as controlling elements. McClintock found that the insertion of an element adjacent to a specific gene could alter the pattern of gene expression by inactivating the gene or by changing the time at which the gene was activated during development. Finally, mobile elements could also respond to trans-acting substances, and thus they were receptive to control by the products of other genetic factors. McClintock (1956, 1965) argued that complex circuits of gene expression could have evolved from the initial interaction of a mobile element with resident genes. Subsequent genetic events would further stabilize the interaction, leading to a new control circuit that included parts of the mobile

Mobile Genetic Elements

ISBN 0-12-638680-3

element associated with the gene. Recent work on bacterial systems (Calos and Miller, 1980; Kleckner, 1981) and in the eukaryotes (Shapiro and Cordell, 1982) has led to the molecular definition of many mobile elements that have properties similar to those described by McClintock in maize. Phase variation was one of the first systems in bacteria to be recognized to have the genetic characteristics that McClintock ascribed to controlling elements. Subsequent analysis of the molecular details of flagellar phase transition indicates that this system evolved from the association of a component of a mobile element with a specific chromosomal gene.

II. The Genetics of Phase Variation

The phenomenon of flagellar phase variation was discovered by Andrewes (1922) in *Salmonella typhimurium*. Andrewes found that a small population of cells derived from a single colony isolated from a culture of *S. typhimurium* had one predominant serotype corresponding to its flagellar antigen. However, if the clone was grown and the population expanded, a second different serotype could be detected. When a cell was selected that had the second serotype, it was found to give rise on further growth to the first serotype. Thus, individual *Salmonella* had the capacity to alternate between two flagellar antigens or phases (see Fig. 1). This antigenic dimorphism was used in the Kauffman-White scheme for the identification and classification of strains of *Salmonella* (Kauffmann, 1951). The transition from one phase to the other occurred with a small probability per bacterial division that ranged from 10^{-5} to 10^{-3} (Stocker, 1949). Furthermore, in a given strain, the probability of phase transition in each direction was genetically determined and could be different. With the discovery of transducing phage, it became possible to analyze the genetic basis of phase variation. Lederberg and Edwards (1953) showed that the two phases corresponded to the gene products of separable genetic loci, *H1* and *H2*. Each gene controlled the synthesis of a specific flagellar structural protein. In the different phases, the bacteria elaborated different forms of the flagellar filament protein (flagellin). The inclusion of antibody in the agar of motility plates selectively immobilized cells with a particular serotype. Expression of a series of alleles at each genetic locus could thus be distinguished with specific antisera. Finally, genetic techniques were used to analyze the nature of the factors that regulated phase transitions. By using bacteriophage transduction from donor strains that were expressing either the *H1* or the *H2* locus into recipients in either phase, Lederberg and Iino (1956) showed that the control of flagellin gene expression resided in the *H2* locus. The *H2* locus could exist in either the "on" or the "off" state, and the state of the *H2* locus determined whether the *H1* gene was expressed or not. If the recipient was in the *H2* (off) state, expression of the transduced donor *H1* serotype was

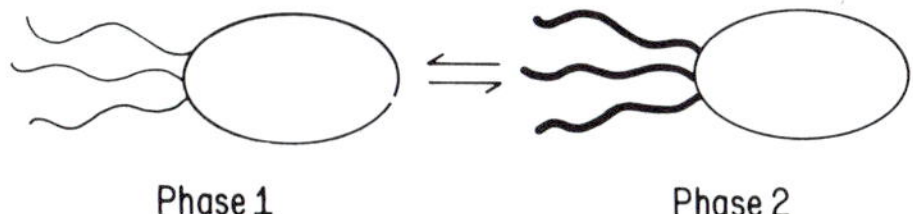

Fig. 1 Phase variation of flagellar serotypes. *Salmonella* bacteria alternately express either a phase 1 or a phase 2 serotype. Alternation between serotypes occurs at a frequency of 10^{-3}–10^{-5} per bacterium per generation. The flagellar serotypes correspond to flagellar filaments polymerized from a single protein called *flagellin*. The flagellar filaments are "propellers" anchored to and rotated by a complex organelle located in the cell envelope. The serotypes are encoded by two unlinked flagellin genes called *H1* and *H2*. When the *H1* gene is transcribed (*H1* on, *H2* off) the phase 1 serotype is expressed, and when the *H2* gene is transcribed (*H1* off, *H2* on) the phase 2 serotype is expressed. A variety of alleles and corresponding serotypes are characteristic of the *H1* and *H2* loci. Variation is thought to enable pathogenic *Salmonella* to evade host immunological defenses.

observed. If this recipient was in the *H2* (on) state, expression of the *H1* serotype was prevented. These observations implied that *H1* gene expression was repressed when the *H2* locus was active. However, the most remarkable finding in this study was that the "state" of expression of the *H2* gene in the donor could be transferred by genetic transduction techniques to the recipient. Thus, the state of the *H2* gene was itself a heritable trait. Lederberg and Iino suggested that the state of the gene could be determined by a genetic modification, and they compared it to the DS elements described by McClintock in maize. Furthermore, some strains of *Salmonella* were found to undergo phase transition at a very low frequency, and this phenotype was associated with a genetic alteration linked to the *H2* gene (Iino, 1961, 1969) called *vH2*$^{-}$. Thus, not only was the state of *H2* expression inherited, but the frequency of transition was also genetically determined. Subsequent genetic analysis of the system has served to clarify some of the factors involved in the regulation of flagellar phase transition (see reviews by Iino, 1969, 1977; Silverman and Simon, 1977). Abortive transduction was used to show (Pearce & Stocker, 1967) that a trans-acting substance encoded by a region adjacent to the *H2* locus could act to repress the activity of the *H1* gene. It was shown (Fujita *et al.*, 1973) that the trans-acting substance corresponded to the product of a gene that was named *rh1* (repressor of *H1*) and that behaved as if it were cotranscribed with the *H2* gene. Furthermore, assays for *H1*-specific mRNA suggested that the *rh1* gene product repressed *H1* gene transcription (Suzuki and Iino, 1975). Finally, studies involving cloned segments of the DNA corresponding to the *H2* and *rh1* regions demonstrated that the appearance of a 16,000 MW polypeptide was correlated with the presence of a region of DNA immediately adjacent to the *H2* gene (Silverman *et al.*, 1979b). The presence of the peptide was also correlated with *rh1* repressor activity. Thus, when the *H2* gene was in the on state, *H2* flagellin and the *rh1* gene product were formed, and the *rh1* gene product repressed *H1* gene activity. When the *H2* gene was in the off state, the repressor was not made and the *H1* gene was expressed. Further work also showed that the *rh1* gene product was able to repress synthesis from the *hag* gene, which is the *E. coli* homolog of the *H1* gene in *Salmonella*

(Makela, 1964). When the *H2* operon was transferred into *Escherichia coli*, phase variation between the *hag* and *H2* gene products could be observed (Enomoto and Stocker, 1975).

The fact that the state of the *H2* gene could be transmitted along with the DNA that carried the genetic information suggested that the change of state was a result of a change in the DNA structure. These kinds of local, yet relatively stable, heritable changes directly associated with a specific gene are interesting because they have the properties that one would expect for elements that are involved in regulating gene expression during differentiation. It therefore became important to understand the molecular nature of this modification. Phase variation has been postulated to be the result of a variety of different local gene rearrangements (Dawson and Smith-Keary, 1963; Stocker and Makela, 1978). On the other hand, it could also result from a specific chemical modification of DNA and the establishment of a heritable pattern of DNA modification (Holliday and Pugh, 1975).

III. The Molecular Basis of Phase Variation

The nature of the molecular change that determined the state of the *H2* gene was found by cloning the DNA that corresponded to this gene and comparing DNA structures adjacent to the gene in the on and off states (Zieg *et al.*, 1977). When DNA corresponding to the genes in these two states was denatured and allowed to renature, a fraction of the renatured molecules was found to show a region of nonhomologous DNA corresponding to approximately 1000 bp (Fig. 2). Sequences in this region were inverted with respect to each other. The sequences were, therefore, not able to hybridize and formed a characteristic inversion loop. Analysis of the heteroduplexes by electron microscopy and analysis of the products of restriction enzyme digestion were all consistent with the interpretation that the region could invert (Zieg *et al.*, 1978a). Furthermore, deletion mutations were constructed that removed a portion of the DNA in this invertible region and resulted in the elimination of the ability of the DNA to invert (Silverman *et al.*, 1979a). This also fixed the flagellar phase and eliminated phase transition. Specific fusions were constructed that juxtaposed the inversion region to structural genes that lacked promoters. It was found that the inversion region contained a promoter that read out of the inversion and could activate adjacent genes. When the DNA was in one orientation, the promoter transcribed genes that were fused to one side of the invertible sequence; when the DNA was in the opposite orientation, the promoter read in the opposite direction (Zieg *et al.*, 1978b; Silverman *et al.*, 1979a). Gene expression encoded by this promoter was regulated by factors that control flagellin gene expression. The invertible region adjacent to the *H2* gene acted as a controlling element. It behaved like a genetic

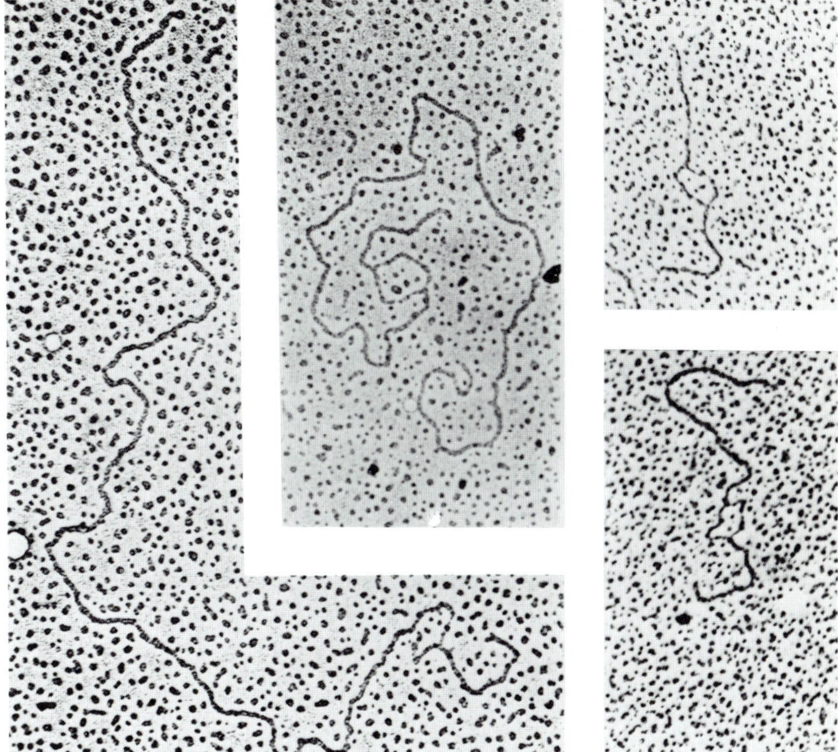

Fig. 2 Heteroduplex analysis of cloned DNA molecules. Electron micrographs show restriction endonuclease fragments from the *H2* gene region of *Salmonella* that were denatured and allowed to reanneal. The source of DNA included molecules representing both the *H2* (on) and *H2* (off) states. When single-stranded DNA from *H2* (on) reanneals with single-stranded DNA from *H2* (off), a duplex molecule is formed that contains a ''bubble'' of nonhomologous DNA approximately 1000 bp in length. This loop structure resulted because the sequences in this region were inverted with respect to each other and, therefore, were not able to hybridize.

flip-flop switch, and in one orientation connected the flagellar promoter to the *H2* operon and in the other orientation disconnected the promoter from the operon (Fig. 3).

Genetic studies were undertaken to map the cis- and trans-acting elements encoded by the invertible region. Tn*5* insertions and a variety of specific deletions were used to mutagenize the region. On the basis of this work (Silverman and Simon, 1980), it was possible to define three cis-acting elements and one trans-acting element that were components of the phase transition switch. One of the cis-acting elements corresponded to a region encompassing 100 bp within the inversion region. When this was deleted, it did not markedly affect the frequency of inversion, However, the *H2* gene was no longer expressed. This sequence was required in cis for the expression of the adjacent *H2* operon, and it corresponded

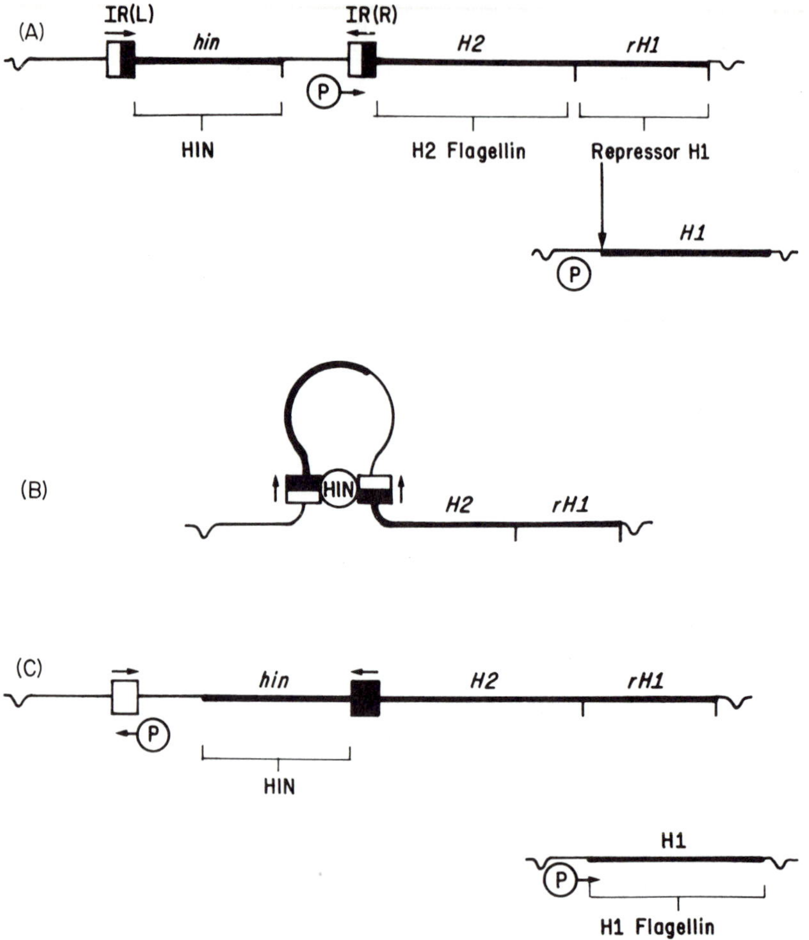

Fig. 3 Genetic rearrangement controls flagellar phase transition. Inversion of a region of DNA of approximately 1000 bp adjacent to the *H2* operon alternately couples and uncouples a promoter element (B). When the promoter is coupled to the operon *H2*, flagellin is synthesized as well as the product of the *rh1* gene that repressed *H1* expression (A; *H2* on, *H1* off). When the promoter is uncoupled from the *H2* operon, no *rh1* gene product is synthesized, and the *H1* gene is expressed (C; *H2* off, *H1* on). IR(L) and IR(R) describe the cis-acting sites in inverted repeat configurations in which a reciprocal recombination event resulting in inversion takes place. The *hin* gene product is encoded by a sequence within the inversion region and is required for inversion. The fate of DNA sequences rearranged by inversion can be followed by referring to the blackened areas of the IR boxes. Note that IR(L) and IR(R) are defined for the *H2* (on) position.

to the flagellar promoter region. Adjacent to this promoter region was another cis-acting site. Deletion of this site, *IR(R)*, completely eliminated inversion. A comparable site was found on the opposite side of the invertible segment; this was termed *IR(L)*. These two cis-acting sites are the termini of the invertible region and are absolutely required for inversion (Fig. 3). Another region corresponding to approximately two-thirds of the 1000-bp sequence was found to be necessary for inversion. However, mutations in this region could be complemented in trans by the presence of an intact sequence on another plasmid in the same cell. On the basis of these experiments, it was concluded that this sequence was responsible for the formation of a gene product that was necessary for inversion. The gene was called *hin* because it mediated inversion associated with the *H2* gene locus. In further studies, the nucleotide sequence of this region was determined, including the sequence for all four products of the crossover region where the inversion occurred (Zieg and Simon, 1980). The nucleotide sequence corresponding to each of the genetically defined functions was determined. The minimal sequences that were found to correspond to IR(R) and IR(L) were 14-bp inverted repeat sequences (see Fig. 5). They represent the switching points at which the inversion occurred. A homologous recombination event between these sites resulted in the inversion of the 980-bp sequence between them. In the *H2* (on) configuration, the promoter for *H2* corresponded to a DNA sequence of approximately 100 bp immediately adjacent to IR(R). The *H2* structural gene for flagellin started with an AUG codon at a position 16 bp away from the IR(R) sequence (Silverman *et al.*, 1981). The *hin* gene corresponded to a DNA sequence that included IR(L) and extended to over 620 bp within the invertible region. This sequence encoded the trans-acting function that was called *Hin,* a 23,000 MW polypeptide. The polypeptide has been identified by using specific DNA fragments and a coupled *in vitro* transcription–translation system (Simon *et al.*, 1980). Figure 3 summarizes these results in terms of the events involved in phase transition. Thus, the regulatory mechanism for *H2* gene switching transition is the reversal of polarity of a promoter element by genetic rearrangement. On the basis of the nucleotide sequences of all the crossover products, we conclude that rearrangement involves conservative, site-specific recombination within homologous 14-bp inverted repeats at the termini of the region.

IV. Bacteriophage Mu

A 3000-bp portion of the bacteriophage Mu genome (G-loop) and a homologous region in bacteriophage P1 (C-loop) have been known for a considerable time to undergo inversion (Chow and Bukhari, 1976; Bächi and Arber, 1977; Lee *et al.*, 1974). It is now clear that the manifestation of this inversion is an alternation in the expression of the genes that determine the host range of the phage particle (Kamp *et al.*, 1978; Bukhari and Abrusio, 1978; Van De Putte *et al.*, 1980). The two configurations of the G-loop are referred to as *G(+)* and *G(−)*. The G-loop in Mu contains specific host range functions that are also found in the C-loop of bacteriophage P1. The orientation of the invertible region affects the nature of the gene product encoded by the G-loop. In one orientation, one set of polypeptides, the product of the S and U genes, are expressed, whereas in the opposite

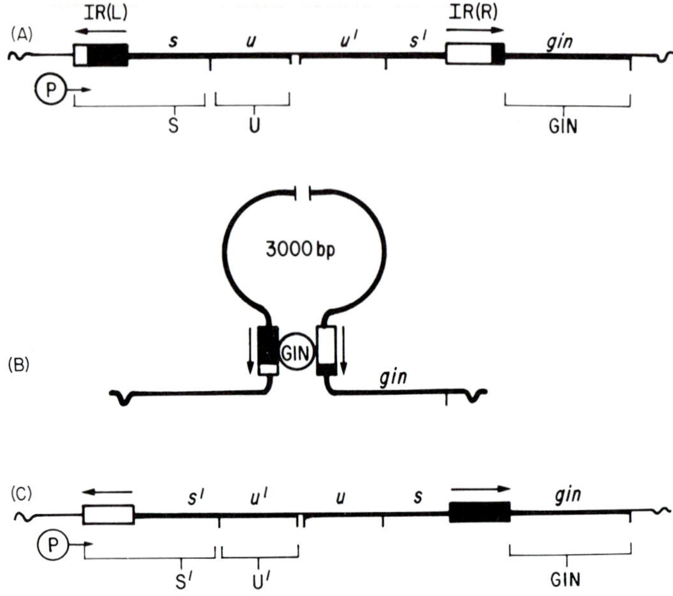

Fig. 4 Mu host range variation is determined by a genetic rearrangement. The host range of the phage is determined by the presence of pairs of gene products in the infectious particle. If gene products S and U are present, the phage will infect *Escherichia coli* (A); if gene product S′ and U′ are present, the phage will infect *Citrobacter* and other genera (C). The alternation in gene product synthesis is controlled by the inversion of a 3000-bp region that encodes an *s, u* operon and an *s*′, *u*′ operon in inverted configuration (B). The promoter and a small part of the coding region lie outside the inversion region. In contrast to phase variation, alternation of expression is by reversible coupling of different coding regions with a static promoter. As in Fig. 3, the cis-acting sites for inversion are marked IR(L) and IR(R). These regions of inverted homology are more extensive than those of *Salmonella* phase variation (34 bp versus 14 bp). The size and shading of IR boxes are related to a comparison of IR sites presented in Fig. 5. The *gin* gene located outside the inversion region encodes a protein required for inversion.

orientation an alternate subset of products determining host ranges S′ and U′ are synthesized (Howe *et al.*, 1979). Inversion of the G-loop is mediated by a gene called *gin,* which encodes a polypeptide of approximately 23,000 MW. This gene is located just outside the inversion region. In the homologous inversion region on bacteriophage P1, a function called *cin* is thought to be necessary to mediate the inversion of the C-loop (Chow *et al.*, 1977; Kamp *et al.*, 1979). Figure 4 schematically describes the regulation of the G-loop function in bacteriophage Mu. In contrast to the phase variation system, variability in Mu host range expression involves the reversible coupling of different coding regions to a static promoter. This is the result of a site-specific homologous recombination event that occurs within 34-bp inverted repeat sequences at both ends of the 3000-bp invertible region (Kahmann and Kamp, 1982). The site-specific recombination event is mediated by the product of the *gin* gene.

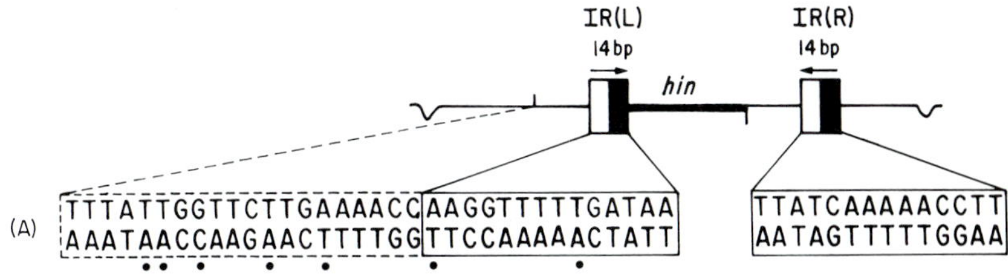

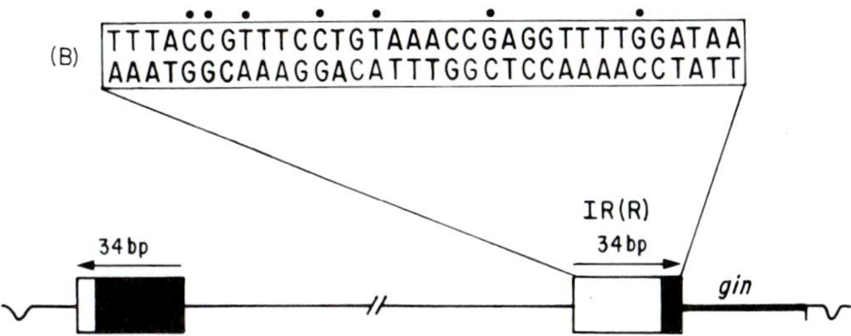

Fig. 5 Comparison of the DNA sequences of sites with inverted repeat (IR) homology from the *Salmonella* (A) and Mu (B) inversion systems. Inversion occurs by homologous recombination across these sites, and these sites probably represent the minimal sequence necessary for the genetically defined cis-acting functions for inversion. Note that the extent of IR homology is greater for the Mu repeats (34 bp) than for the *Salmonella* repeats (14 bp). The *Salmonella* IR appear to be a subset of the Mu IR. The IR(L) region of *Salmonella* actually has extensive homology with the entire IR(R) of Mu. Dots mark base pairs that do not show sequence correspondence. The cis-acting sites of *Salmonella* and Mu not only show sequence similarity but are also functionally equivalent because the *gin* and *hin* activities can be interchanged (see text).

The *Salmonella* and Mu phage inversion systems appear upon superficial examination to be quite dissimilar. However, more detailed analysis reveals striking functional similarities between the two systems. Kahmann and Kamp (1982) determined the DNA base sequence at the IR(R) and IR(L) termini of the G-loop in bacteriophage Mu. These consist of 34-bp inverted repeat sequences. They are homologous with an extended sequence that includes the IR(L) sequence adjacent to the *hin* gene in the *Salmonella* phase variation system. There are only 7-bp mismatches between the 34 bases at the end of the *hin* region and the 34-bp inverted repeat of Mu. Even closer sequence alignment (only 2-bp mismatches) is obtained when comparison is made between the 14-bp IR(L) of *Salmonella* and the homologous region in phage Mu (see Fig. 5). The *gin, hin,* and *cin* genes also encode homologous functions. (*1*) The *gin* and *cin* genes can provide a trans-acting function that mediates phase transition in vH2$^-$ strains of *Salmonella* (Kutsukake and Iino, 1980a,b). The DNA that encodes the vH2$^-$ stabilizing function has been cloned and studied by specific hybridization with the wild-type *H2* region. vH2$^-$ corresponds to an alteration in the *hin* gene sequence that inactivates Hin (Szekely and Simon, 1981). (*2*) The *gin* and *cin* genes can provide a trans-acting function that complements *hin*-defective switching mutants (Silverman *et al.*, 1981). (*3*) The *hin* gene in trans complements mutants of bacteriophage Mu that are *gin*-defective (Kutsukake and Iino, 1980b; Kamp and Kahmann, 1982). Thus, the *gin* and *cin* genes can replace *hin* function, and the *hin* gene can mediate G-loop inversion in bacteriophage Mu. Hybridization studies and measurements of the stability of hybrids between sequences corresponding to the *hin* and *gin* genes suggest that these sequences have approximately 30% mismatch. Comparison of the DNA sequences of the *gin* gene also shows marked homology between *gin* and *hin* (D. Kamp, personal communication). Almost all of the homology between bacteriophage Mu and the *H2* region of *Salmonella* appears to be confined to the regions encoding the *hin* and *gin* genes (Szekely and Simon, 1982).

V. Some Other Site-Specific Recombinational Systems

The *hin, gin,* and *cin* genes mediate conservative, site-specific recombination between homologous inverted repeat sequences leading to inversion. We would predict that these gene products would also have the ability to mediate specific recombination between directly repeated sites leading to deletion or cointegrate formation. Thus, if the IR(R) and IR(L) sites are recloned onto separate compatible plasmids in the presence of an intact *hin* gene, cointegrate plasmids should form by site-specific recombination between the directly repeated configurations of the target sequences. This reaction has been observed (Scott and Simon, 1982). Furthermore, the reverse situation, that is, resolution or deletion mediated by the same kind of reaction, has also been observed. Presumably, the *gin* and *cin* genes are capable of similar activities.

These systems bear striking similarities to other site-specific recombination systems found associated with different transposons and viruses (see Fig. 6). Thus, for example, the TnA family of transposons (Heffron *et al.*, 1979) is found in general to include the following components: (*1*) specific sites at the end of each transposon that are required for transposition; (2) a gene that encodes a high molecular weight polypeptide that is required for transposition and presumably acts at these sites; and (*3*) another gene called *tnpR* (Tn*3*) or *resolvase* (in γδ). The product of the *tnpR* gene apparently is involved in two functions. It acts as a repressor and as a recombinase. It binds to adjacent specific sites, and it represses its own synthesis as well as that of the transposase (Chou *et al.*, 1979; Gill *et al.*, 1979). Its second function is to mediate site-specific recombination (Heffron *et al.*, 1981; Sherratt *et al.*, 1981; Kitts *et al.*, 1982). During transposition, cointegrate molecules are formed. Cointegrates include two copies of the

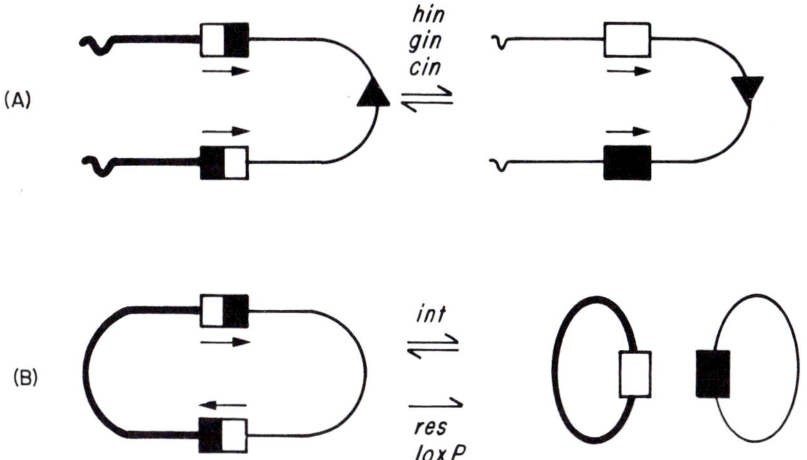

Fig. 6 Site-specific recombination systems. Those systems [*Salmonella* phase variation (*hin*), mu and P1 host range variation (*gin, cin*)] characterized by recombination sites in inverted repeat configuration (A) invert the intervening DNA. Arrows next to boxes show polarity of repeated DNA sequences. Those systems [transposon Tn*3* (*res*), phage λ (*int*), P1 (*loxP*)] characterized by recombination sites in direct repeat configuration (B) result in deletion (excision or resolution) of intervening DNA. As noted in the text, the consequence of the recombination reaction can often be altered by changing the configuration of the cis-acting sites. *Hin* function will cause deletion of intervening DNA if the cis-acting sites are arranged in direct repeat configuration.

transposon in direct repeat configuration. Each copy of the transposon contains a sequence corresponding to the *tnpR* gene and the DNA sequence adjacent to *tnpR* called *IRS*. The *tnpR* gene product resolves the cointegrate structure by mediating site-specific recombination between directly repeated IRS sequences. Finally, the *tnpR* gene product has also been reported to mediate inversion *in vivo* when the IRS site exists on the same molecule in inverted repeat configuration (Heffron *et al.*, 1981). Work on this system has advanced rapidly because overproducers of the *tnpR* gene product resolvase were prepared, and the MW 23,000 protein was isolated (Reed, 1981). Reed and Grindley (1981) showed that resolvase could bind at the IRS site and a number of sites adjacent to IRS, and that it could mediate the cointegrate resolution reaction *in vitro*. Interestingly, they have not been able to demonstrate the inversion reaction *in vitro*. The results with *tnpR* suggest that the purified protein alone, in the absence of any auxiliary proteins, is sufficient to mediate homologous site-specific recombination between directly repeated sequences. However, it is possible that an auxiliary protein is required for the inversion reaction because this reaction is observed *in vivo* but not *in vitro* with the purified protein. Resolvase or *tnpR* functions are ubiquitous. They are part of the TnA class of transposons including Tn*1*, Tn*3*, and γδ (Sherratt *et al.*, 1982).

There is no detectable complementation activity when resolvase is provided in trans to either the phase variation system or the Mu system (M. Silverman and

M. Simon, unpublished results; D. Kamp, personal communication). Resolvase does not appear to mediate inversion in these systems. Nor will it mediate cointegrate formation at the sites recognized by the *hin* and *gin* genes. Furthermore, the *hin* and *gin* gene products do not resolve cointegrates carrying the IRS site (R. Reed, personal communication). On the other hand, it is possible to detect a familial relationship between the resolvase and the *hin* gene. Comparisons of the amino acid sequences deduced from the nucleotide sequence indicated that there was 36% identity in the amino acid sequence corresponding to *hin* and *tnpR* (Simon *et al.*, 1980). Both of these polypeptides, which function to mediate site-specific recombination, are derived from a common ancestor. Whereas homology can be detected between these genes at the amino acid level, experiments using Southern hybridization to detect homology at the nucleic acid sequence level even under low stringency hybridization conditions have not been successful. There are a variety of other homologous, site-specific recombination systems that occur in association with bacteriophage. The classic *int-xis* system in λ mediates bacteriophage integration and excision from the chromosome (Nash *et al.*, 1981). Another similar system is the *lox-cre* system associated with bacteriophage P1 (Sternberg *et al.*, 1981). We have not been able to detect any amino acid homology between *int* and either the TnA resolvase systems or the *hin, gin, cin* family. On the other hand, it is still possible that these systems bear some relationship to each other. They may have evolved and become so specialized that we can no longer detect these relationships at the level of amino acid sequence and must perhaps look at the tertiary structure of these proteins before we can determine how they are related. The sequences at which the recombinases act are all relatively AT-rich. Yet, there is no readily discernible element of specificity that all of the sequences have in common. However, even where specific relationships are known to exist, it is not always easy to discern the sequence elements that are required for site-specific recombination. Thus, for example, Nash (1981) recently compared the sequences of a variety of secondary attachment sites for λ, and there was no simple extensive consensus sequence that was apparently responsible for the specificity of these sites.

We have not exhausted the list of site-specific recombinational functions in bacteria that may influence gene expression. Other systems exist, but they are not as well defined as the ones that we have thus far described. Using the *Salmonella hin* gene as a probe, the genomes of *E. coli* and *Salmonella* have been scanned for related sequences. In *Salmonella* only the *H2*-associated *hin* gene was detected, and no sequence homologous to *hin* was found in *E. coli*. Yet, it is apparent that in *E. coli* an oscillatory switching phenomenon similar to flagellar phase variation exists. Expression of type I pili in *E. coli* switches on and off at frequencies similar to those of *H1* and *H2* variation (Eisenstein, 1981). The molecular details have not yet been described. Many examples of oscillating phenotypes in bacteria and higher organisms exist. Much of the evidence is anecdotal and imprecise, and the very instability of the phenotypes probably discouraged investigation.

Site-specific recombinational functions are not necessarily overt and easy to find because they mediate events that occur at relatively low frequencies. There may, therefore, be a variety of systems that play important roles in gene expression and in the evolution and maintenance of genome structure that have not as yet been defined. Furthermore, each of these systems may have evolved to function in a very specific context. Thus, both the protein and the site at which it functions have undergone specialization and adaptation. With respect to cointegrate resolution, sites adjacent to the IRS site may also be involved in the resolvase-mediated reaction because the resolvase is able to bind at a number of additional sites. Furthermore, mutations at some of these adjacent sites have been shown to affect the rate at which cointegrate resolution occurs *in vivo* (Heffron *et al.*, 1981; Reed and Grindley, 1981). In the phase variation system there is also some difference between the regions flanking sites at which the inversion occurs. For example, the sequences adjacent to the IR(L) site showed extensive homology with the corresponding 34-bp repeat sequence in Mu. On the other hand the sequence adjacent to IR(R) shows much less homology with Mu (Kamp and Kahmann, 1982). Furthermore, the 20-bp sequence adjacent to IR(L) does play a role in the inversion reaction. Variants of the inversion segment have been prepared that carried only the IR(R) inverted repeat on both sides of the 970-bp sequence. These variants are able to invert. However, the inversion occurs at frequencies that are 10–50 times lower than those found when the full 34-bp sequences, IR(L), appeared on both sides of the invertible region (Scott and Simon, 1982).

It is difficult to distinguish operationally between the effect of changes in specific sequences at the site of the recombination event, the effect of changes in the amount of enzyme that is available for the reaction, and the role of other physiological factors in determining the specific rate of the inversion reaction. In 1949, Stocker reported studies that were designed to measure the frequency of flagellar phase transition *in vivo*. He devised an assay that used flagellar phase-specific antisera to immobilize clones that were expressing one antigenic type. By measuring changes in the ratio of immobilized to motile clones, he could estimate the relative rates of switching from one phase to the other. He found that the frequency of phase transition varied with bacterial strains. Furthermore, there was a bias in the frequency of switching so that the rate of transition from phase 2 on to phase 2 off was anywhere from two- to as much as 10-fold higher than the frequency of switching in the opposite direction. He could also estimate the rate at which the culture approached an equilibrium state. In many of these experiments, more than 300 generations of growth were required to achieve apparent equilibrium. The product of the *gin* gene on bacteriophage Mu apparently mediates inversion in both directions with relatively equal frequency, because phage obtained after induction of a lysogen appears to carry equal proportions of the G(+) and G(−) orientations (Bukhari and Abrusio, 1978). The *tnpR* product resolvase, on the other hand, is apparently much more efficient in resolving cointegrate structures than in catalyzing the reverse reaction, that is, the fusion of plasmids each of which carries a single copy of the recognition sequence. It was possible to show that an F factor that carried only one IRS recognition site could mobilize a plasmid that also carried one site if an intact resolvase gene was present (R. Reed and N. Grindley, personal communication). This reaction, however, is extremely inefficient. Furthermore, direct measurement of the rate of the resolution reaction *in vivo* (Muster and Shapiro, unpublished) indicates that the equilibrium of the cointegrate resolution reaction was more than 10^5:1. The resolvase function appears to have evolved to be specifically efficient in resolving cointegrates (Sherratt *et al.*, 1982), whereas the *hin* and *gin* functions are designed to maintain substantial amounts of both inverted forms in the population. The differences in the relative rates of the forward and back reactions observed in *Salmonella* have not been adequately explained. It is possible that the difference is a function of the intracellular concentration of *hin* protein or of some other factor. In early experiments, when the *H2* locus was cloned onto low copy number plasmids or maintained in lysogens, an apparent bias in the distribution of the inverted sequence was observed (Zieg *et al.*, 1977). However, when an abbreviated form of the *H2* inversion region was cloned onto a multicopy plasmid (Scott and Simon, 1982), inversion was found to proceed rapidly in both directions and to result, in about 20 generations of bacterial growth, in a 50:50 equilibrium mixture of the inverted sequences. Apparent differences in the extent of site-specific recombination reactions have been observed in other systems as a function of gene dosage. In the presence of multicopy plasmids that carry the resolvase gene, it has been possible to demonstrate inversion between sites that are recognized by the enzyme but are inverted with respect to each other (Heffron *et al.*, 1981). However, when the resolvase is supplied *in vivo* by only one or two copies of the gene, the inversion reaction is not readily detected (Reed, 1981).

The amount of gene product available to mediate the site-specific recombination event appears to be an important factor in controlling the extent of the reaction. In the case of *hin* and *gin,* we know very little about the mechanisms that regulate the expression of these genes. It would seem that the proteins have to recognize the 14-bp inverted repeated sequence as substrates, and these and adjacent sequences are also part of the promoter region that regulates gene expression. Thus, we might expect recombinational sites to play an important role in regulating the expression of the *hin* and *gin* genes (Silverman *et al.*, 1981). Thus far, there is no direct evidence concerning the regulation of *hin* gene expression. However, in the case of the *tnpR* gene product, it is clear that it binds to regions that regulate both its synthesis and that of the adjacent transposase gene. It definitely functions as a repressor (Kitts *et al.*, 1982). There must be a concerted, highly sophisticated coevolution between the sequences adjacent to and including the site at which recombination occurs and the protein that recognizes these sequences. Campbell (1981) has discussed these kinds of interactions. The recognition sequences play a number of roles. First, they are recognized in order to initiate the site-specific recombination event. Second, both in the TnA transposon systems and in the phase variation and the Mu or P1 systems, these sequences are part of the promoter region for the synthesis of the recombinase gene product. Thus, they not only control the homology that is required to form the substrates in the recombination reaction but may also affect the rate at which the enzyme is synthesized. Finally, this unit (recognition site and recombinase gene) is found associated with a variety of replicons. It is part of the genetic constitution of bacteriophage (Mu, P1, D108), transposons (Tn*1*, Tn*3*, γδ), and cellular genes. The unit behaves as a "module" that can be incorporated into genetic systems as an integral part of their function. Susskind and Botstein (1978) have discussed the role that genetic modules might play in the evolution of bacteriophage. These same arguments can be extended to understand the evolution of transposons and the interaction of transposons with other gene systems.

VI. Evolution of the Phase Variation System

It is clear that the invertible region of the phase variation system of *Salmonella* is related to functionally similar devices in Mu, P1, and transposon TnA. In addition, the *H2* flagellin gene is closely related to the *H1* gene of *Salmonella* and the *hag* gene of *E. coli*. When the DNA sequences comprising the *H2* locus (invertible region and *H2* structural gene) are compared with the Mu phage sequences and with sequences including and 5′ to the *H1* and *hag* genes (Szekely and Simon, 1982), it becomes apparent that the *H2* region is, in fact, a fusion containing sequences from two different sources. There is marked homology between *H2*, *H1*, and *hag* gene sequences, and this close similarity extends for approximately 100 bp from the N-terminus of these genes into the 5′ noncoding region. At this point, there is an abrupt discontinuity where *H2* locus sequences diverge sharply from the *H1* and *hag* sequences, which continue to resemble each other. Approximately 300 bp from the N-terminus of the *H2* gene, considerable homology with Mu sequences is detected. This region (shown in Fig. 7) of homology with Mu extends for 700 bp and corresponds to the IR(R)-*gin* sequence of Mu. A point approximately 100 bp 5′ from the start of the *H2* gene marks a boundary between transposon-like and *H1*-like sequences. Thus, the IR(R) sequence in the phase variation system is embedded in and probably was derived from *H1*-like sequences. In fact, the sequence—ATCAA—which is part of the 14-bp IR(R) recognition sequence (Fig. 5), is conserved in the same relative position in *H1* and *hag*. On the other hand, the IR(L)-*hin* region of *Salmonella* has extensive homology with the IR(R)-*gin* region of Mu (see Fig. 5) and was probably derived from a *gin*-like sequence.

These sequence relationships suggest a process for the evolution of the phase variation system. We have outlined some hypothetical steps in the evolutionary

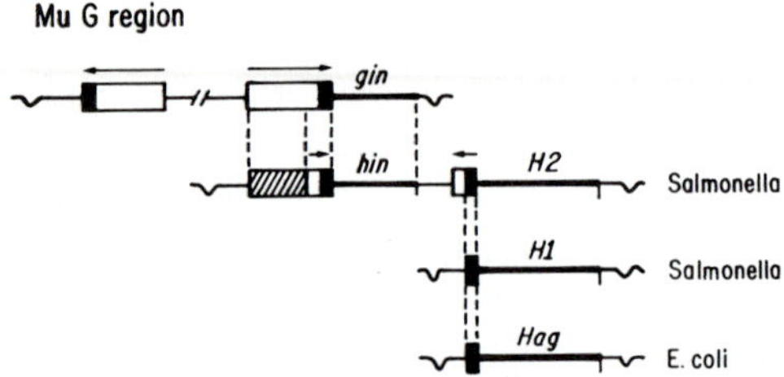

Fig. 7 Sequence relationships. At least two different sources contributed to the DNA sequence of the *H2* region. The *H2* structural gene and a region approximately 100 bp N-terminal to the start of *H2* has a high degree of DNA homology with the *H1* region and with the *hag* region, the homolog of *H1* in *E. coli*. This region extends inside the inversion region and includes the *H2* gene promoter. Past this boundary, about 100 bp left of IR(R), the DNA sequence resembles that of the IR(R)-*gin* region of Mu. IR boxes are not drawn in proportion to genetic scale.

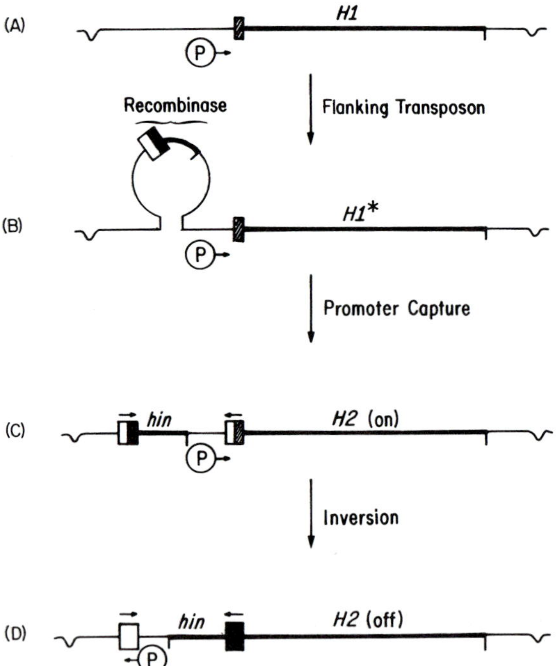

Fig. 8 Evolutionary considerations. To account for the bipartite origin of sequences in the *H2* gene switch, we hypothesize that a mobile genetic element similar to the TnA class of transposons inserted adjacent to an ancestral *H1* gene (A,B). Genetic alteration of a site between the *H1** promoter and the *H1** gene led to a functional recombination site similar to but in inverted configuration to the recombination site carried by the recombinase (C). The recombinase function (*hin*) acting on the pairs of recombination sites inverts the region containing the *H2* gene promoter (D). The recombinase that originally was an integral part of a transposon now functions to regulate an unrelated gene system.

process in Fig. 8. The IR(L)-*hin* recombinase could be derived from a transposon-like unit that inserted adjacent to an ancestral *H1* gene. A low-frequency recombinational event at a site adjacent to that of the flagellin gene could capture the promoter sequence of that gene. Subsequent repair or recombinational processes similar to those of gene conversion could lead to homogenization of the sequence immediately adjacent to the recombination site, with the resultant generation of a functional IR(L)–IR(R) site pair. The consequence of such a promoter capture would be on–off switching of flagellin synthesis mediated by the recombinase function. Thus, we imagine that the phase variation system developed as a result of a number of steps, including the insertion of a mobile element adjacent to an ancestral *H1* gene and the capture of the *H1* promoter by a recombinational function that was part of the mobile element. We have not explicitly addressed the need for a duplicative event that resulted in two flagellin genes in the same cell or the question of the origin of the *rh1* gene.

VII. Functional Significance

It is clear, from the examination of the oscillatory switches described here, that recombinational processes are used to regulate gene expression. What applications do prokaryotes make of these devices? With the *Salmonella* phase variation and Mu phage host range variation, the organism's ability to adapt to changing circumstances is apparently enhanced. In a span of time that is relatively short compared to the evolutionary time scale, an altered identity or life-style could be expressed. The switch in phenotype does not necessarily require a stimulus or inducing event. A small portion of a population could be thought to be pre-adapted to environmental change. Life in the prokaryotic world would surely be subject to variations that are both abrupt and extreme. On the other hand, it is possible that there are events that induce or affect the rate at which switching occurs. In the transposon TnA systems, cointegrate resolution apparently occurs frequently and efficiently. On the other hand, both in the phase variation system and the Mu system, switching of the inversion region occurs at relatively slow rates. In the phase variation system, the fastest observed rate of switching involves one event in 100 bacteria per generation. It is possible that there are host factors in addition to the gene product itself that can increase or decrease the frequency of site-specific recombination and that can enhance either the inversion or the resolution reaction.

As far as prokaryotes are concerned, recombinational switching may be particularly useful for regulating the expression of characteristics that determine how an organism interfaces with its environment. Thus, one might expect to find that properties such as cell motility, permeability, and adhesion are regulated by recombinational devices. It is not clear at this time if utilization of this mechanism of genetic regulation is a general phenomenon. However, observations of unstable oscillating phenotypes, especially with respect to cell surface properties, are common (Lambden *et al.*, 1980; Swanson, 1978). Bacteria isolated from natural environments or as agents of disease readily lose adhesive or virulence properties in a laboratory environment. These may reappear if the cells are passaged in the original environment. Information for these properties is not lost but rather stored in a cryptic form. Recombinational switching may explain how this cryptic information is turned on.

Acknowledgment

The authors' research on phase variation has been supported by grants from the National Science Foundation.

References

Andrewes, F. W. (1922). Studies in group-agglutination. I. The *Salmonella* group and its antigenic structure. *J. Pathol. Bacteriol.* **25,** 515–521.

Bächi, B., and Arber, W. (1977). Physical mapping of BglII, BamHI, EcoRI, HindIII and PstI restriction fragments of bacteriophage PI DNA. *Mol. Gen. Genet.* **153,** 311–324.

Bukhari, A. I., and Abrusio, L. (1978). The invertible segment of bacteriophage Mu DNA determines the adsorption properties of Mu particles. *Nature (London)* **271,** 575–577.

Calos, M. P., and Miller, J. H. (1980). Transposable elements. *Cell* **20,** 579–596.

Campbell, A. (1981). Some general questions about movable elements and their implications. *Cold Spring Harbor Symp. Quant. Biol.* **45,** 1–9.

Chou, J., Lemaux, P., Casadaban, M., and Cohen, S. N. (1979). Transposition protein of Tn3: identification and characterization of an essential repressor controlled gene product. *Nature (London)* **282,** 801–806.

Chow, L. T., and Bukhari, A. I. (1976). The invertible DNA segments of coliphages Mu and P1 are identical. *Virology* **74,** 242–248.

Chow, L. T., Kahmann, R., and Kamp, D. (1977). Electron microscopic characterization of DNA's of non-defective deletion mutants of bacteriophage Mu. *J. Mol. Biol.* **113,** 591–605.

Dawson, G., and Smith-Keary, P. F. (1963). Episomic control of mutation in *Salmonella typhimurium*. *Heredity* **18,** 1–20.

Eisenstein, B. (1981). Phase variation of type I frimbriae in *E. coli* is under transcriptional control. *Nature (London)* **214,** 337–339.

Enomoto, M., and Stocker, B. A. D. (1975). Integration, at *hag* or elsewhere, of *H2* (phase-2 flagellin) genes transduced from *Salmonella* to *Escherichia coli*. *Genetics* **81,** 595–614.

Fujita, H., Yamaguchi, S., and Iino, T. (1973). Studies on H–O variants in *Salmonella* in relation to phase variation. *J. Gen. Microbiol.* **76,** 127–134.

Gill, P. E., Heffron, F., and Falkow, S. (1979). Identification of the protein encoded by the transposable element Tn3 which is required for its transposition. *Nature (London)* **282,** 797–801.

Heffron, F., McCarthy, B. J., Ohtsubo, H., and Ohtsubo, E. (1979). DNA sequence analysis of the ransposon Tn3: three genes and three sites involved in transposition of Tn3. *Cell* **18,** 1153–1163.

Heffron, F., Kostriken, R., Morita, C., and Parker, R. (1981). Tn3 encodes a site-specific recombination system: identification of essential sequences, genes, and the actual site of recombination. *Cold Spring Harbor Symp. Quant. Biol.* **45,** 259–268.

Holliday, R., and Pugh, J. E. (1975). DNA modification mechanisms and gene activity during development. *Science (Washington, D.C.)* **187,** 226–230.

Howe, M. M., Schumm, J. W., and Taylor, A. L. (1979). The S and U genes of bacteriophage Mu are located in the invertible G segment of Mu DNA. *Virology* **92,** 108–124.

Iino, T. (1961). A stabilizer of antigenic phase in *Salmonella abortus-equi*. *Genetics* **46,** 1465–1469.

Iino, T. (1969). Genetics and chemistry of bacterial flagella. *Bacteriol. Rev.* **33,** 454–475.

Iino, T. (1977). Genetics of structure and function of bacterial flagella. *Annu. Rev. Genet.* **11,** 161–182.

Kahmann, R., and Kamp, D. (1982). Substrates for site-specific recombination: the inverted repeat sequences flanking the G segment of bacteriophage Mu. *Cell* (unpublished).

Kamp, D., and Kahmann, R. (1982). The relationship of two invertible segments in bacteriophage Mu and *Salmonella typhimurium* DNA. *Mol. Gen. Genet.* (unpublished).

Kamp, D., Kahmann, R., Zipser, D., Broker, T. R., and Chow, L. T. (1978). Inversion of the G DNA segment of phage Mu controls phage infectivity. *Nature (London)* **271,** 577–580.

Kamp, D., Chow, L. T., Broker, T. R., Kwoh, D., Zipser, D., and Kahmann, R. (1979). Site-specific recombination in phage Mu. *Cold Spring Harbor Symp. Quant. Biol.* **43,** 1159–1167.

Kauffmann, F. (1951). "Enterobacteriaceae." Munksgaard, Copenhagen.

Kitts, P. A., Hamond, A., and Sherratt, D. J. (1982). Inter-replicon transposition of TnI/3 occurs in two sequential genetically separable steps. *Nature (London)* **295,** 626–628.

Kleckner, N. (1981). Transposable elements in prokaryotes. *Annu. Rev. Genet.* **15,** 341–404.

Kutsukake, K., and Iino, T. (1980a). A *trans*-acting factor mediates inversion of a specific DNA segment in flagellar phase variation of *Salmonella*. *Nature (London)* **284,** 479–481.

Kutsukake, K., and Iino, T. (1980b). Inversions of specific DNA segments in flagellar phase variation in *Salmonella* and inversion systems of bacteriophages P1 and Mu. *Proc. Natl. Acad. Sci. USA* **77,** 7338–7341.

Lambden, P. R., Robertson, J. N., and Watt, P. J. (1980). Biological properties of two distinct pilus types produced by isogenic variants of *Neisseria gonorrhoeae*. *J. Bacteriol.* **141,** 393–396.

Lederberg, J., and Edwards, P. R. (1953). Serotypic recombination in *Salmonella*. *J. Immunol.* **71,** 232–240.

Lederberg, J., and Iino, T. (1956). Phase variation in *Salmonella*. *Genetics* **41,** 743–757.

Lee, H. J., Ohtsubo, E., Deonier, R. C., and Davidson, N. (1974). Electron microscope heteroduplex studies of sequence relations among plasmids of *Escherichia coli*. *J. Mol. Biol.* **89,** 585–597.

Makela, P. H. (1964). Genetic homologies between flagellar antigens of *Escherichia coli* and *Salmonella abony*. *J. Gen. Microbiol.* **35,** 503–510.

McClintock, B. (1941). The stability of broken ends of chromosomes in *Zea mays*. *Genetics* **26,** 234–282.

McClintock, B. (1951). Chromosomal organization and genic expression. *Cold Spring Harbor Symp. Quant. Biol.* **16,** 13–47.

McClintock, B. (1956). Controlling elements and the gene. *Cold Spring Harbor Symp. Quant. Biol.* **21,** 197–216.

McClintock, B. (1965). The control of gene action in maize. *Brookhaven Symp. Biol.* **18,** 162–184.

Nash, H. A. (1981). Integration and excision of bacteriophage: the mechanism of conservative site specific recombination. *Annu. Rev. Genet.* **15,** 143–167.

Nash, H. A., Mizuuchi, K., Enquist, L. W., and Weisberg, R. (1981). Strand exchange in integrative recombination: genetics, biochemistry, and models. *Cold Spring Harbor Symp. Quant. Biol.* **45,** 417–428.

Pearce, U. B., and Stocker, B. A. D. (1967). Phase variation of flagellar antigens in *Salmonella:* abortive transduction studies. *J. Gen. Microbiol.* **49,** 335–347.

Reed, R. (1981). Transposon-mediated site-specific recombination: a defined *in vitro* system. *Cell* **25,** 713–719.

Reed, R. R., and Grindley, N. D. F. (1981). Transposon-mediated site-specific recombination *in vitro:* DNA cleavage and protein–DNA linkage at the recombination site. *Cell* **25,** 721–728.

Scott, T. N., and Simon, M. (1982). Genetic analysis of the mechanism of the *Salmonella* phase variation site-specific recombination system. *Mol. Gen. Genet.* (unpublished).

Shapiro, J. A., and Cordell, B. (1982). Eukaryotic mobile and repeated genetic elements. *Biol. Cell.* (in press).

Sherratt, D., Arthur, A., and Burke, M. (1981). Transposon-specified, site-specific recombination systems. *Cold Spring Harbor Symp. Quant. Biol.* **45,** 275–282.

Sherratt, D., Arthur, A., and Dyson, P. (1982). Site-specific recombination. *Nature (London)* **294,** 608–610.

Silverman, M., and Simon, M. (1977). Bacterial flagella. *Annu. Rev. Microbiol.* **31,** 397–419.

Silverman, M., and Simon, M. (1980). Phase variation: genetic analysis of switching mutants. *Cell* **19,** 845–854.

Silverman, M., Zieg, J., Hilmen, M., and Simon, M. (1979a). Phase variation in *Salmonella:* genetic analysis of a recombinational switch. *Proc. Natl. Acad. Sci. USA* **76,** 391–395.

Silverman, M., Zieg, J., and Simon, M. (1979b). Flagellar-phase variation: isolation of the *rhl* gene. *J. Bacteriol.* **137,** 517–523.

Silverman, M., Zieg, J., Mandel, G., and Simon, M. (1981). Analysis of the functional components of the phase variation system. *Cold Spring Harbor Symp. Quant. Biol.* **45,** 17–26.

Simon, M., Zieg, J., Silverman, M., Mandel, G., and Doolittle, R. (1980). Phase variation: evolution of a controlling element. *Science (Washington, D.C.)* **209,** 1370–1374.

Sternberg, N., Hamilton, D., Austin, S., Yarmolinsky, M., and Hoess, R. (1981). Site-specific recombination and its role in the life cycle of bacteriophage P1. *Cold Spring Harbor Symp. Quant. Biol.* **45,** 297–309.

Stocker, B. A. D. (1949). Measurement of the rate of mutation of flagellar antigenic phase in *Salmonella typhimurium. J. Hyg.* **47,** 398–413.

Stocker, B. A. D., and Makela, P. H. (1978). Genetics of the (gram-negative) bacterial surface. *Proc. R. Soc. London Ser. B* **202,** 5–30.

Susskind, M. M., and Botstein, D. (1978). Molecular genetics of bacteriphage P22. *Microbiol. Rev.* **42,** 385–413.

Suzuki, H., and Iino, T. (1975). Absence of messenger ribonucleic acid specific for flagellin in nonflagellate mutants of *Salmonella. J. Mol. Biol.* **95,** 549–546.

Swanson, J. (1978). Studies on gonococcus infection. XII. Colony color and opacity variants of gonococci. *Infect. Immun.* **19,** 320–331.

Szekely, E., and Simon, M. (1981). Homology between the inverible deoxyribonucleic acid sequence that controls flagellar-phase variation in *Salmonella* and deoxyribonucleic acid sequences in other organisms. *J. Bacteriol.* **148,** 829–836.

Szekely, E., and Simon, M. (1982). Evolution of phase variation in *Salmonella*. *Nature (London)* (unpublished).

Van De Putte, P., Cramer, S., and Giphart-Gassler, M. (1980). Invertible DNA determines host specificity of bacteriophage Mu DNA. *Nature (London)* **286,** 218–222.

Zieg, J., and Simon, M. (1980). Analysis of the nucleotide sequence of an invertible controlling element. *Proc. Natl. Acad. Sci. USA* **77,** 4196–4200.

Zieg, J., Silverman, M., Hilmen, M., and Simon: M. (1977). Recombinational switch for gene expression. *Science (Washington, D.C.)* **196,** 170–172.

Zieg, J., Hilmen, M., and Simon, M. (1978a). Regulation of gene expression by site-specific inversion. *Cell* **15,** 237–244.

Zieg, J., Silverman, M., Hilmen, M., and Simon, M. (1978b). The mechanism of phase variation. *In* "The Operon" (J. H. Miller and W. S. Reznikoff, eds.), pp. 411–423. Cold Spring Harbor Lab., Cold Spring Harbor, New York.

CHAPTER 13

Mating-Type Genes of *Saccharomyces cerevisiae*

JAMES E. HABER

Mobile Genetic Elements

ISBN 0-12-638680-3

I. Introduction

The mating-type genes of the yeast *Saccharomyces cerevisiae* have been studied extensively both because of their role in controlling a large variety of cell functions and because of the unusual way in which these genes are themselves regulated. The problem in presenting a review of the remarkable progress made in this field is simply that an investigation of one aspect of the problem frequently presupposes knowledge about another facet. Therefore, a brief general review of the major features of this system is in order before examining any one area in great detail. The two codominant alleles of the yeast mating type locus, *MAT***a** and *MAT*α, are believed to be themselves regulatory genes that control a large number of different cell properties. In most laboratory strains (heterothallic strains) mating-type alleles are essentially stable, although rare conversions of *MAT***a** to *MAT*α or vice versa can occur (Fig. 1). In homothallic strains, howev-

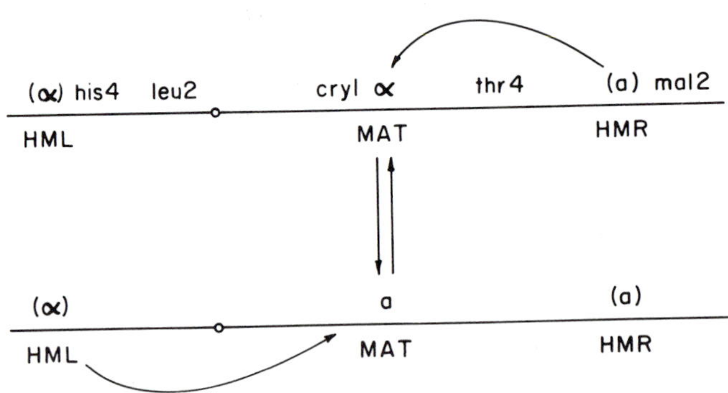

Fig. 1 Mating-type conversions in the yeast *Saccharomyces cerevisiae*. Conversions of the mating-type (*MAT*) locus on chromosome III of yeast occur by transposition of **a**- or α-specific sequences that are found at two unexpressed copies of mating-type information, *HML* and *HMR*, located at the two ends of the same chromosome. In most strains, *HML* carries α information (*HML*α) and *HMR* contains **a** information (*HMR***a**). Several other markers on chromosome III that will be discussed in later sections of the chapter are also noted in their approximate positions. The distance between *MAT* and *HML* or *MAT* and *HMR* are both greater than 50 cM; thus, all three loci segregate essentially independently in meiosis.

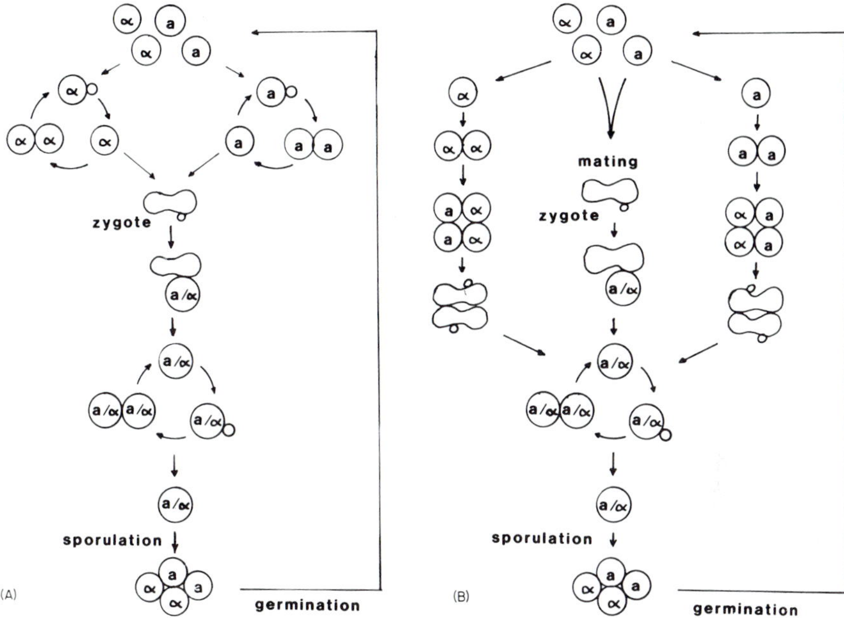

Fig. 2 Heterothallic and homothallic life cycles of *Saccharomyces cerevisiae*. In heterothallic strains (A) haploid spores can either germinate and grow vegetatively in a stable haploid phase or mate with a cell of the opposite mating type, producing a zygote. The zygote, in turn, gives rise to *MAT***a**/*MAT*α diploid cells that can grow vegetatively, but will also undergo meiosis and sporulation when placed under appropriate nitrogen-starvation conditions. When asci containing four haploid spores are returned to a vegetative-growth medium, the entire cycle will be recapitulated. In homothallic strains (B), haploid spores may follow several alternative cell cycles. As in heterothallic strains, cells or spores of opposite mating type can directly form a zygote and give rise to *MAT***a**/*MAT*α diploids. Alternatively, a haploid cell of either mating type can undergo several rounds of mitotic growth, during which one-half of the haploid cells become switched to the opposite mating type. Cells of the opposite mating type, in close proximity, then form zygotes that, in turn, give rise to *MAT***a**/*MAT*α diploids.

er, a haploid cell may switch from one *MAT* allele to the other as frequently as every cell division. These conversions of mating type are not merely phenotypic but represent the physical replacement of one DNA sequence by another. These changes in mating type depend on the presence elsewhere on the same chromosome of intact, but unexpressed, copies of mating type information. The conversion of one mating-type allele to the other results from a transposition event in which a copy of the unexpressed information at either *HML* or *HMR* is used to replace the DNA sequence initially at *MAT*. In most laboratory strains, an unexpressed copy of α information is found on the left arm of chromosome III at the locus *HML*α, whereas an unexpressed copy of **a** information is found near the end of the right arm, at *HMR***a** (Fig. 1). Current research on the yeast mating-type locus has therefore fallen into three major areas. First, there have been extensive studies of the genetic information expressed by *MAT***a** and *MAT*α in controlling a wide variety of cell functions. Second, there has been a major effort to understand how the mating-type information at the *MAT* locus is expressed whereas the same genetic information is not expressed at either *HML* or *HMR*. Third, a great deal of effort has been devoted to understanding the mechanism by which yeast mating-type genes are transposed and also how such transpositions are genetically regulated. It will become obvious that success in understanding any one of these areas has depended on progress in the other two.

II. The Mating-Type Locus

A. Regulation of Cell Type by *MAT* Alleles

The mating-type locus of the yeast *S. cerevisiae* was one of the first genetic loci in yeast to be studied. Lindegren and Lindegren (1943) isolated stable haploids that could be distinguished by their ability to mate with each other and form nonmating diploids. The two mating types, designated **a** and α, were determined by two alleles of a single genetic locus, *MAT***a** and *MAT*α. These two alleles were apparently codominant, because the *MAT***a**/*MAT*α diploids formed by conjugation had a distinctly different phenotype from either haploid; diploids were nonmating. Over the past four decades, it has become clear that the mating type genes directly or indirectly control a number of very different aspects of cell physiology, ranging from mating per se to the ability of diploid cells to undergo differentiation (meiosis and sporulation). Diploids homozygous for either *MAT***a** or *MAT*α can be distinguished from *MAT***a**/*MAT*α cells by a variety of criteria, summarized in Table I. These differences include:

1. *MAT***a**/*MAT***a** and *MAT*α/*MAT*α homozygous diploids mate with cells of the opposite mating type, as do their haploid counterparts, whereas *MAT***a**/*MAT*α diploids are nonmating (Roman *et al.,* 1955).
2. When placed under nitrogen starvation conditions in the presence of acetate, *MAT***a**/*MAT*α diploids will undergo meiosis and form an ascus containing four haploid spores, two *MAT*α and two *MAT***a.** *MAT***a**/*MAT***a** and *MAT*α/*MAT*α diploids, although they exhibit a number of physiological responses similar to those of *MAT***a**/*MAT*α diploids, do not initiate meiotic DNA synthesis and do not enter meiosis (Roman and Sands, 1953; Roth and Lusnak, 1970; Esposito *et al.*, 1969; Petersen *et al.*, 1979; Wright and Dawes, 1979; Pearson and Haber, 1980; Kraig and Haber, 1980, 1981; Zubenko and Jones, 1981).
3. *MAT***a**/*MAT*α diploids have a pattern of budding that is different from that of cells homozygous for mating type. In *MAT***a** and *MAT*α homozygotes, after one cell buds and forms a daughter, the next pair of buds generally appears near

Table I Effect of *MAT* Alleles on Diploid Phenotypes

Condition	Genotype		
	*MAT***a**/*MAT*α	*MAT***a**/*MAT***a**	*MAT*α/*MAT*α
Mating	Nonmating	Mates with α cells	Mates with **a** cells
Sporulation conditions	Meiosis, sporulation	No meiosis or sporulation	No meiosis or sporulation
Budding	Polar	Medial	Medial
Radiation resistance	Greater	—	—
Spontaneous mitotic recombination	Greater	—	—
Mating hormone secretion	None	Secretes **a** factor	Secretes α factor
Ty1 transcription	Low	High	High
Homothallic (*HO*) mating type switching	No	Yes	Yes

the median of the first cell division; however, in *MAT***a**/*MAT*α cells, the two new buds emerge at the opposite ends of the two parent cells (polar) (Hicks *et al.*, 1977b; Strathern *et al.*, 1979a).

4. Sensitivity to ionizing radiation and to radiomimetic chemicals is distinctly different among the different cell types. *MAT***a**/*MAT*α cells are approximately twice as resistant to UV light, X rays, or the mutagen MMS than are *MAT***a** or *MAT*α diploids (Livi and MacKay, 1980).

5. Spontaneous mitotic recombination is about twofold greater in *MAT***a**/*MAT*α diploids than in **a** or α diploids (Friis and Roman, 1968).

6. The mating-type alleles also apparently interact with several cell cycle functions. Reed (1980) has isolated several temperature-sensitive cell cycle mutations that cause *MAT***a** or *MAT*α cells to stop growing at 34°C but arrest *MAT***a**/*MAT*α cells only at 36°C. Similarly, J. H. McCusker (personal communication) has found a "clumpy" cell phenotype that is expressed in *MAT***a** or *MAT*α cells but not in *MAT***a**/*MAT*α diploids.

7. Other mating functions are also controlled by the *MAT* alleles. To arrest cells at the end of the G_1 stage of the cell cycle (in order to facilitate conjugation of cells at the same stage of the cell cycle), *MAT*α cells secrete an oligopeptide mating hormone (α factor) that is not secreted by either *MAT***a** or *MAT***a**/*MAT*α cells (Duntze *et al.*, 1970; Bucking-Throm *et al.*, 1973; Stotzler *et al.*, 1976a,b). Similarly, *MAT***a** cells secrete **a** mating factor that is not produced by the other two cell types (Betz *et al.*, 1977).

8. Mating-type alleles control the level of transcription of at least some of the copies of the yeast transposable element, TY1 (Elder *et al.*, 1981; see also Chapter 7). Transcription of TY1 sequences is approximately 20-fold greater in *MAT***a** or *MAT*α cells than in *MAT***a**/*MAT*α diploids. This same mating-type effect is seen in a variety of mutations selected as apparent up or down promoter

mutations for several metabolic genes (cytochrome *c* (*CYC7*), arginase (*CAR-GA*), and urea amidolyase (*DUR1*) (Errede *et al.*, 1980). In the cases of *CYC7-H2*, it has been demonstrated that the regulatory mutations are in fact insertions of a Ty1-like element and that the mating type effects on transcription of these metabolic genes are in fact mediated by the transposon (Errede *et al.*, 1980).

9. Finally, the homothallism gene, *HO*, which is essential for the efficient switching of mating-type alleles, is also under mating-type control. *MAT***a** and *MAT*α haploids or diploids will continue to switch mating-type alleles, whereas *MAT***a**/*MAT*α diploids (or other strains in which both mating-type alleles are expressed) no longer exhibit high-frequency conversions of the *MAT* alleles. Recently, R. Jensen and I. Herskowitz (personal communication) have cloned this gene and shown by Northern gel analysis that the *HO* gene is transcribed in *MAT***a** or *MAT*α cells but not in *MAT***a**/*MAT*α cells. Analysis of DNA and RNA transcripts from heterothallic (*ho*) strains also suggests that the recessive *ho* allele does not contain significant deletions or insertions relative to *HO* and is probably inactive because of one or more mutations (R. Jensen and I. Herskowitz, personal communication).

That the mating-type alleles affect such a broad spectrum of functions has led to a general supposition that the genes at the *MAT* locus code for regulatory functions that interact with a variety of genes controlling aspects of mating, initiation of meiosis, recombination, bud emergence, and the like. This assumption has been supported by a detailed genetic analysis of the *MAT* locus, which has revealed a single cistron in *MAT***a** and two cistrons in *MAT*α (see Section II,B). Furthermore, a number of mutations at other loci have been discovered that affect some, but not all, of the various *MAT*-specific functions. For example, MacKay and Manney (1974a,b) and Hartwell (1980) have identified a number of mutations that prevent normal conjugation of either *MAT***a** or *MAT*α cells. In addition to at least six "sterile" complementation groups that render both *MAT***a** and *MAT*α cells unable to conjugate, at least one **a**-specific and three α-specific sterile (*ste*) mutations have been demonstrated. Two of the α-specific sterile mutations also affect other cell functions, such as the uptake of dTMP (*tup1;* Wickner, 1974; Lemontt *et al.*, 1980) or the expression of killer toxin (*kex2;* Leibowitz and Wickner, 1976). Some of these mutations may affect the synthesis or secretion of the mating factors.

B. a and α Gene Functions

1. MUTATIONS IN *MAT*α

MacKay and Manney (1974a,b) first identified mutations at the mating-type locus that altered the mating and sporulation ability of *MAT*α strains. Three independently isolated mutations made *MAT*α cells sterile (able to mate only rarely with either *MAT***a** or *MAT*α strains). Two of these mutations, originally designated *VN33* and *VC73*, have been studied extensively. Strathern (1977) demonstrated that the two mutations complement each other to give an α-mating diploid and therefore define two complementation groups. *mat*α*1-5* (originally VN33) renders the cell nearly nonmating; however, a *mat*α*1-5*/*MAT***a** diploid is nonmating and able to sporulate (MacKay and Manney, 1974b; Strathern *et al.*, 1981). Thus, *mat*α*1* mutants can still provide α information necessary to abolish mating in diploids and to permit meiosis and spore formation.[1] In contrast, *mat*α*2-1* (originally VC73) is a sterile mutation that does not provide the α information necessary for *mat*α*2-1*/*MAT***a** diploids to sporulate (MacKay and Manney, 1974b; Strathern *et al.*, 1981). In addition, *mat*α*2-1MAT***a** diploids have a weak **a**-mating phenotype (Haber and George, 1979). Additional alleles of both cistrons have been obtained, with essentially the properties described for the first two mutations. Another of MacKay and Manney's (1974a,b) mutations proved to be a *mat*α*1* mutation (*mat*α*1-2;* originally strain VC2) (Strathern, 1977), and an ochre mutation of *mat*α*1* was also isolated by D. C. Hawthorne (personal communication). MacKay and Manney (1974a,b) also described another *mat*α*2* mutation (*mat*α*2-4;* from strain VP1) (Strathern, 1977). In addition, Tatchell *et al.* (1981) have used *in vitro* mutagenesis (inserting synthetic *Xho*I restriction endonuclease recognition sites) at a variety of locations in a cloned *MAT*α segment to identify a large number of additional mutations. All of these mutations could be classified into the two previously defined complementation groups. However, two mutations in the *mat*α*2* region appear to separate the sterile-mating defect from the sporulation deficiency. One insertion at the 3′-end of the coding sequence of *MAT*α*2* is unable to complement the *mat*α*2-1* mutation with respect to mating but does provide α information for sporulation of a *mat*α*2-1*/*MAT***a** diploid. Sequence analysis suggests that this mutation removes the normal termination codon and extends the *MAT*α*2* polypeptide. The second unusual mutation, an apparent frameshift nearer the amino terminus, does not alter the sterile mating but does provide the information necessary for a *mat*α*2*/*MAT***a** or a *MAT***a**/*MAT***a** diploid to sporulate. These two mutations have not yet been thoroughly investigated.

2. MUTATIONS IN *MAT***a**

A single genetic complementation group has been identified in *MAT***a.** The first mutation, designated *mat***a***, was found by Kassir and Simchen (1976) in a search for sporulation-defective mutations. The *mat***a*** mutation does not alter the cell's ability to mate with *MAT*α strains, but it does not contribute the **a**

[1]Recessive mutations in yeast are designated by lowercase letters. Thus, *mat*α*1-5* is a recessive allele of the *MAT*α*1* cistron. For overall simplicity, when both *MAT*α cistrons are wild-type, the locus is designated *MAT*α. When a cell carries a *mat*α*1* mutation but has a wild-type *MAT*α*2* cistron, it will be designated simply as *mat*α*1*.

functions characteristic of *MAT***a**/*MAT*α diploids. Instead of being nonmating and able to sporulate, *mat***a***/*MAT*α diploids are α-mating and asporogenous. *mat***a*** is also recessive to *MAT***a.** For example, a *mat***a***/*MAT***a** diploid will mate with a *MAT*α/*MAT*α strain to form a nonmating, sporulating tetraploid (Haber and George, 1979). The fact that *mat***a*** cells are **a**-mating is consistent with Strathern's (1977) observation that *mat*α*1 mat*α*2* double mutants (lacking all *MAT*α functions) behave identically to *mat***a***. Thus, strains with no functional *MAT* genes are constitutively **a**-like in mating behavior. This suggests that **a**-mating, in contrast to other **a**-specific functions, is not regulated by the *MAT***a** locus per se.

Additional *mat***a***1* mutations have been isolated by two quite different approaches. Tatchell *et al.* (1981) have used *in vitro* mutagenesis of a cloned *MAT***a** gene to insert a synthetic *Xho*I restriction endonuclease recognition site at random locations in the region. Autonomously replicating plasmids with *in vitro* deletion-inserts were transformed into a *mat***a***/*MAT*α strain to look for plasmids that failed to complement the *mat***a***1* mutation. Four *mat***a***1* frameshift mutations were identified and shown by DNA sequence and restriction endonuclease analysis to lie within the apparent coding region of the *MAT***a***1* transcript.

An alternative approach has been developed by Klar (Klar and Fogel, 1979; Klar, 1980). Several mutations have been identified that permit the expression of the normally silent mating-type information at *HML* and *HMR* (see Section IV,F). Consequently, an *HML*α *MAT*α *HMR***a** strain carrying the *mar1-1* mutation will become nonmating, as both α and **a** information is expressed in the haploid cell. A mutation in *HMR***a** to create an *hmr***a***1* mutation will render the cell α-mating (*HML*α *MAT*α *hmr***a** *mar1*), a phenotype that is easily selected from a background of nonmating cells. In practice, a nonmating *HML*α *MAT*α *HMR***a** *mar1* strain is mixed with a *MAT***a** Mar$^+$ strain carrying complementing auxotrophic markers. The rare wild-type diploids that appear result from conjugation of the *MAT***a** Mar$^+$ strain with α-mating mutants derived from the *HML*α *MAT*α *HMR***a** *mar1* strain. A number of independent mutations of this sort were selected, some of which were shown to be nonsense mutations by the restoration of nonmating in *mar1* strains carrying the temperature-sensitive *amber* suppressor, *SUP4-3* (Klar, 1980). Although these mutations were isolated at *hmr***a***1*, transposition of these sequences to *MAT* in the absence of the *mar1-1* mutation demonstrated that they were equivalent to *mat***a***1* mutations (Klar, 1980).

Although no additional *MAT***a** cistrons have been identified, it is possible that mutations will be isolated that will discriminate between mating and sporulation functions. The chromosomal rearrangement *SAD1*, containing a fusion between *HMR***a** and *MAT*α (Hopper and MacKay, 1980; Kassir and Herskowitz, 1980; J. B. Hicks, personal communication; see Section V,B), does not express **a** information to make either a *MAT*α haploid or a *MAT*α/*MAT*α diploid nonmating. Nevertheless, this sequence does provide information necessary to allow a *MAT*α/*MAT*α diploid to sporulate. Sporulation is quite efficient, with as many as

80% of the cells completing ascospore formation. Northern blot analysis probed with a *MAT***a** probe fails to show significant transcription from this region even when cells are placed under sporulation conditions (Kraig, 1980). These results suggest that the level of expression of *MAT***a** sequences necessary to allow meiosis and sporulation is one to two orders of magnitude less than that necessary to make a *MAT***a**/*MAT*α cell nonmating.

Other evidence argues that the levels of expression of *MAT***a** and *MAT*α must be similar, but not necessarily stoichiometric, to make a cell nonmating. Gunge and Nakatomi (1972) showed, for example, that tetraploids with genotypes *MAT***a**/*MAT*α/*MAT*α/*MAT*α or *MAT*α/*MAT***a**/*MAT***a**/*MAT***a** were nonmating. Similarly, a diploid of genotype *HML*α/*HML*α *mat***a***1*/*MAT*α *HMR***a**/*HMR*α and homozygous for the *cmt* mutation that turns on *HML* and *HMR* is also nonmating (Haber and George, 1979).

C. Physical Characterization of the Mating-Type Genes

Recombinant DNA plasmids carrying mating type genes were first isolated by Hicks *et al.* (1979) and then by Nasmyth and Tatchell (1980). In both cases, pBR322 plasmids carrying a yeast origin of replication and the selectable yeast gene *LEU2* were used. Hicks *et al.* (1979) surveyed a plasmid bank of *Bam*H1 fragments and isolated a transformant that complemented the recessive *mat***a*** mutation. A mitotically unstable clone expressing α information was characterized and shown to contain the normally unexpressed *HML*α gene. The plasmid contained sequences homologous to three different regions of the yeast genome. One of these regions could be shown to contain the *MAT* locus, by a comparison of Southern blots of *Hind*III digests of DNA from isogenic strains carrying either *MAT***a** or *MAT*α. The *MAT*α restriction fragment was approximately 100 bp larger than the equivalent *MAT***a** fragment. This same 100-bp difference can be found between *HML*α and *HML***a** and between the **a** and α alleles of *HMR;* these differences could then be used to assign the *Hind*III restriction fragments of *HML* and *HMR* (Fig. 3). One other useful difference was noted between **a** and α sequences: There is a unique *Bgl*II site in *MAT***a** (and in *HML***a** and *HMR***a**) that is missing in *MAT*α, *HML*α, and *HMR*α.

Nasmyth and Tatchell (1980) isolated a plasmid carrying *MAT*α that complemented a *mat*α*1* mutation. Subsequently, they screened other plasmids for those containing homologous sequences and isolated plasmids containing *HML*α, *MAT***a,** and *HMR***a.** Heteroduplex analysis of these plasmids established the boundaries of the mating-type region (i.e., the extent of homology between *HML*α and *MAT***a,** between *HMR***a** and *MAT*α, and between *HML*α and *HMR***a**). In addition, heteroduplexes between *MAT***a** and *MAT*α confirmed that there was a region of nonhomologous sequences distinguishing the two *MAT* alleles. A

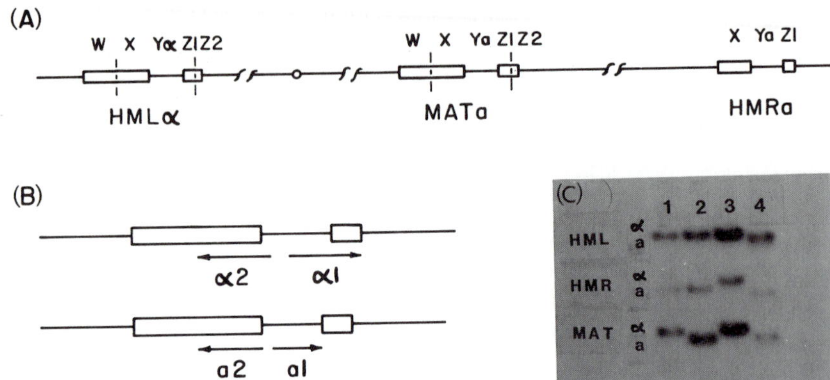

Fig. 3 Structure of mating-type information at *HML*, *MAT*, and *HMR* on chromosome III. (A) The three loci containing **a** or α information lie in the same orientation. **a**- or α-specific sequences are completely nonhomologous. These Y**a** (642 bp) and Yα (747 bp) sequences are surrounded by sequences shared by *HML*, *MAT*, and *HMR*. All three loci share two flanking regions, *X* (704 bp) and *Z*1 (239 bp). In addition, *HML* and *MAT* share two additional sets of sequences: *W* (723 bp) and *Z*2 (88 bp). (B) Transcription and genetic complementation groups of *MAT*α and *MAT***a.** Both *MAT*α transcripts are about 800 bases; the two *MAT***a** transcripts are about 600 and 450 bases. Although both genes are transcribed into two RNAs, there is only one identifiable complementation group in *MAT***a.** (C) Distinctions between **a** and α alleles at *HML*, *MAT*, and *HMR* in Southern blots of genomic DNA digested with *Hind*III gives rise to three fragments: *HML*, *HMR*, and *MAT*. The approximately 100-bp difference between **a** and α alleles at each locus can be seen in blots of the following four strains: 1, *HML*α *MAT*α *HMR***a**; 2, *HML*α *MAT***a** *HMR***a**; 3, *HML*α *MAT*α *HMR*α; 4, *HML***a** *MAT***a** *HMR***a**. The Southern blot was prepared by B. Weiffenbach.

substitution loop was found in the heteroduplexes, with α-specific sequences being about 750 bp long and **a** sequences being about 650 bp long.

Similar studies were carried out by Strathern *et al.* (1979b, 1980), who also established that *HML*, *MAT*, and *HMR* were in the same orientation on chromosome III (Fig. 3). The orientation of mating-type sequences and their rearrangements will be discussed in detail in Section IV,B.

The heteroduplexes formed between *MAT***a** and *HML*α showed two homologous regions flanking the nonhomologous **a** and α sequences. The longer region is approximately 1400 bp, and the shorter region is about 320 bp. In contrast, the homology between *HMR* and either *MAT* or *HML* is substantially less, with about two regions of 700 bp and 230 bp shared by all three regions. The regions of homology shared by all three regions have been designated *X* and *Z*1, flanking the unique sequences *Y***a** or *Y*α. The additional regions, shared by *HML* and *MAT*, are designated *W* and *Z*2 (Fig. 3).

The regions of *MAT***a** and *MAT*α that are transcribed have been established. Northern blots and RNA–DNA heteroduplex analysis have shown that there are two transcripts derived from *MAT*α and two from *MAT***a** (Nasmyth *et al.*, 1981; Klar *et al.*, 1981a). Neither heteroduplex analysis nor S1 cleavage has given evidence of intervening sequences. The two *MAT*α transcripts are bidirectionally transcribed from the unique *Y*α region and extend into the homologous regions shared by *MAT***a** and *MAT*α. Both transcripts are about 800 bp long. From the location of *in vitro* mutations caused by the insertion of *Xho*I linkers, Tatchell *et al.* (1981) concluded that the *MAT*α*1* message is transcribed rightward from *Y*α into *Z*1 and the *MAT*α*2* message extends leftward from *Y*α into *X*. The two *MAT***a** transcripts are also bidirectionally transcribed. One mRNA is transcribed entirely within the *Y***a** region, and from the location of *mat***a***1* mutations have been shown to code for *MAT***a***1* (Tatchell *et al.*, 1981). The second RNA, transcribed leftward, is transcribed entirely within the *X* region from near the *Y***a** junction toward *W*.

DNA sequences of the entire *MAT***a** and *HML*α loci (*W, X, Y, Z*1 and *Z*2) have been determined, as have the sequences for *MAT*α and most of *HMR***a** (Astell *et al.*, 1981; M. Smith, personal communication). Inspection of the DNA sequence confirms that there are two major coding sequences that can be found in *MAT*α. In the case of *MAT***a,** the *MAT***a***1* transcript contains an open reading frame that, in turn, corresponds to a protein of 148 amino acids. In the middle of this sequence (at codon 44) is a UGA codon that would be expected to be a protein synthesis termination codon. Nevertheless, the presence of *mat***a***1* mutations on both sides of this codon has been confirmed by DNA sequencing (Tatchell *et al.*, 1981; K. Nasmyth, personal communication). There is no evidence that the UGA lies in an intervening sequence that is spliced out of the mature *MAT***a***1* transcript. It is therefore possible that two distinctly sized proteins may be synthesized from the *MAT***a***1* message. In contrast, there is no obvious gene product from the second *MAT***a** transcript. This transcript has an open reading frame that is identical to that found for the 3′-end of the *MAT*α*2* transcript; however, there is no in-frame AUG codon that would be expected to initiate translation. In view of the observation that there is only one *mat***a** complementation group and that *in vitro* mutations made in the second transcript region have no obvious phenotype (Tatchell *et al.*, 1981), it is likely that this transcript does not code for a functional protein.

D. Interaction between *MAT* Genes and Other Genes Affecting Mating-Type Control

Strathern *et al.* (1981) have argued that the mating type genes must have a variety of regulatory functions. On the basis of the phenotypes of various *mat* mutations, they proposed the regulatory scheme shown in Fig. 4. Certain **a**-mating functions must be expressed when neither *MAT***a** nor *MAT*α is expressed. The expression of *MAT*α must both turn off some **a**-specific functions and turn on other α-specific genes. Furthermore, at least one *MAT*α cistron must share with *MAT***a** the regulation of **a**/α-specific functions. Strathern *et al.* (1981) argued the *MAT*α*1* must be a positive regulator of α-specific functions, whereas

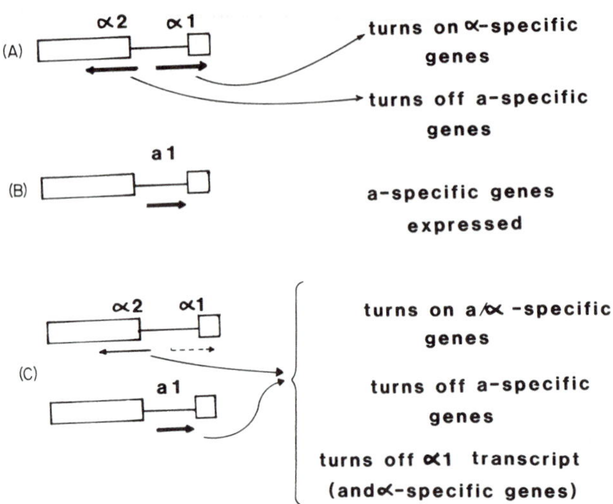

Fig. 4 Mating-type control of cell type. (A) In a *MAT*α haploid cell, the cell is α-mating because the product of the α*1* cistron positively regulates α-specific genes, whereas the product of the α*2* cistron represses **a**-specific functions. (B) In *MAT***a** haploids, **a**-specific functions are expressed, although there is no genetic evidence that a functional *MAT***a***1* cistron is required for this expression. (C) In *MAT***a**/*MAT*α diploid strains, the products of both the *MAT*α*2* and the *MAT***a***1* cistron are required together to turn on **a**/α-specific genes. In addition, these two cistrons act in concert to repress the level of transcription of the *MAT*α cistrons, especially *MAT*α*1*. Finally, the product of the *MAT*α*2* cistron continues to turn off the expression of **a**-specific genes. This model is based primarily on the proposal by Strathern *et al.*

*MAT*α*2* must act both as a negative regulator of **a**-specific genes and as a regulator of **a**/α-specific functions. *MAT***a** must share in regulating **a**/α-specific functions and in turning off α-specific functions. Evidence for some aspects of this scheme has come from studies of the regulation of transcription of the *MAT* genes themselves and from the interactions of *MAT* mutations with other unlinked mutations.

The concomitant expression of both **a** and α cistrons leads to a nonmating phenotype in both diploids and haploids. This change can also be seen in Northern blots of RNA transcripts probed with sequence from the *MAT* locus. Both Klar *et al.* (1981a) and Nasmyth *et al.* (1981b) have shown that the levels of transcription of both *MAT*α cistrons are regulated in *MAT***a**/*MAT*α diploids or in haploids carrying mutations that allow expression of both *HML*α and *HMR***a.** The level of *MAT*α transcripts is reduced more than 10-fold in a *MAT***a**/*MAT*α nonmating cell, in comparison to a *MAT*α strain. The repression of *MAT*α*1* is significantly greater than that of *MAT*α*2*. The level of transcription of the two *MAT***a** transcripts is not significantly affected in *MAT***a**/*MAT*α cells.

The autoregulation of *MAT*α transcripts depends on a functional *MAT***a** and a functional *MAT*α*2* gene. Previous genetic studies had shown that *MAT*α/*mat***a***1* strains are α-mating, whereas *mat*α*2*/*MAT***a** strains are weakly **a**-mating. Neither of these diploids is able to sporulate, in contrast to nonmating *mat*α*1*/*MAT***a** diploids, which are able to sporulate. Thus, diploids that carry mutations in *mat*α*2* or *mat***a***1* cistrons fail to establish the conditions necessary for diploids to be both nonmating and able to sporulate. The lack of regulation can be seen at the level of expression of *MAT*α transcripts. Neither *mat*α*2*/*MAT***a** nor *MAT*α/*mat***a***1* diploids repress the transcription of *MAT*α genes; however, transcription of *MAT*α is reduced in *mat*α*1*/*MAT***a** diploids (Klar *et al.*, 1981a; Nasmyth *et al.*, 1981b). These results reinforce the conclusions of Strathern *et al.* (1981) that the *MAT*α*1* cistron must be a positive regulator of α-specific gene functions the expression of which is repressed in *MAT*α/*MAT***a** diploids.

Recently, V. L. MacKay (personal communication) has provided evidence that at least one other chromosomal gene is required for the normal regulation of *MAT*α transcripts. The *tup1* mutation is a pleiotropic mutation that affects such diverse properties as the uptake of dTMP, acquisition of canavanine resistance, and the level of expression of iso-2-cytochrome *c* (Wickner, 1974; Lemontt *et al.*, 1980; B. Errede and F. Sherman, personal communication). In addition, this mutation is an α-specific sterile mutation. *MAT*α *tup1* strains mate very weakly with *MAT***a** strains, and the resulting diploid, if homozygous for *tup1*, is unable to sporulate (Wickner, 1974). V. L. MacKay (personal communication) hypothesized that *tup1 MAT*α strains behaved identically to strains carrying a *mat*α*2* mutation. MacKay tested this hypothesis by constructing a *mat*α*1 MAT*α*2 tup1* strain (i.e., a strain with a mutant α*1* cistron and a normal α2 cistron) that proved to have an **a**-mating type, analogous to what is found in a *mat*α*1 mat*α*2* double mutant. Thus, it appears that the *TUP1* gene product is necessary for the proper expression of the *MAT*α*2* gene product. This analysis has been extended to a direct observation of the effect of *tup1* on *MAT*α transcripts (V. L. MacKay, personal communication). MacKay's results show that *MAT*α*2* transcription is greatly reduced in such strains, but *MAT*α*1* transcription is much less affected.

A reciprocal effect is seen in the interaction between another α-specific sterile mutation, *ste3*, and the expression of the *MAT*α*1* gene product (Sprague *et al.*, 1981b). A *MAT*α *ste3* strain mates very weakly with *MAT***a** strains, but the resulting diploid, if homozygous for *ste3*, is able to sporulate. The phenotypes of this mutant thus resemble the properties of *MAT*α*1* mutations. More direct evidence that *MAT*α*1* is affected by *ste3* comes from the construction of a *MAT*α*1 mat*α*2 ste3* strain that is **a**-mating rather than nonmating and, again, is the phenotype expected of a strain that fails to express either *MAT*α*1* or *MAT*α*2* gene products. It is not yet known if the *ste3* mutation directly interferes with the transcription of the *MAT*α*1* cistron. It should also be pointed out that two other α-specific sterile mutations, *ste13* and *kex2*, do not interact with *mat*α*2* mutations (Sprague *et al.*, 1981b). Furthermore, the *kex2* mutation, which, like *tup1* mutations, inhibits sporulation of *MAT***a**/*MAT*α diploids, does not behave like *tup1* in its interaction with *MAT*α*2*. A *kex2 mat*α*1 MAT*α*2* strain is sterile (nonmating) rather than **a**-mating, which would be expected if *kex2* prevented expression of *MAT*α*2* directly (J. E. Haber, unpublished).

The regulatory model proposed by Strathern *et al.* (1981) makes a number of other predictions about the interaction of mating type mutations and other mutations that interact with the *MAT* locus. One such prediction is that certain **a**-specific functions should be turned off in *MAT*α/*MAT***a** diploids and that these functions would continue to be expressed in a *mat*α*2* haploid. One such function is the "barrier" phenotype of *MAT***a** cells. *MAT***a** cells secrete a diffusible substance that inactivates α-factor. A growing streak of *MAT***a** cells placed between a diffusing source of α-factor and individual *MAT***a** cells will prevent the arrest of growth of the individual *MAT***a** cells (Hicks and Herskowitz, 1976a). Similarly, *MAT***a** cells incubated in liquid culture with α-factor will become resistant and resume growth (Chan, 1977). A single genetic locus responsible for the barrier phenotype has been identified by Sprague *et al.* (1981a) by the isolation of a *bar1-1* mutation that also renders *MAT***a** cells especially sensitive to the presence of α-factor. The α-specific sterile mutation, *ste3*, described above, also causes *MAT*α cells to produce barrier factor.

One might also predict that mutations can be isolated that would bypass certain aspects of *MAT* locus regulation. One such class of mutations that bypass normal mating-type control of meiosis in sporulation has been described (*csp*, Hopper and Hall, 1975a; *rme*, Kassir and Simchen, 1976; *scm*, Gerlach, 1974). These recessive mutations, when homozygous, allow both *MAT***a**/*MAT***a** and *MAT*α/*MAT*α diploids to sporulate, albeit at a relatively low efficiency. Rine *et al.* (1981) have shown that the sporulation of *rme*/*rme* diploids homozygous for eith *MAT***a** or *MAT*α does not depend on the presence of unexpressed copies of opposite mating-type information at either *HML* or *HMR*. Thus, these mutations appear to bypass the normal *MAT***a**/*MAT*α-dependent regulation of meiosis in sporulation. The *rme* and *csp* mutations may very well be identical, as they were isolated from closely related parent strains and apparently do not complement each other (Rine *et al.*, 1981). The *rme* mutation has been mapped close to the centromere of chromosome VII (Rine *et al.*, 1981).

III. Mating-Type Conversions

A. Hemothallic and Heterothallic Strains

The laboratory strains of *S. cerevisiae* developed by Lindegren and others exhibit stable mating types and could be propagated as haploids indefinitely. However, rare "mutations" of the mating-type locus were observed by Hawthorne (1963a), who showed that when two *MAT*α strains were mated with each other, he recovered rare diploids in which one of the two *MAT*α strains had been changed to *MAT***a.** Similar "forced matings" between two *MAT***a** strains also yield diploids in which one of the two *MAT***a** loci is converted to a *MAT*α allele. In heterothallic strains (i.e., those exhibiting an essentially stable haploid phase with a stable mating type), the frequency of such changes from *MAT***a** to *MAT*α or vice versa is very low, on the order of 10^{-6}.

In contrast, some *Saccharomyces* species exhibit a strong tendency to self-diploidize. In these homothallic strains, a single haploid spore of one mating type can germinate and grow into a colony of cells of both mating types. Cells of opposite mating types can then conjugate, eventually to form a colony consisting almost exclusively of nonmating *MAT***a**/*MAT*α diploids (Fig. 2). The difference between homothallic and heterothallic strains resides primarily in a single dominant genetic locus originally designated *D* (for "diploidization") by Winge and Roberts (1949) and now designated *HO* (Harashima *et al.,* 1974).

The homothallic conversion of cells of one mating type to cells of the opposite mating type has been shown to be a genetic rather than an epigenetic process. When a homothallic diploid is sporulated, the four haploid ascopores contain two *MAT***a** and two *MAT*α segregants. These can be recovered by spore-to-cell matings with heterothallic strains; subsequent genetic analysis has shown that the mating-type alleles generated by homothallic conversion can be recovered as stable heterothallic segregants (Takano and Oshima, 1970b).

The process of homothallic switching of mating types can be followed directly by microscopic inspection of germinating spores. Hicks and Herskowitz (1976b) showed that, in the presence of α-factor, *MAT***a** spores would germinate but not divide. Rather, they form elongated "schmoos" that can be readily distinguished from normally dividing cells. Thus, when four haploid homothallic (*HO*) spores from one meiotic ascus are dissected on nutrient agar containing α-factor, two of the four segregants form the elongated cell shape characteristic of arrested *MAT***a** cells and the other two (*MAT*α) segregants germinate and begin cell division. By subsequent separation of mother and daughter cells, Hicks and Herskowitz (1976b) showed that, at the four-cell stage, two of the four cells derived from a single spore often ceased growing and exhibited the elongated cell shape characteristic of *MAT***a** cells in the presence of α-factor (Fig. 5).

The pattern of mating-type switching is not random; rather, it obeys strict lineage rules. As documented by Hicks and Herskowitz (1976b) and Strathern and Herskowitz (1979), mating-type switching always occurs in pairs of cells. Furthermore, it is always the mother cell and its second daughter that switch mating types, whereas the first daughter and its new daughter cell do not (Fig. 5). Apparently, no cell is able to undergo mating-type switching until it has first undergone a complete cell division; no mating-type conversions are found after the first cell division of a spore and its daughter cell. However, once a cell has divided, it may again switch mating type without an intervening cell division.

The efficiency of mating-type switching is quite high. Strathern and Herskowitz (1979) showed that 75% of the time, a *MAT*α spore will give rise to

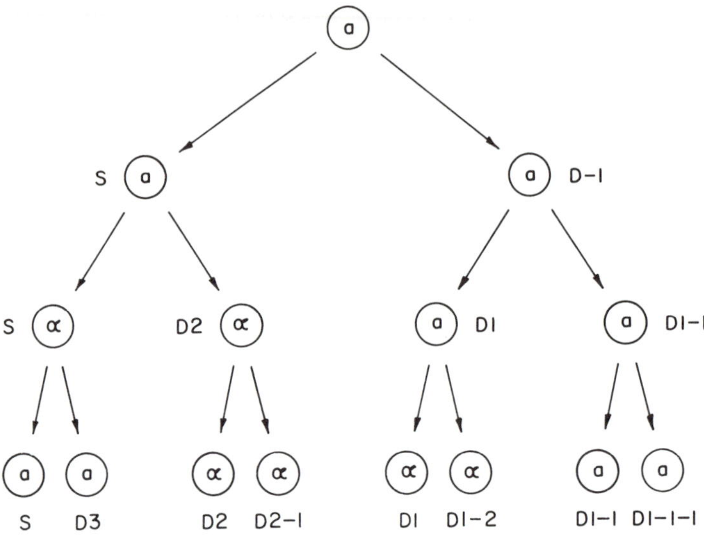

Fig. 5 Cell lineage pattern of mating-type switching. In this example, a spore of mating type **a** (designated S) gives rise after the first cell division to a first daughter (D-1). Both are of the same original-spore mating type. In the second generation, however, two of the four spores (those derived from the S cell) may switch mating type, whereas the two mitotic offspring of the D-1 cell do not. In subsequent generations, any cell that has previously undergone a mitotic division may again switch mating type.

two *MAT***a** and two *MAT*α cells at the four-cell stage. Approximately the same frequency of switching to the opposite mating type is observed if one follows the switching of the progeny of the first daughter cell of such lineages. The frequency of mating-type switching is somewhat strain-dependent, ranging from about 50% in some strains to nearly 90% in others (J. E. Haber, unpublished observations; A. J. S. Klar, personal communication).

Studies with other *Saccharomyces* species identified two other genetic loci that were essential for normal mating-type conversions. Takano and Oshima (1967) found in *S. oviformis* a variant that could switch from *MAT*α to *MAT***a** but was unable to switch from *MAT***a** to *MAT*α. This allelic variant was shown to lie on the left arm of chromosome III, more than 50 cM (centimorgans) from the mating type locus (Harashima and Oshima, 1976; Klar *et al.*, 1980a). An equivalent genetic variant that prevented *MAT*α cells from switching to *MAT***a** but did not prevent *MAT***a** cells from switching to *MAT*α was identified from *S. norbensis* by Santa Maria and Vidal (1970), and was shown to lie more than 50 cM distal to the *MAT* locus on the right arm of chromosome III (Harashima and Oshima, 1976). The initial genetic nomenclature for these two loci (*hm***a** and *hm*α) has been changed in more recent work (see Herskowitz and Oshima, 1981 for a more detailed historical review). In this review, the current nomenclature will be used throughout, even in discussing early experiments.

B. A Transposition Model for Yeast Mating-Type Conversions

In 1971, Oshima and Takano postulated that the homothallic conversion of one mating type to the other involved a transposition and insertion of a ''controlling element,'' analogous to the mechanism proposed by McClintock (1956) in maize. In a standard wild-type strain, the transposition of a controlling element from the locus on the left arm of chromosome III (*HML*) would convert a *MAT***a** cell to *MAT*α, whereas the transposition and insertion of a controlling element from a locus more distal on the right arm of chromosome III (*HMR*) would convert a *MAT*α to *MAT***a.** This proposal was then modified by Harashima *et al.* (1974) to account for observations that different alleles at *HML* and *HMR* appeared to be equivalent in enabling *MAT***a** to switch to *MAT*α and vice versa (Naumov and Tolstorukov, 1973; Harashima *et al.*, 1974). A more explicit version of a controlling-element model was offered in 1977 by Hicks *et al.* (1977). In this model (Fig. 1), both *HML* and *HMR* carry unexpressed copies of either **a** or α information. In the commonly studied strains, *HML* contains α information (*HML*α) and *HMR* contains **a** information (*HMR***a**), although alternative alleles (*HML***a** and *HMR*α) can also be found. Thus, in order for a cell to switch from *MAT***a** to *MAT*α, it must contain either *HML*α or *HMR*α. Similarly, in order for a cell to switch from *MAT*α to *MAT***a,** it must contain either *HMR***a** or *HML***a.** This ''cassette model'' has been confirmed by a variety of genetic and biochemical experiments.

C. Transposition Studies Confirming the Cassette Model

One fundamental aspect of the cassette model is that the replacement of genetic information at the mating-type locus is a nonreciprocal event. That is, when a cell switches from *MAT***a** to *MAT*α, the copy of mating-type information at *MAT* is not simply exchanged in position with the silent copy at *HML*α. If reciprocal exchanges were the rule, a cell of genotype *HML*α *MAT***a** *HMR***a** would be converted to a cell of genotype *HML***a** *MAT*α *HMR***a.** If this *MAT*α cell then mated with a *MAT***a** cell of the original genotype, the resulting diploid would no longer be homozygous for *HML*α and *HMR***a.** In subsequent genetic crosses, one-half of the *MAT***a** segregants should then contain *HML***a** and *HMR***a** and should no longer be able to switch to the opposite mating type (Fig. 6A). Experiments of this sort have clearly shown that such reciprocal exchanges do not generally occur. Thus, the transposition of mating-type information is inherently nonreciprocal, and the information initially at the mating-type locus is presumed to be discarded in some fashion (Fig. 6B).

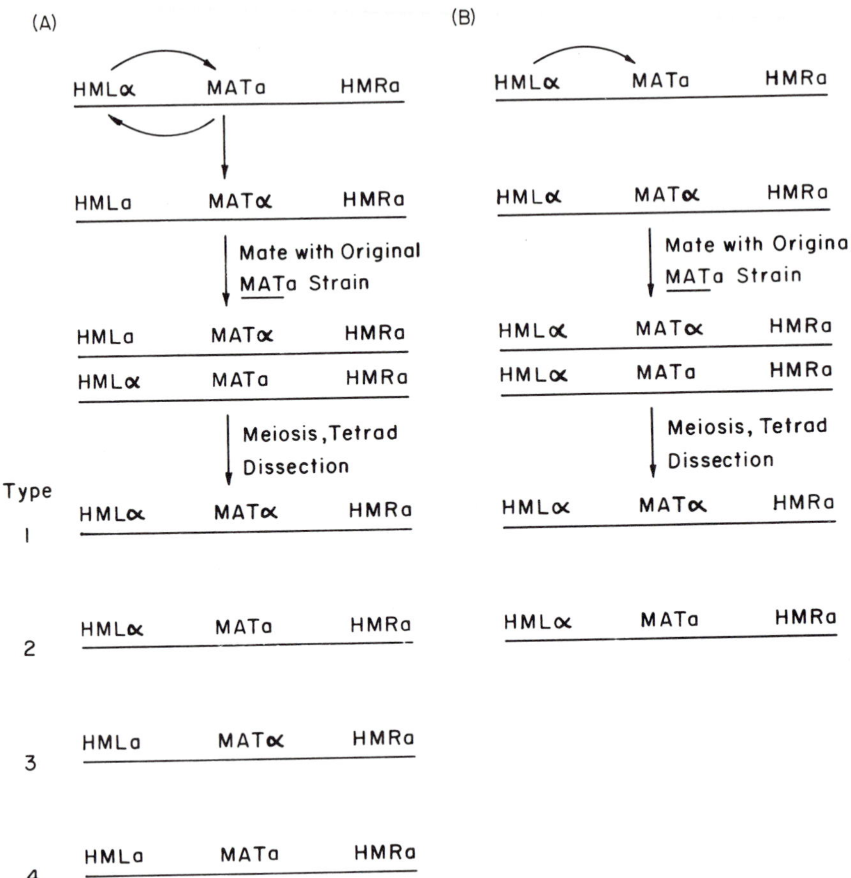

Fig. 6 Two alternative views of mating-type switching. In the model shown in (A), the switching from *MAT***a** to *MAT*α is accompanied by a reciprocal exchange of silent copy information (i.e., *HML*α to *HML***a**). If this were the case, subsequent analysis of meiotic segregants from the homothallic diploids would be expected to show that 25% of the segregants (type 4) are unable to switch from *MAT***a** to *MAT*α, as the independently segregating *HML***a** and *MAT***a** alleles should be found in the same spore. In contrast, if mating-type switching is nonreciprocal (B), only two types of meiotic segregants would be expected from the meiotic diploid, and there would be no cases in which *MAT***a** segregants could not subsequently switch to *MAT*α. The early work of I. Takano and Y. Oshima and their colleagues demonstrated that mating-type switching was indeed nonreciprocal, as shown in (B).

1. "Healing" Experiments

The cassette model predicts that a mutation within the mating-type locus should be "healed" by a series of mating-type switchings. As shown by Hicks and Herskowitz (1976), a mutation within the *mat*α*1* cistron could first be converted to a normal *MAT***a** allele and then to a normal *MAT*α allele either by homothallic mating-type conversions or by much less frequent heterothallic mat-

(A)

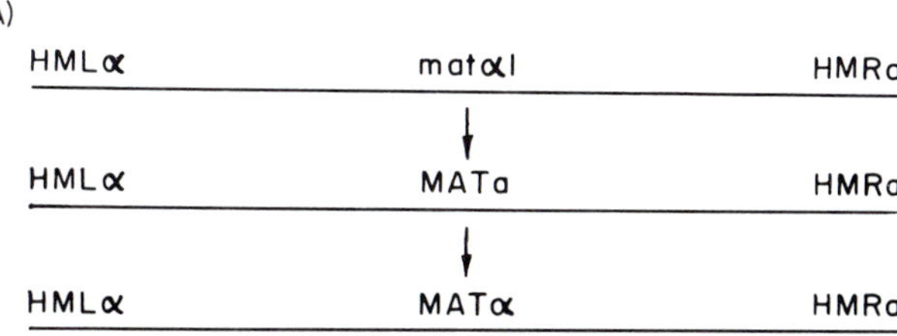

(B)

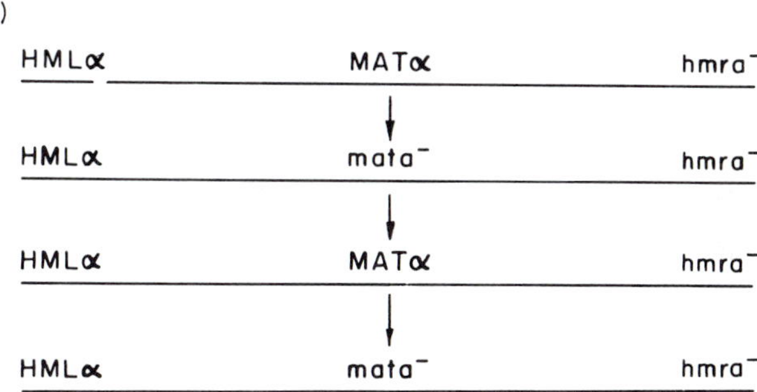

Fig. 7 Healing and wounding of mutant alleles during mating-type switching. (A) A mutation, initially at *MAT*, such as *mat*α*1*, is lost when the cell undergoes mating-type switches to *MAT***a** and subsequently to *MAT*α. The original *MAT*α mutation is apparently excised and is eventually replaced by wild-type information derived by transposition from *HML*α information. In contrast, when the mutant information is not at *MAT*, but at one of the silent loci (e.g., *HMR***a**), mutant information can be repeatedly introduced to the *MAT* locus by transposition. An example of such "wounding" is illustrated in (B).

ing-type conversions (Fig. 7A). Thus, a cell carrying the *mat*α*1* mutation must also carry an unexpressed copy of normal α information that can replace the mutant sequence. Mutations within the *mat*α*2* cistron and in *mat***a***1* have also been shown to be healed in the same way (Strathern *et al.*, 1979a; Klar *et al.*, 1979c). The experiments by Klar *et al.* (1979c) on the healing of the *mat***a***1* mutation were carried out for several generations after the original healing had occurred. In this way, they were able to conclude that the original mutant *mat***a***1* allele did not reappear in subsequent rounds of switching in any of 217 pedigrees that were followed. In addition to the healing of mutations in the structural genes found at the mating-type locus, several cis-acting mutations that interfere with mating type switching may also be healed (see Section V,A,1).

2. "WOUNDING" EXPERIMENTS

Another prediction of the cassette hypothesis is that a mutation at *HML* or *HMR* should be able to be introduced to the mating-type locus. Indeed, such "woundings" should occur every time the cell uses the mutant *HML* or *HMR* cassette in the transposition event. For example, Klar and Fogel (1979) showed that mutations at *hmr***a***1* can be transposed to *MAT* (Fig. 7B). Spores initially of genotype *HO HML*α *MAT*α *hmr***a***1* will switch from *MAT*α to *mat***a***1*, then back to *MAT*α, back to *mat***a***1*, and so on indefinitely because no diploids that are nonmating are formed to turn off the homothallic switching process, Klar (1980) has also shown that some of the mutations generated at *HMR***a** were nonsense mutations, and these same suppressible nonsense mutations can be transposed to the mating-type locus. In addition, Kushner *et al*. (1979) showed that a mutation in *HML*α (*hml*α*-66*) could be repeatedly introduced to replace *MAT***a** by *mat*α. Because *mat*α*1* confers on a haploid cell a sterile or nonmating phenotype, a homothallic cell of genotype *HO hml*α*-66 MAT***a** *HMR***a** will give rise to a colony consisting of both **a**-mating *MAT***a** and sterile *mat*α haploid cells. Additional examples of this type of experiment have been published by Oshima and Takano (1981).

3. EQUIVALENCE OF *HML* AND *HMR* ALLELES

One fundamental assumption of the cassette hypothesis was the postulated equivalence between different alleles of *HML* and *HMR* (Naumov and Taolstorukov, 1973). Tests of these equivalence relationships have now been carried out. Klar and Fogel (1977) demonstrated that *HML***a** and *HMR***a** were equivalent and that *HML***a** and *HML*α were codominant (i.e., both alleles could provide mating type information in a diploid). Additional studies by Arima and Takano (1979) and Harashima and Oshima (1980) showed that *HML*α and *HMR*α were also equivalent.

D. Physical Demonstration of Mating-Type Switching

The availability of cloned DNA sequences homologous to *MAT*, *HML*, and *HMR* (Hicks *et al*., 1979) also made it possible to demonstrate that the switching of *MAT***a** to *MAT*α or of *MAT*α to *MAT***a** was accompanied by a physical change in DNA structure. Because **a**-specific sequences are approximately 100 bp shorter than α-specific sequences, a change from *MAT***a** to *MAT*α is accompanied by a shift in the size of a *Hind*III restriction fragment containing the *MAT* region (Fig. 3C). During such transpositions there is no change in the size of restriction bands containing *HML* or *HMR*. These observations can also be confirmed by using probes complementary to the **a**- or α-specific sequences.

The extent of the transposed region is not yet known. Clearly, the transposition must replace all of the **a**-specific or α-specific sequences (the *Y* region), but it has not yet been determined how much of the homologous regions (*X* and *Z*) to the left or to the right of specific sequences are also transposed. At least some of the sequences in the *X* region of homology must be transposed. The mutation *mat*α*2-4* has been shown by genetic analysis to lie within the region of homology shared by *MAT***a** and *MAT*α (Sprague *et al.*, 1981a). This mutation can also be healed by transposition events (Strathern *et al.*, 1979a). Similarly, a DNA sequence variant in the *Z* region was also shown to appear after switching *MAT*α to *MAT***a** (Astell *et al.*, 1981).

DNA sequence analysis has yet to reveal any distinctive features in the DNA sequence (such as direct or inverted repeated sequences in appropriate positions) that might be used to infer the boundaries of the transposition event. However, as we will discuss in a later section, it is likely that the right-hand boundary of the transposed segment lies near the *Y*/*Z*1 boundary.

IV. Expression of the Silent Genes without Transposition

A. Mutations that Allow Expression of *HML* and *HMR*

Haploid cells normally express only the mating-type information that is found at the *MAT* locus. However, a variety of mutations have now been described that allow the expression of mating-type information at *HML* and *HMR* as well as at *MAT*. The mutations *mar1* (Klar *et al.*, 1979b) and *cmt* (Haber and George, 1979) both cause haploid cells carrying *HML*α and *HMR***a** to become nonmating, as if the cells were *MAT***a**/*MAT*α diploids. The nonmating phenotype depends on the presence of at least one allele of each mating type at either *HML, MAT,* or *HMR*. Because an *HML*α *mat***a***1 HMR*α *cmt* haploid is α-mating, it can be mated with an *HML***a** *mat***a***1 HMR***a** *cmt* strain. The resulting diploid, having no functional mating-type information at the *MAT* locus, is nevertheless nonmating and able to sporulate because of the expression of **a** and α information from the normally silent loci (Haber and George, 1979).

A third mutation, *sir1-1* (Rine *et al.*, 1979), has a somewhat different phenotype. A *sir1-1 HML*α *mat*α*1-5 HMR***a** strain exhibits a normal α-mating type, instead of the nearly nonmating sterile phenotype found in the absence of the *sir1-1* mutation. This suppression of the *mat*α*1-5* mutation depends on the presence of the *HML*α allele. The *sir1-1* mutation also apparently permits some expression of the **a** information at *HMR***a,** because a diploid homozygous for *sir1-1* of genotype *HML*α/*HML*α *MAT*α/*mat***a***1 HMR***a**/*HMR***a** is α-mating but able to sporulate. Thus, *sir1-1* does allow the expression of **a** information sufficient to permit meiosis in a *MAT*α/*mat***a***1* diploid. This allele seems to be unusual, because other *sir1* alleles produce a nonmating phenotype analogous to the *mar* and *cmt* mutations (J. Rine, personal communication).

A variety of additional mutations of this same type have now been described. J. Rine (personal communication) has undertaken an extensive analysis of these mutations and has identified over 50 mutations that can be grouped into at least four complementation groups generally classified as *sir* (silent information regulation) mutants. Complementation analysis has shown that the *mar1* mutation falls in the *sir2* complementation group, whereas *cmt* is an allele of *sir3*. In addition, Rine has shown that one temperature-sensitive sterile mutation isolated by MacKay and Manney (1974a) and two temperature-sensitive sterile mutations isolated by Hartwell (1980) are also *sir* mutations.

B. DNA Sequences Involved in Controlling the Expression of *HML* and *HMR* Alleles

Copies of mating-type information at *HML* and *HMR* apparently contain all of the coding information necessary for full mating-type regulation. One explanation for the fact that these sequences are normally not transcribed is that they contain negative regulatory sequences adjacent to the coding regions that are not found at *MAT*. This view was reinforced by observations that chromosomal rearrangements joining sequences at the mating-type locus to sequences at either *HML* or *HMR* resulted in the expression of the normally silent mating-type information at these loci. For example, Hawthorne (1963a), while studying the rare matings between two *MAT*α strains, recovered a diploid containing a large recessive lethal deletion that appeared to remove nearly all of the right arm of chromosome III between the mating-type locus and the *MAL2* locus near the right end of the chromosome. Subsequent analysis (Strathern *et al.*, 1980; Haber *et al.*, 1980c) has demonstrated that such "Hawthorne deletions" result from a reciprocal recombination event between homologous sequences shared by *MAT* and *HMR* (Fig. 8A). The resulting *MAT/HMR***a** fusion results in the expression of the normally silent **a** sequence at *HMR***a.** Similar fusions have been generated between *MAT* and *HML* to form a deficiency ring chromosome lacking sequences distal to *MAT* and distal to *HML* (Fig. 8B). Such a fusion has been found by Strathern *et al.* (1980) from the forced mating of two *MAT***a** strains, leading to a diploid containing one intact chromosome III (carrying *MAT***a**) and a deficiency ring chromosome in which a *MAT/HML*α fusion expresses the normally silent α information at *HML*α. The ring chromosome generated by this procedure has been physically isolated and characterized (Strathern *et al.*, 1979b). Similar fusions between *MAT* and *HML***a** have also been obtained (Haber *et al.*, 1980c). That one can generate such fusions between *MAT* and *HML* or between *MAT* and *HMR* demonstrates that all loci have their homologous sequences oriented in the same direction (Strathern *et al.*, 1980).

The fusions between *MAT* and *HML* or *MAT* and *HMR* argue that there must be sequences to the left of *HML* or *HMR* that are normally used to prevent the transcription of mating-type information at these loci. A study of other chro-

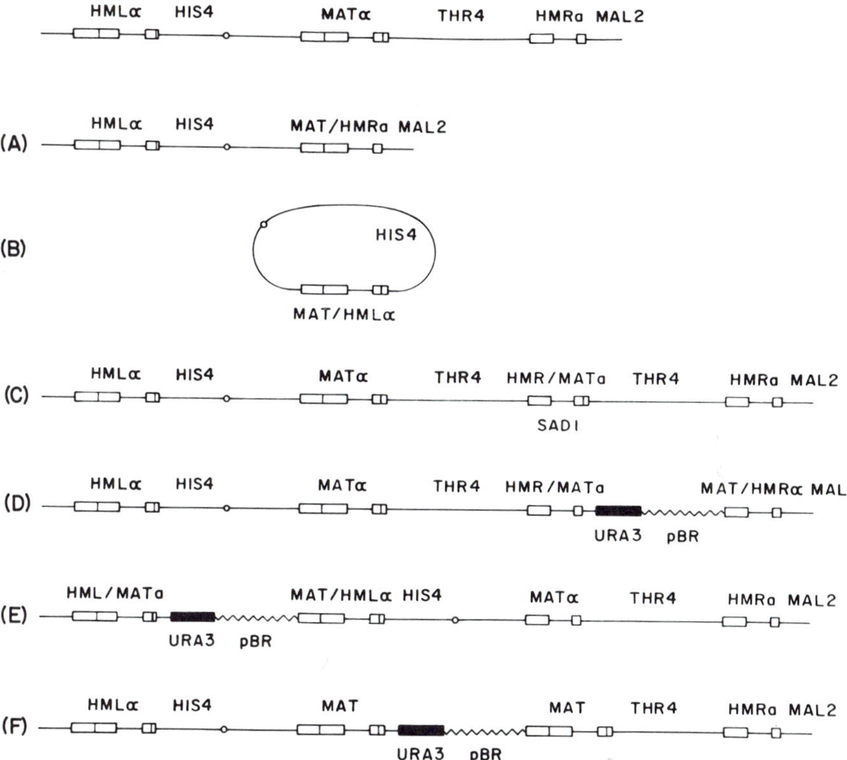

Fig. 8 Fusions of *MAT* sequences with those at *HML* or *HMR*. These hybrid sequences exhibit varying levels of expression of mating-type genes. (A) A Hawthorne deletion joining *MAT* and *HMR***a** causes a recessive lethal deletion of much of the right arm of chromosome III. The *MAT/HMR***a** fusion expresses **a** information and makes a [*MAT/HMR***a**]/*MAT*α diploid nonmating and able to sporulate. (B) A circular chromosome resulting from a *MAT/HML*α fusion is also deleted of many essential genes and is a recessive lethal chromosome. The [MAT/HMLα]/*MAT***a** diploid is nonmating and able to sporulate. (C) The *SAD1* chromosomal rearrangement results from an unequal crossing-over of chromosome III, resulting in a large tandem duplication of part of the right arm. The *HMR/MAT***a** fusions expresses **a** information only very weakly, leaving the *MAT*α haploid α-mating but providing **a** information for sporulation. (D) Similar fusions of *MAT* and *HMR* sequences can be generated by the integration of a *MAT***a**-containing plasmid at *HMR*α. The *HMR/MAT***a** fusion behaves identically to the *SAD1* rearrangement, but the *MAT/HMR*α fusion is only weakly expressed, perhaps because of the adjacent pBR322 sequences. In an analogous *MAT/HMR***a** construction, the **a** sequences are expressed as strongly as the equivalent fusion in (A). In this case, a *MAT*α haploid is nonmating. (E) Equivalent *MAT/HML*α and *HML/MAT***a** fusions created by insertion of a *MAT*-containing plasmid at *HML*α. The *HML/MAT***a** fusion expresses **a** information and makes a *MAT*α haploid nonmating. (F) A tandem duplication of *MAT* genes created by integrating a *MAT*-containing plasmid at *MAT*. Both *MAT***a**-pBR322-*MAT*α and *MAT*α-pBR322-*MAT***a** arrangements are nonmating, as both **a** and α information are expressed. The adjacent pBR322 sequences do not interfere with expression.

mosomal rearrangements suggests that there must also be sequences to the right of *HMR* and *HML* that are essential for the lack of transcription of the two silent copy genes. The chromosomal rearrangement *SAD1*, discussed earlier (Section II), was apparently the result of an unequal crossing-over between *HMR***a** and *MAT*α, yielding **a**-mating information flanked on the left by sequences normally found at *HMR* and on the right by sequences normally found at *MAT* (Fig. 8C). The expression of **a** information from this chromosomal rearrangement is sufficient to allow a *MAT*α/*MAT*α diploid to sporulate but is not sufficient to make such a cell nonmating. The very low level of expression of **a** information from *SAD1* is also evident from direct studies of the level of transcription from this region (Kraig, 1980; E. Kraig, D. T. Rogers, and J. E. Haber, unpublished). The level of transcription of the two *MAT***a** transcripts from *SAD1* is less than a few percent of that found from a normal *MAT***a** locus. The very low level of transcription was also found even under sporulation conditions (in which the level of expression from *SAD1* is sufficient to allow meiosis to occur). Nevertheless, the level of expression from *SAD1* must be substantially greater than from a normal *HMR***a** sequence, which does not provide sufficient mating-type information to allow a *MAT*α/*MAT*α cell to sporulate.

Additional information about sequences that are essential for the normal regulation of the silent genes has been obtained from the studies using recombinant DNA molecules. Using an integrating vector including the *Eco*R1-*Hind*III restriction fragment carrying *MAT***a** and the selectable *URA3* yeast gene in pBR322, Haber and Rogers (1982, and unpublished) integrated *MAT* sequences at *HML* and *HMR*, thus creating fusions similar to those generated by chromosomal rearrangements (Fig. 8D). For example, the insertion of *MAT***a** sequences at *HMR*α can generate two tandem fusion arrangements, one of which is essentially analogous to *SAD1* (*HMR*/*MAT*) and the other of which is equivalent to the Hawthorne deletion (*MAT*/*HMR*). Both genetic and biochemical experiments have shown that the *HMR*/*MAT***a** fusion has all of the same properties as those exhibited by the *SAD1* rearrangement; that is, the sequences are not transcribed sufficiently to make a *MAT*α cell a nonmater, but they are expressed sufficiently to allow a *MAT*α/*mat***a***1* diploid to sporulate. A similar insertion of *MAT***a** at *HML*α has created a hybrid sequence that had not previously been generated by *in vivo* chromosomal rearrangements (Fig. 8E). The properties of this fusion are somewhat different from those found for *HMR*/*MAT***a.** In the *HML*/*MAT***a** fusion, the level of expression of **a** information is sufficient to make a *MAT*α cell a nonmater as well as to provide **a** information for sporulation. Thus, the degree of regulation of expression from sequences to the left of *HML* and from the left of *HMR* are not exactly the same. One possible explanation for the difference is that sequences unique to *HML* are approximately 700 bp farther away from the apparent origin of transcription of **a**-specific genes than are the equivalent regulatory sequences unique to *HMR*. Thus, the mechanism by which regulation is achieved must act over a greater distance in the *HML*/*MAT* fusion than in the *HMR*/*MAT* fusion, and consequently may be less efficient.

The regulation of silent genes has also been studied on autonomously replicating plasmids. It is possible to generate *in vitro* mutations and reintroduce them into yeast cells to determine which sequences are essential for silent-gene regulation. The analysis of the regions surrounding *HMR***a** that are necessary for normal silent-gene regulation has been carried out by generating small deletions at random sites in the *HMR***a** plasmid. The work of K. A. Nasmyth and J. A. Abraham (personal communication) indicates that there are two sets of essential sequences necessary for the regulation of *HMR***a.** One major site lies in the sequences to the left of *HMR* 200–300 bp to the left of the region shared by *MAT* and *HMR;* a second, less extensive, region appears to lie to the right of the homologous sequences shared by *MAT* and *HMR*. Small deletions in the homology region *X* do not alter silent gene regulation.

Similarly, J. R. Broach and J. B. Hicks (personal communication) have shown that insertions of the bacterial transposon Tn*5* in an *HML*α plasmid will alter silent-gene regulation. For example, insertions of Tn*5* into the *W* region of *HML*α or in the sequences to the left of *W* lead to constitutive expression of α-mating information. Some of these mutations may directly alter a site necessary for regulation; however, those insertions into the homology region that is identical at *MAT* and *HML* must alter the silent-gene regulation in some other way. It is quite likely that the insertion of a 5.5-kb element into this region moves the sites used for regulation of transcription too far away to be effective.

C. Changes in Chromatin Structure during Expression of *HML* and *HMR*

Recently, Nasmyth (1982) has provided compelling evidence that the state of expression of the silent genes is correlated with a profound change in chromatin structure. Using limited micrococcal nuclease and DNase I digestions of yeast chromatin, Nasmyth has shown that there is a major DNase I-hypersensitive site that is found in the mating-type region but is not found at *HML* or *HMR* when these genes are unexpressed. However, in the presence of the *mar1* mutation, a DNase I-hypersensitive site can also be found at *HML* and *HMR* in the same position. This major site is found very close to the boundary between the **a** or α specific sequences (*Y***a** or *Y*α) and the *Z*1 region of homology. There is also significant change in the apparent order (phasing) of nucleosomes when these normally untranscribed regions are transcribed. As we will see later, this change in the state of the chromatin also seems to be correlated with the role of the silent-gene copies in acting as donors in homothallic mating-type transpositions.

K. A. Nasmyth (personal communication) has also examined the change in the state of chromatin structure of *HML* and *HMR* by looking directly at the extent of superhelicity of plasmids containing *HML* or *HMR*. Nasmyth has shown that these plasmids are significantly more negatively supercoiled when they are extracted from Mar^+ strains (i.e., where *HML* and *HMR* are not actively transcribed) than from *mar1* strains, where the silent genes are expressed. There is no change in equivalent plasmids containing *MAT* sequences.

It should also be pointed out that the changes in chromatin structure observed by Nasmyth are found as a consequence of changes in *MAR* gene activity but are not directly affected by the mating phenotype of the cell. Even though an *MAT***a**/*MAT*α diploid shows a greatly reduced level of transcription of *MAT*α*1* and *MAT*α*2* sequences, these changes are not manifest in a change of chromatin structure, either at *MAT* or at *HML* or *HMR*.

V. Mechanism of Mating-Type Switching

A. Mutations Affecting Homothallic Mating-Type Switching

In wild-type homothallic strains, the efficiency of mating-type switching is high enough so that within a few generations after the germination of a haploid spore the growing colony is composed of essentially an equal number of *MAT***a** and *MAT*α cells. These cells will conjugate to form a colony consisting of nonmating *MAT***a**/*MAT*α diploids. Mutations that reduce the efficiency of mat-

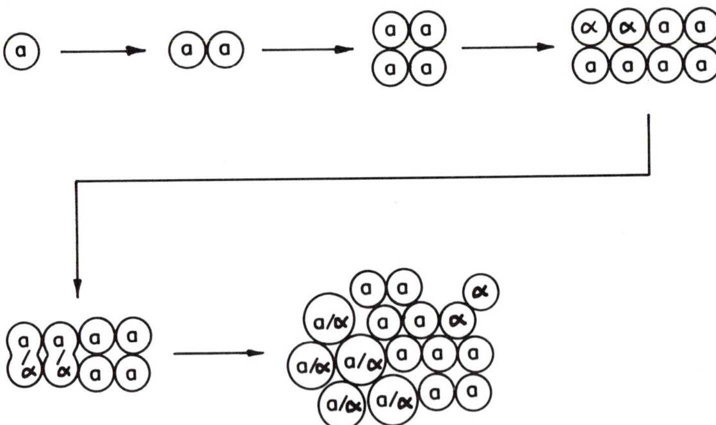

Fig. 9 Cell types appearing in colonies where *MAT* conversions are inefficient. If mating-type switching does not occur frequently, most of the cells arising from a *MAT***a** spore will also be *MAT***a** haploids. When occasional switches to *MAT*α do occur, these cells may mate to form some *MAT***a**/*MAT*α diploids. Some *MAT*α cells that have not yet mated will also be found. The colony, when crossed with mating-type tester strains, will exhibit an asymmetrical (**a**>α) mating phenotype, reflecting the relative frequencies of *MAT***a** and *MAT*α cells.

ing-type switching lead to colonies that still contain a significant number of haploid cells of the original spore mating type (Haber and Garvik, 1977). Under conditions in which mutations significantly reduce, but do not eliminate, mating-type switching, the colony derived from a single spore will consist of a mixture of cell types; haploid cells of the original spore mating type, a much smaller proportion of haploid cells of the opposite mating type (that were presumably generated in the last few cell divisions), and some nonmating *MAT***a**/*MAT*α diploid cells resulting from the conjugation of haploids of the predominant mating type with the relatively rare haploids of the opposite mating type (Fig. 9). This mixture of cells gives rise to a distinctive unequal, dual mating type (**a** > α) or (α > **a**). The proportion of nonmating cells in such slow-switching colonies serves as a reliable index of the relative efficiency of mating-type switching (Davidow and Haber, 1981). Approximately 75% of wild-type homothallic cells that have previously divided will switch to the opposite mating type (Strathern and Herskowitz, 1979). A reduction in the efficiency as little as fivefold will

Table II Mutations and Other Conditions Reducing the Efficiency of Homothallic Mating-Type Switching

	Characteristics	Reference[a]
Mutations		
inc	*MAT***a**-*inc* and *MAT*α-*inc* "inconvertible" mutations are cis-acting and lie within the transposable part of the *MAT* locus	1–3
stk	"Stuck" mutations are cis-acting and closely linked to *MAT*, but lie outside the transposable sequences. All three *stk* mutations affect the switching of *MAT***a** but affect *MAT*α switching only slightly	4
HO-1	Mutations at *HO* reduce the efficiency of switching	2
swi, csm	"Switch" or "control of switching mating type," mutations are unlinked to *MAT*, *HML, HMR*, or *HO*. *swi1* is centromere-linked on chromosome XVI	3, 5
rad52, rad51, rad54	X-ray-sensitive mutations that are lethal in homothallic cells undergoing switching. In *rad52*, the lethal event is a double-stranded chromosome break	6–8
Other conditions		
*HML*α, *HMR***a**	"Reversed" cassettes are inherently less efficient as donors in transposition	9
Translocation of *HML*α	Translocation of a donor to another chromosome markedly reduced the efficiency of switching	10

[a] 1, Takano *et al.* (1973); 2, Mascioli and Haber (1980); 3, Oshima and Takano (1980); 4, Haber *et al.* (1980a); 5, Haber and Garvik (1977); 6, Malone and Esposito (1980); 7, J. Game (personal communications); 8, Weiffenbach and Haber (1981); 9, A. J. S. Klar, J. B. Hicks, and Strathern (1982); 10, Haber *et al.* (1981a).

generate colonies with weak (**a** > α) or (α > **a**) mating phenotypes (Haber *et al.*, 1981a; Davidow and Haber, 1981).

Various kinds of mutations that interfere with mating-type switching have been isolated and are summarized in Table II.

1. MUTATIONS AT *MAT*

Cis-acting mutations that prevent normal mating-type switching have been found both in *MAT***a** and *MAT*α. Takano *et al.* (1973) first identified an "inconvertible" variant of *MAT*α. This *MAT*α-*inc* variant was derived from the related species *S. diastaticus*. This mutation reduces the efficiency of mating type switching to less than one in 10^3 (Takano and Arima, 1979; Strathern *et al.*, 1979a). A similar *MAT***a**-*inc* mutation that reduces the efficiency of switching by approximately 20- to 50-fold was isolated by Mascioli and Haber (1980). Both of these mutations appear to lie within the *MAT* sequences that are replaced by homothallic mating-type switching; that is, they can be healed (Takano *et al.*, 1973; Mascioli and Haber, 1980). For example, the *MAT***a**-*inc* mutation can be converted at low frequency to a normal *MAT*α allele. Subsequently, *MAT*α can be switched to a normal *MAT***a** allele that is no longer defective in mating-type conversion. More recently, a large number of both *MAT***a**-*inc* and *MAT*α-*inc* alleles have been isolated by Oshima and Takano (1980).

Two *inc* mutations have now been cloned and their sequences determined (Fig. 10). In each case, the *inc* mutations appeared to arise by a single base-pair substitution in the *Z*1 region, close to the unique **a** or α information (D. Rogers, M. Zoller, D. Russell, J. E. Haber, and M. Smith, unpublished). The wild-type sequence from the *Y*/*Z* border (CGCAACA . . .) is not found anywhere else in the *MAT* locus. The relation of the site identified by these mutations to the mechanism of mating-type switching will be discussed in more detail in a later section.

A second class of cis-acting mutations at or near *MAT* has also been defined. These "stuck" (*stk*) mutations differ from the *inc* mutations in two respects (Haber *et al.*, 1980a). First, these mutations are not healed by rounds of mating-type conversion; that is, an *stk MAT***a** can switch to an *stk MAT*α and then back to an *stk MAT***a.** This suggests that the *stk* mutations lie outside of those sequences that are normally replaced by mating-type transposition. The second distinctive difference about these *stk* mutations is that they appear to affect the switching of *MAT***a** to *MAT*α much more significantly than they affect the switching of *MAT*α to *MAT***a.** This may reflect a difference in interaction of *MAT* with *HML*α and *HMR***a.** *HML*α contains sequences homologous with the mating-type locus (*W* and *Z*2) that are not found at *HMR*. Thus, a mutation in the *W* or *Z*2 regions of the *MAT* locus might affect transpositions involving *HML*α (i.e., switches of *MAT***a** to *MAT*α) but not transpositions involving *HMR***a** (switches of *MAT*α to *MAT***a**). Three different *stk* mutations have been characterized. All of them map close to the mating-type locus (much less than 1 cM away). The *stk1* mutation appears to lie to the left of the *MAT***a***1* cistron (Haber *et al.*, 1980a).

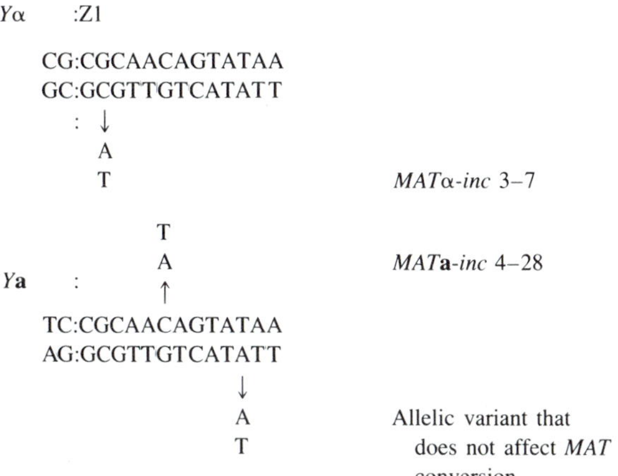

Fig. 10 Single base-pair changes in the *MAT* locus near the *Y*/Z1 border that prevent efficient *MAT* conversions. Two *inc* mutations initially isolated by Oshima and Takano (1980) were cloned and sequenced (D. Rogers, M. Zoller, D. Russell, J. E. Haber, and M. Smith, unpublished). The *MAT***a**-*inc* sequence differs from the *MAT***a** sequence published by Astell *et al.* (1981) at two locations: at the base pair indicated and at the A–T pair 4 bases to the right. However other *MAT***a** sequences from other strains have also contained the T–A pair found here and in *MAT*α. The original *MAT***a** in strain XP8-4A that was cloned and sequenced (Nasmyth and Tatchell, 1980; Astell *et al.*, 1981) apparently contains an allelic variation derived from *HMR***a** of strain XT1172-S245C and is unrelated to the *inc* mutation.

One interesting feature of both *inc* and *stk* mutations is that they promote "illegal" transpositions of mating-type information (Haber *et al.*, 1980b). During normal transpositions of mating-type information, the sequences at *HML*α and *HMR***a** are unaffected; however, when normal transpositions are prevented by *inc* or *stk* mutations, the donor loci are themselves subject to switching (Fig. 11). Thus, an *HO HML*α *stk1 MAT***a** *HMR***a** strain will become completely unable to switch by the conversion of *HML*α to *HML***a** (Haber *et al.*, 1980). In the case of strains carrying *MAT*α-*inc* or *MAT***a**-*inc*, it is possible to show that these illegal transpositions move the mutant mating-type sequences to the donor locus (Mascioli and Haber, 1980; Haber *et al.*, 1980b). In one such experiment, *HML***a** was replaced by *HML*α-*inc*. Subsequently, a strain of genotype *HO HML*α-*inc MAT***a** *HMR***a** was constructed. The switching of *MAT***a** to *MAT*α-*inc* occurred with wild-type efficiency. When the α-*inc* sequence was again at *MAT*, subsequent mating-type switches were prevented. These wounding experiments demonstrated that the *inc* mutation affects the removal of mating-type information from *MAT* but does not affect the transposition of such *inc*-containing sequences from the donor locus.

The availability of an *inc* mutation at *HML*α-*inc* has also made it possible to show that the rare conversions of *HML*α to *HML***a** apparently occur by the same mechanism as the frequent conversions of *MAT*α to *MAT***a.** Whereas *HML* is

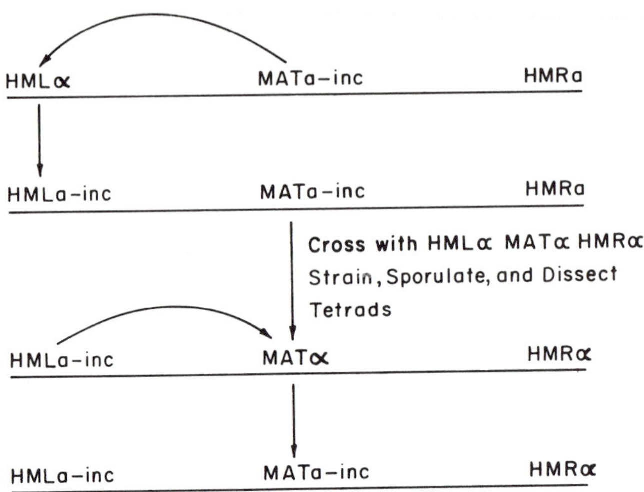

Fig. 11 Illegal transposition of mating-type alleles. Cis-acting mutations such as *MAT***a**-*inc*, *MAT*α-*inc*, and *stk1* prevent the normal replacement of the *MAT* allele. Under these conditions, approximately 1% of the cells undergo a replacement of silent-copy information. In the case illustrated, the *MAT***a**-*inc* allele illegally transposed to replace *HML*α with *HML***a**-*inc*. Subsequent genetic analysis demonstrates that *HML***a**-*inc* actually contains the original *inc* mutation, as this sequence can be transposed back to *MAT*, where it is once again an inconvertible allele. These observations demonstrate that the *inc* mutation prevents the "excision" of *MAT* sequences but does not prevent their transposition.

converted by illegal transposition to *HML***a** in about 1% of cell divisions in an *HO HML*α *stk1 MAT***a** *HMR***a** strain, there were no conversions of an equivalent *HO HML*α-*inc stk1 MAT***a** *HMR***a** strain (Haber *et al.*, 1980b). More recently, Klar *et al.* (1981b) confirmed these observations while following conversions of *HML*α or *HML*α-*inc* that occur at high efficiency in *HO* strains in which the *HML* locus is expressed (see Section V,G).

2. MUTATIONS AT *HML* OR *HMR*

Mutations at *HML* or *HMR* that prevent mating-type switching have generally proven to be actual conversions of *HML*α to *HML***a** or *HMR***a** to *HMR*α (Oshima and Takano, 1980; Klar *et al.*, 1981c). Thus, for example, an *HO HML*α *MAT***a** *HMR***a** strain can occasionally become changed to one of genotype *HO HML***a** *MAT***a** *HMR***a**, so that no switches to the opposite mating type are then possible. Such conversions in wild-type strains occur with a frequency of approximately 10^{-5}. These rare conversions of *HML*α can use either *MAT***a** or *HMR***a** as donor sequence (Haber *et al.*, 1980b; Klar *et al.*, 1981c). To date, no cis-acting mutations at either *HML* or *HMR* have been described. That such mutations exist at *MAT* but not at *HML* or *HMR* is an important fact that must be included in any mechanism of mating-type conversion.

One possible exception to this generalization is a mutation at *HML* described by Oshima and Takano (1981). Some of the properties of this mutation argue that it is a cis-acting mutation at *HML* unable to transpose a normal α sequences to replace *MAT***a**; however, it is more likely that this mutation is an *α1* or *α2* (sterile) mutation that can transpose normally but also contains an *inc* mutation that impairs its subsequent removal from *MAT*. The isolation of *matα1* deletions that are simultaneously *inc* mutations (Weiffenbach and Haber, 1981, unpublished) and the fact that *HMLα-inc* mutations can transpose to *MAT* (Haber *et al.*, 1980b) make this alternative interpretation likely.

3. MUTATIONS AT *HO*

Several mutations of the *HO* gene that reduce mating-type switching have been described (Mascioli and Haber, 1980; Oshima and Takano, 1980; Weiffenbach and Haber, 1981). One temperature-sensitive allele of *HO* has been isolated (B. Weiffenbach and J. E. Haber, unpublished) that almost completely abolishes mating-type switching at 36°C but allows inefficient switching of mating-type alleles at 25°C.

4. "SWITCH" MUTATIONS

Evidence that other gene products are necessary for mating-type switching comes from the isolation of several mutations unlinked to *HO, MAT, HML,* or *HMR* that affect mating type switching. The "switch" mutation *swi1-1* (Haber and Garvik, 1977) reduces the efficiency of mating-type switching to approximately 5% of normal efficiency. This mutation, which maps near *rad1* on the left arm of chromosome XVI, appears to interfere with mating-type switching without interfering with other cellular processes, including meiotic recombination. Other mutations with similar phenotype (*csm;* control of switching mating type) have also been isolated (Oshima and Takano, 1980). M. Stern and I. Herskowitz (personal communication) have placed a larger number of such mutations into four complementation groups.

5. MUTATIONS AFFECTING RADIATION SENSITIVITY AND RECOMBINATION

Malone and Esposito (1980) discovered that the X-ray-sensitive mutation *rad52* prevented normal mating-type switching. This mutation has also been shown to abolish meiotic recombination and substantially to reduce mitotic recombination (Game *et al.*, 1980; Prakash *et al.*, 1980). Malone and Esposito (1980) found that *HO rad52 MATα* spores died within a few cell divisions after germination. In contrast, they found that *HO rad52 MAT***a** cells, although unable to switch to *MATα*, did not die. More recently, Weiffenbach and Haber (1981) have shown that *rad52* is lethal to homothallic spores of both mating types, and that the particular *MAT***a** allele used by Malone and Esposito is a variant that is able to survive in the presence of *HO* and *rad52*. Weiffenbach and Haber (1981) have also shown that any of the mutations that slow down or abolish mating-type switching (i.e., *inc, stk, HO-1,* and *swi1*) prevent the lethality of the *rad52* mutation. On the other hand, cells that cannot switch to their opposite mating type because they contain silent-copy genes of the same mating type information (*HO rad52 HML*α *MAT*α *HMR*α) do not survive (R. E. Malone, personal communication). These observations suggest that all of the slow-switching mutations that have been isolated must act prior to the point at which the *RAD52* gene product is required during mating type conversion.

The nature of the lethal event in *HO rad52* cells has been shown to be a double-stranded chromosome break occurring at or very close to the mating-type locus in cells that attempt to switch mating type (Weiffenbach and Haber, 1981). For example, in an *HO rad52 swi1 MAT*α cell, the colony survives, but at every generation a few cells attempt to switch mating type. These mating-type switches give rise to **a**-mating cells, but subsequent genetic analysis has shown that these **a**-mating cells do not contain a functional *MAT***a** locus; rather, they exhibit an **a**-like mating phenotype by virtue of having lost both *MAT*α*1* and *MAT*α*2* function. Diploids formed by the mating of these **a**-like cells with a *MAT*α*-inc* strain are hemizygous for all markers distal to the mating-type locus, but are still heterozygous for markers proximal to the mating-type locus and on the other arm of chromosome III (Fig. 12).

Similar **a**-like cells can also be found in Rad$^+$ cells. In *HO HML*α *MAT*α *HMR*α colonies, a small portion of the cells also appear to be **a**-mating and prove to have properties identical to those generated by the *RAD52* mutation (J. B. Hicks and J. N. Strathern, personal communication; J. E. Haber, unpublished). Such chromosome breaks are also found in very rare matings between two heterothallic *MAT*α strains (McCusker and Haber, 1981). The broken chromosomes generated by a break at the mating-type locus are mitotically unstable and are generally lost, although a variety of healing events have been described (McCusker and Haber, 1981; Weiffenbach and Haber, 1981).

The appearance of **a**-like cells in an *HO HML*α *MAT*α *HMR*α strain appears to correlate with the formation of two minor *Hin*dIII restriction fragments that are smaller than the *MAT* fragment but that hybridize to *MAT*-specific probes (J. N. Strathern, J. B. Hicks, and A. J. S. Klar, personal communication). From the sizes of the restriction fragments, it appears that they reflect a double-stranded break in the *Z*1 region very close to the boundary between the unique **a** or α sequences (*Y***a** or *Y*α) and *Z*1 (J. N. Strathern, J. B. Hicks, and A. J. S. Klar, personal communication). This site is very similar to the one identified by K. A. Nasmyth (personal communication) as a DNase I-hypersensitive site in the chromatin of the mating-type region. The lethal double-stranded break generated in *rad52* strains may also be at this same location, but no direct demonstration has yet been conducted.

Weiffenbach and Haber (1981, unpublished) have also found that a small

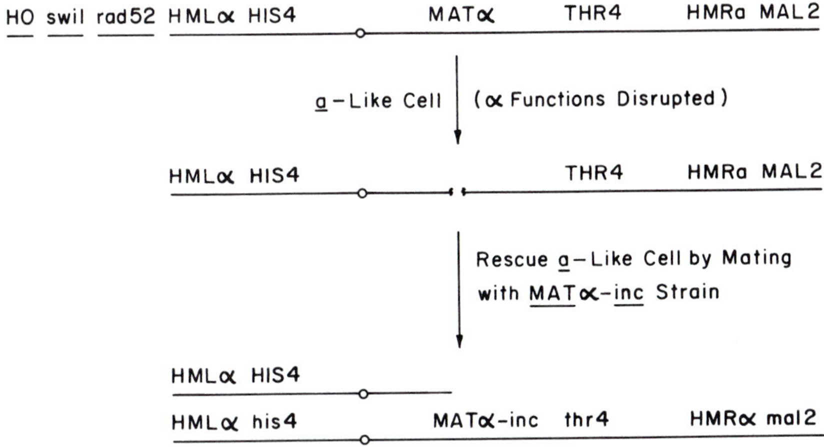

Fig. 12 Creation of lethal double-stranded chromosome breaks at the *MAT*α locus in homothallic cells containing the *rad52* mutation (Weiffenbach and Haber, 1981). In *HO swil MAT*α strains, about 5% of the cells attempt to switch mating type. In the presence of the *rad52* mutation, these attempts to switch mating type are lethal because of the generation of a double-stranded chromosome break. The cell containing the broken chromosome is transiently viable, although the distal portion of chromosome III, lacking a centromere, is lost. The chromosome break also disrupts the expression of α information and the cell becomes **a**-like (i.e., is an **a**-mater but does not express bona fide *MAT***a** functions). The transiently viable cell can be rescued by mating with a Rad$^+$ *MAT*α*-inc* strain. Recessive markers distal to *MAT*α*-inc* are expressed because that segment of the broken homolog was lost. The remaining part of the broken chromosome is also mitotically unstable.

proportion of *HO rad52 MAT*α cells become resistant to the lethality of the *rad52* mutation by the formation of small deletions within the mating-type region (Fig. 13). These deletions render the cells unable to switch and simultaneously cause mutations within the *MAT*α*1* cistron. The smallest of these *mat*α*1-inc* deletions is significantly less than 50 bp, and the largest extends over 750 bp. (The largest deletion removes not only most of the *MAT*α*1* cistron but also part of the *MAT*α*2* cistron, leading to an **a**-like mating cell that cannot undergo homothallic mating-type switching.) All of these deletions remove an *Hha*I site that lies very close to the *Y*α/Z1 border. J. N. Strathern (personal communication) has also created a small *in vitro* deletion in this same region and shown that a plasmid containing *MAT*α is unable to switch. These experiments indicate that a double-stranded chromosome break near the *Y*/Z1 border arises in at least some cells attempting to switch mating-type genes, and that the generation of this break depends on DNA sequences near the *Y*/Z1 border.

Similar but much less extensive studies of other radiation-sensitive genes have also been carried out. J. C. Game (personal communication) has shown that both *rad54-3* and *rad51*, two other mutations that are X-ray-sensitive and deficient in meiotic recombination, are also lethal in homothallic spores attempting to switch mating type. In contrast, two other X-ray-sensitive mutations that also affect

(A)

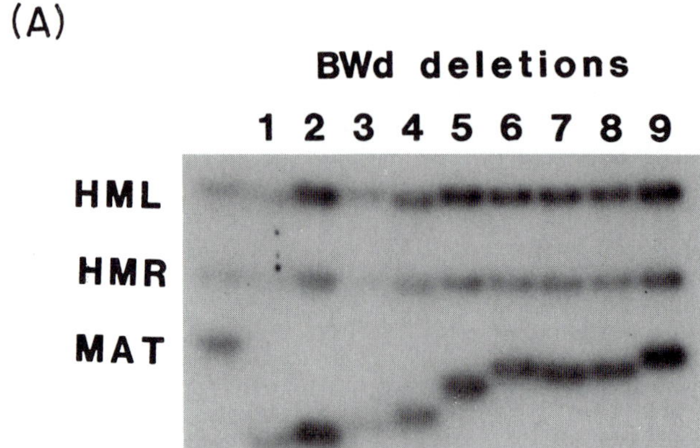

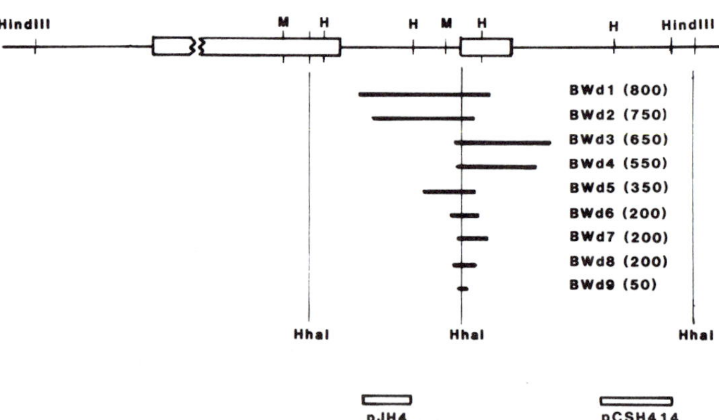

Fig. 13 Formation of deletions in *HO swi1 rad52 MAT*α cells. Nearly all cells that attempt *MAT* conversion in *rad52* strains suffer a lethal double-stranded chromosome break. A small number of cells become *rad52*-resistant by the formation of small deletions within the *MAT*α locus that simultaneously prevent switching and cause *mat*α*1* mutations. These deletions are variable in size and can be seen when DNA from these strains are digested with *Hind*III and the Southern blot is probed with labeled *MAT***a** probe (A). The *HML* and *HMR* bands are unchanged, whereas the *MAT*-containing bands are smaller. Further restriction mapping showed that all of the deletions removed an *HhaI* site very near the *Y*α/*Z* border and were nearly coincident with the position of two *inc* point mutations that have been sequenced (B; see Fig. 10).

meiotic recombination, *rad50* and *rad57*, do not have any apparent effect on mating-type switching. Similarly, none of 14 UV-sensitive mutations (*rad1–rad10, rad14, rad16, rad19, rad22*) have any apparent effect on mating-type switching (J. E. Haber, unpublished).

5. Effect of Other Mutations and Nutritional Conditions on Mating-Type Switching

The ability to switch mating type does not depend on the phenotypic expression of a haploid mating type. For example, a *mat*α*1* mutation causes a cell to be nonmating (sterile); however, such a mutation can be switched to *MAT***a** with wild-type efficiency in both homothallic and heterothallic cells (Hicks and Herskowitz, 1976b; Strathern and Herskowitz, 1979). In addition, cells carrying the nonspecific sterile mutation *ste4*, although they are unable to conjugate, do switch *MAT* alleles (J. B. Hicks, personal communication). The *ste4* mutation does not turn on silent copy genes but interferes with some step in the conjugation of haploid cells.

Efficient mating-type switching also takes place when cells are germinated on relatively poor nutrient medium or in the presence of 25 mg/ml ethidium bromide, which interferes with the replication of mitochondrial DNA and renders cells petite within a few generations (J. E. Haber, unpublished). Similarly, mating-type switching occurs in haploid spores carrying the *petS* mutation that makes cells petite and also causes the loss of other cytoplasmic elements, such as the double-stranded killer RNA (J.E. Haber, unpublished). Mating-type conversions apparently do not occur in cells the growth of which has been arrested at the end of the G1 stage of the cell cycle either by α-factor or at stationary phase (J. N. Strathern, personal communication).

B. Initiation of Mating-Type Switching Events

Both genetic and biochemical experiments have suggested that the events that initiate the transposition of mating-type alleles occur at the *MAT* locus rather than at *HML* or *HMR*. Several experiments have shown that, once activated, a *MAT* allele may be converted by any homologous sequence with which it is able to pair, including another *MAT* locus. For example, in homothallic diploids initially of genotype *HML*α*/HML*α *mat***a***-1/MAT*α*-inc HMR***a***/HMR***a**, one finds not only transposition of **a** information from *HMR***a**, to make a *MAT***a***/MAT*α*-inc* diploid, but also diploids that have become homozygous for *MAT*α*-inc* (Haber *et al.*, 1980b; J. E. Haber, unpublished). More than 5% of the diploids recovered from such an experiment result from the conversion of *mat***a***1* to *MAT*α*-inc*. In these cases, one *MAT* allele has been used to convert the other. Such events would not be expected if mating-type switching was initiated by some replication event at *HMR***a**. Similarly, in transformed haploid strains carrying a tandem duplication of mating-type genes (see Section V,B), one *MAT* allele can frequently be used as the donor to convert the other in homothallic strains.

It should also be recalled that all of the cis-acting mutations that prevent mating-type switching are found at or immediately adjacent to that *MAT* locus rather than at *HML* or at *HMR*. The *MAT***a**-*inc* and *MAT*α-*inc* mutations not only prevent cells from switching efficiently but also prevent the lethal effects of the *rad52* mutation (presumably a double-stranded break within the *MAT* locus itself) (Weiffenbach and Haber, 1981). Furthermore, we have shown that the *inc* mutations themselves, when transposed to *HML* or *HMR*, are perfectly capable of acting as efficient *donors* of mating-type information, even though they prevent further mating-type switching once they have been transposed back to the *MAT* locus (Haber *et al.*, 1980b).

A direct biochemical indication of this activation event comes from the observation of a double-stranded chromosome break occurring within the *MAT* locus in a small fraction of cells attempting to switch mating type (J. N. Strathern, J. B. Hicks, and A. J. S. Klar, personal communication). This double-stranded chromosome break occurs very close to the *Y*/*Z*1 border and is found in both *HML***a** *MAT***a** *HMR***a** and *HML*α *MAT*α *HMR*α homothallic strains in which mating-type switching continues indefinitely. J. N. Strathern (personal communication) have also found a similar double-stranded break in *HO HML*α *MAT*α *hmr***a***1* strains in which homothallic switching from *MAT*α to *mat***a***1* occurs repeatedly. A possibly identical double-stranded break is found in homothallic strains carrying the *rad52* mutation (Weiffenbach and Haber, 1981). These data do not definitively show that the initial event at the mating-type locus is actually a double-stranded chromosome break (as opposed to an initial single-stranded event), but they do argue that some event occurs at *MAT* that does not occur at *HML* or *HMR*.

Direct evidence for the formation of a site-specific, double-stranded cleavage of *MAT* DNA has recently been obtained by R. Kostriken and F. Heffron (personal communication). Cell extracts from *HO* cells contained a site-specific endonuclease that cleaved *MAT* DNA near the *Y*/*Z* border. This activity was not found in extracts from *ho* cells. The cleavage generated a 4-bp, 3′-staggered cut 3 bp from the *Y*/*Z* border:

```
CGCAACA   GG
GCG      TTGTC
```

This site coincides with the location of the two *inc* mutations that have been sequenced (Fig. 10). Thus, the activation of *MAT* switching appears to involve the formation of a double-stranded cleavage at *MAT* very close to the *Y*/*Z* border.

C. Evidence for Physical Pairing between the Donor Locus and *MAT*

Mating-type switching appears to require homologous pairing between *MAT* and either *HML* and *HMR*. We have found that a significant fraction of mating-type switching events are accompanied by an intrachromosomal recombination event joining *MAT* and *HMR* or *MAT* and *HML* (Haber *et al.*, 1980c). A recombination event joining *MAT* and *HMR* would lead to the formation of a large, lethal deletion of several essential genes lying between *MAT* and *HMR*. (The reciprocal recombination product would be an acentric circular fragment that would be lost during mitosis.) Such deletions can, however, be rescued either by mating these cells with another haploid strain (Hawthorne, 1963a; McCusker and Haber, 1981) or by carrying out homothallic switching experiments in diploid strains (Haber *et al.*, 1980c). In homothallic diploids initially of genotype *HMLα/HMLα mat***a***1/MATα-inc HMR***a***/HMRα*, homothallic mating-type switching will replace *mat***a***1* with *MAT***a** (the other chromosome in the diploid contains the *MATα-inc* mutation that is not able to undergo mating-type conversion). Approximately 1% of the nonmating homothallic diploids recovered from such experiments contain a recessive lethal deletion resulting from the recombinational joining of *MAT* and *HMR***a** (Fig. 14). The high frequency of these recombination events depends on the presence of the *HO* allele. Similar fusions of *mat***a***1* with *HML***a** (to generate a deficiency ring chromosome) were also found in similar experiments (Haber *et al.*, 1980c). It should be pointed out that these recombination events occurred not only between *mat***a***1* and either *HML***a** or *HMR***a** (i.e., during homologous replacement of mating-type information) but also during the switching of *mat***a***1* to *MATα*. When a *mat***a***1/MAT***a***-inc* diploid was used, approximately 1% of the conversions of *mat***a***1* to a functional α allele were accompanied by recombination events joining *MAT* and *HMR* or *MAT* and

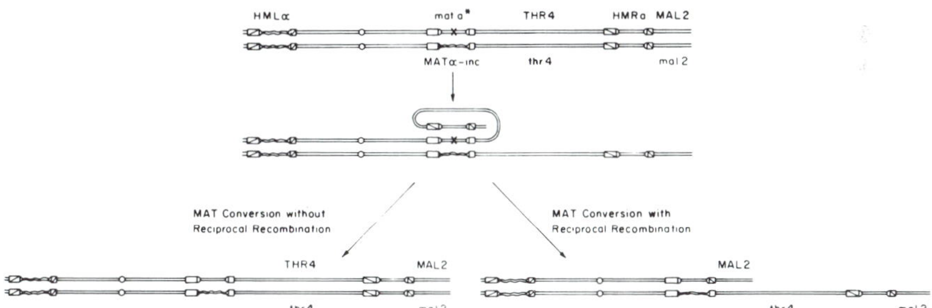

Fig. 14 Reciprocal recombination accompanying *MAT* switching. A *ho/HO mat***a***1/MATα-inc* diploid, constructed by conjugation, is initially α-mating and will switch to *MAT***a***/MATα-inc*. Normal switching of *mat***a***1* to *MAT***a** occurs as a nonreciprocal transposition event, leaving both homologs of chromosome III unaltered. However, about 1–2% of the time, conversions of *mat***a***1* are accompanied by a reciprocal recombination event joining homologous sequences at *MAT* and *HMR*. This recombination event results in a deletion of all the sequences between *mat***a***1* and *HMR***a** and the uncovering of recessive markers on the opposite homolog. The *MAL2* gene distal to *HMR***a** is not deleted. The other product of reciprocal recombination, a circle lacking a centromere and carrying the deleted sequences, is lost. The formation of *MAT/HMR* deletions during mating-type switching argues that the donor (*HMR* or *HML*) and recipient (*MAT*) sequences must physically pair during the transposition event.

HML (Haber *et al.*, 1980c). These observations argue that during homothallic switching there is homologous pairing between *MAT* and both *HML* and *HMR* and suggest that such pairing may be an essential feature of mating-type switching.

D. Evidence of Heteroduplex Formation during Mating-Type Conversions

During *MAT* switching, DNA from both the donor locus and the *MAT* locus must be present simultaneously in a heteroduplex. Klar *et al.* (1980b) have used a homothallic diploid initially of genotype *HML*α/*HML*α *mat***a***1*-1/*MAT*α-*inc* *hmr***a***1-2*/*HMR*α in which the two **a** mutations (at *MAT* and *HMR*) were not identical. Although many transposition events simply replaced one mutant **a** allele with the other, a significant fraction of transposition events resulted in the formation of a wild-type *MAT* allele that could have been generated only by an intragenic recombination event between the two mutant **a** sequences (Fig. 15). These observations argue that, at least in the case of homologous mating-type replacement (i.e., **a** sequences pairing with and replacing **a** sequences), some sort of heteroduplex structure must be generated. There is as yet no experimental evidence that such heteroduplexes are formed when the donor and recipient contain nonhomologous sequences. If such heteroduplex regions are formed during heterologous mating-type switching, only the *W* and *Z* sequences flanking Y**a** and Yα could form a stable heteroduplex.

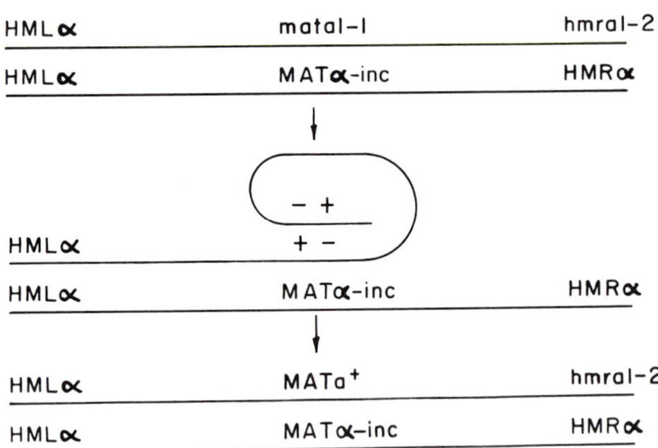

Fig. 15 Evidence of heteroduplex formation during mating-type conversion (based on Klar *et al.*, 1980b). A homothallic *mat***a***1*-1/*MAT*α-*inc* diploid has an α-mating type because the *mat***a***1* allele is recessive. This α-mating diploid will undergo mating-type switching. Only the *mat***a***1-1* allele will undergo conversion, as the *MAT*α-*inc* allele cannot switch. The transposition of **a** information from another **a**1 mutant (*hmr***a***1-2*) frequently results in the formation of wild-type *MAT***a** information. Wild-type **a** information from two **a***1* mutants must have been formed by intragenic recombination.

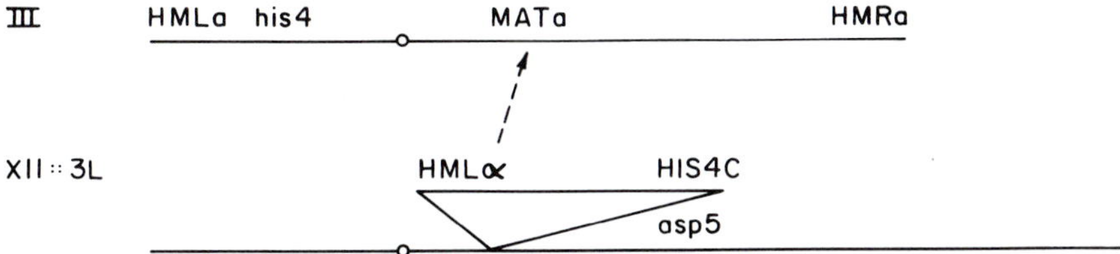

Fig. 16 Effect of interchromosomal interactions between *HML*α and *MAT***a.** Mating-type switching is significantly reduced when the donor (*HML*α) is translocated to chromosome XII. (Haber *et al.*, 1981a.)

E. "Topology" of Mating-Type Switching

All three copies of mating-type information (*HML, MAT, HMR*) are oriented in the same direction on chromosome III. The evidence that mating-type switching involves an intrachromosomal recombination event, in which sequences at *HML* or *HMR* must pair directly with those at *MAT* (Haber *et al.*, 1980c), suggests further that the efficiency of mating-type switching might be severely impaired if such pairings were *inter*chromosomal rather than *intra*chromosomal. Indeed, the efficiency of mating-type switching is reduced approximately 100-fold in an *HO HML***a** *MAT***a** *HMR***a** strain that also carries *HML*α as part of a large insertional translocation of much of the left arm of chromosome III to chromosome XII (Fig. 16) (Haber *et al.*, 1981a). A similar conclusion has been reached by J. Margolskee and I. Herskowitz (personal communication) using a reciprocal translocation between chromosomes I and III that separates *HML*α from *MAT***a** and *HMR***a.** On the other hand, the efficiency of switching is only slightly affected if transposition occurs in trans from a homologous chromosome in a diploid (Klar and Fogel, 1977; Rine and Herskowitz, 1980). The relatively efficient switching in an interchromosomal transposition between wholly homologous chromosomes may indicate that mitotic chromosomes are frequently synapsed.

Transformation and integration of plasmids containing the mating-type locus also make it possible to ask whether efficiency of mating-type switching is affected by the orientation of the mating-type locus relative to *HML* and *HMR*. L. S. Davidow and J. E. Haber (unpublished) have constructed plasmids containing *MAT***a** and *LEU2* in such a way that the mating-type region can be integrated either in the same orientation or in opposite orientation at the *LEU2* locus. These transformed strains also carried a double mutation, *mat***a***1-inc,* at the normal *MAT* locus, so that the expression and switching of mating type must occur at the newly integrated site. The efficiency of mating-type switching when *MAT* was integrated near *LEU2* was reduced approximately threefold, regardless of the orientation of the *MAT* sequences relative to those at *HML* or *HMR*. These transformed strains also make it possible to ask whether a change in the orientation of the *MAT* locus relative to the centromere has any effect on the pedigree of mating-type switching. However, even when the *MAT* locus is inverted relative to the centromere (and *HML* and *HMR*), the pedigree of mating-type switching remains the same: mother cells that have previously divided will switch to produce two cells of opposite mating type, whereas "inexperienced" cells that have not previously divided do not switch mating type. A similar conclusion has been reached by A. J. S. Klar and D. C. Hawthorne (personal communication), who have used a pericentric inversion of chromosome III that inverts both *HML* and the centromere relative to *MAT* and *HMR*. Again, there is no effect on the pedigree of mating-type switching in this inversion.

F. Selection of *HML* or *HMR* as Donor

During normal mating-type switching, approximately 75% of the eligible cells switch to the opposite mating type. Thus, the selection of *HML* or *HMR***a** is not random; rather, a *MAT* cell must usually derive its information from *HMR***a,** just as a *MAT***a** cell must preferentially use *HML*. The selection of *HML*α or *HMR***a** does not, however, depend on the actual sequence of mating-type information at the *MAT* locus. For example, it is possible to construct homothallic diploids carrying *mat***a***1*/*MAT*α*-inc* and homozygous for *HML***a** and *HMR***a.** Such cells, which have an α-mating phenotype, nevertheless switch *mat***a***1* to *MAT***a** with an efficiency that is essentially identical to that found in the reciprocal experiment involving a homothallic diploid carrying *mat***a***1*/*MAT***a***-inc* and homozygous for *HMR*α and *HML*α (Haber *et al.*, 1980c). Thus, the *mat***a***1* allele could be switched efficiently to either *MAT***a** or *MAT*α. It is therefore possible that it is the cell's *phenotype* rather than its genotype that determines whether a cell will use a donor containing **a** or α information.

One surprising aspect of mating-type switching is that it is apparently much less efficient in homothallic cells carrying *HML***a** and *HMR*α (i.e., in which the positions of the silent mating-type genes are reversed from the more frequent arrangement that has been studied). The apparently slow switching of a *MAT*α cell that must derive **a** information from *HML***a** or of a *MAT***a** cell that must derive its α information from *HMR*α is, however, only apparent. Klar *et al.* (1982) have recently demonstrated that under these circumstances most mating-type switches go undetected, as *MAT*α switches to another copy of *MAT*α rather than to *MAT***a.** This conclusion has been substantiated by the use of **a** or α alleles at *HMR* and *HML* that are genetically or biochemically distinct from those found at *MAT*. A *MAT***a** cell interacts selectively with sequences at *HML* (regardless of whether it contains α or **a** information), and *MAT*α preferentially pairs with *HMR*.

Klar *et al.* (1982) have concluded that there must be some sort of competition in the selection of *HMR* or *HML* as donor, because under conditions in which there is only a single copy of mating-type information available for switching, both *MAT***a** and *MAT*α will switch equally efficiently from the single donor locus. This experiment has been carried out in two ways. First, Klar *et al.* have generated a haploid-viable ring chromosome in which a recombination event between *HML* and *HMR* has occurred. The resulting ring chromosome contains only a single donor locus consisting of the *X* region from *HMR* and the *Z1/Z2* regions from *HML* flanking unexpressed α information. Using this single donor locus containing either **a** or α sequences, both *MAT*α and *MAT***a** switch efficiently to the opposite mating type. In addition, Klar *et al.* (1982) have deleted nearly all of the *HMR* locus, so that a linear chromosome III contains only *HML* as a donor. In a haploid containing only *HML***a** and *MAT*α, homothallic mating-type switching occurs in over 80% of the eligible cells. This is in sharp contrast to homothallic strains containing *HML***a** *MAT*α *HMR*α, in which only 7% of the switches were conversions from *MAT*α to *MAT***a**. Thus, the removal of *HMR* sequences led to a significant increase in the use of *HML* in the switching of *MAT*α.

The selection of *HMR* as the donor locus in cells that have an α phenotype and of *HML* in cells that have an **a** phenotype must depend on the recognition of sequences that are not identical between the two donor loci. At least some of this recognition may depend on the fact that *HML* shares considerably more homologous sequences with the mating-type locus than does *HMR* (the *X* and *Z2* regions). It is possible that these *stk* mutations (which affect the switching of *MAT***a** much more than that of *MAT*α) interfere with the association of the additional homologous sequences shared by *HML*α and *MAT***a** but do not interfere with the association of *HMR***a** with *MAT*α (which do not share these additional homologous sequences).

G. Distinction between Donor and Recipient Loci

In wild-type homothallic strains, the switching of mating-type alleles is extremely directional: The *MAT* locus is switched, whereas those of *HML* and *HMR* are not. The normal directionality of mating-type switching can be abolished by mutations that permit expression of the normally silent *HML* and *HMR* loci (Klar *et al.*, 1981b). The effect of mutations such as *mar1* or *cmt* on mating-type switching cannot be studied in normal strains that carry functional copies of both **a** and α information. In the presence of the *mar*-type mutations, homothallic haploids carrying at least one copy of both **a** and α information are nonmating and homothallic mating-type switching is repressed (Haber and George, 1979; Klar *et al.*, 1981b). However, by using strains carrying normal α information but with mutations in all copies of **a** information, it is possible to demonstrate that frequent conversions of mating-type information can occur at *HML*, *MAT*, and *HMR* when a *mar* mutation is present. For example, Klar *et al.*, (1981b) has shown that haploid homothallic spores initially of genotype *HML*α *MAT*α *hmr***a***1* can give rise to the **a**-mating cells with genotype *hml***a***1* *mat***a***1* *hmr***a***1*.

Klar *et al.* (1981b) have carried out additional experiments in homothallic diploid cells homozygous for *mar1*. In a zygote formed by mating a heterothallic *mar1 HML*α *mat***a***1 hmr***a***1* strain with an *HO mar1 hml***a***1 mat***a***1 hmr***a***1* strain, the switching of *HML*α to *hml***a***1* can be followed by the response of progeny cells to α factor. In such diploids, the conversion of *HML*α to *hml***a***1* occurs at the four-cell stage in approximately 50% of the cases examined. There were, however, a significant number of unusual switching events noted in these studies, both in switching *HML*α and in equivalent studies of the switching of *HMR*α in such diploids. Approximately 10% of the time, the pattern of switching violated the standard pedigree rules that have been established for switching at *MAT*. There were cases in which daughter cells, rather than mother cells, switched at the second generation; in other cases, only one of a pair a cells appeared to have switched mating type. No such aberrations were found in equivalent studies in which the only α allele was at *MAT*.

Similar experiments have also been carried out by A. M. Comeau and J. E. Haber (unpublished), using the *cmt* mutation to express *HML* and *HMR*. However, unlike the results reported by Klar *et al.* (1981b), a high proportion of the switching events in *HO cmt hml***a** *MAT*α *hmr***a** cells gave rise to **a**-mating cells that were only transiently viable. These **a**-mating cells could be rescued by mating with a *CMT MAT*α*-inc* strain, in much the same way that transiently viable *HO swi1 rad52* cells could be rescued (Weiffenbach and Haber, 1981). Further analysis has indicated that at least some of the lethality associated with homothallic switching in *cmt* strains arose from frequent reciprocal recombination between *MAT*α and *hml***a** to form a circular deficiency chromosome. Thus, the simultaneous expression of the "silent copy genes," as well as *MAT*, also changed the type of recombination events normally found during switching.

In addition to mutations such as *mar1* and *cmt* that allow the expression of *HML* and *HMR*, various chromosomal rearrangements joining *MAT* with *HML* or *HMR* also lead to the expression of mating-type genes (see Section IV,B). These rearrangements have also been shown to be switched in homothallic strains.

As described earlier, the chromosomal rearrangement *SAD1* and an equivalent fusion between *HMR* and *MAT***a** generated by the insertion of a *MAT***a** plasmid at *HMR* are expressed sufficiently to provide sporulation information but not sufficiently to make *MAT*α cells nonmating. Recent studies (J. E. Haber and P. C. Thorburn, unpublished) have shown that the **a** information in the *HMR/MAT***a** fusion can be used as a donor to switch *MAT*α. In addition, this sequence is itself switched from *HMR/MAT***a** to *HMR/MAT*α in approximately 1–5% of the cells examined. Similarly, the α sequences in a *MAT/HMR*α or *MAT/HML*α sequence created by the insertion of a *MAT* plasmid at *HMR* or *HML* are also not expressed strongly enough to make *MAT***a** cells nonmating; however, these weakly expressed sequences can be switched to *MAT/HMR***a** or *MAT/HML***a.** Thus, when mating-type sequences are expressed, either by chromosomal rearrangement or by the recessive *mar* mutations, these sequences become actively switched by the homothallic switching mechanism. All of these data argue that the normal directionality of mating-type switching is a consequence of the lack of expression of *HML* or *HMR*.

H. Homothallic Conversions of Tandem Duplications of Yeast Mating-Type Genes

An alternative approach to studying the mechanism of *MAT* conversions has been to follow switching events in haploid strains containing a tandem duplication of mating-type genes. Such strains have been constructed by the transformation and integration of a pBR322 plasmid containing both the *Eco*R1-*Hind*III *MAT***a** fragment and the selectable *URA3* gene (Haber and Rogers, 1982).

An equivalent strain containing a tandem duplication of *MAT*α genes has also been obtained. When these heterothallic transformants were crossed with homothallic *ura3* spores, meiotic segregants carrying both *HO* and the tandem duplication of the *MAT* genes were obtained. The apparent efficiency of mating-type switching is very high, as all of the homothallic segregants were nonmating. These nonmating segregants were of several types. A few percent were haploid strains in which one or the other *MAT* locus had switched to the opposite mating type, to yield *MAT***a**-*URA3*-pBR322-*MAT*α or *MAT*α-*URA3*-pBR322-*MAT***a** haploid nonmating strains. In the vast majority of cases, the tandem duplication was converted to the opposite mating type, which, following conjugation, yielded nonmating diploid cells. Approximately 10% of the time, the nonmating diploids proved to be of genotype

$$\frac{MAT\mathbf{a}\text{-URA3-pBR322-}MAT\mathbf{a}}{MAT\alpha\text{-URA3-pBR322-}MAT\alpha}$$

These represent cases in which both *MAT* alleles had switched. By far the most frequent event, however, was the formation of diploids of genotype

$$\frac{MAT\mathbf{a}\text{-URA3-pBR322-}MAT\mathbf{a}}{MAT\alpha}$$

The switching of the tandem duplication to a single *MAT* locus of opposite mating type appears to represent cases in which the switching of mating-type alleles was either preceded or accompanied by the excision of the tandem duplication. Pedigree studies have shown that, in fact, the loss of the tandem duplication occurs in the same cell division as the switching of mating-type alleles, and therefore accompanies the conversion event (J. E. Haber and D. T. Rogers, unpublished).

These results can be understood if one assumes that the entire tandem duplication can be treated as a single mating-type locus during homothallic mating-type switching. One example of how this may occur is shown in Fig. 17, where the donor *HML*α locus pairs with homologous regions of two different *MAT***a** genes.

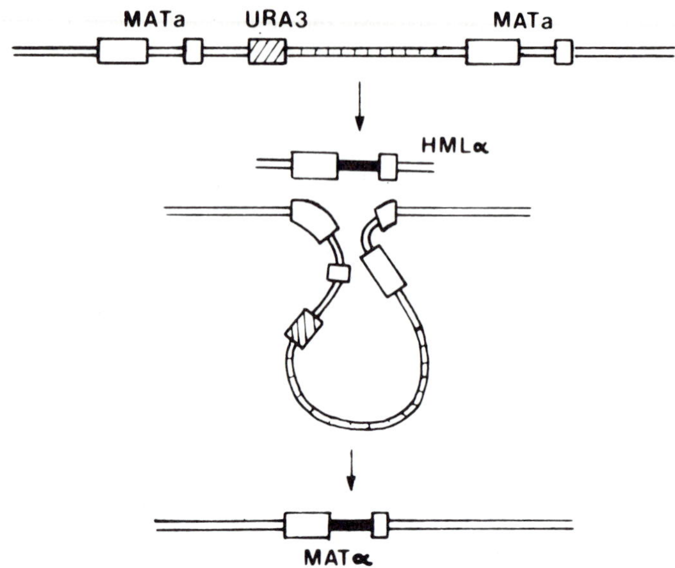

Fig. 17 Conversion of a tandem duplication of *MAT* genes. A tandem duplication of *MAT***a** was created by transformation and integration by homologous recombination of a *MAT***a** plasmid carrying pBR322 and the yeast *URA3* gene. The predominant conversion that has been observed results in the replacement of the entire tandem duplication by a single copy of opposite mating-type information. Pedigree studies have shown that this conversion occurs within a single cell division. Such a conversion may occur after pairing of the *HML*α donor locus with the distal portions of the two *MAT***a** gene regions. A mitotic gene conversion would then replace a large (7-kb) substitution loop with a small α-specific sequence.

The pairing of the outside homologous regions with *HML* in this way generates a very large region of nonhomology between *HML*α and the tandem duplication. Instead of a substitution loop of approximately 700 bp, the nonhomologous region in the tandem duplication is over 7000 bp. Nevertheless, this tandem duplication can be efficiently switched to a single copy of the opposite mating type.

It should be stressed that this excision of the tandem duplication of *MAT* genes occurs at high frequency only in homothallic cells. In heterothallic strains of genotype *HML***a** (*MAT***a**-*URA*3-pBR322-*MAT***a**) *HMR***a,** the *URA3* gene is lost, with a frequency of less than 10^{-3}. In contrast, homothallic strains of this genotype lose *URA*3 so frequently that a colony derived from an *HO HML***a** (*MAT***a**-*URA*3-pBR322-*MAT***a**) *HMR***a** spore is almost completely Ura$^-$. The few Ura$^+$ cells that remain in the colony appear to contain mutations that prevent efficient mating-type switching. This observation has led to a powerful new method to select mutations defective in switching, as it relies on a simple nutritional selection of Ura$^+$ papillae to find new mutations (J. E. Haber and N. Rudin, unpublished).

We have also shown that a tandem duplication in which the right *MAT* locus contains the *mat***a***1-inc* mutation and the left *MAT* locus carries *MAT*α [*MAT*α-*URA3*-pBR322-(*mat***a***1-inc*)], the *MAT*α locus is able to switch to *MAT***a**, but excision of the tandem duplication is almost completely prevented. This result suggests that the excision of the plasmid requires normal mating-type switching to be possible at the right *MAT* locus. In contrast, in a tandem duplication where *mat***a***1-inc* is on the left [(*mat***a***1-inc*)-*URA3*-pBR322-*MAT*α], switching to a single copy of *MAT***a** does occur. These observations are consistent with the scheme shown in Fig. 17. Our studies using tandem duplications have also given one other surprising result: the selection of *HML* or *HMR* as the "appropriate" donor can apparently be overridden by placing an "inappropriate" donor closer to the *MAT* locus. Based on the results of Klar *et al*. (1982) (Section V,F), one would expect that in an *HML*α [*MAT*α-*URA3*-pBR322-(*mat***a***1-inc*)] *HMR***a** homothallic strain, the donor of **a** information would be *HMR***a** (lacking the *W* and *Z2* homology regions). In fact, the *mat***a***1-inc* sequence (containing both *W* and *Z2*, as well as additional homologous flanking sequences) was used as the donor more than 90% of the time to create a [(*mat***a***1-inc*)-URA3-pBR322-(*mat***a***1-inc*)] strain that could not switch again (J. E. Haber, unpublished). Thus, a sequence very close to *MAT*α that resembles *HML* more than *HMR* in the extent of its shared homology with *MAT* was preferentially used as the donor.

I. Transpositions of Tandem Duplications of *MAT* Information

A similar approach has been used to investigate whether or not the mating-type switching system can lead to the transposition of a tandem duplication of mating-type genes as well as a single *MAT* locus. For these experiments, tandem duplications at *HMR* were created by the integration of a plasmid containing *MAT***a** and *URA3* (Haber and Rogers, 1982). As described earlier (Section IV,B), the integration of *MAT***a** at *HMR***a** creates a tandem duplication in which at least one of the two **a** regions is expressed. In a strain that also carries *MAT*α, the transformants will therefore be nonmating. Heterothallic conversions of *MAT*α to *MAT***a** (or to a tandem duplication of **a** information) were recovered by crossing these nonmating haploid transformants with a *MAT*α strain. Two different types of transformants were examined in this way. In one case, a single integration of the *MAT***a** plasmid at *HMR***a** was used; and among 20 independent switches of the *MAT* locus, all resulted from a transposition of a single **a** sequence. In contrast, when transpositions were examined from a nonmating strain that contained two copies of the *MAT***a** plasmid integrated at *HMR***a** (Fig. 18), two of 10 independent conversions of α to **a** resulted from a transposition of a tandem duplication of **a** information flanking both *URA3* and pBR322. Subsequent genetic and biochemical analysis showed that these transpositions occurred without

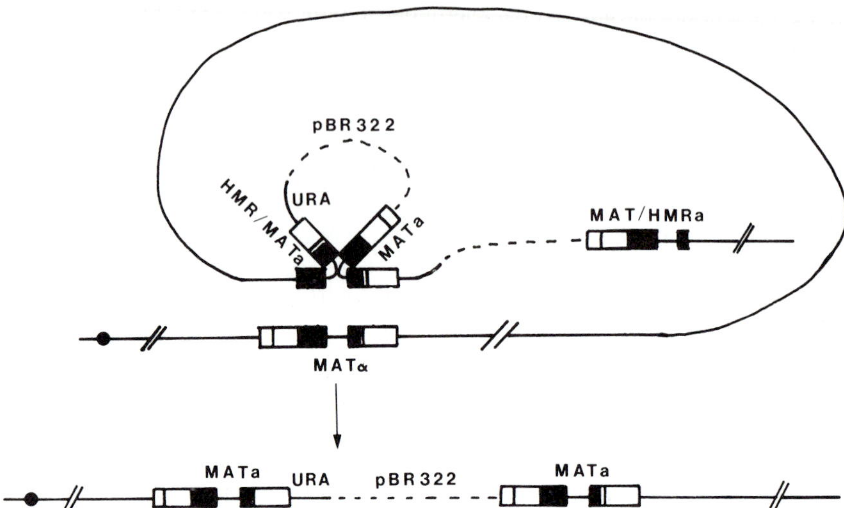

Fig. 18 A model for the transposition of a tandem duplication of mating-type genes from *HMR* :: (*MAT***a**-URA3-pBR322)$_2$ to replace *MAT*α. A gene conversion event in which sequences at *HMR* are used to replace nonhomologous sequences at *MAT* first requires the homologous pairing of these regions, as shown. Instead of the normal 0.7-kb region of nonhomology transferred between *HMR***a** and *MAT*α, the pairing of the outside homologous regions of the tandem duplication at *HMR***a** :: (*MAT***a**-*URA3*-pBR322)$_2$ leads to the insertion of a 7.9-kb region of DNA not homologous to *Y*α.

loss or alteration of the integrated sequences at *HMR* and therefore represent a bona fide transposition of a tandem duplication of mating-type genes. That such transpositions of a tandem duplication occurred in the double integrant and not in the single integrant may reflect the fact that more extensive pairing between homologous sequences is possible in the second case. The results can be explained by assuming that homologous sequences at the ends of a tandem duplication are used to pair with the sequences at *MAT* and that a large substitution loop is then used as a template for generating new mating-type sequences (Fig. 18).

J. Transpositions of Mating-Type Genes Containing Large Insertions

A. J. S. Klar (personal communication) has used an insertion of the TY1 transposable element at *HMR***a** as a donor of mating-type information. His results show that this insert, in the **a**-specific portion of *HMR***a,** is transposed along with normal *HMR***a** sequences during mating-type switching. These results again suggest that mating-type switching can accommodate nonhomologous sequences that are not normally found in the mating-type genes.

K. Relation of Mating-Type Switching to Other Mitotic Recombination Events

The transposition of mating-type genes appears to involve many of the same steps as other mitotic gene conversion events in yeast. What distinguishes mating-type switching is that it is highly site-specific, is under the genetic control of the *HO* locus, and occurs at an extremely high frequency. We have asked whether the ''induction'' of all of the enzymatic machinery necessary to accomplish mating-type switching also increases the mitotic recombination at other unlinked loci. In a *mat***a***1*/*MAT*α-*inc* homothallic diploid, essentially every cell will undergo a conversion event to replace *mat***a***1* by *MAT***a.** We have constructed such diploids by mating a heterothallic *mat***a***1* haploid with a homothallic *MAT*α-*inc* haploid and selecting zygotes. The diploids were also heteroallelic for *lys2-1*/*lys2-2*, *his4-39*/*his4-864*, and *leu2-1*/*leu2-27* and were heterozygous for *can1*/+. After three to four generations (during which time most cells had switched to become *MAT***a**/*MAT*α-*inc*), the frequency of intragenic Lys$^+$, His$^+$, or Leu$^+$ recombinants were measured and compared with the frequency of those arising from the mating of a normal *MAT***a** heterothallic strain with the homothallic *MAT*α-*inc* diploid. In the latter case, no mating-type switching would occur, because the *MAT***a**/*MAT*α-*inc* diploid expresses both functional **a** and α genes, so that *HO* is not expressed. There was no significant difference in the proportion of prototrophic colonies or in the fraction of canavanine-resistant colonies when these two diploid populations were compared (J. E. Haber and L. B. Rowe, unpublished). Thus, there is no evidence that the site-specific recombination events for mating-type switching lead to an increased level of recombination elsewhere in the genome.

One mutation that affects mitotic recombination has been examined for its effect on mating-type switching. The mutation *spo11* has been found by Esposito and Wagstaff (1981) to show an increased level of mitotic recombination. Their work suggests that the *spo11* mutation specifically increases reciprocal recombination. An examination of homothallic strains carrying *spo11* as well as the *swi1* mutation (to slow down the frequency of switching) indicates that there is no increase in the formation of so-called Hawthorne deletions (reciprocal recombinations between *MAT* and *HMR***a**) when *MAT*α switches to *MAT***a** (J. E. Haber, unpublished). Thus, *spo11* does not appear to have any significant effect on mating-type switching.

VI. Possible Mechanisms of Transposition of Yeast Mating-Type Genes

The extensive genetic and molecular analysis of *MAT* switching has ruled out several early and provocative models. Both Hawthorne (1963b) and Brown

(1976) offered versions of a flip-flop mechanism in which transcription at *MAT* could be directed to copy either **a** or α sequences that were both present at *MAT*. Demonstrations that mutations now known to be within coding sequences could be healed made these mechanisms unlikely, and subsequent DNA sequence analysis has proven that there are unexpressed complete copies of mating-type genes at *HML* and *HMR* that can physically replace the **a** or α sequence at *MAT*.

The data reviewed in Section V suggest that the transposition of yeast mating-type genes involves an intrachromosomal mitotic recombination event that depends on pairing between the homologous sequences at *MAT* and either *HML* or *HMR*. Generally, this recombination event is not reciprocal, but leads to the transfer of sequences from *HML* or *HMR* to *MAT* without altering the donor sequences. Some transposition events are, however, accompanied by reciprocal exchange of flanking sequences. These characteristics of mating-type switching are strongly reminiscent of both meiotic and mitotic gene-conversion events.

A. An Analogy: Meiotic and Mitotic Gene Conversion

Gene conversion is a well-studied form of nonreciprocal recombination found in both mitotic and meiotic cells. Most attention has been focused on meiotic gene conversion, where one recovers tetrads in which two alleles of a locus do not segregate to 2:2 but rather to 1:3 or 3:1. Such events are presumed to arise by mismatch repair of heteroduplex DNA. The formation of such a heteroduplex is believed to be an essential feature of all meiotic recombination, and approximately 50% of meiotic gene conversions are accompanied by a reciprocal exchange of flanking markers. The remaining gene conversions occur without reciprocal recombination (Fogel *et al.*, 1978).

A variety of molecular mechanisms have been proposed to account for meiotic gene-conversion events. The extensive data from yeast meiosis in particular have led to the suggestion that only a single heteroduplex region is formed initially (Meselson and Radding, 1976). The formation of a heteroduplex can be envisioned to occur in a variety of ways (Meselson and Radding, 1976; Resnick, 1976; Stahl, 1979). All of these mechanisms presume that a single strand of DNA "invades" and displaces a strain of the same polarity in another DNA molecule. There is good, although indirect, evidence that the invading strand is preferentially used as a template in mismatch repair of the heteroduplex (thus leading to a gene conversion) (Savage and Hastings, 1981).

It should be pointed out that meiotic gene conversions can occur even when one homolog contains a deletion or an insertion (Fink and Styles, 1974; Fogel *et al.*, 1978). Meiotic gene conversions of *MAT* alleles (where both *MAT* regions carry a nonhomologous sequence) have also been documented (Klar *et al.*, 1979a). Thus, the normal heterothallic meiotic cell contains all of the enzymatic components necessary to replace nonhomologous sequences at *MAT*.

Mitotic gene conversion may differ from meiotic events in several respects. First, recombination events in general appear to occur at the G_1 stage of the cell cycle (i.e., before DNA replication), although meiotic recombination occurs after chromosomal replication (Esposito, 1978). Second, there are several mutations that affect recombination differently in meiosis than in mitosis (Malone *et al.*, 1980; Malone and Esposito, 1981). Third, only about 10% of all gene conversion events in diploids are accompanied by a reciprocal exchange of flanking markers (Esposito and Wagstaff, 1981). Intrachromosomal gene conversions have also been studied in haploids by following recombination between tandem duplications of either the *his4* or *leu2* genes in which the two copies carry different alleles (Jackson and Fink, 1981; Klein and Petes, 1981). Here, too, about 5% of intragenic recombinations that produce His^+ or Leu^+ cells are accompanied by a reciprocal recombination event that excises the intervening plasmid sequences.

It is also important to note that recent studies of the effect of the *rad52* mutation on mitotic recombination have suggested that reciprocal recombination and nonreciprocal mitotic gene conversions may not derive from the same sequence of events (as had been argued for meiotic gene conversion) (Jackson and Fink, 1981; Orr-Weaver *et al.*, 1981). Their studies showed that *rad52* prevented intrachromosomal conversions occurring without reciprocal exchange of flanking markers but did not interfere significantly with reciprocal recombination events. It is striking that these same effects of *rad52* were seen in *MAT* conversions (Weiffenbach and Haber, 1981).

In the broadest outlines, at least, *MAT* switching appears to be quite similar to other intrachromosomal gene conversions (Klein and Petes, 1981; Jackson and Fink, 1981). What distinguishes *MAT* conversions are their very high frequency in homothallic strains. It is possible that this highly efficient switching depends on one or two facilitating genes, but that otherwise *MAT* conversions could depend on all the other functions necessary for intrachromosomal mitotic gene conversion. One such facilitating event may be the site-specific nucleolytic cleavage (activation?) of the *MAT* locus. A second (as yet undemonstrated) would be a protein-mediated specific pairing of the homologous regions shared by *MAT* and *HML* or *HMR*.

Another useful way to think about *MAT* switching may be its analogy to the stimulation of recombination between plasmids and a homologous chromosomal site by double-stranded DNA breaks (and even gaps) (Orr-Weaver *et al.*, 1981). In their experiments, an internal segment of a cloned yeast gene was excised and the plasmid not religated. When this broken, gapped DNA was used in a transformation experiment, it was repaired by a gene-conversion event involving an intact homologous sequence on the chromosome. The initiation of a double-stranded break very close to the *Y/Z* border at *MAT* creates a somewhat analogous situation in which homologous sequences are interrupted not by a gap but by nonhomologous sequences.

Recently, we have obtained further evidence to support the idea that a double-stranded break will serve to establish the direction of gene conversion (J. E. Haber and M. Resnick, unpublished). We have followed mitotic gene conversion events in zygotes arising by the mating of two haploids carrying different alleles of *lys2*. Prior to mating and zygote formation, one or the other haploid was irradiated by γ rays to produce random double-stranded breaks. More than 85% of the Lys$^+$ colonies arose by a conversion event that changed the *lys2* allele of the irradiated parent, leaving the *lys2* allele of the unirradiated parent unchanged. These results are analogous to the high degree of directionality found at the *MAT* locus, where the DNA sequences near the double-stranded break are converted.

B. A Mitotic Gene-Conversion Mechanism for *MAT* Switching

A model for the transposition of *MAT* alleles can be constructed by taking into account the experimental observations that have been reviewed. The several experiments that suggest direct physical pairing between *MAT* and *HML* or *HMR* make it unlikely that transposition involves a diffusible copy of *HML* or *HMR* that integrates at *MAT*. Similarly, the observation of a heteroduplex formed between *mat***a***1* and *hmr***a***1* sequences argues against the complete excision of *MAT* sequences prior to the insertion of new mating-type sequences.

One version of a site-specific mitotic gene conversion model is depicted in Fig. 19. In this model, as in the meiotic gene-conversion mechanisms of Meselson and Radding (1976) or Stahl (1979), a single DNA strand from *HML* or at *HMR* is generated and displaces a strand of the same polarity at *MAT*.

This model supposes that mating type switching is actually initiated by a double-stranded cut in *MAT* DNA near the *Y*/*Z*1 border (Fig. 19A). Because *Y***a** and *Y*α are nonhomologous, only the DNA to the right of the double-stranded cut can invade the donor locus (Fig. 19B). This incoming strand can be used to initiate DNA synthesis in *HMR* (or *HML*) proceeding from *Z*1 toward *W*, with the displacement of a donor strand. The progression of DNA replication across the *Y* region into *W* eventually generates DNA on both sides of the unique *Y* region that can form a stable heteroduplex at *MAT* (Fig. 19C). Subsequent mismatch repair, favoring the incoming strand, will then result in the switching of mating type genes (Fig. 19D). To facilitate the inclusion of the incoming strand, including the nonhomologous region, it is likely that one or both of the strands of the *Y* region are removed by an exonuclease.

In the model shown in Fig. 19, the generation of a single DNA strand from *HML* or *HMR* is not initiated at these loci, but rather begins with a nucleolytic cleavage at *MAT*. Such an activation step is similar to that suggested by Stahl (1979) and is also similar to the model of Orr-Weaver *et al.* (1981) for a mechanism of mitotic recombination initiated by a double-stranded break. As pointed out before, the events that *initiate MAT* switching seem to occur at *MAT*

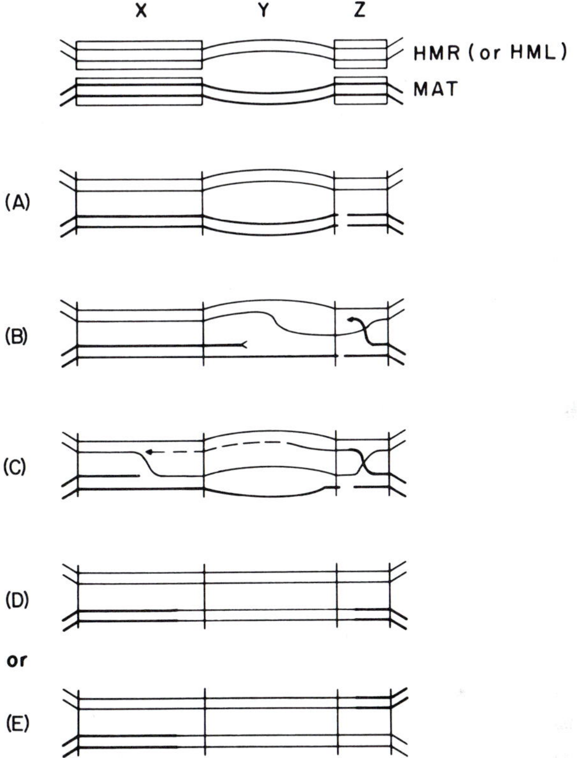

Fig. 19 A mitotic gene-conversion mechanism for the switching of yeast mating-type genes. (A) Mating-type switching is initiated by a site-specific nuclear cleavage in the *Z* region of the *MAT* locus. The site has been identified by the location of two *inc* point mutations. Mating-type switching first requires the homologous pairing of sequences at both *MAT* and *HMR* or *HML*. In this example, a double-stranded chromosome break is pictured, although subsequent events could also be accommodated by a single-stranded chromosome cleavage at this site. (B) A single strand at the *MAT* locus may then invade and displace a strand of the donor locus. It should be noted that only the strand in the *Z* locus is able to make such a displacement, as the DNA sequences in the *Y* locus are normally not homologous between *MAT* and the donor locus. (C) Continued displacement of a strand from the donor locus that can form a heteroduplex with the *MAT* locus may occur by the initiation of DNA replication on the invading *MAT* strand, using the donor locus as a template. (D or E) Finally, the preferential mismatch repair using the invading DNA strand will result in the conversion of the *MAT* locus without a change in the donor locus. It should also be noted that the heteroduplex structure illustrated in (C) contains a crossover structure (Holiday structure) in the *Z* region that could isomerize during or before mismatch correction. If such an isomerization occurred, mating-type switching would be accompanied by a reciprocal recombination that would join the left-hand portion of the *MAT* locus to the right-hand portion of *HMR* (a so-called Hawthorne deletion) (cf. Fig. 14).

rather than at *HML* or *HMR*. For example, the *α-inc* mutation prevents switching of *MAT*α*-inc* to *MAT***a** (and also *HML*α*-inc* to *HML***a**), but it does not prevent *HML*α*-inc* from acting as an efficient donor to transpose the mutation back to *MAT*. Furthermore, in diploids, the switching of *mat***a***1* can use *MAT*α*-inc* as the donor, so that neither *HML* nor *HMR* is required to initiate switching. These observations, plus the finding of double-stranded DNA breaks at *MAT* (both in Rad$^+$ and *rad52* homothallic strains), argue that an event occurring at *MAT* somehow initiates subsequent replacement of one allele by the other.

The location of the double-stranded DNA breakpoint and the position of both *MAT***a***-inc* and *MAT*α*-inc* point mutants within 10 bp of the *Y*/*Z*1 border suggest that this initiation event involves a site-specific cut in the DNA at this point. It should be remembered that the present data do not yet demonstrate that the initial event at *MAT* is a double-stranded DNA cut. It is still possible that the first step is a single-stranded cut, perhaps followed by the formation of a long, single-stranded gap in the *MAT* region. Such a single-stranded gap could be labile and lead to double-stranded breaks, especially in *rad52* cells. Although *rad52* has been strongly implicated in double-stranded DNA repair (Resnick and Martin, 1976), it is also important to note that in *rad52* cells attempting to undergo meiosis, DNA is recovered with many single-stranded cuts rather than double-stranded ones (Resnick *et al.*, 1981). Thus, *rad52* may affect several enzymatic activities of DNA repair, and it is not yet clear what intermediate fails to be repaired by *rad52* during mating-type switching. That the nicking or cutting of DNA near the *Y*/*Z*1 boundary occurs at *MAT* and not at *HML* or *HMR* must reflect the fact that *MAT* is transcribed whereas *HML* and *HMR* are not (see Section IV). One would expect that similar nucleolytic events will also be found at *HML* or *HMR* in strains in which these loci are also expressed. This site-specific cleavage of *MAT* DNA may provide the basis for the fact that *MAT* is switched, whereas *HML* and *HMR* are not.

One other aspect of this model deserves comment. The switching of a *MAT* allele to the opposite mating type apparently involves the formation of a heteroduplex structure between donor and recipient DNA strands. One must ask how a single strand of DNA that can displace the approximately 650–750 bp of *nonhomologous Y* sequences before forming again as a heteroduplex on the other side. From healing experiments, it is clear that at least part of *X* and all of *Y* must be replaced by a transposition event. We assume that a heteroduplex must be formed both in *W* and in *Z* so that replacement of *MAT* sequences can occur. The formation of a heteroduplex including the *Y* region can obviously be facilitated by the excision of at least one strand of the *MAT Y* region, so that the incoming strand is not in competition with a resident strand that has the same polarity and is completely homologous to the opposite strand. These considerations raise the possibility that at least one strand of the nonhomologous sequence at *MAT* must be removed before or during DNA strand transfer from *HML* or *HMR* and the formation of a heteroduplex. This constraint is not, however, different from what is necessary for meiotic gene conversion of *MAT* alleles in heterothallic strains.

The extent of the *Z* region that must be transferred (or mismatch repaired if there are allelic variants) is not known. We do know that α-*inc* and **a**-*inc* mutations can both be healed (when they are at *MAT*) and can be transposed (e.g., from *HML* to *MAT*). In addition, a base-pair variant 11 bp from the *Y/Z* border was apparently transposed from *HMR***a** to replace *MAT*α in a pair of strains studied by Astell *et al.* (1981).

The details of the formation of the heteroduplex shown in Fig. 19 are one of several possible models. It is possible that both strands of the *MAT* Y region are "chewed back" into the X region, where a second strand invasion of *MAT* DNA into the homologous donor could occur, in a manner similar to that proposed by Orr-Weaver *et al.* (1981). One might also imagine that the initiation of DNA replication at *HMR* or *HML* would ultimately be resolved by the switching of the DNA polymerase back to the *MAT* strand, with yet another replication event restoring the donor locus.

A gene conversion model is consistent with the experimental observations that have been reviewed. The mechanism presumes that switching depends on intrachromosomal pairing of sites at *HML* or *HMR* with homologous regions at *MAT*. Consistent with this picture is the fact that switching is severely impaired if it requires *inter*chromosomal pairing between *HML*α and *MAT***a.** The model also accounts for the heteroduplex formation observed in the generation of wild-type **a** sequences from two different **a**1 mutants at *MAT* and *HMR*. In addition, the resolution of a "Holliday" structure (Fig. 19C) after an isomerization would lead to a reciprocal recombination event joining *MAT* to *HML* or *HMR* accompanying the gene-conversion event (Fig. 14). A gene-conversion model can also account for the efficient switching of a tandem duplication of *MAT* genes to a single *MAT* gene of opposite mating type (Fig. 17).

C. Unanswered Questions

Studies of the expression and transposition of *MAT* genes have raised a number of questions that have not yet yielded to experimental observation. For example, how does the cell "select" *HML* or *HMR* to act as donor? The evidence to date suggests that this selection is mediated by the cell's mating phenotype rather than its genotype. A *mat***a***1* sequence can switch efficiently to *MAT***a** or to *MAT*α, depending on the initial phenotype of the cell. This suggests that there should be a genetically determined way to select (by promoting specific pairing of homologous sequences?) *HML* or *HMR* to pair with *MAT*. No mutations have yet been identified that alter the preference of donor selection. The *stk* mutants may, however, represent sequences at *MAT* that may be part of the selection of donors. Similarly, no mutants have been described that alter the regulation of the *HO* locus (e.g., one might find *MAT***a**/*MAT*α diploids in which *HO* functions are still active).

One is also confronted with the fascinating question of how the asymmetric cell lineage of *MAT* conversions occur. Only a cell that has previously divided can switch mating type. There is the possibility that old DNA strands might be preferentially inherited by mother cells and that these chromatids may be modified (methylated?) to allow them to be switched, whereas the newer DNA strands would not (Holliday and Pugh, 1975). The experiments in which *MAT* sequences were inverted relative to the centromere and donor loci (Section V) seem to rule out such models. However, Klar *et al.* (1981b) also looked at switching in *mat***a***1*/*MAT*α*-inc* zygotes formed from cells in which *HO* had been on in both parents for many generations. Thus, 25% of the zygotes should have been formed between two *HO* cells each of which was already a mother. No switches were observed at the first cell division. Thus, a cytoplasmic determinant was not demonstrated. It is of course possible that zygote formation "erases" distinctions between mother and daughter cells. These results leave open the question of how lineages are established.

D. Relation of Mating-Type Switching to Other Transposing Systems

It appears that the mechanism of mating-type switching differs significantly from the mechanisms proposed for other transposable elements. DNA sequence analysis in the *W* and *Z* regions does not indicate that transpositions of *MAT* genes produce the duplication of a small number of nucleotides, as has been found for all of the insertions of bacterial transposons or in the transpositions of eukaryotic elements such as the yeast TY1 or *Drosophila copia* elements (see Chapters 7 and 8). Furthermore, it appears that mating-type switching involves a pairing of extensive homologous sequences and is far more site-specific in the insertions of transposable elements into essentially nonhomologous sequences. Nor does mating-type switching appear to involve the kind of recombination described in bacterial phase variation or in the joining of sequences in the mammalian immune system.

It seems likely that the ability to control differentiating cell types by the replacement of cassettes (Herskowitz *et al.*, 1977) is too good an idea to have been used only in *Saccharomyces cerevisiae*. In fact, homothallic mating type conversion of a very distantly related yeast, *Schizosaccharomyces pombe*, appears from genetic studies to switch by a related mechanism (Egel, 1977; Egel and Gutz, 1981; Beach *et al.*, 1982). There is, however, no obvious homology between *MAT* DNA of *Saccharomyces cerevisiae* and that of *Schizosaccharomyces pombe* (A. J. S. Klar and R. Egel, personal communication). The transposition of a silent copy gene to a site where it is expressed may also occur in the trypanosome surface glycoprotein genes (Bernards *et al.*, 1981; see also Chapter 14). A possibly similar mechanism may also act in the generation of diversity in the H2 locus of the mouse immune system (J. Seidman; personal communication).

One suspects that as other developmental systems are understood in more detail, mechanisms similar to that found in *MAT* switching will be described. Are some of the lineage mutants in *Caenorhabditis elegans* (Chalfie *et al.*, 1981) examples of *inc* or *swi*-like mutations in which the switch from one cell type to the other has been prevented by inefficient transposition of a new cassette?

Although other more complex examples may emerge, the prospects of understanding the yeast *MAT* system in greater detail will continue to absorb the energies of a number of laboratories.

Acknowledgments

I am grateful to Ira Herskowitz, Jim Hicks, Amar Klar, Vivian MacKay, Kim Nasmyth, Jasper Rine, Michael Smith, George Sprague, Jeff Strathern, Isamu Takano, and Kelley Tatchell for communicating the results of unpublished experiments. I am also indebted to Anne Comeau, Lance Davidow, Chris Ilgen, Barbara Garvik, Jeanne George, Ellen Kraig, John McCusker, Deborah Mascioli, Susan Raposa, Dave Rogers, Lucy Rowe, Norah Rudin, Walter Savage, Sue Stewart, Pat Thorburn, and Barbara Weiffenbach for their many contributions, published and unpublished, from my lab. Peter Cherbas and Bob Schleif offered many thoughtful suggestions about this review. Phyllis Bishop, Cathy Halainen, Ruth Ladd, and a wonderful word processor made it possible. Some of the work reported in this paper was supported by USPHS grant GM20056 and NSF grants PCM 11749 and PCM 8110633.

References

Arima, K., and Takano, I. (1979). Evidence for co-dominance of the homothallic genes, *HMα/hmα* and *HMa/hma*, in Saccharomyces yeasts. *Genetics* **93,** 1–12.

Astell, C. R., Ahlstrom-Jonasson, L., Smith, M., Tatchell, K., Nasmyth, K. A., and Hall, B. D. (1981). The sequence of the DNAs coding for the mating type loci of *Saccharomyces cerevisiae*. *Cell* **27,** 15–23.

Beach, D., Nurse, P., and Egel, R. (1982). Molecular rearrangement of mating-type genes in fission yeast. *Nature (London)* **296,** 682–683.

Bernards, A., Van der Ploeg, L. H. T., Frasch, A. C. C., and Borst, P. (1981). Activation of trypanosome surface glycoprotein genes involves a duplication–transposition leading to an altered 3′ end. *Cell* **27,** 497–505.

Betz, R., MacKay, V. L., and Duntze, W. (1977). **a**-Factor from *Saccharomyces cerevisiae:* partial characterization of a mating hormone produced by cells of mating type **a.** *J. Bacteriol.* **132,** 462–472.

Brown, W. S. (1976). A cross-over shunt model for alternate potentiation of yeast mating-type alleles. *J. Genet.* **62,** 81–91.

Bucking-Throm, E., Duntze, W., Hartwell, L. H., and Manney, T. R. (1973). Reversible arrest of haploid cells at the initiation of DNA synthesis by a diffusible sex factor. *Exp. Cell Res.* **76,** 99–110.

Chalfie, M., Horvitz, H. R., and Sulston, J. E. (1981). Mutations that lead to reiterations in the cell lineages of *C. elegans*. *Cell* **24,** 59–69.

Chan, R. K. (1977). Recovery of *Saccharomyces cerevisiae* mating-type **a** cells from G1 arrest by α factor. *J. Bacteriol.* **130,** 766–774.

Davidow, L. S., and Haber, J. E. (1981). Relations between the efficiency of homothallic switching of yeast mating type genes and the distribution of cell types. *Mol. Cell. Biol.* **1,** 1120–1124.

Duntze, W., MacKay, V., and Manney, T. (1970). *Saccharomyces cerevisiae:* a diffusable sex factor. *Science (Washington, D.C.)* **168,** 1472–1473.

Egel, R. (1977). "Flip-flop" control and transposition of mating-type genes in fission yeast. *In* "DNA Insertion Elements, Plasmids and Episomes" (A. Bukhari, J. Shapiro, and S. Adhya, eds.), pp. 447–455. Cold Spring Harbor Lab., Cold Spring Harbor, New York.

Egel, R., and Gutz, H. (1981). Gene activation by copy transposition in mating type-switching of a homothallic fission yeast. *Curr. Genet.* **3,** 5–12.

Elder, R. T., St. John, T. P., Stinchcomb, D. T., and Davis, R. W. (1981). Studies on the transposable element TY1 of yeast. I. RNA homologous to Ty1. *Cold Spring Harbor Symp. Quant. Biol.* **45,** 581–591.

Errede, B., Cardillo, T. S., Sherman, F., DuBois, E., DesChamps, J., and Wiame, J. M. (1980). Mating signals control expression of mutations resulting from the insertion of a transposable repetiative element adjacent to diverse yeast gene. *Cell* **22,** 427–436.

Esposito, M. (1978). Evidence that spontaneous mitotic recombination occurs at the two-strand stage. *Proc. Natl. Acad. Sci. USA* **75,** 4436–4441.

Esposito, M., and Wagstaff, J. (1981). Mechanisms of mitotic recombination. *In* "The Molecular Biology of the Yeast *Saccharomyces cerevisiae*" (J. Strathern, E. W. Jones, and J. R. Broach, eds.), pp. 341–370. Cold Spring Harbor Lab., Cold Spring Harbor, New York.

Esposito, M., Esposito, R., Arnaud, M., and Halvorson, H. (1969). Acetate utilization and macromolecular synthesis during sporulation of yeast. *J. Bacteriol.* **100,** 180–186.

Fink, G. R., and Styles, C. (1974). Gene conversion of deletions in the *his4* region of yeast. *Genetics* **77,** 231–244.

Fogel, S., Mortimer R., Lusnak, K., and Tavares, E. (1978). Meiotic gene conversion: a signal of the basic recombination event in yeast. *Cold Spring Harbor Symp. Quant. Biol.* **42,** 1325–1341.

Friis, J., and Roman, H. (1968). The effect of the mating-type alleles on intragenic recombination in yeast. *Genetics* **59,** 33–36.

Game, J. C., Zamb, T. J., Braun, R. J., Resnick, M., and Roth, R. M. (1980). The role of radiation (rad) genes in meiotic recombination in yeast. *Genetics* **94,** 51–68.

Gerlach, W. L. (1974). Sporulation in mating-type homozygotes of *Saccharomyces cerevisiae*. *Heredity* **32,** 241–249.

Gunge, N., and Nakatomi, Y. (1972). Genetic mechanisms of rare matings of the yeast *Saccharomyces cerevisiae* heterozygous for mating-type. *Genetics* **70,** 41–58.

Haber, J. E., and Garvik, B. (1977). A new gene affecting the efficiency of mating type interconversions in homothallic strains of *Saccharomyces cerevisiae*. *Genetics* **87,** 33–50.

Haber, J. E., and George, J. P. (1979). A mutation that permits the expression of normally silent copies of mating-type information in *Saccharomyces cerevisiae*. *Genetics* **92,** 13–35.

Haber, J. E., and Rogers, D. T. (1982). Transposition of a tandem duplication of yeast mating-type genes. *Nature (London)* **296,** 768–770.

Haber, J. E., Mascioli, D. W., and Rogers, D. T. (1980a). Illegal transposition of mating type genes in yeast. *Cell* **20,** 519–528.

Haber, J. E., Rogers, D. T., and McCusker, J. (1980b). Homothallic conversions of yeast mating-type genes occur by intrachromosomal recombination. *Cell* **22,** 277–289.

Haber, J. E., Savage, W. T., Raposa, S. M., Weiffenbach, B., and Rowe, L. B. (1980c). Genetic identification of sequences adjacent to the yeast mating type locus that are essential for transposition of new mating type alleles. *Proc. Natl. Acad. Sci. USA* **77,** 2824–2828.

Haber, J. E., Rowe, L. B., and Rogers, D. T. (1981a). Transposition of yeast mating type genes from two translocations of the left arm of chromosome III. *Mol. Cell. Biol.* **1,** 1106–1123.

Haber, J. E., Weiffenbach, B., Rogers, D. T., McCusker, J., and Rowe, L. B. (1981b). Chromosomal rearrangements accompanying yeast mating-type switching: evidence for a gene-conversion model. *Cold Spring Harbor Symp. Quant. Biol.* **45,** 991–1002.

Harashima, S., and Oshima, Y. (1976). Mapping of the homothallism genes, *HM*α and *HM***a**, in *Saccharomyces* yeasts. *Genetics* **84,** 437–451.

Harashima, S., and Oshima, Y. (1980). Functional equivalence and co-dominance of homothallism genes, *HM*α/*HM*α and *HM***a**/*HM***a**, in *Saccharomyces* yeasts. *Genetics* **95,** 819–831.

Harashima, S., Nogi, Y., and Oshima, Y. (1974). The genetic system controlling homothallism in *Saccharomyces* yeasts. *Genetics* **77,** 639–650.

Hartwell, L. H. (1980). Mutants of *S. cerevisiae* unresponsive to cell division control by polypeptide mating hormones. *J. Cell Biol.* **85,** 811–822.

Hawthorne, D. C. (1963a). A deletion in yeast and its bearing on the structure of the mating type locus. Genetics **48,** 1727–1729.

Hawthorne, D. C. (1963b). Directed mutation of the mating type alleles as an explanation of homothallism in yeast. *In* "Genetics Today. Proceedings of the 11th International Congress of Genetics" (S. J. Geerts, ed.), Vol. 1, pp. 34–35. Pergamon, Oxford.

Herskowitz, I., and Oshima, Y. (1981). *In* "The Molecular Biology of the Yeast *Saccharomyces cerevisiae*" (J. N. Strathern, E. W. Jones, and J. Broach, eds.), pp. 181–209. Cold Spring Harbor Lab., Cold Spring Harbor, New York.

Herskowitz, I., Strathern, J. N., Hicks, J. B., and Rine, J. (1977). Mating type interconversion in yeast and its relationship to development in higher eucaryotes. *In* "Proceedings of the 1977 ICN–UCLA Symposium: Molecular Approaches to Eucaryotic Genetic Systems" (G. Wilcox, J. Abelson, and C. F. Fox, eds.), pp. 193–215. Academic Press, New York.

Hicks, J. B., and Herskowitz, I. (1976a). Evidence for a new diffusible element of mating pheromones in yeast. *Nature (London)* **260,** 246–248.

Hicks, J. B., and Herskowitz, I. (1976b). Interconversion of yeast mating types. I. Direct observations of the action of homothallism (*HO*) gene. *Genetics* **83,** 245–258.

Hicks, J. B., and Herskowitz, I. (1977). Interconversion of yeast mating types. II. Restoration of mating ability to sterile mutants in homothallic and heterothallic strains. *Genetics* **85,** 373–393.

Hicks, J. B., Strathern, J. N., and Herskowitz, I. (1977a). The cassette model of mating-type interconversion. *In* "DNA Insertion Elements, Plasmids, and Episomes" (A. I. Bukhari, J. A. Shapiro, and S. L. Adhya, eds.), pp. 457–462. Cold Spring Harbor Lab., Cold Spring Harbor, New York.

Hicks, J. B., Strathern, J. N., and Herskowitz, I. (1977b). Interconversion of yeast mating types. III. Action of the homothallism (*HO*) gene in cells homozygous for the mating type locus. *Genetics* **85,** 395–405.

Hicks, J. B., Strathern, J. N., and Klar, A. J. S. (1979). Transposable mating type genes in *Saccharomyces cerevisiae*. *Nature (London)* **282,** 478–483.

Holliday, R., and Pugh, J. E. (1975). DNA modification mechanisms and gene activity during development. *Science (Washington, D.C.)* **187,** 226–232.

Hopper, A. K., and Hall, B. D. (1975a). Mating type and sporulation in yeast. I. Mutations which alter mating-type control over sporulation. *Genetics* **80,** 41–59.

Hopper, A. K., and Hall, B. D. (1975b). Mutation of a heterothallic strain to homothallism. *Genetics* **80,** 77–85.

Hopper, A. K., and MacKay, V. L. (1980). Control of sporulation in yeast: *SAD*—a mating-type specific, unstable alteration that uncouples sporulation from mating-type control. *Mol. Gen. Genet.* **180,** 301–314.

Hopper, A., Magee, P., Welch, S., and Hall, B. (1974). Macromolecule synthesis and breakdown in relation to sporulation and meiosis in yeast. *J. Bacteriol.* **119,** 619–628.

Jackson, J. A., and Fink, G. R. (1981). Gene conversion between duplicated genetic elements in yeast. *Nature (London)* **292,** 306–311.

Kassir, Y., and Herskowitz, I. (1980). A dominant mutation (*SAD*) bypassing the requirement for the **a** mating type locus in yeast sporulation. *Mol. Gen. Genet.* **180,** 315–322.

Kassir, Y., and Simchen, G. (1976). Regulation of mating and meiosis in yeast by the mating-type region. *Genetics* **82,** 187–206.

Klar, A. J. S. (1980). Interconversion of yeast cell types by transposable genes. *Genetics* **95,** 631–648.

Klar, A. J. S., and Fogel, S. (1977). The action of homothallism genes in *Saccharomyces* diploids during vegetative growth and the equivalence of *hm*a and *HM*α loci functions. *Genetics* **85,** 407–416.

Klar, A. J. S., and Fogel, S. (1979). Activation of mating type genes by transposition in *Saccharomyces cerevisiae*. *Proc. Natl. Acad. Sci. USA* **76,** 4539–4543.

Klar, A. J. S., Fogel, S., and Lusnak, K. (1979a). Gene conversion of the mating-type locus in *Saccharomyces cerevisiae*. *Genetics* **92,** 777–782.

Klar, A. J. S., Fogel, S., and MacLeod, K. (1979b). *MAR*1—a regulator of *HM*a and *hM*α loci in *Saccharomyces cerevisiae*. *Genetics* **93,** 37–50.

Klar, A. J. S., Fogel, S., and Radin, D. N. (1979c). Switching of a mating-type *a* mutant allele in budding yeast *Saccharomyces cerevisiae*. *Genetics* **92,** 759–776.

Klar, A. J. S., McIndoo, J., Hicks, J. B., and Strathern, J. N. (1980a). Precise mapping of the homothallism genes, *HML* and *HMR*, in yeast *Saccharomyces cerevisiae*. *Genetics* **96,** 315–320.

Klar, A. J. S., McIndoo, J., Strathern, J. N., and Hicks, J. B. (1980b). Evidence for a physical interaction between the transposed and the substituted sequences during mating type gene transposition in yeast. *Cell* **22,** 291–298.

Klar, A. J. S., Strathern, J. N., Broach, J. R., and Hicks, J. B. (1981a). Regulation of transcription in expressed and unexpressed mating type cassettes of yeast. *Nature (London)* **289,** 239–244.

Klar, A. J. S., Strathern, J. N., and Hicks, J. B. (1981b). A position-effect control for gene transposition: state of expression of yeast mating-type genes affects their ability to switch. *Cell* **25,** 517–524.

Klar, A. J. S., Hicks, J. B., and Strathern, J. N. (1981c). Irregular transpositions of mating-type genes in yeast. *Cold Spring Harbor Symp. Quant. Biol.* **45,** 983–1002.

Klar, A. J. S., Hicks, J. B., and Strathern, J. N. (1982). Directionality of yeast mating-type interconversion. *Cell* **28,** 551–561.

Klein, H. L., and Petes, T. D. (1981). Intrachromosomal gene conversion in yeast. *Nature (London)* **289,** 144–148.

Kraig, E. (1980). Messenger RNA and protein metabolism during sporulation of *Saccharomyces cerevisiae*. Ph.D. Dissertation, Brandeis Univ., Waltham, Massachusetts.

Kraig, E., and Haber, J. (1980). Messenger ribonucleic acid and protein metabolism during sporulation of *Saccharomyces cerevisiae*. *J. Bacteriol.* **144,** 1098–1112.

Kraig, E., and Haber, J. (1981). Effect of environment and genotype on the synthesis of abundant protein during sporulation of *Saccharomyces cerevisiae*. *Spores* **8,** 298–304.

Kushner, P. J., Blair, L. C., and Herskowitz, I. (1979). Control of yeast cell types by mobile genes—a test. *Proc. Natl. Acad. Sci. USA* **76,** 5264–5268.

Leibowitz, M. J., and Wickner, R. B. (1976). A chromosomal gene required for killer plasmid expression, mating, and spore maturation in *Saccharomyces cerevisiae*. *Proc. Natl. Acad. Sci. USA* **73,** 2061–2065.

Lemontt, J. F., Fugit, D. R., and MacKay, V. L. (1980). Pleiotropic mutations at the *TUP1* locus that affect the expression of mating-type-dependent functions in *Saccharomyces cerevisiae*. *Genetics* **94,** 899–920.

Lindegren, C. C., and Lindegren, G. (1943). A new method for hybridizing yeast. *Proc. Natl. Acad. Sci. USA* **29,** 306–308.

Livi, G. P., and MacKay, V. L. (1980). Mating-type regulation of methyl methanesulfonate sensitivity in *Saccharomyces cerevisiae*. *Genetics* **95,** 259–271.

MacKay, V. L., and Manney, T. R. (1974a). Mutations affecting sexual conjugation and related processes in *Saccharomyces cerevisiae*. I. Isolation and phenotype characterization of nonmating mutants. *Genetics* **76,** 255–271.

MacKay, V. L., and Manney, T. R. (1974b). Mutations affecting sexual conjugation and related processes in *Saccharomyces cerevisiae*. II. Genetic analysis of nonmating mutants. *Genetics* **76,** 273–288.

Malone, R. E., and Esposito, R. E. (1980). The *RAD52* gene as required for homothallic interconversion of mating types and spontaneous mitotic recombination in yeast. *Proc. Natl. Acad. Sci. USA* **77,** 503–507.

Malone, R. E., and Esposito, R. E. (1981). Recombinationless meiosis in *Saccharomyces cerevisiae*. *Mol. Cell. Biol.* **1,** 891–901.

Malone, R. E., Golin, J. E., and Esposito, M. S. (1980). Mitotic versus meiotic recombination in *Saccharomyces cerevisiae*. *Curr. Genet.* **1,** 241–242.

Mascioli, D. W., and Haber, J. E. (1980). A cis-acting mutation within the *MAT*a locus of *Saccharomyces cerevisiae* that prevents efficient homothallic mating type switching. *Genetics* **94,** 341–360.

McClintock, B. (1956). Controlling elements and the gene. *Cold Spring Harbor Symp. Quant. Biol.* **21,** 197–216.

McCusker, J. H., and Haber, J. E. (1981). Evidence of chromosomal breaks near the mating type locus of *S. cerevisiae* that accompany *MAT*α × *MAT*α matings. *Genetics* (unpublished).

Meselson, M., and Radding, C. (1975). A general model for genetic recombination. *Proc. Natl. Acad. Sci. USA* **72,** 358–361.

Nasmyth, K. A. (1982). Regulation of yeast mating-type chromatin structure by *SIR:* an action at a distance affecting both transcription and transposition. *Cell* **30,** 567–578.

Nasmyth, K. A., and Tatchell, K. (1980). The structure of transposable yeast mating type loci. *Cell* **19,** 753–764.

Nasmyth, K. A., Tatchell, K., Hall, B. D., Astell, C., and Smith, M. (1981). The physical mapping of transcripts at the mating type loci in *Saccharomyces cerevisiae*. *Nature (London)* **289,** 244–250.

Naumov, G. I., and Tolstorukov, I. I. (1973). Comparative genetics of yeast. X. Reidentification of mutators of mating types in Saccharomyces. *Genetika (Moscow)* **9,** 82–91. [Sov. Genet. (Engl. transl.)].

Orr-Weaver, T. L., Szostak, J. W., and Rothstein, R. J. (1981). Yeast transformation: a model system for the study of recombination. *Proc. Natl. Acad. Sci. USA* **78,** 6354–6358.

Oshima, Y., and Takano, I. (1971). Mating types in *Saccharomyces:* their convertibility and homothallism. *Genetics* **67,** 327–335.

Oshima, Y., and Takano, I. (1972). Genetic controlling system for homothallism and a novel method for breeding triploid cells in *Saccharomyces*. *In* "Fermentation Technology Today: Proceedings of 4th International Fermentation Symposium" (G. Terui, ed.), p. 847–852. Soc. Fermentation Technology, Osaka, Japan.

Oshima, T., and Takano, I. (1980). Mutants showing heterothallism from a homothallic strain of *Saccharomyces cerevisiae*. *Genetics* **94,** 841–857.

Oshima, T., and Takano, I. (1981). Mating-type differentiation by transposition of controlling elements in *Saccharomyces cerevisiae*. *Genetics* **97,** 531–549.

Pearson, N. J., and Haber, J. E. (1980). Changes in the regulation of ribosomal protein synthesis during vegetative growth and sporulation in yeast. *J. Bacteriol.* **143,** 1411–1419.

Petersen, J. G. L., Kielland-Brandt, M. C., and Nilsson-Tillgren, T. (1979). Protein patterns of yeast during sporulation. *Carlsberg Res. Commun.* **44,** 149–162.

Prakash, S., Prakash, L., Burke, W., and Montelone, B. A. (1980). Effects of the *RAD52* gene on recombination in *Saccharomyces cerevisiae*. *Genetics* **94,** 31–50.

Reed, S. I. (1980). The selection of *Saccharomyces cerevisiae* mutants defective in the start event of cell division. *Genetics* **95,** 561–577.

Resnick, M. A. (1976). Repair of double-strand breaks in DNA; a model involving recombination. *J. Theor. Biol.* **59,** 97–106.

Resnick, M. A., and Martin, P. (1976). The repair of double-stranded breaks in the nuclear DNA of *Saccharomyces cerevisiae* and its genetic control. *Mol. Gen. Genet.* **143,** 119–129.

Resnick, M. A., Kasimos, J. N., Game, J. C., Braun, R. J., and Roth, R. M. (1981). Changes in DNA during meiosis in a repair-deficient mutant (*rad52*) of yeast. *Science (Washington, D.C.)* **212,** 543–545.

Rine, J., and Herskowitz, I. (1980). The *trans* action of *HMRa* in mating type interconversion. *Mol. Gen. Genet.* **180,** 99–105.

Rine, J. D., Strathern, J. N., Hicks, J. B., and Herskowitz, I. (1979). A suppressor of mating type locus mutations in *Saccharomyces cerevisiae:* evidence for and identification of cryptic mating type loci. *Genetics* **93,** 877–901.

Rine, J., Sprague, G. F., Jr., and Herskowitz, I. (1981). *rme1* mutation of *Saccharomyces cerevisiae:* map position and bypass of mating type locus control of sporulation. *Mol. Cell. Biol.* **1,** 958–960.

Roman, H., and Sands, S. M. (1953). Heterogeneity of clones of *Saccharomyces* derived from haploid ascospores. *Proc. Natl. Acad. Sci. USA* **39,** 171–179.

Roman, H., Phillips, M. M., and Sands, S. M. (1955). Studies of polyploid *Saccharomyces cerevisiae*. I. Tetraploid segregation. *Genetics* **40,** 546–561.

Roth, R.; and Lusnak, K. (1970). DNA synthesis during yeast sporulation: genetic control of an early development event. *Science (Washington, D.C.)* **168,** 493–494.

Santa Maria, J., and Vidal, D. (1970). Segregacion anormal del "mating type" en *Saccharomyces*. *Inst. Nac. Invest. Agron. (Madrid)* **30,** 1–21.

Savage, E. A., and Hastings, P. J. (1981). Marker effects and nature of the recombination event at the *his1* locus of *Saccharomyces cerevisiae*. *Curr. Genet.* **3,** 37–47.

Sprague, G. F., Jr., and Herskowitz, I. (1981). Control of yeast cell type by the mating type locus. I. Identification and control of expression of the *a*-specific gene *BAR1*. *J. Mol. Biol.* **153,** 305–321.

Sprague, G. F., Jr., Rine, J., and Herskowitz, I. (1981a). Homology and non-homology at the yeast mating type locus. *Nature (London)* **289,** 250–252.

Sprague, G. F., Jr., Herskowitz, I., and Rine, J. (1981b). Control of yeast cell type by the mating type locus. II. Genetic interactions between *MAT*α and unlinked α-specific *STE* genes. *J. Mol. Biol.* **153,** 323–335.

Stahl, F. W. (1979). "Genetic Recombination." Freeman, San Francisco, California.

Stotzler, D., and Duntze, W. (1976a). Isolation and characterization of four related peptides exhibiting α-factor activity from *Saccharomyces cerevisiae*. *Eur. J. Biochem.* **65,** 257–262.

Stotzler, D., Kiltz, H.-H., and Duntze, W. (1976b). Primary structure of α-factor peptides from *Saccharomyces cerevisiae*. *Eur. J. Biochem.* **69,** 397–400.

Strathern, J. N. (1977). Regulation of cell type in *Saccharomyces cerevisiae*. Ph.D. Dissertation, Univ. of Oregon, Eugene.

Strathern, J. N., and Herskowitz, I. (1979). Asymmetry and directionality in production of new cell types during clonal growth: the switching pattern of homothallic yeast. *Cell* **17,** 371–381.

Strathern, J. N., Blair, L. C., and Herskowitz, I. (1979a). Healing of *mat* mutations and control of mating type interconversion by the mating type locus in *Saccharomyces cerevisiae*. *Proc. Natl. Acad. Sci. USA* **77,** 2829–2833.

Strathern, J. N., Newlon, C. S., Herskowitz, I., and Hicks, J. B. (1979b). Isolation of a circular derivative of yeast chromosome III: implications for the mechanism of mating type interconversion. *Cell* **18,** 309–319.

Strathern, J., Spatola, E., McGill, C., and Hicks, J. B. (1980). Structure and organization of transposable mating type cassettes in *Saccharomyces* yeasts. *Proc. Natl. Acad. Sci. USA* **77,** 2839–2843.

Strathern, J. N., Hicks, J. B., and Herskowitz, I. (1981). Control of cell type in yeast by the mating type locus: the α1–α2 hypothesis. *J. Mol. Biol.* **147,** 357–373.

Takahashi, T. (1958). Complementary genes controlling homothallism in Saccharomyces. *Genetics* **43,** 705–714.

Takano, I., and Arima, K. (1979). Evidence of insensitivity of the α-*inc* allele to the function of the homothallic genes in Saccharomyces yeasts. *Genetics* **57,** 875–885.

Takano, I., and Oshima, Y. (1967). An allele specific and a complementary determinant controlling homothallism in *Saccharomyces oviformis*. *Genetics* **57,** 875–885.

Takano, I., and Oshima, Y. (1970a). Allelism tests among various homothallism-controlling genes and gene systems in Saccharomyces. *Genetics* **64,** 229–238.

Takano, I., and Oshima, Y. (1970b). Mutational nature of an allele-specific conversion of the mating type by the homothallic gene *HO*α in Saccharomyces. *Genetics* **65,** 421–427.

Takano, I., Kusumi, T., and Oshima, Y. (1973). An α mating-type allele insensitive to the mutagenic action of the homothallic gene system in *Saccharomyces diastaticus*. *Mol. Gen. Genet.* **126,** 19–28.

Tatchell, K., Nasmyth, K., Hall, B. D., Astell, C. R., and Smith, M. (1981). *In vitro* mutation analysis of the mating type locus in yeast cell. *Cell* **27,** 27–35.

Weiffenbach, B., and Haber, J. E. (1981). Homothallic mating type switching generates lethal chromosome breaks in *rad52* strains of *Saccharomyces cerevisiae*. *Mol. Cell. Biol.* **1,** 522–534.

Wickner, R. B. (1974). Mutants of *Saccharomyces cerevisiae* that incorporate deoxythmydine-5′-monophosphate into deoxyribonucleic acid *in vivo*. *J. Bacteriol.* **117,** 252–260.

Winge, O., and Roberts, C. (1949). A gene for diploidization in yeast. *C. R. Trav. Lab. Carlsberg Ser. Physiol.* **24,** 341–346.

Wright, J. R., and Dawes, I. W. (1979). Sporulation-specific protein changes in yeast. *FEBS Lett.* **104,** 183–186.

Zubenko, G. S., and Jones, E. W. (1981). Protein degradation, meiosis and sporulation in proteinase-deficient mutants of *Saccharomyces cerevisiae*. *Genetics* **97,** 45–64.

CHAPTER 14

Antigenic Variation in Trypanosomes

P. BORST

Mobile Genetic Elements

ISBN 0-12-638680-3

I. Introduction

Trypanosomes are unicellular flagellates (Fig. 1A) that belong to the order Kinetoplastida, which contains some of the most successful parasitic protozoa

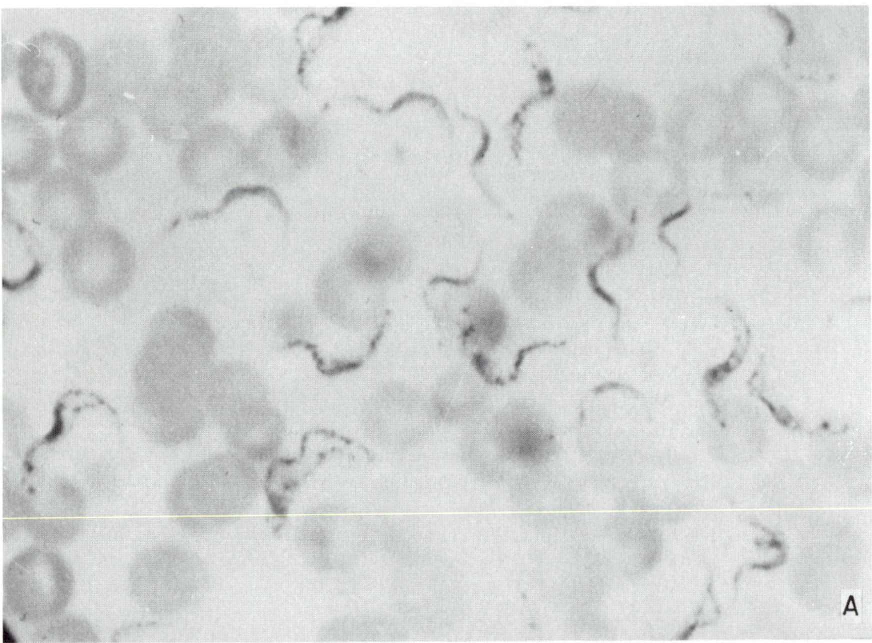

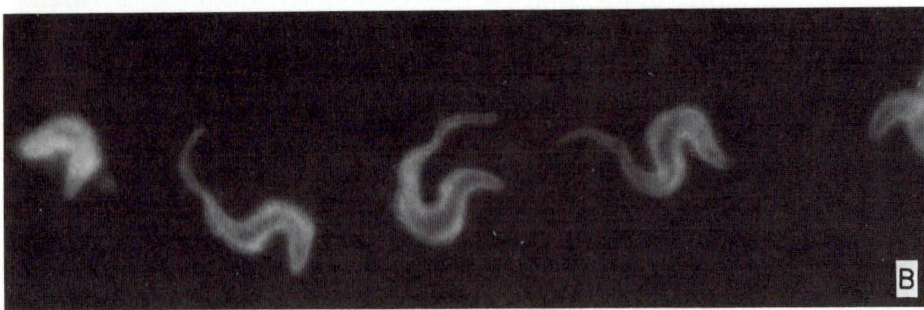

Fig. 1 *Trypanosoma brucei*. (A) *T. brucei* in a blood smear from a heavily infected rat (Giemsa stain). (B) *T. brucei* treated with a specific rabbit anti-VSG antibody, followed by fluorescein isothiocyanate-conjugated sheep antirabbit antibody. Note the homogeneous coating of the trypanosomes with antibody. (C) Diagrammatic representation of the life cycle of *T. brucei*. The slender

known. Most representatives of this order escape the host defense in the usual fashion by hiding in host cells. The African trypanosomes, however, have the unique ability to survive for long periods in the plasma and tissue spaces of the vertebrate host, in spite of a massive antibody response against the parasite's surface antigens. The trypanosome does this by varying the composition of its surface coat. The antigenic properties of the coat are due to a single glycoprotein—the variant surface glycoprotein, or VSG, which covers the entire surface of the trypanosome, probably in the form of a monolayer. A single trypanosome can produce more than 100 VSG variants so different in amino acid sequence that they do not have (surface) antigenic determinants in common. By switching from one VSG to the next, a small fraction of the trypanosomes continues to evade immunodestruction by the host.

For decades, the mechanism of antigenic variation at the gene level has been the subject of intense speculation and, up to 1978, even the rapid accumulation of mutations in one or a few VSG genes was considered a viable alternative (Seed, 1978). The recombinant DNA technique made it possible to analyze the VSG genes directly, and this analysis has shown that trypanosomes contain a large number (around 10^3) of VSG genes and that these genes are present, whether they are expressed or not. Antigenic variation is, therefore, based on the

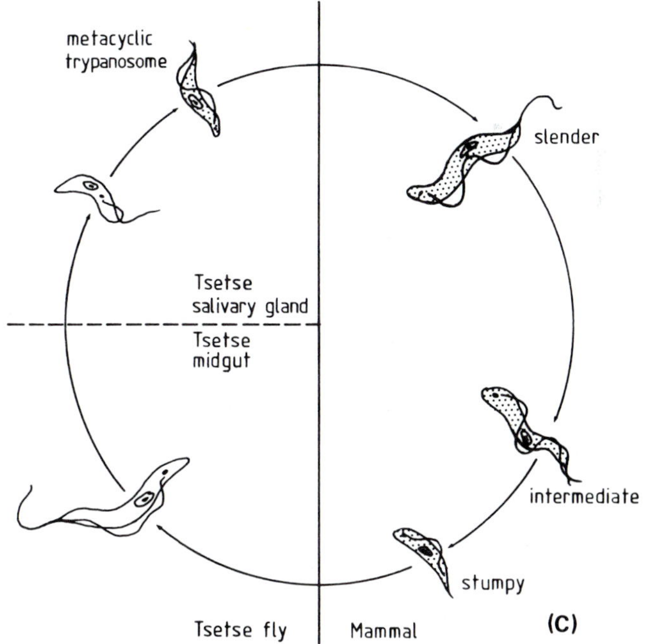

forms make up the bulk of the dividing parasites in the mammalian bloodstream. The stumpy forms are preadapted to life in the insect. Trypanosomes with surface coat are mottled. [(A, B) photographs by Dr. G. A. M. Cross; (C) simplified from Vickerman (1974).]

differential expression of a large repertoire of VSG genes. These experiments have eliminated all models in which the trypanosome economizes on gene usage by recombining coding segments at the DNA or RNA level or by hypermutation of a limited number of genes.

An important contribution of this early analysis of VSG genes was the discovery of three types of DNA rearrangements in trypanosomes:

1. The activation of some VSG genes requires a duplication–transposition of these genes. This is an unambiguous example of a developmental gene transposition comparable to that involved in the mating-type switch in yeast. It provides a simple explanation for the tight intergenic exclusion observed for VSG genes, that is, that only one is expressed at a time. This duplication–transposition will be the main topic of this chapter.
2. Deletions and insertions frequently occur in a region downstream of a particular subset of VSG genes. Activation of these genes does not seem to involve a simple duplication–transposition; how the expression of these genes is controlled is still a matter of speculation. I shall discuss here the nature of these downstream rearrangements and possible ways in which they could contribute to the control of the expression of these genes.
3. Comparison of closely related trypanosome isolates has shown that they may differ substantially in their VSG gene repertoire. These differences are the result of gene duplications and deletions and the rapid sequence evolution of a subset of VSG genes. These gene rearrangements must be important for the evolution of the VSG gene repertoire, and this will be considered at the end of this chapter.

There are several recent reviews that provide introductions to the general biology of antigenic variation in trypanosomes. The life cycle and immunology are simply and clearly described for nonparasitologists by Turner and Cordingley (1981), Marcu and Williams (1981), and Englund *et al.* (1982). The biochemistry of surface antigens has been reviewed by Cross (1977, 1978, 1979) and Borst and Cross (1982); and the recent work on DNA by Englund *et al.* (1982) and Borst and Cross (1982). The older literature on antigenic variation has been summarized by Vickerman (1974, 1978), Gray and Luckins (1976), and Doyle (1977). The general biology and the medical and economic importance of trypanosomes are described in detail in the beautiful monograph by Hoare (1972), as well as in the two volumes on "Biology of the Kinetoplastida" edited by Lumsden and Evans (1976, 1979).

With such ample coverage of the basic facts available, the initial sections of this chapter, which summarize these facts, will be limited to the minimum required for an understanding of the DNA rearrangements. In discussing the rearrangements, the emphasis will be on the mechanism and the evolutionary aspects, their use to the trypanosome, and the possible significance of such rearrangements for protozoa in general. There is some overlap between this chapter and the review I recently wrote with Cross (Borst and Cross, 1982). This is noticeable in sections II,A–C, III, IV,A, B, and D, and V,C.

II. The Basic Biology of Trypanosomes and Antigenic Variation

A. Taxonomy and Life Cycle

The African trypanosomes that show antigenic variation belong to the genus *Trypanosoma*. The species most intensively studied belong to the subgenus *Trypanozoon,* which includes *Trypanosoma brucei, T. evansi,* and *T. equiperdum*. Three variants of *T. brucei* are known: *T. brucei brucei, T. brucei gambiense,* and *T. brucei rhodesiense*. The latter two cause sleeping sickness in man and animals; the former does not infect man because it is lysed by human serum. On the basis of enzyme polymorphisms and DNA analysis, the overall differences between these three *T. brucei* variants are small, and the ones that infect man can be considered host-range mutants of *T. brucei brucei* that have acquired resistance to the lytic action of human serum (see Borst *et al.*, 1981b).

The life cycle of *T. brucei* is complex, as shown by the diagram in Fig. 1C. The tsetse fly (*Glossina*) becomes infected by sucking blood from an infected mammal. The trypanosomes multiply at first in the insect midgut, where they lose their protective surface coat. Later, they migrate to the salivary glands of the fly, where they reacquire their coat. When the fly takes another blood meal, some 10^4 infective trypanosomes are discharged into the mammalian host with the fly's saliva.

In the mammalian host, a relapsing parasitemia develops. After a period of trypanosome multiplication, the host develops antibodies against the surface coat of the predominant variant antigen types (VATs) present, and these are destroyed. A small fraction of the trypanosomes has switched coat, however, and escapes destruction. After 7–10 days a new parasitemia develops, followed by immune destruction of the major VATs present. In this way, a trypanosome infection may persist for months, showing an oscillating parasitemia in which serologically distinct parasites appear at 7–10-day intervals. In African game animals adapted to *T. brucei,* the blood level of trypanosomes remains low, and the chronic infection is tolerated and may eventually be overcome. In imported cattle or in humans, on the other hand, untreated infections are usually fatal. The multiplication in the insect is not required for the maintenance of active antigenic variation. This is shown by the effective antigenic variation displayed by *T. evansi* (passively transmitted by flies) and *T. equiperdum* (passively transmitted between horses as a venereal disease).

B. Genetics, Ploidy, and Nuclear DNA Content

Both in the fly and in mammals, *T. brucei* exists only outside cells and multiplies by binary fission. Direct evidence for a sexual stage is missing, and at present there is no known mechanism for genetic exchange between trypanosomes, although there are indirect indications that this must exist in nature (Tait, 1980).

Because trypanosomes have a primitive mitotic apparatus and no recognizable condensed chromosomes in any phase of their cell cycle, the question of ploidy has to be studied by indirect means. The haploid genome size of *T. brucei* nuclear DNA has been estimated by renaturation experiments at 3.7×10^4 kb, with 32% of the genome being in middle- or high-repetitive DNA (Borst *et al.*, 1980). The complexity of the trypanosome genome is, therefore, about six times that of *Escherichia coli*. The DNA per nucleus in bloodstream forms, determined by absorption photometry or fluorescence emission photometry of Feulgen-stained cells, is 0.091 pg or 8.2×10^4 kb (Borst *et al.*, 1982). This makes the trypanosome diploid. A similar conclusion was reached by a different route by Tait (1980). Studying enzyme polymorphisms, he found that for two-subunit enzymes the relative proportion of homodimers and heterodimers was exactly that predicted for a diploid organism with efficient means of genetic exchange.

C. The Nature of the Surface Coat and the Structure of VSGs

The surface coat makes up 7–10% of the total cell protein and consists mainly of a single protein—the VSG—that appears to cover the entire external surface of the trypanosome as a monolayer (see Fig. 1B; Cross, 1977; Holder and Cross, 1981; Borst and Cross, 1982). Isolated VSGs consist of a single polypeptide of about 55,000 MW with the glycosyl chains generally attached to the C-terminal half. VSGs from different VATs differ radically in amino acid sequence, in carbohydrate content (7–17%, w/w), and in the positions where carbohydrate is attached. In the surface coat, VSG molecules are probably tightly packed, with their C-terminal part oriented toward the membrane. The consequence is that only the antigenic determinants in the N-terminal half of the protein are accessible to host antibodies. Indeed, there is an immunogenic carbohydrate side chain attached to the C-terminal amino acid of many different VSGs that induces antibodies that cross-react with all these VSGs but not with the corresponding intact trypanosomes.

Trypanosomes do not continuously shed their coat, and VSG–membrane association appears to be stable so long as the cells are viable. The mode of attachment of VSG to membrane is unclear. There is no evidence for a hydrophobic domain inserted into the lipid layer, and in the mature VSGs that have been (partially) sequenced, a stretch of hydrophobic amino acids at the C-terminus is conspicuously absent (Boothroyd *et al.*, 1980; Pays *et al.*, 1981b; Rice-Ficht *et al.*, 1981). Possibly the C-terminal carbohydrate is involved, because this has been found in all 11 VSGs analyzed.

D. Switching of Coats

Switching of coats is not induced by external factors; the order in which VATs appear during chronic infections is programmed at the switching level. Switching is not induced by antibody because it occurs in immunocompromised animals or in culture (Doyle *et al.*, 1979). Other facts as well (see Turner and Cordingley, 1981) indicate that a low rate of switching is an intrinsic property of trypanosomes.

Whether VSGs are produced in a defined order is still controversial. Because this is important for the mechanism of switching, I shall consider the available data in some detail.

The appearance of VATs in the infective cycle is clearly nonrandom, and this is manifested at two levels:

1. In a chronic infection in mammals, different VATs appear in an imprecisely predictable order. Some VATs always appear very early, others early, and still others semilate or late in infection, but within—say—the early group, the order varies in different infections (Van Meirvenne *et al.*, 1975a,b, 1977; Capbern *et al.*, 1977; Kosinski, 1980; Miller and Turner, 1981). When trypanosomes are transferred by syringe in the course of a chronic infection to a fresh animal, the program is "reset," that is, the very early VATs reappear (Capbern *et al.*, 1977). This indicates that in a chronic infection early VATs continuously reappear as a result of switching, but these are not detected later in the infection because they are immediately eliminated by antibody.
2. The metacyclic population that develops in the salivary glands of the fly is heterogeneous but dominated by a limited number of VATs (sometimes only one). These VATs differ from the very early ones found in a chronic infection started with trypanosomes from another mammal by syringe. The metacyclic repertoire is strain-specific and not influenced by the nature of the VAT population ingested by the fly (Le Ray *et al.*, 1977, 1978; Barry *et al.*, 1979; Hudson *et al.*, 1980; Hajduk and Vickerman, 1981).

The interpretion of these results is complicated by the fact that the growth rate of a trypanosome can depend on the VSG made; there are even competition effects between different VATs of the same stock (Seed *et al.*, 1977; Seed, 1978; Miller and Turner, 1981). Nevertheless, the programmed appearance of VATs cannot be explained by growth effects alone for two reasons. First, different trypanosome strains may express the same set of VSG genes in different order (Van Meirvenne *et al.*, 1975b, 1977); likewise a certain VAT may be dominant in the metacyclic population of one trypanosome strain, but not in another (Jenni, 1977; Barry *et al.*, 1979). Second, in a given strain, some VATs tend to appear after others with high probability (Miller and Turner, 1981). Both these phenomena would seem to require some kind of programming at the switching level. I shall return in Section IV,E to the possible mechanisms for this programming.

Only a few years ago, the picture of antigenic switches in trypanosomes looked considerably less complex than it is now known to be. The metacyclic poulation was thought to contain only a single VAT, and in the mammalian host VATs were thought to follow each other in a largely fixed and predictable order. The present complexity, however, makes much more sense if one considers the selective pressure the trypanosome encounters in its natural habitat. In the tsetse belt of Africa, most of the game animals are infected with *T. brucei* (or its relatives), and for a *T. brucei* strain to start a successful superinfection, extensive VAT heterogeneity in the initial inoculation would maximize the chance of introducing at least one VAT that the host has not yet encountered. Once the trypanosome has gained access, the situation changes. Programmed switching and growth competition should now drastically limit the number of VATs expressed at high levels; otherwise the repertoire would be rapidly exhausted. The order of VAT appearance in the mammal should not be too rigorously programmed, however. If VAT 5 can appear only after VAT 4 has been expressed, the trypanosome would be very vulnerable to specific or nonspecific host defense measures against early VSGs, or to the effect of a prior infection of the host with another strain of trypanosomes. So, a sloppy order of VSG gene expression seems a sensible strategy for survival. In a later section, I shall argue that developmental gene rearrangements are optimally suited to producing such a sloppy order of gene expression.

E. Trypanosomes in the Laboratory

It is not always obvious to outsiders how one can do experiments on antigens that undergo continuous variation. Two factors are important. First, the rate of switching is low, especially in trypanosome populations that have been transferred from rodent to rodent by syringe. Second, it is possible to clone single trypanosomes in irradiated mice and grow them up to populations containing less than one in 10^5 heterotypes (i.e., VATs other than the one cloned). Such populations can be stored in liquid N_2 and used to infect rats. Rodent-adapted trypanosomes multiply very rapidly and can kill a rat in 3–4 days. In the terminal phase, up to 20% of the blood volume consists of trypanosomes, and these usually contain less than 1% heterotypes, providing ample homogeneous material for experiments.

Trypanosomes can also be grown in culture. Coated infective trypanosomes can be obtained only in low yield over a feeder layer of mammalian cells. Trypanosomes with properties similar to those found in the insect gut can be grown in good yield in axenic culture at 27°C. These trypanosomes do not contain a surface coat and probably do not synthesize any VSG.

F. Nomenclature

To describe the phenomenon of antigenic variation, parasitologists have developed a rather peculiar nomenclature that requires some explanation. *VATs* are trypanosomes that express a particular VSG. Populations of trypanosomes, each of which can express the same VAT repertoire, are called a *serodeme*. A mixture of trypanosomes isolated from an animal in nature is called a primary *isolate*. Such isolates can contain several different trypanosome strains, because in nature mixed infections are common. Populations derived from the primary isolate are called *stocks* when they have not been characterized and *strains* when they have. In practice, only populations derived from a single cloned trypanosome can be called strains, but in view of the confusion generated by the word stock, as defined here, I have avoided its use in this chapter.

Primary isolates are designated by locality of isolation/year of isolation/lab code/number—for instance, SERENGETI/58/EATRO/1716 (EATRO, East African Trypanosomiasis Research Organization). The specific trypanosome population used is added in square brackets—for instance, [Stabilate GUP 310] (GUP, Glasgow University Protozoology).

A VAT expressed by a majority of the trypanosomes in a clone is called a VAT *homotype;* the minor VATs also present are called *heterotypes*. VATs are often designated by codes such as AnTat 1.1 or ILTat 1.3. AnTat 1.1 means Antwerp *Trypanozoon* (the subgenus to which *T. brucei* belongs) antigen type, serodeme 1, VAT 1. ILTat means ILRAD *Trypanozoon* antigen type. Problems arise when the same VAT is reexpressed in a serodeme, as it is often not known whether the same allele is expressed. Formally, such reexpressors should get a new clone number, but to avoid confusion they are usually designated by the letters a, b, c, and so on; for instance, the reexpressed AnTat 1.1 is called AnTat 1.1b.

A final note of caution is needed about the use of laboratory strains for the study of antigenic variation. This is unavoidable to get a ready and reproducible source of trypanosomes, but it is obvious that syringe passage in rodents in the laboratory selects for rapid growth and high virulence, properties that are disadvantageous in the natural habitat of the parasite, because they will rapidly eliminate suitable hosts. It is possible, therefore, that some of the unusual phenomena observed with the early VATs in the laboratory (see Section V) do not occur in that form in nature.

III. VSG Messenger RNAs and VSG Genes

A. The Sequence of VSG Messenger RNAs

VSGs are encoded in prominent poly(A)-tailed messenger RNAs (mRNAs). Complementary DNA (cDNA) clones containing partial copies of these mRNAs have been readily obtained by several routes (see the literature in Borst and Cross, 1982). The sequence elements present in a VSG mRNA were first identified for VSG 117 of the 427 strain of *T. brucei* (Boothroyd *et al.*, 1980, 1981).

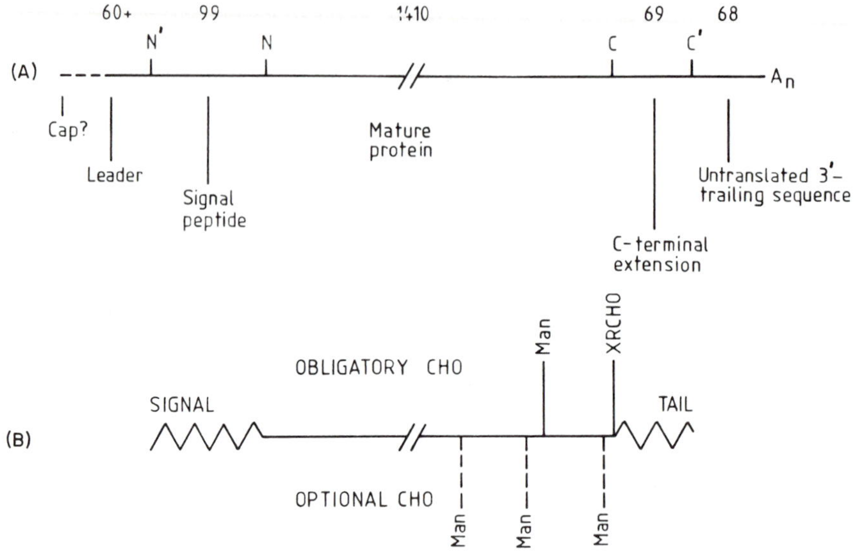

Fig. 2 Schematic drawing of a VSG (precursor) molecule (B) and its mRNA (A). The wavy lines indicate the N-terminal (leader) and C-terminal (tail) extensions that are present in the VSG precursor but absent in the mature VSG molecule. The numbers over the mRNA refer to the size in nucleotides of the mRNA of VSG 117, strain 427 of *Trypanosoma brucei* (see text). The exact length of the leader sequence is not known, but it exceeds 66 nucleotides.

Because the entire 117 VSG protein has been sequenced, the positions corresponding to the N- and C-termini of the mature VSG could be unambiguously assigned in the cDNA sequence. The main features of the mRNA and the corresponding precursor and mature proteins are shown in Fig. 2.

The most interesting feature of the VSG mRNA sequence is a C-terminal extension beyond the C-terminus of the mature protein (Boothroyd *et al.*, 1980). This extension has been detected in the cDNAs of six additional VSG mRNAs. In all cases, the sequence is hydrophobic and ends in a run of eight hydrophobic amino acids with Leu-Leu-Phe or Leu-Leu-Leu as C-terminus. Five of the seven C-terminal extensions are 23 amino acids long and homologous; the other two are only 17 amino acids long and rather different.

The hydrophobic tail is absent from VSGs released from the membrane by mild procedures in the presence of a cocktail of protease inhibitors, and its absence from mature VSGs does not appear to be an isolation artifact. Boothroyd *et al.* (1980, 1981) have proposed that the hydrophobic tail may initially serve to anchor precursor VSGs to internal membranes during intracellular transport and may be cleaved from the VSG in a final processing step after the VSG has attained its position in the surface coat. If this is correct, the extension would only have a signal function to route the VSG to the cell membrane. The large differences in size and sequence between the C-terminal extensions of some of these VSGs suggest that the requirements for the C-terminal signal are as loose as those for N-terminal signals.

B. Structure, Organization, and Relatedness of VSG Genes

Using cloned DNA complementary to several VSG mRNAs as probes, the organization of VSG genes has been studied by conventional techniques. Initial studies were done by hybridizing the cDNA probes to Southern blots of re-

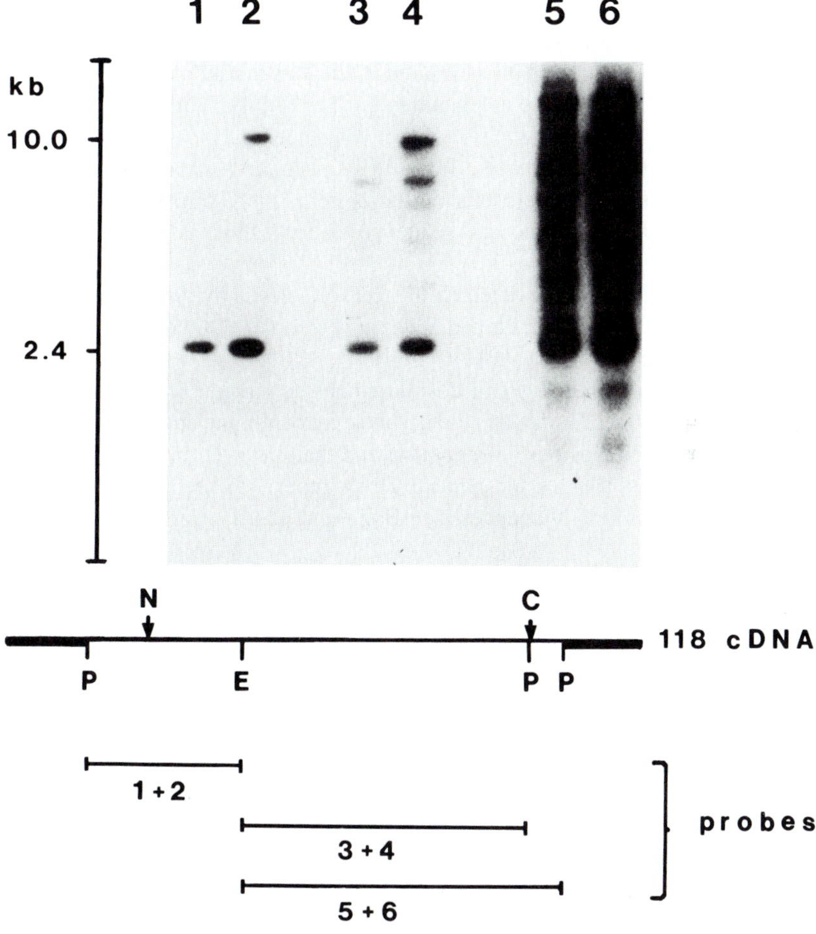

Fig. 3 Autoradiogram showing hybridization of different VSG 118 cDNA subprobes to blots of size-fractionated PstI digests of nuclear DNA from a heterologous variant (lanes 1, 3, and 5) and variant 118 itself (lanes 2, 4, and 6). (From Bernards *et al.*, 1981.)

striction digests of nuclear DNA. More recently, individual VSG genes and their flanking sequences have been cloned as recombinant DNA in *E. coli* and sequenced. The main findings are:

1. The gene for each VSG is present in each trypanosome of a strain whether the gene is expressed or not (Hoeijmakers *et al.*, 1980).
2. Each cDNA probe sees a family of VSG genes that are related in sequence (Hoeijmakers *et al.*, 1980; Pays *et al.*, 1981a). The degree of homology between family members increases from 5′ to 3′ (Frasch *et al.*, 1980; Borst *et al.*, 1981a), as illustrated by the blot in Fig. 3. The results suggest that the rate of acceptance of mutations in the gene decreases from 5′ to 3′ (Frasch *et al.*, 1980).

Although it is clear that the sequences corresponding to the C-terminal extension and the 3′-untranslated sequence of the VSG mRNA give rise to the most extensive cross-hybridization (cf. Fig. 3), the protein-coding sequence itself makes a clear contribution. This is limited in the 118 family of the 427 strain of *T. brucei* (Fig. 3) but is more extensive in the 117 family (Van der Ploeg *et al.*, 1982), and it has also been seen by direct sequence analysis of cDNAs of the ILTat serodeme (Rice-Ficht *et al.*, 1981). This peculiar gradient of change may result from the functional demands on the protein, the C-terminus being more restricted in permissible amino acid sequences because of its association with the membrane.

3. No introns have been detected in two VSG genes by analysis of hybrids between VSG mRNA and the cloned genes by R-looping (Borst *et al.*, 1981a) or S_1 nuclease (Bernards *et al.*, 1981).
4. An unpublished analysis of cosmid clones containing 35-kb segments of nuclear DNA by L. H. T. Van der Ploeg (personal communication) indicates that there are 10^3 VSG genes per trypanosome and that these are highly clustered in the genome. About 10% of the nuclear DNA appears to be taken up by VSG genes and their flanking sequences (see Borst and Cross, 1982).

IV. The Expression of Some VSG Genes Controlled by a Duplication–Transposition

A. Expression of Some VSG Genes Accompanied By the Appearance of an Extra Copy of This Gene

By hybridizing blots of nuclear DNA from several variants with a probe for one VSG gene, Hoeijmakers *et al.* (1980) discovered that the expression of some VSG genes is accompanied by a duplication of the gene and the transposition of the duplicate into a different DNA segment. The extra copy, called the *ex-*

pression-linked extra copy, or *ELC*, disappears when the gene is switched off (Pays *et al.*, 1981a) and reappears (in modified form) when the gene is turned on (Pays *et al.*, 1981b; Bernards *et al.*, 1981; Michels *et al.*, 1982).

Progress in the analysis of ELCs has been slow for two reasons. First, there is a strong selection against ELC-containing fragments when trypanosome DNA is cloned as recombinant DNA in *E. coli*, irrespective of whether plasmid, λ, or cosmid vectors are used (L. H. T. Van der Ploeg, personal communication). Hence, no ELC clones have been obtained for any of the *T. brucei* strains studied. Second, the presence of a family of related genes hybridizing to the same cDNA probe makes the construction of restriction maps of genes in total nuclear DNA laborious.

Figure 4 shows the physical maps of ELC genes and corresponding basic copy (BC) genes now available for the 427 strain of *T. brucei*. Comparison of homologous BC and ELC shows that the flanking sequences are radically different. The ELC is, therefore, the consequence of a duplication–transposition, rather than a duplication–inversion or a duplication followed by an insertion of a promoter.

The distribution of restriction sites around the ELC gene in Fig. 4 is highly unusual and characteristic. The 117 and 118 genes seem to have been transposed into the same (or a homologous) expression site. In both cases, the transposed segment starts 1–2 kb upstream of the gene (Van der Ploeg *et al.*, 1982) and ends within the 3′-end of the gene (see below).

In the ELC the transposed segment is flanked by long DNA segments that are not cut by any of the restriction endonucleases tested, including TaqI (TCGA). It seems likely that these "barren" regions are AT-rich and/or contain short repeats that are sufficiently homogeneous to lack restriction sites.

Two independently isolated trypanosome clones expressing the 118 gene contain ELCs that differ in the sequence surrounding the expressed copy. The 118a and 118b ELC maps in Fig. 4 look similar, however, in that 118b can be derived from 118a by inserting 10 kb of DNA upstream of the transposed segment and deleting 2.7 kb downstream. Two additional trypanosome clones making VSG 118—118c and 118d—appear to contain ELCs that represent additional variations on this theme. All four ELCs look similar, but the length of the barren regions differs in all four (P. A. M. Michels, personal communication). Although we cannot exclude the possibility that there are four (or more) expression sites, it seems more likely that there is a single expression site and that the (repetitive?) flanking regions change in size by duplications or deletions of repeats. If genes could be inserted in different positions in this expression site, this could contribute to the size variations observed (see Section IV,C).

The barren region downstream of the gene ends at a point where all enzymes tested seem to cut the DNA at the same position. This point may be the end of a chromosome, because it is preferentially attacked by exonuclease BAL-31 in large DNA (T. De Lange, personal communication).

From tentative restriction maps of the DNA from the AnTat stock of *T. brucei*, Pays *et al.* (1981b) have also concluded that two VSG genes may be transposed

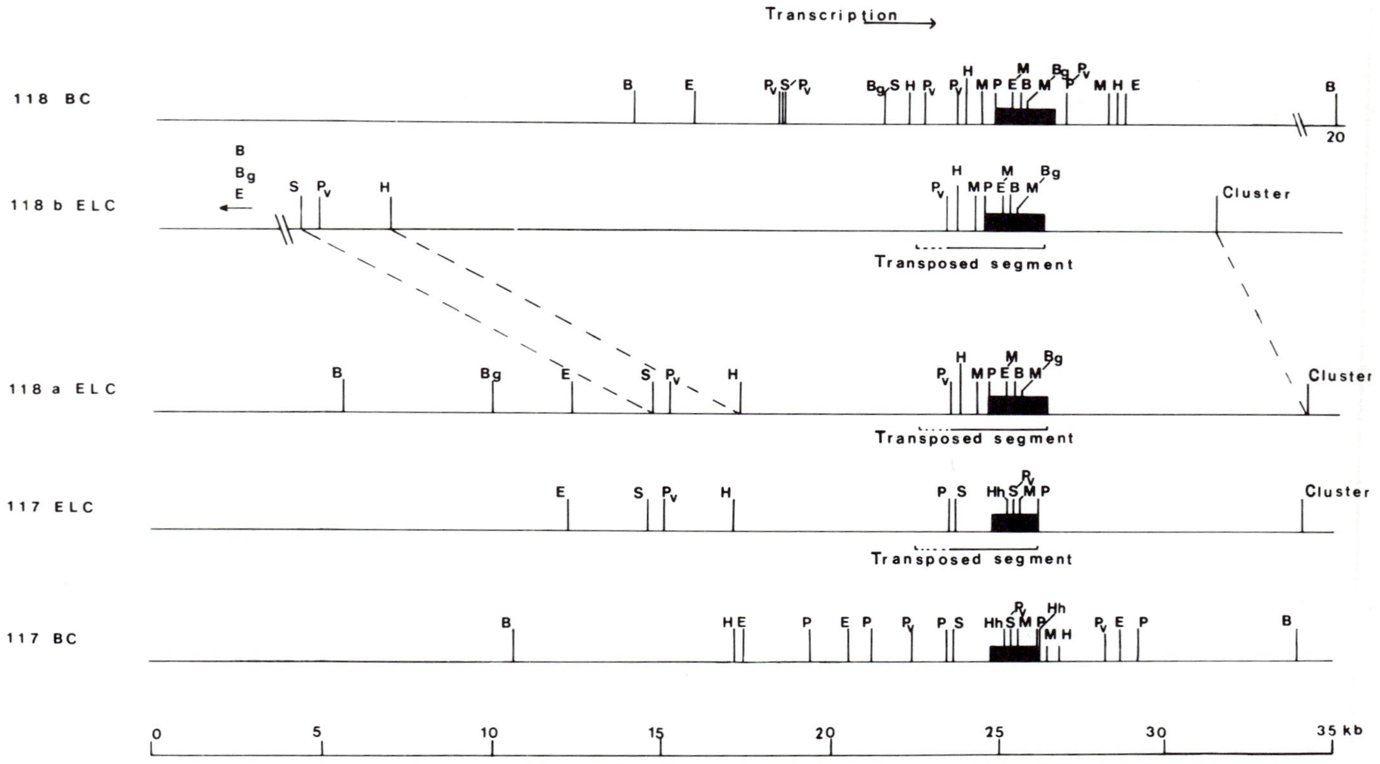

Transcription →
118 BC
B E Pv S Pv Bg S H Pv Pv H M P E M B M Bg P Pv M H E B
20
118 b ELC
B Bg E ←
S Pv H
Pv H M P E M B M Bg
Cluster
Transposed segment
118 a ELC
B Bg E S Pv H
Pv H M P E M B M Bg
Cluster
Transposed segment
117 ELC
E S Pv H
P S Hh S Rv M P
Cluster
Transposed segment
117 BC
B H E P E P Pv P S Hh S Rv M P Hh M H Pv E P B
0 5 10 15 20 25 30 35 kb

into a similar expression site. In that case, the location of the transposed segment in the expression site appears to be the same for 1.1 and the second version of this gene—1.1b—and only 0.2–0.6 kb are cotransposed with the gene. Whether the expression site of the AnTat stock is similar to that of the 427 strain cannot be decided from the limited mapping data of Pays *et al.* (1981b).

B. The Duplication–Transposition of a BC Gene into the Expression Site Accompanied by a Recombination that Leads to a Replacement of the 3′-End of the Gene

Sequence analysis of cloned DNA complementary to a series of VSG mRNAs (see Fig. 5) led to the remarkable finding that the 3′-regions of these mRNAs show a high degree of homology even in cases in which the remainder of the mRNA sequence lacks detectable homology by hybridization, as in the 117–221 couple. This suggested to us that it might be involved in the recombination process presumably required for transposing the BC gene into the expression site. Proof that this is indeed so now rests on the following evidence (see Bernards *et al.*, 1981):

1. As shown in Fig. 6, the 3′-end of the 117 BC gene differs from the corresponding mRNA sequence (deduced from cloned cDNA) by multiple point mutations, deletions, and insertions. Similar differences have been found for the 118 BC gene and 118 cDNA (A. Bernards, personal communication).
2. The 118 BC gene lacks a BspI site present 87 bp upstream of the poly(A) tail of the 118 mRNA. This site is present in the ELC gene, as deduced from nuclear blotting experiments.
3. The two independently isolated trypanosome clones that express the 118 gene (118a and 118b; see Fig. 4) produce mRNAs with different 3′-ends (Michels *et al.*, 1982).

A remarkable feature of the end of the 117 BC gene, shown in Fig. 6, is that it looks functional. Despite multiple differences with the 117 cDNA clone, it has a continuous reading frame for a 3′-terminal protein extension, and the nontranslated trailing sequence has the blocks of homology shared by all VSG mRNAs sequenced thus far (Fig. 5). The same situation has been found for the 118 BC gene (A. Bernards, personal communication). This suggests that crossovers be-

---

Fig. 4 Restriction enzyme cleavage sites in and around the BC and ELC genes for VSGs 117 and 118 in *Trypanosoma brucei,* strain 427. The 118b ELC is from a recent reexpression of this gene (Michels *et al.*, 1982). The ''cluster'' at the 3′-end of the ELC maps indicates a small region where at least 15 restriction enzymes appear to cut; it may be the end of a chromosome (see text). Abbreviations: B, *Bam*HI; Bg, *Bgl*II; E, *Eco*RI; Hh, *Hha*I; H, *Hind*III; M, *Msp*I; P, *Pst*I; Pv, *Pvu*II; S, *Sal*I. (From Van der Ploeg *et al.*, 1982.)

```
Homology      . . . . . # ## #. . . . #. . * * . . . . ## . * #* . * . . . . . . # . . . . ** . . * . . . . * #* ## . . . #* * * * #* * . * * #
AnTat 1.1     TCCTCTATTCTACTAACAA AGAACTTCGCCCTCAGCGTG GTTTCTGCTGCGTTAGTGGCACTGCTGTTCTAA
AnTat 1.8     TCCTCTATTCTAGTAAACA AACAATTAGCCCTCAGCGTG GTTTCTGCTGCATTTGCGGCCTTGCTTTTTTAA
117           TCCTCTATTCTAGTAACCA AGAAATTCGCCCTCACCGTG GTTTCTGCTGCATTTGTGGCCTTGCTTTTTTAA
118           GGTAGTTTTCTAACAAGCA AACAATTCGCCTTCAGCGTG GTTTCTGCTGCATTTGTGGCCTTGCTTTTCTAA
221           AACACCACAGGAAGCAGCA ATTCTTTTGTCATTAGCAAG ACCCCTCTTTGGCTTGCAGTTTTGCTTTTTTAA
IlTat 1.3     TCCACTTTTATCCTAAACA AACAATTCGCCCTCAGCGTG GTTTCTGCTGCATTTGCGGCCTTGCTTTTTTAA
IlTat 1.1     AATTCTTTTGTCATTCATA AGGCACCA. . . CTT. . . . . . . TTTCT. . . . . . . . TGCGTTTTTGCTTTTTTAA
117BC-1       TCCTCTATTCTAGTAACCA AAAAATTCGCCCTCA. . GTG. . . . CTGCTGCGTTTGCGGCCTTGCTTTTTTAG
```

```
Homology      . . . . . . . . . . . . . . . . . . . ##* * * * . . . . . . . . . . . . . . # * * ###. . . . . . +++++. . . . . . * * #* #* * * . . . . . . . . . . . . . . . . . . * * * * * * * * * * * * * #* * . . . . . . . . . . . . . . . . .
AnTat 1.1     ACA. . . . . . . . CCTTTCTT CCCCTCT. . . . CTTT. . . AAAATTTTCC. . . . . . . . . . . . . . . TTGCTACTTGAAAAACTT. . . . . . . CTGATATATTTTAACACCTTT. . . . . . . . . . . . . . (A)
AnTat 1.81    TTAGCC. . . . . . . . . . . . . CCCCTCTTT. . CT. . . . AAAAATTTTCCCCCTGCTAAAA . . TTTTGCTACT. . . . . . . . . . . . . . . . . CTGATATATTTTAACACCT. . . . . . . . . . . . . . . . (A)
AnTat 1.8r    TTAGCC. . . . . . . . . . . . . CCCCTCTTT. . CT. . . . AAAAATTTTCCCCCTGCTAAAA . . TTTTGCTACT. . . . . . . . . . . . . . . . . CTGATATATTTTAACACCT. . . AAAAAATTCCCCG(A)
117           . . . . . . . . . . . . . . . TTTTCCCCCTCTTTTTCTT. . . . AAAAATTC. . . . . . . . . . . . . . . . . . TTGCTACTTGAAAA . C. TC. . . . . . CTGATATATTTTAACAC. . . . . . . GCAAATT. . . (A)
118           AA. . . . . . . . CAAATTTTT CCCCTCAATTTT. . . . . . . AAAATTT. . . . . . . . . . . . . . . . . . TTGCTACTTGAAAAACTTT. . . . . . CTGATATATTTTAACAC. . . . . . . . GTAAGTT. . . (A)
221           TTTCCCCCCTCAAATTTCC CCCCTCCTTTT. . . . . . . . . AAAATTTTCC. . . . . . . . . . . . . . . . TTGCTACTTGAAAA . CTTT. . . . . . TTGATATATTTTAACACCAAAACC. . . . . . . . . . . . (A)
IlTat 1.3     . . . . TCCCTTT. . . . . . . . CCCCTCTTTTTCCATGCTAAAAATTCT. . . . . TGCTAAAA AATTTTGCTACTTGAAAA . CTTT. . . . . . CTGATATATTTTAACACCT. . . . . . . . . . . . . . . . (A)
IlTat 1.1     TTTTTCCCTTTAAAATTTT   CCTCCTTTAAAAAC. . .   AAATTTT. . . . . . . . . . . . . . . . . . . TT  CTACTTGAAAA . CTTCTGATATTTGATATATTTTAACACCTTT. . . . . . . . . . . . . . . (A)
117BC-1       CATAACG. . . . . . . . . TTCCCCCTCAACTTTTTTCCTTAAAGAACT. . . . . . . . . . . . . . . . . TTGCCACT. . . . . . . . . . . . . . . . . . CTGATATATTTTAATACCT. . . . . . . . . . . . .
```

Fig. 5 Nucleotide sequences corresponding to the VSG precursor "tail" and 3′-untranslated regions of cloned cDNA from seven trypanosome variants, aligned to maximize their homology. For comparison, the terminal sequence of one BC gene (trypanosome variant 117) is also included. The AnTat sequences are from Matthyssens *et al.* (1981); 117, 118, 221, and 117-BC-1 are from Boothroyd *et al.* (1981), Majumder *et al.* (1981), and Bernards *et al.* (1981); ILTat sequences are from Rice-Ficht *et al.* (1981). VSGs from 117 and AnTat 1.8 are antigenically indistinguishable (Cross, 1977). AnTat 1.81 and 1.8r are two independent clones differing at their extreme 3′-ends. An asterisk (*) indicates perfect homology between all sequences; # indicates only one sequence preventing *; + indicates a repeat of part of a homologous sequence. The sequences are printed in two blocks corresponding to the precursor "tail" (upper) and untranslated sequences (lower).

```
117 cDNA     GAT  TCC  TCT  ATT  CTA  GTA  ACC  AAG  AAA  TTC  GCC  CTC  ACC  GTG  GTT  TCT  GCT  GCA  TTT   GTG  GCC  TTG  CTT
117 BC DNA   GAT  TCC  TCT  ATT  CTA  GTA  ACC  AAA  AAA  TTC  GCC  CTC  A. .  GTG  . . .  .CT  GCT  GCG  TTT   GCG  GCC  TTG  CTT
VSG 117      Asp* Ser  Ser  Ile  Leu  Val  Thr  Lys  Lys  Phe  Ala  Leu  Thr  Val  Val  Ser  Ala  Ala  Phe   Val  Ala  Leu  Leu
BC 117       Asp  Ser  Ser  Ile  Leu  Val  Thr  Lys  Lys  Phe  Ala  Leu  S    erA  . . .  la   Ala  Ala  Phe   Ala  Ala  Leu  Leu
117 cDNA     TTT  TAA T. . . . TTTCCCCCTC. . . . . TTTTTCTTAAAAATTCTTGCTACTTGAAAACTC CTGATATATTTTAACACGCAAATTAn
117 BC DNA   TTT  TAG CATAACGTTCCCCCTCAACTTTTTTCCTTAAAGAACTTTGC. . . . . . . . . . ACT. CTGATATATTTTAATACCTTGTATT
VSG 117      Phe  End
BC 117       Phe  End
```

Fig. 6 Comparison of the 3′-terminal DNA sequences of the 117 BC gene and VSG 117 cDNA and the amino acid sequence deduced from the DNA sequence. The sequences are aligned to maximize the homology; differences between the two sequences are underlined. The glycosylated Asp residue that forms the mature C-terminus of VSG 117 is designated Asp*. (From Bernards *et al.*, 1981.)

tween the BC gene and the expression site during ELC formation could occur anywhere along the 3′-end, leading to a variable contribution of the BC gene to the ELC. Evidence compatible with this hypothesis has recently been obtained with the second trypanosome clone expressing the 118 VSG gene (118b in Fig. 4). This clone produces an mRNA containing all of the 3′-end of the BC gene (P. A. M. Michels and A. Y. C. Liu, personal communication), and the crossover may, therefore, have occurred within the terminal homology block of 14 bp present in all VSG mRNAs sequenced (see Fig. 5). However, we cannot yet exclude the possibility that the 118 BC has recombined with an identical sequence in the expression site in the generation of the 118b ELC. In that case, the position of the crossover could have been farther upstream.

C. Transposition Mechanisms

It is clear from the previous sections that the formation of an ELC involves a replicative transposition. There are three prototypes of such an event: the transposition of a transposable element; a direct gene conversion, as seen in the mating type switch in yeast; and the transposition of a DNA copy of an RNA transcript of a gene. To see whether the formation of an ELC imitates any of these prototypes or follows a novel course, one would require information on the state of the expression site before the transposition, on the (edges of the) transposed segment, and on the ELC formed. Preferably, this information should be available for several genes and several reexpressions of the same gene, but this is not yet available. What is known can be summarized in four points:

1. The last 150 nucleotides at the 3′-end of VSG genes have substantial homology. This is obvious from the blot in Fig. 3 and from a direct comparison of the ends of the 117 and 118 BC genes (A. Bernards, personal communication). This end is replaced in part during the generation of the ELC. Different VSG mRNAs have similar but different ends (Fig. 5). The ends on VSG BC genes and on the mRNAs look as if they come from the same collection. This is compatible with a model in which, during the duplication–transposition, the incoming gene pairs with the preceding gene in the expression site and (partially) exchanges ends. This is testable, but the experiment is difficult and has not been done. Other possibilities remain. For instance, the expression site might contain a large number of "ends" that could all act as targets for the incoming gene. It is not obvious how such an array of ends could remain stable and heterogeneous.

Although the homology blocks at the 3′-end of the transposed segment are unambiguous, they are short, the longest common block being 14 nucleotides (Fig. 5). From the hybridization analysis of cloned BC genes, it is also unlikely that the homology between different BC genes extends beyond the 3′-end of the transposed sequence (L. H. T. Van der Ploeg, personal communication). It

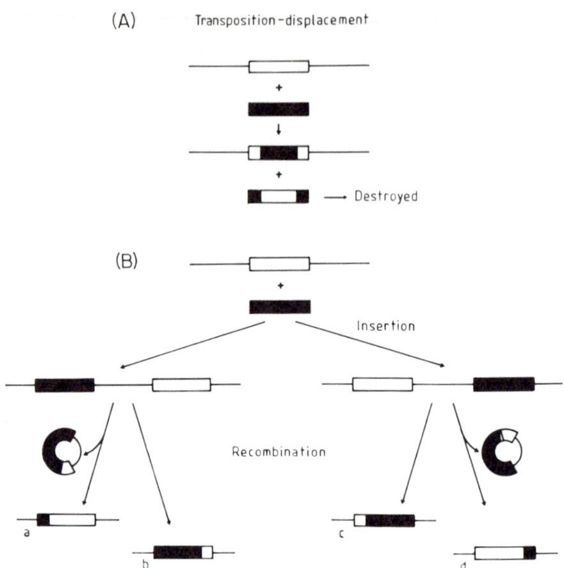

Fig. 7 Speculative models for the formation of ELCs. (A) Transposition–displacement. (B) Transposition followed by internal recombination. Because of space limitations, only two of the four circles excised in model (B) are shown.

follows that the transposition cannot involve long blocks of homology, as in the mating type switch in yeast.

2. The fragment that contains the 5′-edge of the transposed segment of the 118 gene hybridizes with the analogous breakpoint fragment of the 117 gene and with many bands in nuclear DNA blots. The exact breakpoint is not known; nor is it known whether this repetitive element is present adjacent to the breakpoint. For the remainder of this discussion I shall assume, however, that this is the case and that this 5′-repetitive element is common to the transposed segments of all VSG genes. The hybridization between the 5′-edge fragments of the 117 and 118 BC genes is removed by stringent washing, and the homology is, therefore, limited, as it is at the 3′-edge of the transposed segments. The 3′- and 5′-edge fragments have no detectable homology. The transposed segment is, therefore, not bordered by straight or inverse repeats long enough to be detected by hybridization. Short repeats could be present.

3. The expression site contains only a single gene.
4. An ELC that is switched off (usually) disappears.

These facts can be accommodated by two types of models illustrated in Fig. 7. In model A the incoming gene directly displaces the resident one in the expression site, as in gene conversion. In model B the incoming gene is transposed somewhere into the expression site, followed by a recombination that throws out either the old gene (resulting in switching) or the new one (no switch). If the initial transposition in model B does not change the ends of the transposed segment, the internal recombination will alter only the 3′-end if the incoming gene ends up in front of the resident one in the expression site (case b). If it is inserted behind the resident gene, it will retain its 3′-end (case c). It is also possible, however, that the ends are altered already during the initial transposition if the expression site contains tandem arrays of 3′-end sequences, a possibility raised in the preceding paragraph.

From the data presented under points 1 and 2 above, it is clear that long segments of homology (larger than 50 bp) are not present between the incoming VSG gene and the expression site. Hence, transposition cannot involve heteroduplex formation over longer stretches, as required in standard homologous recombination systems. Several plausible alternatives remain, however. In model B the new gene could enter the expression site in the same way as a bacterial transposable element with a high target specificity. A variant of this transposition mode could operate in model A, although a single-strand aggression of the incoming strand at both edges of the preceding gene, followed by gene conversion, is easier to envisage. In view of the limited homology, such single-strand aggression would probably require stabilization by a specific protein. The situation is reminiscent of the limited homology used in the recombinations that create functional immunoglobulin chain genes (Sakano *et al.*, 1979; Kataoka *et al.*, 1981). Obviously, the duplication and transposition do not have to be linked but could be separated, for instance, if reverse transcription is used in the duplication. An advantage of such a mechanism is that it does not involve direct interactions between genes that are far apart in the genome; nor does it require temporary cleavage of DNA strands, with the attendant risk of irreparable chromosome breakage.

From this brief discussion, it should be clear that the limited data available are compatible with virtually all mechanisms used in nature for replicative transpositions. I favor a gene conversion model because gene conversion provides the simplest way to displace the preceding gene from the expression site. Obviously, this gene conversion would be very different from the one thought to be involved in the mating-type switch in yeast, because only limited homology is available for the incoming strand.

D. How Does Transposition of a VSG Gene Activate Its Transcription?

By RNA blot experiments, Hoeijmakers *et al.* (1980) found that VSG gene expression is controlled at the transcription level. DNA complementary to VSG 117 mRNA hybridized only to RNA from trypanosomes making VSG 117 and not to trypanosomes making 121 or 221 (Hoeijmakers *et al.*, 1980). No transcripts other than mature mRNA were detected in these experiments. Also because no introns were found in the BC genes, it appeared that the mature mRNA was the primary transcript of the VSG gene. Recent experiments, summarized in part by Bernards *et al.* (1981), indicate, however, that the situation is more complex. Comparison of the 5′-ends of the cDNA and the BC gene showed that the sequence corresponding to the 5′-end of the mRNA is not contiguous with the remainder of the mRNA sequence in the BC gene. About 35 bp are missing. Where is this leader sequence coming from? Using DNA probes from the cotransposed 1-kb segment in front of the 118 gene, three additional transcripts have been identified. These transcripts are found only in a trypanosome variant that makes VSG 118 and are present at low concentrations. From the map position of these RNAs, we infer that VSG gene transcription may start beyond the transposed segment. The expression site may, therefore, provide the promoter required to initiate transcription of VSG genes. This speculative scheme also implies that the initial transcript is spliced to yield a mature mRNA with a 5′-terminus coming from outside the transposed segment.

It should be stressed that this transcription model is based on limited data. Pulse-chase experiments, verifying that the minor RNAs are precursor RNAs, are lacking. The mapping of the minor RNAs relative to the 5′-end of the transposed segment is also not precise enough to determine whether they extend beyond this end or not. The possibility remains, therefore, that the transposition creates a new promoter at the breakpoint of the transposed segment or activates a promoter within this segment.

Granted these uncertainties, the model makes sense. Promoter addition provides a simple and effective way of controlling VSG gene expression. The postulated splicing step creates flexibility for the gene transposition, which may be useful if such a large number of genes have to use the same expression site. Splicing but not trimming allows the retention of the 5′-triphosphate end of the precursor RNA, which may be useful for capping (whether VSG mRNAs carry caps is not known). Splicing can accommodate intervening sequences of various lengths, and these may be present because the results of Pays *et al.* (1981b) suggest that the cotransposed segment of the AnTat 1.1 gene is smaller than 0.6 kb, that is, much smaller than the cotransposed segments of the 117 and 118 genes (Fig. 4). By combining splicing with promoter addition, the trypanosome adds a eukaryotic finesse to a prokaryotic method for controlling gene expression. We return to this point in a later section.

Treatment of isolated trypanosome nuclei with DNase I leads to a preferential loss of one of the fragments derived from the putative ELC in the 1.1 variant of the Antwerp strain (Pays *et al.*, 1981c). The hypersensitivity is lost when the nuclei are first washed in high salt (E. Pays, personal communication) under conditions known to abolish the DNase hypersensitivity of transcribed genes in isolated vertebrate chromatin. Because no complete physical map of the ELC gene of the 1.1 variant is yet available, it is not known how far this DNase I sensitivity extends beyond the gene itself.

E. What Determines the Order of VSG Gene Expression?

In a chronic infection, different VSG genes appear in an imprecisely predictable order. I have argued in Section II,D that this order must be determined, in part, by genetic programming. This program could be based on gene location, gene homology, or affinity for a protein involved in the duplication–transposition. For simplicity, I shall call this protein the *switchase,* although it is not known whether such an enzyme is involved.

Programming based on gene location requires that all VSG genes have a definable position relative to each other and to the expression site, that is, that they be on one chromosome. This is not impossible because VSG genes appear clustered and restricted to 10% of the genome (Section III,B). Order would be imposed by the distance to the expression site, for instance, because transposition requires pairing of the BC genes and the expression site. It is also possible, however, that each transposition would be initiated by a switchase sliding down the DNA, starting at the expression site. Again, the chance of transposition of a VSG gene would depend on the distance between the gene and the expression site.

A simple switching program based on homology predicts that genes are selected on the basis of their degree of homology to the gene already present in the expression site. This is intrinsically implausible, because it would strongly favor reexpression of the preceding gene in the sequence. Moreover, this programming mode predicts that there would be a rather high homology between the 3′-end of a VSG ELC (mRNA) and the BC gene from which it is derived. The results in Fig. 5 do not support this prediction.

A program based on affinity of the BC genes for the switchase could also explain the existence of a sloppy order. Unfortunately, this alternative model generates no predictions that can be simply tested before the transposition system is reconstructed *in vitro*.

None of these models provides a simple explanation for the observation of Miller and Turner (1981) that some VSG genes tend to be always expressed in sequence. This would seem to require cotransposition of adjacent genes to the expression site. The gene closest to the promoter would be expressed, and an internal recombination in the expression site would activate the cotransposed adjacent gene. At least in the ELCs mapped in Amsterdam (see Fig. 4) and by Pays *et al.* (1981b), this does not seem to occur because the transposed segment is too small to accommodate more than one VSG gene. The situation may be different, however, for the genes studied by Miller and Turner (1981).

How would the trypanosome activate a separate subset of VSG genes in the salivary gland of the fly to yield the metacyclic trypanosome population? Immunological characterization of the VATs involved suggests that the metacyclic VATs are also expressed in chronic infections in vertebrates; the metacyclic program, therefore, differs from that operating in the vertebrate but uses in part the same genes. This could be due to the use of a different expression site or a different switchase. The environment in the salivary gland of the tsetse fly and in the vertebrate bloodstream is rather different, however. This could have drastic effects on the *milieu intérieur* of the trypanosome and could also affect the relative growth rates of different VATs. It is, therefore, not yet clear whether the differences in the metacyclic and vertebrate VAT repertoires are programmed at the genetic level or due to secondary effects.

F. What Triggers Switching?

Because the data are still scarce, it is a little early to speculate about the switching mechanism. Three facts seem well established:

1. Switching is not induced by antibody or any other external agent.
2. When the trypanosome enters a mammalian host through a fly bite, the switching rate is initially high and then decreases. This is not observed when the trypanosomes are transmitted by syringe from an infected mammal to a fresh one. The rate of switching, therefore, shows memory. This could, for instance, be due to overproduction of an enzyme involved in the duplication-transposition when the trypanosome is located in the salivary gland of the fly.
3. Switching is sufficiently frequent to allow a single cloned and mammal-adapted trypanosome to start a chronic relapsing infection in a new host. It is infrequent enough to allow the isolation from mammals of trypanosome populations containing less than 1 in 10^5 heterotypes, as judged from experiments in which a preimmunized host is infected with a known dose of trypanosomes (see Doyle, 1977).

These facts demonstrate that switching in trypanosomes is not an event like the mating type switch in yeast, which occurs with high probability during every other cell division (in cells that have budded at least once) (Strathern and Herskowitz, 1979). It seems likely that it is an infrequent event controlled by an enzyme present at very low concentration, as in the flip-flop control of phase variation in *Salmonella* or in the transposition of transposable elements in bacteria in general.

Shortly after the switch, the trypanosome should be producing two types of VSGs at the same time, temporarily leading to production of zebra trypanosomes. This raises a question. Are the "old" VSG molecules gradually diluted out during growth and division, segregated out into one of the progeny, or preferentially shed? The answer is not known and is hard to find because trypanosomes cannot be made to switch simultaneously to the same new antigen. It seems likely, however, that old antigens are not simply diluted out because this would make the switched trypanosome vulnerable to antibodies against the old antigen for several generations after switching. The marked and unexplained tendency of homologous VSGs to form dimers in solution (Auffret and Turner, 1981), and possibly higher oligomers on the trypanosome membrane, may provide a mechanism for segregating old VSGs into separate domains that could either be shed or segregated out into one daughter cell during division.

G. Why Is a Transposition Mechanism Used to Control VSG Gene Expression?

There are three precedents for gene rearrangements associated with the control of gene expression: phase variation in prokaryotes (Zieg *et al.*, 1978), the mating type switch in yeast (Hicks *et al.*, 1979; Klar *et al.*, 1981; Nasmyth *et al.*, 1981), and the synthesis of antibody genes (Early and Hood, 1981). Whether the gene rearrangements that lead to a functional antibody L- or H-chain gene are really a prerequisite for transcription is not certain, and the detection of transcripts from variable-region genes that are not or are incorrectly rearranged makes even this doubtful. In phase variation and the mating type switch, however, gene rearrangement triggers the start of gene transcription, and in both cases it is generally accepted that this complex mechanism for the control of gene expression has evolved to allow the alternating expression of two sets of genes, A and B. In phase variation, the switch from A to B is effected by the inversion of a DNA segment that makes or breaks the link between gene and promoter; in the mating type switch, a single expression site can be filled with genetic cassette A or B. In both cases, the expression of gene set A prevents expression of set B, and vice versa. In both examples, therefore, gene rearrangement seems to have evolved to solve a switching problem. This makes it plausible that the duplication–transposition mechanism for VSG gene control has also arisen in response to the far more complex switching problem of the trypanosomes.

Other explanations for the adoption of the transposition mechanism to control VSG gene expression are equally speculative but less attractive. Trypanosomes might use this mechanism for the control of all protein-coding genes, or it may be more important for ensuring the exclusive expression of a single gene. Further experiments should allow the testing of these alternatives, and I shall disregard them in the remainder of this chapter.

If transposition control has evolved to solve the switching problem, the question remains of why other solutions were excluded. Why not inversion control, as in antigenic variation in bacteria? If all VSG genes are arranged in a row next to a single promoter, inversion might bring any of these under the control of this promoter and, if nearby genes have a better chance of being at the edge of the inverting segment than faraway genes, this would also result in a preferred order of appearance. This mechanism would require, however, that all VSG genes be in one DNA molecule, and this might put a ceiling on the number of VSG genes. Inversion would presumably involve a simultaneous double-stranded break and, intuitively, this seems a risky mode of gene switching. Finally, continued inversion would scramble gene order, unless a gene is first inverted back to its original position, before the next is inverted into the expression site. This would make the mechanism hopelessly complex.

The expression of gene families can also be tightly controlled without gene rearrangement. However, an important feature of such developmentally regulated gene families is that genes are switched on in a precise order and that subsequent switch-off is in practice irreversible. In this case, therefore, the order of genes on a chromosome could help to limit gene expression to a subset of genes by position effects that would progress over the chromosome. Such spatial ordering of genes according to the temporal order of expression in development has been observed for globin genes (Bernards and Flavell, 1980; Efstratiadis *et al.*, 1980).

The problem for the trypanosome is more complex because the expression of VSG genes must be very tightly controlled; the gene family is very large; the order of expression is not precise; and switch-off is not irreversible. These problems are neatly solved by the duplication–transposition mechanism, and none of the other mechanisms considered here can match it in terms of precision, simplicity, and versatility.

H. The Evolutionary Origin of Antigenic Variation

Little is known about the evolutionary origin of antigenic variation. DNA complementary to VSG mRNAs of *T. brucei* weakly hybridizes to DNA from other Kinetoplastida genera (A. C. C. Frasch, personal communication), but the sequences involved have not been isolated and characterized. An elaborate system of antigenic variation is also present in some free-living, nonparasitic protozoa (see Section VI), and this suggests that the ability to alter surface properties may be widespread throughout the protozoa and merely perfected by trypanosomes.

Transposition control of VSG gene expression may have arisen early or late in the development of antigenic variation. Turner and Cordingley (1981), who also provide an interesting discussion of the evolutionary origin of the surface coat and *T. brucei* in general, favor a model in which the members of a large family of VSG genes acquire the ability to transpose by insertion of terminal repetitive elements. I find this late acquisition of transposition less likely because one would expect that repertoire expansion would be advantageous only when an efficient system for mutual exclusion was already available. I also think that this model underplays the complexity of transposition control, which involves not only a replicative transposition but also a displacement that has to be precise. Such complex systems may not often arise in evolution. I therefore expect that transposition control was acquired early, and I am impressed with the fact that the only other system known to involve a precise duplication–transposition control—the mating-type switch in yeast—also affects cell surface properties. It is, therefore, possible that antigenic variation is an offshoot of mating-type control. The analysis of antigenic variation and mating-type control in other protozoa should throw light on this problem.

V. VSG Gene Rearrangements Not Linked to Expression

A. Some VSG Genes Expressed without Obvious Duplication–Transposition

An outstanding feature of antigenic variation is that the expression of one VSG gene seems to prevent the expression of all others. VSG gene expression is mutually exclusive. The duplication–transposition system discussed in the preceding section would be perfectly adapted to carry out this task if three conditions were met:

1. Genes activated by transposition are not activated by other means.
2. Only one expression site is functional in a trypanosome clone.
3. All VSG genes are activated by transposition into the same expression site.

The limited facts available are compatible with condition 1. Most facts are also compatible with condition 2, with one puzzling exception: Hoeijmakers *et al.* (1980) found that a band associated with the ELC of their gene 121 shortened by 4 kb, but did not disappear when this gene was switched off and gene 221 switched on. In later switches this phenomenon of the persistent ELC was not reproduced, and it has not been explained. The possibility that two expression sites are available, with some kind of allelic exclusion, cannot be dismissed in this diploid organism.

The most baffling finding, however, is that condition 3 is apparently not met, because not all VSG genes seem to show the simple transposition activation illustrated in Fig. 4. Three of the four VSG genes studied by the ILRAD group (Williams *et al.*, 1981; Young *et al.*, 1982) and one of the VSG genes studied in Amsterdam (Borst *et al.*, 1981a) show no sign of expression-linked duplication. I shall call these genes NTA (nontransposition-activated) genes to distinguish them from the ones that are transposed (TA genes).

The known NTA genes have three properties in common:

1. NTA genes are present in multiple copies or, more precisely, are part of a set of isogenes that are too closely related to be distinguished by stringent hybridization.
2. Members of the NTA isogene groups show frequent DNA rearrangements 3′ of the gene; these rearrangements have no clear relation to the expression of these genes.
3. NTA genes are expressed early in infection.

There are two principally different ways to explain the activation of the NTA genes:

1. NTA genes are activated by a duplication–transposition, but the ELC is not recognized as such in nuclear DNA blots. There are three versions of this explanation:

a. The transposed segment is so large that its transposition is not detectable by alteration of outside cleavage sites in blot analysis. One would still expect a clear doubling of the hybridizing bands, however, and this is not seen.

b. The gene is duplicated (long) before it is switched on, and the switch-on and switch-off are brought about by rearrangements in the transposed segment without direct loss of the gene. If the transposed segment is long and contains several genes, the duplication may not be easily detected (see also Young *et al.*, 1982). This explanation is vague enough to be compatible with the available evidence. It should be noted, however, that this farfetched explanation still results in a mode of gene control that is basically different from that outlined in Section IV and will require additional hypothetical control mechanisms (see below).

c. A third alternative has recently been put forward by A. Bernards (personal communication). From the fact that the flanking regions of the ELC in the expression site vary in size in different reexpressions of the 118 gene (see Section IV,A), he infers that these regions are intrinsically unstable. This instability is somehow promoted when an NTA gene is in the expression site. This results in such frequent changes in the size of the flanking regions that the ELC-containing fragments become too heterogeneous to be detected in blots of nuclear DNA. This "ghost ELC" should, of course, be detectable by using restriction endonucleases that cut within the transposed segment. If the gene is switched on, the fragments from within the transposed segment should double in relative amount. This analysis is complicated, however, by the presence of multiple NTA isogenes. A further complication is that the 3′-flanking sequence of the NTA gene studied by our group (the 221 gene) resembles the barren regions that flank an ELC in the expression site (see Section V,B). If this resemblance is not spurious, such genes may enter the expression site by recombination of flanking sequences within the expression site and, therefore, without alterations in the 3′-end of the gene. Catching a ghost ELC is, therefore, not so simple.

2. NTA genes are not activated by duplication–transposition but by other mechanisms. The problem then shifts to the organization of mutual exclusion of VSG gene expression, and I have been unable to devise a plausible model that solves this problem. Two possibilities can be considered:

a. The activation of an NTA gene involves the transposition of a DNA segment from the expression site to a region downstream of the BC gene. This recombination activates the BC gene *in situ,* for instance, by introducing an "enhancer region" (Banerji *et al.*, 1981) that promotes access of RNA polymerase. This transposition must lead to the elimination of the gene present in the expression site to explain the exclusion of TA genes by the activation of an NTA gene. Because the same NTA gene can be turned on again and again by transferring the trypanosomes into fresh rabbits, all the steps in the NTA gene activation must be reversible, that is, the segment transposed from the expression site to the NTA gene must be transferred back again at some time to inactivate this gene and to allow the use of the expression site for another gene.

b. There is a regulatory locus downstream of the NTA genes that controls the expression of the gene lying 5′ of it. This locus is unstable and changes often from off to on or back. In the on position it activates the adjacent gene and blocks transcription of TA genes in the expression site. This model implies that there could be simultaneous expression of different NTA genes. This has not been observed, but if such patchwork trypanosomes have an unstable coat, they might be rapidly eliminated. Another implication is that the activation of an NTA gene blocks the expression of the gene in the expression site but does not empty this site. We have observed in practice, however, that the 118 ELC is lost when NTA gene 221 is switched on (Michels *et al.*, 1982). Unfortunately, this experiment is not conclusive because there is no probe for the regions surrounding the ELC. The possibility cannot be excluded, therefore, that the 118 gene has been replaced by another gene in the expression site.

In summary, it is clear from this somewhat laborious discussion that the ghost ELC hypothesis provides the only simple and plausible explanation for the puzzling results obtained with the NTA genes. Although it may seem a little facile to come up with a ghost ELC when no real ELC can be found, this hypothesis is not entirely without circumstantial evidence. First, Pays *et al.* (1981b, p. 4230) have had difficulties in mapping restriction sites downstream of the AnTat 1.1 ELC and have raised the possibility "that the 3′ neighborhood of the ELC of the gene is variable, even within a cloned population of trypanosomes." Second, the 3′-ends of the cDNAs arising from NTA genes 221 and ILTat 1.3 are similar to the 3′-ends coming from TA genes (Fig. 5). If these ends are essential for duplication-transposition, as argued in Section IV, one would not expect them to be preserved in the mRNAs copied from NTA genes unless these genes also go through a duplication–transposition. Finally, it seems fair to stress that my view on this issue is not shared by Williams and his associates. In their latest paper they conclude, on the basis of an extensive discussion of all the evidence, that "The control of expression by coupled duplication and transposition is not sufficient to account for the selection of a single VSG gene for expression" (Young *et al.*, 1982, p. 803). The point should be settled before this book is published (see Note Added in Proof).

B. Rearrangements Downstream of VSG Genes

In the preceding section, I described a class of NTA genes that do not give a detectable ELC when activated. These genes are also characterized by a region downstream of the gene where deletions and insertions occur that are not obviously related to expression. Five genes of this type have thus far been identified in the ILTat serodeme (Young *et al.*, 1982) and three in Amsterdam (Borst *et al.*, 1981a; A. Bernards, unpublished observations). The study of these genes is complicated by the fact that they are all part of a set of closely related isogenes that are difficult or impossible to distinguish in nuclear blotting experiments. A further complication is that these genes are selected against when nuclear DNA is cloned as recombinant DNA in *E. coli,* and none of them has been obtained with both flanking regions in clone banks (J. R. Young and L. H. T. Van der Ploeg, personal communications). Despite these complications, detailed restriction endonuclease cleavage maps have been constructed for the ILTat 1.2 gene and the 221 gene of the 427 strain; these maps are shown in Fig. 8. The most remarkable aspect of these maps, especially that of the 221 gene, is that the 3′-region of the

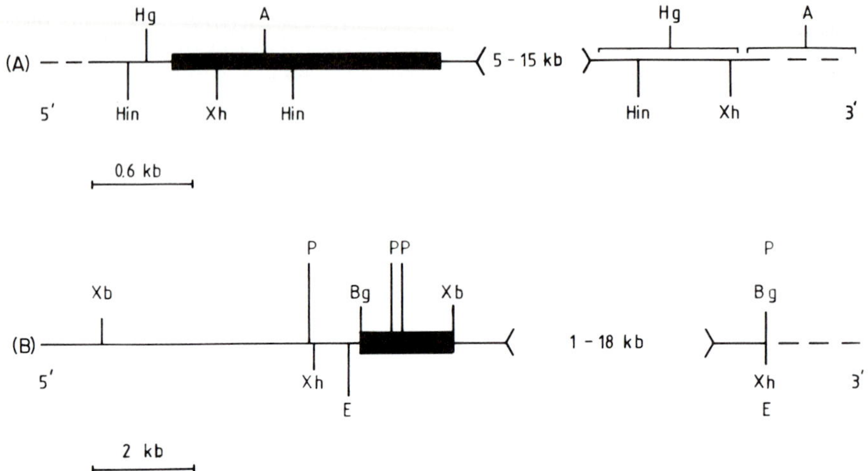

Fig. 8 DNA rearrangements downstream of VSG genes. (A) The genomic context of ILTat 1.2 is redrawn in simplified form from Williams *et al.* (1981). (B) The results for 221 are unpublished data from A. Bernards (personal communication) and are based on nine trypanosome variants in which the variable region has six different sizes. Abbreviations of the restriction endonucleases: A, *Ava*II; Bg, *Bgl*II; E, *Eco*RI; Hg, *Hgi*AI; Hin, *Hind*III; P, *Pst*I; Xb, *Xba*I; Xh, *Xho*I. The black boxes represent the sequence corresponding to the mRNA (incomplete at both sides in ILTat 1.2).

gene resembles the corresponding region in the ELC gene maps in Fig. 4. The variable region is poor in restriction sites and ends in a cluster of sites. Although this may be a coincidence, it is also possible that the 221 gene may, in fact, be flanked by a copy of the 3′-part of the expression site. This would explain the size variations of this region in different trypanosome clones and the selection against it in recombinant DNA cloning experiments, whereas the 5′-flanking region of the 221 gene can be recovered from such banks (T. De Lange, personal communication).

VSG genes with a segment of DNA from the expression site attached to it could easily arise as a side product of the transposition pathway outlined in Fig. 7B. Such copies might add on to the ends of chromosomes and might be preferentially activated, that is, they may be early genes, because of a more extensive homology with the expression site.

C. The Role of Gene Rearrangements and Selective Mutagenesis in the Evolution of VSG Genes: Disposable Genes

The evolution of VSG genes has been analyzed by comparing VSG genes in related trypanosome strains (Frasch *et al.*, 1980, 1982). Nuclear DNA was digested with restriction endonucleases, and the conservation of recognition sites in and around VSG genes was determined by blot analysis. The 117 gene was found to be largely conserved in 12 out of 13 strains. This gene is probably identical to the 1.8 gene studied in Antwerp, and the corresponding VSG is found largely unaltered in trypanosome strains throughout Africa. Conservation of this gene is, therefore, expected and has also been observed by Pays *et al.* (1981b,d) for additional strains. On the other hand, the 221 gene present in the strain studied in Amsterdam was found to be absent in each of the 12 other strains analyzed, even though some of the 221-related genes appeared to be rather conserved. The 118 gene was present in seven of the 13 strains analyzed, but in this case four isogenes could be distinguished in different strains, differing by 5–8% in sequence, as calculated from multiple restriction site changes.

Most of the strains analyzed by Frasch *et al.* (1980, 1982) are closely related, as judged from enzyme polymorphisms and kinetoplast DNA sequence comparisons (Borst *et al.*, 1981b). It is clear, therefore, that the large differences in the VSG gene repertoire are due to the rapid evolution of some VSG genes. Although we have sampled only a tiny fraction of the repertoire, a similar pattern of conserved and nonconserved genes is also seen with the many related genes hybridizing at relaxed stringency. I therefore think that *T. brucei* strains may differ substantially in their total VSG gene repertoires. This can account for the large differences in expressed VAT repertoires previously observed (Van Meirvenne *et al.*, 1975b).

The most remarkable feature of the rapid evolution of VSG genes is its selectivity. Whereas some genes (such as 117) are conserved, others are not. The coexistence of conserved and rapidly evolving genes for the same function in one genome can be explained in several ways. I favor the hypothesis that trypanosomes have developed a mechanism for speeding up the evolution of a subset of their VSG genes (see Frasch *et al.*, 1982). In this hypothesis there is a limited set of stable VSG genes. Gene copies of this set can be transposed into a specialized region of the genome in which sequence evolution is rapid. Genes are lost from this region at the same rate as they are transported into it. An obvious source of additional VSG gene copies is provided by the duplication–transposition that controls VSG gene expression, because every time a new ELC is formed, the old one is displaced from the expression site (Fig. 7). This discarded ELC is usually degraded, because it is not found back as an extra copy in DNA blotting experiments. It may have a finite chance, however, of inserting elsewhere into the DNA.

Returning to the argument presented in the preceding section, an evolutionary scenario could be envisaged in which new VSG genes arise as NTA genes. After a period of rapid mutational alterations, they would either be lost again or end up in the collection of relatively stable VSG genes.

Hypermutagenesis will increase the rate of production of nonfunctional genes. If these genes can still be switched on, the trypanosome carrying them will be eliminated. In large populations of rapidly multiplying organisms, this should not be a serious disadvantage. Mutations that lead to nonfunctional genes that can no longer be switched on pose more serious problems. These pseudogenes cannot be eliminated by simple means, and their accumulation will inflate the genome.

In principle, there are three ways in which the trypanosome could cope with this problem:

1. Drastic contraction and expansion of the repertoire. Trypanosomes expanding a basic set of genes with too many dead ones in it would be eliminated. Such contraction–expansion cycles could be effected by deletion–duplication, gene conversion, or master-slave replication mechanisms. There is no evidence that the gene repertoire ever contracts, but the data are scarce.
2. Gene conversion spreading preferentially certain genes through the repertoire. If gene conversion would preferentially use the gene that has just been displaced from the expression site by switching, this would be a means of spreading functional genes, rather than dead ones, through the population. A segmental gene conversion has recently been invoked to explain the apparent ability of mammals to limit the number of nonfunctional Ig V genes (Baltimore, 1981b). Why functional genes would have a better chance of spreading in that case is not clear, however.
3. Gene-rectifying mechanisms work specifically on the sequences that are involved in switching. This is an attractive hypothesis, because in this way most genes can be regularly switched on and checked unless all the switch sequences go flat and the trypanosome is eliminated. Gene conversion might be the simplest mechanism of rectification.

What could be the selective pressure to create new VSG genes, once the repertoire is large enough for a trypanosome strain to maintain itself for years in a vertebrate? In my opinion, the driving force must be the competition between trypanosomes for a limited number of vertebrate hosts. A trypanosome carrying new VATs has a much better chance of establishing a superinfection of a chronically infected host, which should be a major advantage if most of the vertebrate population is infected.

A final point concerns the observations mentioned in Section III,B, which indicate that 5′-halves of VSG genes evolve more rapidly than 3′-halves. This could, in principle, also be due to a more selective local hypermutagenesis, as found for the V gene parts of Ig genes (Baltimore, 1981b). There is no evidence on this point, and it will not be easy to distinguish between selective hypermutagenesis and a higher rate of fixation of mutations in the 5′-half of VSG genes. The second alternative is simpler and seems more plausible.

VI. Antigenic Variation in Protozoa Other than Trypanosomes

Antigenic variation occurs in parasitic protozoa other than trypanosomes, for example, in *Plasmodium* (malaria) and *Babesia* (see Turner, 1980), but also in large, free-living ciliates such as *Paramecium* and *Tetrahymena*. No studies have been done with these organisms at the DNA level, however, and it is therefore not known whether DNA rearrangements are involved. Nevertheless, I shall briefly consider the ciliates here. On the one hand, antigenic variation in these organisms shares the two main features of antigenic variation in trypanosomes: mutual intergenic exclusion and reversibility. On the other hand, the presence of a highly polyploid macronucleus in ciliates makes it unlikely that a simple transposition system could control gene expression in this case. I shall only discuss *Paramecium* here, because its genetic and biochemical analysis is the most advanced.

With its length of about 100 μm, *Paramecium aurelia* is nearly 10 times as long as trypanosomes and is one of the largest unicellular organisms. More than 75 years ago, Rössle made the first antisera against *Paramecium* and showed that these immobilized the paramecia by agglutination of the cilia that they use for locomotion. Clumping together of individuals by agglutination, as with trypanosomes, does not occur. The antigens responsible for this phenomenon are called *immobilization* or *i-antigens* (*I-Ags*). The work of Jollos and Sonneborn established that one strain can alternatively express several surface antigens. In the following decennia, the phenomenology, genetics, and biochemistry of antigenic variation were worked out by Sonneborn (1974), Beale (1954), Preer (1969), Finger (1974), and Sommerville (1970).

Genetic analysis has shown the presence of at least 12 unlinked loci that code for I-Ags in *Paramecium*. I-Ags specified by different loci share few if any antigenic determinants. Several alleles are known at most loci, and in a heterozygote either one allele (allelic exclusion; Capdeville, 1979a,b) or both alleles present can be expressed. However, there is no simultaneous expression of more than one locus at a time; mutual intergenic exclusion is, therefore, usually complete.

Biochemical analyses indicate that the I-Ags constitute about 10% of the total protein of a *Paramecium* and that I-Ag molecules form a surface coat that covers the entire surface of the organism (see Ramanathan *et al.*, 1981). The I-Ags are very large proteins (about MW 250,000), and I-Ags from different loci differ radically in amino acid composition and peptide maps. The function of I-Ags in *Paramecium* is unknown. Rather than serving as an exchangeable coat to elude predators, they are usually thought to function as a selective barrier that helps to control the access of substances to the membrane. The mRNAs for the I-Ags are about 8000 nucleotides long, contain a poly(A) tail, are so prominent as to be directly visible in ethidium bromide-stained agarose gels, and are present only when the corresponding I-Ag is made (Preer, 1981).

So far, this picture of antigenic variation in *Paramecium* is very reminiscent of the situation in trypanosomes, and the experimental results—notably the intergenic exclusion—would seem to be readily explainable by a transposition-activation model (Brown, 1981). There are a number of complications, however, that are not so easy to account for by this model. The most formidable of these is the presence of a macronucleus.

Ciliates are very large unicellular organisms, and to feed the enormous cytoplasmic volume with sufficient RNA, they use nuclear polyploidy. In the simplest ciliates, multiple and equivalent nuclei are used for this purpose. In the more complex ciliates, nuclear dualism has developed in which a diploid micronucleus responsible for genetic continuity gives rise to a polyploid macronucleus that is used for the bulk of the RNA synthesis (see Raikov, 1976). The organization and genetic complexity of this macronucleus show considerable variation among ciliates. In the *Hypotrichida* the macronucleus is composed of gene-sized molecules and contains only a small fraction of the gene complement of the micronucleus (Lawn *et al.*, 1978). In *Paramecium* the macronuclear DNA is long, most or all of the micronucleus genes are represented, and the ploidy is about 10^3 (Cummings, 1975; McTavish and Sommerville, 1980).

It is clear that in such a highly polyploid system duplication-transposition control of gene expression will not be simple. The concerted transposition of 10^3 genes to 10^3 expression sites seems an improbable event. Activation of a single micronuclear gene, followed by amplification and transposition, is conceptually simpler, but there is no evidence for any genetic activity of micronuclei in vegetative cells.

A further complication is that in *Paramecium* the antigen produced is controlled in part by environmental conditions. For instance, *P. primaurelia,* stock 60, will quantitatively switch from I-Ag S to I-Ag D when the temperature is raised from 15° to 35°C (see Beale, 1954). This is clearly very different from the situation in trypanosomes, in which switching is not induced and one can never predict with certainty the next antigen to be produced.

In summary, it is clear that ciliates, like trypanosomes, show antigenic variation and intergenic exclusion of I-Ag expression, but the mechanism involved in the ciliates is certainly not a simple replica of the duplication-transposition found in trypanosomes. Obviously, the phenomenon is very complex, and it may be a consolation that other attempts to explain it in molecular terms have also "not been very fruitful," as Finger (1974, p. 133) puts it.

There is no reported evidence for other regulatory or mutagenic forms of DNA rearrangement in ciliates, apart from the formation of macronuclei already mentioned. *Paramecium* can, however, undergo mating type switches (resulting in "selfers") that could be analogous to the mating type switches of fungi. Although the interpretation of these events is complicated by the segregation of the genetic information in the macronuclei, Preer (1969) concludes that changes in gene expression are also involved. This may be another area where transposition control would seem a reasonable working hypothesis.

VII. Summary and Outlook

African trypanosomes are completely covered by a protective surface coat containing a single glycoprotein, the VSG; this is the only antigenic structure in intact trypanosomes exposed to host antibodies. One trypanosome can produce more than 100 VSGs that differ so much in amino acid sequence that they have no antigenic determinants in common. By regularly switching the VSG produced, a fraction of the trypanosomes continues to escape immunodestruction in the vertebrate host.

In a chronic infection, VSG genes are switched on in an imprecisely predictable order, but never more than one at a time (intergenic exclusion). This unusual genetic behavior can be simply explained by a modified cassette mechanism, as first pointed out by Hibner and So (quoted by Strathern and Herskowitz, 1979). In this mechanism, VSG genes are activated by a duplicative transposition into a single expression site that can accommodate only one VSG gene at a time.

The evidence that such a developmental gene transposition controls the expression of many VSG genes is now conclusive. The transposition involves a recombination that leads to a replacement of the 3′-end of the gene. This recombination appears to depend on conserved sequences in the gene segment corresponding to the 3′-untranslated region of VSG mRNA. There is suggestive evidence that the transposition does not activate the VSG gene by breaking a negative control circuit, but rather by promoter addition. Transposition does not require large segments of sequence homology between expression site and transposed segment, in contrast to the situation found in the mating type switch in yeast. Displacement of the preceding gene in the expression site may occur by gene conversion, but other mechanisms are neither excluded nor implausible.

There are several other major gaps in our understanding of the transposition control of VSG gene expression. Some VSG genes seem to be activated without the appearance of a transposed copy, and this makes no sense. The number of expression sites available for VSG gene expression is not known, and all attempts to clone an expression site as recombinant DNA in *E. coli* have thus far failed. The molecular basis for the ordered appearance of VSGs in a chronic infection is still unknown. With the techniques now available and the increasing number of laboratories working on antigenic variation, these problems should be solved in the next 2 years.

In addition to developmental gene transpositions, VSG genes in trypanosomes are involved in mutagenic gene rearrangements. The sequences around the expression site undergo frequent length variations, and the same phenomenon has been observed for a region downstream of some VSG genes. Both rearrangements may be due to the presence of (similar?) repetitive sequences. Comparison of VSG genes in closely related trypanosome strains has shown the existence of a subset of unstable VSG genes; these duplicate, evolve rapidly by point mutations, and disappear by deletion. The possibility that a mutagenic process is involved remains open.

Although antigenic variation is known to occur in other parasitic and free-living protozoa (such as *Paramecium* and *Tetrahymena*), the molecular basis of this process has not yet been studied at the DNA level. No experiments have been reported on other phenomena that might involve gene transposition, such as mating type switches, or on the presence of movable genetic elements in protozoa. Because protozoa represent the most diverse and versatile phylum of the eukaryotic world, more interest in their DNA seems warranted.

Acknowledgments

I am indebted to my students and colleagues for allowing me to quote their unpublished observations and for their critical comments. The experimental work received financial support from the World Health Organization (grant T16/181/T7/34) and from the Foundation for Fundamental Biological Research (BION), which is subsidized by The Netherlands Organization for the Advancement of Pure Research (ZWO).

References

Auffret, C. A., and Turner, M. J. (1981). Variant specific antigens of *Trypanosoma brucei* exist in solution as glycoprotein dimers. *Biochem. J.* **193,** 647–650.

Baltimore, D. (1981a). Somatic mutation gains its place among the generators of diversity. *Cell* **26,** 295–296.

Baltimore, D. (1981b). Gene conversion: some implications for immunoglobulin genes. *Cell* **24,** 592–594.

Banerji, J., Rusconi, S., and Schaffner, W. (1981). Expression of a β-globin gene is enhanced by remote SV40 DNA sequences. *Cell* **27,** 299–308.

Barry, J. D., Hajduk, S. L., Vickerman, K., and Le Ray, D. (1979). Detection of multiple variable antigen types in metacyclic populations of *T. brucei*. *Trans. R. Soc. Trop. Med. Hyg.* **73,** 205–208.

Beale, G. H. (1954). The genetics of *Paramecium aurelia*. *Cambridge Monogr. Exp. Biol.*, No. 2.

Bernards, R., and Flavell, R. A. (1980). Physical mapping of the globin gene deletion in hereditary persistance of foetal haemoglobin (HPFG). *Nucleic Acids Res.* **8,** 1521–1534.

Bernards, A., Van der Ploeg, L. H. T., Frasch, A. C. C., Borst, P., Boothroyd, J. C., Coleman, S., and Cross, G. A. M. (1981). Activation of trypanosome surface glycoprotein genes involves a duplication–transposition leading to an altered 3′ end. *Cell* **27,** 497–505.

Boothroyd, J. C., Cross, G. A. M., Hoeijmakers, J. H. J., and Borst, P. (1980). A variant surface glycoprotein of *Trypanosoma brucei* is synthesized with a hydrophobic carboxy-terminal extension absent from purified glycoprotein. *Nature (London)* **288,** 624–626.

Boothroyd, J. C., Paynter, C. A., Cross, G. A. M., Bernards, A., and Borst, P. (1981). Variant surface glycoproteins of *Trypanosoma brucei* are synthesized with cleavable hydrophobic sequences at the carboxy and amino termini. *Nucleic Acids Res.* **9,** 4735–4743.

Borst, P., and Cross, G. A. M. (1982). The molecular basis for trypanosome antigenic variation. *Cell* **29,** 291–303.

Borst, P., Fase-Fowler, F., Frasch, A. C. C., Hoeijmakers, J. H. J., and Weijers, P. J. (1980). Characterization of DNA from *Trypanosoma brucei* and related trypanosomes by restriction endonuclease digestion. *Mol. Biochem. Parasitol.* **1,** 221–246.

Borst, P., Frasch, A. C. C.,Bernards, A., Van der Ploeg, L. H. T., Hoeijmakers, J. H. J., Arnberg, A. C., and Cross, G. A. M. (1981a). DNA rearrangements involving the genes for variant antigens in *Trypanosoma brucei*. *Cold Spring Harbor Symp. Quant. Biol.* **45,** 935–943.

Borst, P., Fase-Fowler, F., and Gibson, W. C. (1981b). Quantitation of genetic differences between *Trypanosoma brucei gambiense, rhodesiense* and *brucei* by restriction enzyme digestion of kinetoplast DNA. *Mol. Biochem. Parasitol.* **3,** 117–131.

Borst, P., Van der Ploeg, M., Van Hoek, J. F. M., Tas, J., and James, J. (1982). On the DNA content and ploidy of trypanosomes. *Mol. Biochem. Parasitol.* **6,** 13–23.

Brown, D. D. (1981). Gene expression in eukaryotes. *Science (Washington, D.C.)* **211,** 667–674.

Capbern, A., Giroud, C., Baltz, T., and Mattern, P. (1977). *Trypanosoma equiperdum:* étude des variations antigéniques au cours de la trypanosome expérimentale du lapin. *Exp. Parasitol.* **42,** 6–13.

Capdeville, Y. (1979a). Regulation of surface antigen expression in *Paramecium primaurelia*. II. Role of the surface antigen itself. *J. Cell. Physiol.* **99,** 383–393.

Capdeville, Y. (1979b). Intergenic and interallelic exclusion in *Paramecium primaurelia:* immunological comparisons between allelic and non-allelic surface antigens. *Immunogenetics (NY)* **9,** 77–96.

Cross, G. A. M. (1977). Isolation, structure and function of variant-specific surface antigens. *Ann. Soc. Belge Med. Trop.* **57,** 389–399.

Cross, G. A. M. (1978). Antigenic variation in trypanosomes. *Proc. R. Soc. London Ser. B* **202,** 55–72.

Cross, G. A. M. (1979). Immunological aspects of antigenic variation in trypanosomes. *J. Gen. Microbiol.* **113,** 1–11.

Cummings, D. J. (1975). Studies on macronuclear DNA from *Paramecium aurelia. Chromosoma* **53,** 191–208.

Doyle, J.J. (1977). Antigenic variation in the salivarian trypanosomes. *In* "Immunity to Blood Parasites of Animals and Man" (L. H. Miller, J. A. Pino, and J. J. McKelvey, eds.), pp. 31–63. Plenum, New York.

Doyle, J. J., Hirumi, H., Hirumi, K., Lupton, E. N., and Cross, G. A. M. (1979). Antigenic variation in clones of animal-infective *Trypanosoma brucei* derived and maintained *in vitro*. *Parasitology* **80,** 359–369.

Early, P., and Hood, L. (1981). Allelic exclusion and nonproductive immunoglobulin gene rearrangements. *Cell* **24,** 1–3.

Efstratiadis, A., Posakony, J. W., Maniatis, T., Lawn, R. M., O'Connell, C., Spriyz, R. A., DeRiel, J. K., Forget, B. G., Weissman, S. M., Slighton, J. L., Biechl, A. E., Smithies, O., Baralle, F. E., and Shoulders, C. C. (1980). The structure and evolution of the human β-globin gene family. *Cell* **21,** 653–668.

Englund, P. T., Hajduk, S. L., and Marini, J. C. (1982). The molecular biology of trypanosomes. *Annu. Rev. Biochem.* **51,** 695–726.

Finger, I. (1974). Surface antigens of *Paramecium aurelia. In "Paramecium,* a Current Survey" (W. J. Van Wagtendonk, ed.), pp. 131–162. Elsevier, Amsterdam.

Frasch, A. C. C. Bernards, A., Van der Ploeg, L. H. T., Borst, P., Hoeijmakers, J. H. J., Van den Burg, J., and Cross, G. A. M. (1980). The genes for the variable surface glycoproteins of *Trypanosoma brucei*. *In* "The Biochemistry of Parasites and Host–Parasite Relationships: The Host–Invader Interplay" (H. Van den Bossche, ed.), pp. 235–239. North-Holland, Amsterdam.

Frasch, A. C. C., Borst, P., and Van den Burg, J. (1982). Rapid evolution of genes coding for variant surface glycoproteins in trypanosomes. *Gene* **17,** 197–211.

Gray, A. R., and Luckins, A. G. (1976). Antigenic variation in salivarian trypanosomes. *In* "Biology of the Kinetoplastida" W. H. R. Lumsden and D. A. Evans, eds.), Vol. 1, pp. 493–542. Academic Press, New York/London.

Hajduk, S. L., and Vickerman, K. (1981). Antigenic variation in cyclically-transmitted *Trypanosoma brucei*. Variable antigen type composition of the first parasitaemia in mice bitten by trypanosome-infected *Glossina morsitans*. *Parasitology* **83,** 609–621.

Hicks, J., Strathern, J. N., and Klar, A. J. S. (1979). Transposable mating type genes in *Saccharomyces cerevisiae*. *Nature (London)* **282,** 478–483.

Hoare, C. A. (1972). "The Trypanosomes of Mammals: A Zoological Monograph." Blackwell, Oxford.

Hoeijmakers, J. H. J., Frasch, A. C. C., Bernards, A., Borst, P., and Cross, G. A. M. (1980). Novel expression-linked copies of the genes for variant surface antigens in trypanosomes. *Nature (London)* **284,** 78–80.

Holder, A. A., and Cross, G. A. M. (1981). Glycopeptides from variant surface glycoproteins of *Trypanosoma brucei*. C-terminal location of antigenically cross-reacting carbohydrate moieties. *Mol. Biochem. Parasitol.* **2,** 135–150.

Hudson, K. M., Taylor, A. E. R., and Elce, B. J. (1980). Antigenic changes in *T. brucei* on transmission by tsetse fly. *Parasite Immunol.* **2,** 57–69.

Jenni, L. (1977). Comparisons of antigenic types of *Trypanosoma (T.) brucei* strains transmitted by *Glossina m. morsitans*. *Acta Trop.* **34,** 35–41.

Kataoka, T., Miyata, T., and Honjo, T. (1981). Repetitive sequences in class-switch recombination regions of immunoglobulin heavy chain genes. *Cell* **23,** 357–368.

Klar, A. J. S., Strathern, J. N., Broach, J. R., and Hicks, J. B. (1981). Regulation of transcription in expressed and unexpressed mating-type cassettes of yeast. *Nature (London)* **289,** 239–244.

Kosinski, R. J. (1980). Antigenic variation in trypanosomes: a computer analysis of variant order. *Parasitology* **80,** 343–357.

Lawn, R. M., Heumann, J. M., Herrick, G., and Prescott, D. M. (1978). The gene-size DNA molecules in Oxytricha. *Cold Spring Harbor Symp. Quant. Biol.* **42,** 483–492.

LeRay, D., Barry, J. D., Easton, C., and Vickerman, K. (1977). First tsetse fly transmission of the 'AnTat' serodeme of *Trypanosoma brucei*. *Ann. Soc. Belge Med. Trop.* **57,** 369–381.

LeRay, D., Barry, J. D., and Vickerman, K. (1978). Antigenic heterogeneity of metacyclic forms of *Trypanosoma brucei*. *Nature (London)* **273,** 300–302.

Lumsden, W. H. R., and Evans, D. A. (1976). "Biology of the Kinetoplastida," Vol. 1. Academic Press, New York/London.

Lumsden, W. H. R., and Evans, D. A. (1979). "Biology of the Kinetoplastida," Vol. 2. Academic Press, New York/London.

Majumder, H. K., Boothroyd, J. C., and Weber, H. (1981). Homologous 3′-terminal regions of mRNAs for surface antigens of different antigenic variants of *Trypanosoma brucei*. *Nucleic Acids Res.* **9,** 4745–4753.

Marcu, K. B., and Williams, R. O. (1981). Microbial surface elements: the case of variant surface glycoprotein (VSG) genes of African trypanosomes. *In* "Genetic Engineering" (J. K. Setlow and A. Hollaender, eds.), Vol. 3, pp. 129–155. Plenum, New York.

Matthyssens, G., Michiels, F., Hamers, R., Pays, E., and Steinert, M. (1981). Two variant surface glycoproteins of *Trypanosoma brucei* have a conserved C-terminus. *Nature (London)* **293,** 230–233.

McTavish, C., and Sommerville, J. (1980). Macronuclear DNA organization and transcription of *Paramecium primaurelia*. *Chromosoma* **78,** 147–164.

Michels, P. A. M., Bernards, A., Van der Ploeg, L. H. T., and Borst, P. (1982). Characterization of the expression-linked gene copies of variant surface glycoprotein 118 in two independently isolated clones of *Trypanosoma brucei*. *Nucleic Acids Res.* **10,** 2353–2366.

Miller, E. N., and Turner, M. J. (1981). Analysis of antigenic types appearing in first relapse populations of clones of *Trypanosoma brucei*. *Parasitology* **82,** 63–80.

Nasmyth, K. A., Tatchell, K., Hall, B. D., Astell, C., and Smith, M. (1981). A position effect in the control of transcription at yeast mating type loci. *Nature (London)* **289,** 244–250.

Pays, E., Van Meirvenne, N., Le Ray, D., and Steinert, M. (1981a). Gene duplication and transposition linked to antigenic variation in *Trypanosoma brucei. Proc. Natl. Acad. Sci. USA* **78,** 2673–2677.

Pays, E., Lheureux, M., and Steinert, M. (1981b). Analysis of the DNA and RNA changes associated with the expression of isotypic variant-specific antigens of trypanosomes. *Nucleic Acids Res.* **9,** 4225–4238.

Pays, E., Lheureux, M., and Steinert, M. (1981c). The expression-linked copy of surface antigen gene in *Trypanosoma* is probably the one transcribed. *Nature (London)* **292,** 365–367.

Pays, E., Lheureux, M., Vervoort, T., and Steinert, M. (1981d). Conservation of a variant-specific surface antigen gene in different trypanosome species and sub-species. *Mol. Biochem. Parasitol.* **4,** 349–357.

Preer, J. R., Jr. (1969). Genetics of the protozoa. *In* "Research in Protozoology" (T.-T. Chen, ed.), Vol. 3, pp. 129–278. Pergamon, Oxford.

Preer, J. R., Jr. (1981). mRNAs for the immobilization antigens of *Paramecium. Proc. Natl. Acad. Sci. USA* **78,** 6776–6778.

Raikov, I. B. (1976). Evolution of macronuclear organization. *Annu. Rev. Genet.* **10,** 413–440.

Ramanathan, R., Adoutte, A., and Dute, R. R. (1981). Biochemical studies of the excitable membrane of *Paramecium tetraurelia.* V. Effects of proteases on the ciliary membrane. *Biochim. Biophys. Acta* **641,** 349–365.

Rice-Ficht, A. C., Chen, K. K., and Donelson, J. E. (1981). Sequence homologies near the C-termini of the variable surface glycoproteins of *Trypanosome brucei. Nature (London)* **294,** 53–57.

Sakano, H., Hüppi, K., Heinrich, G., and Tonegawa, S. (1979). Sequences at the somatic recombination sites of immunoglobulin light-chain genes. *Nature (London)* **280,** 288–294.

Seed, J. R. (1978). Competition among serologically different clones of *Trypanosoma brucei gambiense in vivo. J. Protozool.* **25,** 526–529.

Seed, J. R., Kemp, W. M., and Brown, R. A. (1977). Symposium on experimental parasitology of host-parasite interfaces. *TJS Spec. Publ.,* No. 2, 1–13.

Sommerville, J. (1970). Serotype expression in *Paramecium. In* "Advances in Microbial Physiology" (A. H. Rose and J. F. Wilkinson, eds.), Vol. 4, pp. 131–178. Academic Press, New York/London.

Sonneborn, T. M. (1974). *Paramecium aurelia. Handb. Genet.* **2,** 469–594.

Strathern, J. N., and Herskowitz, I. (1979). Asymmetry and directionality in production of new cell types during clonal growth: the switching pattern of homothallic yeast. *Cell* **17,** 371–381.

Tait, A. (1980). Evidence for diplody and mating in trypanosomes. *Nature (London)* **287,** 536–538.

Turner, M. J. (1980). Antigenic variation. *In* "The Molecular Basis of Microbial Pathogenicity" (H. Smith, J. J. Skehel, and M. J. Turner, eds.), pp. 133–158. Verlag Chemie, Weinheim.

Turner, M. J., and Cordingley, J. S. (1981). Evolution of antigenic variation in the salivarian trypanosomes. *In* "Molecular and Cellular Aspects of Microbial Evolution" (M. J. Carlile, J. F. Collins, and B. E. B. Moseley, eds.), pp. 313–347. Cambridge Univ. Press, Cambridge.

Van der Ploeg, L. H. T., Bernards, A., Rijsewijk, F. A. M., and Borst, P. (1982). Characterization of the DNA duplication–transposition that controls the expression of two genes for variant surface glycoproteins in *Trypanosoma brucei. Nucleic Acids Res.* **10,** 593–609.

Van Meirvenne, N., Janssens, P. G., and Magnus, E. (1975a). Antigenic variation in syringe passaged populations of *Trypanosoma (Trypanozoon) brucei.* I. Rationalization of the experimental approach. *Ann. Soc. Belge Med. Trop.* **55,** 1–23.

Van Meirvenne, N., Janssens, P. G., Magnus, E., Lumsden, W. H. R., and Herbert, W. J. (1975b).

Antigenic variation in syringe passaged populations of *Trypanosoma (Trypanozoon) brucei*. II. Comparative studies on two antigenic-type collections. *Ann. Soc. Belge Med. Trop.* **55,** 25–30.

Van Meirvenne, N., Magnus, E., and Vervoort, T. (1977). Comparisons of variable antigenic types produced by trypanosome strains of the sub-genus *Trypanozoon*. *Ann. Soc. Belge Med. Trop.* **57,** 409–423.

Vickerman, K. (1974). Antigenic variation in African trypanosomes. *In* "Parasites in the Immunized Host: Mechanism of Survival," pp. 53–80. Associated Scientific Publishers, Amsterdam.

Vickerman, K. (1978). Antigenic variation in trypanosomes. *Nature (London)* **273,** 613–617.

Williams, R. O., Young, J. R., Majiwa, P. A. O., Doyle, J. J., and Shapiro, S. Z. (1981). Contextural genomic rearrangements of variable-antigen genes in *Trypanosoma brucei*. *Cold Spring Harbor Symp. Quant. Biol.* **45,** 945–949.

Young, J. R., Donelson, J. E., Majiwa, P. A. O., Shapiro, S. Z., and Williams, R. O. (1982). Analysis of genomic rearrangements associated with two variable antigen genes of *Trypanosoma brucei*. *Nucleic Acids Res.* **10,** 803–819.

Zieg, J., Hillmen, M., and Simon, M. (1978). Regulation of gene expression by site-specific inversion. *Cell* **15,** 237–244.

Note Added in Proof

Since this chapter was written, many additional papers have been published and relevant new results obtained in three areas.

1. The analysis of the 5′-end of VSG mRNAs has been extended to four independent trypanosome clones that make VSG 118, two that make VSG 117, and one that makes VSG 221. All have the same 35-nucleotide sequence at their 5′-end, and none has this sequence encoded contiguously with the remainder of the gene (Van der Ploeg *et al.*, 1982; Boothroyd and Cross, 1982; Borst *et al.*, 1983a). There are about 200 copies of this 35-bp mini-exon sequence in the genome, and these are spread over one or several repetitive areas lacking EcoRI, HindIII, and KpnI sites (Borst *et al.*, 1983a). Although these have not yet been linked to the VSG gene expression site, I like the idea that there might be an area upstream of the ELC with many promoter–mini-exon modules, each able to act as the start of pre-mRNA synthesis. One (randomly picked) mini-exon in cloned DNA has been sequenced and found to contain the complete 35-nucleotide sequence. The intron has the conventional GT/AG borders (Borst *et al.*, 1983a).

2. Further analysis of the NTA genes (the VSG genes that do not seem to duplicate when activated) has confirmed that these genes do not duplicate, not even in ghost form, and that they all lie at the end of a (different) chromosome (Williams *et al.*, 1982; Borst *et al.*, 1983a), like the expression site (De Lange and Borst, 1982) for the TA genes. These observations and the important finding that one of the NTA genes, the 221 gene, contains the same 35-nucleotide sequence spliced onto the end of its mRNA as that of the TA genes (see 1) have led to a simple chromosome-end-exchange model for the activation of NTA genes (Borst *et al.*, 1983a,b). This model postulates that the 221 gene enters the expression site, because a reciprocal translocation exchanges the ends on which ELC and 221 gene lie. The ELC is thus disconnected from the promoter–mini-exon area, and the 221 is placed downstream of this area. The reciprocal translocation is thought to involve homologous sequences and to be freely reversible. This model explains all known facts, even the "lingering" 121 ELC observed when a 121 trypanosome variant gave rise to a 221 variant (see p. 645). The proof for the actual translocation is still lacking, however, and the model remains a speculation at present.

3. Rice-Ficht *et al.* (1982) have published experiments that they have interpreted as meaning that the ELC is made by a mutagenic polymerase that introduces 4% point mutations. We have not seen anything resembling this in our strain, and we think that these widely publicized results are either due to (partial) gene conversion or analysis of the wrong BC gene (Borst *et al.*, 1983a).

Allen, G., Gurnett, L. P., and Cross, G. A. M. (1982). Complete amino acid sequence of a variant surface glycoprotein (VSG 117) from *Trypanosoma brucei*. *J. Mol. Biol.* **157,** 527–546.

Boothroyd, J. C., Paynter, C. A., Coleman, S. L., and Cross, G. A. M. (1982). Complete nucleotide sequence of complementary DNA coding for a variant surface glycoprotein from *Trypanosoma brucei*. *J. Mol. Biol.* **157,** 547–556.

Boothroyd, J. C., and Cross, G. A. M. (1982). Transcripts coding for variant surface glycoproteins of *Trypanosoma brucei* have a short, identical exon at their 5′-end. *Gene* **20,** 279–287.

Borst, P., Bernards, A., Van der Ploeg, L. H. T., Michels, P. A. M., Liu, A. Y. C., De Lange, T., Sloof, P., Veeneman, G. H., Tromp, M., and Van Boom, J. (1983a). DNA rearrangements controlling the expression of genes for variant surface antigens in trypanosomes. *Consequences DNA Structure Genome Arrangement, John Innes Symp. Biol., 5th* (in press).

Borst, P., Bernards, A., Van der Ploeg, L. H. T., Michels, P. A. M., Liu, A. Y. C., and De Lange, T. (1983b). Gene rearrangements controlling the expression of genes for variant surface antigens in trypanosomes. *Proc.—ICN-UCLA Symp. Tumor Viruses Differentiation* (in press).

De Lange, T., and Borst, P. (1982). Genomic environment of the expression-linked extra copies of genes for surface antigens of *Trypanosoma brucei* resembles the end of a chromosome. *Nature (London)* **299,** 451–453.

Majiwa, P. A. O., Young, J. R., Englund, P. T., Shapiro, S. Z., and Williams, R. O. (1982). Two distinct forms of surface antigen gene rearrangements in *Trypanosoma brucei*. *Nature (London)* **297,** 514–516.

McConnel, J., Gurnett, A. M., Cordingley, J. S., Walker, J. E., and Turner, M. J. (1981). Biosynthesis of *Trypanosoma brucei* variant surface glycoprotein. I. Synthesis, size, and processing of an N-terminal signal peptide. *Mol. Biochem. Parasitol.* **4,** 225–242.

McConnel, J., Cordingley, J. S., and Turner, M. J. (1982). The biosynthesis of *Trypanosoma brucei* variant surface glycoproteins—*in vitro* processing of signal peptide and glycosylation using heterologous rough endoplasmic reticulum vesicles. *Mol. Biochem. Parasitol.* **6,** 161–174.

Pays, E., Lheureux, M., and Steinert, M. (1982). Structure and expression of a *Trypanosoma brucei gambiense* variant specific antigenic gene. *Nucl. Acids Res.* **10,** 3149–3163.

Rice-Ficht, A. C., Chen, K. K., and Donelson, J. E. (1982). Point mutations during generation of expression-linked extra copy of trypanosome surface glycoprotein gene. *Nature (London)* **298,** 676–679.

Van der Ploeg, L. H. T., Liu, A. Y. C., Michels, P. A. M., De Lange, T., Borst, P., Majumder, H. K., Weber, H., Veeneman, G. H., and Van Boom, J. (1982a). RNA splicing is required to make the messenger RNA for a variant surface antigen in trypanosomes. *Nucl. Acids Res.* **10,** 3591–3604.

Van der Ploeg, L. H. T., Valerio, D., De Lange, T., Bernards, A., Borst, P., and Grosveld, F. G. (1982b). An analysis of cosmid clones of nuclear DNA from *Trypanosoma brucei* shows that the genes for variant surface glycoproteins are clustered in the genome. *Nucl. Acids Res.* **10,** 5905–5923.

Williams, R. O., Young, J. R., and Majiwa, P. A. O. (1982). Genomic environment of *Trypanosoma brucei* VSG genes: presence of a minichromosome. *Nature (London)* **299,** 417–421.

Index

T

V

Y

Z